		III A	IV A	V A	VI A	VII A	0
		P_1	P_2	P_3	P_4	P_5	S_2
							2 He Helium 4.00260
		5 $He+2g$ B Boron 10.81	6 $He+2g$ C Carbon 12.011	7 $He+2/3$ N Nitrogen 14.0067	8 O Oxygen 15.9994	9 F Fluorine 18.998403	10 Ne Neon 20.179 $2P_6$
		13 Al Aluminum 26.98154	14 Si Silicon 28.0855	15 P Phosphorus 30.97376	16 S Sulfur 32.06	17 Cl Chlorine 35.453	18 Ar Argon 39.948 $3P_6$

	I B	II B							
28 Ni Nickel 58.69	29 Cu Copper 63.546	30 Zn Zinc 65.38	31 Ga Gallium 69.72	32 Ge Germanium 72.59	33 As Arsenic 74.9216	34 Se Selenium 78.96	35 Br Bromine 79.904	36 Kr Krypton 83.80 $4P_6$
46 Pd Palladium 106.42	47 Ag Silver 107.868	48 Cd Cadmium 112.41	49 In Indium 114.82	50 Sn Tin 118.69	51 Sb Antimony 121.75	52 Te Tellurium 127.60	53 I Iodine 126.9045	54 Xe Xenon 131.29 $5P_6$
78 Pt Platinum 195.08	79 Au Gold 196.9665	80 Hg Mercury 200.59	81 Tl Thallium 204.383	82 Pb Lead 207.2	83 Bi Bismuth 208.9804	84 Po Polonium (209)[a]	85 At Astatine (210)[a]	86 Rn Radon (222)[a] $6P_6$

metals ← → nonmetals

63 Eu Europium 151.96	64 Gd Gadolinium 157.25	65 Tb Terbium 158.9254	66 Dy Dysprosium 162.50	67 Ho Holmium 164.9304	68 Er Erbium 167.26	69 Tm Thulium 168.9342	70 Yb Ytterbium 173.04	71 Lu Lutetium 174.967
95 Am Americium (243)[a]	96 Cm Curium (247)[a]	97 Bk Berkelium (247)[a]	98 Cf Californium (251)[a]	99 Es Einsteinium (252)[a]	100 Fm Fermium (257)[a]	101 Md Mendelevium (258)[a]	102 No Nobelium (259)[a]	103 Lr Lawrencium (260)[a]

Chemistry

C H E M I S T R Y Fifth Edition

Charles E. Mortimer

Muhlenberg College

Wadsworth Publishing Company • Belmont, California • A Division of Wadsworth, Inc.

To: J.S.M. and C.E.M.[II]

ISBN 0-534-01184-5

A study guide has been specially designed to help students master the concepts presented in this textbook. Order from your bookstore.

Cover photos:
Pamela S. Roberson (first and third photos from top)
Werner H. Müller/Peter Arnold, Inc. (second photo)
The Image Bank West/Gabe Palmer (fourth photo)

Chemistry Editor: Jack Carey
Production Editor: Hal Humphrey
Designer: Adriane Bosworth
Technical Illustrators: J & R Art Services
Photo Researchers: Roberta Spieckerman Associates

Printed in the United States of America
3 4 5 6 7 8 9 10—87 86 85 84

Library of Congress Cataloging in Publication Data

Mortimer, Charles E.
  Chemistry.

  (Wadsworth series in chemistry)
  Includes index.
  1. Chemistry. I. Title. II. Series.
QD31.2.M65 1983   540   82–11133
ISBN 0–534–01184–5

PREFACE

Sometimes it seems as though every product that is advertised on television is said to be *new and improved*. I hope that this trite phrase describes this edition, but at the same time, I believe that the character of the text that has made it successful through the previous four editions has been retained. This book was written to explain chemistry, not just present chemical facts. Consequently, each concept continues to be explained as fully as is necessary for understanding, simplified where necessary, but never distorted.

Probably the most prominent change made in the preparation of the fifth edition is the alteration in the sequence of topics. *Stoichiometry*, which is central to the understanding of all chemical concepts, is introduced early (Chapter 2). As a result, the use of stoichiometric principles can be expanded and reinforced throughout the entire course. Furthermore, this early placement facilitates the design of a coordinated laboratory program (for which a section on "Reactions in Solution" has been introduced).

Thermochemistry (Chapter 3) follows stoichiometry, underscoring the fact that chemistry is a science that is concerned with both energy and matter and that both are amenable to quantitative treatment. The early discussion of thermochemistry prepares the way for the use of energy concepts (such as lattice energy, ionization energy, and bond energy) in the development of later topics.

A new chapter on *Reactions in Aqueous Solution* (Chapter 11) is introduced after the chapter on *Solutions* (Chapter 10), which it logically follows. The discussion of these reactions, which constitute a high proportion of all chemical reactions investigated, lays the groundwork for later discussions (notably, ionic equilibria, acids and bases, electrochemistry, and descriptive chemistry).

Electrochemistry (Chapter 18) is postponed until after *Thermodynamics* (Chapter 17) and *Equilibrium* (Chapters 13 through 16) have been discussed. In this way, the principles of thermodynamics (particularly Gibbs free energy) and equilibrium (notably, equilibrium expressions) can be used to develop electrochemical concepts (electromotive force, electrode potentials, the Nernst equation). Oxidation numbers and oxidation-reduction reactions are discussed in Chapter 11, earlier than in the formal treatment of electrochemistry.

The descriptive chemistry of the *Nonmetals* (Chapters 19 through 22) appears later in the book so that it continues to follow the treatment of electrochemistry. A discussion of descriptive chemistry, if it is to be more than superficial, demands an understanding of electrochemical concepts—electrolysis and electrode potentials. The chemistry of oxygen and hydrogen is now incorporated into the treatment of the nonmetals rather than appear as a separate chapter (old Chapter 8). The descriptive chemistry of the nonmetals is followed by a chapter on *Metals* (Chapter 23) and industrial aspects are emphasized in both.

The chapter on *Organic Chemistry* (Chapter 26) together with a new chapter on *Biochemistry* (Chapter 27), which logically follows, are placed at the end of the book. The chapter on *Nuclear Chemistry* (Chapter 25) has been moved forward to accommodate this placement.

Several chapters have been divided to make the text more flexible and to facilitate the preparation of a course outline: *Stoichiometry* (Chapter 2) and *Thermochemistry* (Chapter 3) are now separate chapters; the *Ionic Bond* (Chapter 5) is now covered in a separate chapter from the *Covalent Bond* (Chapters 6 and 7); *Kinetics* (Chapter 12) and *Equilibrium* (Chapter 13) appear separately; and the descriptive chemistry of the *Nonmetals* has been divided into four parts (Chapters 19 through 22).

Many changes have been made in the text itself in the interest of keeping it up-to-date and improving the clarity and usefulness of the presentation. For example, sections on air pollution, the corrosion of iron, and industrial uses of the nonmetals have been added.

A number of features appear in this edition to help the student. *Summaries* are given at the end of each chapter. They provide the student with a quick check-list of the concepts covered in the chapter.

Key terms are listed and defined at the end of each chapter. Students will find these glossaries useful as an aid in studying the material of the chapter, as a quick reference for future work, and as a help in solving the chapter-end problems that follow these lists. Important *new terms* continue to be set in color type at the point in the text where they are first introduced and defined.

Boxes are used to set off step-by-step directions for the solution of basic types of problems. Students will find this boxed material useful for initial assignments and also for reference in later work.

Chapter-end problems, totaling about 1200, are grouped according to type. Many of these problems are new. Answers for approximately half the problems, those that are color-keyed, are given in the appendix. The more difficult problems are marked with asterisks.

Examples, designed to illustrate how to solve chemical problems, are used throughout the text. The number of these examples has been increased, particularly in the early chapters.

Photographs, of scientists and subjects of chemical interest, are used to enliven the text. They accompany the *figures*, which augment and amplify the discussion.

Notes on mathematical operations appear in the appendix. These notes include discussions on the use of exponents, scientific notation, common and natural logarithms, and the quadratic formula.

The following supplemetary items are available: study guide, solutions manual, answer booklet, instructor's manual with test items, and transparency masters.

I sincerely thank the following persons for their comments and suggestions:

David L. Adams, North Shore Community College
John E. Bauman Jr., University of Missouri
Paul A. Barks, North Hennepin Community College
Neil R. Coley, Chabot College
John DeKorte, Northern Arizona University
Geoffry Davies, Northeastern University
Phil Davis, University of Tennessee
Lawrence Epstein, University of Pittsburgh
Patrick Garvey, Des Moines Area Community College
Peter J. Hansen, Northwestern College

Larry C. Hall, Vanderbilt University
David W. Herlocker, Western Maryland College
Delwin Johnson, St. Louis Community College at Forest Park
George B. Kauffman, California State University, Fresno
Robert P. Lang, Quincy College
Lester R. Morss, Argonne National Laboratory
John Maurer, University of Wyoming
William McCurdy, Ohio State University
Robert C. Melucci, Community College of Philadelphia
Lucy T. Pryde, Southwestern College
Fred H. Redmore, Highland Community College
Lewis Radonovich, University of North Dakota
Roland R. Roskos, University of Wisconsin
Larry Thompson, University of Minnesota
James A. Weiss, Penn State University, Scranton Campus

The Wadsworth staff have been congenial, helpful, and efficient in the production of this book. I appreciate their efforts and express my gratitude. My thanks go especially to: Jack Carey, Chemistry Editor; Hal Humphrey, Production Editor; Adriane Bosworth, Designer; and Harriet Serenkin.

Suggestions for the improvement of this edition will be welcomed.

Charles E. Mortimer

BRIEF CONTENTS

DETAILED CONTENTS

INTRODUCTION

Chemistry may be defined as the science that is concerned with the characterization, composition, and transformations of matter. This definition, however, is far from adequate. The interplay between the branches of modern science causes the boundaries between them to be so indistinct that it is almost impossible to stake out a field and say "this is chemistry." Not only do the interests of scientific fields overlap, but concepts and methods find universal application. Moreover, this definition fails to convey the spirit of chemistry, for it, like all science, is a vital, growing enterprise, not an accumulation of knowledge. It is self-generating; the very nature of each new chemical concept stimulates fresh observation and experimentation that lead to progressive refinement as well as to the development of other concepts. In the light of scientific growth, it is not surprising that a given scientific pursuit frequently crosses artificial, human-imposed boundaries.

Nevertheless, there is a common, if somewhat vague, understanding of the province of chemistry, and we must return to our preliminary definition; a fuller understanding should emerge as this book unfolds. Chemistry is concerned with the composition and the structure of substances and with the forces that hold the structures together. The physical properties of substances are studied since they provide clues for structural determinations, serve as a basis for identification and classification, and indicate possible uses for specific materials. The focus of chemistry, however, is probably the *chemical reaction*. The interest of chemistry extends to every conceivable aspect of these transformations and includes such considerations as detailed descriptions of how and at what rates reactions proceed, the conditions required to bring about desired changes and to prevent undesired changes, the energy changes that accompany chemical reactions, the syntheses of substances that occur in nature and of those that have no natural counterparts, and the quantitative mass relations between the materials involved in chemical changes.

1.1 The Development of Modern Chemistry

Modern chemistry, which emerged late in the eighteenth century, took hundreds of years to develop. The story of its development can be divided roughly into five periods:

1. Practical arts (— to 600 B.C.). The production of metals from ores, the manufacture of pottery, brewing, baking, and the preparation of medicines, dyes, and drugs are ancient arts. Archaeological evidence proves that the inhabitants of ancient Egypt and Mesopotamia were skilled in these crafts, but how and when they developed are not known.

These arts, which are chemical processes, became highly developed during this period. The development, however, was *empirical*, that is, based on practical experience alone without reference to underlying chemical principles. The Egyptian metalworkers knew how to obtain copper by heating malachite ore with charcoal. They did not know, nor did they seek to know, why the process worked and what actually occurred in the fire.

2. Greek (600 B.C. to 300 B.C.). The philosophical aspect (or theoretical aspect) of chemistry began in classical Greece about 600 B.C. The foundation of Greek science was the search for principles through which an understanding of nature could be obtained. Two theories of the Greeks became very important in the centuries that followed:

a. A concept that all terrestial substances are composed of four elements (earth, air, fire, and water) in various proportions originated with Greek philosophers of this period.

b. A theory that matter consists of separate and distinct units called **atoms** was proposed by Leucippus and extended by Democritus in the fifth century B.C.

Plato proposed that the atoms of one element differ in shape from the atoms of another. Furthermore, he believed that atoms of one element could be changed (or **transmuted**) into atoms of another by changing the shape of the atoms.

The concept of transmutation is also found in Aristotle's theories. Aristotle (who did not believe in the existence of atoms) proposed that the elements, and therefore all substances, are composed of the same primary matter and differ only in the forms that this primary matter assumes. To Aristotle, the form included not only the shape but also the qualities (such as color and hardness) that distinguish one substance from others. He proposed that changes in form constantly occur in nature and that all material things (animate and inanimate) grow and develop from immature forms to adult forms. (Throughout the middle ages, it was believed that minerals could grow and that mines would be replenished after minerals were removed from them.)

3. Alchemy (300 B.C. to 1650 A.D.). The philosophical tradition of ancient Greece and the craft tradition of ancient Egypt met in Alexandria, Egypt (the city founded by Alexander the Great in 331 B.C.), and **alchemy** was the result of the union. The early alchemists used Egyptian techniques for the handling of materials to investigate theories concerned with the nature of matter. Books written in Alexandria (the oldest known works on chemical topics) contain diagrams of chemical apparatus and descriptions of many laboratory operations (for example, distillation, crystallization, and sublimation).

The philosphical content of alchemy incorporated elements of astrology and mysticism into the theories of the earlier Greeks. A dominant interest of the alchemists was the transmutation of base metals, such as iron and lead, into the noble metal, gold. They believed that a metal could be changed by changing its qualities (particularly its color) and that such changes occur in nature—that metals strive for the perfection represented by gold. Furthermore, the alchemists believed that these changes could be brought about by means of a very small amount of a powerful transmuting agent (later called the *philosopher's stone*).

In the seventh century A.D., the Arabs conquered the centers of Hellenistic civilization (including Egypt in 640 A.D.), and alchemy passed into their hands.

The Alchemist, painted by the Flemish artist David Teniers in 1648. *Fisher Scientific Company.*

Greek texts were translated into Arabic and served as the foundation for the work of Arab alchemists. The Arabs called the philospher's stone *aliksir* (which was later corrupted into *elixir*). Arab alchemists believed that this substance could not only ennoble metals by transmuting them into gold but also could ennoble life by curing all diseases. For centuries afterward, the two principal goals of alchemy were the transmutation of base metals into gold and the discovery of an *elixir of life* that could make man immortal by preventing death.

In the twelfth and thirteenth centuries, alchemy was gradually introduced into Europe by the translation of Arabic works into Latin. Most of the translations were made in Spain where, after the Islamic conquest in the eighth century, a rich Moorish culture was established and flourished.

A school of iatrochemistry, a branch of alchemy concerned with medicine, flourished in the sixteenth and seventeenth centuries. On the whole, however, European alchemists added little that was new to alchemical theory. Their work is important because they preserved the large body of chemical data that they received from the past, added to it, and passed it on to later chemists.

Alchemy lasted until the seventeenth century. Gradually the theories and attitudes of the alchemists began to be questioned. The work of Robert Boyle, who published *The Sceptical Chymist* in 1661, is noteworthy. Although Boyle believed that the transmutation of base metals into gold might be possible, he severely criticized alchemical thought. Boyle emphasized that chemical theory should be derived from experimental evidence.

4. Phlogiston (1650 to 1790). Throughout most of the eighteenth century, the phlogiston theory dominated chemistry. This theory, which was later shown to be erroneous, was principally the work of Georg Ernst Stahl. **Phlogiston** (a "fire principle") was assumed to be a constituent of any substance that could undergo combustion.

Upon combustion, a substance was thought to lose its phlogiston and be reduced to a simpler form. Air was believed to function in a combustion by carrying off the phlogiston as it was released. Whereas we would think of the combustion of wood in the following terms:

$$\text{wood} + \text{oxygen gas (from air)} \longrightarrow \text{ashes} + \text{oxygen-containing gases}$$

according to the phlogiston theory,

$$\text{wood} \longrightarrow \text{ashes} + \text{phlogiston (removed by air)}$$

Wood, therefore, was believed to be a compound composed of ashes and phlogiston. Readily combustible materials were thought to be rich in phlogiston.

The phlogiston theory interpreted **calcination** in a similar way. The formation of a metal oxide (called a calx) by heating a metal in air is called a calcination:

$$\text{metal} + \text{oxygen gas (from air)} \longrightarrow \text{calx (metal oxide)}$$

According to the phlogiston theory, a metal is assumed to be a compound composed of a calx and phlogiston. Calcination, therefore, was thought to be the loss of phlogiston by a metal:

$$\text{metal} \longrightarrow \text{calx} + \text{phlogiston (removed by air)}$$

The phlogiston theory was extended to explain many other chemical phenomena. The preparation of certain metals, for example, can be accomplished by heating the metal oxide with carbon:

$$\text{calx (metal oxide)} + \text{carbon} \longrightarrow \text{metal} + \text{carbon monoxide gas}$$

In a process of this type, the carbon (supposedly rich in phlogiston) was thought to replace the phlogiston lost by calcination:

$$\text{calx} + \text{phlogiston (from carbon)} \longrightarrow \text{metal}$$

One difficulty inherent in the phlogiston theory was never adequately explained. When wood burns, it supposedly *loses* phlogiston and the resulting ashes weigh *less* than the original piece of wood. On the other hand, in calcination, the *loss* of phlogiston is accompanied by an increase in weight since the calx (a metal oxide) weighs *more* than the original metal. The adherents of the phlogiston theory recognized this problem, but throughout most of the eighteenth century the importance of weighing and measuring was not realized.

5. Modern chemistry (1790 —). The work of Antoine Lavoisier in the late eighteenth century is generally regarded as the beginning of modern chemistry. Lavoisier deliberately set out to overthrow the phlogiston theory and revolutionize chemistry. He relied on the results of quantitative experimentation (he used the chemical balance extensively) to arrive at his explanations of a number of chemical phenomena.

Antoine Lavoisier, 1743–1794.
Smithsonian Institution.

The **law of conservation of mass** states that there is no detectable change in mass during the course of a chemical reaction. In other words, the total mass of all materials entering into a chemical reaction equals the total mass of all the products of the reaction. This law is implicit in earlier work, but Lavoisier stated it explicitly and used it as the cornerstone of his science. To Lavoisier, therefore, the phlogiston theory was impossible.

The roles that gases play in reactions proved to be a stumbling block to the development of chemical theory. When the law of conservation of mass is applied to a combustion or to a calcination, the masses of the gases used or produced in these reactions must be taken into account. The correct interpretation of these processes, therefore, had to wait until chemists identified the gases involved and developed methods to handle and measure gases. Lavoisier drew upon the results of other scientists' work with gases to explain these reactions.

In interpreting chemical phenomena, Lavoisier used the modern definitions of elements and compounds (see Section 1.2). The phlogiston theory regarded a metal as a *compound* composed of a calx and phlogiston. Lavoisier showed that a metal is an *element* and that the corresponding calx is a *compound* composed of the metal and oxygen from the air.

In his book *Traité Elémentaire de Chimie* (*Elementary Treatise on Chemistry*), published in 1789, Lavoisier used essentially modern terminology. The present-day language of chemistry is based on the system of nomenclature that Lavoisier helped to devise.

The achievements of scientists since the 1790s are described throughout this book. More has been learned about chemistry in the two centuries following Lavoisier than in the twenty centuries preceding him. Chemistry has gradually developed five principal branches (these divisions, however, are arbitrary and the classification is subject to criticism):

a. Organic chemistry. The chemistry of most of the compounds of carbon. At one time it was assumed that these compounds could only be obtained from plant or animal life or derived from other compounds that had been obtained from living material.

b. Inorganic chemistry. The chemistry of all the elements except carbon. Some simple carbon compounds (for example, carbon dioxide and the carbonates) are traditionally classified as inorganic compounds.

c. Analytical chemistry. The identification of the composition, both qualitative and quantitative, of substances.

d. Physical chemistry. The study of the physical principles that underlie the structure of matter and chemical transformations.

e. Biochemistry. The chemistry of living systems, both plant and animal.

1.2 Elements, Compounds, and Mixtures

Matter, the material of which the universe is composed, may be defined as anything that occupies space and has mass. **Mass** is a measure of quantity of matter. A body that is not being acted upon by some external force has a tendency to remain at rest or, when it is in motion, to continue in uniform motion in the same direction. This property is known as **inertia.** The mass of a body is proportional to the inertia of the body.

The mass of a body is invariable; the weight of a body is not. **Weight** is the gravitational force of attraction exerted by the earth on a body; the weight of a given body varies with the distance of that body from the center of the earth. The weight of a body is directly proportional to its mass as well as to the earth's gravitational attraction. At any given place, therefore, two objects of equal mass have equal weights.

The ancient Greeks originated the concept that all matter is composed of a limited number of simple substances called **elements.** The Greeks assumed that all terrestrial matter is derived from four elements: *earth, air, fire,* and *water.* Since heavenly bodies were thought to be perfect and unchangeable, celestial matter was assumed to be composed of a different element, the *ether,* which later came to be known as *quintessence* (from Latin, meaning *fifth element*). This Greek theory dominated scientific thought for centuries.

In 1661, Robert Boyle proposed an essentially modern definition of an element in his book *The Sceptical Chymist:* "I now mean by Elements . . . certain Primitive and Simple, or perfectly unmingled bodies; which not being made of any other bodies, or of one another, are the Ingredients of which all those call'd perfectly mixt Bodies are immediately compounded, and into which they are ultimately resolved."

Boyle made no attempt to identify specific substances as elements. He did, however, emphasize that the proof of the existence of elements, as well as the identification of them, rested on chemical experimentation.

Boyle's concept of a chemical element was firmly established by Antoine Lavoisier in the following century. Lavoisier accepted a substance as an element if it could not be decomposed into simpler substances. Furthermore, he showed that a compound is produced by the union of elements. Lavoisier correctly identified 23 elements (although he incorrectly included light, heat, and several simple compounds in his list).

At the present time, 106 elements are known. Of these, 85 have been isolated from natural sources; the remainder have been prepared by nuclear reactions (Section 25.5).

Each element is assigned a one- or two-letter **chemical symbol** that has been decided upon by international agreement. Whereas the name of an element may differ from one language to another, the symbol does not. Nitrogen, for example, is called *azoto* in Italian and *stickstoff* in German, but the symbol for nitrogen is N in any language. These symbols are listed in the table of the elements that appears inside the back cover of this book.

Most of the symbols correspond closely to the English names of the elements. Some of them, however, do not. The symbols for some of the elements have been assigned on the basis of their Latin names; these elements are listed in Table 1.1. The symbol for tungsten, W, is derived from the German name for the element, *wolfram.*

The 15 most abundant elements in the earth's crust, bodies of water, and atmosphere are listed in Table 1.2. This classification includes those parts of the universe from which we can obtain the elements. The earth consists of a core (which is probably composed of iron and nickel) surrounded successively by a mantle and a thin crust. The crust is about 20 to 40 miles thick and constitutes only about 1% of the earth's mass.

If the entire earth were considered, a list different from that of Table 1.2 would result, and the most abundant element would be iron. On the other hand, the most abundant element in the universe as a whole is hydrogen, which is thought to constitute about 75% of the total mass of the universe.

Table 1.1 Symbols of elements derived from Latin names

English Name	Latin Name	Symbol
antimony	stibium	Sb
copper	cuprum	Cu
gold	aurum	Au
iron	ferrum	Fe
lead	plumbum	Pb
mercury	hydrargyrum	Hg
potassium	kalium	K
silver	argentum	Ag
sodium	natrium	Na
tin	stannum	Sn

Table 1.2 Abundance of the elements (earth's crust, bodies of water, and atmosphere)

Rank	Element	Symbol	Percent by Mass
1	oxygen	O	49.2
2	silicon	Si	25.7
3	aluminum	Al	7.5
4	iron	Fe	4.7
5	calcium	Ca	3.4
6	sodium	Na	2.6
7	potassium	K	2.4
8	magnesium	Mg	1.9
9	hydrogen	H	0.9
10	titanium	Ti	0.6
11	chlorine	Cl	0.2
12	phosphorus	P	0.1
13	manganese	Mn	0.1
14	carbon	C	0.09
15	sulfur	S	0.05
—	all others	—	0.56

Whether an element finds wide commercial use depends not only upon its abundance but also upon its accessibility. Some familiar elements (such as copper, tin, and lead) are not particularly abundant but are found in nature in deposits of ores from which they can be obtained readily. Other elements that are more abundant (such as titanium, rubidium, and zirconium) are not widely used because either their ores are widespread in nature or the extraction of the elements from their ores is difficult or expensive.

Compounds are substances that are composed of two or more elements in fixed proportions. The law of definite proportions (first proposed by Joseph Proust in 1799) states that a pure compound always consists of the same elements combined in the same proportion by mass. The compound *water*, for example, is always formed from the elements *hydrogen* and *oxygen* in the proportion 11.19% hydrogen to 88.81% oxygen. Over ten thousand inorganic compounds are known, and over one million organic compounds have been either synthesized or isolated from natural sources. Compounds have properties that are different from the properties of the elements of which they are composed.

An element or a compound is called a **pure substance.** All other kinds of matter are mixtures. **Mixtures** consist of two or more pure substances and have variable compositions. The properties of a mixture depend upon the composition of the mixture and the properties of the pure substances that form the mixture. There are two types of mixtures. A **heterogeneous mixture** is not uniform throughout but consists of parts that are physically distinct. A sample containing iron and sand, for example, is a heterogeneous mixture. A **homogeneous mixture** appears uniform throughout and is usually called a **solution.** Air, salt dissolved in water, and a silver-gold alloy are examples of a gaseous, a liquid, and a solid solution, respectively.

The classification of matter is summarized in Figure 1.1. From the figure we see that the only type of *heterogeneous matter* is the heterogeneous mixture. The classification *homogeneous matter*, however, includes homogeneous mixtures and pure substances (elements and compounds).

A physically distinct portion of matter that is uniform throughout in composition and properties is called a **phase.** Homogeneous materials consist of only one phase. Heterogeneous materials consist of more than one phase. The phases of heterogeneous mixtures have distinct boundaries and are usually easily discernible.

In the heterogeneous mixture *granite*, for example, it is possible to identify pink feldspar crystals, colorless quartz crystals, and shiny black mica crystals. When the number of phases in a sample is being determined, all portions of the same kind are counted as a single phase. Granite, therefore, is said to consist of three phases. The relative proportions of the three phases of granite may vary from sample to sample.

Figure 1.1 notes that both types of mixtures can be separated into their components by *physical means*, but that compounds can be separated into their constituent elements only by *chemical means*. Changes in state (such as the melting of a solid and the vaporization of a liquid), as well as changes in shape or state of subdivision, are examples of **physical changes**—changes that do not involve the production of new chemical species. Physical means (such as filtration and distillation) may be used to separate the components of a mixture, but a substance that was not present in the original mixture is never produced by these means. **Chemical changes,** on the other hand, are transformations in which substances are converted into other substances.

1.3 The Metric System

The metric system of measurement is used in all scientific studies. As a result of a treaty signed in 1875, metric conventions are established and modified when necessary by international agreement. From time to time, an international group, the General Conference of Weights and Measures, meets to ratify improvements in the metric system. The currently approved **International System of Units** (*Le Système International d'Unitès*, officially abbreviated **SI**) is a modernization and simplification of an old system that developed from one proposed by the French

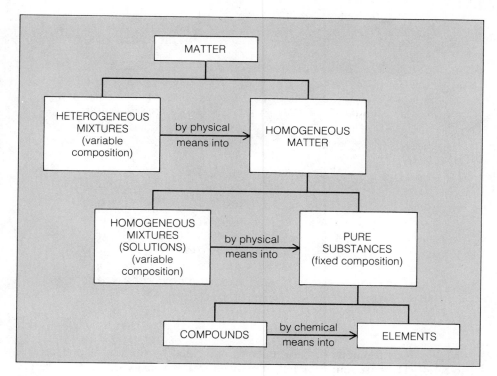

Figure 1.1 Classification of matter

Academy of Science in 1790. Lavoisier was a member of the committee that formulated the original system.

The International System is founded on seven **base units** and two **supplementary units** (see Table 1.3 and the appendix). The selection of primary standards for the base units is arbitrary. For example, the primary standard of mass, the kilogram, is defined as the mass of a cylinder of platinum-irridium alloy that is kept at the International Bureau of Weights and Measures at Sèvres, France. Throughout the years, the primary standards for some base units have been changed when new standards appeared to be superior to old ones.

Table 1.3	Base units and supplementary units of the International System of Units		
	Measurement	Unit	Symbol
Base units	length	meter	m
	mass	kilogram	kg
	time	second	s
	electric current	ampere	A
	temperature	kelvin	K
	amount of substance	mole	mol
	luminous intensity	candela	cd
Supplementary units	plane angle	radian	rad
	solid angle	steradian	sr

Table 1.4	Prefixes used to modify unit terms in the metric system		
Prefix	Abbreviation	Factor	
tera-	T-		$1\ 000\ 000\ 000\ 000\ \times$ or 10^{12}
giga-	G-		$1\ 000\ 000\ 000\ \times$ or 10^{9}
mega-	M-		$1\ 000\ 000\ \times$ or 10^{6}
kilo-	k-		$1\ 000\ \times$ or 10^{3}
hecto-	h-		$100\ \times$ or 10^{2}
deka-	da-		$10\ \times$ or 10
deci-	d-	$0.1\ \times$	or 10^{-1}
centi-	c-	$0.01\ \times$	or 10^{-2}
milli-	m-	$0.001\ \times$	or 10^{-3}
micro-	μ-	$0.000\ 001\ \times$	or 10^{-6}
nano-	n-	$0.000\ 000\ 001\ \times$	or 10^{-9}
pico-	p-	$0.000\ 000\ 000\ 001\ \times$	or 10^{-12}
femto-	f-	$0.000\ 000\ 000\ 000\ 001\ \times$	or 10^{-15}
atto-	a-	$0.000\ 000\ 000\ 000\ 000\ 001\ \times$ or 10^{-18}	

Multiples or fractions of base units are indicated by the use of prefixes (see Table 1.4). The base unit of length, the meter (m), is usually not used to record the distances between cities. A larger unit, the kilometer (km), is more convenient. The kilometer is equal to 1000 meters, and the name for this unit is obtained by adding the prefix *kilo-* (which means $1000 \times$) to the name of the base unit:

$$1 \text{ km} = 1000 \text{ m}$$

The centimeter (cm) is a smaller unit than the meter. The prefix *centi-* means $0.01 \times$, and 1 centimeter is 0.01 meter:

$$1 \text{ cm} = 0.01 \text{ m}$$

Note that the name for the base unit for mass, the kilogram, contains a prefix. The names of other units of mass are obtained by substituting other prefixes for the prefix *kilo-*. The name of no other base unit contains a prefix.

Other SI units, called **derived units,** are obtained from the base units by algebraic combination. Examples are the SI unit for volume, which is the cubic meter (abbreviated m^3) and the SI unit for velocity, which is the meter per second (abbreviated m/s or $m \cdot s^{-1}$).

Some derived units are given special names. The SI unit for force, for example, is the newton, N. This unit is derived from the base units for mass (the kilogram, kg), length (the meter, m), and time (the second, s). The **newton** is the force that gives a mass of 1 kg an acceleration of 1 m/s^2 (see Section 3.1):

$$1 \text{ N} = 1 \text{ kg} \cdot \text{m/s}^2$$

The current terminology of the International System has been developed since 1960. Some units that were defined prior to this time do not conform to SI rules and are not SI units. The use of some of these units is, however, permitted.

The liter, for example, which is defined as 1 cubic decimeter (and hence is 1000 cm^3), may be used in addition to the official SI unit of volume, the cubic meter. Certain other units that are not a part of SI are to be retained for a limited period of time. The standard atmosphere (atm, a unit of pressure) falls into this category. The use of still other units that are outside the International System is discouraged. For example, the International Committee of Weights and Measures considers it preferable to avoid the use of the calorie as an energy unit.

Not all scientists have adopted SI units, but use of the system appears to be growing. Strict adherence to the International System, however, poses a problem since it eliminates some units that previously have been used widely. Since much of the data found in the chemical literature has been recorded in units that are not SI units, one must be familiar with both the old and the new units.

1.4 Significant Figures

Every measurement is uncertain to some extent. Suppose, for example, that we wish to measure the mass of an object. If we use a platform balance, we can determine the mass to the nearest 0.1 g. An analytical balance, on the other hand, is capable of giving results correct to the nearest 0.0001 g. The exactness, or precision, of the measurement depends upon the limitations of the measuring device and the skill with which it is used.

The precision of a measurement is indicated by the number of figures used to record it. The digits in a properly recorded measurement are **significant figures.** These figures include all those that are known with certainty plus one more, which is an estimate.

Suppose that a platform balance is used and the mass of an object is determined to be 12.3 g. The chances are slight that the actual mass of the object is exactly 12.3 g, no more nor less. We are sure of the first two figures (the 1 and the 2); we known that the mass is greater than 12 g. The third figure (the 3), however, is somewhat inexact. At best, it tells us that the true mass lies closer to 12.3 g than to either 12.2 g or 12.4 g. If, for example, the actual mass were 12.28 . . . g or 12.33 . . . g, the value would be correctly recorded in either case as 12.3 g *to three significant figures.*

If, in our example, we add a zero to the measurement, we indicate a value containing *four significant figures* (12.30 g), which is incorrect and misleading. This value indicates that the actual mass is between 12.29 g and 12.31 g. We have, however, no idea of the magnitude of the integer of the second decimal place since we have determined the value only to the nearest 0.1 g. The zero does not indicate that the second decimal place is unknown or undetermined. Rather, it should be interpreted in the same way that any other figure is (see, however, rule 1 that follows). Since the uncertainty in the measurement lies in the 3, this digit should be the last significant figure reported.

On the other hand, we have no right to drop a zero if it is significant. A value of 12.0 g that has been determined to the precision indicated should be recorded that way. It is incorrect to record 12. g for this measurement since 12. g indicates a precision of only *two* significant figures instead of the *three* significant figures of the measurement.

The following rules can be used to determine the proper number of significant figures to be recorded for a measurement.

Modern analytical balances capable of giving results to the nearest 0.1 mg. *Left:* A mechanical single-pan balance. *Right:* An electronic, digital read-out balance that can be interfaced with other equipment. *Sauter Division of Mettler Instrument Corporation.*

1. *Zeros used to locate the decimal point are not significant.* Suppose that the distance between two points is measured as 3 cm. This measurement could also be expressed as 0.03 m since 1 cm is 0.01 m:

$$3 \text{ cm} = 0.03 \text{ m}$$

Both values, however, contain only *one* significant figure. The zeros in the second value, since they merely serve to locate the decimal point, are not significant. The precision of a measurement cannot be increased by changing units.

Zeros that arise as a part of a measurement are significant. The number 0.0005030 has four significant figures. The zeros after 5 are significant. Those preceding the numeral 5 are not significant since they have been added only to locate the decimal point.

Occasionally, it is difficult to interpret the number of significant figures in a value that contains zeros, such as 600. Are the zeros significant, or do they merely serve to locate the decimal point? This type of problem can be avoided by using scientific notation (see Appendix C.2). The decimal point is located by the power of 10 employed; the first part of the term contains only significant figures. The value 600, therefore, can be expressed in any of the following ways depending upon how precisely the measurement has been made:

6.00×10^2 (three significant figures)

6.0×10^2 (two significant figures)

6×10^2 (one significant figure)

2. *Certain values, such as those that arise from the definition of terms, are exact.* For example, by definition, there are *exactly* 1000 ml in 1 liter. The value 1000 may be considered to have an infinite number of significant figures (zeros) following the decimal point.

Values obtained by counting may also be exact. The H_2 molecule, for example, contains exactly 2 atoms, not 2.1 or 2.3. Other counts, however, are inexact. The population of the world, for example, is estimated and is not derived from an actual count.

3. *At times, the answer to a calculation contains more figures than are significant.* The following rules should be used to round off such a value to the correct number of digits.

    a. If the figure following the last number to be retained is less than 5, all the unwanted figures are discarded and the last number is left unchanged:

        3.6247 is 3.62 to three significant figures

    b. If the figure following the last number to be retained is greater than 5, or 5 with other digits following it, the last figure is increased by 1 and the unwanted figures are discarded:

        7.5647 is 7.565 to four significant figures

        6.2501 is 6.3 to two significant figures

    c. If the figure following the last figure to be retained is 5 and there are only zeros following the 5, the 5 is discarded and the last figure is increased by 1 if it is an odd number or left unchanged if it is an even number. In a case of this type, the last figure of the rounded-off value is always an even number. Zero is considered to be an even number:

        3.250 is 3.2 to two significant figures

        7.635 is 7.64 to three significant figures

        8.105 is 8.10 to three significant figures

The idea behind this procedure, which is arbitrary, is that on the average as many values will be increased as are decreased.

The number of significant figures in the answer to a calculation depends upon the numbers of significant figures in the values used in the calculation. Consider the following problem. If we place 2.38 g of salt in a container that has a mass of 52.2 g, what will be the mass of the container plus salt? Simple addition gives 54.58 g. But we cannot know the mass of the two together any more precisely than we know the mass of one alone. The result must be rounded off to the nearest 0.1 g, which gives 54.6 g.

4. *The result of an* addition *or* subtraction *should be reported to the same number of decimal places as that of the term with the least number of decimal places.* The answer for the addition

$$
\begin{array}{r}
161.032 \\
5.6 \\
32.4524 \\
\hline
199.0844
\end{array}
$$

should be reported as 199.1 since the number 5.6 has only one digit following the decimal point.

5. *The answer to a* multiplication *or* division *is rounded off to the same number of significant figures as is possessed by the least precise term used in the calculation.* The result of the multiplication

$$152.06 \times 0.24 = 36.4944$$

should be reported as 36 since the least precise term in the calculation is 0.24 (two significant figures).

1.5 Chemical Calculations

Units should be indicated as an integral part of all measurements. It makes little sense to say that the length of an object is 5.0. What does this value mean: 5.0 cm, 5.0 m, 5.0 ft? Careful use of units will simplify problem solving and reduce the probability of making errors.

The unit labels that are included in the terms employed in a calculation should undergo the same mathematical operations as the numbers. In any calculation, the units that appear in both numerator and denominator are canceled, and those remaining appear as a part of the answer. If the answer does not have the unit sought, a mistake has been made in the way that the calculation has been set up.

Many problems may be solved by the use of one or more **conversion factors.** A factor of this type is derived from an equality and is designed to convert a measurement from one unit into another. Suppose, for example, that we wish to calculate the number of centimeters in 5.00 inches (in.). The conversion factor that we need to solve the problem is derived from the exact relation

$$2.54 \text{ cm} = 1.00 \text{ in.}$$

If we divide both sides of this equation by 1.00 in., we obtain

$$\frac{2.54 \text{ cm}}{1.00 \text{ in.}} = 1$$

The factor (2.54 cm/1.00 in.) is equal to 1 since the numerator and the denominator are equivalent.

Our problem can be stated in the following way:

$$? \text{ cm} = 5.00 \text{ in.}$$

Multiplication by the conversion factor that we derived solves the problem

$$? \text{ cm} = 5.00 \text{ in.} \left(\frac{2.54 \text{ cm}}{1.00 \text{ in.}} \right) = 12.7 \text{ cm}$$

Since the factor is equal to 1, this operation does not change the value of the given quantity. Notice that the *inch* labels cancel, which leaves the answer in the desired unit, *centimeters.*

A second conversion factor can be derived from the relation

$$2.54 \text{ cm} = 1.00 \text{ in.}$$

by dividing both sides of the equation by 2.54 cm:

$$1 = \frac{1.00 \text{ in.}}{2.54 \text{ cm}}$$

This factor, which is also equal to 1, is the reciprocal of the factor previously derived and can be used to convert centimeters to inches. For example, the number of inches that equals 20.0 cm can be found in the following way:

$$? \text{ in.} = 20.0 \text{ cm} \left(\frac{1.00 \text{ in.}}{2.54 \text{ cm}} \right) = 7.87 \text{ in.}$$

A single equality that relates two units, therefore, can be used to derive two conversion factors. The factors are reciprocals of one another. In the solution to a problem, the correct factor to use is the one that will lead to the cancellation of the unit, or units, that must be eliminated. Notice that this unit should be found in the denominator of the factor.

The solution to some problems requires the use of several factors. If we wish to find the number of centimeters in 0.750 ft, we can state the problem in the following way:

$$? \text{ cm} = 0.750 \text{ ft}$$

Since 1.00 ft = 12.0 in., we derive the conversion factor (12.0 in./1.00 ft), which is of course equal to 1. Multiplication by this factor converts feet into inches but does not complete the solution:

$$? \text{ cm} = 0.750 \text{ ft} \; \frac{12.0 \text{ in.}}{1.00 \text{ ft}}$$

The factor needed to convert inches into centimeters is (2.54 cm/1.00 in.), and thus

$$? \text{ cm} = 0.750 \text{ ft} \; \frac{12.0 \text{ in.}}{1.00 \text{ ft}} \left(\frac{2.54 \text{ cm}}{1.00 \text{ in.}} \right) = 22.9 \text{ cm}$$

Example 1.1

If Jules Verne had used SI units, what title would he have given his book *Twenty Thousand Leagues under the Sea*? Express the answer in the SI unit that will give the smallest number that is greater than 1. One league is 3.45 miles; 1 mile is 1609 m.

Solution

First we convert leagues into meters. The conversion is accomplished by the use of two factors derived from the data given:

$$? \text{ m} = 20,000 \text{ league} \left(\frac{3.45 \text{ mile}}{1 \text{ league}} \right) \left(\frac{1609 \text{ m}}{1 \text{ mile}} \right) = 111,000,000 \text{ m} = 1.11 \times 10^8 \text{ m}$$

Notice that the factors successively convert leagues to miles and then miles to meters. A given factor converts the units in the *denominator* of the factor to the units in the *numerator* of the factor.

Next, we change the units of the answer from the base unit *meter* to the SI unit that will satisfy the requirement stated in the problem. From Table 1.4 we note that a megameter (Mm) is 10^6 meters and a gigameter (Gm) is 10^9 meters. The magnitude of our answer (10^8) is between the two. In order to get an answer that is greater than 1, we convert to megameters:

$$? \text{ Mm} = 1.11 \times 10^8 \text{ m} \left(\frac{1 \text{ Mm}}{10^6 \text{ m}} \right) = 1.11 \times 10^2 \text{ Mm} = 111 \text{ Mm}$$

or, *One Hundred and Eleven Megameters under the Sea.* (Notice that the radius of the earth is only 6.37 Mm!)

Factors can be derived from percentages. Consider, for example, the percentages used to express the composition of the alloy used to make the American five-cent piece. The "nickel" is actually 75.0% copper and 25.0% nickel, by mass. Six factors, counting reciprocals, can be derived from this data.

Since a *percentage* is the number of parts *per hundred*, it is convenient to use exactly one hundred mass units of the alloy in the derivation of the factors. In 100.0 g of alloy there would be 75.0 g of copper and 25.0 g of nickel. Thus,

1. 100.0 g alloy = 75.0 g Cu
2. 100.0 g alloy = 25.0 g Ni
3. 75.0 g Cu = 25.0 g Ni

Each of these relations will yield two factors—one the inverse of the other. The factor required for the solution of a problem can be derived from that relation that involves the pertinent units.

Example 1.2

How many grams of nickel must be added to 50.0 g of copper to make the coinage alloy previously described?

Solution

The problem is stated in the following way:

$$? \text{ g Ni} = 50.0 \text{ g Cu}$$

To solve the problem, we need a conversion factor that relates *g Cu* (in the denominator) to *g Ni* (in the numerator). Relation 3 given previously can yield such a factor; it is (25.0 g Ni/75.0 g Cu). The solution is

$$? \text{ g Ni} = 50.0 \text{ g Cu} \left(\frac{25.0 \text{ g Ni}}{75.0 \text{ g Cu}} \right) = 16.7 \text{ g Ni}$$

Example 1.3

Sterling silver is an alloy consisting of 92.5% Ag and 7.5% Cu. How many kilograms of sterling silver can be made from 3.00 kg of pure silver?

Solution

The problem is stated in the following way:

? kg sterling = 3.00 kg Ag

It is clear that we need a factor in which the unit in the denominator is *kg Ag*. We can derive the desired factor from the percentage of silver in sterling silver. Since sterling silver is 92.5% Ag by mass, 100.0 kg of sterling silver contains 92.5 kg of silver:

100.0 kg sterling = 92.5 kg Ag

The factor we need, therefore, is (100.0 kg sterling/92.5 kg Ag). Notice that the label *kg Ag* appears in the denominator of this factor and will cancel the unit of the given quantity:

$$? \text{ kg sterling} = 3.00 \text{ kg Ag} \left(\frac{100.0 \text{ kg sterling}}{92.5 \text{ kg Ag}} \right) = 3.24 \text{ kg sterling}$$

Frequently, information is given in the form of ratios. The cost per unit item, the distance traveled per unit time, and the number of items per unit mass are examples. The word *per* implies division, and the number in the denominator is 1 (exactly) unless specified otherwise. A speed of 50 kilometers per hour is 50 km/1 hr.

The numerator and denominator of such a ratio are equivalent:

50 km = 1 hr

These ratios may be used, therefore, as conversion factors—either in the form in which they are given (50 km/1 hr) or in the inverted form (1 hr/50 km).

If a ratio is desired as the answer to a problem, the calculation is set up by using the numerator of the ratio as the quantity desired and the denominator of the ratio as the quantity given. If, for example, we wish to find a rate of travel in *kilometers per hour*, we state the problem in the following way:

? km = 1 hr

Example 1.4

What is the speed of a car (in km/hr) if it travels 16 km in 13 min?

Solution

The problem is stated by writing

$? \text{ km} = 1 \text{ hr}$

The only information that is given in the problem and that can be used to derive a conversion factor is

$16 \text{ km} = 13 \text{ min}$

Since the answer is to be expressed in *km*, the factor that has this unit in the numerator is the one that should be used (16 km/13 min).

The unit in the denominator of the factor (min), however, will not cancel the unit of the given quantity (hr). This cancellation can be brought about by using another factor, one derived from

$60 \text{ min} = 1 \text{ hr}$

The factor derived from this equality is (60 min/1 hr) since the unit in the denominator (hr) will cancel the unit of the given quantity. Use of both factors gives

$$? \text{ km} = 1 \text{ hr} \left(\frac{60 \text{ min}}{1 \text{ hr}} \right) \left(\frac{16 \text{ km}}{13 \text{ min}} \right) = 74 \text{ km}$$

The car traveled at the rate of 74 km/hr.

Density is one type of ratio that is frequently used in chemistry. The **density** of a substance is the mass per unit volume of that substance:

$$\text{density} = \frac{\text{mass}}{\text{volume}}$$

Density is usually expressed in g/cm³. For pure liquids or solutions, the units usually employed are g/ml. Since one liter is 1000 cm³ and one liter contains 1000 ml, 1 cm³ equals 1 ml, and g/cm³ is equivalent to g/ml.

Example 1.5

The mass of the earth is 5.976×10^{24} kg and the volume of the earth is 1.083×10^{21} m³. What is the mean density of the earth?

Solution

The problem can be solved by finding the number of grams contained in one cubic centimeter:

$? \text{ g} = 1 \text{ cm}^3$

Since the volume is given in the problem in terms of *m³*, the *cm³* in our set-up must be changed to *m³*. By taking the third power of both sides of the equation

$1 \text{ cm} = 10^{-2} \text{ m}$

we derive

$$1 \text{ cm}^3 = 10^{-6} \text{ m}^3$$

which we use to convert the *cm³* units in our set-up to *m³*:

$$? \text{ g} = 1 \text{ cm}^3 \left(\frac{10^{-6} \text{ m}^3}{1 \text{ cm}^3} \right)$$

The problem states that

$$1.083 \times 10^{21} \text{ m}^3 = 5.976 \times 10^{24} \text{ kg}$$

and the factor we derive from this relation is used next to find the mass in *kg*:

$$? \text{ g} = 1 \text{ cm}^3 \left(\frac{10^{-6} \text{ m}^3}{1 \text{ cm}^3} \right) \left(\frac{5.976 \times 10^{24} \text{ kg}}{1.083 \times 10^{21} \text{ m}^3} \right)$$

The solution is completed by converting *kg* into *g*:

$$1 \text{ kg} = 1000 \text{ g} = 10^3 \text{ g}$$

$$? \text{ g} = 1 \text{ cm}^3 \left(\frac{10^{-6} \text{ m}^3}{1 \text{ cm}^3} \right) \left(\frac{5.976 \times 10^{24} \text{ kg}}{1.083 \times 10^{21} \text{ m}^3} \right) \left(\frac{10^3 \text{ g}}{1 \text{ kg}} \right) = 5.518 \text{ g}$$

The mean density of the earth is 5.518 g/cm³. (In comparison, the density of water is 1.00 g/cm³.)

Example 1.6

The mean density of the moon is 3.341 g/cm³ and the mass of the moon is 7.350×10^{25} g. What is the volume of the moon?

Solution

Density relates mass and volume. We are given the mass of the moon and asked to find the volume. We state the problem as

$$? \text{ cm}^3 = 7.350 \times 10^{25} \text{ g}$$

The factor that we use to solve the problem is the reciprocal of the density (1 cm³/3.341 g). In this way, the *g* units will cancel:

$$? \text{ cm}^3 = 7.350 \times 10^{25} \text{ g} \left(\frac{1 \text{ cm}^3}{3.341 \text{ g}} \right) = 2.200 \times 10^{25} \text{ cm}^{3*}$$

---

* Cancellation will not be indicated in future examples.

Summary

The topics that have been discussed in this chapter are

1. The development of chemistry from its roots in the practical arts of ancient civilizations and the theories of the ancient Greeks.

2. The classification of matter into pure substances (elements and compounds) and mixtures. Since chemistry is the study of the composition, properties, and transformations of matter, this classification is of central importance.

3. The assignment of chemical symbols to the elements.

4. The abundance of the elements.

5. The metric system of measurement.

6. The use of significant figures to indicate the precision of measurements.

7. A method of calculating that employs conversion factors.

8. Calculations involving percentages, rates, and densities.

Key Terms

Some of the more important terms introduced in this chapter are listed below. Definitions for terms not included in this list may be located in the text by use of the index.

Chemical symbol (Section 1.2)   A one- or two-letter abbreviation assigned by international agreement to each element.

Chemistry (Introduction)   The science that is concerned with the characterization, composition, and transformations of matter.

Compound (Section 1.2)   A pure substance that is composed of two or more elements in fixed proportions and that can be chemically decomposed into these elements.

Conversion factor (Section 1.5)   A ratio in which the numerator and denominator are equivalent quantities expressed in different units. A conversion factor is equal to 1 and is used in a calculation to convert the units of a measurement into other units.

Density (Section 1.5)   Mass per unit volume.

Element (Section 1.2)   A pure substance that cannot be decomposed into simpler substances.

Law of conservation of mass (Section 1.1)   There is no detectable change in mass during the course of a chemical reaction.

Law of definite proportions (Section 1.2)   A pure compound always consists of the same elements combined in the same proportions by mass.

Mass (Section 1.2)   A measure of quantity of matter.

Matter (Section 1.2)   Anything that occupies space and has mass.

Metric system (Section 1.3)   A decimal system of measurement that is used in all scientific studies.

Mixture (Section 1.2)   A sample of matter that consists of two or more pure substances, does not have a fixed composition, and may be decomposed into its components by physical means.

Phase (Section 1.2)   A physically distinct portion of matter that is uniform throughout in composition and properties.

SI unit (Section 1.3)   A unit that is used in the International System of Units (*Le Système International d'Unités*).

Significant figures (Section 1.4)   Digits in a measurement that indicate the precision of the measurement. These figures include all those that are known with certainty plus one more, which is an estimate.

Solution (Section 1.2)   A mixture of two or more pure substances that is uniform throughout (homogeneous).

Substance (Section 1.2)   An element or a compound. Substances have fixed compositions and properties.

Weight (Section 1.2)   The gravitational force of attraction exerted by the earth on a body.

Problems*

1.1 Compare and contrast: **(a)** law of conservation of mass, law of definite proportions; **(b)** mixture, compound; **(c)** heterogeneous mixture, homogeneous mixture; **(d)** physical changes, chemical changes; **(e)** organic chemistry, biochemistry.

1.2 Give the names of the elements for which the symbols are: **(a)** Sr, **(b)** Sb, **(c)** Al, **(d)** Au, **(e)** Ag, **(f)** Si, **(g)** Hg, **(h)** He,   Na, **(j)** Ne, **(k)** Ca, **(l)** Cd.

1.3 Give the symbols for the following elements: **(a)** tin, **(b)** titanium, **(c)** phosphorus, **(d)** potassium, **(e)** copper, **(f)** cobalt, **(g)** iron, **(h)** iodine, **(i)** chlorine, **(j)** chromium, **(k)** magnesium, **(l)** manganese, **(m)** lithium, **(n)** lead.

1.4 Determine the number of significant figures in each of the following: **(a)** 500.0, **(b)** 500, **(c)** 0.05, **(d)** 10.072, **(e)** 0.03040, **(f)** 6,000, **(g)** 6003, **(h)** 0.6542.

1.5 Perform the following calculations and report the answer to the proper number of significant figures: **(a)** $123.4 + 12.34 + 1.234$, **(b)** $123.4/12.34$, **(c)** $6.524 - 5.624$, **(d)** $5.0 + 0.005$, **(e)** $16.0 \times 18.75 \times 0.375$, **(f)** $1.0625/505$.

1.6 Perform the following calculations and record the answer to the proper number of significant figures: **(a)** $6.50 \times 10^{-7} - 5.603 \times 10^{-8}$, **(b)** $6.50 \times 10^7 + 5.603 \times 10^8$, **(c)** $(5.5 \times 10^{-5})^2$, **(d)** $(3.52 \times 25)/91.75$, **(e)** $13.6 + 0.03$, **(f)** $156.2/0.62$.

1.7 **(a)** How many centimeters are in 1 kilometer? **(b)** How many kilograms are in 1 milligram? **(c)** How many nanoseconds are in 10 milliseconds? **(d)** How many terameters are in 100 micrometers?

1.8 The liter is defined as 1 cubic decimeter. **(a)** How many liters are in 1 cubic meter? **(b)** How many cubic meters are in 1 liter?

1.9 The ångstrom unit (Å), which is defined as 10^{-10} m, is not an SI unit. **(a)** How many nanometers are equal to 1 Å? **(b)** How many picometers are equal to 1 Å? **(c)** The radius of the chlorine atom is 0.99 Å. What is the distance in nanometers and in picometers?

1.10 How many meters tall is a horse that stands 15 hands? One hand is 4.00 inches, and 1 inch is 2.54 cm.

1.11 One furlong is defined as one-eighth of a mile. How many kilometers are there in a six-furlong race? The following relations are exact: 1 inch = 2.54 cm, 12 inches = 1 foot, 5280 feet = 1 mile. Give answer to three significant figures.

1.12 A day in Venus is 1.01×10^7 seconds long. How many hours is this? How many earth days?

1.13 A tun consists of four hogsheads, one hogshead is 0.500 butt, one butt is 126 gallons, one gallon is 3.785 liters, one liter is 1.00 dm^3. How many m^3 are equal to 1.00 tun?

1.14 The mean distance of the earth from the sun is 1.496×10^8 km, which is defined as one astronomical unit (au). The mean radius of the orbit of the moon around the earth is 0.002570 au. The mean radius of the earth at the equator is 6378 km. The distance from the earth to the moon is the equivalent of how many trips around the circumference of the earth at the equator?

1.15 Convert the measurements of the following recipe (for plain layer cake) into milliliters: 1 cup sugar, $\frac{1}{4}$ cup butter, 2 eggs (which cannot be converted), 1 teaspoon vanilla, $1\frac{2}{3}$ cup flour, $\frac{1}{2}$ tablespoon baking powder, $\frac{1}{4}$ teaspoon salt, and $\frac{1}{2}$ cup milk. Given: 1 pint = 473.2 ml, 1 pint = 2 cups, 1 cup = 16 tablespoons, 1 tablespoon = 3 teaspoons.

---

* The more difficult problems are marked with asterisks. The appendix contains answers to color-keyed problems.

1.16 If the metric system is adopted for everyday use, fabric will be sold in meters (not yards), milk in liters (not quarts), and meat in kilograms (not pounds). Given the following: 1 inch = 2.54 cm, 1 pint = 473.2 ml, and 1 pound = 453.6 g, what percentage increase does each of the following changes represent? **(a)** fabric: 1.00 meter instead of 1.00 yard; **(b)** milk: 1.00 liter instead of 1.00 quart; **(c)** meat: 1.00 kilogram instead of 2.00 pound.

1.17 Pure gold is 24 karat (also written carat, but abbreviated K). **(a)** A ring is made of 14 K gold. What percentage of the metal is gold? **(b)** A gold alloy used in dental work contains 92% Au. How is this alloy rated in terms of karats of gold?

1.18 A form of white gold called platinum gold is 60.0% Au and 40.0% Pt. **(a)** How many grams of platinum must be used to make 0.500 kg of alloy? **(b)** How many grams of the alloy can be made from 0.500 kg of platinum?

1.19 How many grams of zinc must be used with 1.95 kg of copper to make a type of brass that is 35.0% Zn and 65.0% Cu?

1.20 A 200-pound person contains 3.6 g of iron. What is the percentage of iron in the body? One pound is 453.6 g.

*1.21 A white brass consists of 60.0% copper, 15.0% nickel, and 25.0% zinc. **(a)** How many grams of the alloy can be made from 1500 g of Cu, 360 g of Ni, and 500 g of Zn? **(b)** How many grams of each of the pure metals would be left over?

1.22 One knot is 1.1508 miles per hour. What is this speed in cm/s? (The following relations are exact: 1 inch = 2.54 cm, 12 inches = 1 foot, 5280 feet = 1 mile.)

1.23 The speed limit on a highway is 55 miles per hour. Convert this value to kilometers per hour. (The following relations are exact: 1 inch = 2.54 cm, 12 inches = 1 foot, 5280 feet = 1 mile.)

1.24 The equator of Mars is 2.13×10^4 km long. The rotational velocity of Mars about its axis is 0.240 km/s at the equator. How many hours constitute a day on Mars?

1.25 The earth rotates around its axis with a velocity such that a person standing on the equator is moving at a rate of 1039 miles per hour. **(a)** What is this speed in m/s? **(b)** Since it takes 24.0 hours for the earth to make a complete revolution, what is the circumference of the earth at the equator (in m)? **(c)** What is the radius of the earth at the equator (in m)?

1.26 Under certain conditions, the speed of a hydrogen molecule is 1.84×10^3 m/s. In one second, however, the molecule undergoes 1.40×10^{10} collisions with other molecules. On the average, what is the distance traveled between collisions **(a)** in meters? **(b)** in micrometers?

1.27 The density of diamond is 3.51 g/cm³. What is the volume of a 1.00 carat diamond? (One carat is 0.200 g.)

1.28 The density of diamond is 3.51 g/cm³ and of graphite is 2.22 g/cm³. Both substances are pure carbon. What volume would 10.0 g of carbon occupy **(a)** in the form of diamond? **(b)** in the form of graphite?

1.29 According to estimates, 1.0 g of seawater contains 4.0 pg of Au. If the total mass of the oceans is 1.6×10^{12} Tg, how many grams of gold are present in the oceans of the earth?

*1.30 A very long tube with a cross-sectional area of 1.00 cm² is filled with mercury to a height of 76.0 cm. At what height would water stand in this tube if it were filled with a mass of water equal to that of the mercury? (The density of mercury is 13.60 g/cm³ and the density of water is 1.00 g/cm³.)

*1.31 A quarter-pound stick of butter measures $1\frac{5}{16}$ inch by $1\frac{5}{16}$ inch by $4\frac{11}{16}$ inch. **(a)** What is the density of butter in g/cm³? **(b)** Will butter float or sink in water (density = 1.00 g/cm³)?

*1.32 The mass of Mars is 6.414×10^{14} Tg and the density of Mars is 3.966 g/cm³. What is the radius of Mars? Express the answer in the SI unit that will give the smallest number greater than 1.

*1.33 The density of Venus is 0.9524 times that of earth, and the volume of Venus is 0.8578 times that of earth. The mass of earth is 5.975×10^{27} g. What is the mass of Venus?

STOICHIOMETRY

Alfred North Whitehead, the philosopher and mathematician, wrote, "All science as it grows towards perfection becomes mathematical in its ideas." Modern chemistry began when Lavoisier and chemists of his time recognized the importance of careful measurement and began to ask questions that could be answered quantitatively. Stoichiometry (derived from the Greek *stoicheion*, meaning "element," and *metron*, meaning "to measure") is the branch of chemistry that deals with the quantitative relationships between elements and compounds in chemical reactions. The atomic theory of matter is basic to this study.

2.1 Dalton's Atomic Theory

Credit for the first atomic theory is usually given to the ancient Greeks, but this concept may have had its origins in even earlier civilizations. Two theories prevailed among the Greeks. Aristotle (fourth century B.C.) believed that matter is continuous and, hence, hypothetically can be divided endlessly into smaller and smaller particles. The atomic theory of Leucippus and Democritus (fifth century B.C.) held that the subdivision of matter would ultimately yield atoms, which could not be further divided. The word *atom* is derived from the Greek word *atomos*, which means "uncut" or "indivisible."

The theories of the ancient Greeks were based on abstract thought, not on planned experimentation. For approximately two thousand years the atomic theory remained mere speculation. The existence of atoms was accepted by Robert Boyle in his book *The Sceptical Chymist* (1661) and by Isaac Newton in his books *Principia* (1687) and *Opticks* (1704). John Dalton, however, proposed an atomic theory, which he developed in the years 1803 to 1808, that is a landmark in the history of chemistry.

Many scientists of the time believed that all matter consists of atoms, but Dalton went further. Dalton made the atomic theory quantitative by showing that it is possible to determine the relative masses of the atoms of different elements. The principal postulates of Dalton's theory are

1. Elements are composed of extremely small particles called atoms. All atoms of the same element are alike, and atoms of different elements are different.

2. The separation of atoms and the union of atoms occur in chemical reactions. In these reactions, no atom is created or destroyed, and no atom of one element is converted into an atom of another element.

John Dalton, 1766–1844.
Smithsonian Institution.

3. A chemical compound is the result of the combination of atoms of two or more elements in a simple numerical ratio.

Dalton believed that all atoms of a given element have equal atomic masses. Today we know that many elements consist of several types of atoms that differ in mass (see the discussion of isotopes in Section 4.6). All atoms of the same element, however, react chemically in the same way. We can, therefore, work with Dalton's theory by using an average mass for the atoms of each element. In most calculations, no mistake is made by proceeding as if an element consists of only one type of atom with an average mass.

Dalton derived the quantitative aspects of his theory from the laws of chemical change. His second postulate accounts for the **law of conservation of mass,** which states that there is no detectable change in mass during the course of a chemical reaction. Since chemical reactions consist of the separating and joining of atoms and since atoms are neither created nor destroyed in these processes, the total mass of all the materials entering into a chemical reaction must equal the total mass of all the products of the reaction.

The third postulate of Dalton's theory explains the **law of definite proportions,** which states that a pure compound always contains the same elements combined in the same proportions by mass. Since a given compound is the result of the combination of atoms of two or more elements in a fixed ratio, the proportions by mass of the elements present in the compound are also fixed.

On the basis of his theory, Dalton proposed a third law of chemical combination, the **law of multiple proportions.** This law states that when two elements, A and B, form more than one compound, the amounts of A that are combined in these compounds with a fixed amount of B are in a small whole-number ratio. This law follows from Dalton's view that the atoms in a compound are combined in a fixed proportion. For example, carbon and oxygen form two compounds: carbon dioxide and carbon monoxide. In carbon dioxide, *two* atoms of oxygen are combined with one atom of carbon, and in carbon monoxide, *one* atom of oxygen is combined with one atom of carbon. When the two compounds are compared, therefore, the masses of oxygen that combine with a fixed mass of carbon stand in the ratio of 2:1. The experimental verification of the law of multiple proportions was strong support for Dalton's theory.

2.2 Atomic Weights

A very important aspect of Dalton's work is his attempt to determine the relative masses of atoms. Water is a compound that consists of 88.8% oxygen and 11.2% hydrogen by mass. Dalton incorrectly assumed that one atom of oxygen is combined with one atom of hydrogen in water. On this basis, the mass of a single oxygen atom and the mass of a single hydrogen atom would stand in the ratio of 88.8 to 11.2, which is approximately 8 to 1. The arbitrary assignment of a mass of 1 to the hydrogen atom would give a relative mass of approximately 8 to the oxygen atom.*

The formulation that Dalton used for water is incorrect. Actually, one atom of oxygen is combined with *two* atoms of hydrogen. The oxygen atom, therefore, has a mass that is approximately 8 times the mass of *two* hydrogen atoms. If one

---

* Since the figures that Dalton used for the percent composition of water were very inaccurate, Dalton actually proposed a relative mass of 7 for the oxygen atom.

hydrogen atom is assigned a mass of 1, two hydrogen atoms would have a combined mass of 2. Thus on this scale, the oxygen atom has a relative mass of approximately 8 times 2, or 16.

The relative mass of the carbon atom can be derived in the following way. Carbon monoxide consists of approximately 4 parts by mass of oxygen and 3 parts by mass of carbon. One oxygen atom is combined with one carbon atom in this compound, and therefore the masses of the oxygen atom and the carbon atom must stand in the ratio of approximately 4 to 3. Based on the value of 16 that we derived for the oxygen atom, the approximate relative mass of the carbon atom is 12.

Notice that two types of information are needed to apply Dalton's method successfully: the combining ratio by mass of the elements in a compound and the combining ratio by numbers of atoms of these elements. The atomic combining ratios were not available to Dalton and the chemists of his time. Many years passed before a way around this difficulty was discovered and a correct set of relative atomic masses was developed (see Section 4.7). At the present time, they are determined by the use of instruments known as mass spectrometers (see Section 4.6), rather than by means of chemical analysis.

Even though Dalton made errors in assigning relative atomic masses, he must be given credit for introducing the concept and recognizing its importance. The relative masses of atoms serves as the cornerstone of chemical stoichiometry. These values are called **atomic weights,** a term that is not literally correct (since it refers to masses, not weights), but a term that is sanctioned by long usage.

Any relative atomic weight scale must be based on the arbitrary assignment of a value to one atom that is chosen as a standard. Dalton used the hydrogen atom as his standard and assigned it a value of 1. In later years, chemists used naturally occurring oxygen as the standard and set its atomic weight equal to exactly 16. Today, a particular type of carbon atom, the carbon-12 atom, is employed as the standard and assigned a mass of exactly 12 (see Section 4.7).

An element may occur in nature as a mixture of various types of atoms that have identical chemical properties but that differ slightly in mass. With very few exceptions, a mixture of this type has a constant composition. The atomic weight of such an element is an average value that takes into account the masses of these types of atoms and their relative abundances in nature (see Section 4.7).

Several types of carbon atoms occur in nature. The carbon-12 atom, which is employed as the standard for atomic weights, is the most abundant one. When the percentages and masses of all the types of carbon atoms are taken into account, the average relative mass for naturally occurring carbon is 12.011, which is recorded as the atomic weight of carbon. About three-quarters of the atomic weights of the elements are average values that take into account the several types of atoms that make up the element. The remainder give the relative mass of a single type of atom. A table of modern atomic weights appears inside the back cover of this book.

2.3 Formulas

The chemical symbols that are assigned to the elements (Section 1.2) are used to write formulas that describe the atomic composition of compounds. The formula of water is H_2O, which indicates that there are two atoms of hydrogen for every one atom of oxygen in the compound. The subscripts of the formula indicate the relative number of atoms of each type that are combined. If a symbol carries no

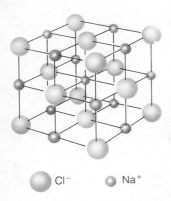

Cl^- Na^+

Figure 2.1 Sodium chloride crystal lattice

subscript, the number 1 is assumed. The formula of sulfuric acid is H_2SO_4, which indicates that the combining ratio of this compound is two atoms of hydrogen to one atom of sulfur to four atoms of oxygen.

A **molecule** is a particle formed from two or more atoms. Some (but not all) compounds occur in molecular form. In these cases, the formula gives the number of atoms of each type in a single molecule of the compound. Formulas of this type are sometimes called **molecular formulas.** Both water and sulfuric acid are molecular in nature, and both H_2O and H_2SO_4 are molecular formulas.

The molecular formula of hydrogen peroxide, H_2O_2, indicates that there are two atoms of hydrogen and two atoms of oxygen in a molecule of hydrogen peroxide. Notice that the ratio of hydrogen atoms to oxygen atoms (2 to 2) is not the *simplest* whole-number ratio (which is 1 to 1). A formula that is written using the simplest whole-number ratio is called the **simplest formula,** or **empirical formula.** The *molecular* formula of hydrogen peroxide is H_2O_2; the *empirical* formula is HO. At times the data at hand are sufficient to derive only an empirical formula.

For some molecular compounds, the molecular and empirical formulas are identical; examples are H_2O, H_2SO_4, CO_2, and NH_3. For many molecular compounds, however, the molecular and empirical formulas are different. The molecular formulas

$$N_2H_4 \qquad B_3N_3H_6 \qquad C_6H_6$$

correspond to the empirical formulas

$$NH_2 \qquad BNH_2 \qquad CH$$

Notice that the atomic ratio for an empirical formula can be obtained by reducing the atomic ratio of the molecular formula to the lowest possible set of whole numbers.

Some compounds are not composed of molecules. Sodium chloride, for example, is made up of **ions**—particles that bear positive or negative charges (Chapter 5). The sodium and chloride ions, which are derived from sodium and chlorine atoms, form a crystal. A sodium chloride crystal is composed of large numbers of these ions, which are held together by the attractions between the dissimilar charges of the ions (see Figure 2.1).

In the crystal, there is one sodium ion (Na^+) for every one chloride ion (Cl^-). The formula of the compound is NaCl. The formula does not describe a molecule nor does it indicate that the ions are paired, since no ion in the crystal can be considered as belonging exclusively to another. Rather, the formula gives the *simplest* ratio of atoms of each type required to produce the compound. The formula NaCl, therefore, is an empirical formula.

The formulas of ionic compounds are derived from the formulas of their ions. Since a crystal is electrically neutral, the total charge of the positive ions must equal the total charge of the negative ions. Consider barium chloride. The formula of the barium ion is Ba^{2+}, and the formula of the chloride ion is Cl^-. The two ions

$$Ba^{2+} \qquad and \qquad 2\,Cl^- \qquad form \qquad BaCl_2$$

The arrangement of ions in a $BaCl_2$ crystal is different from that in the NaCl crystal. The ions in each of these crystals are present in the ratio indicated by their formulas (*one* Ba^{2+} to *two* Cl^- in $BaCl_2$, and *one* Na^+ to *one* Cl^- in NaCl).

The formulas of ionic compounds such as $BaCl_2$ and $NaCl$ are empirical formulas. The formulas of a few ionic compounds, however, can be reduced to simpler terms. Sodium peroxide is such a compound. In sodium peroxide, two sodium ions (Na^+) are present for every one peroxide ion (O_2^{2-}):

$$2\,Na^+ \quad \text{and} \quad O_2^{2-} \quad \text{form} \quad Na_2O_2$$

This formula can be reduced to NaO, which is the *empirical* formula of sodium peroxide. What do we call the formula Na_2O_2? It seems odd to call Na_2O_2 a molecular formula since Na_2O_2 is an ionic compound and molecules of Na_2O_2 do not exist. For this reason, some chemists prefer to call all molecular formulas **true formulas.** The problem encountered with the formula of sodium peroxide is not common. The formulas of most ionic compounds are empirical formulas, and the atomic ratios they indicate cannot be reduced.

Formulas are also used to designate the atomic composition of the molecules of which some elements are composed. A number of elements occur in nature as **diatomic molecules**—molecules that contain two atoms joined together. These elements, together with their molecular formulas, are listed in Table 2.1. Some elements are composed of molecules that are formed from more than two atoms. Sulfur molecules, for example, consist of eight atoms and have the molecular formula S_8. The molecular formula of the phosphorus molecules is P_4. The *empirical formulas* of *all* elements, of course, consist of only the symbols for the elements.

Table 2.1 Elements that occur in nature as diatomic molecules	
Element	Formula
hydrogen	H_2
nitrogen	N_2
oxygen	O_2
fluorine	F_2
chlorine	Cl_2
bromine	Br_2
iodine	I_2

2.4 The Mole

The atomic weight of fluorine is 19.0 and of hydrogen is 1.0, which means that an atom of fluorine is 19 times heavier than an atom of hydrogen.* If we take 100 fluorine atoms and 100 hydrogen atoms, the mass of the collection of fluorine atoms will be 19 times the mass of the collection of hydrogen atoms. The masses of any two samples of fluorine and hydrogen that contain the same number of atoms will stand in the ratio of 19.0 to 1.0, which is the ratio of their atomic weights.

Now suppose that we take 19.0 g of fluorine and 1.0 g of hydrogen, which are values in grams numerically equal to the atomic weights of the elements. Since the masses of the samples stand in the ratio of 19.0 to 1.0, the samples must contain the same number of atoms. In fact, a sample of any element that has a mass in grams numerically equal to the atomic weight of the element will contain this same number of atoms.

This number is called **Avogadro's number,** named in honor of Amedeo Avogadro, who first interpreted the behavior of gases in chemical reactions in terms of the number of reacting molecules (see Section 8.8). The value of Avogadro's number has been experimentally determined; to six significant figures it is 6.02205×10^{23}. The amount of a substance that contains Avogadro's number of elementary units is called a **mole** (abbreviated mol), which is an SI base unit. The mole is defined as the amount of substance that contains as many elementary entities as there are atoms in exactly 12 g of carbon-12.

---

* In order to simplify the discussion, we have rounded off these atomic weights to the first figure following the decimal point.

Thus, a sample of an element that has a mass in grams numerically equal to the atomic weight of the element is a mole of atoms of the element and contains Avogadro's number of atoms. The atomic weight of beryllium, for example, is 9.01218. A 9.01218 g sample of beryllium is a mole of beryllium atoms and contains 6.02205×10^{23} beryllium atoms (Avogadro's number).

The atomic weights used to solve a problem should be expressed to the proper number of significant figures. The data given in the statement of a problem determine how precise the answer to the problem should be. The atomic weights used in the solution should be expressed to the number of significant figures that reflects this precision.

Example 2.1

How many moles of aluminum are there in 125 g of Al?

Solution

Notice that the answer should be expressed to three significant figures. First, we state the problem in the following way:

? mol Al = 125 g Al

Next, we derive a conversion factor to solve the problem. To three significant figures the atomic weight of Al is 27.0; therefore,

1 mol Al = 27.0 g Al

We use the conversion factor that has the unit *g Al* in the denominator since this is the unit that must be eliminated:

$$? \text{ mol Al} = 125 \text{ g Al} \left(\frac{1 \text{ mol Al}}{27.0 \text{ g Al}} \right) = 4.63 \text{ mol Al}$$

Example 2.2

How many grams of gold constitute 0.2500 mol of Au?

Solution

The answer should be expressed to four significant figures. We state the problem:

? g Au = 0.2500 mol Au

To four significant figures the atomic weight of Au is 197.0; therefore,

1 mol Au = 197.0 g Au

The unit that must be eliminated in this problem is *mol Au*, and the conversion factor must have this unit in the denominator:

$$? \text{ g Au} = 0.2500 \text{ mol Au} \left(\frac{197.0 \text{ g Au}}{1 \text{ mol Au}} \right) = 49.25 \text{ g Au}$$

Example 2.3

How many carbon atoms are there in a 1.000 carat diamond? Diamond is pure carbon and one carat is exactly 0.2 g.

Solution

Since the value 0.2 g, which arises from the definition of the carat, is exact, it does not limit the number of significant figures in the answer. The precision of the answer is limited by the value 1.000 carat (four significant figures). The problem is

$$? \text{ atoms C} = 0.2000 \text{ g C}$$

We derive a conversion factor from the atomic weight of C (to four significant figures):

$$1 \text{ mol C} = 12.01 \text{ g C}$$

with the unit g C in the denominator so that this unit will cancel:

$$? \text{ atoms C} = 0.2000 \text{ g C} \left(\frac{1 \text{ mol C}}{12.01 \text{ g C}} \right)$$

At this point, the calculation would give an answer expressed in mol C. Using Avogadro's number (to four significant figures), we derive a conversion factor from

$$1 \text{ mol C} = 6.022 \times 10^{23} \text{ atoms C}$$

with the unit *mol C* in the denominator so that it will cancel. Multiplication by this factor completes the solution:

$$? \text{ atoms C} = 0.2000 \text{ g C} \left(\frac{1 \text{ mol C}}{12.01 \text{ g C}} \right) \left(\frac{6.022 \times 10^{23} \text{ atoms C}}{1 \text{ mol C}} \right)$$
$$= 1.003 \times 10^{22} \text{ atoms C}$$

What about compounds? The **formula weight** of a substance is the sum of the atomic weights of all the atoms in the formula of the substance. The formula weight of H_2O, for example, can be calculated as follows:

$$
\begin{aligned}
2(\text{atomic weight H}) = 2(1.0) &= 2.0 \\
\text{atomic weight O} &= 16.0 \\
\hline
\text{formula weight } H_2O &= 18.0
\end{aligned}
$$

The formula weight of $BaCl_2$ is

$$
\begin{aligned}
\text{atomic weight Ba} &= 137.3 \\
2(\text{atomic weight Cl}) = 2(35.5) &= 71.0 \\
\hline
\text{formula weight } BaCl_2 &= 208.3
\end{aligned}
$$

If the formula in question pertains to a molecular substance and is a molecular formula, the corresponding formula weight may also be called a molecular weight.

A **molecular weight** is the sum of the atomic weights of the atoms that constitute a *molecule*. The formula weight of H_2O is also the molecular weight of the substance since the formula is a description of the composition of the water molecule. In the case of $BaCl_2$, however, the formula weight is *not* a molecular weight since $BaCl_2$ is an ionic compound and molecules of $BaCl_2$ do not exist. How to distinguish between molecular and ionic substances is a topic that will be discussed in later chapters.

A mole consists of Avogadro's number of entities. A mole of a molecular substance consists of Avogadro's number of molecules. For these substances, a sample that has a mass in grams numerically equal to the molecular weight is a mole of the substance and contains Avogadro's number of *molecules*. A mole of H_2O, therefore, has a mass of 18.0 g and contains 6.02×10^{23} H_2O molecules. Since there are two atoms of H and one atom of O in a molecule of H_2O, a mole of H_2O molecules (18.0 g) contains two moles of H atoms (2.0 g) and one mole of O atoms (16.0 g).

When the mole designation is used, the type of entity being measured *must* be specified. A mole of H *atoms* contains 6.02×10^{23} H atoms and to three significant figures has a mass of 1.01 g; a mole of H_2 *molecules* contains 6.02×10^{23} H_2 molecules and has a mass of 2.02 g. For fluorine,

$$1 \text{ mol F} = 6.02 \times 10^{23} \text{ F atoms} = 19.0 \text{ g fluorine}$$

$$1 \text{ mol F}_2 = 6.02 \times 10^{23} \text{ F}_2 \text{ molecules} = 38.0 \text{ g fluorine}$$

What about ionic substances? The designation "1 mol $BaCl_2$" means that the sample contains Avogadro's number of $BaCl_2$ formula units—the entity specified. One mole of $BaCl_2$, therefore, has a mass of 208.3 g, the formula weight of $BaCl_2$. In reality, one mole of $BaCl_2$ contains

$$1 \text{ mol Ba}^{2+} = 6.02 \times 10^{23} \text{ Ba}^{2+} \text{ ions} = 137.3 \text{ g barium*}$$

$$2 \text{ mol Cl}^- = 2(6.02 \times 10^{23}) \text{ Cl}^- \text{ ions} = 2(35.5) \text{ g Cl}^- = 71.0 \text{ g chlorine}$$

which together make up

$$1 \text{ mol BaCl}_2 = 6.02 \times 10^{23} \text{ BaCl}_2 \text{ units} = 208.3 \text{ g BaCl}_2$$

2.5 Derivation of Formulas

Data from the chemical analysis of a compound are used to derive the empirical formula of the compound. The analysis gives the proportions by mass of the elements that make up the compound. The simplest or empirical formula indicates the atomic proportions of the compound—the relative numbers of atoms of various types that make up the compound.

Since a mole of atoms of one element contains the same number of atoms as a mole of atoms of any other element, the ratio by moles is the same as the ratio by atoms. The number of moles of each element present in a sample of the compound is readily obtained from the mass of each element present. The simplest

---

* For all practical purposes, an ion has the same mass as the atom from which it is derived.

1. If the data are given in terms of percentage composition, base the calculation on a 100.0 g sample of the compound. In this instance, the number of grams of each element present in the sample will be numerically equal to the percentage of that element present in the compound. There is no need to find percentages if the data are given in terms of the number of grams of each element present in a sample.

2. Convert the number of grams of each element present in the sample to the number of moles of *atoms* of each element. The conversion factors needed are derived from the fact that 1 mol of *atoms* of an element (numerator) is an *atomic* weight in grams (denominator).

3. Divide each of the values obtained in step 2 by the smallest value. If every number obtained in this way is not a whole number multiply *each number* by the same simple integer in such a way that whole numbers will result.

4. A ratio by moles of atoms is the same as a ratio by atoms. The whole numbers obtained in step 3 are the subscripts of the empirical formula.

whole-number ratio by moles (which is the same as the ratio by atoms) is used to write the empirical formula. The procedure is illustrated in the examples that follow.

Example 2.4

What is the empirical formula of a compound that contains 43.6% P and 56.4% O?

Solution

We assume, for convenience, that we have a sample that has a mass of 100.0 g. On the basis of the percentage composition, this sample contains 43.6 g P and 56.4 g O.

Next, we use the method illustrated in Example 2.1 to find the number of moles of P atoms and O atoms in these quantities. To three significant figures, the atomic weight of P is 31.0 and of O is 16.0:

$$? \text{ mol P} = 43.6 \text{ g P} \left(\frac{1 \text{ mol P}}{31.0 \text{ g P}} \right) = 1.41 \text{ mol P}$$

$$? \text{ mol O} = 56.4 \text{ g O} \left(\frac{1 \text{ mol O}}{16.0 \text{ g O}} \right) = 3.53 \text{ mol O}$$

The ratio by atoms is the same as the ratio by moles of atoms. There are, therefore, 1.41 atoms of P for every 3.53 atoms of O in the compound. We need, however, the simplest *whole-number* ratio in order to write the formula. By dividing the two values by the smaller value, we get

$$\text{for P,} \frac{1.41}{1.41} = 1.00 \qquad \text{for O,} \frac{3.53}{1.41} = 2.50$$

We still do not have a whole-number ratio, but we can get one by multiplying each of these values by 2. Hence, the simplest whole-number ratio is 2 to 5, and the empirical formula is P_2O_5.

Example 2.5

Caffeine, which occurs in coffee, tea, and kola nuts, is a stimulant for the central nervous system. A 1.261 g sample of pure caffeine contains 0.624 g C, 0.065 g H, 0.364 g N, and 0.208 g O. What is the empirical formula of caffeine?

Solution

The results of a chemical analysis are usually reported in terms of percentages. However, any mass ratio can be converted into a mole ratio and, in this form, used to derive an empirical formula. There is no need to convert the data given in this example into percentages. We calculate the number of moles of each element present in the sample:

$$? \text{ mol C} = 0.624 \text{ g C} \left(\frac{1 \text{ mol C}}{12.0 \text{ g C}} \right) = 0.0520 \text{ mol C}$$

$$? \text{ mol H} = 0.065 \text{ g H} \left(\frac{1 \text{ mol H}}{1.0 \text{ g H}} \right) = 0.065 \text{ mol H}$$

$$? \text{ mol N} = 0.364 \text{ g N} \left(\frac{1 \text{ mol N}}{14.0 \text{ g N}} \right) = 0.0260 \text{ mol N}$$

$$? \text{ mol O} = 0.208 \text{ g O} \left(\frac{1 \text{ mol O}}{16.0 \text{ g O}} \right) = 0.0130 \text{ mol O}$$

Division of each of these values by the smallest value (0.0130) gives the ratio

$$4 \text{ mol C}: \quad 5 \text{ mol H}: \quad 2 \text{ mol N}: \quad 1 \text{ mol O}$$

and the empirical formula of caffeine, therefore, is $C_4H_5N_2O$.

The molecular formula of a compound can be derived from the empirical formula if the molecular weight of the compound is known.

Example 2.6

What is the molecular formula of the oxide of phosphorus that has the empirical formula P_2O_5 (derived in Example 2.4) if the molecular weight of this compound is 284?

Solution

The value obtained by adding the atomic weights indicated by the empirical formula P_2O_5 is 142. If we divide this formula weight into the actual molecular weight, we get

$$\frac{284}{142} = 2$$

There are, therefore, twice as many atoms of each kind present in a molecule as is indicated by the empirical formula. The molecular formula is P_4O_{10}.

Example 2.7

The molecular weight of caffeine is 194 and the empirical formula of caffeine is $C_4H_5N_2O$. What is the molecular formula of caffeine?

Solution

The formula weight indicated by $C_4H_5N_2O$ is 97. Since the molecular weight is twice this value, the molecular formula of caffeine is $C_8H_{10}N_4O_2$.

Example 2.8

Glucose, a simple sugar, is a constituent of human blood and tissue fluids and is utilized by cells as a principal source of energy. The compound contains 40.0% C, 6.73% H, and 53.3% O and has a molecular weight of 180.2. What is the molecular formula of glucose?

Solution

The most convenient way to solve the problem is to calculate the number of moles of each element present in a mole of glucose. We first determine the number of grams of each element in a mole of glucose (180.2 g). Since the compound contains 40.0% C, there are 40.0 g of C in 100 g of glucose, and we use the factor (40.0 g C/100 g glucose):

$$? \text{ g C} = 1 \text{ mol glucose} \left(\frac{180.2 \text{ g glucose}}{1 \text{ mol glucose}}\right)\left(\frac{40.0 \text{ g C}}{100 \text{ g glucose}}\right) = 72.1 \text{ g C}$$

In like manner, the number of grams of H and of O can be found:

$$? \text{ g H} = 1 \text{ mol glucose} \left(\frac{180.2 \text{ g glucose}}{1 \text{ mol glucose}}\right)\left(\frac{6.73 \text{ g H}}{100 \text{ g glucose}}\right) = 12.1 \text{ g H}$$

$$? \text{ g O} = 1 \text{ mol glucose} \left(\frac{180.2 \text{ g glucose}}{1 \text{ mol glucose}}\right)\left(\frac{53.3 \text{ g O}}{100 \text{ g glucose}}\right) = 96.0 \text{ g O}$$

Next, we determine the number of moles of atoms that each of these values represents:

$$? \text{ mol C} = 72.1 \text{ g C} \left(\frac{1 \text{ mol C}}{12.0 \text{ g C}} \right) = 6.00 \text{ mol C}$$

$$? \text{ mol H} = 12.1 \text{ g H} \left(\frac{1 \text{ mol H}}{1.01 \text{ g H}} \right) = 12.0 \text{ mol H}$$

$$? \text{ mol O} = 96.0 \text{ g O} \left(\frac{1 \text{ mol O}}{16.0 \text{ g O}} \right) = 6.00 \text{ mol O}$$

These values are the number of moles of atoms of each element present in a mole of glucose molecules. They are also the number of atoms of each type present in a molecule of glucose. The molecular formula, therefore, is $C_6H_{12}O_6$.

The problem may also be solved by first determining the empirical formula from the analytical data (it is CH_2O) and then using the molecular weight to derive the molecular formula.

2.6 Percentage Composition of Compounds

The percentage composition of a compound is readily calculated from the formula of the compound. The subscripts of the formula give the number of moles of each element in a mole of the compound. From this information and from the atomic weights of the elements, we can obtain the number of grams of each element contained in a mole of the compound. The percentage of a given element is 100 times the mass of the element divided by the mass of a mole of the compound. The process is illustrated in Example 2.9.

Example 2.9

What is the percentage of Fe in Fe_2O_3 calculated to four significant figures?

Solution

One mole of Fe_2O_3 contains

$$\begin{array}{ll} 2 \text{ mol Fe} = 2(55.8) \text{ g Fe} = & 111.6 \text{ g Fe} \\ 3 \text{ mol O} = 3(16.0) \text{ g O} = & \underline{48.0 \text{ g O}} \\ & 159.6 \text{ g} \end{array}$$

The sum of the masses, 159.6 g, is the mass of one mole of Fe_2O_3. The percentage of Fe in Fe_2O_3 is

$$\frac{111.6 \text{ g Fe}}{159.6 \text{ g Fe}_2\text{O}_3} \times 100\% = 69.92\% \text{ Fe in Fe}_2\text{O}_3$$

The percentage composition of a compound is frequently determined by chemical analysis. These data can then be used to find the empirical formula of the compound. Example 2.10 is an illustration of a method used for the analysis of organic compounds.

Example 2.10

Nicotine is a compound that contains carbon, hydrogen, and nitrogen. If a 2.50 g sample of nicotine is burned in oxygen, 6.78 g of CO_2, 1.94 g of H_2O, and 0.432 g of N_2 are the products of the combustion. What is the percentage composition of nicotine?

Solution

Note that we will work to three significant figures. We first calculate the quantity of each element present in the 2.50 g sample of nicotine. The carbon in the sample formed 6.78 g of CO_2. We ask, therefore,

$$? \text{ g C} = 6.78 \text{ g CO}_2$$

The conversion factor that we use to solve this problem is the fraction that we would use to find the percentage of C in CO_2. Since 1 mol CO_2 (44.0 g CO_2) contains 1 mol C (12.0 g C),

$$12.0 \text{ g C} = 44.0 \text{ g CO}_2$$

We derive the conversion factor (12.0 g C/44.0 g CO_2):

$$? \text{ g C} = 6.78 \text{ g CO}_2 \left(\frac{12.0 \text{ g C}}{44.0 \text{ g CO}_2} \right) = 1.85 \text{ g C}$$

The same procedure is used to find the number of grams of hydrogen in the sample of nicotine. The hydrogen of the nicotine formed 1.94 g H_2O. In 1 mol of H_2O (18.0 g) there are 2 mol of H atoms (2.02 g). Therefore,

$$? \text{ g H} = 1.94 \text{ g H}_2O \left(\frac{2.02 \text{ g H}}{18.0 \text{ g H}_2O} \right) = 0.218 \text{ g H}$$

In a combustion such as the one described, the nitrogen does not combine with oxygen but is evolved as N_2. Hence, the sample contained 0.432 g N.

The quantity of each element present in the 2.50 g sample is used to determine the percentage composition of nicotine:

$$\frac{1.85 \text{ g C}}{2.50 \text{ g nicotine}} \times 100\% = 74.0\% \text{ C in nicotine}$$

$$\frac{0.218 \text{ g H}}{2.50 \text{ g nicotine}} \times 100\% = 8.72\% \text{ H in nicotine}$$

$$\frac{0.432 \text{ g N}}{2.50 \text{ g nicotine}} \times 100\% = 17.3\% \text{ N in nicotine}$$

These data can be used to find the empirical formula of nicotine, which is C_5H_7N.

Some simple stoichiometric problems can be solved by use of the proportions derived from formulas.

Example 2.11

Silver sulfide, Ag_2S, occurs as the mineral argentite, which is an ore of silver. How many grams of silver are theoretically obtainable from 250.0 g of an impure ore that is 70.00% Ag_2S?

Solution

The problem can be stated as follows:

? g Ag = 250.0 g ore

If we take 100 g of the ore, we will get 70.00 g Ag_2S since the ore is 70.00% Ag_2S. Notice that the number 100 is exact (it arises from the definition of percent); the number 70.00 is not. Therefore,

70.00 g Ag_2S = 100 g ore

and the factor (70.00 g Ag_2S/100 g ore) can be derived:

$$? \text{ g Ag} = 250.0 \text{ g ore} \left(\frac{70.00 \text{ g Ag}_2\text{S}}{100 \text{ g ore}} \right)$$

The *g ore* labels cancel, and at this point we have an answer in *g Ag_2S.*

The solution to the problem can be completed by use of the same factor that would be used to find the percentage of Ag in Ag_2S. From the formula Ag_2S, we derive

2 mol Ag = 1 mol Ag_2S

2(107.9) g Ag = 247.9 g Ag_2S

215.8 g Ag = 247.9 g Ag_2S

Therefore,

$$? \text{ g Ag} = 250.0 \text{ g ore} \left(\frac{70.00 \text{ g Ag}_2\text{S}}{100 \text{ g ore}} \right) \left(\frac{215.8 \text{ g Ag}}{247.9 \text{ g Ag}_2\text{S}} \right) = 152.3 \text{ g Ag}$$

2.7 Chemical Equations

Chemical equations are representations of reactions in terms of the symbols and formulas of the elements and compounds involved. The reactants are indicated on the left and the products on the right. An arrow is used instead of the customary equal sign of the algebraic equation; it may be considered as an abbreviation for the word *yields.*

Chemical equations report the results of experimentation. One of the goals of chemistry is the discovery and development of principles that make it possible to predict the products of chemical reactions; careful attention will be given to any such generalizations. All too often, however, the products of a particular set of reactants must be memorized, and any prediction is subject to modification if experiment dictates. What may appear reasonable on paper is not necessarily what occurs in the laboratory.

The first step in writing a chemical equation is to ascertain the products of the reaction in question. Carbon disulfide, CS_2, reacts with chlorine, Cl_2, to produce carbon tetrachloride, CCl_4, and disulfur dichloride, S_2Cl_2. To represent this, we write

$$CS_2 + Cl_2 \longrightarrow CCl_4 + S_2Cl_2$$

This equation is not quantitatively correct because it violates the law of conservation of mass. There must be as many atoms of each element, combined or uncombined, indicated on the left side of an equation as there are on the right side. Although one carbon atom and two sulfur atoms are indicated on both the left side and the right side of the equation, two chlorine atoms (one Cl_2 molecule) appear on the left and six chlorine atoms appear on the right. The equation can be balanced* by indicating that three molecules of chlorine should be used for the reaction. Thus,

$$CS_2 + 3\,Cl_2 \longrightarrow CCl_4 + S_2Cl_2$$

The simplest types of chemical equations are balanced by trial and error, as the following examples will illustrate. When steam is passed over hot iron, hydrogen gas and an oxide of iron that has the formula Fe_3O_4 are produced. Thus,

$$Fe + H_2O \longrightarrow Fe_3O_4 + H_2$$

It is tempting to substitute another oxide of iron, FeO, for Fe_3O_4, since this would immediately produce a balanced equation. Such an equation, however, would be without value; experiment indicates that Fe_3O_4, not FeO, is a product of the reaction. Balancing an equation is never accomplished by altering the formulas of the products of the reaction. In the equation for the reaction of iron and steam, three atoms of Fe and four molecules of H_2O are needed to provide the iron and oxygen atoms required for the formation of Fe_3O_4:

$$3\,Fe + 4\,H_2O \longrightarrow Fe_3O_4 + H_2$$

The equation is now balanced except for hydrogen, which may be balanced as follows:

$$3\,Fe + 4\,H_2O \longrightarrow Fe_3O_4 + 4\,H_2$$

In much the same manner we can balance an equation for the complete combustion of ethane (C_2H_6) in oxygen. The products of this reaction are carbon dioxide and water:

$$C_2H_6 + O_2 \longrightarrow CO_2 + H_2O$$

---

* Strictly speaking, the expression is not an "equation" until it is balanced.

To balance the two carbon atoms of C_2H_6, the production of two molecules of CO_2 must be indicated, and the six hydrogen atoms of C_2H_6 require that three molecules of H_2O be produced:

$$C_2H_6 + O_2 \longrightarrow 2\,CO_2 + 3\,H_2O$$

Only the oxygen remains unbalanced; there are seven atoms of oxygen on the right and only two on the left. In order to get seven atoms of oxygen on the left, we would have to take $3\frac{1}{2}$, or $\frac{7}{2}$, molecules of O_2:

$$C_2H_6 + \tfrac{7}{2}O_2 \longrightarrow 2\,CO_2 + 3\,H_2O$$

Customarily, equations are written with whole number coefficients. By multiplying the entire equation by two, we get

$$2\,C_2H_6 + 7\,O_2 \longrightarrow 4\,CO_2 + 6\,H_2O$$

2.8 Problems Based on Chemical Equations

A chemical equation can be interpreted in several different ways. Consider, for example, the equation

$$2\,H_2 + O_2 \longrightarrow 2\,H_2O$$

On the simplest level this equation shows that hydrogen reacts with oxygen to produce water. On the atomic–molecular level, it states that

$$2 \text{ molecules } H_2 + 1 \text{ molecule } O_2 \longrightarrow 2 \text{ molecules } H_2O$$

It can also be read as

$$2 \text{ mol } H_2 + 1 \text{ mol } O_2 \longrightarrow 2 \text{ mol } H_2O$$

since 1 mol of H_2, 1 mol of O_2, and 1 mol of H_2O all contain the same number of molecules (Avogadro's number).

The last interpretation is the one that enables us to solve stoichiometric problems. The coefficients of the chemical equation give the ratios, by moles, in which the substances react and are produced. Since 2 mol of H_2 react with 1 mol of O_2, 10 mol of H_2 would require 5 mol of O_2 for reaction. Since 2 mol of H_2 produce 2 mol of H_2O, the reaction of 10 mol of H_2 would produce 10 mol of H_2O.

Example 2.12

Determine the number of moles of O_2 that are required to react with 5.00 mol of C_2H_6 according to the equation

$$2\,C_2H_6 + 7\,O_2 \longrightarrow 4\,CO_2 + 6\,H_2O$$

Solution

The problem can be stated as follows:

$$? \text{ mol } O_2 = 5.00 \text{ mol } C_2H_6$$

The stoichiometric relationship derived from the coefficients of the chemical equation is

$$2 \text{ mol } C_2H_6 = 7 \text{ mol } O_2$$

From this relationship, we can derive the conversion factor that we need to solve the problem. Since this ratio must have the units mol C_2H_6 in the denominator, it is (7 mol O_2/2 mol C_2H_6). The solution is

$$? \text{ mol } O_2 = 5.00 \text{ mol } C_2H_6 \ \frac{7 \text{ mol } O_2}{2 \text{ mol } C_2H_6} = 17.5 \text{ mol } O_2$$

Example 2.13

Chlorine can be prepared by the reaction

$$MnO_2 + 4\,HCl \longrightarrow MnCl_2 + Cl_2 + 2\,H_2O$$

(a) How many grams of HCl are required to react with 25.0 g of MnO_2? (b) How many grams of Cl_2 are produced by the reaction?

Solution

The problem is

$$? \text{ g HCl} = 25.0 \text{ g } MnO_2$$

The stoichiometric ratio to be derived from the chemical equation will be expressed in moles. Therefore, we convert *g MnO_2* into *mol MnO_2*. The molecular weight of MnO_2 is 86.9:

$$? \text{ g HCl} = 25.0 \text{ g } MnO_2 \left(\frac{1 \text{ mol } MnO_2}{86.9 \text{ g } MnO_2} \right)$$

The chemical equation gives the relation

$$1 \text{ mol } MnO_2 = 4 \text{ mol HCl}$$

from which we derive the conversion factor (4 mol HCl/1 mol MnO_2):

$$? \text{ g HCl} = 25.0 \text{ g } MnO_2 \left(\frac{1 \text{ mol } MnO_2}{86.9 \text{ g } MnO_2} \right) \left(\frac{4 \text{ mol HCl}}{1 \text{ mol } MnO_2} \right)$$

At this point, the calculation would give the number of moles of HCl required. We must, therefore, convert *mol HCl* into *g HCl* to get our answer. The molecular weight of HCl is 36.5:

$$? \text{ g HCl} = 25.0 \text{ g } MnO_2 \left(\frac{1 \text{ mol } MnO_2}{86.9 \text{ g } MnO_2} \right) \left(\frac{4 \text{ mol HCl}}{1 \text{ mol } MnO_2} \right) \left(\frac{36.5 \text{ g HCl}}{1 \text{ mol HCl}} \right)$$

$$= 42.0 \text{ g HCl}$$

(b) The same procedure is used to solve this problem. Grams of MnO_2 is converted into moles of MnO_2. The mole relation from the chemical equation

$$1 \text{ mol } MnO_2 = 1 \text{ mol } Cl_2$$

is used to find the number of moles of Cl_2 produced. Finally, moles of Cl_2 is converted into grams of Cl_2:

$$? \text{ g } Cl_2 = 25.0 \text{ g } MnO_2 \left(\frac{1 \text{ mol } MnO_2}{86.9 \text{ g } MnO_2}\right)\left(\frac{1 \text{ mol } Cl_2}{1 \text{ mol } MnO_2}\right)\left(\frac{71.0 \text{ g } Cl_2}{1 \text{ mol } Cl_2}\right)$$

$$= 20.4 \text{ g } Cl_2$$

Example 2.14

The amount of carbon monoxide in a sample of a gas can be determined by the reaction

$$I_2O_5 + 5\,CO \longrightarrow I_2 + 5\,CO_2$$

If a gas sample liberates 0.192 g of I_2, how many grams of CO were present in the sample?

Solution

The relation, by moles, of the two substances of interest is obtained from the chemical equation. It is

$$5 \text{ mol } CO = 1 \text{ mol } I_2$$

We also need to know

$$1 \text{ mol } CO = 28.0 \text{ g } CO$$

$$1 \text{ mol } I_2 = 254 \text{ g } I_2$$

Conversion factors derived from these three relations are needed to solve the problem. The solution is

$$? \text{ g } CO = 0.192 \text{ g } I_2 \left(\frac{1 \text{ mol } I_2}{254 \text{ g } I_2}\right)\left(\frac{5 \text{ mol } CO}{1 \text{ mol } I_2}\right)\left(\frac{28.0 \text{ g } CO}{1 \text{ mol } CO}\right) = 0.106 \text{ g } CO$$

In some problems, quantities are given for two or more reactants. Suppose, for example, that we are asked how much H_2O can be prepared from 2 mol of H_2 and 2 mol of O_2. The chemical equation

$$2\,H_2 + O_2 \longrightarrow 2\,H_2O$$

states that 2 mol of H_2 will react with only 1 mol of O_2. In the problem, however, 2 mol of H_2 and 2 mol of O_2 are given. More O_2 has been supplied than can be used. When all the H_2 has been consumed, the reaction will stop. At this point, 1 mol of O_2 will have been used and 1 mol of O_2 will remain unreacted.

The amount of H_2 supplied limits the reaction and determines how much H_2O will be formed. Hydrogen, therefore, is called the **limiting reactant.**

Whenever the quantities of two or more reactants are given in a problem, we must determine which one limits the reaction before the problem can be solved.

Example 2.15

How many moles of H_2 can theoretically be prepared from 4.00 mol of Fe and 5.00 mol of H_2O? The chemical equation for the reaction is

$$3\,Fe + 4\,H_2O \longrightarrow Fe_3O_4 + 4\,H_2$$

Solution

The first step is to determine which reactant limits the reaction. The chemical equation read in terms of moles specifies 3 mol Fe. The amount of Fe given in the problem is 4.00 mol Fe, which is

$$\frac{4.00 \text{ mol Fe}}{3 \text{ mol Fe}} = 1.33$$

times the amount specified in the chemical equation. The chemical equation specifies 4 mol H_2O and the amount given in the problem is 5.00 mol H_2O, which is

$$\frac{5.00 \text{ mol } H_2O}{4 \text{ mol } H_2O} = 1.25$$

times the amount specified in the chemical equation. The H_2O, therefore, limits the extent of the reaction since a smaller *proportionate* amount has been supplied (1.25 is a smaller number than 1.33). Since 1.25 times the amount of H_2O specified in the equation has been supplied, only 1.25 times the amount of Fe specified in the equation can react. The remainder will be left over. The problem is solved on the basis of the H_2O supplied (the limiting reactant):

$$? \text{ mol } H_2 = 5.00 \text{ mol } H_2O \left(\frac{4 \text{ mol } H_2}{4 \text{ mol } H_2O} \right) = 5.00 \text{ mol } H_2$$

Example 2.16

How many grams of N_2F_4 can theoretically be prepared from 4.00 g of NH_3 and 14.0 g of F_2? The chemical equation for the reaction is

$$2\,NH_3 + 5\,F_2 \longrightarrow N_2F_4 + 6\,HF$$

Solution

The first step is to determine which reactant limits the reaction. We find the number of moles of each reactant present before reaction. The molecular weight of NH_3 is 17.0 and of F_2 is 38.0:

$$? \text{ mol NH}_3 = 4.00 \text{ g NH}_3 \left(\frac{1 \text{ mol NH}_3}{17.0 \text{ g NH}_3} \right) = 0.235 \text{ mol NH}_3$$

$$? \text{ mol F}_2 = 14.0 \text{ g F}_2 \left(\frac{1 \text{ mol F}_2}{38.0 \text{ g F}_2} \right) = 0.368 \text{ mol F}_2$$

If we read the chemical equation in terms of moles, we note that

$$2 \text{ mol NH}_3 = 5 \text{ mol F}_2$$

We compare the number of moles supplied with these quantities. The problem specified 0.235 mol of NH_3, which is

$$\frac{0.235 \text{ mol NH}_3}{2 \text{ mol NH}_3} = 0.118$$

of the amount stated in the relation derived from the chemical equation. The 0.368 mol F_2 is

$$\frac{0.368 \text{ mol F}_2}{5 \text{ mol F}_2} = 0.0736$$

of the amount given in the relation from the chemical equation. The F_2, therefore, is the limiting reactant since a smaller *proportionate* amount has been supplied

Calculations Based on Chemical Equations

1. State the problem. Indicate the substance sought (using grams as the desired unit), an equal sign, and the mass of the substance given (in grams).

2. Enter the factor that will convert the mass of the substance given into moles of substance given. The conversion factor is derived from the fact that 1 mol of a substance (numerator) is a molecular weight in grams (denominator).

3. Enter the conversion factor that is derived from the coefficients of the chemical equation and that relates the number of moles of substance sought (numerator) to the number of moles of substance given (denominator).

4. Enter the factor that will convert the number of moles of substance sought to grams of substance sought. The molecular weight of the substance sought in grams (numerator) is 1 mol of the substance sought (denominator).

5. Carry out the mathematical operations indicated to obtain the answer. All units should cancel except grams of the substance sought.

If more than one quantity is given in the problem:

1. Calculate the number of moles of each of the given reactants (see preceding step 2).

2. Divide each of these values by the coefficient of the chemical equation that pertains to the substance being considered.

3. The smallest number obtained in step 2 pertains to the reactant that limits the extent of the reaction. Use the quantity of this reactant to solve the problem in the way previously outlined.

(0.0736 is a smaller number than 0.118). The problem is solved on the basis of the amount of F_2 supplied:

$$? \text{ g } N_2F_4 = 0.368 \text{ mol } F_2$$

The relation between the quantity of F_2 employed and the quantity of N_2F_4 produced is derived from the chemical equation. It is

$$5 \text{ mol } F_2 = 1 \text{ mol } N_2F_4$$

The molecular weight of N_2F_4 is 104:

$$1 \text{ mol } N_2F_4 = 104 \text{ g } N_2F_4$$

The solution to the problem is

$$? \text{ g } N_2F_4 = 0.368 \text{ mol } F_2 \left(\frac{1 \text{ mol } N_2F_4}{5 \text{ mol } F_2} \right) \left(\frac{104 \text{ g } N_2F_4}{1 \text{ mol } N_2F_4} \right) = 7.65 \text{ g } N_2F_4$$

Frequently, the quantity of a product actually obtained from a reaction is less than the amount calculated. It may be that part of the reactants do not react, or that part of the reactants react in a way different from that desired (side reactions), or that not all of the product is recovered. The **percent yield** relates the amount of product that is actually obtained (the actual yield) to the amount that theory would predict (the theoretical yield):

$$\text{percent yield} = \frac{\text{actual yield}}{\text{theoretical yield}} \times 100\%$$

Example 2.17

If 4.80 g of N_2F_4 is obtained from the experiment described in Example 2.16, what is the percent yield?

Solution

The theoretical yield is the result of the calculation of Example 2.16, 7.65 g N_2F_4. The actual yield is 4.80 g N_2F_4. Therefore, the percent yield is

$$\frac{4.80 \text{ g } N_2F_4}{7.65 \text{ g } N_2F_4} \times 100\% = 62.7\%$$

2.9 Stoichiometry of Reactions in Solution

Many chemical reactions are carried out in aqueous solutions. The quantities of reactants for a reaction of this type are usually stated in terms of solution concentrations. The amount of a substance dissolved in a given volume of solution is called the **concentration** of the solution.

The **molarity**, M, of a solution is the number of moles of a substance (called a solute) dissolved in one liter of solution. Notice that the definition is based on *one liter*. The value stated as the molarity of a solution pertains to the amount of solute that would be present in exactly one liter of the solution. If a sample of the solution is less (or more) than one liter, the number of moles of solute in the sample is proportionately less (or more) than the numerical value of the molarity.

Notice also that the definition of molarity is based on one liter of *solution* and not on one liter of solvent (which is usually water). Because the definition is made in this way, it is relatively simple to determine the number of moles of solute in a measured volume of a solution on the basis of the molar concentration of the solution. One liter (which is 1000 ml) of a 3.0 M solution contains 3.0 mol of solute, one-half liter (which is 500 ml) contains 1.5 mol of solute, one-quarter liter (which is 250 ml) contains 0.75 mol of solute, and so on.

Example 2.18

How many moles of $AgNO_3$ are present in 25.0 ml of 0.600 M $AgNO_3$ solution?

Solution

The problem is

? mol $AgNO_3$ = 25.0 ml $AgNO_3$ sol'n

Since the concentration of $AgNO_3$ in the solution is 0.600 M,

0.600 mol $AgNO_3$ = 1000 ml $AgNO_3$ sol'n

from which we derive a conversion factor to solve the problem:

$$? \text{ mol } AgNO_3 = 25.0 \text{ ml } AgNO_3 \text{ sol'n} \left(\frac{0.600 \text{ mol } AgNO_3}{1000 \text{ ml } AgNO_3 \text{ sol'n}} \right)$$

$$= 0.0150 \text{ mol } AgNO_3$$

Example 2.19

How many grams of NaOH are required to prepare 0.250 liter of a 0.300 M solution of NaOH?

Solution

The solution is 0.300 M, and therefore

0.300 mol NaOH = 1 liter NaOH sol'n

We find the number of moles of NaOH required to prepare 0.250 liter of solution:

$$? \text{ mol NaOH} = 0.250 \text{ liter NaOH sol'n} \left(\frac{0.300 \text{ mol NaOH}}{1 \text{ liter NaOH sol'n}} \right)$$

$$= 0.0750 \text{ mol NaOH}$$

The formula weight of NaOH to three significant figures is 40.0. Therefore,

$$40.0 \text{ g NaOH} = 1 \text{ mol NaOH}$$

The number of grams of NaOH needed is

$$? \text{ g NaOH} = 0.0750 \text{ mol NaOH} \left(\frac{40.0 \text{ g NaOH}}{1 \text{ mol NaOH}} \right) = 3.00 \text{ g NaOH}$$

Example 2.20

How many milliliters of 0.750 M NaOH are required to react with 50.0 ml of 0.150 M H_2SO_4 according to the following equation?

$$H_2SO_4 + 2 NaOH \longrightarrow Na_2SO_4 + 2 H_2O$$

Solution

We find the number of moles of H_2SO_4 in the sample:

$$? \text{ mol } H_2SO_4 = 50.0 \text{ ml } H_2SO_4 \text{ sol'n} \left(\frac{0.150 \text{ mol } H_2SO_4}{1000 \text{ ml } H_2SO_4 \text{ sol'n}} \right)$$

$$= 0.00750 \text{ mol } H_2SO_4$$

From the equation, we see that

$$2 \text{ mol NaOH} = 1 \text{ mol } H_2SO_4$$

Therefore,

$$? \text{ mol NaOH} = 0.00750 \text{ mol } H_2SO_4 \left(\frac{2 \text{ mol NaOH}}{1 \text{ mol } H_2SO_4} \right) = 0.0150 \text{ mol NaOH}$$

Finally, we find the volume of 0.750 M NaOH solution that contains 0.0150 mol of NaOH:

$$? \text{ ml NaOH sol'n} = 0.0150 \text{ mol NaOH} \left(\frac{1000 \text{ ml NaOH sol'n}}{0.750 \text{ mol NaOH}} \right)$$

$$= 20.0 \text{ ml NaOH sol'n}$$

The problem could have been solved in one step as illustrated in the following example.

Example 2.21

A soda mint tablet contains $NaHCO_3$ as an antacid. One tablet requires 34.5 ml of 0.138 M HCl solution for complete reaction. Determine the number of grams of $NaHCO_3$ that one tablet contains:

$$NaHCO_3 + HCl \longrightarrow NaCl + H_2O + CO_2$$

Solution

$$? \text{ g } NaHCO_3 = 34.5 \text{ ml HCl sol'n} \left(\frac{0.138 \text{ mol HCl}}{1000 \text{ ml HCl sol'n}} \right) \left(\frac{1 \text{ mol } NaHCO_3}{1 \text{ mol HCl}} \right) \left(\frac{84.0 \text{ g } NaHCO_3}{1 \text{ mol } NaHCO_3} \right)$$

$$= 0.400 \text{ g } NaHCO_3$$

The first factor (derived from the molarity of the HCl solution) is used to find the number of moles of HCl in the sample of HCl solution. The second factor (derived from the coefficients of the chemical equation) converts this number of moles of HCl into the number of moles of $NaHCO_3$ that will react with it. The last factor (derived from the formula weight of $NaHCO_3$) converts *moles* of $NaHCO_3$ into *grams* of $NaHCO_3$.

Additional discussion of solutions and examples that pertain to them is found in Section 10.6.

Summary

The topics that have been discussed in this chapter are

1. Dalton's atomic theory, which provided the foundation for chemical stoichiometry.

2. The assignment of atomic weights to the elements, the basis of chemical stoichiometry.

3. The types of chemical formulas that are used to describe compounds, and their interpretations.

4. The mole, a unit consisting of Avogadro's number of entities that permits calculations involving realistic quantities of elements and compounds.

5. The assignment of molecular weights and formula weights.

6. How formulas can be derived from experimental data.

7. The derivation from chemical formulas of the per-centage composition of compounds and the use of these values in some simple calculations.

8. An introduction to the chemical equation.

9. The solution of problems based on chemical equations, including (a) the calculation of the theoretical quantity of a reactant needed for—or a substance produced by—a chemical reaction, (b) a method to determine the limiting reactant when amounts of two or more reactants are given in the problem, and (c) the calculation of the percent yield of a chemical process.

10. The use of molarity to express the concentration of a substance in solution.

11. Calculations that involve reactions that take place in solution.

Key Terms

Some of the more important terms introduced in this chapter are listed below. Definitions for terms not included in this list may be located in the text by use of the index.

Actual yield (Section 2.8) The amount of product actually obtained from a chemical reaction.

Atom (Section 2.1) The smallest particle of an element that retains the properties of the element.

Atomic weight (Section 2.2) The average mass of the atoms of an element, relative to the mass of a carbon-12 atom taken as exactly 12.

Avogadro's number (Section 2.4) The number of entities in one mole; 6.02205×10^{23}.

Chemical equation (Section 2.7) A representation of a chemical reaction in terms of the symbols and formulas of the elements and compounds involved.

Concentration (Section 2.9) The amount of a substance dissolved in a given quantity of solution or solvent.

Diatomic molecule (Section 2.3) A molecule consisting of two atoms.

Empirical formula (Section 2.3) A formula for a compound that is written using the simplest whole-number ratio of atoms present in the compound; also called the **simplest formula**.

Formula weight (Section 2.4) The sum of the atomic weights of the atoms in a formula.

Ion (Section 2.3) A particle made up of an atom or a group of atoms that bears either a positive or a negative charge.

Limiting reactant (Section 2.8) The reactant that, based on the chemical equation, is supplied in the smallest stoichiometric amount and hence limits the quantity of product that can be obtained from a chemical reaction.

Molarity (Section 2.9) The number of moles of a substance (called a solute) dissolved in one liter of solution.

Mole (Section 2.4) The amount of substance that contains the same number of elementary entities as there are atoms in exactly 12 g of carbon-12; a collection of Avogadro's number of units.

Molecular formula (Section 2.3) A chemical formula for a molecular substance that gives the number and type of each atom present in a molecule of the substance.

Molecular weight (Section 2.4) The sum of the atomic weights of the atoms that constitute a molecule.

Molecule (Section 2.3) A particle formed from two or more atoms.

Percent yield (Section 2.8) 100% times the actual yield divided by the theoretical yield.

Stoichiometry (Introduction) The quantitative relationships between the elements and compounds in chemical reactions.

Theoretical yield (Section 2.8) The maximum amount of product that can be obtained from a chemical reaction, as calculated by use of stoichiometric theory on the basis of the chemical equation for the reaction.

Problems*

Dalton's Theory, Atomic Weight

2.1 State the law of conservation of mass and the law of definite proportions. How do they differ? How does Dalton's theory account for them?

2.2 Compare and contrast the law of definite proportions and the law of multiple proportions. Use the compounds NO and NO_2 in your discussion.

2.3 Explain why relative atomic weights have no units.

2.4 Methane has the formula CH_4 and is 75.0% carbon. Show how these data can be used to assign a relative atomic mass to the carbon atom on the basis of the mass of the hydrogen atom being set equal to 1.00.

2.5 List the seven elements that occur in nature as diatomic molecules.

The Mole, Avogadro's Number

2.6 How many moles and how many molecules are present in 50.0 g of **(a)** H_2, **(b)** H_2O, **(c)** H_2SO_4?

---

* The more difficult problems are marked with asterisks. The appendix contains answers to color-keyed problems.

2.7 How many atoms are present in each of the samples described in Problem 2.6?

2.8 Only one type of aluminum atom occurs in nature. To four significant figures, what is the mass (in grams) of one Al atom?

2.9 One atom of an element has a mass of 9.786×10^{-23} g. What is the atomic weight of the element?

2.10 The international prototype of the kilogram is a cylinder of an alloy that is 90.000% platinum and 10.000% iridium. **(a)** How many moles of Pt and how many moles of Ir are present in the cylinder? **(b)** How many atoms of each kind are present?

2.11 One ounce (avdp) is 28.350 g. **(a)** How many moles and how many atoms of Au are present in 1.0000 ounce of Au? **(b)** If gold sells for $650.00 an ounce, how many atoms can you buy for a dollar?

*__2.12__ The distance from the earth to the sun is 1.496×10^8 km. Suppose that the atoms in 1.000 mol were enlarged into spheres 1.000 cm in diameter. If these spheres were arranged in a line touching one another, would they reach to the sun?

*__2.13__ Pure gold is 24 karat. If a 14.0 karat gold alloy consists of 14.0 parts by mass of Au and 10.0 parts by mass of Cu, how many Cu atoms are present in the alloy for every one atom of Au?

Formulas

2.14 Determine the molecular formulas of the compounds for which the following empirical formulas and molecular weights pertain: **(a)** HBS_2, 227.81; **(b)** $NaSO_2$, 174.10; **(c)** V_3S_4, 281.06; **(d)** $NaPO_3$, 815.69; **(e)** CH_2, 56.11.

2.15 Determine the molecular formulas of the compounds for which the following empirical formulas and molecular weights pertain: **(a)** COS, 60.07; **(b)** B_5H_4, 232.33; **(c)** S_2N, 156.25; **(d)** NSF, 195.20; **(e)** $PNCl_2$, 579.43.

2.16 What is the empirical formula of a compound that is 7.40% Li, 11.53% B, and 81.07% F?

2.17 Quinine is 74.05% C, 7.46% H, 9.86% O, and 8.63% N. What is the empirical formula of quinine?

2.18 Putrescine, a product of decaying flesh, is 54.50% C, 13.72% H, and 31.78% N. What is the empirical formula of putrescine?

2.19 Hydroxyl apatite, an important constituent of bones and teeth, is 39.895% Ca, 18.498% P, 41.406% O, and 0.201% H. What is the empirical formula of hydroxyl apatite?

2.20 Aspirin is 60.00% C, 4.48% H, and 35.52% O. What is the empirical formula of aspirin?

2.21 L-Dopa, a drug used in the treatment of Parkinson's disease, is 54.82% C, 5.62% H, 7.10% N, and 32.46% O. What is the empirical formula of L-Dopa?

2.22 The molecular weight of citric acid is 192.13 and the compound is 37.51% C, 58.29% O, and 4.20% H. What is the molecular formula of citric acid?

2.23 The molecular weight of saccharin is 183.18 and the compound is 45.90% C, 2.75% H, 26.20% O, 17.50% S, and 7.65% N. What is the molecular formula of saccharin?

2.24 A sample of a compound that contains only C and H was burned in oxygen and 9.24 g of CO_2 and 3.15 g of H_2O were obtained. **(a)** How many moles of C atoms and how many moles of H atoms did the sample contain? **(b)** What is the empirical formula of the compound? **(c)** What was the mass of the sample that was burned?

2.25 A sample of a compound that contains only C, H, and S was burned in oxygen, and 15.84 g of CO_2, 3.24 g of H_2O, and 5.77 g of SO_2 were obtained. **(a)** How many moles of C atoms, moles of H atoms, and moles of S atoms did the sample contain? **(b)** What is the empirical formula of the compound? **(c)** What was the mass of the sample that was burned?

*__2.26__ A 7.61 g sample of p-aminobenzoic acid (PABA, a compound used in sunscreen products) was burned in oxygen and 17.1 g of CO_2, 3.50 g of H_2O, and 0.777 g of N_2 were obtained. The compound contains carbon, hydrogen, nitrogen, *and oxygen.* **(a)** How many moles of C atoms, moles of H atoms, and moles of N atoms did the sample contain? **(b)** What mass of C, H, and N did the sample contain? **(c)** Based on the mass of the original sample, what mass of O did the sample contain? **(d)** How many moles of O atoms did the sample contain? **(e)** What is the empirical formula of PABA?

2.27 Upon heating 7.50 g of a hydrate $CoCl_2 \cdot xH_2O$ in a vacuum, the water was driven off and 4.09 g of anhydrous $CoCl_2$ remained. What is the value of x in the formula $CoCl_2 \cdot xH_2O$?

2.28 Upon heating 6.45 g of a hydrate $CuSO_4 \cdot xH_2O$ in a vacuum, the water was driven off and 4.82 g of anhydrous $CuSO_4$ remained. What is the value of x in the formula $CuSO_4 \cdot xH_2O$?

*__2.29__ A 6.29 g sample of a compound that contains vanadium and chlorine is dissolved in water. The addition of a water-soluble silver salt precipitates AgCl, which is insoluble in water. The process yields 17.19 g of AgCl. What is the empirical formula of the chloride of vanadium?

Percentage Composition

2.30 To three significant figures, what percentage of nickel carbonyl, $Ni(CO)_4$, is nickel?

2.31 To four significant figures, what percentage of the mineral witherite, $BaCO_3$, is barium?

2.32 To four significant figures, what percentage of zircon, $ZrSiO_4$, is zirconium?

2.33 What mass of zinc is theoretically obtainable from 1.25 kg of sphalerite ore that is 75.0% ZnS?

2.34 What mass of copper is theoretically obtainable from 10.0 kg of chalcocite ore that is 25.0% Cu_2S?

2.35 How many grams of xenon and of fluorine are theoretically needed to make 1.000 g of XeF_4?

2.36 How many grams of lithium and of nitrogen are theoretically needed to make 5.000 g of Li_3N?

2.37 A 1.74 g sample of a compound that contains only C and H was burned in oxygen, and 5.28 g of CO_2 and 2.70 g of H_2O were obtained. What is the percentage composition of the compound?

*2.38 Cholesterol is a compound that contains carbon, hydrogen, *and oxygen*. The combustion of a 9.50 g sample of the compound yields 29.20 g of CO_2 and 10.18 g of H_2O. What is the percentage composition of the compound?

*2.39 The mineral hematite is Fe_2O_3. Hematite ore contains unwanted material, called gangue, in addition to Fe_2O_3. If 1.000 kg of ore contains 0.5920 kg of Fe, what percentage of the ore is Fe_2O_3?

*2.40 Sulfur-containing compounds are an undesirable component of some oils. The amount of sulfur in an oil can be determined by oxidizing the S to sulfate, SO_4^{2-}, and precipitating the sulfate ion as barium sulfate, $BaSO_4$, which can be collected, dried, and weighed. From an 8.25 g sample of an oil, 0.929 g of $BaSO_4$ was obtained. What is the percentage of sulfur in the oil?

Chemical Equations

2.41 Balance the following chemical equations:
(a) $V_2O_5 + H_2 \longrightarrow V_2O_3 + H_2O$
(b) $B_2O_3 + C \longrightarrow B_4C + CO$
(c) $Bi + O_2 \longrightarrow Bi_2O_3$
(d) $CaC_2 + H_2O \longrightarrow Ca(OH)_2 + H_2C_2$
(e) $Ba(NO_3)_2 + H_2SO_4 \longrightarrow BaSO_4 + HNO_3$

2.42 Balance the following chemical equations:
(a) $NO_2 + H_2O \longrightarrow HNO_3 + NO$
(b) $Al_2S_3 + H_2O \longrightarrow Al(OH)_3 + H_2S$
(c) $SiCl_4 + Si \longrightarrow Si_2Cl_6$
(d) $(NH_4)_2Cr_2O_7 \longrightarrow N_2 + H_2O + Cr_2O_3$
(e) $Ca_3N_2 + H_2O \longrightarrow Ca(OH)_2 + NH_3$

2.43 Gasohol is a mixture of gasoline and ethyl alcohol. **(a)** Write the chemical equation for the combustion of octane (C_8H_{18}, a component of gasoline) in O_2. (The products of the reaction are CO_2 and H_2O.) **(b)** Write the chemical equation for the combustion of ethyl alcohol (C_2H_6O) in O_2. (The products of the reaction are CO_2 and H_2O.)

Problems Based on Chemical Equations

2.44 Determine the number of grams of H_3PO_4 that can be made from 100.0 g of P_4O_{10}:

$$P_4O_{10} + 6H_2O \longrightarrow 4H_3PO_4$$

2.45 Use the equation

$$2NaNH_2 + N_2O \longrightarrow NaN_3 + NaOH + NH_3$$

(a) to determine the number of grams of $NaNH_2$ and of N_2O that are required to make 5.00 g of NaN_3. **(b)** How many grams of NH_3 are produced?

2.46 Pure, dry NO gas can be made by the following reaction:

$$3KNO_2 + KNO_3 + Cr_2O_3 \longrightarrow 4NO + 2K_2CrO_4$$

How many grams of each of the reactants are needed to make 2.50 g of NO?

2.47 Determine the number of grams of HI that will be produced by adding 3.50 g of PI_3 to excess water:

$$PI_3 + 3H_2O \longrightarrow 3HI + H_3PO_3$$

2.48 A 13.38 g sample of a material that contains some As_4O_6 requires 5.330 g of I_2 for reaction according to the chemical equation

$$As_4O_6 + 4I_2 + 4H_2O \longrightarrow 2As_2O_5 + 8HI$$

(a) What mass of As_4O_6 reacted with the I_2 supplied? **(b)** What percentage of the sample is As_4O_6? **(c)** What percentage of the sample is As?

2.49 A 6.55 g sample of a mixture of Na_2SO_3 and Na_2SO_4 was dissolved in water and heated with solid sulfur. The Na_2SO_4 does not react, but the Na_2SO_3 reacts as follows:

$$Na_2SO_3 + S \longrightarrow Na_2S_2O_3$$

and 1.23 g of S dissolved to form $Na_2S_2O_3$. What percentage of the original mixture was Na_2SO_3?

2.50 How many grams of NH_4SCN can be made from 5.00 g of CS_2 and 4.00 g of NH_3? The equation for the reaction is

$$CS_2 + 2NH_3 \longrightarrow NH_4SCN + H_2S$$

2.51 How many grams of OF_2 can be made from 1.60 g of F_2 and 1.60 g of NaOH? The equation is

$$2F_2 + 2NaOH \longrightarrow OF_2 + 2NaF + H_2O$$

2.52 Determine the number of grams of B_2H_6 that can be made from 3.204 g of $NaBH_4$ and 5.424 g of BF_3 by the following reaction:

$$3NaBH_4 + 4BF_3 \longrightarrow 3NaBF_4 + 2B_2H_6$$

2.53 Determine the number of grams of SF_4 that can be made from 4.00 g of SCl_2 and 2.00 g of NaF by the following reaction:

$$3SCl_2 + 4NaF \longrightarrow SF_4 + S_2Cl_2 + 4NaCl$$

2.54 (a) How many grams of $OP(NH_2)_3$ should be secured from the reaction of 7.00 g of $OPCl_3$ and 5.00 g of NH_3? The equation is

$$OPCl_3 + 6NH_3 \longrightarrow OP(NH_2)_3 + 3NH_4Cl$$

(b) If 3.50 g of $OP(NH_2)_3$ were isolated, what was the percentage yield?

2.55 (a) How many grams of Ti metal are required to react with 3.513 g of $TiCl_4$? The equation for the reaction is

$$3TiCl_4 + Ti \longrightarrow 4TiCl_3$$

(b) How many grams of $TiCl_3$ should be produced by the reaction?

(c) If 3.000 g of $TiCl_3$ are isolated as the product of the reaction, what is the percentage yield?

2.56 (a) How many grams of NaN_3 should be secured from the reaction of 3.50 g of $NaNH_2$ and 3.50 g of $NaNO_3$? The equation is

$$3\,NaNH_2 + NaNO_3 \longrightarrow NaN_3 + 3\,NaOH + NH_3$$

(b) If 1.20 g of NaN_3 are isolated, what is the percentage yield?

*2.57 A mixture of sodium oxide, Na_2O, and barium oxide, BaO, that weighs 5.00 g is dissolved in water. This solution is then treated with dilute sulfuric acid, H_2SO_4, which converts the oxides to sulfates. Barium sulfate, $BaSO_4$, precipitates from the solution, but sodium sulfate, Na_2SO_4, is soluble and remains in solution. The $BaSO_4$ is collected by filtration and is found to weigh 3.43 g when dried. What percentage of the original sample of mixed oxides is BaO?

*2.58 A 10.50 g sample of a mixture of calcium carbonate, $CaCO_3$, and calcium sulfate, $CaSO_4$, is heated to decompose the carbonate:

$$CaCO_3 \longrightarrow CaO + CO_2$$

The CO_2 gas escapes and the $CaSO_4$ is not decomposed by heating. The final mass of the sample is 7.64 g. What percentage of the original mixture is $CaCO_3$?

*2.59 A 9.90 g sample of a mixture of $CaCO_3$ and $NaHCO_3$ is heated and the compounds decompose:

$$CaCO_3 \longrightarrow CaO + CO_2$$

$$2\,NaHCO_3 \longrightarrow Na_2CO_3 + CO_2 + H_2O$$

The decomposition of the sample yielded 2.86 g of CO_2 and 0.900 g of H_2O. What percentage of the original mixture is $CaCO_3$?

Reactions in Solution

2.60 How many grams of H_2SO_4 are needed to prepare 375 ml of a 6.00 M solution of H_2SO_4?

2.61 How many grams of KIO_3 are needed to prepare 5.000 liter of a 0.1000 M solution of KIO_3?

2.62 How many grams of $NaOH$ are needed to prepare 0.250 liter of a 1.50 M solution of $NaOH$?

2.63 How many milliliters of 3.00 M H_3PO_4 are required to react with 28.8 ml of 5.00 M KOH? The equation for the reaction is

$$H_3PO_4 + 3\,KOH \longrightarrow K_3PO_4 + 3\,H_2O$$

2.64 How many milliliters of 0.500 M $AgNO_3$ are needed to react with 25.0 ml of 0.750 M Na_2CrO_4? The equation for the reaction is

$$Na_2CrO_4 + 2\,AgNO_3 \longrightarrow Ag_2CrO_4 + 2\,NaNO_3$$

2.65 How many milliliters of 0.150 M $KMnO_4$ are needed to react with 15.0 ml of 0.250 M $FeCl_2$? The equation is

$$5\,FeCl_2 + KMnO_4 + 8\,HCl \longrightarrow$$
$$5\,FeCl_3 + MnCl_2 + KCl + 4\,H_2O$$

2.66 How many grams of solid CaO are needed to react with 50.0 ml of 0.600 M HCl? The equation for the reaction is

$$CaO + 2\,HCl \longrightarrow CaCl_2 + H_2O$$

2.67 How many grams of I_2 are required to react with 45.0 ml of 0.500 M $Na_2S_2O_3$? The equation is

$$2\,Na_2S_2O_3 + I_2 \longrightarrow 2\,NaI + Na_2S_4O_6$$

2.68 (a) How many grams of Na_2CO_3 are present in an impure sample of the compound if 35.0 ml of 0.250 M HCl are required to react with it? The equation for the reaction is

$$Na_2CO_3 + 2\,HCl \longrightarrow 2\,NaCl + CO_2 + H_2O$$

(b) If the sample weighed 1.25 g, what percentage of the material is Na_2CO_3?

THERMOCHEMISTRY

C H A P T E R

3

In the course of a chemical reaction, energy is either liberated or absorbed. Calculations relating to these energy changes are as important as those concerned with the masses of reacting substances. Thermochemistry is the study of the heat released or absorbed by chemical and physical changes. In succeeding chapters, calculations involving these energy changes will be frequently encountered. In this chapter, this type of calculation will be introduced.

3.1 Energy Measurement

It is common to think of force as the application of physical strength—as pushing. If the effects of friction are neglected, a body in motion remains in motion at a constant velocity, and a body at rest stays at rest (its velocity is zero). If these bodies are pushed, there will be a change in their velocities. The increase in velocity per unit time is called the acceleration.

Suppose, for example, that we have a body moving at a velocity of 1 m/s. Assume that this body is acted on by a constant force, that is, given a sustained push. The body will move faster and faster. At the end of 1 second it may be moving at the rate of 2 m/s. At the end of 2 s, its speed may be 3 m/s. If the body picks up speed at the rate of one meter per second in a second, its acceleration is said to be 1 m/s^2.

A force that gives a *one gram* body an acceleration of 1 m/s^2 is not so large as a force that gives a *one kilogram* body the same acceleration. The magnitude of a force (F), therefore, is proportional to the mass of the body (m) as well as to the acceleration (a) that the force produces:

$$F = ma \tag{3.1}$$

The SI unit of force is called the newton (symbol, N) and is derived from the base units of mass (the kilogram), length (the meter), and time (the second):

$$F = ma$$
$$1 \text{ N} = (1 \text{ kg})(1 \text{ m/s}^2)$$
$$1 \text{ N} = 1 \text{ kg} \cdot \text{m/s}^2$$

Work (W) is defined as the force times the distance through which the force acts (d):

$$W = Fd \qquad\qquad (3.2)$$

In the International System, the unit of work is the joule (symbol, J). The **joule** is defined as the work done when a force of one newton acts through a distance of one meter:

$$W = Fd$$
$$1 \text{ J} = (1 \text{ N})(1 \text{ m})$$
$$= 1 \text{ N} \cdot \text{m}$$
$$= 1 \text{ kg} \cdot \text{m}^2/\text{s}^2$$

Energy may be defined as the capacity to do work. There are many forms of energy, such as heat energy, electrical energy, and chemical energy. When one form of energy is converted into another form, energy is neither created nor destroyed. The SI unit of work, the joule, is the unit used for all energy measurements, including heat measurements. The unit is named in honor of James Joule (1818–1889), a student of John Dalton, who demonstrated that a given quantity of work always produces the same quantity of heat.

3.2 Temperature and Heat

Most liquids expand as the temperature increases. The mercury thermometer is designed to use the expansion of mercury to measure temperature. The thermometer consists of a small bulb sealed to a tube that has a narrow bore (called a capillary tube). The bulb and part of the tube contain mercury, the space above the mercury is evacuated, and the upper end of the tube is sealed. When the temperature increases, the mercury expands and rises in the capillary tube.

The **Celsius temperature scale,** named for Anders Celsius, a Swedish astronomer, is employed in scientific studies and is part of the International System. The scale is based on the assignment of 0°C to the normal freezing point of water and 100°C to the normal boiling point of water. When a thermometer is placed in a mixture of ice and water, the mercury will stand at a height that is marked on the tube as 0°C. When the thermometer is placed in boiling water under standard atmospheric pressure, the mercury will rise to a position that is marked 100°C. The tube is marked between these two fixed points to indicate 100 equal divisions, each of which represents one degree. The thermometer is calibrated below 0°C and above 100°C by marking off degrees of the same size. The Celsius scale was formerly called the **centigrade scale,** derived from the Latin words *centum* (a hundred) and *gradus* (a degree).

On the **Fahrenheit temperature scale** (named for G. Daniel Fahrenheit, a German instrument maker) the normal freezing point of water is 32°F and the normal boiling point of water is 212°F. Since there are 100 Celsius degrees and 180 (212 minus 32) Fahrenheit degrees between these two fixed points, 5 Celsius degrees equal 9 Fahrenheit degrees.

The Fahrenheit temperature scale is not used in scientific work. Conversion of a temperature from the Fahrenheit scale (t_F) to the Celsius scale (t_C) can be accomplished in the following way:

1. Subtract 32 from the Fahrenheit reading. The value obtained tells how many Fahrenheit degrees the temperature is above the freezing point of water.

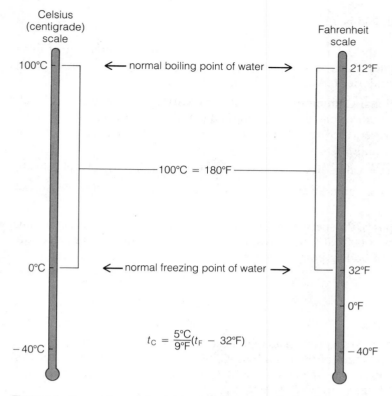

Celsius
(centigrade)
scale

Fahrenheit
scale

100°C ← normal boiling point of water → 212°F

100°C = 180°F

0°C ← normal freezing point of water → 32°F

0°F

$$t_C = \frac{5°C}{9°F}(t_F - 32°F)$$

−40°C −40°F

Figure 3.1 Comparison of the Celsius (centigrade) and Fahrenheit temperature scales

2. Since 5 Celsius degrees equal 9 Fahrenheit degrees, 5/9 of the value obtained is the number of Celsius degrees above the freezing point of water, which is 0°C.

Hence,

$$t_C = \frac{5°C}{9°F}(t_F - 32°F) \tag{3.3}$$

In Figure 3.1, the two temperature scales are compared. The thermodynamic temperature scale, called the Kelvin scale, is described in Section 8.3.

Temperature is defined as the degree of hotness. Heat, on the other hand, is a form of energy. In the past, chemists have customarily measured heat in calories. The **specific heat** of a substance is defined as the amount of heat required to raise the temperature of 1 g of the substance by 1°C. The calorie was originally defined in terms of the specific heat of water. The one-degree temperature interval had to be specified, however, since the specific heat of water changes slightly as the temperature changes. For many years, the **calorie** was defined as the amount of heat required to raise the temperature of 1 g of water from 14.5°C to 15.5°C.

Very precise determinations of heat energy, in joules, can be made by electrical measurements. The joule, therefore, is a better primary standard of heat than the specific heat of water. The calorie is now defined by its joule equivalent:

1 cal = 4.184 J (exactly)

Several points should be noted:

1. The joule and the calorie are relatively small units for measurement of thermochemical values. Such values are frequently reported in kilojoules (a kJ is 1000 J) and kilocalories (a kcal is 1000 cal).

2. The International Committee of Weights and Measures recommends that all energy measurements be based on the joule and that the calorie no longer be used. In the past, however, thermochemical values have customarily been recorded in calories and kilocalories.

    a. To convert a value given in *calories to joules*, multiply by (4.184 J/1 cal).

    b. To convert a value given in *kilocalories to kilojoules*, multiply by (4.184 kJ/1 kcal).

3. For our purposes, the specific heat of water can be considered to be a constant, 4.184 J/(g°C) or 1.000 cal/(g°C), over any temperature interval between the freezing point and boiling point of water.

3.3 Calorimetry

The **heat capacity** (C) of a given mass of a substance is the amount of heat required to raise the temperature of the mass by 1°C. Specific heat is the heat capacity of *one gram* of a substance—the amount of heat required to raise the temperature of 1 g of the substance by 1°C. Therefore,

$$C = (\text{mass})(\text{specific heat}) \qquad (3.4)$$

Since the specific heat of water is 4.184 J/(g°C), the heat capacity of 500 g of water is

$$
\begin{aligned}
C &= [500 \text{ g}][4.18 \text{ J/(g°C)}] \\
&= 2090 \text{ J/°C} \\
&= 2.09 \text{ kJ/°C}
\end{aligned}
$$

This sample absorbs 2.09 kJ of heat for each degree that the temperature increases. Twice this amount of heat would be required to raise the temperature by 2°C. In general,

$$q = C(t_2 - t_1) \qquad (3.5)$$

where q is the heat absorbed by the sample, C is the heat capacity of the sample, t_2 is the final temperature, and t_1 is the initial temperature. The heat absorbed by a 500-g sample of water when it is heated from 20.00°C to 25.00°C can be calculated in the following way:

$$
\begin{aligned}
q &= C(t_2 - t_1) \\
&= (2.09 \text{ kJ/°C})(25.00°C - 20.00°C) \\
&= (2.09 \text{ kJ/°C})(5.00°C) \\
&= 10.4 \text{ kJ}
\end{aligned}
$$

A device called a **calorimeter** is used to measure the heat changes that accompany chemical reactions. A calorimeter consists of a vessel, in which the reaction is conducted, submerged in a weighed quantity of water in a well-insulated container. The reaction is run using known quantities of reactants, and the heat evolved by the reaction increases the temperature of the water and the calorimeter. The amount of heat liberated by the reaction can be calculated from the increase in temperature if the total heat capacity of the calorimeter and its contents is known.

Example 3.1

A bomb type of calorimeter (Figure 3.2) is used to measure the heat evolved by the combustion of glucose, $C_6H_{12}O_6$:

$$C_6H_{12}O_6(s) + 6O_2(g) \longrightarrow 6CO_2(g) + 6H_2O(l)$$

A 3.00-g sample of glucose is placed in the bomb, which is then filled with oxygen gas under pressure. The bomb is placed in a well-insulated calorimeter vessel that is filled with 1.20 kg of water. The initial temperature of the assembly is 19.00°C. The reaction mixture is ignited by the electrical heating of a wire within the bomb. The reaction causes the temperature of the calorimeter and its contents to increase to 25.50°C. The heat capacity of the calorimeter is 2.21 kJ/°C. The molecular weight of glucose is 180. How much heat is evolved by the combustion of 1 mol of glucose?

Solution

Since 1200 g of water is employed and the specific heat of water is 4.18 J/(g°C), the heat capacity of the water in the calorimeter, C_{H_2O}, is

$$C = (mass)(specific\ heat)$$

$$C_{H_2O} = [1200\ g][4.18\ J/(g°C)]$$
$$= 5016\ J/°C = 5.02\ kJ/°C$$

The heat capacity of the calorimeter, C_{cal}, is 2.21 kJ/°C. The total heat capacity, C_{total}, is

$$C_{total} = C_{H_2O} + C_{cal}$$
$$= 5.02\ kJ/°C + 2.21\ kJ/°C$$
$$= 7.23\ kJ/°C$$

Thus, 7.23 kJ of heat is needed to raise the temperature of the assembly by 1°C. The amount of heat *absorbed* by the calorimeter and the water is

$$q = C(t_2 - t_1)$$
$$= (7.23\ kJ/°C)(25.50°C - 19.00°C)$$
$$= (7.23\ kJ/°C)(6.50°C)$$
$$= 47.0\ kJ$$

This quantity (47.0 kJ) is also the amount of heat *evolved* by the combustion of 3.00 g of glucose. Therefore,

$$47.0\ kJ = 3.00\ g\ C_6H_{12}O_6$$

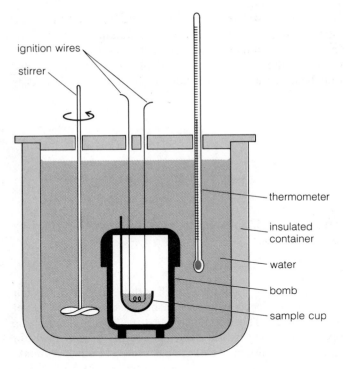

Figure 3.2 Bomb calorimeter

For a mol of glucose (180 g of glucose), the quantity of heat evolved is

$$? \text{ kJ} = 180 \text{ g } C_6H_{12}O_6 \left(\frac{47.0 \text{ kJ}}{3.00 \text{ g } C_6H_{12}O_6} \right) = 2.82 \times 10^3 \text{ kJ}$$

3.4 Thermochemical Equations

If a reaction that produces a gas is run in a closed container, the pressure inside the container will increase. Most reactions, however, are run in containers that are open to the atmosphere. For these reactions, the pressure is constant whether gases are produced or not.*

---

* Reactions run in a calorimeter bomb may or may not cause the pressure inside the bomb to change. The equation for the reaction described in Example 3.1 is

$$C_6H_{12}O_6(s) + 6 O_2(g) \longrightarrow 6 CO_2(g) + 6 H_2O(l)$$

Notice that 6 mol of gas (O_2 gas) are consumed and 6 mol of gas (CO_2 gas) are produced. The pressure inside the bomb, therefore, does not change when the reaction occurs.

If a reaction produces more moles of gas than it uses, the pressure inside a calorimeter bomb builds up. If such a reaction occurs in a container that is open to the atmosphere, the gases that are produced escape. The pressure in this case remains constant and equal to the pressure of the atmosphere. For such a reaction, the energy change measured when the pressure is allowed to change is slightly different from the energy change measured at constant pressure. In a case of this type a correction is applied to the value obtained by use of a bomb calorimeter (see Section 17.2).

There are no appreciable pressure effects for many reactions, including those that do not involve gases (for example, reactions run in solution) and those in which the number of moles of gas used equals the number of moles of gas produced.

The heat liberated or absorbed by reactions that are conducted under constant pressure can be related to a property that is called **enthalpy** and is given the symbol H. Every pure substance is assumed to have an enthalpy (which is also called a heat content). A given set of reactants, therefore, has a definite total enthalpy, $H_{reactants}$. The corresponding set of products also has a definite total enthalpy, $H_{products}$. The heat of reaction is the difference between these enthalpies and is therefore given the symbol ΔH. The upper case Greek delta, Δ, is used to indicate a difference:

$$\Delta H = H_{products} - H_{reactants} \tag{3.6}$$

Reactions that liberate heat are called **exothermic reactions**. For reactions of this type, the products have a lower enthalpy than the reactants; ΔH is a negative value. When the reaction occurs, the products replace the reactants in the system. The enthalpy of the reaction system declines (ΔH is negative), and the difference is given off as heat (see Figure 3.3).

Reactions that absorb heat are called **endothermic reactions**. For reactions of this type, the enthalpy of the products is higher than the enthalpy of the reactants, and ΔH is positive. When the reaction occurs, heat must be supplied in order to raise the enthalpy of the system (see Figure 3.4).

The enthalpies of chemical substances depend upon temperature and pressure. By convention, ΔH values are usually reported for reactions run at 25°C and standard atmospheric pressure (Section 8.2). If any other conditions are employed, they are noted.

Thermochemical data may be given by writing a chemical equation for the reaction under consideration and listing beside it the ΔH value for the reaction *as it is written*. The appropriate value of ΔH is the one required when the equation is read in molar quantities. Contrary to usual practice, fractional coefficients may be used to balance the chemical equation. A fractional coefficient simply indicates a fraction of a mole of a substance. Thus,

$$H_2(g) + \tfrac{1}{2}O_2(g) \longrightarrow H_2O(l) \qquad \Delta H = -286 \text{ kJ}$$

When 1 mol of hydrogen gas reacts with $\tfrac{1}{2}$ mol of oxygen gas to produce 1 mol of liquid water, 286 kJ of heat is evolved.

The state of each substance in the reaction must be indicated in the equation. A designation, such as (g) for gas, (s) for solid, (l) for liquid, or (aq) for "in aqueous solution," is placed after each formula. The need for this convention can be demonstrated by comparing the following equation with the preceding one:

$$H_2(g) + \tfrac{1}{2}O_2(g) \longrightarrow H_2O(g) \qquad \Delta H = -242 \text{ kJ}$$

Notice that 44 kJ less heat is liberated by the second reaction than by the first reaction. This quantity of heat is used to convert 1 mol of $H_2O(l)$ to 1 mol of $H_2O(g)$.

When an equation is reversed, the sign of ΔH is changed. A reaction that is endothermic in one direction is exothermic in the opposite direction:

$$\tfrac{1}{2}H_2(g) + \tfrac{1}{2}I_2(s) \longrightarrow HI(g) \qquad \Delta H = +25.9 \text{ kJ}$$

$$HI(g) \longrightarrow \tfrac{1}{2}H_2(g) + \tfrac{1}{2}I_2(s) \qquad \Delta H = -25.9 \text{ kJ}$$

If the coefficients of the substances in a chemical equation are multiplied by a factor, the ΔH value must be multiplied by the same factor. For example, if the

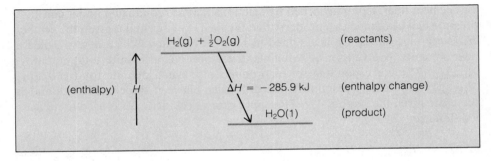

Figure 3.3 Enthalpy diagram for an exothermic reaction

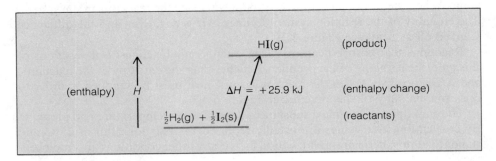

Figure 3.4 Enthalpy diagram for an endothermic reaction

last equation is multiplied through by 2, the ΔH value must also be multiplied by 2:

$$2\,HI(g) \longrightarrow H_2(g) + I_2(s) \qquad \Delta H = 2(-25.9\text{ kJ}) = -51.8\text{ kJ}$$

In like manner, the coefficients of an equation and the ΔH value can be divided by the same number.

The conventions for writing thermochemical equations can be summarized as follows:

1. For exothermic reactions (the reaction system loses heat), ΔH is negative. For endothermic reactions (the reaction system absorbs heat), ΔH is positive.

2. Unless otherwise noted, ΔH values are given for reactions run at 25°C and standard atmospheric pressure.

3. Designations such as (g), (s), (l), and (aq) are placed behind the formulas in the equation to indicate the physical state of each substance.

4. The coefficients of the substances of the chemical equation indicate the number of moles of each substance involved in the reaction (fractions may be used), and the ΔH value given corresponds to these quantities of materials.

5. If the coefficients in the chemical equation are multiplied or divided by a factor, the ΔH value must be multiplied or divided by the same factor.

6. When a chemical equation is reversed, the sign but not the magnitude of the ΔH value is changed.

Thermochemical problems are solved in much the same way that simple stoichiometric problems are solved.

Example 3.2

The thermite reaction is highly exothermic:

$$2\,Al(s) + Fe_2O_3(s) \longrightarrow 2\,Fe(s) + Al_2O_3(s) \qquad \Delta H = -848\,kJ$$

How much heat is liberated when 36.0 g of Al reacts with excess Fe_2O_3?

Solution

The equation and ΔH value show that

$$-848\,kJ = 2\,mol\,Al$$

Since the atomic weight of Al is 27.0,

$$? \,kJ = 36.0\,g\,Al \left(\frac{1\,mol\,Al}{27.0\,g\,Al}\right)\left(\frac{-848\,kJ}{2\,mol\,Al}\right) = -565\,kJ$$

3.5 The Law of Hess

The basis of many thermochemical calculations is the **law of constant heat summation**, which was established experimentally by G. H. Hess in 1840. This **law of Hess** states that the change in enthalpy for any chemical reaction is constant, whether the reaction occurs in one step or in several steps. Thermochemical data, therefore, may be treated algebraically.

Consider, for example, the reaction of graphite with oxygen that produces carbon dioxide gas:

$$C(graphite) + O_2(g) \longrightarrow CO_2(g) \qquad \Delta H = -393.5\,kJ$$

This transformation also can occur in two steps: the reaction of graphite with O_2 that forms CO followed by the reaction of the CO with O_2 that forms CO_2. Addition of the equations for the steps gives a result that is identical with the equation for the direct reaction (see Figure 3.5).

$$\begin{array}{ll}
C(graphite) + \tfrac{1}{2}O_2(g) \longrightarrow CO(g) & \Delta H = -110.5\,kJ \\
CO(g) + \tfrac{1}{2}O_2(g) \longrightarrow CO_2(g) & \Delta H = -283.0\,kJ \\
\hline
C(graphite) + \quad O_2(g) \longrightarrow CO_2(g) & \Delta H = -393.5\,kJ
\end{array}$$

Since thermochemical data can be treated algebraically, it is possible to derive an enthalpy of reaction from measurements made on other reactions. Suppose, for example, that the following thermochemical equations are given:

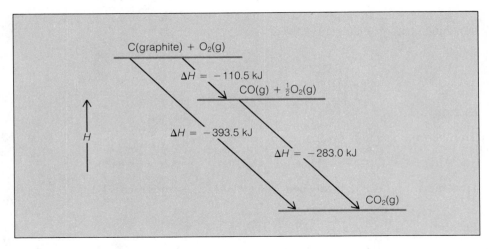

Figure 3.5 Enthalpy diagram to illustrate the law of Hess

$$C(\text{graphite}) + O_2(g) \longrightarrow CO_2(g) \qquad \Delta H = -393.5 \text{ kJ} \qquad (3.7)$$

$$H_2(g) + \tfrac{1}{2}O_2(g) \longrightarrow H_2O(l) \qquad \Delta H = -285.9 \text{ kJ} \qquad (3.8)$$

$$CH_4(g) + 2O_2(g) \longrightarrow CO_2(g) + 2H_2O(l) \qquad \Delta H = -890.4 \text{ kJ} \qquad (3.9)$$

These relations can be used to find the ΔH for the reaction in which methane, CH_4, is prepared from carbon and hydrogen. This enthalpy change cannot be measured directly:

$$C(\text{graphite}) + 2H_2(g) \longrightarrow CH_4(g) \qquad \Delta H = ?$$

Since 1 mol of C(graphite) appears on the left of equation (3.7) and also on the left of the desired equation, equation (3.7) is written as previously given:

$$C(\text{graphite}) + O_2(g) \longrightarrow CO_2(g) \qquad \Delta H = -393.5 \text{ kJ} \qquad (3.7)$$

Two moles of $H_2(g)$ appear on the left of the desired equation and only 1 mol of $H_2(g)$ appears on the left of equation (3.8). Equation (3.8), therefore, is multiplied through by 2, and the ΔH value is multiplied by 2:

$$2H_2(g) + O_2(g) \longrightarrow 2H_2O(l) \qquad \Delta H = -571.8 \text{ kJ} \qquad (3.10)$$

One mole of $CH_4(g)$ appears on the *right* side of the desired equation. Therefore, equation (3.9) is reversed, and the sign of the ΔH value is changed:

$$CO_2(g) + 2H_2O(l) \longrightarrow CH_4(g) + 2O_2(g) \qquad \Delta H = +890.4 \text{ kJ} \qquad (3.11)$$

Equations (3.7), (3.10), and (3.11) are added. Terms common to both sides of the final equation ($2O_2$, CO_2, and $2H_2O$) are canceled.

$$C(\text{graphite}) + 2H_2(g) \longrightarrow CH_4(g) \qquad \Delta H = -74.9 \text{ kJ}$$

The resulting ΔH value is the enthalpy of reaction that was sought.

Example 3.3

Given the following thermochemical equations:

$$4\,NH_3(g) + 3\,O_2(g) \longrightarrow 2\,N_2(g) + 6\,H_2O(l) \qquad \Delta H = -1531\ kJ \qquad (3.12)$$

$$N_2O(g) + H_2(g) \longrightarrow N_2(g) + H_2O(l) \qquad \Delta H = -367.4\ kJ \qquad (3.13)$$

$$H_2(g) + \tfrac{1}{2}O_2(g) \longrightarrow H_2O(l) \qquad \Delta H = -285.9\ kJ \qquad (3.14)$$

Find the value of ΔH for the reaction

$$2\,NH_3(g) + 3\,N_2O(g) \longrightarrow 4\,N_2(g) + 3\,H_2O(l)$$

Solution

Since the desired equation has 2 mol of NH_3 on the left, we divide equation (3.12) by 2 and the ΔH value by 2. We multiply equation (3.13) and the corresponding ΔH value by 3 so that the coefficient of N_2O in the final equation will be 3. To eliminate the $3\,H_2(g)$ added to the left by the last step, we reverse equation (3.14) and multiply it by 3; the ΔH value is multiplied by 3 and its sign changed:

$$2\,NH_3(g) + \tfrac{3}{2}O_2(g) \longrightarrow N_2(g) + 3\,H_2O(l) \qquad \Delta H = -765.5\ kJ$$

$$3\,N_2O(g) + 3\,H_2(g) \longrightarrow 3\,N_2(g) + 3\,H_2O(l) \qquad \Delta H = -1102.2\ kJ$$

$$3\,H_2O(l) \longrightarrow 3\,H_2(g) + \tfrac{3}{2}O_2(g) \qquad \Delta H = +857.7\ kJ$$

The equations and ΔH values are added. Terms common to both sides of the final equation ($\tfrac{3}{2}O_2$, $3\,H_2$, and $3\,H_2O$) are canceled:

$$2\,NH_3(g) + 3\,N_2O(g) \longrightarrow 4\,N_2(g) + 3\,H_2O(l) \qquad \Delta H = -1010.0\ kJ$$

3.6 Enthalpies of Formation

The enthalpy of an element or compound depends upon the temperature and pressure. If we wish to compare ΔH values, the conditions under which they have been measured must be identical. The **standard state** of a substance is the state in which the substance is stable at 1 atm pressure and 25°C. The symbol ΔH° is used to indicate standard enthalpy changes, which apply to reactions involving only materials in their standard states.

The **standard enthalpy of formation**, ΔH_f°, of a compound is the enthalpy change for the reaction in which 1 mol of the compound in its standard state is made from its elements in their standard states. Enthalpy of formation, therefore, is a specific type of enthalpy change. The ΔH values for the reactions shown in Figures 3.3 and 3.4 are in reality the ΔH_f° values of $H_2O(l)$ and $HI(g)$, respectively:

$$H_2(g) + \tfrac{1}{2}O_2(g) \longrightarrow H_2O(l) \qquad \Delta H_f^\circ = -285.9\ kJ$$

$$\tfrac{1}{2}H_2(g) + \tfrac{1}{2}I_2(s) \longrightarrow HI(g) \qquad \Delta H_f^\circ = +25.9\ kJ$$

Enthalpies of formation are either measured directly or calculated from other thermochemical data by applying the law of Hess. The result of the calculation of this type given in Section 3.5,

$$C(\text{graphite}) + 2\,H_2(g) \longrightarrow CH_4(g) \qquad \Delta H_f^\circ = -74.9 \text{ kJ}$$

is the enthalpy of formation of $CH_4(g)$. Some enthalpies of formation are listed in Table 3.1.

Table 3.1 Enthalpies of formation at 25°C and 1 atm

Compound	ΔH_f° (kJ/mol)	Compound	ΔH_f° (kJ/mol)
$H_2O(g)$	−241.8	$COCl_2(g)$	−223.
$H_2O(l)$	−285.9	$SO_2(g)$	−296.9
$HF(g)$	−269.	$CO(g)$	−110.5
$HCl(g)$	−92.30	$CO_2(g)$	−393.5
$HBr(g)$	−36.2	$NO(g)$	+90.37
$HI(g)$	+25.9	$NO_2(g)$	+33.8
$H_2S(g)$	−20.2	$HNO_3(l)$	−173.2
$HCN(g)$	+130.5	$NH_4NO_3(s)$	−365.1
$NH_3(g)$	−46.19	$NaCl(s)$	−411.0
$PH_3(g)$	+9.25	$MgO(s)$	−601.83
$CH_4(g)$	−74.85	$CaO(s)$	−635.5
$C_2H_6(g)$	−84.68	$Ca(OH)_2(s)$	−986.59
$C_2H_4(g)$	+52.30	$CaCO_3(s)$	−1206.9
$C_2H_2(g)$	+226.7	$Ca_3P_2(s)$	−504.17
$C_6H_6(l)$	+49.04	$BaO(s)$	−588.1
$CH_3OH(g)$	−201.2	$BaCO_3(s)$	−1218.
$CH_3OH(l)$	−238.6	$Al_2O_3(s)$	−1669.8
$CH_3NH_2(g)$	−28.	$Fe_2O_3(s)$	−822.2
$NF_3(g)$	−113.	$AgC(s)$	−127.0
$CF_4(g)$	−913.4	$HgBr_2(s)$	−169.
$CHCl_3(l)$	−132.	$ZnO(s)$	−348.0

The enthalpy change for a reaction can be calculated from the enthalpies of formation of the compounds involved in the reaction. For example, the enthalpy change for the reaction

$$C_2H_4(g) + H_2(g) \longrightarrow C_2H_6(g) \qquad \Delta H = ?$$

can be calculated from the enthalpies of formation of ethylene, $C_2H_4(g)$, and ethane, $C_2H_6(g)$:

$$2\,C(\text{graphite}) + 2\,H_2(g) \longrightarrow C_2H_4(g) \qquad \Delta H_f^\circ = +52.30 \text{ kJ} \qquad (3.15)$$

$$2\,C(\text{graphite}) + 3\,H_2(g) \longrightarrow C_2H_6(g) \qquad \Delta H_f^\circ = -84.68 \text{ kJ} \qquad (3.16)$$

The reverse of equation (3.15) indicates a transformation in which $C_2H_4(g)$ breaks down into its elements. The appropriate value of the enthalpy change for the reversed reaction is $-\Delta H_f^\circ(C_2H_4)$ or −52.30 kJ. The elements from the decomposition of $C_2H_4(g)$ plus an additional mole of $H_2(g)$ can be imagined to form

$C_2H_6(g)$; equation (3.16) is written as shown. When these two equations are added, the desired thermochemical expression is obtained:

$$\begin{array}{ll} C_2H_4(g) \longrightarrow 2\,C(\text{graphite}) + 2\,H_2(g) & -\Delta H_f^\circ = -\ 52.30 \text{ kJ} \\ 2\,C(\text{graphite}) + 3\,H_2(g) \longrightarrow C_2H_6(g) & \Delta H_f^\circ = -\ 84.68 \text{ kJ} \\ \hline C_2H_4(g) + H_2(g) \longrightarrow C_2H_6(g) & \Delta H^\circ = -136.98 \text{ kJ} \end{array}$$

The ΔH of the reaction is, therefore, $\Delta H_f^\circ(C_2H_6) - \Delta H_f^\circ(C_2H_4)$.

In general, a ΔH° value for a reaction may be obtained by subtracting the sum of the enthalpies of formation of the reactants from the sum of the enthalpies of formation of the products:

$$\Delta H^\circ = \sum \Delta H_f^\circ(\text{products}) - \sum \Delta H_f^\circ(\text{reactants}) \tag{3.17}$$

The uppercase Greek sigma, Σ, indicates a sum. By reversing the sign of $\Sigma \Delta H_f^\circ$ (reactants), we indicate a process in which the reactants are broken down into the elements. The formation of the products from these elements is indicated by the term $\Sigma \Delta H_f^\circ$(products).

Two factors frequently cause trouble in the use of this approach for the calculation of ΔH° values.

1. Enthalpies of formation are given in kilojoules per *mole*. The chemical equation that corresponds to a value listed in Table 3.1 pertains to the formation of only *one* mole of the compound. If more than one mole (or less than one mole) of the compound is involved in the reaction being studied, the ΔH_f° value must be multiplied by the number of moles involved.

2. The enthalpy of formation of an *element* in its standard state is zero (the enthalpy change when an element is prepared from itself). No terms for elements are added into the sums $\Sigma \Delta H_f^\circ$(products) and $\Sigma \Delta H_f^\circ$(reactants).

Consider the reaction

$$2\,NH_3(g) + 3\,Cl_2(g) \longrightarrow N_2(g) + 6\,HCl(g) \qquad \Delta H^\circ = ?$$

The enthalpy change for the reaction can be calculated in the following way:

$$\begin{aligned} \Delta H^\circ &= \sum \Delta H_f^\circ(\text{products}) - \sum \Delta H_f^\circ(\text{reactants}) \\ &= 6\Delta H_f^\circ(HCl) - 2\Delta H_f^\circ(NH_3) \\ &= 6(-92.30 \text{ kJ}) - 2(-46.19 \text{ kJ}) \\ &= -553.80 \text{ kJ} + 92.38 \text{ kJ} = -461.42 \text{ kJ} \end{aligned}$$

This calculation may be checked by the addition of the appropriate thermochemical equations. Since 6 mol of HC(g) are prepared, the equation for the formation of HCl(g) is multiplied through by 6 and the enthalpy of formation of HCl(g) is multiplied by 6. Two moles of $NH_3(g)$ are *consumed* in the reaction. The equation for the *formation* of $NH_3(g)$ is multiplied through by 2 and *reversed;* the value of ΔH_f° is multiplied by 2 and the sign changed. These two thermochemical expressions are added. No separate equations are introduced for the elements that are involved in the reaction (Cl_2 and N_2):

$$
\begin{array}{lll}
3\,H_2(g) + 3\,Cl_2(g) \longrightarrow 6\,HCl(g) & \Delta H^\circ = & 6\Delta H_f^\circ = -553.80 \text{ kJ} \\
2\,NH_3(g) \longrightarrow N_2(g) + 3\,H_2(g) & \Delta H^\circ = -2\Delta H_f^\circ = & +92.38 \text{ kJ} \\
\hline
2\,NH_3(g) + 3\,Cl_2(g) \longrightarrow N_2(g) + 6\,HCl(g) & \Delta H^\circ = & -461.42 \text{ kJ}
\end{array}
$$

The terms "$3\,H_2(g)$" cancel in the addition. Notice that "$3\,Cl_2(g)$" and "$N_2(g)$" appear in the final equation even though no special provision was made for their introduction.

Example 3.4

Use enthalpies of formation to calculate the ΔH° of the reaction

$$Fe_2O_3(s) + 3\,CO(g) \longrightarrow 2\,Fe(s) + 3\,CO_2(g)$$

Solution

The values needed can be found in Table 3.1

$$
\begin{aligned}
\Delta H^\circ &= \sum \Delta H_f^\circ(\text{products}) - \sum \Delta H_f^\circ(\text{reactants}) \\
&= 3\Delta H_f^\circ(CO_2) - [\Delta H_f^\circ(Fe_2O_3) + 3\Delta H_f^\circ(CO)] \\
&= 3(-393.5 \text{ kJ}) - [(-822.2 \text{ kJ}) + 3(-110.5 \text{ kJ})] \\
&= -1180.5 \text{ kJ} + 1153.7 \text{ kJ} = -26.8 \text{ kJ}
\end{aligned}
$$

Example 3.5

Given the following data:

$$CH_4(g) + 2\,O_2(g) \longrightarrow CO_2(g) + 2\,H_2O(l) \qquad \Delta H^\circ = -890.4 \text{ kJ}$$

the enthalpy of formation of $CO_2(g)$ is -393.5 kJ/mol, the enthalpy of formation of $H_2O(l)$ is -285.9 kJ/mol. Calculate the enthalpy of formation of $CH_4(g)$.

Solution

In this case, the value of ΔH° for a reaction is known and a ΔH_f° value for one of the reactants is sought.

$$
\begin{aligned}
\Delta H^\circ &= \sum \Delta H_f^\circ(\text{products}) - \sum \Delta H_f^\circ(\text{reactants}) \\
\Delta H^\circ &= \Delta H_f^\circ(CO_2) + 2\Delta H_f^\circ(H_2O) - \Delta H_f^\circ(CH_4) \\
-890.4 \text{ kJ} &= (-393.5 \text{ kJ}) + 2(-285.9 \text{ kJ}) - \Delta H_f^\circ(CH_4) \\
-890.4 \text{ kJ} &= -965.3 \text{ kJ} - \Delta H_f^\circ(CH_4) \\
\Delta H_f^\circ(CH_4) &= -74.9 \text{ kJ}
\end{aligned}
$$

Chapter 3   Thermochemistry

3.7 Bond Energies

Atoms are held together in molecules by chemical bonds (see Chapter 6). The energy required to *break* the bond that holds two atoms together in a diatomic molecule is called the **bond dissociation energy.** These values are reported in kilojoules per mole of bonds. In the following equations, which illustrate this process, dashes are used to represent the bonds between atoms; H_2, for example, appears as H—H:

$$\text{H—H}(g) \longrightarrow 2\,\text{H}(g) \qquad \Delta H = +435 \text{ kJ}$$

$$\text{Cl—Cl}(g) \longrightarrow 2\,\text{Cl}(g) \qquad \Delta H = +243 \text{ kJ}$$

$$\text{H—Cl}(g) \longrightarrow \text{H}(g) + \text{Cl}(g) \qquad \Delta H = +431 \text{ kJ}$$

Each of these ΔH values is *positive*, which indicates that energy is *absorbed* in each process. The bond in the H_2 molecule is the strongest of the three. It takes the most energy to pull the atoms of the H_2 molecule apart.

If one of these equations is reversed, the sign of the ΔH value must be changed:

$$\text{H}(g) + \text{Cl}(g) \longrightarrow \text{H—Cl}(g) \qquad \Delta H = -431 \text{ kJ}$$

When a bond forms, energy is *released*—the same amount that is *required* to break the bond.

Bond energies may be used to determine ΔH values. Consider the reaction

$$\text{H}_2(g) + \text{Cl}_2(g) \longrightarrow 2\,\text{HCl}(g) \qquad \Delta H = 2\Delta H_f^\circ = -184.6 \text{ kJ}$$

The ΔH for this reaction is twice the enthalpy of formation of HCl(g) since the equation indicates the formation of two moles of HCl(g). We can derive this ΔH

value from bond energies in the following way. The enthalpy change is the sum of the energy *required* to break 1 mol of H—H bonds, the energy *required* to break 1 mol of Cl—Cl bonds, and the energy *evolved* by the formation of 2 mol of H—Cl bonds.

$$H—H(g) \longrightarrow 2 H(g) \qquad\qquad \Delta H = +435 \text{ kJ}$$

$$Cl—Cl(g) \longrightarrow 2 Cl(g) \qquad\qquad \Delta H = +243 \text{ kJ}$$

$$2 H(g) + 2 Cl(g) \longrightarrow 2 H—Cl(g) \qquad \Delta H = 2(-431)\text{kJ} = -862 \text{ kJ}$$

The sum of these equations is

$$H—H(g) + Cl—Cl(g) \longrightarrow 2 H—Cl(g) \qquad \Delta H = -184 \text{ kJ}$$

A molecule that contains more than two atoms, such as H_2O, is called a **polyatomic molecule.** Molecules of this type contain more than one bond. There are, for example, two H—O bonds in the H_2O molecule. The ΔH for

$$H—O—H(g) \longrightarrow 2 H(g) + O(g) \qquad \Delta H = +926 \text{ kJ}$$

refers to a process in which *two* moles of H—O bonds are broken. The **average bond energy** of the H—O bond, therefore, is $+926 \text{ kJ}/2$ mol, or $+463 \text{ kJ/mol}$.

In the H_2O molecule, the H—O bonds are equivalent. If the bonds were broken one at a time, however, the ΔH values would not be the same.

$$H—O—H(g) \longrightarrow H(g) + O—H(g) \qquad \Delta H = +501 \text{ kJ}$$

$$O—H(g) \longrightarrow O(g) + H(g) \qquad \Delta H = +425 \text{ kJ}$$

The average of the ΔH values for the steps is $+463 \text{ kJ/mol}$, which is the average bond energy. In general, the second bond of a molecule such as H_2O is easier to break than the first. The fragment remaining after one H has been removed (O—H) is not as stable as the original molecule (H—O—H).

The bond energy of a given type of bond is not the same in all molecules containing that bond. The H—O bond energy in H—O—H is different from the H—O bond energy in H—O—Cl. The values listed in Table 3.2 for diatomic molecules are bond dissociation energies. The other values listed are average bond energies, and each of these values is an average derived from a large number of cases. Since average bond energies are approximations, a ΔH value obtained by use of these values must be regarded as an estimate.

In some molecules, two atoms are bonded together by more than one bond. Two nitrogen atoms, for example, can be joined by a single bond (N—N), a double bond (N=N), or a triple bond (N≡N), depending on the molecule. Multiple bonds are indicated in Table 3.2.

Example 3.6

Use average bond energies to calculate the value of ΔH for the reaction

$$\begin{array}{c} H \\ | \\ 2 H—N—H(g) + 3 Cl—Cl(g) \longrightarrow N≡N(g) + 6 H—Cl(g) \end{array}$$

Solution

We can imagine the reaction to take place by a series of steps. Energy is absorbed (ΔH is positive) when a bond is broken, and energy is evolved (ΔH is negative) when a bond is formed.

Six N—H bonds are broken:

$$2\ \overset{\displaystyle H}{\underset{\displaystyle |}{H-N-H}}(g) \longrightarrow 2\,N(g) + 6\,H(g) \qquad \Delta H = 6(+389)\ \text{kJ} = +2334\ \text{kJ}$$

Three Cl—Cl bonds are broken:

$$3\,\text{Cl—Cl}(g) \longrightarrow 6\,\text{Cl}(g) \qquad \Delta H = 3(+243)\ \text{kJ} = +729\ \text{kJ}$$

One N≡N bond is formed:

$$2\,N(g) \longrightarrow N{\equiv}N(g) \qquad \Delta H = -941\ \text{kJ}$$

Six H—Cl bonds are formed:

$$6\,H(g) + 6\,\text{Cl}(g) \longrightarrow 6\,\text{H—Cl}(g) \qquad \Delta H = 6(-431)\ \text{kJ} = -2586\ \text{kJ}$$

The sum of these steps is the answer to the problem:

$$2\,NH_3(g) + 3\,Cl_2(g) \longrightarrow N_2(g) + 6\,HCl(g) \qquad \Delta H = -464\ \text{kJ}$$

In Section 3.6, enthalpies of formation were used to calculate the value of ΔH for this reaction. The value obtained in this way (-461 kJ) is a more reliable value than the one derived from bond energies (-464 kJ).

Table 3.2 Average bond energies[a]

Bond	Average Bond Energy (kJ/mol)	Bond	Average Bond Energy (kJ/mol)
H—H	435	P—H	318
H—F	565	N—Cl	201
H—Cl	431	P—Cl	326
H—Br	364	C—C	347
H—I	297	C=C	619
F—F	155	C≡C	812
Cl—Cl	243	C—H	414
Br—Br	193	C—O	335
I—I	151	C=O	707
O—O	138	C—F	485
O₂[b]	494	C—Cl	326
O—H	463	C—N	293
O—F	184	C=N	616
O—Cl	205	C≡N	879
N—N	159	S—H	339
N=N	418	S—S	213
N≡N	941	S—Cl	276
N—H	389		

[a] Reactants and products in gaseous state.
[b] Double bond of molecular oxygen.

Summary

The topics that have been discussed in this chapter are

1. The meaning of the term *energy* and the definition of the *joule*, the SI unit used for all energy measurments.

2. The Celsius and Fahrenheit temperature scales.

3. The definition of the calorie and the interconversion of calories and joules.

4. The measurement of the heat absorbed or released by a chemical reaction.

5. The concept of enthalpy, H, and enthalpy change, ΔH.

6. How to write thermochemical equations (chemical equations with an indication of the corresponding enthalpy change for the reaction) and the stoichiometry of thermochemical systems.

7. The law of Hess, which is the justification for the manipulation of thermochemical equations to yield new ΔH values.

8. Enthalpies of formation and the use of these values to determine enthalpy changes.

9. Bond energies and their use in the determination of approximate ΔH values.

Key Terms

Some of the more important terms introduced in this chapter are listed below. Definitions for terms not included in this list may be located in the text by use of the index.

Bond energy (Section 3.7)   The energy required to break a bond between two atoms in a molecule in the gaseous state.

calorie, cal (Section 3.2)   The amount of heat required to raise the temperature of 1 g of water from 14.5 to 15.5°C; 1 cal = 4.184 J (exactly).

Calorimeter (Section 3.3)   A device used to measure the heat transferred in chemical reactions and physical changes.

Celsius temperature scale (Section 3.2)   A temperature scale based on the assignment of 0°C to the normal freezing point of water and 100°C to the normal boiling point of water.

Endothermic reaction (Section 3.4)   A chemical reaction in which heat is absorbed.

Energy (Section 3.1)   The capacity to do work.

Enthalpy, H (Section 3.4)   The heat content of a sample of matter; for a reaction run at constant pressure, the change in enthalpy, ΔH, is the heat transferred (liberated or absorbed).

Enthalpy of formation, ΔH_f (Section 3.6)   For a given compound, the enthalpy change for a reaction in which 1 mol of the compound is prepared from its elements. If the elements and the compound are in their standard states, the value is called the standard enthalpy of formation, ΔH_f.

Exothermic reaction (Section 3.4)   A chemical reaction in which heat is liberated.

Fahrenheit temperature scale (Section 3.2)   A temperature scale on which the normal freezing point of water is 32°F and the normal boiling point of water is 212°F.

Joule, J (Section 3.1)   The SI unit for all energy measurements; 1 kg·m²/s².

Heat capacity (Section 3.3)   The amount of heat required to raise the temperature of a given mass by 1°C.

Law of Hess, law of constant heat summation (Section 3.5)   The change in enthalpy for any chemical reaction is constant, whether the reaction occurs in one step or in several steps.

Polyatomic molecule (Section 3.7)   A molecule that contains more than two atoms.

Specific heat (Section 3.2)   The amount of heat required to raise the temperature of 1 g of a substance by 1°C.

Standard state (Section 3.6)   The state of a substance in which the substance is stable at 1 atm pressure and 25°C.

Temperature (Section 3.2)   Degree of hotness.

Thermochemistry (Introduction)   Study of the energy changes that accompany chemical and physical changes.

Problems*

Heat Measurements, Calorimetry

3.1 Normal body temperature is 98.6°F. What is normal body temperature in degrees Celsius?

3.2 What temperature in degrees Celsius corresponds to 0°F?

3.3 What temperature in degrees Celsius corresponds to −40°F?

3.4 A thermostat is set at 68°F. What is the setting in degrees Celsius?

3.5 What is the heat capacity of 325 g of water?

3.6 How many kilojoules of heat are required to raise the temperature of 1.50 kg of water from 22.00°C to 25.00°C?

3.7 How many kilojoules of heat are liberated by a reaction if the heat produced by the reaction raises the temperature of 1.75 kg of water from 23.00°C to 42.00°C?

3.8 What is the specific heat of ethyl alcohol if 129 J of heat is required to raise the temperature of 15.0 g of ethyl alcohol from 22.70°C to 26.20°C?

3.9 The specific heat of lead is 0.129 J/(g°C). How many joules of heat are required to raise the temperature of 207 g of lead from 22.25°C to 27.65°C?

3.10 The specific heat of diethyl ether is 2.33 J/(g°C). If 113 J of heat raises the temperature of a 12.5 g sample of diethyl ether to 27.35°C, what was the initial temperature of the sample?

3.11 The specific heat of nickel is 0.444 J/(g°C). If 50.0 J of heat is added to a 32.3 g sample of nickel at 23.25°C, what is the final temperature of the sample?

3.12 If 95.5 J of heat raises the temperature of a sample of gold from 21.50°C to 29.35°C, what is the mass of the sample? The specific heat of gold is 0.132 J/(g°C).

3.13 A 1.45 g sample of acetic acid, $HC_2H_3O_2$, was burned in excess oxygen in a calorimeter. The calorimeter contained 0.750 kg of water and had a heat capacity of 2.67 kJ/°C. The temperature of the calorimeter and its contents increased from 24.32°C to 27.95°C. What quantity of heat would be liberated by the combustion of 1.00 mol of acetic acid?

3.14 A 2.30 g sample of quinone, $C_6H_4O_2$, was burned in excess oxygen in a calorimeter. The calorimeter contained 1.00 kg of water and had a heat capacity of 3.27 kJ/°C. The temperature of the calorimeter and its contents increased from 19.22°C to 27.07°C. What quantity of heat would be liberated by the combustion of 1.00 mol of quinone?

3.15 The combustion of 1.00 mol of glucose, $C_6H_{12}O_6$, liberates 2.82×10^3 kJ of heat. If 1.25 g of glucose is burned in a calorimeter containing 0.950 kg of water, and the temperature of the assembly increases from 20.10°C to 23.25°C, what is the heat capacity of the calorimeter?

3.16 The combustion of 1.00 mol of sucrose, $C_{12}H_{22}O_{11}$, liberates 5.65×10^3 kJ of heat. A calorimeter that has a heat capacity of 1.23 kJ/°C contains 0.600 kg of water. How many grams of sucrose should be burned to raise the temperature of the calorimeter and its contents from 23.00°C to 27.00°C?

Thermochemical Equations

3.17 Indicate whether each of the following reactions is *exothermic* or *endothermic*:

(a) $Br_2(l) + Cl_2(g) \longrightarrow 2\,BrCl(g) \quad \Delta H = +29.4\,kJ$

(b) $NH_3(g) + HCl(g) \longrightarrow NH_4Cl(s) \quad \Delta H = -176\,kJ$

(c) $N_2O_4(g) \longrightarrow 2\,NO_2(g) \quad \Delta H = +58.0\,kJ$

(d) $CS_2(l) + 3\,Cl_2(g) \longrightarrow CCl_4(l) + S_2Cl_2(l)$
$$\Delta H = -112\,kJ$$

3.18 The combustion of 1.000 g of benzene, $C_6H_6(l)$, in $O_2(g)$ liberates 41.84 kJ of heat and yields $CO_2(g)$ and $H_2O(l)$. Write the thermochemical equation for the combustion of *one mole* of $C_6H_6(l)$.

3.19 The combustion of 1.000 g of ethyl alcohol, $C_2H_5OH(l)$, liberates 29.69 kJ of heat and yields $CO_2(g)$ and $H_2O(l)$. Write the thermochemical equation for the combustion of *one mole* of $C_2H_5OH(l)$.

3.20 Hydrazine, $N_2H_4(l)$, is used in rocket fuel. The thermochemical equation for the combustion of hydrazine is

$$N_2H_4(l) + O_2(g) \longrightarrow N_2(g) + 2\,H_2O(l)$$
$$\Delta H = -622.4\,kJ$$

What quantity of heat is liberated by the combustion of 1.000 g of $N_2H_4(l)$?

3.21 Glucose, $C_6H_{12}O_6(s)$, is converted into ethyl alcohol, $C_2H_5OH(l)$, in the fermentation of fruit juice to produce wine:

$$C_6H_{12}O_6(s) \longrightarrow 2\,C_2H_5OH(l) + 2\,CO_2(g)$$
$$\Delta H = -67.0\,kJ$$

What quantity of heat is liberated when a liter of wine containing 95.0 g of $C_2H_5OH(l)$ is produced?

3.22 Given the thermochemical equation

$$2\,NaN_3(s) \longrightarrow 2\,Na(s) + 3\,N_2(g) \quad \Delta H = +42.7\,kJ$$

---

* The more difficult problems are marked with asterisks. The appendix contains answers to color-keyed problems.

What is the value of ΔH for the preparation of 0.150 kg of $N_2(g)$?

3.23 Given the thermochemical equation

$$2\,NH_3(g) + 3\,N_2O(g) \longrightarrow 4\,N_2(g) + 3\,H_2O(l)$$
$$\Delta H = -1010\,kJ$$

(a) What quantity of heat is liberated by the reaction of 50.0 g of $N_2O(g)$ with excess $NH_3(g)$? **(b)** What quantity of heat is liberated by the reaction that produces 50.0 g of $N_2(g)$?

Law of Hess

3.24 Given:
(a) $H_2S(g) + \frac{3}{2}O_2(g) \longrightarrow H_2O(l) + SO_2(g)$
$$\Delta H = -562.6\,kJ$$
(b) $CS_2(l) + 3\,O_2(g) \longrightarrow CO_2(g) + 2\,SO_2(g)$
$$\Delta H = -1075.2\,kJ$$

Calculate the value of ΔH for the reaction

$$CS_2(l) + 2\,H_2O(l) \longrightarrow CO_2(g) + 2\,H_2S(g)$$

3.25 Given:
(a) $2\,NH_3(g) + 3\,N_2O(g) \longrightarrow 4\,N_2(g) + 3\,H_2O(l)$
$$\Delta H = -1010\,kJ$$
(b) $4\,NH_3(g) + 3\,O_2(g) \longrightarrow 2\,N_2(g) + 6\,H_2O(l)$
$$\Delta H = -1531\,kJ$$

Calculate the value of ΔH for the reaction

$$N_2(g) + \tfrac{1}{2}O_2(g) \longrightarrow N_2O(g)$$

3.26 Given:
(a) $2\,NF_3(g) + 2\,NO(g) \longrightarrow N_2F_4(g) + 2\,ONF(g)$
$$\Delta H = -82.9\,kJ$$
(b) $NO(g) + \tfrac{1}{2}F_2(g) \longrightarrow ONF(g)\quad \Delta H = -156.9\,kJ$

(c) $Cu(s) + F_2(g) \longrightarrow CuF_2(s)\quad \Delta H = -531.0\,kJ$

Calculate the value of ΔH for the reaction

$$2\,NF_3(g) + Cu(s) \longrightarrow N_2F_4(g) + CuF_2(s)$$

3.27 Given:
(a) $FeO(s) + H_2(g) \longrightarrow Fe(s) + H_2O(g)$
$$\Delta H = +24.7\,kJ$$
(b) $3\,FeO(s) + \tfrac{1}{2}O_2(g) \longrightarrow Fe_3O_4(s)\quad \Delta H = -317.6\,kJ$

(c) $H_2(g) + \tfrac{1}{2}O_2(g) \longrightarrow H_2O(g)\quad \Delta H = -241.8\,kJ$

Calculate the value of ΔH for the reaction

$$3\,Fe(s) + 4\,H_2O(g) \longrightarrow Fe_3O_4(s) + 4\,H_2(g)$$

3.28 Given:
(a) $BCl_3(g) + 3\,H_2O(l) \longrightarrow H_3BO_3(s) + 3\,HCl(g)$
$$\Delta H = -112.5\,kJ$$
(b) $B_2H_6(g) + 6\,H_2O(l) \longrightarrow 2\,H_3BO_3(s) + 6\,H_2(g)$
$$\Delta H = -493.4\,kJ$$
(c) $\tfrac{1}{2}H_2(g) + \tfrac{1}{2}Cl_2(g) \longrightarrow HCl(g)\quad \Delta H = -92.3\,kJ$

Calculate the value of ΔH for the reaction

$$B_2H_6(g) + 6\,Cl_2(g) \longrightarrow 2\,BCl_3(g) + 6\,HCl(g)$$

3.29 Given:
(a) $OF_2(g) + H_2O(l) \longrightarrow O_2(g) + 2\,HF(g)$
$$\Delta H = -276.6\,kJ$$
(b) $SF_4(g) + 2\,H_2O(l) \longrightarrow SO_2(g) + 4\,HF(g)$
$$\Delta H = -827.5\,kJ$$
(c) $S(s) + O_2(g) \longrightarrow SO_2(g)\quad \Delta H = -296.9\,kJ$

Calculate the value of ΔH for the reaction

$$2\,S(s) + 2\,OF_2(g) \longrightarrow SO_2(g) + SF_4(g)$$

3.30 Given:
(a) $OSCl_2(l) + H_2O(l) \longrightarrow SO_2(g) + 2\,HCl(g)$
$$\Delta H = +10.3\,kJ$$
(b) $PCl_3(l) + \tfrac{1}{2}O_2(g) \longrightarrow OPCl_3(l)\quad \Delta H = -325.1\,kJ$

(c) $P(s) + \tfrac{3}{2}Cl_2(g) \longrightarrow PCl_3(l)\quad \Delta H = -306.7\,kJ$

(d) $4\,HCl(g) + O_2(g) \longrightarrow 2\,Cl_2(g) + 2\,H_2O(l)$
$$\Delta H = -202.6\,kJ$$

Calculate the value of ΔH for the reaction

$$2\,P(s) + 2\,SO_2(g) + 5\,Cl_2(g) \longrightarrow$$
$$2\,OSCl_2(l) + 2\,OPCl_3(l)$$

3.31 Given:
(a) $2\,ClF_3(g) + 2\,NH_3(g) \longrightarrow N_2(g) + 6\,HF(g) + Cl_2(g)$
$$\Delta H = -1195.6\,kJ$$
(b) $N_2H_4(l) + O_2(g) \longrightarrow N_2(g) + 2\,H_2O(l)$
$$\Delta H = -622.4\,kJ$$
(c) $4\,NH_3(g) + 3\,O_2(g) \longrightarrow 2\,N_2(g) + 6\,H_2O(l)$
$$\Delta H = -1530.6\,kJ$$

Calculate the value of ΔH for the reaction

$$3\,N_2H_4(l) + 4\,ClF_3(g) \longrightarrow$$
$$3\,N_2(g) + 12\,HF(g) + 2\,Cl_2(g)$$

*__3.32__ Given:
(a) $NH_3(g) + HNO_3(l) \longrightarrow NH_4NO_3(s)$
$$\Delta H = -145.7\,kJ$$
(b) $NH_4NO_3(s) \longrightarrow N_2O(g) + 2\,H_2O(l)$
$$\Delta H = -125.2\,kJ$$
(c) $3\,NO(g) \longrightarrow N_2O(g) + NO_2(g)\quad \Delta H = -155.8\,kJ$

(d) $4\,NH_3(g) + 5\,O_2(g) \longrightarrow 4\,NO(g) + 6\,H_2O(l)$
$$\Delta H = -1169.2\,kJ$$

(e) $NO(g) + \tfrac{1}{2}O_2(g) \longrightarrow NO_2(g)\quad \Delta H = -56.6\,kJ$

Calculate the value of ΔH for the reaction

$$3\,NO_2(g) + H_2O(l) \longrightarrow 2\,HNO_3(l) + NO(g)$$

*__3.33__ Given:
(a) $2\,NH_3(g) + 3\,N_2O(g) \longrightarrow 4\,N_2(g) + 3\,H_2O(l)$
$$\Delta H = -1010\,kJ$$
(b) $N_2O(g) + 3\,H_2(g) \longrightarrow N_2H_4(l) + H_2O(l)$
$$\Delta H = -317\,kJ$$

(c) $2 NH_3(g) + \frac{1}{2} O_2(g) \longrightarrow N_2H_4(l) + H_2O(l)$
$$\Delta H = -143 \text{ kJ}$$

(d) $H_2(g) + \frac{1}{2} O_2(g) \longrightarrow H_2O(l) \quad \Delta H = -286 \text{ kJ}$

Calculate the value of ΔH for the reaction

$$N_2H_4(l) + O_2(g) \longrightarrow N_2(g) + 2 H_2O(l)$$

Enthalpies of Formation

3.34 Write thermochemical equations that correspond to the following standard enthalpies of formation: **(a)** $AgCl(s)$, -127 kJ/mol; **(b)** $NO_2(g)$, $+33.8$ kJ/mol; **(c)** $CaCO_3(s)$, -1206.9 kJ/mol.

3.35 Write thermochemical equations that correspond to the following standard enthalpies of formation: **(a)** $HCN(g)$, $+130.5$ kJ/mol; **(b)** $CS_2(l)$, $+87.86$ kJ/mol; **(c)** $NH_4NO_3(s)$, -365.1 kJ/mol.

3.36 Use enthalpies of formation (Table 3.1) to calculate the value of $\Delta H°$ for the reaction

$$2 H_2S(g) + 3 O_2(g) \longrightarrow 2 H_2O(l) + 2 SO_2(g)$$

3.37 Use enthalpies of formation (Table 3.1) to calculate the value of $\Delta H°$ for the reaction

$$Fe_2O_3(s) + 3 H_2(g) \longrightarrow 3 Fe(s) + 3 H_2O(g)$$

3.38 Use enthalpies of formation (Table 3.1) to calculate the value of $\Delta H°$ for the reaction

$$2 NH_3(g) + 2 CH_4(g) + 3 O_2(g) \longrightarrow$$
$$2 HCN(g) + 6 H_2O(l)$$

3.39 Use enthalpies of formation (Table 3.1) to calculate the value of $\Delta H°$ for the reaction

$$NH_3(g) + 3 F_2(g) \longrightarrow NF_3(g) + 3 HF(g)$$

3.40 **(a)** Write the chemical equation for the combustion of one mole of methyl alcohol, $CH_3OH(l)$, in $O_2(g)$. The products of the reaction are $CO_2(g)$ and $H_2O(l)$. **(b)** Use enthalpies of formation (Table 3.1) to calculate the value of $\Delta H°$ for the reaction.

3.41 **(a)** Write the chemical equation for the combustion of one mole of benzene, $C_6H_6(l)$, in $O_2(g)$. The products of the reaction are $CO_2(g)$ and $H_2O(l)$. **(b)** Use enthalpies of formation (Table 3.1) to calculate the value of $\Delta H°$ for the reaction.

3.42 **(a)** Write the thermochemical equation for the combustion of one mole of hydrazine, $N_2H_4(l)$, in $O_2(g)$. The products of the reaction are $N_2(g)$ and $H_2O(l)$, and the value of $\Delta H°$ for the reaction is -622.4 kJ. **(b)** Use your answer together with values from Table 3.1 to calculate the enthalpy of formation of hydrazine.

3.43 **(a)** Write the thermochemical equation for the combustion of one mole of urea, $CO(NH_2)_2(s)$, in $O_2(g)$. The products of the reaction are $N_2(g)$, $CO_2(g)$, and $H_2O(l)$, and the value of $\Delta H°$ for the reaction is -632 kJ. **(b)** Use your answer together with values from Table 3.1 to calculate the standard enthalpy of formation of urea.

3.44 Given the thermochemical equation

$$2 ClF_3(g) + 2 NH_3(g) \longrightarrow N_2(g) + 6 HF(g) + Cl_2(g)$$
$$\Delta H° = -1195.6 \text{ kJ}$$

use this equation and values from Table 3.1 to calculate the standard enthalpy of formation of $ClF_3(g)$.

3.45 Given the thermochemical equation

$$CaCO_3(s) + 2 NH_3(g) \longrightarrow CaCN_2(s) + 3 H_2O(l)$$
$$\Delta H° = +90.1 \text{ kJ}$$

use this equation and values from Table 3.1 to calculate the standard enthalpy of formation of $CaCN_2(s)$.

3.46 Given the thermochemical equation

$$CaC_2(s) + 2 H_2O(l) \longrightarrow C_2H_2(g) + Ca(OH)_2(s)$$
$$\Delta H° = -125.3 \text{ kJ}$$

use this equation and values from Table 3.1 to calculate the standard enthalpy of formation of $CaC_2(s)$.

Bond Energies

3.47 Use average bond energies (Table 3.2) to calculate the enthalpy of formation of $HCl(g)$. Compare the answer to the value given in Table 3.1.

3.48 Use average bond energies (Table 3.2) to calculate the enthalpy of formation of $N_2H_4(g)$:

$$N{\equiv}N(g) + 2 H{-}H(g) \longrightarrow \underset{\displaystyle \overset{|}{H} \quad \overset{|}{H}}{H{-}N{-}N{-}H}(g)$$

3.49 For $C(\text{graphite}) \rightarrow C(g)$, $\Delta H = +717$ kJ/mol. Use this fact together with average bond energies from Table 3.2 to find the enthalpy of formation of $HCN(g)$:

$$\frac{1}{2} H{-}H(g) + C(\text{graphite}) + \frac{1}{2} N{\equiv}N(g) \longrightarrow$$
$$H{-}C{\equiv}N(g)$$

3.50 For $C(\text{graphite}) \rightarrow C(g)$, $\Delta H = +717$ kJ/mol. Use this fact together with average bond energies from Table 3.2 to find the enthalpy of formation of $CH_3OH(g)$:

$$C(\text{graphite}) + 2 H{-}H(g) + \frac{1}{2} O_2(g) \longrightarrow$$

$$\underset{\displaystyle \overset{|}{H}}{\overset{\displaystyle \overset{H}{|}}{H{-}C{-}O{-}H}}(g)$$

3.51 Given the thermochemical equation

$$XeF_2(g) + H_2(g) \longrightarrow 2 HF(g) + Xe(g)$$
$$\Delta H = -430 \text{ kJ}$$

use this equation together with average bond energies from Table 3.2 to calculate the average bond energy of the Xe—F bond in $XeF_2(g)$.

3.52 The standard enthalpy of formation of $ClF_5(g)$ is -254.8 kJ/mol. Use this value together with average bond energies from Table 3.2 to calculate the average bond energy of the Cl—F bond in $ClF_5(g)$.

3.53 Use average bond energies (Table 3.2) to calculate ΔH for the reaction:

$$
\underset{\underset{\text{H}}{|}}{\text{H}}\text{C}=\underset{\underset{\text{H}}{|}}{\text{C}}\text{—H(g)} + \text{H—H(g)} \longrightarrow \text{H}\text{—}\underset{\underset{\text{H}}{|}}{\overset{\overset{\text{H}}{|}}{\text{C}}}\text{—}\underset{\underset{\text{H}}{|}}{\overset{\overset{\text{H}}{|}}{\text{C}}}\text{—H(g)}
$$

3.54 Use average bond energies (Table 3.2) to calculate ΔH for the reaction:

$$
\text{H}\text{—}\underset{\underset{\text{H}}{|}}{\overset{\overset{\text{H}}{|}}{\text{C}}}\text{—O—H(g)} + \text{H—Cl(g)} \longrightarrow
$$

$$
\text{H}\text{—}\underset{\underset{\text{H}}{|}}{\overset{\overset{\text{H}}{|}}{\text{C}}}\text{—Cl(g)} + \text{H—O—H(g)}
$$

3.55 Use average bond energies (Table 3.2) to calculate ΔH for the reaction:

$$
4\,\text{H—Cl(g)} + \text{O}_2(g) \longrightarrow
$$
$$
2\,\text{H—O—H(g)} + 2\,\text{Cl—Cl(g)}
$$

3.56 Use average bond energies (Table 3.2) to calculate ΔH for the reaction:

$$
\text{F—O—F(g)} + \text{H—O—H(g)} \longrightarrow
$$
$$
\text{O}_2(g) + 2\,\text{H—F(g)}
$$

3.57 Use average bond energies (Table 3.2) to calculate ΔH for the reaction:

$$
4\,\text{H—}\underset{\underset{\text{H}}{|}}{\overset{\overset{}{}}{\text{N}}}\text{—H(g)} + 3\,\text{O}_2(g) \longrightarrow
$$

$$
2\,\text{N}\equiv\text{N(g)} + 6\,\text{H—O—H(g)}
$$

3.58 (a) Use average bond energies (Table 3.2) to calculate ΔH for the reaction:

$$
\text{H—C}\equiv\text{N(g)} + 2\,\text{H—H(g)} \longrightarrow \text{H}\text{—}\underset{\underset{\text{H}}{|}}{\overset{\overset{\text{H}}{|}}{\text{C}}}\text{—}\underset{\underset{\text{H}}{|}}{\overset{\overset{\text{H}}{|}}{\text{N}}}\text{(g)}
$$

(b) Calculate ΔH for the reaction using enthalpies of formation from Table 3.1. How well do the values agree?

3.59 (a) Use average bond energies (Table 3.2) to calculate ΔH for the reaction:

$$
\text{H}\text{—}\underset{\underset{\text{H}}{|}}{\overset{\overset{\text{H}}{|}}{\text{C}}}\text{—H(g)} + 4\,\text{F—F(g)} \longrightarrow
$$

$$
\text{F}\text{—}\underset{\underset{\text{F}}{|}}{\overset{\overset{\text{F}}{|}}{\text{C}}}\text{—F(g)} + 4\,\text{H—F(g)}
$$

(b) Calculate ΔH for the reaction using enthalpies of formation from Table 3.1. How well do the values agree?

ATOMIC STRUCTURE

Modern atomic theory has provided a foundation upon which modern chemistry has been built. An understanding of atomic structure and the ways in which atoms interact is central to an understanding of chemistry.

Individual atoms cannot be weighed, measured, or examined directly. Indirect evidence, therefore, has been used to develop the atomic theory. Some nineteenth-century scientists (Wilhelm Ostwald, for example) did not believe that atoms exist. Others (Michael Faraday, for example) thought of the atomic theory as a convenient concept that may or may not represent reality. Still others (Ludwig Boltzmann, for example) believed in the physical existence of atoms.

Whether atoms are real or imaginary was an open question as late as 1904. In that year a debate, attended by noted international scientists, was held at the St. Louis World's Fair. Ostwald argued against the existence of atoms and Jacobus van't Hoff and Boltzmann defended the other side.

4.1 The Electron

In Dalton's theory and in the theories of the Greeks, atoms were regarded as the smallest possible components of matter. Toward the end of the nineteenth century, it began to appear that the atom itself might be composed of even smaller particles. This change in viewpoint was brought about by experiments with electricity.

In 1807–8, the English chemist Humphry Davy discovered five elements (potassium, sodium, calcium, strontium, and barium) by using electricity to decompose compounds. This work led Davy to propose that elements are held together in compounds by attractions that are electrical in nature.

In 1832–33, Michael Faraday ran an important series of experiments on chemical electrolysis, in which compounds are decomposed by electricity. Faraday studied the relationship between the amount of electricity used and the amount of compound decomposed. He formulated the laws of chemical electrolysis (Section 18.4). On the basis of Faraday's work, George Johnstone Stoney proposed in 1874 that units of electrical charge are associated with atoms. In 1891, Stoney suggested that these units be called electrons.

Attempts to pass an electric current through a vacuum led to the discovery of cathode rays by Julius Plücker in 1859. Two electrodes are sealed in a glass tube from which air is almost completely removed. When a high voltage is impressed across these electrodes, rays stream from the negative electrode, called the cathode.

These rays are negatively charged, travel in straight lines, and cause the walls opposite the cathode to glow.

In the latter part of the nineteenth century, cathode rays were extensively studied. The results of the experiments of many scientists led to the conclusion that the rays are streams of fast-moving, negatively charged particles. These particles were eventually called electrons as suggested by Stoney. The electrons, which originate from the metal of which the cathode is composed, are the same no matter what material is employed for the electrode.

Since unlike charges attract, the streams of electrons that constitute a cathode ray are attracted toward the positive plate when two oppositely charged plates are placed on either side of them (Figure 4.1, path c). The rays, therefore, are deflected from their usual straight-line path in an electric field. The degree of deflection varies

1. *Directly* with the size of the charge of the particle, e. A particle with a high charge is deflected more than one with a low charge.

2. *Inversely* with the mass of the particle, m. A particle with a large mass is deflected less than one with a small mass.

The ratio of charge to mass, e/m, therefore, determines the extent to which the electrons are deflected from a straight-line path in an electric field. Electrons are deflected in a magnetic field also. In this case, however, the deflection is at right angles to the applied field (Figure 4.1, path a).

In 1897, Joseph J. Thomson determined the value of e/m for the electron by studying the deflections of cathode rays in electric and magnetic fields. The value is

$$e/m = -1.7588 \times 10^8 \text{ C/g}$$

The *coulomb* (C) is the SI unit of electric charge. One coulomb is the quantity of charge that passes a given point in an electric circuit in one second when the current is one ampere.

The first precise measurement of the charge of the electron was made by Robert A. Millikan in 1909. In Millikan's experiment (Figure 4.2), electrons are produced by the action of X rays on the molecules of which air is composed. Very small drops of oil pick up electrons and acquire electric charges. The oil drops are allowed to settle between two horizontal plates, and the mass of a single drop is determined by measuring its rate of fall.

When the plates are charged, the rate of fall of the drop is altered because the negatively charged drop is attracted to the positive plate. Measurement of the rate of fall in these circumstances enables calculation of the charge on the drop. Since a given drop can pick up one or more electrons, the charges calculated in this way are not identical. They are, however, all simple multiples of the same value, which is assumed to be the charge on a single electron:

$$e = -1.6022 \times 10^{-19} \text{ C}$$

The mass of the electron can be calculated from the value of e/m and the value of e:

$$m = \frac{e}{e/m} = \frac{-1.6022 \times 10^{-19} \text{ C}}{-1.7588 \times 10^8 \text{ C/g}} = 9.1096 \times 10^{-28} \text{ g}$$

Joseph J. Thomson, 1856–1940. *Burndy Library, courtesy AIP Niels Bohr Library.*

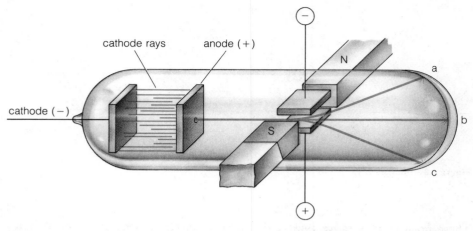

cathode rays   anode (+)

cathode (−)

N

a

c

b

S

c

−

+

Figure 4.1 Deflections of cathode rays. (a) In a magnetic field. (b) With magnetic and electric fields balanced; no net deflection. (c) In an electric field.

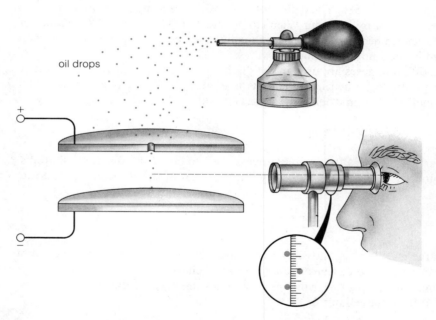

oil drops

+

−

Figure 4.2 Millikan's determination of the charge on the electron

4.2 The Proton

If one or more electrons are removed from a neutral atom or molecule, the residue has a positive charge equal to the sum of the negative charges of the electrons removed. If one electron is removed from an atom of neon (symbol, Ne), a Ne^+ ion results; if two are removed, a Ne^{2+} ion results; and so forth. Positive particles of this type (positive ions) are formed in an electric discharge tube when cathode rays rip electrons from the atoms or molecules of the gas present in the tube.

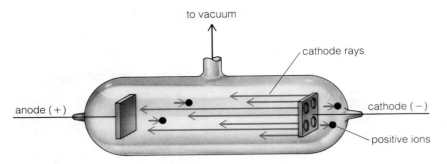

Figure 4.3 Positive rays

These positive ions move *toward* the negative electrode (the cathode). If holes have been bored in this electrode, the positive ions pass through them (see Figure 4.3). The electrons of the cathode rays, since they are negatively charged, move in the opposite direction.

These streams of positive ions, called **positive rays,** were first observed by Eugen Goldstein in 1886. The deflections of positive rays in electric and magnetic fields were studied by Wilhelm Wien (1898) and J. J. Thomson (1906). Values of e/m were determined for the positive ions by using essentially the same method employed in the study of cathode rays.

When different gases are used in the discharge tube, different types of positive ions are produced. When hydrogen gas is used, a positive particle results that has the smallest mass (and hence the largest e/m value) of any positive ion observed:

$$e/m = +9.5791 \times 10^4 \text{ C/g}$$

These particles, now called **protons,** are assumed to be a component of all atoms. The proton has a charge equal in magnitude to that of the electron but opposite in sign:

$$e = +1.6022 \times 10^{-19} \text{ C}$$

This charge is called the **unit electrical charge.** The proton is said to have a unit positive charge and the electron, a unit negative charge.

The mass of the proton, which is 1836 times the mass of the electron, can be calculated from these data:

$$m = \frac{e}{e/m} = \frac{+1.6022 \times 10^{-19} \text{ C}}{+9.5791 \times 10^4 \text{ C/g}} = 1.6726 \times 10^{-24} \text{ g}$$

4.3 The Neutron

Since atoms are electrically neutral, a given atom must contain as many electrons as protons. To account for the total masses of atoms, Ernest Rutherford in 1920 postulated the existence of an uncharged particle. Since this particle is uncharged, it is difficult to detect and characterize. In 1932, however, James Chadwick published the results of his work, which established the existence of the **neutron.** He

Table 4.1 Subatomic particles

| Particle | Mass | | Charge[b] |
	Grams	Unified Atomic Mass Units[a]	
Electron	9.109535×10^{-28}	0.0005485803	1−
Proton	1.672649×10^{-24}	1.007276	1+
Neutron	1.674954×10^{-24}	1.008665	0

[a] The unified atomic mass unit (u) is 1/12 the mass of a ^{12}C atom (Section 4.7).

[b] The unit charge is 1.60219×10^{-19} C.

was able to calculate the mass of the neutron from data on certain nuclear reactions (see Chapter 25) in which neutrons are produced. By taking into account the masses and energies of all particles used and produced in these reactions, Chadwick determined the mass of the neutron, which is very slightly larger than that of the proton. The neutron has a mass of 1.6750×10^{-24} g and the proton, 1.6726×10^{-24} g.

The properties of the electron, proton, and neutron are summarized in Table 4.1. At the present time, many other subatomic particles have been identified. Their relationship to atomic structure, however, is not completely understood. For the study of chemistry, atomic structure is adequately explained on the basis of the electron, proton, and neutron.

4.4 The Nuclear Atom

Certain atoms are unstable combinations of the subatomic particles. These atoms spontaneously emit rays and in this way change into atoms with a different chemical identity. This process, called **radioactivity,** was discovered by Henri Becquerel in 1896. In subsequent years, Ernest Rutherford explained the nature of the three types of rays emitted by radioactive substances that occur in nature (see Table 4.2). Other types of rays have now been identified; these rays, however, result from the disintegration of atoms that do not occur in nature, but have been made by nuclear reactions.

Ernest Rutherford, 1871–1937. *American Institute of Physics, Niels Bohr Library.*

Alpha rays consist of particles that carry a 2+ charge and have a mass approximately four times that of the proton. These α particles are ejected from the radioactive substance at speeds around 16,000 km/s (approximately 0.05 times the speed of light). When α particles were first studied, the neutron had not yet been discovered. We now know that an α particle consists of two protons and two neutrons. **Beta rays** are streams of electrons that travel at approximately 130,000 km/s (approximately 0.4 times the speed of light). **Gamma radiation** is essentially a highly energetic form of light. Gamma rays are uncharged and are similar to X rays.

In 1911, Rutherford reported the results of experiments in which α particles were used to investigate the structure of the atom. A beam of α particles was directed against a very thin (about 0.0004-cm thick) foil of gold, platinum, silver, or copper. The large majority of the α particles went directly through the foil in use. Some, however, were deflected from their straight-line path, and a few recoiled back toward their source (Figure 4.4).

Table 4.2 Types of radioactive emissions			
Ray	Symbol	Composition	Charge of Component
Alpha	α	particles containing 2 protons and 2 neutrons	2+
Beta	β	electrons	1−
Gamma	γ	very short wavelength electromagnetic radiation	0

Rutherford explained the results of these experiments by proposing that a **nucleus** exists in the center of the atom. Most of the mass and all of the positive charge of the atom are concentrated in the nucleus. The electrons (which occupy most of the total volume of the atom) are outside the nucleus (**extranuclear**) and in rapid motion around it. The nucleus is now believed to contain protons and neutrons, which together account for the mass of the nucleus. Since an atom is electrically neutral, the total positive charge of the nucleus (from the protons that it contains) equals the total negative charge of all the electrons. The number of electrons, therefore, is equal to the number of protons.

It is important to grasp the scale of this model of the atom. If a nucleus of an atom were the size of a tennis ball, the atom would have a diameter of over one mile. Since most of the volume of an atom is empty space, most α particles pass directly through the target foils. The comparatively light electrons do not deflect the heavier, fast-moving α particles. A close approach of an α particle (which is positively charged) to a nucleus (which is also positively charged) results in the repulsion of the α particle and deflection from its straight-line path. In a very few instances, when an α particle scores a direct hit on a nucleus, the particle is reflected back toward its source.

The composition and stability of the nucleus are discussed in Chapter 25.

4.5 Atomic Symbols

An atom is identified by two numbers: the atomic number and the mass number. The **atomic number**, Z, is the number of unit positive charges on the nucleus. Since the proton has a $1+$ charge, the atomic number is equal to the number of protons in the nucleus of the atom. An atom is electrically neutral. The atomic number, therefore, also indicates the number of extranuclear electrons in an uncombined atom.

The **mass number**, A, of an atom is the total number of protons and neutrons (collectively called **nucleons**) in the nucleus of the atom. The number of neutrons can be calculated by subtracting the atomic number (the number of protons) from the mass number (the number of protons and neutrons taken together):

$$\text{number of neutrons} = A - Z \tag{4.1}$$

The mass *number* is the total *number* of nucleons in a nucleus and not the mass of the nucleus. The mass number, however, is a whole-number approximation of the atomic mass in atomic mass units (u) since the mass of the proton as well as

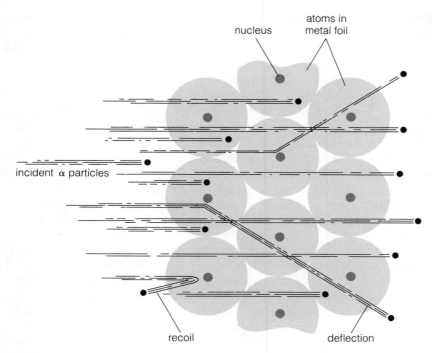

Figure 4.4 Deflections and recoil of α particles by nuclei of metal foil in Rutherford's experiment. (Not to scale.)

the mass of the neutron is approximately equal to 1 u and the mass of the electron is negligible.

An atom of an element is designated by the chemical symbol for the element with the atomic number of the element placed at the lower left and the mass number placed at the upper left*:

$$^A_Z Symbol$$

The symbol $^{35}_{17}Cl$ designates an atom of chlorine that has 17 protons (Z) and 18 neutrons ($A - Z$) in the nucleus and 17 extranuclear electrons (Z). An atom of sodium with the symbol $^{23}_{11}Na$ has 11 protons and 12 neutrons in the nucleus and 11 electrons in motion around this nucleus.

4.6 Isotopes

All atoms of a given element have the same atomic number. Some elements, however, consist of several types of atoms that differ from one another in mass number. Atoms that have the same atomic number but different mass numbers are called isotopes.

---

* The other two corners are reserved for other designations: the upper right for the charge if the atom has lost or gained electrons and become an ion, the lower right corner for the number of atoms present in a molecule or a formula unit.

Two isotopes of chlorine occur in nature: $^{35}_{17}Cl$ and $^{37}_{17}Cl$. The atomic compositions of these isotopes are:

$^{35}_{17}Cl$	17 protons	18 neutrons	17 electrons
$^{37}_{17}Cl$	17 protons	20 neutrons	17 electrons

Both these atoms have 17 protons and 17 electrons, but $^{35}_{17}Cl$ has 18 neutrons and $^{37}_{17}Cl$ has 20 neutrons. Isotopes, therefore, differ in the number of neutrons in the nucleus, which means that they differ in atomic mass.

The chemical properties of an atom depend principally upon the numbers of protons and electrons that the atom contains, which are determined by the atomic number. Isotopes of the same element, therefore, have very similar (in most cases, indistinguishable) chemical properties. Some elements exist in nature in only one isotopic form (for example, sodium, beryllium, and fluorine). Most elements, however, have more than one natural isotope—tin has 10.

The **mass spectrometer** is used to determine the types of isotopes present in an element, the exact atomic masses of these isotopes, and the relative amount of each isotope present. The essential features of the instrument are shown in Figure 4.5. Positive ions are produced from vaporized material by bombardment with electrons. These ions are attracted toward a negatively charged slit. They are accelerated by the attraction and pass through the slit with a high velocity.

The beam of ions is then passed through a magnetic field. Charged particles follow a circular path in a magnetic field. The higher the charge on the particle (e), the more it is deflected from a straight-line path. The mass of a particle (m), on the other hand, is inversely related to the degree of deflection. The larger the mass of a particle, the less it is deflected from a straight-line path. The radius of the path of a positive ion in the magnetic field, therefore, depends upon the value of e/m.

All the ions that pass through the final slit have the same value of e/m. Ions with other values of e/m can be made to pass through this slit by adjusting the magnetic field strength or the voltage used to accelerate the ions. Consequently, each type of ion present can be made to pass through the slit separately. The detector measures the intensity of each ion beam, which depends upon the relative amount of each isotope present in the sample.

Isobars are atoms of different elements that have the same mass number but different atomic numbers. Thus, $^{36}_{16}S$ and $^{36}_{18}Ar$ are isobars. The total number of nucleons in each of the isobars is the same, but the numbers of protons and neutrons are different:

$^{36}_{16}S$	16 protons	20 neutrons	16 electrons
$^{36}_{18}Ar$	18 protons	18 neutrons	18 electrons

Notice that both isobars have a total of 36 nucleons. Isobars are not alike chemically since chemical characteristics depend principally upon the numbers of protons and electrons, which are determined by the atomic number.

4.7 Atomic Weights

The **atomic mass unit** (for which the SI symbol is u) is defined as one-twelfth the mass of the $^{12}_{6}C$ atom. The assignment of exactly 12 u to the mass of this isotope of carbon is arbitrary. Other standards have been used in the past.

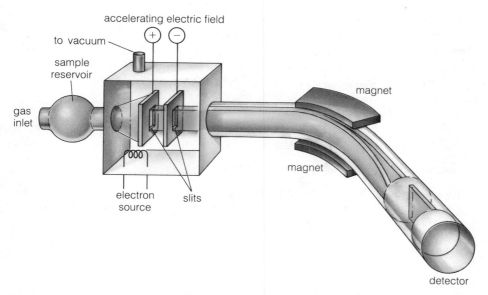

accelerating electric field

to vacuum

sample
reservoir

gas
inlet

electron
source

slits

magnet

magnet

detector

Figure 4.5   Essential features of a mass spectrometer

The masses of the proton, neutron, and electron based on this scale are given in Table 4.1. The mass of an atom, however, cannot be calculated from these values. With the exception of $_1^1H$ (the nucleus of which consists of a single proton), the sum of the masses of the particles that make up a nucleus is always larger than the actual mass of the nucleus.

Einstein has shown that matter and energy are equivalent. These mass differences, in terms of energy, account for what is called the **binding energy** of the nucleus. If it were possible to pull a nucleus apart, the binding energy would be the energy required to do the job. The reverse process, the condensation of nucleons into a nucleus, would release the binding energy with an attendant decrease in mass. Binding energy is discussed in Section 25.6.

Atomic masses are determined by use of the mass spectrometer. Most elements occur in nature as mixtures of isotopes. In these cases, the instrument can be used to determine the relative amount of each isotope present in the element as well as the atomic mass of each isotope. The data for chlorine show that the element consists of 75.77% $_{17}^{35}Cl$ atoms (mass, 34.969 u) and 24.23% $_{17}^{37}Cl$ atoms (mass, 36.966 u). Any sample of chlorine from a natural source consists of these two isotopes in this proportion.

The atomic weight of the element chlorine is the weighted average of the atomic masses of the natural isotopes. We *cannot* find the average by adding the masses of the isotopes and dividing by 2. The value obtained in this way would be correct only if the element consisted of equal numbers of atoms of the two isotopes. Instead, the *weighted* average is found by multiplying the atomic mass of each isotope by its fractional abundance and adding the values obtained. A fractional abundance is the decimal equivalent of the percent abundance:

	(*abundance*)	(*mass*)	
$_{17}^{35}Cl$	(0.7577)	(34.969 u) =	26.496 u
$_{17}^{37}Cl$	(0.2423)	(36.966 u) =	8.957 u
			35.453 u

The accepted value for chlorine is 35.453 ± 0.001 u.

There is no atom of chlorine with a mass of 35.453 u, but it is convenient to think in terms of one. In most calculations, no mistake is made by assuming that a sample of an element consists of only one type of atom with the average mass, the atomic weight. This assumption is valid because there is a huge number of atoms, in even a small sample of matter. There are, for example, more atoms in a drop of water than there are people on earth.

Example 4.1

What is the atomic weight of magnesium to four significant figures? The element consists of 78.99% $^{24}_{12}$Mg atoms (mass, 23.99 u), 10.00% $^{25}_{12}$Mg atoms (mass, 24.99 u), and 11.01% $^{26}_{12}$Mg atoms (mass, 25.98 u).

Solution

$$
\begin{array}{ll}
& \textit{(abundance)} \quad \textit{(mass)} \\
^{24}_{12}\text{Mg} & (0.7899)(23.99 \text{ u}) = 18.95 \text{ u} \\
^{25}_{12}\text{Mg} & (0.1000)(24.99 \text{ u}) = 2.50 \text{ u} \\
^{26}_{12}\text{Mg} & (0.1101)(25.98 \text{ u}) = \underline{2.86 \text{ u}} \\
& \text{atomic weight Mg} = 24.31 \text{ u}
\end{array}
$$

Example 4.2

Carbon occurs in nature as a mixture of $^{12}_{6}$C and $^{13}_{6}$C. The atomic mass of $^{12}_{6}$C is exactly 12 u, by definition, and the atomic mass of $^{13}_{6}$C is 13.003 u. The atomic weight of carbon is 12.011 u. What is the atom percent of $^{12}_{6}$C in natural carbon?

Solution

The equation to determine the atomic weight of carbon is

$$
\text{(abundance } ^{12}_{6}\text{C)(mass } ^{12}_{6}\text{C)} + \text{(abundance } ^{13}_{6}\text{C)(mass } ^{13}_{6}\text{C)}
$$
$$
= \text{atomic weight C}
$$

If we let x equal the abundance of $^{12}_{6}$C, then $(1 - x)$ is the abundance of $^{13}_{6}$C. Therefore,

$$
(x)(12.000) + (1 - x)(13.003) = 12.011
$$
$$
12.000x + 13.003 - 13.003x = 12.011
$$
$$
-1.003x = -0.992
$$
$$
x = 0.989
$$

Atoms of $^{12}_{6}$C constitute 98.9% of the total number.

Very small amounts of $^{14}_{6}$C also occur in nature. The amount is so small, however, that the atomic weight of carbon can be calculated while neglecting this isotope.

4.8 Electromagnetic Radiation

Radio waves, infrared waves, visible light, and X rays are types of electromagnetic radiation. Electromagnetic radiation travels through space in a wave motion (Figure 4.6). The following terms are used to describe these waves.

1. The wavelength, λ (lambda), is the distance between two similar points on two successive waves (such as the distance from crest to crest or trough to trough).

2. The amplitude, a, of a wave is the height of a crest (or the depth of a trough). The intensity (or brightness) of the radiation is proportional to the square of the amplitude, a^2.

3. In a vacuum, all waves, regardless of wavelength, travel at the same speed, 2.9979×10^8 m/s. This speed is called the speed of light and is given the symbol c.

4. The frequency of the radiation, v (nu), is the number of waves that pass a given spot in a second. For a given type of radiation, the wavelength times the number of waves per second (the frequency) equals the distance traveled per second (the speed of light):

$$\lambda v = c \tag{4.2}$$

and therefore,

$$v = \frac{c}{\lambda} \tag{4.3}$$

The electromagnetic spectrum is shown in Figure 4.7. Radio waves have very long wavelengths, infrared waves (radiant heat) have moderate wavelengths, and γ rays (from radioactive decay) have extremely short wavelengths. White light (visible light) consists of radiation with wavelengths in the approximate range of 4×10^{-7} m to 7.5×10^{-7} m (which is 400 nm to 750 nm).*

The wave theory successfully interprets many properties of electromagnetic radiation. Other properties, however, require that such radiation be considered as consisting of particles. In 1900, Max Planck proposed the quantum theory of radiant energy. Planck suggested that radiant energy could be absorbed or given off only in definite quantities, called quanta. The energy of a quantum, E, is proportional to the frequency of the radiation, v:

$$E = hv \tag{4.4}$$

The proportionality constant, h, is Planck's constant, 6.6262×10^{-34} J $\cdot$ s.

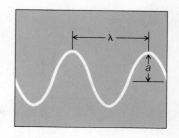

Figure 4.6 Wavelength, λ, and amplitude, a, of a wave

---

* In the past, wavelengths have been measured in ångstrom units (Å), 10^{-10} m. This unit is not a part of the International System. The International Committee of Weights and Measures recommends that the nanometer (nm), 10^{-9} m, be used instead:

$1 \text{ Å} = 10^{-10}$ m $= 10^{-8}$ cm

$1 \text{ nm} = 10^{-9}$ m $= 10^{-7}$ cm

Therefore,

$1 \text{ Å} = 0.1$ nm   and $1 \text{ nm} = 10$ Å

The range 400 nm to 750 nm corresponds to 4000 Å to 7500 Å.

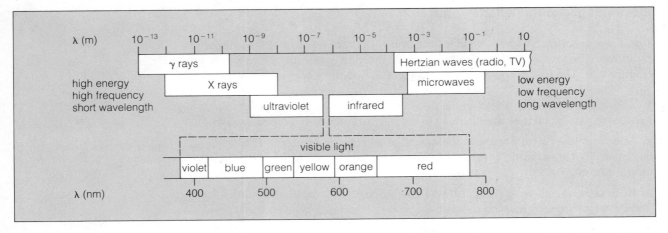

Figure 4.7 Electromagnetic radiation. (Note that the approximate ranges of electromagnetic radiations are plotted on a logarithmic scale in the upper part of the diagram. The spectrum of visible light is not plotted in this way.)

Since E and v are directly proportional, high energy radiation has a high frequency. A high frequency means that a large number of waves pass a spot in one second. The wavelength of high energy radiation, therefore, must be short. On the other hand, low energy radiation has a low frequency and a long wavelength. In 1905, Albert Einstein proposed that Planck's quanta are discontinuous bits of energy, which were later named **photons.**

4.9 Atomic Spectra

When a ray of light is passed through a prism, the ray is bent, or refracted. The amount that a wave is refracted depends upon its wavelength. A wave with a short wavelength is bent more than one with a long wavelength. Since ordinary white light consists of waves with all the wavelengths in the visible range, a ray of white light is spread out into a wide band called a **continuous spectrum.** The spectrum is a rainbow of colors with no blank spots—violet merges into blue, blue into green, and so on.

When gases or vapors of a chemical substance are heated in an electric arc or a Bunsen flame, light is emitted. If a ray of this light is passed through a prism, a **line spectrum** is produced (Figure 4.8). This spectrum consists of a limited number of colored lines, each of which corresponds to a different wavelength of light. The line spectrum of each element is unique.

The frequencies that correspond to the lines in the visible region of the hydrogen spectrum are given by the equation:

$$v = \frac{c}{\lambda} = (3.289 \times 10^{15}/s)\left(\frac{1}{2^2} - \frac{1}{n^2}\right) \qquad n = 3, 4, 5, \ldots \qquad (4.5)$$

where n is an integer equal to, or greater than, 3.

This relationship, proposed by J. J. Balmer in 1885, was derived from experimental observations and was not based on any theory of atomic structure. The

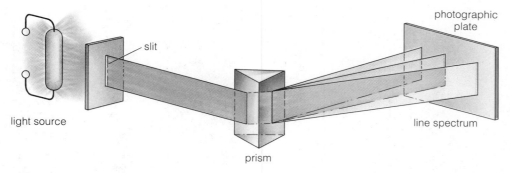

Figure 4.8 The spectroscope

Niels Bohr, 1885–1962.
*American Institute of Physics,
Niels Bohr Library.*

series of spectral lines in the visible region that is described by the Balmer equation is called the **Balmer series.**

In 1913, Niels Bohr proposed a theory for the electronic structure of the hydrogen atom that explained the line spectrum of this element. The hydrogen atom contains one electron and a nucleus that consists of a single proton. Bohr's theory includes the following points.

1. The electron of the hydrogen atom can exist only in certain spherical **orbits** (which are also called **energy levels** or **shells**). These shells are arranged concentrically around the nucleus. Each shell is designated by a letter (K, L, M, N, O, . . .) or a value of n (1, 2, 3, 4, 5, . . .).

2. The electron has a definite energy characteristic of the orbit in which it is moving. The K level ($n = 1$), the shell closest to the nucleus, has the smallest radius. An electron in the K level has the lowest possible energy. With increasing distance from the nucleus (K, L, M, N, O; $n = 1, 2, 3, 4, 5$), the radius of the shell and the energy of an electron in the shell increase. The electron cannot have an energy that would place it between the permissible shells.

3. When the electrons of an atom are as close to the nucleus as possible (for hydrogen, one electron in the K shell), they are in the condition of lowest possible energy, called the **ground state.** When the atoms are heated in an electric arc or Bunsen flame, electrons absorb energy and jump to outer levels, which are higher energy states. The atoms are said to be in **excited states.**

4. When an electron falls back to a lower level, it emits a definite amount of energy. The energy difference between the high energy state and low energy state is emitted in the form of a quantum of light. The light quantum has a characteristic frequency (and wavelength) and produces a characteristic spectral line. In spectral studies, many atoms are absorbing energy at the same time that many others are emitting it. Each spectral line corresponds to a different electron transition.

Bohr derived an equation for the energy that an electron would have in each orbit, E_{orbit}. This equation can be simplified to

$$E_{orbit} = -\frac{(2.179 \times 10^{-18} \text{ J})}{n^2} \qquad n = 1, 2, 3, \ldots \qquad (4.6)$$

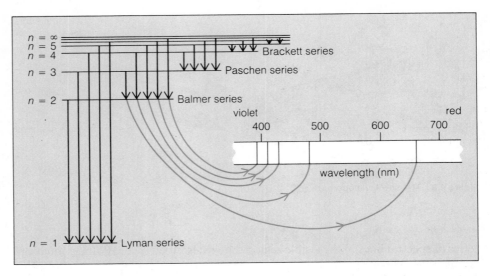

Figure 4.9 The relation between some electron transitions of the hydrogen atom and the spectral lines of the visible region

We shall let the energy of the electron in an outer level (n_o) be indicated by E_o and the energy of the electron in an inner level (n_i) be indicated by E_i. When the electron falls from the outer level to the inner level, $(E_o - E_i)$ is given off as a photon of light. According to Planck's equation, the energy of a photon is equal to $h\nu$. Therefore,

$$h\nu = E_o - E_i \qquad \text{(for } E_o \rightarrow E_i) \qquad (4.7)$$

$$h\nu = \frac{(-2.179 \times 10^{-18} \text{ J})}{n_o^2} - \frac{(-2.179 \times 10^{-18} \text{ J})}{n_i^2}$$

$$h\nu = (2.179 \times 10^{-18} \text{ J})\left(\frac{1}{n_i^2} - \frac{1}{n_o^2}\right) \qquad (4.8)$$

Since $h = 6.626 \times 10^{-34}$ J·s,

$$\nu = \left(\frac{2.179 \times 10^{-18} \text{ J}}{6.626 \times 10^{-34} \text{ J·s}}\right)\left(\frac{1}{n_i^2} - \frac{1}{n_o^2}\right)$$

$$\nu = (3.289 \times 10^{15}/\text{s})\left(\frac{1}{n_i^2} - \frac{1}{n_o^2}\right) \qquad (4.9)$$

The lines produced by electron transitions to the $n = 2$ level from higher levels are described by the equation

$$\nu = (3.289 \times 10^{15}/\text{s})\left(\frac{1}{2^2} - \frac{1}{n_o^2}\right) \qquad n = 3, 4, 5, \ldots \qquad (4.5)$$

This equation is the same as the equation derived by Balmer from experimental data.

The relationship between some of the electron transitions of the hydrogen atom and the spectral lines is illustrated in Figure 4.9. Since electron transitions to the $n = 1$ level (**Lyman series**) release more energy than those to the $n = 2$ level

(Balmer series), the wavelengths of the lines of the Lyman series are *shorter* than those of the Balmer series. The lines of the Lyman series occur in the ultraviolet region. On the other hand, the lines of the Paschen series, which represent transitions to the $n = 3$ level, occur at wavelengths *longer* than the Balmer series. The Paschen lines appear in the infrared region.

The Bohr theory is highly successful in interpreting the spectrum of hydrogen. If fails, however, to explain the spectra of atoms that contain more than one electron. Bohr's model of the atom, therefore, ultimately had to be modified (see Section 4.11).

Example 4.3

What are the frequency and wavelength of the line in the hydrogen spectrum that corresponds to an electron transition from the $n = 3$ level to the $n = 2$ level?

Solution

$$\nu = (3.289 \times 10^{15}/s)\left(\frac{1}{n_i^2} - \frac{1}{n_o^2}\right)$$

$$= (3.289 \times 10^{15}/s)\left(\frac{1}{2^2} - \frac{1}{3^2}\right)$$

$$= (3.289 \times 10^{15}/s)\left(\frac{1}{4} - \frac{1}{9}\right)$$

$$= 0.4568 \times 10^{15}/s = 4.568 \times 10^{14}/s$$

Since

$$\nu\lambda = c$$

$$\lambda = \frac{c}{\nu}$$

$$= \frac{2.998 \times 10^8 \text{ m/s}}{4.568 \times 10^{14}/s}$$

$$= 6.563 \times 10^{-7} \text{ m} = 656.3 \text{ nm}$$

Note that the conversion from meters to nanometers can be accomplished in the following way:

$$? \text{ nm} = 6.563 \times 10^{-7} \text{ m}\left(\frac{1 \text{ nm}}{10^{-9} \text{ m}}\right) = 6.563 \times 10^2 \text{ nm} = 656.3 \text{ nm}$$

4.10 Atomic Number and the Periodic Law

Early in the nineteenth century, chemists became interested in the chemical and physical similarities that exist between elements. In 1817 and 1829, Johann W. Döbereiner published articles in which he examined the properties of sets of

Dmitri Mendeleev, 1834–1907. *Smithsonian Institution.*

elements that he called triads (Ca, Sr, Ba; Li, Na, K; Cl, Br, I; and S, Se, Te). The elements of each set have similar properties, and the atomic weight of the second element of a set is approximately equal to the average of the atomic weights of the other two elements of the set.

In following years, many chemists attempted to classify the elements into groups on the basis of similarities in properties. In the years 1863–66, John A. R. Newlands proposed and developed his "law of octaves." Newlands stated that when the elements are listed by increasing atomic weight, the eighth element is similar to the first, the ninth to the second, and so forth. He compared this relationship to octaves of musical notes. Unfortunately, the actual relationship is not so simple as Newlands supposed. His work seemed forced and was not taken seriously by other chemists.

The modern periodic classification of the elements stems from the works of Julius Lothar Meyer (1869) and, in particular, Dmitri Mendeleev (1869). Mendeleev proposed a periodic law: when the elements are studied in order of increasing atomic weight, similarities in properties recur periodically. Mendeleev's table listed the elements in such a way that similar elements appeared in vertical columns, called groups (see Figure 4.10).

In order to make similar elements appear under one another, Mendeleev had to leave blanks for undiscovered elements in his table. On the basis of his system, he predicted the properties of three of the missing elements. The subsequent discovery of scandium, gallium, and germanium, each of which was found to have properties much like those predicted by Mendeleev, demonstrated the validity of the periodic system. The existence of the noble gases (He, Ne, Ar, Kr, Xe, and Rn) was unforeseen by Mendeleev. Nevertheless, after their discovery in the years 1892–98, these elements readily fitted into the periodic table.

The plan of the periodic table required that three elements (K, Ni, and I) be placed out of the order determined by increasing atomic weight. Iodine, for example, should be element number 52 on the basis of atomic weight. Instead, iodine was arbitrarily made element number 53 so that it would fall in a group of the table with chemically similar elements (F, Cl, and Br). Subsequent study of the periodic classification convinced many that some fundamental property other than atomic weight is the cause of the observed periodicity. It was proposed that this fundamental property is in some way related to atomic number, which at that time was only a serial number derived from the periodic system.

The work of Henry G. J. Moseley in the years 1913 and 1914 solved the problem. When high-energy cathode rays are focused on a target, X rays are produced (Figure 4.11). This X radiation can be resolved into its component wavelengths, and the line spectra thus obtained can be recorded photographically. Different X-ray spectra result when different elements are used as targets; each spectrum consists of only a few lines.

Moseley studied the X-ray spectra of 38 elements with atomic numbers between 13 (aluminum) and 79 (gold). Using a corresponding spectral line for each element, he found that there is a linear relationship between the square root of the frequency of the line and the atomic number of the element (Figure 4.12). In other words, the square root of the frequency of the spectral line increases by a constant amount from element to element when the elements are arranged by increasing atomic number.

Moseley was able, therefore, to assign the correct atomic number to any element on the basis of its X-ray spectrum. In this way he settled the problem involving the classification of elements that have atomic weights out of line with those of their neighbors (K, Ni, and I). He also stated that there should be 14 elements in the

Henry G.J. Moseley, 1887–1915. *American Institute of Physics, Niels Bohr Library, W. F. Meggers Collection.*

Chapter 4   Atomic Structure

Period	Group I		II		III		IV		V		VI		VII		VIII			(0)
	a	b	a	b	a	b	a	b	a	b	a	b	a	b	a		b	
1	H 1.0																He 4.0	
2	Li 6.9		Be 9.0			B 10.8		C 12.0		N 14.0		O 16.0		F 19.0				Ne 20.2
3	Na 23.0		Mg 24.3			Al 27.0		Si 28.1		P 31.0		S 32.1		Cl 35.5				Ar 39.9
4	K 39.1		Ca 40.1		Sc 45.0		Ti 47.9		V 50.9		Cr 52.0		Mn 54.9		Fe 55.8 Co 58.9 Ni 58.7			
		Cu 63.5		Zn 65.4		Ga 69.7		Ge 72.6		As 74.9		Se 79.0		Br 79.9				Kr 83.8
5	Rb 85.5		Sr 87.6		Y 88.9		Zr 91.2		Nb 92.9		Mo 95.9		Tc		Ru 101.1 Rh 102.9 Pd 106.4			
		Ag 107.9		Cd 112.4		In 114.8		Sn 118.7		Sb 121.8		Te 127.6		I 126.9				Xe 131.3
6	Cs 132.9		Ba 137.3		La* 138.9		Hf 178.5		Ta 180.9		W 183.9		Re 186.2		Os 190.2 Ir 192.2 Pt 195.1			
		Au 197.0		Hg 200.6		Tl 204.4		Pb 207.2		Bi 209.0		Po		At				Rn
7	Fr		Ra		Ac**													

*	Ce 140.1	Pr 140.9	Nd 144.2	Pm	Sm 150.4	Eu 152.0	Gd 157.3	Tb 158.9	Dy 162.5	Ho 164.9	Er 167.3	Tm 168.9	Yb 173.0	Lu 175.0
**	Th 232.0	Pa	U 238.0	Np	Pu	Am	Cm	Bk	Cf	Es	Fm	Md	No	Lr

Figure 4.10 Periodic table based on the Mendeleev table of 1871. (Elements that were not known in 1871 appear in the colored squares.)

series from $_{58}$Ce to $_{71}$Lu (found at the bottom of the chart in Figure 4.10), and he established that these elements should follow lanthanum in the periodic table. Moseley's diagrams indicated that at that time four elements before number 79 (gold) remained to be discovered (numbers 43, 61, 72, and 75). On the basis of Moseley's work, the **periodic law** was redefined: the chemical and physical properties of the elements are periodic functions of *atomic number*.

Moseley's atomic numbers agreed roughly with the nuclear charges that Rutherford had calculated on the basis of the α-particle scattering experiments. Moseley proposed, therefore, that the atomic number, Z, is the number of units of positive charge of the atomic nucleus. He said, "There is in the atom a fundamental quantity, which increases by regular steps as we pass from one element to the next. This quantity can only be the charge on the central positive nucleus."

X rays are electromagnetic radiations that have much shorter wavelengths (and consequently higher frequencies and energies) than those of visible light (Section 4.8). The X-ray spectrum of an element is believed to arise from certain electron

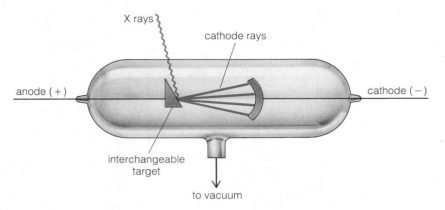

Figure 4.11   X-ray tube

transitions within the target atoms of the element (Section 4.8). In the X-ray tube, the cathode rays rip electrons from the inner shells of the target atoms. X rays are produced when outer electrons fall back into these vacancies. Since an electron transition to the K level of an atom from a higher level is one in which a relatively large amount of energy is released, the frequency of the resulting radiation is high. The corresponding wavelength, therefore, is short and characteristic of X rays.

The frequency of the radiation released by an electron transition also depends upon the charge on the nucleus of the atom. The amount of energy released is directly proportional to the square of the nuclear charge (Z^2). The higher the charge, the more energy released and the shorter the wavelength of the radiation emitted. Moseley's observations reflect this relationship.

At the present time, the most popular form of the periodic table is the long form that is found inside the front cover of this book. The **periods** consist of those elements that are arranged in horizontal rows in the table. The organization is such that elements of similar chemical and physical properties (called **groups** or **families**) appear in vertical columns. In this table, the set of elements (the lanthanides) that appears at the bottom should actually appear in the body of the chart after lanthanum in proper order by atomic number so that the sixth period contains 32 elements. For convenience in reproduction this arrangement is not used. The chart should be vertically cut, the sections separated, and the lanthanides inserted in their proper positions. For the actinides, which appear below the lanthanides at the bottom of the periodic table, the same considerations pertain. They should be inserted in the seventh period (which is as yet incomplete) after actinium.

The first period consists of only two elements, hydrogen and helium. Subsequent periods have 8, 8, 18, 18, and 32 elements. With the exception of the first period, each complete period begins with an **alkali metal** (group I A)—a highly reactive, light, silvery metal—and ends with a **noble gas** (group 0)—a colorless gas of low reactivity. The element before the noble gas of each complete period (except for the first period) is a **halogen** (group VII A)—a very reactive nonmetal.

A general pattern exists for each period after the first. Starting with an alkali metal, the properties change from element to element—the metallic properties fade and are gradually replaced by nonmetallic characteristics. After a highly reactive nonmetal (a halogen) is reached, each complete period ends with a noble gas.

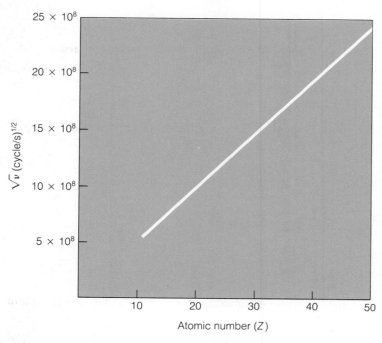

Figure 4.12 Relationship between frequency of characteristic X-ray lines and atomic number

4.11 Wave Mechanics

Bohr regarded the electron as a charged particle in motion, and he assumed that the electron in a hydrogen atom can possess only certain definite quantities of energy. Other aspects of the theory, however, he derived from the laws of classical physics that pertain to the behavior of charged particles. It soon became evident that this approach was inadequate and that a new one was needed.

In order to predict the path of a moving body, we must know both its position and velocity at the same time. Werner Heisenberg's uncertainty principle (1926) states that it is impossible to determine simultaneously the exact *position* and exact *momentum* of a body as small as the electron. The more precisely we try to determine one of these values, the more uncertain we are of the other.

We see objects by noting interference with the light rays used to illuminate them. Radiation with an extremely short wavelength would be needed to locate an object as small as the electron. Radiation that has a short wavelength has a high frequency and is very energetic (Section 4.8). When it strikes the electron, the impact causes the direction of motion and speed of the electron to change. The attempt to locate the electron changes the momentum of the electron drastically.

Photons with longer wavelengths are less energetic and would have a smaller effect on the momentum of the electron. Because of their longer wavelength, such photons, however, would not indicate the position of the electron very precisely. It appears, therefore, that an exact description of the path of an electron in a Bohr orbit is not possible.

In the same way that light has both a wave and a particle-like character, matter has a dual nature. In 1924, Louis de Broglie proposed that electrons and other particles have wave properties. The energy of a photon of light, E, is equal to its frequency, v, times Planck's constant, h:

Louis de Broglie, 1892–.
*American Institute of Physics,
Niels Bohr Library.*

$$E = hv \qquad (4.4)$$

Since $v = c/\lambda$, where c is the speed of light and λ is the wavelength (Section 4.8), we can substitute c/λ for v:

$$E = h\frac{c}{\lambda} \qquad (4.10)$$

Using Einstein's equation, $E = mc^2$, where m is the effective mass of the photon, we substitute mc^2 for E:

$$mc^2 = h\frac{c}{\lambda}$$

This equation is solved for λ, the wavelength:

$$\lambda = \frac{h}{mc} \qquad (4.11)$$

According to de Broglie, a similar equation can be used to assign a wavelength to an electron:

$$\lambda = \frac{h}{mv} \qquad (4.12)$$

Erwin Schrödinger, 1887–1961. *California Institute of Technology Archives.*

where m is the mass of the electron and v is its velocity. This postulate has been confirmed by a variety of experimental data. In 1926, Erwin Schrödinger used de Broglie's relation to develop an equation that describes the electron in terms of its wave character.

The **Schrödinger equation** is the basis of **wave mechanics.** The equation is written in terms of a **wave function,** ψ (psi), for an electron. When the equation is solved for the electron in the hydrogen atom, a series of wave functions is obtained. Each wave function corresponds to a definite energy state for the electron and pertains to a region in which the electron may be found. The wave function of an electron describes what is called an **orbital** (so named to distinguish it from the orbit of Bohr).

The intensity of a wave is proportional to the square of its amplitude. The wave function, ψ, is an amplitude function. At any position in space, the value of ψ^2 for a very small volume is proportional to the electron charge density. The charge of the electron can be assumed to be spread out into a charge cloud by the rapid motion of the electron. The cloud is denser in some regions than others. The probability of finding the electron in a given region is proportional to the density of the charge cloud at that spot. The probability is high in a region where the cloud is dense. This interpretation does not attempt to describe the path of the electron, it merely predicts where an electron is likely to be found.

For an electron in the $n = 1$ state of the hydrogen atom, the charge cloud has the greatest density near the nucleus and becomes thinner as the distance from the nucleus increases (Figure 4.13). More information about this probability distribution can be obtained from the curves of Figure 4.14. In curve (a), ψ^2 is plotted against distance from the nucleus. The probability of finding the electron in a small volume segment is greatest near the nucleus and approaches zero as the distance from the nucleus increases.

Figure 4.13 Cross section of the charge cloud of an electron in the $n = 1$ state of the hydrogen atom

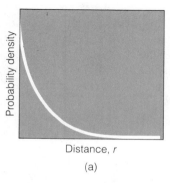
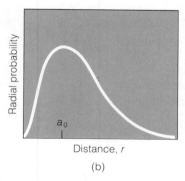

Figure 4.14 Probability curves for an electron in the $n = 1$ state of the hydrogen atom. (a) Probability of finding the electron per unit volume versus distance from the nucleus. (b) Probability of finding the electron at a given distance versus distance from the nucleus.

Curve (b) is a radial probability curve. The *total* probability of finding the electron at a given distance from the nucleus is plotted against distance. Imagine a group of very thin spherical shells arranged one after the other with the center of each shell at the center of the nucleus. What is the probability of finding the electron in each of these shells? The probability of finding the electron in a small volume segment is greatest near the nucleus. A shell close to the nucleus, however, contains fewer of these volume segments than one farther out. The radial probability takes both factors into account.

The curve shows a maximum at a distance a_0. The total probability of finding the electron at all points of distance r from the nucleus is greatest when r is equal to a_0. This value is the same as the value determined by the Bohr theory for the radius of the $n = 1$ shell. In the Bohr theory, a_0 is the distance at which the electron is *always* found in the $n = 1$ shell. In wave mechanics, a_0 is the distance at which the electron is likely to be found most often.

It is not possible to draw a shape that bounds a region in which the probability of finding the electron is 100%. A surface can be drawn, however, that connects points of equal probability and that encloses a volume in which the probability of finding the electron is high (for example, 90%). Such a representation, called a **boundary surface diagram,** is shown in Figure 4.15 for the electron in the $n = 1$ state of the hydrogen atom.

4.12 Quantum Numbers

In wave mechanics, the electron distribution of an atom containing a number of electrons is divided into *shells*. The shells, in turn, are thought to consist of one or more *subshells*, and the subshells are assumed to be composed of one or more *orbitals*, which the *electrons* occupy. Each electron of an atom is identified by a combination of four quantum numbers, which loosely indicate shell, subshell, orbital, and electron.

The **principal quantum number,** n, corresponds approximately to the n introduced by Bohr. It identifies the shell, or level, to which the electron belongs. These shells are regions where the probability of finding an electron is high. The value of n is a positive integer:

$$n = 1, 2, 3, \ldots$$

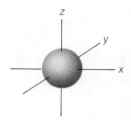

Figure 4.15 Boundary surface representation of an electron in the $n = 1$ state of the hydrogen atom. (Volume encloses 90% of the electron density. Nucleus is at the origin.)

Table 4.3 Subshell notations		
n	l	Spectroscopic Notation
1	0	1s
2	0	2s
2	1	2p
3	0	3s
3	1	3p
3	2	3d
4	0	4s
4	1	4p
4	2	4d
4	3	4f

The larger the value of n, the farther the shell is from the nucleus.

Each shell consists of one or more subshell, or sublevels. The number of subshells in a principal shell is equal to the value of n. There is only one subshell in the $n = 1$ shell, there are two in the $n = 2$ shell, three in the $n = 3$ shell, and so on. Each subshell in a shell is assigned a subsidiary quantum number, l. The values of l for the sublevels of a shell are determined by the shell's value of n. There is one value of l for each term in the series:

$$l = 0, 1, 2, 3, \ldots, (n - 1) \tag{4.13}$$

When $n = 1$, the only value of l is 0, and there is only one subshell. When $n = 2$, there are two subshells that have l values of 0 and 1, respectively. When $n = 3$, the three subshells have l values of 0, 1, and 2.

Other symbols are used at times to denote subshells. A letter is used to represent each value of l in the following way:

$$l = 0, 1, 2, 3, 4, 5, \ldots$$

$$\text{notation} = s, p, d, f, g, h, \ldots$$

The first four notations are the initial letters of adjectives formerly used to identify spectral lines: sharp, principal, diffuse, and fundamental. For l values higher than 3, the letters proceed alphabetically—g, h, i, and so on. Combining the principal quantum number with one of these letters gives a convenient way to designate a subshell. The subshell with $n = 2$ and $l = 0$ is called the $2s$ subshell. The subshell with $n = 2$ and $l = 1$ is called the $2p$ subshell. Table 4.3 contains a summary of subshell notation for the first four shells.

Each subshell consists of one or more orbitals. The number of orbitals in a subshell is given by the equation

$$\text{number of orbitals} = 2l + 1 \tag{4.14}$$

In any $l = 0$ subshell, for example, there is $2(0) + 1 = 1$ orbital. In any $l = 1$ subshell, there are $2(1) + 1 = 3$ orbitals. In any $l = 2$ subshell, there are $2(2) + 1 = 5$ orbitals. In other words,

$$\text{notation} = s, p, d, f, g, \ldots$$
$$l = 0, 1, 2, 3, 4, \ldots$$
$$\text{number of orbitals} = 1, 3, 5, 7, 9, \ldots$$

An s subshell consists of one orbital, a p subshell consists of three orbitals, a d subshell consists of five orbitals, and so on.

Each orbital within a given subshell is identified by a magnetic orbital quantum number, m_l. For any subshell, the values of m_l are given by the terms in the series:

$$m_l = +l, +(l - 1), \ldots, 0, \ldots, -(l - 1), -l$$

Thus, for $l = 0$, the only permitted values of m_l is 0 (one s orbital). For $l = 1$, m_l can be $+1$, 0, and -1 (three p orbitals). For $l = 2$, m_l can be $+2$, $+1$, 0, -1, and -2 (five d orbitals). Notice that the values of m_l are derived from l, and that the values of l are derived from n.

Chapter 4 Atomic Structure

Each orbital in an atom, therefore, is identified by a set of values for n, l, and m_l. An orbital described by the quantum numbers $n = 2$, $l = 1$, and $m_l = 0$, is an orbital in the p subshell of the second shell—a $2p$ orbital. The quantum numbers for the orbitals of the first four shells are given in Table 4.4.

Table 4.4	The orbitals of the first four shells			
Shell n	Subshell l	Orbital m_l	Subshell Notation	Number of Orbitals per Subshell
1	0	0	1s	1
2	0	0	2s	1
	1	$+1, 0, -1$	2p	3
3	0	0	3s	1
	1	$+1, 0, -1$	3p	3
	2	$+2, +1, 0, -1, -2$	3d	5
4	0	0	4s	1
	1	$+1, 0, -1$	4p	3
	2	$+2, +1, 0, -1, -2$	4d	5
	3	$+3, +2, +1, 0, -1, -2, -3$	4f	7

In Section 4.11, the electron charge cloud of the $1s$ orbital was discussed (see Figures 4.13, 4.14, and 4.15). Each of the three parts of Figure 4.16 pertains to the charge cloud of the $2s$ orbital. Figure 4.16 (a) is a radial probability curve for the $2s$ orbital. The curve shows two places where the probability of finding an electron is relatively high: one is close to the nucleus and the other is farther out. In the charge cloud of the $2s$ orbital, therefore, there are two regions where the electron density is relatively high. They appear in Figure 4.16(b), which is a cross section of the electron density of the orbital. The boundary surface diagram of the $2s$ orbital, however, appears the same as that for the $1s$ orbital, except for size. Compare Figure 4.16(c) with Figure 4.15. All s orbitals are spherical.

Boundary surface diagrams for the three $2p$ orbitals are shown in Figure 4.17. In these diagrams, the nucleus is at the origin. The electron density of a p orbital is not spherical. Instead, each p orbital consists of two sections, called lobes, that are on either side of a plane that passes through the nucleus. The shapes of the three orbitals are identical. They differ, however, in the way that the lobes are directed. Since the lobes may be considered to lie along the x, y, or z axis, they are given the designations $2p_x$, $2p_y$, and $2p_z$.

A magnetic field has no effect on the energy of an s electron since an s orbital is spherical. No matter how we turn a sphere, it always looks the same with regard to a fixed frame of reference. A spherical orbital (an s orbital) in a magnetic field presents only one aspect toward the lines of force.

The p orbitals are not spherical. Each p sublevel consists of three orbitals that differ in their orientation. Each p orbital presents a different aspect toward the lines of force of a magnetic field. These p orbitals are identical in terms of energy. In the absence of the magnetic field, no distinction between electrons that occupy different p orbitals is noted. If spectral studies are run using a magnetic field, however, some of the lines of the spectrum are split into several lines. This effect, called the **Zeeman effect,** disappears when the magnetic field is removed.

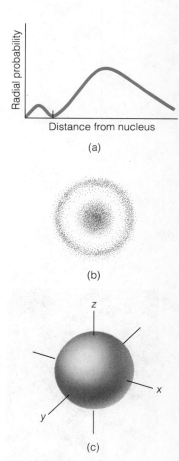

Figure 4.16 Diagrams for the $2s$ orbital. (a) Radial probability distribution. (b) Cross section of electron charge cloud. (c) Boundary surface diagram.

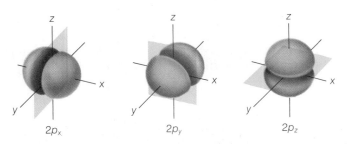

Figure 4.17 Boundary surface diagrams for the 2p orbitals

The m_l value of an orbital is associated with the orientation of the orbital in relation to a direction fixed by a magnetic field.

Boundary surface diagrams of the five $3d$ orbitals are shown in Figure 4.18. The shape of the d_{z^2} orbital is different from the others, but they are all equivalent in terms of energy.

The first three quantum numbers (n, l, and m_l) arise from solutions to the Schrödinger wave equation. A fourth quantum number, the **magnetic spin quantum number,** m_s, is necessary to describe an electron completely. An electron has magnetic properties that are like those of a charged particle that is spinning on an axis. A spinning charge generates a magnetic field and an electron has a magnetic field associated with it that can be described in terms of an apparent spin. The magnetic spin quantum number of an electron can have one of two values:

$$m_s = +\tfrac{1}{2} \text{ or } -\tfrac{1}{2}$$

Two electrons that have different m_s values (one $+\frac{1}{2}$ and the other $-\frac{1}{2}$) are said to have *opposed spins*. The spin magnetic moments of these two electrons cancel each other. Each orbital can hold two electrons with opposed spins.

Each electron, therefore, may be described by a set of four quantum numbers:

1. n gives the shell and the relative average distance of the electron from the nucleus.

2. l gives the subshell and the shape of the orbital for the electron. Each orbital of a given subshell is equivalent in energy in the absence of a magnetic field.

3. m_l designates the orientation of the orbital.

4. m_s refers to the spin of the electron.

The **exclusion principle** of Wolfgang Pauli states that no two electrons in the same atom may have identical sets of all four quantum numbers. Even if two electrons have the same values for n, l, and m_l they will differ in their m_s values. This situation indicates that two electrons are paired in a single orbital. Two electrons paired in a $1s$ orbital, for example, have (n, l, m_l, m_s) quantum sets of $(1, 0, 0, +\frac{1}{2})$ and $(1, 0, 0, -\frac{1}{2})$. According to the exclusion principle, therefore, an orbital may hold no more than two electrons.

Quantum numbers for the orbitals of the first four shells are given in Table 4.4. The quantum number sets for the electrons in these orbitals can be obtained by indicating an m_s value (either $+\frac{1}{2}$ or $-\frac{1}{2}$) with the set of n, l, m_l values that denote the orbital.

The maximum number of electrons that a shell can hold is given by $2n^2$. Each orbital can hold two electrons. The maximum number of electrons in a shell,

Wolfgang Pauli, 1900–1958.
CERN, courtesy AIP Niels Bohr Library.

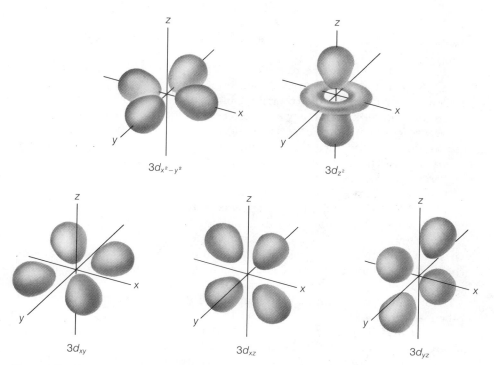

$3d_{x^2-y^2}$ $3d_{z^2}$

$3d_{xy}$ $3d_{xz}$ $3d_{yz}$

Figure 4.18   Boundary surface diagrams for the 3d orbitals

therefore, is equal to twice the number of orbitals in the shell. The maximum number of electrons for a subshell can be calculated by multiplying the number of orbitals in the subshell by 2. The capacities of the shells and subshells from $n = 1$ to $n = 4$ are given in Table 4.5.

Table 4.5 Maximum number of electrons for the subshells of the first four shells			
Subshell Notation	Orbitals per Subshell	Electrons per Subshell	Electrons per Shell ($2n^2$)
1s	1	2	2
2s	1	2	8
2p	3	6	
3s	1	2	18
3p	3	6	
3d	5	10	
4s	1	2	32
4p	3	6	
4d	5	10	
4f	7	14	

4.13 Orbital Filling and Hund's Rule

The way electrons are arranged in an atom is called the **electron configuration** of the atom. For the first 18 elements, the ground-state electron configurations can be derived by assuming that the electrons occupy the shells by increasing value of n, and within a shell by increasing value of l. The situation for elements with atomic numbers higher than 18 is slightly more complicated and is discussed in Section 4.14.

Two ways to indicate the electron configuration of an atom are shown in Table 4.6. In the **orbital diagrams,** each orbital is indicated by a dash and an electron is represented by an arrow either pointing up, ↑, to represent one direction of electron spin, or pointing down, ↓, to represent the opposite direction (m_s can be either $+\frac{1}{2}$ or $-\frac{1}{2}$). In the **electronic notations,** the electron configuration of an atom is summarized in a slightly different way. The symbols $1s$, $2s$, $2p$, and so on, are used to indicate *subshells*, and superscripts are added to indicate the number of electrons in each subshell.

The single electron of a hydrogen atom occupies a $1s$ orbital ($n = 1$, $l = 0$, $m_l = 0$). In the orbital diagram that appears in Table 4.6, one arrow is shown in the blank for the $1s$ orbital. The electronic notation for the H atom is $1s^1$.

The helium atom has two electrons with opposite spins in the $1s$ orbital, and two arrows, pointing in opposite directions, are shown in the blank for the $1s$ orbital in the orbital diagram. The electronic notation for the He atom is $1s^2$. Note that the $n = 1$ shell of the helium atom is filled.

The lithium atom has a pair of electrons in the $1s$ orbital plus one electron in the $2s$ orbital ($n = 2$, $l = 0$, $m_l = 0$). The electronic notation for the Li atom $1s^2 2s^1$. The next atom, beryllium, has electron pairs in both the $1s$ and $2s$ orbitals; the electronic notation for the Be atom is $1s^2\,2s^2$.

The boron atom has five electrons. Two electrons, with spins paired, occupy the $1s$ orbital, another pair of electrons occupies the $2s$ orbital, and the fifth electron is present in a $2p$ orbital. The $2p$ subshell ($n = 2$, $l = 1$) consists of three orbitals (with m_l values of $+1$, 0 and -1). Since the three $2p$ orbitals are of equal energy, the fifth electron of boron can occupy any one of the three. In the orbital diagram for boron that appears in Table 4.6, an arrow is shown in one of the $2p$ orbitals, but these orbitals are not identified by m_l values. The electronic notation for the B atom is $1s^2\,2s^2\,2p^1$.

The electron configuration of the sixth element, carbon, can be derived from the configuration of boron by indicating an additional electron. Questions arise, however, concerning the placement of this sixth electron of carbon. Does the sixth electron belong in the $2p$ orbital that already holds one electron, or does it belong in another orbital? What is the spin orientation of the sixth electron?

Table 4.6 Electron configurations of the first ten elements

	Orbital Diagram					Electronic Notation
	1s	2s	2p			
$_1$H	↑					$1s^1$
$_2$He	↑↓					$1s^2$
$_3$Li	↑↓	↑				$1s^2\,2s^1$
$_4$Be	↑↓	↑↓				$1s^2\,2s^2$
$_5$B	↑↓	↑↓	↑			$1s^2\,2s^2\,2p^1$
$_6$C	↑↓	↑↓	↑	↑		$1s^2\,2s^2\,2p^2$
$_7$N	↑↓	↑↓	↑	↑	↑	$1s^2\,2s^2\,2p^3$
$_8$O	↑↓	↑↓	↑↓	↑	↑	$1s^2\,2s^2\,2p^4$
$_9$F	↑↓	↑↓	↑↓	↑↓	↑	$1s^2\,2s^2\,2p^5$
$_{10}$Ne	↑↓	↑↓	↑↓	↑↓	↑↓	$1s^2\,2s^2\,2p^6$

Chapter 4   Atomic Structure

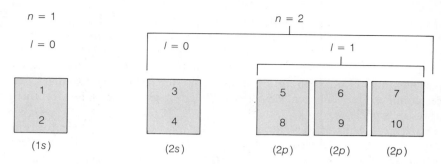

Figure 4.19 Order in which the orbitals of the $n = 1$ and $n = 2$ shells are filled

Hund's rule of maximum multiplicity provides the answers. Hund's rule states that the electrons are distributed among the orbitals of a subshell in a way that gives the maximum number of unpaired electrons with parallel spins. The term *parallel spins* means that all the unpaired electrons spin in the same direction—all the m_s values of these electrons have the same sign.

In carbon, therefore, each of the $2p$ electrons occupies a separate orbital, and these two electrons have the same spin orientation. These two unpaired electrons are clearly shown in the orbital diagram for carbon given in Table 4.6. The distinction, however, is not apparent in the electronic notation, $1s^2 \, 2s^2 \, 2p^2$.

Notice that all the superscripts in this notation are even numbers. Such a situation does *not* mean that all the electrons are paired in orbitals. Customarily, the electronic notation gives the electron configuration by *subshells* (not by *orbitals*). The unpaired electrons can be shown, however, by using a separate designation for each orbital. The notation for carbon would be $1s^2 \, 2s^2 \, 2p^1 \, 2p^1$. This device is not necessary if we remember that there are three $2p$ orbitals and that Hund's rule requires that the two electrons occupy separate orbitals.

Figure 4.19 illustrates Hund's rule. The orbitals are represented by squares. The order of filling is shown by numbers entered into these squares. Since electrons are negatively charged and repel each other, they spread out and occupy the $2p$ orbitals singly before they begin to pair. The electron configurations for B, C, N, O, F, and Ne given in Table 4.6 illustrate Hund's rule. The five orbitals of a d subshell and the seven orbitals of an f subshell are filled in the same way—electrons successively occupy each orbital of the subshell singly and only after each orbital holds one electron does electron pairing occur. Hund's rule has been confirmed by magnetic measurements.

The number of unpaired electrons in an atom, ion, or molecule can be determined by magnetic measurements. **Paramagnetic substances** are drawn into a magnetic field. Substances that contain unpaired electrons are paramagnetic. The magnetic moment depends upon the number of unpaired electrons present. Two effects contribute to the paramagnetism of an atom: the spin of the unpaired electrons and the orbital motion of these electrons. The effect of electron spin is the greater of the two, and in many cases, the effect of orbital motion is negligible.

Diamagnetic substances are weakly repelled by a magnetic field. A material is diamagnetic if all of its electrons are paired. Diamagnetism is a property of all matter, but it is obscured by the stronger paramagnetic effect if unpaired electrons are present.*

---

* Ferromagnetic substances, such as iron, are strongly attracted into a magnetic field. Ferromagnetism is a form of paramagnetism that is shown by only a few solid substances.

In Table 4.7, the electronic notations for the *outer shells* of atoms of the elements of the first three periods are shown. In each atom that contains electrons in inner shells, these shells are complete. Notice the similarity between the configurations of elements of the same group. All of the elements of group I A, for example, have one electron in an *s* orbital in the outer shell. The similarity in electron configuration between elements of a given group accounts for their similarities in properties.

The outermost shells of these atoms are called valence shells, and electrons in them are called valence electrons. All the electrons in the valence shell, regardless of subshell, are counted as valence electrons. For elements that are members of A groups, the number of valence electrons is the same as the group number. The noble gases (group 0) have eight electrons in their valence shell with the exception of helium, which has two.

I A							0
$_1$H $1s^1$	II A	III A	IV A	V A	IV A	VII A	$_2$He $1s^2$
$_3$Li $2s^1$	$_4$Be $2s^2$	$_5$B $2s^22p^1$	$_6$C $2s^22p^2$	$_7$N $2s^22p^3$	$_8$O $2s^22p^4$	$_9$F $2s^22p^5$	$_{10}$Ne $2s^22p^6$
$_{11}$Na $3s^1$	$_{12}$Mg $3s^2$	$_{13}$Al $3s^23p^1$	$_{14}$Si $3s^23p^2$	$_{15}$P $3s^23p^3$	$_{16}$S $3s^23p^4$	$_{17}$Cl $3s^23p^5$	$_{18}$Ar $3s^23p^6$

Table 4.7   Electron configurations of the outer shells of the elements of the first three periods

4.14 Electronic Structures of the Elements

The data of Tables 4.6 and 4.7 indicate a way that the electron configurations of atoms can be derived. We start with the hydrogen atom, which has one electron in a 1*s* orbital. By adding one electron, we get the configuration of an atom of the next element, helium (which is $1s^2$). In this manner, we go from element to element until we derive the configuration of the atom that we desire. This method was first suggested by Wolfgang Pauli and is called the aufbau method (*aufbau* means "building up" in German).

In a few cases, the electron configurations obtained by use of the aufbau method are in error. These errors, however, are small and usually involve only one misplaced electron. The correct electron configurations of the elements are given later, in Table 4.8.

The electron added in going from one element to the next in the aufbau procedure is called the differentiating electron. It makes the configuration of an atom different from that of the atom that precedes it. The differentiating electron is added in each step to the orbital of lowest energy available to it.

All the orbitals of a given subshell have equivalent energies. The energy of any one of the 3*p* orbitals, for example, is the same as the energy of either of the other two 3*p* orbitals. All five 3*d* orbitals are alike in terms of energy. Orbitals that belong to different subshells of the same shell, however, have different energies. For a given value of *n*, the energies increase in the order $s < p < d < f$.

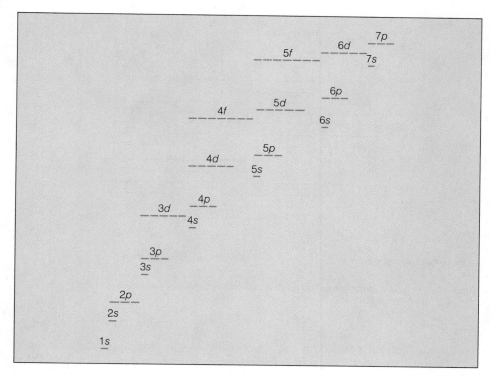

Figure 4.20 Aufbau order of atomic orbitals

In the $n = 3$ shell, for example, the $3s$ orbital has the lowest energy, a $3p$ orbital has an intermediate energy, and a $3d$ orbital has the highest energy. At times, the energies of orbitals from different shells overlap. In some atoms, for example, the $4s$ orbital has a lower energy than a $3d$ orbital.

There is no standard order of orbitals, based on energy, that pertains to all the elements. In the *hypothetical* aufbau process, the character of the atom changes as protons and neutrons are added to the nucleus and more and more electrons are introduced. Fortunately, the orbital energy order varies from element to element in a slow and regular manner. As a result, the aufbau order shown in Figure 4.20 can be derived.

This order pertains *only* to the orbital position that the differentiating electron takes in the aufbau process. Electron configurations can be derived from the diagram shown in Figure 4.20 by successively filling the orbitals starting at the bottom of the chart and proceeding upward. Remember that there are three orbitals in a p sublevel, five in a d, and seven in an f. A given sublevel is filled before electrons are added to the next sublevel.

The periodic table can be used to derive electron configurations. In Figure 4.21, the type of differentiating electron is related to the position of the element in the periodic chart. Notice that the table can be divided into an "s block," a "p block," a "d block," and an "f block." The principal quantum number of the differentiating electron is equal to the period number for A family elements (the "s block" and "p block" elements), the period number minus 1 for the "d block" elements, and the period number minus 2 for the "f block" elements. Use a periodic table (such as the one inside the front cover) to follow the discussion as we derive the electron configurations of the elements.

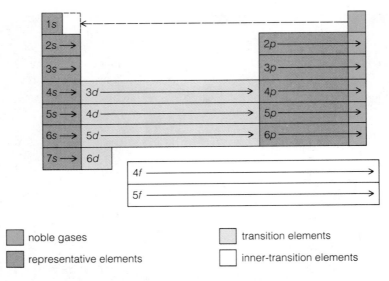

1s			
2s →		2p →	
3s →		3p →	
4s →	3d →	4p →	
5s →	4d →	5p →	
6s →	5d →	6p →	
7s →	6d		

4f →

5f →

■ noble gases   ▪ transition elements

■ representative elements   □ inner-transition elements

Figure 4.21 Type of differentiating electron related to position of the element in the periodic table

The first period consists of only two elements—hydrogen and helium—both of which are "s block" elements. The configuration of hydrogen is $1s^1$, and that of helium is $1s^2$.

The second period begins with lithium ($1s^2\,2s^1$) and beryllium ($1s^2\,2s^2$), for which electrons are added to the $2s$ orbital. For the six elements that complete this period—boron ($1s^2\,2s^2\,2p^1$) to the noble gas neon ($1s^2\,2s^2\,2p^6$)—electrons are gradually added to the three $2p$ orbitals.

The pattern of the second period is repeated in the third. The two "s block" elements are sodium ($1s^2\,2s^2\,2p^6\,3s^1$) and magnesium ($1s^2\,2s^2\,2p^6\,3s^2$). The six "p block" elements go from aluminum ($1s^2\,2s^2\,2p^6\,3s^2\,3p^1$) to the noble gas argon ($1s^2\,2s^2\,2p^6\,3s^2\,3p^6$).

In the discussion of the configurations of the remaining elements, only the outer orbitals will be indicated. The first overlap of orbital energies is observed with potassium ($Z = 19$), the first element of the fourth period. The configuration of potassium is . . . $3s^2\,3p^6\,4s^1$ despite the fact that $3d$ orbitals are vacant. In like manner, calcium ($Z = 20$) has the configuration . . . $3s^2\,3p^6\,4s^2$. Notice that potassium and calcium are "s block" elements.

With the next element, scandium ($Z = 21$), the $3d$ subshell comes into use (. . . $3s^2\,3p^6\,3d^1\,4s^2$). In the series from scandium to zinc, the $3d$ subshell is gradually filled. The configuration of zinc ($Z = 30$) is . . . $3s^2\,3p^6\,3d^{10}\,4s^2$. These "d block" elements are called **transition elements.** They are said to exhibit **inner building** since the last electron is added to the shell ($n = 3$) next to the outermost shell ($n = 4$). The elements from number 21 to 30, which belong to B families, are said to form the first transition series.

With element 31, gallium (. . . $3s^2\,3p^6\,3d^{10}\,4s^2\,4p^1$), the $4p$ subshell begins to be filled. The fourth period ends with krypton ($Z = 36$, . . . $3s^2\,3p^6\,3d^{10}\,4s^2\,4p^6$).

The fifth period starts with rubidium ($Z = 37$, . . . $4s^2\,4p^6\,5s^1$) and strontium ($Z = 38$, . . . $4s^2\,4p^6\,5s^2$) with electrons being added to the $5s$ subshell even though both $4d$ and $4f$ orbitals are vacant. Notice that these two elements are "s block" elements. A second transition series follows, in which electrons are added to the $4d$ subshell. It starts with yttrium ($Z = 39$, . . . $4s^2\,4p^6\,4d^1\,5s^2$) and ends with cadmium ($Z = 48, \ldots 4s^2\ 4p^6\ 4d^{10}\ 5s^2$). The fifth period ends with the series from indium to xenon with electrons being added to the $5p$ subshell. Xenon ($Z = 54$) has the configuration $\ldots 4s^2\ 4p^6\ 4d^{10}\ 5s^2\ 5p^6$. The $4f$ subshell is still vacant at the end of this period.

The sixth period is far more complicated as far as orbital overlap is concerned. The first two elements, cesium ($Z = 55, \ldots 4d^{10}\ 5s^2\ 5p^6\ 6s^1$) and barium ($Z = 56, \ldots 4d^{10}\ 5s^2\ 5p^6\ 6s^2$) have electrons in the $6s$ subshell. Here, we come upon a complication of the aufbau procedure. The $4f$ and $5d$ subshells are close in energy. The next electron (for lanthanum, $Z = 57$) is added to the $5d$ subshell (thus lanthanum is a transition element), but the next electron (for cerium, $Z = 58$) is added to the $4f$ subshell and the electron added for lanthanum falls back into the $4f$ subshell. For La, the configuration is $\ldots 4d^{10}\ 4f^0\ 5s^2\ 5p^6\ 5d^1\ 6s^2$. For Ce, the configuration is $\ldots 4d^{10}\ 4f^2\ 5s^2\ 5p^6\ 5d^0\ 6s^2$. For the elements 58 to 70 (cerium to ytterbium), electrons are added to the $4f$ subshell.

These elements are called **inner-transition elements**. For elements of this type, electron addition occurs in the third subshell ($4f$) from the outermost subshell ($6s$). After the $4f$ subshell has been filled, the next electron is added to the $5d$ subshell. Thus, for lutetium ($Z = 71$) the configuration is $\ldots 4d^{10}\ 4f^{14}\ 5s^2\ 5p^6\ 5d^1\ 6s^2$. This third transition series is completed and the $5d$ subshell filled with element 80, mercury ($\ldots 4d^{10}\ 4f^{14}\ 5s^2\ 5p^6\ 5d^{10}\ 6s^2$). The period ends with the filling of the $6p$ sublevel in elements 81 to 86.

The seventh period is incomplete and includes many elements that do not occur in nature but have been made by nuclear reactions. In general, this period follows the pattern established by the sixth period. Elements 87 and 88 have electrons added to the $7s$ sublevel, for element 89 an electron is added to the $6d$ sublevel, elements 90 to 103 constitute a second inner-transition series and exhibit an electron buildup of the $5f$ sublevel, and the transition elements 104, 105, and 106 have electrons added to the $6d$ sublevel. The actual electron configurations of some seventh-period atoms deviate slightly from the configurations predicted by the aufbau method (Table 4.8).

To determine the electronic configuration of any element, we start with hydrogen and on the basis of the periodic table account for every electron added until the desired element is reached. The method is illustrated by the following examples.

Example 4.4

Write the electronic notation for the electronic configuration of tin ($Z = 50$).

Solution

We trace our way through the periodic table adding terms so as to account for the electron added to each element up to $Z = 50$ (tin). Check the method by using a periodic table:

first period: $1s^2$ (which takes us up to $_2$He)
second period: $2s^2\ 2p^6$ (which takes us up to $_{10}$Ne)
third period: $3s^2\ 3p^6$ (which takes us up to $_{18}$Ar)
fourth period: $4s^2\ 3d^{10}\ 4p^6$ (which takes us up to $_{36}$Kr)
fifth period: $5s^2\ 4d^{10}\ 5p^2$ (which takes us up to $_{50}$Sn)

Table 4.8	Electronic configurations of the elements																					

Element	Z	1s	2s	2p	3s	3p	3d	4s	4p	4d	4f	5s	5p	5d	5f	6s	6p	6d	7s
H	1	1																	
He	2	2																	
Li	3	2	1																
Be	4	2	2																
B	5	2	2	1															
C	6	2	2	2															
N	7	2	2	3															
O	8	2	2	4															
F	9	2	2	5															
Ne	10	2	2	6															
Na	11	2	2	6	1														
Mg	12	2	2	6	2														
Al	13	2	2	6	2	1													
Si	14	2	2	6	2	2													
P	15	2	2	6	2	3													
S	16	2	2	6	2	4													
Cl	17	2	2	6	2	5													
Ar	18	2	2	6	2	6													
K	19	2	2	6	2	6		1											
Ca	20	2	2	6	2	6		2											
Sc	21	2	2	6	2	6	1	2											
Ti	22	2	2	6	2	6	2	2											
V	23	2	2	6	2	6	3	2											
Cr	24	2	2	6	2	6	5	1											
Mn	25	2	2	6	2	6	5	2											
Fe	26	2	2	6	2	6	6	2											
Co	27	2	2	6	2	6	7	2											
Ni	28	2	2	6	2	6	8	2											
Cu	29	2	2	6	2	6	10	1											
Zn	30	2	2	6	2	6	10	2											
Ga	31	2	2	6	2	6	10	2	1										
Ge	32	2	2	6	2	6	10	2	2										
As	33	2	2	6	2	6	10	2	3										
Se	34	2	2	6	2	6	10	2	4										
Br	35	2	2	6	2	6	10	2	5										
Kr	36	2	2	6	2	6	10	2	6										
Rb	37	2	2	6	2	6	10	2	6			1							
Sr	38	2	2	6	2	6	10	2	6			2							
Y	39	2	2	6	2	6	10	2	6	1		2							
Zr	40	2	2	6	2	6	10	2	6	2		2							
Nb	41	2	2	6	2	6	10	2	6	4		1							
Mo	42	2	2	6	2	6	10	2	6	5		1							
Tc	43	2	2	6	2	6	10	2	6	6		1							
Ru	44	2	2	6	2	6	10	2	6	7		1							
Rh	45	2	2	6	2	6	10	2	6	8		1							
Pd	46	2	2	6	2	6	10	2	6	10									
Ag	47	2	2	6	2	6	10	2	6	10		1							
Cd	48	2	2	6	2	6	10	2	6	10		2							
In	49	2	2	6	2	6	10	2	6	10		2	1						
Sn	50	2	2	6	2	6	10	2	6	10		2	2						
Sb	51	2	2	6	2	6	10	2	6	10		2	3						
Te	52	2	2	6	2	6	10	2	6	10		2	4						
I	53	2	2	6	2	6	10	2	6	10		2	5						
Xe	54	2	2	6	2	6	10	2	6	10		2	6						

Element	Z	1s	2s	2p	3s	3p	3d	4s	4p	4d	4f	5s	5p	5d	5f	6s	6p	6d	7s
Cs	55	2	2	6	2	6	10	2	6	10		2	6			1			
Ba	56	2	2	6	2	6	10	2	6	10		2	6			2			
La	57	2	2	6	2	6	10	2	6	10		2	6	1		2			
Ce	58	2	2	6	2	6	10	2	6	10	2	2	6			2			
Pr	59	2	2	6	2	6	10	2	6	10	3	2	6			2			
Nd	60	2	2	6	2	6	10	2	6	10	4	2	6			2			
Pm	61	2	2	6	2	6	10	2	6	10	5	2	6			2			
Sm	62	2	2	6	2	6	10	2	6	10	6	2	6			2			
Eu	63	2	2	6	2	6	10	2	6	10	7	2	6			2			
Gd	64	2	2	6	2	6	10	2	6	10	7	2	6	1		2			
Tb	65	2	2	6	2	6	10	2	6	10	9	2	6			2			
Dy	66	2	2	6	2	6	10	2	6	10	10	2	6			2			
Ho	67	2	2	6	2	6	10	2	6	10	11	2	6			2			
Er	68	2	2	6	2	6	10	2	6	10	12	2	6			2			
Tm	69	2	2	6	2	6	10	2	6	10	13	2	6			2			
Yb	70	2	2	6	2	6	10	2	6	10	14	2	6			2			
Lu	71	2	2	6	2	6	10	2	6	10	14	2	6	1		2			
Hf	72	2	2	6	2	6	10	2	6	10	14	2	6	2		2			
Ta	73	2	2	6	2	6	10	2	6	10	14	2	6	3		2			
W	74	2	2	6	2	6	10	2	6	10	14	2	6	4		2			
Re	75	2	2	6	2	6	10	2	6	10	14	2	6	5		2			
Os	76	2	2	6	2	6	10	2	6	10	14	2	6	6		2			
Ir	77	2	2	6	2	6	10	2	6	10	14	2	6	7		2			
Pt	78	2	2	6	2	6	10	2	6	10	14	2	6	9		1			
Au	79	2	2	6	2	6	10	2	6	10	14	2	6	10		1			
Hg	80	2	2	6	2	6	10	2	6	10	14	2	6	10		2			
Tl	81	2	2	6	2	6	10	2	6	10	14	2	6	10		2	1		
Pb	82	2	2	6	2	6	10	2	6	10	14	2	6	10		2	2		
Bi	83	2	2	6	2	6	10	2	6	10	14	2	6	10		2	3		
Po	84	2	2	6	2	6	10	2	6	10	14	2	6	10		2	4		
At	85	2	2	6	2	6	10	2	6	10	14	2	6	10		2	5		
Rn	86	2	2	6	2	6	10	2	6	10	14	2	6	10		2	6		
Fr	87	2	2	6	2	6	10	2	6	10	14	2	6	10		2	6		1
Ra	88	2	2	6	2	6	10	2	6	10	14	2	6	10		2	6		2
Ac	89	2	2	6	2	6	10	2	6	10	14	2	6	10		2	6	1	2
Th	90	2	2	6	2	6	10	2	6	10	14	2	6	10		2	6	2	2
Pa	91	2	2	6	2	6	10	2	6	10	14	2	6	10	2	2	6	1	2
U	92	2	2	6	2	6	10	2	6	10	14	2	6	10	3	2	6	1	2
Np	93	2	2	6	2	6	10	2	6	10	14	2	6	10	4	2	6	1	2
Pu	94	2	2	6	2	6	10	2	6	10	14	2	6	10	6	2	6		2
Am	95	2	2	6	2	6	10	2	6	10	14	2	6	10	7	2	6		2
Cm	96	2	2	6	2	6	10	2	6	10	14	2	6	10	7	2	6	1	2
Bk	97	2	2	6	2	6	10	2	6	10	14	2	6	10	8	2	6	1	2
Cf	98	2	2	6	2	6	10	2	6	10	14	2	6	10	10	2	6		2
Es	99	2	2	6	2	6	10	2	6	10	14	2	6	10	11	2	6		2
Fm	100	2	2	6	2	6	10	2	6	10	14	2	6	10	12	2	6		2
Md	101	2	2	6	2	6	10	2	6	10	14	2	6	10	13	2	6		2
No	102	2	2	6	2	6	10	2	6	10	14	2	6	10	14	2	6		2
Lr	103	2	2	6	2	6	10	2	6	10	14	2	6	10	14	2	6	1	2

The terms should be rearranged to give the notation in proper sequence:

$$1s^2 \ 2s^2 2p^6 \ 3s^2 \ 3p^6 \ 3d^{10} \ 4s^2 \ 4p^6 \ 4d^{10} \ 5s^2 \ 5p^2$$

Example 4.5

Write the electronic notation for the electronic configuration of neodymium ($Z = 60$).

Solution

first period: $1s^2$
second period: $2s^2 \ 2p^6$
third period: $3s^2 \ 3p^6$
fourth period: $4s^2 \ 3d^{10} \ 4p^6$
fifth period: $5s^2 \ 4d^{10} \ 5p^6$
sixth period: $6s^2 \ 4f^4$

The notation, upon rearrangement, is

$$1s^2 \ 2s \ 2p^6 \ 3s^2 \ 3p^6 \ 3d^{10} \ 4s^2 \ 4p^6 \ 4d^{10} \ 4f^4 \ 5s^2 \ 5p^6 \ 6s^2$$

The notations for lanthanum ($Z = 57$) and the lanthanides ($Z = 58$ to 71) pose a problem. The notation for $_{57}$La ends . . . $4f^0 \ 5s^2 \ 5p^6 \ 5d^1 \ 6s^2$. One might expect that the notation for the next element, $_{58}$Ce, would end . . . $4f^1 \ 5s^2 \ 5p^6 \ 5d^1 \ 6s^2$. Instead, the differentiating $5d$ electron added for $_{57}$La falls back into the $4f$ subshell so that the notation for $_{58}$Ce ends . . . $4f^2 \ 5s^2 \ 5p^6 \ 5d^0 \ 6s^2$. Subsequent notations end . . . $4f^3 \ 5s^2 \ 5p^6 \ 5d^0 \ 6s^2$ (for $_{59}$Pr), . . . $4f^4 \ 5s^2 \ 5p^6 \ 5d^0 \ 6s^2$ (for $_{60}$Nd), and so on.

Example 4.6

Write the electronic notation for the electronic configuration of tungsten ($Z = 74$).

Solution

first period: $1s^2$
second period: $2s^2 \ 2p^6$
third period: $3s^2 \ 3p^6$
fourth period: $4s^2 \ 3d^{10} \ 4p^6$
fifth period: $5s^2 \ 4d^{10} \ 5p^6$
sixth period: $6s^2 \ 4f^{14} \ 5d^4$

Upon rearrangement, the notation is

$$1s^2 \ 2s^2 \ 2p^6 \ 3s^2 \ 3p^6 \ 3d^{10} \ 4s^2 \ 4p^6 \ 4d^{10} \ 4f^{14} \ 5s^2 \ 5p^6 \ 5d^4 \ 6s^2$$

The aufbau order cannot be used to interpret processes that involve the loss of electrons (ionizations). The configuration of the iron atom, Fe, is $1s^2 \ 2s^2 \ 2p^6 \ 3s^2 \ 3p^6 \ 3d^6 \ 4s^2$, and that of the Fe^{2+} ion is $1s^2 \ 2s^2 \ 2p^6 \ 3s^2 \ 3p^6 \ 3d^6$.

Ionization, therefore, results in the loss of the $4s$ electrons even though $3d$ electrons are the last added by the aufbau method. The Fe atom has 26 protons in the nucleus and 26 electrons. The Fe^{2+} ion has 26 protons in the nucleus but only 24 electrons. The order of orbital energies is different in the atom and the ion. In general, the first electrons lost in an ionization are those with the highest value of n and l. Electron notations, therefore, should be written by increasing value of n and not by the hypothetical order of filling.

4.15 Half-filled and Filled Subshells

In Table 4.8 the correct electronic configurations of the elements are listed. The configurations predicted by the aufbau procedure are confirmed by spectral and magnetic studies for most elements. There are a few, however, that exhibit slight variations from the standard pattern. In certain instances, it is possible to explain these variations on the basis of the stability of a filled or half-filled subshell.

The predicted configuration for the $3d$ and $4s$ subshells in the chromium atom (Z, 24) is $3d^4 4s^2$, whereas the experimentally derived configuration is $3d^5 4s^1$. Presumably the stability gained by having one unpaired electron in each of the five $3d$ orbitals (a half-filled subshell) accounts for the fact that the $3d^5 4s^1$ configuration is the one observed. The existence of a half-filled subshell also accounts for the fact that the configuration for the $4d$ and $5s$ subshells of molybdenum (Z, 42) is $4d^5 5s^1$ rather than the predicted $4d^4 5s^2$.

The stability of a half-filled $4f$ subshell is evident in the configuration of gadolinium (Z, 64). The configuration predicted by the aufbau method for this inner-transition element ends . . . $4f^8 5s^2 5p^6 5d^0 6s^2$. The accepted structure of gadolinium, however, ends . . . $4f^7 5s^2 5p^6 5d^1 6s^2$, which contains a half-filled $4f$ subshell and one $5d$ electron.

For copper (Z, 29), the predicted configuration for the last two subshells is $3d^9 4s^2$, whereas the experimentally derived structure is $3d^{10} 4s^1$. The explanation for this deviation lies in the stability of the $3d^{10} 4s^1$ configuration that results from the completed $3d$ subshell. Silver (Z, 47) and gold (Z, 79) also have configurations with completely filled d subshells instead of the $(n - 1)d^9 ns^2$ configurations predicted. In the case of palladium (Z, 46), two electrons are involved—the only case with a difference of more than one electron. The predicted configuration for the last two subshells of palladium is $4d^8 5s^2$; the observed configuration is $4d^{10} 5s^0$.

Half-filled and filled subshells also contribute to the stability of atoms in cases where the aufbau order is followed. Several examples are noted in Sections 5.2 and 5.3. The stability of the noble gases, however, is the most important example. The noble gases have configurations in which all subshells are filled. That these arrangements are very stable is shown by the low order of chemical reactivity of these elements.

Other types of deviations are observed, particularly among elements with high atomic numbers. For our purposes these exceptions are not important. In general, the chemistry of the elements is satisfactorily explained on the basis of the predicted configurations.

4.16 Types of Elements

The classification of the elements into metals and nonmetals is based on certain properties. Metals are good conductors of heat and electricity, have a characteristic luster, are malleable (capable of being pounded into shapes), and are ductile (capable of being drawn into wire). Nonmetals, on the other hand, are very poor conductors of heat and electricity, are not lustrous, and are brittle in the solid state.

About 80% of the known elements are metals. The stepped, diagonal line that appears in the periodic table marks the approximate division between the metals and the nonmetals. The nonmetals are found to the right of this line. The division, however, is not sharp since the elements that appear close to this line (sometimes called metalloids) have properties that are intermediate between metallic and nonmetallic properties.

The elements may also be classified according to their electron configurations.

1. The noble gases. In the periodic table the noble gases are found at the end of each period in group 0. They are colorless monatomic gases, which are chemically unreactive, and diamagnetic. With the exception of helium (which has the configuration $1s^2$), all the noble gases have outer configurations of $ns^2\,np^6$, a very stable arrangement.

2. The representative elements. These elements are found in the A families of the periodic table and include metals and nonmetals. They exhibit a wide range of chemical behavior and physical characteristics. Some of the elements are diamagnetic and some are paramagnetic. The compounds of these elements, however, are generally diamagnetic and colorless. All of their electronic shells are either complete or stable ($ns^2\,np^6$) except the outer shell to which the last electron may be considered as having been added. This outer shell is termed the valence shell: electrons in it are valence electrons. The number of valence electrons for each atom is the same as the group number. The chemistry of these elements depends upon these valence electrons.

3. The transition elements. These elements are found in the B families of the periodic table. They are characterized by inner building—the differentiating electron added by the aufbau procedure is an inner d electron. Electrons from the two outermost shells are used in chemical reactions. All of these elements are metals. Most of them are paramagnetic and form highly colored, paramagnetic compounds.

4. The inner-transition elements. These elements are found at the bottom of the periodic table, but they belong to the sixth and seventh periods after the elements of group III B. The sixth-period series of 14 elements that follows lanthanum is called the lanthanide series. The seventh-period series that follows actinium is known as the actinide series. The differentiating electron in each atom is an f electron. It is added to the shell that is two shells in from the outermost shell. The outer three shells, therefore, may be involved in the chemistry of these elements. All inner-transition elements are metals. They are paramagnetic and their compounds are paramagnetic and highly colored.

Summary

The topics that have been discussed in this chapter are

1. The discovery and characterization of the electron, proton, and neutron.

2. The nuclear structure of the atom.

3. The structure and characterization of isotopes and the symbolism used to designate specific isotopes.

4. The masses of atoms based on the atomic mass unit, u.

5. The atomic weight of an element based on the atomic mass and natural abundance of its natural isotopes.

6. The wave theory and quantum theory of electromagnetic radiation.

7. Atomic spectra and the Bohr theory for the electronic structure of the hydrogen atom.

8. The development and organization of the periodic table.

9. The uncertainty principle, the discovery of the wave properties of the electron, and the description of the electronic distribution of atoms in terms of wave mechanics.

10. The use of quantum numbers to define the electrons present in an atom.

11. Hund's rule to predict the order of orbital occupancy in subshells.

12. The magnetic properties of atoms.

13. The aufbau method for predicting the electronic configurations of atoms and the exceptions to such predictions brought about by the presence of filled and half-filled subshells.

14. The classification of the elements into noble gases, representative elements, transition elements, and inner-transition elements.

Key Terms

Some of the more important terms introduced in this chapter are listed below. Definitions for terms not included in this list may be located in the text by use of the index.

Alpha particle, α (Section 4.4)   A particle that consists of two protons and two neutrons and that is emitted by certain radioactive nuclei.

Atomic mass unit, u (Section 4.7)   A unit of mass equal to one-twelfth the mass of a $^{12}_{6}C$ atom.

Atomic number, Z (Section 4.5)   The number of protons in the nucleus of an atom of an element.

Aufbau method (Section 4.4)   A method of deriving the electron configurations of atoms in which electrons are successively added on the basis of orbital energies until the desired configuration is obtained.

Beta particle, β (Section 4.4)   An electron emitted by certain radioactive nuclei.

Binding energy (Section 4.7)   The energy required by a hypothetical process in which a nucleus is decomposed into nucleons; the energy equivalent of the difference between the sum of the masses of the nucleons of a nucleus and the actual mass of the nucleus.

Diamagnetic substance (Section 4.13)   A substance that is repelled by a magnetic field, behavior exhibited by substances in which all electrons are paired.

Electromagnetic radiation (Section 4.8)   Radiant energy that is propagated through a vacuum at a characteristic speed, c, and interpreted in terms of waves or quanta.

Electron (Section 4.1)   A subatomic particle that has an approximate mass of 0.00055 u, carries a one-unit negative charge, and is found outside the nucleus in an atom.

Electronic configuration (Section 4.13)   The manner in which the electrons are arranged in an atom; may be indicated by an orbital diagram or an electronic notation (see Table 4.6).

Energy level, energy shell (Sections 4.9 and 4.12)   A group of orbitals in an atom that have the same value of n, the principal quantum number.

Excited state (Section 4.9)   A state of an atom in which the electronic configuration gives the atom a higher energy than the ground state.

Exclusion principle of Pauli (Section 4.12)   No two electrons in the same atom may have identical sets of all four quantum numbers.

Frequency, ν (Section 4.8)   The number of waves of electromagnetic radiation that pass a given spot in 1 s; $v = c/\lambda$.

Gamma radiation, γ (Section 4.4)   Very short wavelength electromagnetic radiation emitted by certain radioactive nuclei.

Ground state (Section 4.9)   The state of lowest possible energy of an atom in which all electrons in the atom are as close to the nucleus as possible.

Group, family (Section 4.10)   Elements that appear in vertical columns in the periodic table.

Hund's rule (Section 4.13)   Electrons are distributed among the orbitals of a subshell in a way that gives the maximum number of unpaired electrons with parallel spins.

Inner-transition element (Sections 4.14 and 4.16)   The lanthanide and actinide elements found at the bottom of the periodic table in which the differentiating electron added by the aufbau method is an f electron added to the shell that is two shells in from the outermost shell.

Isobars (Section 4.6)   Atoms of different elements that have the same mass number but different atomic numbers.

Isotopes (Section 4.6)   Atoms of the same element that have the same atomic number but different mass numbers.

Magnetic orbital quantum number, m_l (Section 4.12)   Indicates the orientation of the orbital of the electron to which this value pertains. For a given value of l, m_l may have all the integral values from $+l$ to $-l$ (including 0).

The number of m_l values for each l value is the number of orbitals in that subshell.

Magnetic spin quantum number, m_s (Section 4.12)   Refers to the relative spin of the electron to which the value pertains. Each orbital may hold two electrons of opposed spin ($+\frac{1}{2}$ and $-\frac{1}{2}$).

Mass number, A (Section 4.5)   The number of protons and neutrons, taken together, in the nucleus of an atom.

Mass spectrograph (Section 4.6)   An instrument used to determine the types of isotopes present in an element, the exact masses of these isotopes, and the relative amount of each isotope present.

Metal (Section 4.16)   An element that has a characteristic luster, high thermal and electrical conductivities, and malleability; found in the periodic table to the left of the stepped, diagonal line.

Neutron (Section 4.3)   A subatomic particle that has an approximate mass of 1.0087 u, is uncharged, and is found in the nucleus of the atom.

Noble gas (Section 4.16)   An element in group 0 of the periodic classification. With the exception of helium (which has the electronic configuration $1s^2$), all the noble gases have outer electronic configurations of $ns^2\,np^6$.

Nonmetal (Section 4.16)   An element that is not lustrous, is a poor thermal and electrical conductor, and is brittle in the solid state; found in the periodic table to the right of the stepped, diagonal line.

Nucleon (Section 4.5)   A proton or a neutron, both of which are found in the atomic nucleus.

Nucleus (Section 4.4)   The small dense, positively charged center of an atom that contains the protons and neutrons.

Orbit (Section 4.9)   In the Bohr theory, an allowed state of an electron, characterized by a value of n.

Orbital (Section 4.11)   An energy state for an electron characterized by three quantum numbers: n, l, and m_l. A given orbital may hold two electrons with opposed spins.

Paramagnetic substance (Section 4.13)   A substance that is drawn into a magnetic field, behavior that is characteristic of substances that contain unpaired electrons.

Period (Section 4.10)   Those elements that are arranged in horizontal rows in the periodic table.

Periodic law (Section 4.10)   The chemical and physical properties of the elements are periodic functions of atomic number.

Photon (Section 4.8)   A quantum of radiant energy.

Principal quantum number, n (Section 4.12)   Indicates the energy shell of the electron to which the value pertains. The values of n are positive integers: 1, 2, 3, ...

Proton (Section 4.2)   A subatomic particle that has a mass of approximately 1.0073 u, carries a unit positive charge, and is found in the nucleus of the atom.

Quantum (Section 4.8)   A small definite quantity of radiant energy. Planck's theory assumes that radiant energy is absorbed or emitted in these quanta. The energy of a quantum, E, is directly proportional to the frequency of the radiation, v; and the proportionality constant, h, is Planck's constant (6.6262×10^{-34} J·s).

Radioactivity (Section 4.4)   The spontaneous emission of radioactive rays by an unstable atomic nucleus, which in the process is transformed into a different nucleus.

Representative element (Section 4.16)   An A family element of the periodic table.

Spectrum (Section 4.9)   A pattern of light produced by the dispersal of a light beam into its component wavelengths. Since it consists of all wavelengths, white light produces a continuous spectrum. Light emitted by a substance in an excited state, however, produces a line spectrum in which only certain wavelengths appear.

Speed of light, c (Section 4.8)   The speed at which the waves of all electromagnetic radiation travel in a vacuum; 2.9979×10^8 m/s.

Subshell (Section 4.12)   A division of an electron shell characterized by a particular value of l. The designations $s, p, d, f, \ldots$ are used for $l = 0, 1, 2, 3, \ldots$.

Subsidiary quantum number, l (Section 4.12)   Indicates the type of subshell and the shape of the orbital of the electron to which this quantum number pertains. In a given shell (indicated by n), l may have all the integral values in the series: $0, 1, 2, 3, \ldots (n - 1)$.

Transition element (Sections 4.14 and 4.16)   An element found in a B family in the periodic table. For these elements, the differentiating electron added by the aufbau method is a d electron added to the shell that is next to the outermost shell.

Uncertainty principle (Section 4.11)   It is impossible to determine, simultaneously, the exact position and exact momentum of an electron.

Unit charge (Section 4.2) 1.6022×10^{-19} C. The magnitude of the charge on the proton and electron; the proton has a unit *positive* charge and the electron a unit *negative* charge.

Valence electrons (Section 4.13)   The electrons found in the outermost shell in the ground state of an atom of an A family element.

Wave function, ψ (Section 4.11)   A solution to the Schrödinger wave equation; it describes an orbital. The square of the wave function, ψ^2, at any point is proportional to the electron charge density or the probability of finding the electron at that point.

Wavelength, λ (Section 4.8)   The distance between two similar points on two successive waves of electromagnetic radiation.

Problems*

Nuclear Atom, Atomic Symbols

4.1 J. J. Thomson determined the ratio of charge to mass of an electron (e/m). Why was the method he used unable to yield either value separately?

4.2 Which positive ion is deflected more in an electric field? Why? **(a)** H^+ or Ne^+ **(b)** Ne^+ or Ne^{2+}.

4.3 The proton is thought to have a radius of 1.30×10^{-13} cm and a mass of 1.67×10^{-24} g. **(a)** What is the density of the proton in g/cm^3? Assume the proton to be spherical. The volume of a sphere, V, is given by the formula $V = \frac{4}{3}\pi r^3$, where r is the radius of the sphere. **(b)** A basketball has a radius of 12.0 cm. What would be the mass of a basketball if it had the same density as a proton? Could you lift it?

4.4 Describe the three types of rays emitted by radioactive substances that occur in nature.

4.5 Rutherford used several metal foils in his alpha particle scattering experiments. Compare the number of wide-angle deflections observed for a Cu foil with the number observed for an Au foil of the same thickness.

4.6 The approximate radius of a nucleus, r, is given by the formula $r = A^{1/3}(1.3 \times 10^{-13}$ cm), where A is the mass number of the nucleus. The atomic radius of the $^{27}_{13}$Al atom is approximately 143 pm. If the diameter of an $^{27}_{13}$Al atom were 1.00 km (which is 0.621 mile), what would be the diameter of its nucleus (in cm)?

4.7 Use the data given in Problem 4.6 to calculate the percentage of the total volume of the aluminum atom that is occupied by the nucleus. The volume of a sphere, V, is given by the formula $V = \frac{4}{3}\pi r^3$, where r is the radius of the sphere.

4.8 **(a)** Describe the composition of the $^{138}_{56}$Ba atom. **(b)** Give the symbol that designates the atom that contains 83 protons and 126 neutrons.

4.9 **(a)** Describe the composition of the $^{195}_{78}$Pt atom. **(b)** Give the symbol that designates the atom that contains 30 protons and 34 neutrons

4.10 Complete the following table:

Symbol	Z	A	Protons	Neutrons	Electrons
P	15	31			
Ti		48			
	37	85			37
Hg				122	80
Ce		140		82	
Fe^{3+}				30	
	35			44	36

4.11 Complete the following table:

Symbol	Z	A	Protons	Neutrons	Electrons
K	19	41			
Mn		55			
	40	90			40
Pb				126	82
Xe		132		78	
Se^{2-}				46	
	48			66	46

Isotopes, Atomic Weights

4.12 Compare and contrast: **(a)** isotope, isobar; **(b)** atomic mass, mass number; **(c)** nucleon, neutron; **(d)** atomic number, atomic weight.

4.13 Each capital letter in the following list stands for an atom that contains the number of protons (p) and neutrons (n) listed beside it:

A	26 p, 28 n		D	24 p, 26 n
B	24 p, 30 n		E	26 p, 27 n
C	23 p, 27 n		F	23 p, 30 n

(a) Classify the atoms into groups of isotopes. **(b)** Classify them into groups of isobars. **(c)** Write the atomic symbol, complete with atomic number and mass number, for each atom.

4.14 Which is larger: an atomic mass unit based on the mass of the $^{19}_{9}$F atom set at exactly 19, or the current standard? Only one isotope of fluorine occurs in nature, $^{19}_{9}$F.

4.15 The element vanadium consists of two isotopes: $^{50}_{23}$V, which has an atomic mass of 49.9472 u, and $^{51}_{23}$V, which has an atomic mass of 50.9440 u. The atomic weight of vanadium is 50.9415. What is the percent abundance of each of the two isotopes?

4.16 The element rhenium consists of two isotopes: $^{185}_{75}$Re, which has an atomic mass of 184.953 u, and $^{187}_{75}$Re, which has an atomic mass of 186.956 u. The atomic weight of rhenium is 186.207. What is the percent abundance of each of the two isotopes?

4.17 Lithium occurs in nature as a mixture of $^{6}_{3}$Li atoms (mass, 6.015 u) and $^{7}_{3}$Li atoms (mass, 7.016 u). The atomic weight of lithium is 6.941. What is the percent abundance of each of the two isotopes?

4.18 If an element consists of 60.10% of atoms with a mass of 68.926 u and 39.90% of atoms with a mass of 70.925 u, what is the atomic weight of the element?

* The more difficult problems are marked with asterisks. The appendix contains answers to color-keyed problems.

4.19 If an element consists of 90.51% of atoms with a mass of 19.992 u, 0.27% of atoms with a mass of 20.994 u, and 9.22% of atoms with a mass of 21.990 u, what is the atomic weight of the element?

Electromagnetic Radiation

4.20 Compare and contrast: **(a)** wavelength, frequency; **(b)** wavelength, amplitude; **(c)** line spectrum, continuous spectrum; **(d)** ground state, excited state; **(e)** photons, light ray.

4.21 Which radiation is more energetic: **(a)** infrared radiation or microwaves, **(b)** yellow light or blue light, **(c)** a radiowave or a microwave?

4.22 A compound used in sunscreens, p-aminobenzoic acid (called PABA) absorbs ultraviolet radiation with maximum absorption occurring at 265 nm. What is the corresponding frequency? The speed of light, c, is 3.00×10^8 m/s.

4.23 What is the frequency and energy per quantum of: **(a)** red light with a wavelength of 700 nm and **(b)** violet light with a wavelength of 400 nm? The speed of light, c, is 3.00×10^8 m/s and Planck's constant, h, is 6.63×10^{-34} J·s.

4.24 What is the frequency and energy per quantum of: **(a)** a gamma ray with a wavelength of 1.00 pm and **(b)** a microwave with a wavelength of 0.100 cm? The speed of light, c, is 3.00×10^8 m/s and Planck's constant, h, is 6.63×10^{-34} J·s.

4.25 What is the wavelength and energy per quantum of: **(a)** a radiowave with a frequency of 9.00×10^5/s and **(b)** an X ray with a frequency of 1.20×10^{18}/s? The speed of light, c, is 3.00×10^8 m/s and Planck's constant is 6.63×10^{-34} J·s.

4.26 How many photons are in a 1.00×10^{-16} J signal of light with a wavelength of 500 nm? The speed of light, c, is 3.00×10^8 m/s and Planck's constant, h, is 6.63×10^{-34} J·s.

4.27 Voyager 1 sent back pictures of Saturn from a distance of 8.0×10^6 miles. How long did it take a signal to reach the earth? The speed of light, c, is 3.00×10^8 m/s and 1.000 mile is 1609 m.

***4.28** The photoelectric effect consists of the emission of electrons from the surface of a metal when the metal is irradiated by light. It requires a photon with a minimum energy of 7.58×10^{-19} J to eject an electron from silver. What frequency and wavelength of light correspond to this value? The speed of light, c, is 3.00×10^8 m/s and Planck's constant, h, is 6.63×10^{-34} J·s.

***4.29** The photoelectric effect consists of the emission of electrons from the surface of a metal when the metal is irradiated by light. If light with a wavelength of 400 nm falls on the surface of potassium metal, electrons, each with a kinetic energy of 1.38×10^{-19} J, are ejected. **(a)** What is the energy of a 400 nm photon? **(b)** If 1.38×10^{-19} J of the energy of the incident photon is imparted

to the ejected electron as kinetic energy, how much energy is required to release the electron from the metal? **(c)** What are the minimum frequency and corresponding wavelength of light required to release an electron from potassium? The speed of light, c, is 3.00×10^8 m/s and Planck's constant, h, is 6.63×10^{-34} J·s.

Atomic Spectra

4.30 According to Bohr, what is the source of the light emitted by a substance in a spectroscope?

4.31 What is the wavelength of the spectral line that corresponds to an electron transition from the $n = 4$ level to the $n = 1$ level in the hydrogen atom?

4.32 What is the wavelength of the spectral line that corresponds to an electron transition from the $n = 4$ level to the $n = 3$ level in the hydrogen atom?

4.33 The spectral lines of hydrogen in the visible region correspond to electron transitions to the $n = 2$ level from higher levels. What is the electron transition that corresponds to the 410.2 nm spectral line?

4.34 A handbook identifies the persistent spectral lines of hydrogen as having wavelengths of 121.6 nm, 486.1 nm, and 656.3 nm. **(a)** In what region of the electromagnetic spectrum does each occur? **(b)** Identify the electron transition that corresponds to each line.

***4.35** The Pfund series of spectral lines of hydrogen occurs at wavelengths of from 2.279 μm to 7.459 μm. What are the corresponding electron transitions?

***4.36** The amount of energy required to remove the most loosely held electron from an isolated atom in its ground state is called the first ionization energy of the element under consideration. **(a)** Calculate the frequency of the hydrogen spectral line that corresponds to an electron transition from $n = \infty$ to $n = 1$. **(b)** Calculate the energy of this transition. Planck's constant is 6.626×10^{-34} J·s. **(c)** What is the first ionization energy of hydrogen in kJ/mol?

***4.37** The Bohr theory successfully interprets the spectra of species that have only one electron. For the Li^{2+} ion, the two lines of lowest energy that correspond to electron transitions to the $n = 1$ level have wavelengths of 13.50 nm and 11.39 nm. **(a)** What is the value of K in the relationship: $v = K[(1/n_i^2) - (1/n_o^2)]$? **(b)** What is the wavelength of the next line in the series? **(c)** What are the wavelengths of the two spectral lines that correspond to transitions from the $n = 3$ and $n = 4$ levels to the $n = 2$ level?

***4.38** **(a)** Based on the answer to Problem 4.37a and the equation used in Example 4.3, derive an equation relating the value of K in the equation $v = K[(1/n_i^2) - (1/n_o^2)]$ and the atomic number, Z, of the one-electron species under consideration. **(b)** Predict the value of K for He^+.

Periodic Law

4.39 Compare the elements of **(a)** a period and **(b)** a group.

4.40 What change did Moseley make in Mendeleev's periodic law?

4.41 What is the origin of X rays?

*__4.42__ Moseley found that the frequency, v, of a characteristic line of the X-ray spectrum of an element is related to the atomic number, Z, of the element by the formula $\sqrt{v} = a(Z - b)$ where a is approximately $5.00 \times 10^7/\sqrt{s}$ and b is approximately 1.00. What is the atomic number of an element for which the corresponding line in the X-ray spectrum occurs at a wavelength of 0.164 nm? What is the element?

*__4.43__ What is the wavelength of the line in the X-ray spectrum of $_{24}Cr$ that conforms to the formula given in Problem 4.42?

Quantum Numbers

4.44 List the four quantum numbers, tell what each idenfies, and state the values that each may assume.

4.45 Describe the $n = 4$ level in terms of sublevels, orbitals, and electrons.

*__4.46__ **(a)** What is the de Broglie wavelength of an electron moving at one-tenth the speed of light? **(b)** What is the de Broglie wavelength of a 70.0 kg person jogging at the rate of 2.70 m/s? The mass of an electron is 9.11×10^{-28} g, the speed of light is 3.00×10^8 m/s, Planck's constant is 6.63×10^{-34} J · s, and 1 J is 1 kg · m²/s².

4.47 Sketch boundary-surface diagrams for $1s$, $2p$, and $3d$ orbitals.

4.48 What is the interpretation of the Bohr radius, a_0, in terms of the Bohr theory and in terms of wave mechanics?

4.49 Give the values for all four quantum numbers for each electron in the ground state of the neon atom. Use positive values of m_l and m_s first.

4.50 Each electron in an atom may be characterized by a set of four quantum numbers. For each of the following parts, tell how many different sets of quantum numbers are possible such that each set contains all of the values listed: **(a)** $n = 3$; **(b)** $n = 3$, $l = 3$; **(c)** $n = 3$, $l = 2$; **(d)** $n = 3$, $l = 2$, $m_l = -2$; **(e)** $n = 3$, $l = 0$, $m_l = 0$; **(f)** $n = 3$, $l = 0$, $m_l = +2$; **(g)** $n = 3$, $l = 1$.

4.51 Each electron in an atom may be characterized by a set of four quantum numbers. For each of the following parts, tell how many different sets of quantum numbers are possible such that each set contains all of the values listed: **(a)** $n = 4$, $l = 0$; **(b)** $n = 4$, $l = 0$, $m_l = +3$; **(c)** $n = 4$, $l = 3$, $m_l = 0$; **(d)** $n = 4$, $l = 2$; **(e)** $n = 4$, $l = 4$; **(f)** $n = 4$, $l = 3$; **(g)** $n = 4$.

4.52 In the ground state of $_{37}Rb$: **(a)** how many electrons have $l = 0$ as one of their quantum numbers? **(b)** how many electrons have $m_l = 0$ as one of their quantum numbers? **(c)** how many electrons have $m_l = -1$ as one of their quantum numbers?

4.53 In the ground state of $_{54}Xe$: **(a)** how many electrons

have $l = 1$ as one of their quantum numbers? **(b)** how many electrons have $m_l = +1$ as one of their quantum numbers? **(c)** How many electrons have $m_l = +2$ as one of their quantum numbers?

Electronic Configurations

4.54 Give the orbital diagram for the electronic configuration of $_{26}Fe$ and the corresponding electronic notation.

4.55 State Hund's rule. How have magnetic measurements helped to establish this rule?

4.56 Identify the atoms that have the following ground-state electronic configurations in their outer shell or shells: **(a)** $3s^2\ 3p^5$, **(b)** $3s^2\ 3p^6\ 3d^5\ 4s^2$, **(c)** $3s^2\ 3p^6\ 4s^2$, **(d)** $3s^2\ 3p^6\ 3d^{10}\ 4s^2$, **(e)** $4s^2\ 4p^6$.

4.57 Identify the atoms that have the following ground-state electronic configurations in their outer shell or shells: **(a)** $5s^2\ 5p^6$, **(b)** $4s^2\ 4p^6\ 4d^{10}\ 5s^2$, **(c)** $4s^2\ 4p^6\ 4d^{10}\ 4f^7\ 5s^2\ 5p^6\ 6s^2$, **(d)** $5s^2\ 5p^6\ 5d^7\ 6s^2$, **(e)** $5s^2\ 5p^6\ 6s^2$.

4.58 Which of the atoms listed in Problems 4.56 and 4.57 are paramagnetic?

4.59 Write the notations for the ground-state electronic configurations of the following atoms: **(a)** $_{38}Sr$, **(b)** $_{50}Sn$, **(c)** $_{62}Sm$, **(d)** $_{28}Ni$, **(e)** $_{71}Lu$, **(f)** $_{80}Hg$.

4.60 State the number of unpaired electrons in each of the atoms listed in Problem 4.59. Which are paramagnetic?

4.61 Write the notations for the ground-state electronic configurations of the following atoms: **(a)** $_{52}Te$, **(b)** $_{92}U$, **(c)** $_{23}V$, **(d)** $_{74}W$, **(e)** $_{54}Xe$, **(f)** $_{70}Yb$, **(g)** $_{40}Zr$.

4.62 State the number of unpaired electrons in each of the atoms listed in Problem 4.61. Which are paramagnetic?

4.63 Write the notations for the ground-state electronic configurations of the following ions: **(a)** $_{47}Ag^+$, **(b)** $_{82}Pb^{2+}$, **(c)** $_{21}Sc^{3+}$, **(d)** $_{24}Cr^{3+}$, **(e)** $_{16}S^{2-}$, **(f)** $_{53}I^-$.

4.64 State the number of unpaired electrons in each of the ions listed in Problem 4.63. Which are paramagnetic?

4.65 Describe how the ground-state electronic configurations of the elements of a period compare to one another and how the ground-state electronic configurations of the elements of a group are related.

4.66 Examine the electronic configurations given in Table 4.8. List the elements of the first six periods that have configurations that deviate from those predicted by the aufbau method. In which of these cases can the deviation be ascribed to the presence of a half-filled subshell? a filled subshell?

4.67 Classify each of the following elements as a noble gas, a representative element, a transition element, or an inner-transition element. Also state whether the element is a metal or a nonmetal: **(a)** Na, **(b)** N, **(c)** Ni, **(d)** In, **(e)** Nd, **(f)** Ne.

CHAPTER 5 PROPERTIES OF ATOMS AND THE IONIC BOND

Chemical bonds, which form when atoms combine, are the result of changes in electron distribution. There are three fundamental types of bonding:

1. **Ionic bonding** (Chapter 5) results when electrons are transferred from one type of atom to another. The atoms of one of the reacting elements lose electrons and become positively charged ions. The atoms of the other reactant gain electrons and become negatively charged ions. The electrostatic (plus–minus) attraction between the oppositely charged ions holds them in a crystal.

2. In **covalent bonding** (Chapters 6 and 7) electrons are shared, not transferred. A single covalent bond consists of a pair of electrons shared by two atoms. Molecules are made up of atoms covalently bonded to each other.

3. **Metallic bonding** (Chapter 23) is found in metals and alloys. The metal atoms are arranged in a three-dimensional structure. The outer electrons of these atoms are free to move throughout the structure and are responsible for binding it together.

5.1 Atomic Sizes

How an atom reacts depends upon many factors. Nuclear charge and electron configuration are the most important. The effective size of the atom is also of importance. Determination of this size, however, is a problem. The wave theory predicts that beyond a region of high density, the electron cloud of an atom thins out gradually and ends only at infinity. We cannot isolate and measure a single atom.

It is possible, however, to measure in several ways the distance between the nuclei of two atoms that are bonded together. **Atomic radii** are derived from these bond distances.* The Cl—Cl bond distance in the Cl_2 molecule, for example, is 198 pm.† One-half of this value, 99 pm, is assumed to be the atomic radius of chlorine. In turn, the atomic radius of Cl (99 pm) can be subtracted from the C—Cl bond distance (176 pm) to derive the atomic radius of C (77 pm). The effective

---

* The bond distances of *single* covalent bonds are employed. See Section 6.1.

† Atomic dimensions have been recorded in angstrom units in the past (1 Å = 10^{-10} m). In the International System, these dimensions are recorded in nanometers (1 nm = 10^{-9} m) or picometers (1 pm = 10^{-12} m). The Cl—Cl bond distance is

$$1.98 \text{ Å} = 0.198 \text{ nm} = 198 \text{ pm} = 1.98 \times 10^{-10} \text{ m}$$

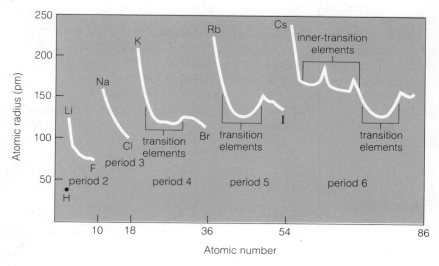

Figure 5.1 Atomic radii of the elements in picometers (1 pm = 10^{-12} m)

size of an atom may vary slightly from bond to bond when the atom is bonded to other types of atoms. Such variations, however, are usually less than a few picometers. Data derived in this way, therefore, can be used in making comparisons.

Atomic radius is plotted against atomic number in Figure 5.1. Values for the noble gases are not available. Two trends should be noted.

1. Within a group of the periodic table, an increase in atomic radius is generally observed from top to bottom. Values for the group I A elements (Li, Na, K, Rb, and Cs) and the group VII A elements (F, Cl, Br, and I) are labeled in Figure 5.1. The increase in atomic radius within each group is clearly evident. As we move from one atom to the next down a group, an additional electron level is employed, and an increase in atomic size follows.

There is, however, also an increase in the number of protons in the nucleus. The resulting increase in nuclear charge is a factor that tends to decrease atomic size. The full nuclear charge, however, is shielded from the outer electrons by electrons that lay between them and the nucleus. The number of shielding electrons increases from atom to atom in a group along with the increase in nuclear charge. As a result, the effective nuclear charge that an outer electron experiences is not the full nuclear charge. The size, therefore, is largely determined by the value of the principal quantum number, n, of the outer electrons.

2. The atomic radii of the representative elements decrease across a period from left to right. Notice the portion of the curve in Figure 5.1 that pertains to the elements of the second period (from Li to F). As we move from one atom to the next, an electron is added to the *same level* ($n = 2$) and a proton is added to the nucleus. The increase in nuclear charge is not accompanied by an equivalent increase in shielding. The shielding of one electron in a level by another electron in the same level is not very effective. As a result, the effective nuclear charge experienced by an electron in the $n = 2$ level increases across the period, these electrons are drawn in toward the nucleus, and the atomic size decreases.

The transition elements and inner-transition elements show some variations from this general pattern. For the transition elements the differentiating electrons

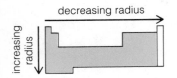

decreasing radius →

increasing radius ↓

Figure 5.2 General trends in atomic radii in relation to the periodic classification

fill inner *d* orbitals. The effect of nuclear charge on outer, size-determining electrons is reduced by the shielding effect of inner electrons. For a transition series, therefore, the gradual buildup of electrons in inner *d* orbitals at first retards the rate of decrease in atomic radius and then, toward the end of the series when the inner *d* subshell nears completion, causes the radius to increase.

The general trends in atomic radii are summarized in Figure 5.2. As a rule, atoms of metals are larger than atoms of nonmetals. The atomic radii of most metals are larger than 120 pm. The atomic radii of most nonmetals are smaller than 120 pm.

5.2 Ionization Energies

The amount of energy required to remove the most loosely held electron from an isolated atom in its ground state is called the **first ionization energy:**

$$A(g) \longrightarrow A^+(g) + e^-$$

The symbol A(g) stands for a gaseous atom of any element.

Recall the conventions regarding the assignment of signs to energy terms that were presented in Section 3.4:

1. When a system *absorbs* energy, the corresponding energy term is given a *positive* sign. A process of this type is called **endothermic.**

2. When a system *evolves* energy, the corresponding energy term is given a *negative* sign. A process of this type is called **exothermic.**

In the determination of ionization energies, energy is *used* to pull the electron away from the atom where it is held by the attraction that the nucleus exerts. Since energy is absorbed, ionization energies have positive signs.

Ionization energies can be derived from the line spectra of atoms (Section 4.9). Figure 4.9 shows that the electronic energy levels converge to a limit. The ionization energy is the energy required to raise an electron to this limit, or it is the energy emitted when an electron falls from this limit back to the ground state. The limit is readily identified from a spectrum since beyond this point, in the direction of shorter wavelengths, the spectrum no longer consists of lines but is continuous. An electron is not restricted to quantized energy states after it has been removed from the atom and hence produces a continuous spectrum.

Ionization energies are given in electron volts (eV) for individual electrons or in kilojoules per mole (kJ/mol) for a mole of electrons.* An ionization energy expressed in kJ/mol refers to the energy required to remove 1 mol of electrons (6.022×10^{23} electrons) from 1 mol of atoms (6.022×10^{23} atoms). Ionization energy is plotted against atomic number in Figure 5.3. We can make the following generalizations:

---

* One electron volt is the kinetic energy acquired by an electron in passing through a potential difference of 1 V in a vacuum:

$$1 \text{ eV} = 1.6022 \times 10^{-19} \text{ J} = 96.487 \text{ kJ/mol}$$

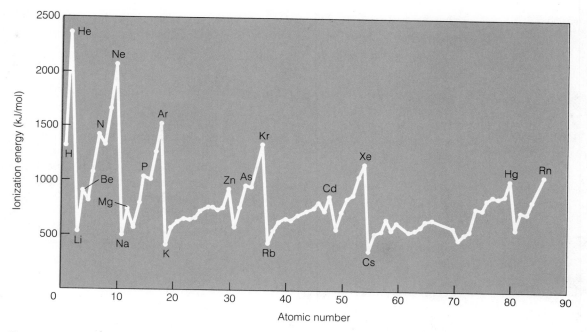

Figure 5.3 First ionization energies of the elements versus atomic number

1. In general, ionization energy *increases* across a period from left to right. Notice the portions of the curve that pertain to the second period elements (from Li to Ne), the third period elements (from Na to Ar), and so on. The ionization energy increases because the atoms become smaller and the effective nuclear charge increases. Removal of the electron becomes more and more difficult.

2. In general, ionization energy *decreases* within a group of representative elements from top to bottom. The elements of group I A (Li, Na, K, Rb, and Cs) and the elements of group 0 (He, Ne, Ar, Kr, Xe, and Rn) are labeled in Figure 5.3. As we go from atom to atom down a group, the nuclear charge increases, but the effect is largely canceled by the increase in the number of shielding electrons in the inner shells. The atoms become larger, and the electron that is removed comes from a higher and higher level. Removal of the electron becomes easier and ionization energy decreases.

The ionization energies of the transition elements do not increase across a period as rapidly as those of the representative elements. The ionization energies of the inner-transition elements remain almost constant. In these series, the differentiating electrons are being added to inner shells. The resulting increase in shielding accounts for the effects noted.

Atoms of metals tend to lose electrons and become positive ions in chemical reactions. Atoms of nonmetals do not usually behave in this way. Metals, therefore, are elements that have comparatively low ionization energies and nonmetals are elements that have comparatively high ones. The ionization energies of most metals are below 1000 kJ/mol and of most nonmetals are above 1000 kJ/mol.

Several features of the curve in Figure 5.3 relate to certain electron configurations. Consider the points on the curve that pertain to

1. The noble gases (He, Ne, Ar, Kr, Xe and Rn), each of which has an $ns^2\,np^6$ configuration in the outer shell (except for He which has the configuration $1s^2$).

2. The elements Be, Mg, Zn, Cd, and Hg, each of which has a *filled s subshell* in the outermost shell (ns^2).

3. The elements N, P, and As, each of which has a *half-filled p subshell* in the outermost shell ($ns^2\,np^3$).

The ionization energy of each of these elements (particularly the noble gases) is higher than the ionization energy of the element that follows it in the periodic table. These three types of electron configuration, therefore, may be said to be comparatively stable, and it is relatively difficult to remove an electron from them. In each case, it is easier to remove the differentiating electron added by the aufbau method to form the next element from an atom of the next element.

We have so far discussed only *first* ionization energies. The second ionization energy of an element refers to the removal of *one* electron from a 1+ ion of the element:

$$A^+(g) \longrightarrow A^{2+}(g) + e^-$$

The *third* ionization energy pertains to the removal of *one* electron from a 2+ ion. Removing an electron, which has a negative charge, from an ion that has a positive charge becomes more and more difficult as the charge on the ion increases. Consequently, the ionization energies increase in the order: first < second < third, and so on. Since ionization energies above the third are extremely large for any element, ions with charges higher than 3+ seldom exist under ordinary conditions.

Ionization energies in kilojoules per mole for the first three elements of the third period are listed in Table 5.1. For each element, the ionization energies in-

Table 5.1	Ionization energies of the third period metals				
		Ionization Energies (kJ/mol)			
Metal	Group	First	Second	Third	Fourth
Na	I A	+496	+4,563	+6,913	+9,541
Mg	II A	+738	+1,450	+7,731	+10,545
Al	III A	+577	+1,816	+2,744	+11,575

crease from the first to the fourth, as expected. Notice, however, that in each case there is a decided jump in the required energy after all the valence electrons have been removed. These points are marked in the table; the number of valence electrons equals the group number. After the valence electrons have been removed, it is necessary to break into the very stable $2s^2\,2p^6$ noble-gas arrangement of the shell beneath the valence shell to remove the next electron.

The trends in first ionization energy are summarized in Figure 5.4. Notice that the most reactive metals are found in the lower left corner of the periodic table. This reactivity in terms of electron loss decreases as we move upward or to the right from this corner of the chart.

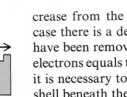

increasing ionization energy

decreasing ionization energy

Figure 5.4 General trends in ionization energy in relation to the periodic classification

5.3 Electron Affinities

The energy change associated with the process in which an electron is added to a gaseous atom in its ground state is called a **first electron affinity**:

$$e^- + A(g) \longrightarrow A^-(g)$$

Notice that an electron affinity pertains to a process in which a *negative* ion is produced from a neutral atom (by electron *gain*). An ionization energy, on the other hand, pertains to a process in which a *positive* ion is produced from a neutral atom (by electron *loss*).

Some electron affinities are listed in Table 5.2. Energy is usually (but not always) *evolved* by these processes. Most first electron affinities, therefore, have *negative* signs.* The first electron affinity of fluorine, for example, is -322 kJ/mol:

$$e^- + F(g) \longrightarrow F^-(g) \qquad \Delta H = -322 \text{ kJ}$$

Some of the values listed in Table 5.2 have positive signs. For example,

$$e^- + Ne(g) \longrightarrow Ne^-(g) \qquad \Delta H = +21 \text{ kJ}$$

A positive sign on an electron affinity value indicates that work must be done (energy absorbed) to force the atom under consideration to accept an additional electron. An electron approaching a neutral atom is attracted by the nucleus of the atom but repelled by the electrons of the atom. If the attraction is greater than the repulsion, energy is released when the negative ion forms. If the repulsion is greater than the attraction, energy is required to form the negative ion.

A small atom should have a greater tendency toward electron gain than a large atom. The added electron is, on the average, closer to the positively charged

Table 5.2   Electron affinities (kJ/mol)[a]

A. Addition of one electron

H −73			B −23	C −123	N 0	O −142	F −322	He (+21)
Li −60	Be (+240)		B −23	C −123	N 0	O −142	F −322	Ne (+29)
Na −53	Mg (+230)		Al −44	Si −120	P −74	S −200	Cl −348	Ar (+35)
K −48	Ca (+156)		Ga (−36)	Ge −116	As −77	Se −195	Br −324	Kr (+39)
Rb −47	Sr (+168)		In −34	Sn −121	Sb −101	Te −190	I −295	Xe (+41)
Cs −46	Ba (+52)		Tl −50	Pb −101	Bi −101	Po (−170)	At (−270)	Rn (+41)

B. Addition of two electrons

O +702	S +332

[a] Values in parentheses are estimated.

---

* In some sources, electron affinity is defined in terms of the *energy released* by these processes. In these sources, electron affinity values are given positive signs if they pertain to processes in which energy is liberated. This sign convention is the opposite of the one employed in this book. *Negative* signs are consistently used to indicate energy *released*, and *positive* signs are used to indicate energy *absorbed*.

nucleus of a small atom. The atomic radii of the elements decrease and the effective nuclear charges increase across a period from left to right. We would expect the electron affinities of the corresponding elements to become larger and larger negative values. This trend is only roughly followed (see Table 5.2.).

Exceptions to this generalization should be noted. In the second period, for example, the values for beryllium (filled $2s$ subshell), nitrogen (half-filled $2p$ subshell), and neon (all subshells filled) are out of line. These elements have relatively stable electron configurations and do not accept additional electrons readily. Similar exceptions may be noted for corresponding elements of the other periods. In all the periods, the atom with the greatest tendency toward electron gain (largest negative value) is the group VII A element (F, Cl, Br, I, and At). The electron configurations of each of these elements is one electron short of a noble-gas configuration.

How do electron affinities vary from element to element within the same periodic group? No pattern that pertains to all the groups seems to exist. In the case of the group VII A elements, the electron affinity of fluorine appears to be out of line (see Table 5.2.). Based on atomic size, fluorine—the smallest atom of the group—might be expected to liberate the most energy when an electron is added. In the case of a small atom, however, the added electron is not only strongly attracted by the nucleus but it is also strongly repelled by the electrons already present in the atom. The electron charge of the valence electrons is more concentrated in a small shell than it is when the *same number* of electrons is placed in a larger shell. For fluorine, it is believed that this stronger repulsion offsets the stronger attraction brought about by small atomic size.

Some **second electron affinities** have been determined. These values refer to processes in which an electron is added to a negative ion. For example,

$$e^- + O^-(g) \longrightarrow O^{2-}(g) \qquad \Delta H = +844 \text{ kJ}$$

Since a negative ion and an electron repel each other, energy is required, not released, by the process. All second electron affinities have positive signs.

Calculations for the production of a multicharged negative ion must take into consideration all pertinent electron affinities. The total effect for any such ion is always endothermic. In the formation of the O^{2-} ion from the O atom, for example, less energy is *released* by the first electron affinity,

$$e^- + O(g) \longrightarrow O^-(g) \qquad \Delta H = -142 \text{ kJ}$$

than is *required* by the second electron affinity,

$$e^- + O^-(g) \longrightarrow O^{2-}(g) \qquad \Delta H = +844 \text{ kJ}$$

The overall process, therefore, is endothermic:

$$2e^- + O(g) \longrightarrow O^{2-}(g) \qquad \Delta H = +702 \text{ kJ}$$

5.4 The Ionic Bond

When a metal reacts with a nonmetal, electrons are transferred from the atoms of the metal to the atoms of the nonmetal, and an **ionic** (or **electrovalent**) **compound** is produced. The atoms that lose electrons become positive ions, called **cations.**

Chapter 5   Properties of Atoms and the Ionic Bond

Atoms that gain electrons become negative ions, called anions. These ions attract one another to form a crystal.

Compounds of the A family elements are often represented by using the symbols of the elements together with dots to indicate valence electrons. Valence electrons are the only electrons that are involved in the chemical reactions of the representative elements.

Consider the reaction of a sodium atom and a chlorine atom. Sodium is in group I A. Its atoms have one valence electron. Chlorine is a member of group VII A. Atoms of chlorine have seven valence electrons. The sodium atom loses one electron; the chlorine atom gains an electron:

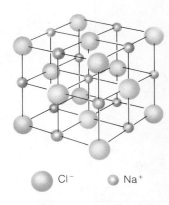

Cl⁻   Na⁺

Figure 5.5 Sodium chloride crystal lattice

$$\text{Na}\cdot \; + \cdot\ddot{\underset{\cdot\cdot}{\text{Cl}}}: \longrightarrow \text{Na}^+ \; + \; :\ddot{\underset{\cdot\cdot}{\text{Cl}}}:^-$$

The sodium ion that forms has a 1+ charge since the sodium nucleus contains 11 protons (11+ charge) and the ion has only 10 electrons (one having been lost). The chloride ion that forms has a 1− charge since the chlorine nucleus contains 17 protons (17+ charge) and the ion has 18 electrons (one having been gained).

In the reaction, the total number of electrons lost by sodium must equal the total number of electrons gained by chlorine. Thus, the number of sodium ions produced is the same as the number of chloride ions produced, and the formula, NaCl, gives the simplest ratio of ions present in the compound (1 to 1). These ions attract one another to form a crystal (Figure 5.5).

In a sodium chloride crystal, no ion may be considered as belonging exclusively to another. Rather, each sodium ion is surrounded by six chloride ions and each chloride ion is surrounded by six sodium ions. The arrangement of ions in the crystal is such that the repulsion of like-charged ions is outweighed by the attraction of oppositely charged ions. The net attraction holds the crystal together.

The complete electronic configurations of the atoms and ions of this reaction are

$$\text{Na}(1s^2 \, 2s^2 \, 2p^6 \, 3s^1) \longrightarrow \text{Na}^+(1s^2 \, 2s^2 \, 2p^6) + e^-$$
$$e^- + \text{Cl}(1s^2 \, 2s^2 \, 2p^6 \, 3s^2 \, 3p^5) \longrightarrow \text{Cl}^-(1s^2 \, 2s^2 \, 2p^6 \, 3s^2 \, 3p^6)$$

The sodium ion has an electronic configuration identical to that of neon, and the chloride ion has the same configuration as argon. The ions may be said to be isoelectronic (the same in electronic configuration) with neon and argon, respectively.

In ionic reactions, most A family elements lose or gain electrons in such a way as to produce ions that are isoelectronic with a noble gas. Most of these noble gas ions have eight electrons in the outermost shell (called an s^2p^6 configuration). A few, however, have the $1s^2$ configuration of helium (Li^+, Be^{2+}, and H^-), which is called an s^2 configuration.

An atom of oxygen (which is a group VI A nonmetal) has six valence electrons. It gains two electrons to attain the s^2p^6 configuration of neon:

$$2e^- + \text{O}(1s^2 \, 2s^2 \, 2p^4) \longrightarrow \text{O}^{2-}(1s^2 \, 2s^2 \, 2p^6)$$

In a reaction between sodium and oxygen, two atoms of sodium are required for every one atom of oxygen since the number of electrons lost must equal the number of electrons gained:

$$2\,\text{Na}\cdot + \ddot{\underset{\cdot\cdot}{\text{O}}}\text{:} \longrightarrow 2\,\text{Na}^+ + \text{:}\ddot{\underset{\cdot\cdot}{\text{O}}}\text{:}^{2-}$$

The simplest ratio of ions present in the product, sodium oxide, is indicated by the formula of the compound, Na_2O.

The formula of an ionic compound can be derived from the formulas of the ions it contains. The *total* positive charge of the cations must equal the *total* negative charge of the anions. The s^2p^6 ion derived from calcium (a group II A metal) has a charge of $2+$:

$$\text{Ca}(1s^2\,2s^2\,2p^6\,3s^2\,3p^6\,4s^2) \longrightarrow \text{Ca}^{2+}(1s^2\,2s^2\,2p^6\,3s^2\,3p^6) + 2e^-$$

The compound formed from Ca^{2+} and Cl^- ions, calcium chloride, has the formula $CaCl_2$. Two Cl^- ions are required for every Ca^{2+} ion. Calcium oxide, which is composed of Ca^{2+} and O^{2-} ions, has the formula CaO. The formula gives the *simplest* ratio of ions.

An atom of Al (a group III A metal) attains a noble-gas configuration by the loss of three electrons. The oxide of aluminum consists of Al^{3+} and O^{2-} ions. In order to balance the charge, two Al^{3+} ions (total charge, $6+$) and three O^{2-} ions (total charge, $6-$) must be indicated in the formula for the compound, Al_2O_3.

5.5 Lattice Energies

The enthalpy change associated with the condensation of gaseous positive and negative ions into a crystal is called a **crystal energy** or a **lattice energy.** The lattice energy of sodium chloride, for example, is -789 kJ/mol:

$$\text{Na}^+(g) + \text{Cl}^-(g) \longrightarrow \text{NaCl}(s) \qquad \Delta H = -789 \text{ kJ}$$

Since energy is always *evolved* in these processes, all lattice energies have *negative* signs. The lattice energy (with the sign changed) of a given crystal may also be viewed as the amount of energy required to separate the ions of that crystal:

$$\text{NaCl}(s) \longrightarrow \text{Na}^+(g) + \text{Cl}^-(g) \qquad \Delta H = +789 \text{ kJ}$$

The importance of the lattice energy may be seen by using a method of analysis developed independently by Max Born and Fritz Haber in 1916. The **Born-Haber cycle** for the preparation of sodium chloride will serve as an example (see Figure 5.6).

The Born-Haber analysis is based on the law of Hess (Section 3.5), which states that the change in enthalpy for any chemical reaction is a constant, no matter whether the reaction is brought about in one step or in several steps. The enthalpy change for the preparation of *one mole* of NaCl(s) *in one step* from Na(s) and $Cl_2(g)$ is the enthalpy of formation of the compound (Section 3.6):

$$\text{Na}(s) + \tfrac{1}{2}\text{Cl}_2(g) \longrightarrow \text{NaCl}(s) \qquad \Delta H_f^\circ = -411 \text{ kJ} \qquad (5.1)$$

We can also *imagine* one mole of NaCl(s) to be prepared from Na(s) and $Cl_2(g)$ by a series of steps. The sum of the ΔH values for these steps must equal, according to the law of Hess, the enthalpy of formation of NaCl(s), which is the ΔH value for the single-step preparation. A list of this series of steps follows.

Cluster of halite (sodium chloride) crystals. Note the cubic structure. *United States Geological Survey.*

Chapter 5 Properties of Atoms and the Ionic Bond

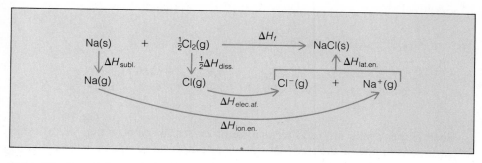

Figure 5.6 Born-Haber cycle for NaCl(s)

1. Crystalline sodium metal is sublimed into gaseous sodium atoms; 108 kJ of energy is *absorbed* per mole of Na (the enthalpy of sublimation):

$$Na(s) \longrightarrow Na(g) \qquad \Delta H_{subl} = +108 \text{ kJ} \qquad (5.2)$$

2. *One-half mole* of gaseous Cl_2 molecules is dissociated into *one mole* of gaseous Cl atoms; 122 kJ of energy is *absorbed* in the process. The enthalpy of dissociation of $Cl_2(g)$, also called the bond energy of the Cl—Cl bond (Section 3.7), is +243 kJ per mole of Cl_2. Since the dissociation of one mole of Cl_2 produces two moles of Cl atoms, and since only one mole of Cl atoms is needed to form one mole of NaCl, only one-half the dissociation energy is needed:

$$\tfrac{1}{2}Cl_2(g) \longrightarrow Cl(g) \qquad \tfrac{1}{2}\Delta H_{diss} = \tfrac{1}{2}(+243 \text{ kJ}) = +122 \text{ kJ} \qquad (5.3)$$

3. The gaseous sodium atoms are ionized into gaseous sodium ions. The amount of energy *required* is the first ionization energy of sodium (Section 5.2):

$$Na(g) \longrightarrow Na^+(g) + e^- \qquad \Delta H_{ion\ en} = +496 \text{ kJ} \qquad (5.4)$$

4. Electrons are added to the gaseous chlorine atoms to produce gaseous chloride ions. The enthalpy change per mole of Cl(g) is the first electron affinity of chlorine (Section 5.3). Energy is *liberated*:

$$Cl(g) + e^- \longrightarrow Cl^-(g) \qquad \Delta H_{elec\ af} = -348 \text{ kJ} \qquad (5.5)$$

This is the first step in which energy is liberated. Even so, the amount liberated is not sufficient to make up for the total amount absorbed in the preceding steps.

5. In this last step, the gaseous ions condense into one mole of crystalline sodium chloride. The corresponding enthalpy change, the lattice energy of NaCl(s), is −789 kJ/mol, which represents energy *liberated*:

$$Na^+(g) + Cl^-(g) \longrightarrow NaCl(s) \qquad \Delta H_{lat\ en} = -789 \text{ kJ} \qquad (5.6)$$

It is clear that most of the energy liberated by the overall reaction comes from this step. It is this step that makes the whole process energetically favorable.

When the thermochemical equations given in steps 1 to 5 [Equations (5.2) to (5.6)] are added, the result is the equation for the enthalpy of formation of NaCl(s):

$$Na(s) + \tfrac{1}{2}Cl_2(g) \longrightarrow NaCl(s) \qquad \Delta H_f^\circ = -411 \text{ kJ} \qquad (5.1)$$

We can check the cycle in the following way:

$$\Delta H_f^\circ = \Delta H_{subl} + \tfrac{1}{2}\Delta H_{diss} + \Delta H_{ion\ en} + \Delta H_{elec\ af} + \Delta H_{lat\ en}$$
$$= +108 \text{ kJ} + 122 \text{ kJ} + 496 \text{ kJ} - 348 \text{ kJ} - 789 \text{ kJ}$$
$$= -411 \text{ kJ}$$

Born-Haber cycles are used to analyze processes to see how they are affected by variations in one of the steps. They may also be used to calculate the enthalpy change for one of the steps, or for the entire process.

Example 5.1

Calculate the lattice energy of $MgCl_2(s)$. For magnesium, the enthalpy of sublimation is $+150$ kJ/mol, the first ionization energy is $+738$ kJ/mol, and the second ionization energy is $+1450$ kJ/mol. For chlorine, the dissociation energy is $+243$ kJ/mol of $Cl_2(g)$, and the first electron affinity is -348 kJ/mol of $Cl(g)$. For $MgCl_2(s)$, the enthalpy of formation is -642 kJ/mol.

Solution

The thermochemical equations for the steps of the cycle must add up to the thermochemical equation for the enthalpy of formation of one mole of $MgCl_2(s)$:

$$Mg(s) + Cl_2(g) \longrightarrow MgCl_2(s) \qquad \Delta H_f^\circ = -642 \text{ kJ}$$

Notice that in this case the positive ion has a $2+$ charge and hence the first and second ionization energies of the metal must be used. In addition, since one mole of $MgCl_2$ is formed from two moles of Cl atoms, we must employ the dissociation energy of one mole of Cl_2 molecules as well as two times the electron affinity of one mole of Cl atoms. The steps are

step	chemical equation	ΔH
sublimation of Mg	$Mg(s) \longrightarrow Mg(g)$	$+150$ kJ
first ionization energy of Mg	$Mg(g) \longrightarrow Mg^+(g) + e^-$	$+738$ kJ
second ionization energy of Mg	$Mg^+(g) \longrightarrow Mg^{2+}(g) + e^-$	$+1450$ kJ
dissociation of Cl_2	$Cl_2(g) \longrightarrow 2\ Cl(g)$	$+243$ kJ *(Continued)*

step	chemical equation	ΔH
first electron affinity for 2 mol of Cl atoms	$2\ Cl(g) + 2e^- \longrightarrow 2\ Cl^-(g)$	$2(-348\ kJ) = -696\ kJ$
lattice energy	$Mg^{2+}(g) + 2\ Cl^-(g) \longrightarrow MgCl_2(s)$	$\Delta H_{lat\ en}$
total	$Mg(s) + Cl_2(g) \longrightarrow MgCl_2(s)$	$+1885\ kJ + \Delta H_{lat\ en}$

The total value of ΔH, however, must equal the enthalpy of formation of $MgCl_2(s)$; therefore,

$$+1885\ kJ + \Delta H_{lat\ en} = -642\ kJ$$

$$\Delta H_{lat\ en} = -2527\ kJ$$

The lattice energy of $MgCl_2(s)$ is -2527 kJ/mol.

5.6 Types of Ions

The driving force of the ionic reaction is the electrostatic attraction of the ions for one another. This attraction results in the liberation of the lattice energy. Some lattice energies are listed in Table 5.3. Notice that more energy is liberated by the formation of crystals that contain ions with charges higher than $1+$ and $1-$ than is liberated by the formation of crystals that contain only $1+$ and $1-$ ions. Ions with high charges attract oppositely charged ions more strongly than ions with only a $1+$ or a $1-$ charge. The four values given in Table 5.3 reflect this fact.

Lattice energy is an important factor in determining the charges that atoms assume in the formation of an ionic crystal:*

1. *Why doesn't Na lose two electrons and become "Na^{2+}"?* The energy required for this process is the sum of the first and second ionization energies of Na (Table 5.1):

$$+496\ kJ/mol + 4563\ kJ/mol = +5059\ kJ/mol$$

The lattice energy of an imaginary "$NaCl_2$" would be far too small to compensate for the energy required by this ionization. The lattice energy of "$NaCl_2$" would probably be approximately the same as that of $MgCl_2$, -2527 kJ/mol (Table 5.3). The removal of an electron beyond the noble-gas configuration requires too much energy.

2. *Since Na cannot lose two electrons in ion formation, how is Mg able to lose two*

---

* The correct thermodynamic function to use to check reaction spontaneity is the change in Gibbs free energy, ΔG, not the change in enthalpy, ΔH (see Section 17.4). This approach is valid, however, for the formation of an ionic crystal.

Table 5.3   Lattice energies

Compound	Component Ions	Lattice Energy (kJ/mol)
NaCl	Na^+, Cl^-	-789
Na_2O	$2Na^+$, O^{2-}	-2570
$MgCl_2$	Mg^{2+}, $2Cl^-$	-2527
MgO	Mg^{2+}, O^{2-}	-3890

electrons to become Mg^{2+}? The Mg atom attains a noble-gas configuration through this loss. The sum of the first and second ionization energies of Mg (Table 5.1) is much less than the sum for Na:

$$+738 \text{ kJ/mol} + 1450 \text{ kJ/mol} = +2188 \text{ kJ/mol}$$

The lattice energy of $MgCl_2$, -2527 kJ/mol (Table 5.3) is more than enough to supply the energy required for this ionization.

3. *Since less energy is required to remove one electron than is required to remove two, why doesn't Mg form a "Mg^+" ion?* In this case, the energy required to form "Mg^+" would be only $+738$ kJ/mol (the first ionization energy of Mg, Table 5.1). On the other hand, the lattice energy of the imaginary "MgCl" would be only about -789 kJ/mol (similar to that of NaCl). The larger lattice energy of $MgCl_2$ (-2527 kJ/mol, brought about by the higher charge on Mg^{2+}) favors the formation of $MgCl_2$.

We can see, therefore, why so many metals form cations with s^2p^6 (noble-gas) configurations. More than three electrons are never lost or gained in ion formation. The energy required to bring about a gain or loss of more than three electrons is not available. Compounds with formulas such as $TiCl_4$, $SnBr_4$, SF_6, PCl_5, and SiO_2 are not ionic.

Energy considerations also favor the formation of negative ions with noble-gas electron configurations. All monatomic anions are noble-gas ions. Atoms of nonmetals add electrons until the noble-gas configuration is reached, and the attainment of this limit is favored:

1. *According to the electron affinity data in Table 5.2, the formation of O^- releases energy (-142 kJ/mol) and the formation of O^{2-} requires energy ($+702$ kJ/mol). Why doesn't the O atom form the "O^-" ion rather than the O^{2-} ion?* The lattice energy of an imaginary "NaO" would probably be only about -789 kJ/mol (similar to that of NaCl). The lattice energy of Na_2O (which contains the O^{2-} ion) is -2570 kJ/mol. On balance, the formation of the noble-gas ion, O^{2-}, is favored.

2. *Why doesn't the Cl atom form a "Cl^{2-}" ion instead of the Cl^- ion?* More energy would be liberated by the formation of the crystal lattice of an imaginary "Na_2Cl" (probably, about -2570 kJ/mol) than is liberated by the formation of the NaCl lattice (-789 kJ/mol). The production of the "Cl^{2-}" ion, however, would require the input of an extremely large amount of energy—more than any lattice energy could supply. The electron added to Cl^- to produce "Cl^{2-}" would be added beyond an s^2p^6 configuration. It would be added to the next higher quantum level, screened from the nuclear charge by a closed s^2p^6 shell, and repelled by the electrons

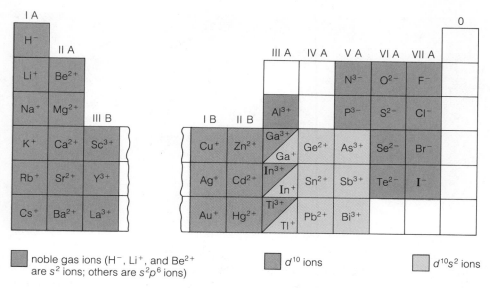

Figure 5.7 Types of ions and their relations to the periodic table

already present in the negatively charged Cl^- ion. The addition would require a large amount of energy. Such additions—beyond a noble-gas configuration—are never observed.

The noble-gas cations and anions and their relation to the periodic table are shown in Figure 5.7.

Certain metals undergo ionic reactions even though they cannot possibly produce s^2p^6 cations. Zinc, for example, would have to lose 12 electrons to get a noble-gas configuration. In its reactions, zinc forms the Zn^{2+} ion by the loss of two electrons:

$$Zn(\ldots 3s^2\ 3p^6\ 3d^{10}\ 4s^2) \rightarrow Zn^{2+}(\ldots 3s^2\ 3p^6\ 3d^{10}) + 2e^-$$

The ionization energies of Zn and the lattice energies of Zn^{2+} compounds favor the formation of this ion. The electron configuration of the Zn^{2+} ion is a stable one. All of the subshells in the outer shell of the Zn^{2+} ion are filled. Other atoms form ions with $ns^2\ np^6\ nd^{10}$ configurations similar to that of Zn^{2+}. They are called d^{10} **ions** and are listed in Figure 5.7.

Tin forms a cation, Sn^{2+}, that will serve as an example of another type of ion:

$$Sn(\ldots 4s^2\ 4p^6\ 4d^{10}\ 5s^2\ 5p^2) \rightarrow Sn^{2+}(\ldots 4s^2\ 4p^6\ 4d^{10}\ 5s^2) + 2e^-$$

This configuration is also a stable one. Ions with similar configurations are called $d^{10}s^2$ **ions.** Nine of them are listed in Figure 5.7. The larger elements of group III A can form both $d^{10}s^2$ ions and d^{10} ions.

In the reactions of the transition metals, inner d electrons may be lost as well as the outer s electrons. The outer s electrons, however, are lost first. Most transition elements cannot form ions with any of the regular configurations (s^2, s^2p^6, d^{10}, or $d^{10}s^2$). A list of some of these from the first transition series is given in Table 5.4. The configuration listed in the table is that of the outer shell of the ion.

Many transition metals form more than one type of cation. Examples are Cu^+ and Cu^{2+}, Cr^{2+} and Cr^{3+}, and Fe^{2+} and Fe^{3+}. More energy is required to

Table 5.4 Some first transition series cations	
Electron Configuration	Example
$3s^2\,3p^6\,3d^1$	Ti^{3+}
$3s^2\,3p^6\,3d^2$	V^{3+}
$3s^2\,3p^6\,3d^3$	Cr^{3+}, V^{2+}
$3s^2\,3p^6\,3d^4$	Cr^{2+}, Mn^{3+}
$3s^2\,3p^6\,3d^5$	Mn^{2+}, Fe^{3+}
$3s^2\,3p^6\,3d^6$	Fe^{2+}, Co^{3+}
$3s^2\,3p^6\,3d^7$	Co^{2+}
$3s^2\,3p^6\,3d^8$	Ni^{2+}
$3s^2\,3p^6\,3d^9$	Cu^{2+}

produce the Fe^{3+} ion than is required to produce the Fe^{2+} ion. The lattice energies of Fe^{3+} compounds, however, are larger than those of Fe^{2+} compounds. These factors balance to the point that it is possible to prepare compounds of both ions.

5.7 Ionic Radius

The distance between the centers of two adjacent ions in a crystal can be determined by X-ray diffraction (Section 9.12). For most crystals, this distance is the sum of the radius of a cation and the radius of an anion. Dividing such a distance to obtain the two radii is a problem.

One solution to the problem is to study a crystal composed of very small cations and large anions, such as lithium iodide (Figure 5.8a). In the LiI crystal, I^- ions are assumed to touch each other. The distance between two I^- ions (d in Figure 5.8a) is divided in half to get the radius of the I^- ion:

$$\text{radius } I^- = 432 \text{ pm}/2 = 216 \text{ pm}$$

In most crystals, the anions do not touch each other. The distance d in Figure 5.8b could not be used in a similar way.

Once the radius of the I^- ion has been found, other ionic radii can be calculated. If the distance between the center of a K^+ ion and an I^- ion is determined (d' in Figure 5.8b), the radius of the K^+ ion can be found by subtracting the I^- radius from the K^+ to I^- distance:

$$d' = \text{radius } K^+ + \text{radius } I^-$$

$$349 \text{ pm} = \text{radius } K^+ + 216 \text{ pm}$$

$$\text{radius } K^+ = 133 \text{ pm}$$

A positive ion is always smaller than the atom from which it is derived (Figure 5.9). The radius of K is 203 pm, and the radius of K^+ is 133 pm. In the formation of the K^+ ion, the loss of an electron represents the loss of the entire $n = 4$ level of the K atom. Furthermore, the protons outnumber the electrons

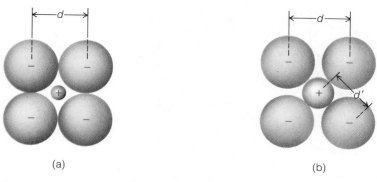

(a)   (b)

Figure 5.8 Determination of ionic radii. (See text for further discussion.)

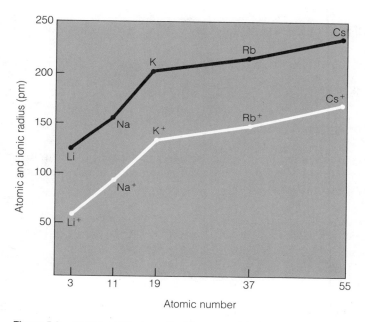

Figure 5.9 Atomic and ionic radii of the group I A elements

in the positive ion. The electrons of the ion are drawn in closer to the nucleus. For similar reasons, a 2^+ ion is larger than a $3+$ ion. For example,

radius Fe $= 117$ pm   radius $Fe^{2+} = 75$ pm   radius $Fe^{3+} = 60$ pm

A negative ion is always larger than the atom from which it is derived (Figure 5.10). The radius of the Cl atom is 99 pm, and the radius of the Cl^- ion is 181 pm. In the formation of the Cl^- ion, the addition of an electron causes an increase in the extent to which the valence electrons repel one another, and the valence level expands. The chloride ion has 18 electrons and 17 protons.

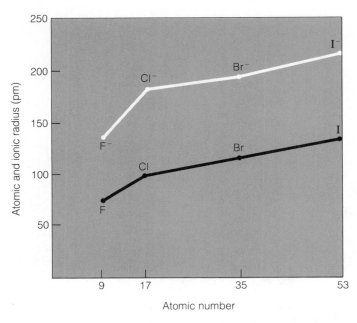

Figure 5.10 Atomic and ionic radii of the group VII A elements

5.8 Nomenclature of Ionic Compounds

The nomenclature (system of naming) of ionic compounds is based on a number of rules. The cation (positive ion) of the compound is named first and the anion (negative ion) is named second.

1. Cations. A monatomic ion is one that is derived from a single atom. Most cations are monatomic ions formed by metal atoms. If the metal forms only one type of cation, the name of the ion is the same as the name of the metal:

Na^+ is the sodium ion
Mg^{2+} is the magnesium ion
Al^{3+} is the aluminum ion

Some metals form more than one type of cation. In these cases, the distinction between the cations may be made by indicating the charge of the cation in its name. The charge is indicated by a Roman numeral in parentheses added to the *English* name of the metal:

Cu^+ is the copper(I) ion and Cu^{2+} is the copper(II) ion
Fe^{2+} is the iron(II) ion and Fe^{3+} is the iron(III) ion

In an older method used to distinguish between two types of ion formed by a metal, the ending of the name of the metal is changed. The *Latin* name of the metal is employed when the symbol for it is derived from the Latin. The ending *-ous* is used in the name of that ion of the pair that has the lower charge, and the ending *-ic* is used to name the ion that has the higher charge:

Cu^+ is the cuprous ion and Cu^{2+} is the cupric ion
Fe^{2+} is the ferrous ion and Fe^{3+} is the ferric ion

Table 5.5 Some common ions

Cations		Anions	
ammonium	NH_4^+	acetate	$C_2H_3O_2^-$
copper(I) or cuprous	Cu^+	bromide	$\cdot Br^-$
lithium	Li^+	chlorate	ClO_3^-
potassium	K^+	chloride	Cl^-
silver	Ag^+	chlorite	ClO_2^-
sodium	Na^+	cyanide	CN^-
		fluoride	F^-
barium	Ba^{2+}	hydroxide	OH^-
cadmium	Cd^{2+}	hypochlorite	ClO^-
calcium	Ca^{2+}	iodide	I^-
chromium(II) or chromous	Cr^{2+}	nitrate	NO_3^-
cobalt(II) or cobaltous	Co^{2+}	nitrite	NO_2^-
copper(II) or cupric	Cu^{2+}	perchlorate	ClO_4^-
iron(II) or ferrous	Fe^{2+}	permanganate	MnO_4^-
lead(II) or plumbous	Pb^{2+}		
magnesium	Mg^{2+}	carbonate	CO_3^{2-}
manganese(II) or manganous	Mn^{2+}	chromate	CrO_4^{2-}
mercury(I) or mercurous*	Hg_2^{2+}	dichromate	$Cr_2O_7^{2-}$
mercury(II) or mercuric	Hg^{2+}	oxide	O^{2-}
nickel(II) or nickelous	Ni^{2+}	peroxide	O_2^{2-}
tin(II) or stannous	Sn^{2+}	sulfate	SO_4^{2-}
zinc	Zn^{2+}	sulfide	S^{2-}
		sulfite	SO_3^{2-}
aluminum	Al^{3+}		
chromium (III) or chromic	Cr^{3+}	arsenate	AsO_4^{3-}
iron(III) or ferric	Fe^{3+}	nitride	N^{3-}
		phosphate	PO_4^{3-}

* A diatomic ion that is given the name mercury(I) because it can be thought of as consisting of two Hg^+ ions.

Notice that the charge is not explicitly given and that the method cannot be used if the metal forms more than two different cations.

A **polyatomic ion** is one that is formed from several atoms held together by covalent bonds (see Section 6.1). There are not many polyatomic cations, but here are two common ones:

NH_4^+ is the ammonium ion
Hg_2^{2+} is the mercury(I) or mercurous ion

The Hg_2^{2+} ion is given the name mercury(I) because it may be considered to be made up of two Hg^+ ions. A list of common cations is given in Table 5.5.

2. Anions. Monatomic anions are formed by atoms of nonmetals. Their names are derived by replacing the usual ending of the name of the nonmetal with the ending -ide:

Cl^- is the chloride ion
O^{2-} is the oxide ion
N^{3-} is the nitride ion

All ions that have names that end with -*ide* are not monatomic, however. A few polyatomic anions have this ending; for example,

CN^- is the cyanide ion
OH^- is the hydroxide ion
O_2^{2-} is the peroxide ion

Many polyatomic anions are known. Common ones are included in the list of anions given in Table 5.5, a list that should be mastered at this point. The system used for naming these anions is discussed in Section 11.6.

The name of an ionic compound consists of the name of the cation followed by the name of the anion (as a separate word):

Fe_2O_3 is iron(III) oxide or ferric oxide
$PbCO_3$ is lead(II) carbonate or plumbous carbonate
Ag_3PO_4 is silver phosphate
$(NH_4)_2S$ is ammonium sulfide
$Cu(CN)_2$ is copper(II) cyanide or cupric cyanide
$Mg(NO_3)_2$ is magnesium nitrate

Summary

The topics that have been discussed in this chapter are

1. The determination of atomic radius, the relationship between atomic radius and position in the periodic table, and the features of atomic structure that influence atomic radius.

2. The determination of ionization energy, periodic trends in the ionization energies of the elements, and the influence of atomic structure on ionization energy.

3. The determination of electron affinity and the correlation of electron affinity to atomic structure and position in the periodic table.

4. The formation of the ionic bond.

5. The Born-Haber treatment of the enthalpy changes involved in the formation of ionic crystals; the importance of lattice energy.

6. Types of ions; what determines the size of the charge that an ion carries.

7. The determination of ionic radius and the relation between ionic radius and atomic radius.

8. Nomenclature of ionic compounds.

Key Terms

Some of the more important terms introduced in this chapter are listed below. Definitions for terms not included in this list may be located in the text by use of the index.

Anion (Section 5.4) A negatively charged ion; an atom or a group of atoms that has gained one or more electrons.

Atomic radius (Section 5.1) An approximation of the radius of an atom based on the division of bond distances.

Bond distance (Section 5.1) The distance between the nuclei of two atoms that are bonded together.

Born–Haber cycle (Section 5.5) A method of analysis of the enthalpy change of a process in which the ΔH

for the entire process is set equal to the sum of the ΔH values for a series of steps that produce the same change.

Cation (Section 5.4) A positively charged ion; an atom or a group of atoms that has lost one or more electrons.

d^{10} ion (Section 5.6) A cation that has an $ns^2 np^6 nd^{10}$ electron configuration in its outer shell (where n is the principal quantum number of this shell).

$d^{10}s^2$ ion (Section 5.6) A cation that has a d^{10} electron configuration plus an additional shell that contains two electrons in an s orbital.

Effective nuclear charge (Section 5.1) The positive charge experienced by an electron that results from the charge of the nucleus decreased by the shielding effect of inner electrons.

Electron affinity (Section 5.3) The energy change associated with the process in which an electron is added to a gaseous atom in its ground state is a first electron affinity. Second and higher electron affinities pertain to processes in which electrons are added to negative ions.

Enthalpy of sublimation (Section 5.5) The enthalpy change associated with a process in which a solid is converted directly into a gas.

Ionic bonding (Section 5.4) The attraction that exists between positive and negative ions and that holds them together into a crystal structure; results from the transfer of electrons.

Ionic radius (Section 5.7) An approximation of the radius of an ion based on the division of the distances between the nuclei of adjacent ions in an ionic crystal.

Ionization energy (Section 5.2) The amount of energy required to remove the most loosely held electron from an isolated atom in its ground state is a first ionization energy. Second and higher ionization energies pertain to processes in which electrons are removed from positive ions.

Isoelectronic (Section 5.4) The same in electronic configuration.

Lattice energy (Section 5.5) The enthalpy change associated with the condensation of gaseous ions into an ionic crystal.

Noble-gas ion (Section 5.6) A cation or an anion that is isoelectronic with a noble gas; an s^2 or an s^2p^6 ion.

s^2 ion (Section 5.6) A cation or anion that has two electrons in an s orbital as the configuration of its outer shell.

s^2p^6 ion (Section 5.6) A cation or anion that has two electrons in an s orbital and six electrons in three p orbitals as the configuration of its outer shell.

Shielding (Section 5.1) The effect brought about by inner electrons in diminishing the nuclear charge experienced by outer electrons.

Problems*

Atomic Radii

5.1 List the factors that influence the size of atoms and give an example to illustrate each one.

5.2 In general, how do the atomic radii of metals and nonmetals compare?

5.3 Explain why a minimum is observed in each curve that is obtained by plotting atomic radius against atomic number for the members of each transition series.

5.4 Which member of each of the following pairs would you predict to be larger: **(a)** Si or S, **(b)** Si or Sn, **(c)** Si or Ga, **(d)** Si or Al, **(e)** Si or Mg, **(f)** Si or F, **(g)** Si or C?

5.5 Which member of each of the following pairs would you predict to be larger: **(a)** Te or Po, **(b)** P or S, **(c)** Sr or Ba, **(d)** Sr or Sb, **(e)** In or Ge, **(f)** Pb or Bi, **(g)** K or Mg?

5.6 Given the following bond distances: Br—Cl, 213 pm; I—Br, 247 pm; and Cl—Cl, 198 pm, what is the I—Cl bond distance?

Ionization Energy

5.7 Discuss the factors that influence ionization energy.

5.8 How do the first ionization energies of metals compare to those of nonmetals?

5.9 Why is the second ionization energy of an element always larger than the first ionization energy?

5.10 Why are the first ionization energies of Be and N out of line in comparison to those of the other members of the second period?

5.11 Which member of each of the following pairs has the higher first ionization energy: **(a)** Rb or Sr, **(b)** Sr or Sn, **(c)** Sn or Sb, **(d)** Sb or Bi, **(e)** Sn or Se, **(f)** Se or S, **(g)** S or Ar?

5.12 Which member of each of the following pairs has the higher first ionization energy: **(a)** O or Ne, **(b)** Ne or Ar, **(c)** O or S, **(d)** S or F, **(e)** N or O, **(f)** Na or Mg, **(g)** Mg or Ca?

Electron Affinity

5.13 The first ionization energy of F is $+1680$ kJ/mol and the first electron affinity of F is -322 kJ/mol. Explain clearly what these values mean.

5.14 List the atomic factors that influence electron affinity.

5.15 Why do some electron affinity values recorded in Table 5.2 have positive signs?

5.16 Explain the following on the basis of electron configuration: **(a)** the electron affinity of S is -200 kJ/mol and of Cl is -348 kJ/mol, **(b)** the electron affinity of Si is -120 kJ/mol and of P is -74 kJ/mol, **(c)** the electron affinity of Li is -60 kJ/mol and of Be is $+240$ kJ/mol.

5.17 Why are all second electron affinities positive values?

5.18 Of all the elements in the third period (Na through Ar): **(a)** which has the largest atomic radius, **(b)** which has the highest first ionization energy, **(c)** which is the most reactive metal, **(d)** which is the most reactive nonmetal, **(e)** which is the least reactive, **(f)** how many are metals?

---

* The appendix contains answers to color-keyed problems.

5.19 Use the following data to calculate the lattice energy of KF. The enthalpy of formation of KF is -563 kJ/mol. The enthalpy of sublimation of K is $+90$ kJ/mol, and the first ionization energy of K is $+415$ kJ/mol. The dissociation energy of F_2 is $+155$ kJ/mol F_2 and the first electron affinity of F is -322 kJ/mol F.

5.20 Use the following data to calculate the lattice energy of $SrCl_2$. The enthalpy of formation of $SrCl_2$ is -828 kJ/mol. The enthalpy of sublimation of Sr is $+164$ kJ/mol, the first ionization energy of Sr is $+549$ kJ/mol, and the second ionization energy of Sr is $+1064$ kJ/mol. The dissociation energy of Cl_2 is $+243$ kJ/mol Cl_2 and the first electron affinity of Cl is -348 kJ/mol Cl.

5.21 Use the following data to calculate the lattice energy of CaS. The enthalpy of formation of CaS is -482 kJ/mol. The enthalpy of sublimation of Ca is $+192$ kJ/mol, the first ionization energy of Ca is $+590$ kJ/mol, and the second ionization energy of Ca is $+1145$ kJ/mol. The enthalpy change for the transformation $S(s) \rightarrow S(g)$ is $+279$ kJ/mol. The first electron affinity of S is -200 kJ/mol S and the second electron affinity of S is $+522$ kJ/mol S^-.

5.22 Use the following data to calculate the enthalpy of formation of MgO. The enthalpy of sublimation of Mg is $+150$ kJ/mol, the first ionization energy of Mg is $+738$ kJ/mol, and the second ionization energy of Mg is $+1450$ kJ/mol. The enthalpy of dissociation of O_2 is $+494$ kJ/mol O_2. The first electron affinity of O is -142 kJ/mol O and the second electron affinity of O is $+844$ kJ/mol O^-. The lattice energy of MgO is -3890 kJ/mol.

5.23 Use the following data to calculate the electron affinity of Br. The enthalpy of formation of RbBr is -389 kJ/mol. The enthalpy of sublimation of Rb is $+86$ kJ/mol and the first ionization energy of Rb is $+403$ kJ/mol. The enthalpy change for the transformation $Br_2(1) \rightarrow 2Br(g)$ is $+224$ kJ/mol Br_2. The lattice energy of RbBr is -666 kJ/mol.

5.24 Consider the lattice energies of KCl, K_2O, and CaO. List the compounds in the order of increasing quantity of energy released. Why did you select the order that you used?

5.25 Of all the steps in a Born-Haber cycle, why is the lattice energy important to the success of the preparation of an ionic compound?

The Ionic Bond, Types of Ions

5.26 Give formulas for the oxides, chlorides, and nitrides of sodium, magnesium, and aluminum.

5.27 Write notations for the ground-state electronic configurations of the following ions: (a) Mg^{2+}, (b) Cr^{2+}, (c) Co^{2+}, (d) Pd^{2+}, (e) Ag^+, (f) I^-.

5.28 (a) State the number of unpaired electrons in each of the ions listed in Problem 5.27. (b) Which of these ions would you expect to be paramagnetic and which diamagnetic?

5.29 Write notations for the ground-state electronic configurations of the following ions: (a) S^{2-}, (b) Cu^+, (c) Cu^{2+}, (d) Sc^{3+}, (e) F^-, (f) Hg^{2+}, (g) Pb^{2+}, (h) Cr^{3+}.

5.30 (a) State the number of unpaired electrons in each of the ions listed in Problem 5.29. (b) which of these ions would you expect to be paramagnetic and which diamagnetic?

5.31 For each of the following, give the formulas of two ions that are isoelectronic with the atom or ion listed: (a) Kr, (b) Zn^{2+}, (c) Zn, (d) O^{2-}, (e) Ca^{2+}.

5.32 For each of the following, give the formulas of two ions that are isoelectronic with the atom or ion listed: (a) Cd^{2+}, (b) Ag^+, (c) Rb^+, (d) Hg, (e) Xe.

5.33 Give examples of s^2 ions, s^2p^6 ions, d^{10} ions, and $d^{10}s^2$ ions.

5.34 Identify the s^2p^6, d^{10}, and $d^{10}s^2$ ions that are included in the following list: Al^{3+}, Ga^{3+}, Sc^{3+}, N^{3-}, Pb^{2+}, K^+, Ba^{2+}, Cu^+, Cd^{2+}, Tl^+, Bi^{3+}.

5.35 Using arguments based on the energy changes involved in ion formation, explain why Na forms Na^+, not Na^{2+}, but Cu forms both Cu^+ and Cu^{2+}.

5.36 Which member of each of the following pairs would you predict to be larger: (a) Cs^+ or Ba^{2+}, (b) S or S^{2-}, (c) S^{2-} or Cl^-, (d) Cr^{2+} or Cr^{3+}, (e) Ag or Ag^+, (f) Cu^{2+} or Ag^+?

5.37 Which member of each of the following pairs would you predict to be larger: (a) Br or Br^-, (b) Cs or Cs^+, (c) O^{2-} or F^-, (d) Au^+ or Au^{3+}, (e) In^{3+} or Tl^+, (f) In^+ or Sn^{2+}?

Nomenclature of Ionic Compounds

5.38 Give formulas for (a) chromium(III) oxide, (b) calcium phosphate, (c) silver dichromate, (d) magnesium chlorate, (e) nickel(III) nitrate, (f) zinc carbonate.

5.39 Give formulas for (a) iron(III) sulfate, (b) copper(I) bromide, (c) barium hydroxide, (d) gold(III) sulfide, (e) lead(II) chromate, (f) ammonium acetate.

5.40 Name the following: (a) $MnSO_4$, (b) $Mg_3(PO_4)_2$, (c) $PbCO_3$, (d) $HgCl_2$, (e) Na_2O_2, (f) $Al_2(SO_4)_3$.

5.41 Name the following: (a) $Ca(ClO_4)_2$, (b) $Co(NO_3)_2$, (c) SnF_2, (d) $KMnO_4$, (e) $FePO_4$, (f) Hg_2I_2.

THE COVALENT BOND

C H A P T E R

6

In the last chapter we discussed the formation and properties of ionic compounds. In this chapter, the covalent bond (in which electrons are shared by the bonded atoms) will be introduced. In addition, we will consider bonds that have a character intermediate between the purely ionic and the purely covalent.

6.1 Covalent Bonding

When atoms of nonmetals interact, molecules are formed which are held together by covalent bonds. Since these atoms are similar in their attraction for electrons (identical when two atoms of the same element are considered), electron transfer does not occur; instead, electrons are shared. A **covalent bond** consists of a pair of electrons (with opposite spins) that is shared by two atoms.

As an example, consider the bond formed by two hydrogen atoms. An individual hydrogen atom has a single electron that is symmetrically distributed around the nucleus in a $1s$ orbital. When two hydrogen atoms form a covalent bond, the atomic orbitals overlap in such a way that the electron clouds reinforce each other in the region between the nuclei, and there is an increased probability of finding an electron in this region. According to the Pauli exclusion principle, the two electrons of the bond must have opposite spins. The strength of the covalent bond comes from the attraction of the positively charged nuclei for the negative cloud of the bond (see Figure 6.1).

The hydrogen molecule can be represented by the symbol H:H or H—H. Although the electrons belong to the molecule as a whole, each hydrogen atom can be considered to have the noble-gas configuration of helium (two electrons in the $n = 1$ level). This consideration is based on the premise that both shared electrons contribute to the stable configuration of each hydrogen atom.

The formula, H_2, describes a discrete unit—a molecule—and hydrogen gas consists of a collection of such molecules. There are no molecules in strictly ionic materials. The formula Na_2Cl_2 is incorrect because sodium chloride is an ionic compound and the simplest ratio of ions in a crystal of sodium chloride is 1 to 1; a molecule of formula Na_2Cl_2 does not exist. For covalent materials, however, a formula such as H_2O_2 can be correct; this formula describes a molecule containing two hydrogen atoms and two oxygen atoms.

The hydrogen molecule can be described as being **diatomic** (containing two atoms). Certain other elements also exist as diatomic molecules. An atom of any group VII A element, for example, has seven valence electrons. By the forma-

Figure 6.1 Representation of the electron distribution in a hydrogen molecule

tion of a covalent bond between two of these atoms, each atom attains an octet configuration characteristic of the noble gases. Thus, fluorine gas consists of F_2 molecules:

$$:\!\overset{\cdot\cdot}{\underset{\cdot\cdot}{F}}\!\cdot \;+\; \cdot\overset{\cdot\cdot}{\underset{\cdot\cdot}{F}}\!: \;\longrightarrow\; :\!\overset{\cdot\cdot}{\underset{\cdot\cdot}{F}}\!:\!\overset{\cdot\cdot}{\underset{\cdot\cdot}{F}}\!:$$

Only the electrons between the two atoms are shared and form a part of the covalent bond (although the molecular orbital theory considers that all of the electrons affect the bonding—see Section 7.4).

More than one covalent bond may form between two atoms. A nitrogen atom (group V A) has five valence electrons:

$$:\!\overset{\cdot}{N}\!\cdot \;+\; \cdot\overset{\cdot}{N}\!: \;\longrightarrow\; :\!N\!:::\!N\!:$$

In the molecule, N_2, six electrons are shared in three covalent bonds (usually called a triple bond). Notice that, as a result of this formulation, each of the nitrogen atoms can be considered to have an octet of electrons.

Nonmetallic elements that exist as diatomic molecules are H_2, F_2, Cl_2, Br_2, I_2, N_2, and O_2. (Oxygen is a special case and will be discussed in Section 7.4). These elements are always indicated in this way in chemical equations.

The electron-dot formulas we have been using are called **valence-bond structures** or **Lewis structures,** named after Gilbert N. Lewis who proposed this theory of covalent bonding in 1916 (more recent theories of covalent bonding are discussed in Chapter 7). The Lewis theory emphasizes the attainment of noble-gas configurations on the part of atoms in covalent molecules. Since the number of valence electrons is the same as the group number for the nonmetals, one might predict that VII A elements, such as Cl, would form one covalent bond to attain a stable octet; VI A elements, such as O and S, two covalent bonds; V A elements, such as N and P, three covalent bonds; and IV A elements, such as C, four covalent bonds. These predictions are borne out in many compounds containing only simple covalent bonds:

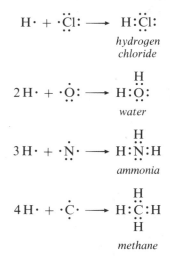

Gilbert N. Lewis, 1875–1946.
Photographer Johan Hagemeyer: Bancroft Library, courtesy AIP Neils Bohr Library

Notice that in these molecules, each hydrogen atom can be considered to have a complete $n = 1$ shell; the other atoms have characteristic noble-gas octets.

The covalent bonding of compounds can also be indicated by dashes; each dash represents one bond, a pair of electrons:

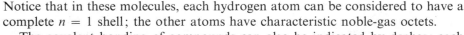

$$:\overset{..}{\underset{..}{Cl}}-\overset{..}{\underset{..}{P}}-\overset{..}{\underset{..}{Cl}}: \qquad :\overset{..}{\underset{..}{Cl}}-\overset{..}{\underset{..}{O}}-\overset{..}{\underset{..}{Cl}}: \qquad$$

phosphorous trichloride   dichlorine oxide   ethane

The following are examples of molecules that contain double and triple bonds:

$$:\overset{..}{\underset{..}{O}}: \;+\; :\overset{.}{\underset{.}{C}}: \;+\; :\overset{..}{\underset{..}{O}}: \;\longrightarrow\; :\overset{..}{O}::C::\overset{..}{O}: \qquad (\text{or } :\overset{..}{O}{=}C{=}\overset{..}{O}:)$$

carbon dioxide

$$2\,H\cdot \;+\; \cdot\overset{.}{C}: \;+\; :\overset{.}{C}\cdot \;+\; 2\,H\cdot \;\longrightarrow\; H:\overset{..}{C}:\overset{..}{C}:H \qquad (\text{or } H{-}\overset{H}{\underset{}{C}}{=}\overset{H}{\underset{}{C}}{-}H)$$

ethylene

$$H\cdot \;+\; \cdot\overset{.}{\underset{.}{C}}: \;+\; :\overset{.}{\underset{.}{C}}\cdot \;+\; \cdot H \;\longrightarrow\; H:C:::C:H \qquad (\text{or } H{-}C{\equiv}C{-}H)$$

acetylene

Notice that in each compound the number of covalent bonds on each atom agrees with the number predicted.

A method for drawing Lewis structures is given in Section 6.3. Before the method is presented, however, the concept of formal charge must be introduced since the method relies upon the use of formal charge to check the validity of its results.

6.2 Formal Charge

In the formation of certain covalent bonds, *both* of the shared electrons are furnished by *one* of the bonded atoms. For example, in the reaction of ammonia with a proton (a hydrogen atom stripped of its electron), the unshared electron pair of the nitrogen atom of NH_3 is used to form a new covalent bond:

$$H:\overset{..}{\underset{H}{N}}:H \;+\; H^+ \;\longrightarrow\; \left[H:\overset{..}{\underset{H}{N}}:H\right]^+$$

A bond formed in this way is frequently called a "coordinate covalent" bond, but it is probably unwise to do so. Labeling a specific bond as a "coordinate covalent" bond implies that it is different from other covalent bonds and has little justification. All electrons are alike no matter what their source. All the bonds in NH_4^+ are identical. It is impossible to distinguish between them.

Notice, however, that the number of covalent bonds on the N atom of NH_4^+ does not agree with the number predicted in the preceding section. Since a nitrogen atom has five valence electrons (group V A), it would be expected to satisfy the octet principle through the formation of three covalent bonds. This prediction is correct for NH_3; it is not correct for NH_4^+.

An answer to the question may be obtained by calculating the formal charges of the atoms in NH_4^+. The **formal charge** of an atom in a molecule is calculated in the following way. The number of valence electrons that an A family atom has is equal to its group number. If all the valence electrons were removed from the atom in question, the resulting ion would have a *positive* charge equal to

$$+(group\ no.)$$

This would be the charge on the atom if it had no valence electrons associated with it. In a molecule, however, the atom has valence electrons associated with it—shared (as covalent bonds) and, in some cases, unshared. If the electron pair of each covalent bond is divided equally between the two atoms that it bonds, the atom in question will get one electron for each covalent bond that it has (a $1-$ charge for each *bond*). In addition, the atom will have a $1-$ charge for each unshared *electron* that it has in the molecule. The total *negative* charge from these sources is

$$-[(no.\ bonds) + (no.\ unshared\ e^-)]$$

The formal charge of the bonded atom, therefore, is given by the formula

$$formal\ charge = +\ (group\ no.) - (no\ bonds) - (no.\ unshared\ e^-) \qquad (6.1)$$

Since the N atom (a group V A atom) in NH_4^+ has four bonds and no unshared electrons, its formal charge is

$$formal\ charge = +5 - 4 - 0 = 1+$$

Each H atom in the NH_4^+ ion has a formal charge of zero:

$$formal\ charge = +1 - 1 - 0 = 0$$

The formal charge of the N atom in NH_4^+ is indicated in the following way:

$$
\begin{array}{c}
\text{H} \\
| \\
\text{H}-\overset{\oplus}{\text{N}}-\text{H} \\
| \\
\text{H}
\end{array}
$$

We can now explain the difference between the predicted and actual number of bonds on the N atom in NH_4^+. A hypothetical N^+ would have four valence electrons and be able to form four covalent bonds. An uncharged N atom, on the other hand, has five valence electrons and can form only three covalent bonds.

A formal charge is, as the name implies, only a formality. In the assignment of formal charges, the assumption is made that the electron pair of any covalent bond is shared equally by the bonded atoms. Such an assumption is usually not true, and formal charges must be interpreted carefully. The electron density on the N atom in NH_4^+ is less than that on the N atom in NH_3, but the actual charge is not a full positive charge because the bonding electrons are not equally shared.

As another example of the way in which formal charges are calculated, consider the $POCl_3$ molecule:

$$
\begin{array}{c}
:\overset{\cdot\cdot}{\text{O}}: \\
| \\
:\overset{\cdot\cdot}{\underset{\cdot\cdot}{\text{Cl}}}-\text{P}-\overset{\cdot\cdot}{\underset{\cdot\cdot}{\text{Cl}}}: \\
| \\
:\overset{\cdot\cdot}{\underset{\cdot\cdot}{\text{Cl}}}:
\end{array}
$$

The formal charge of the O atom is

$$formal\ charge = +(group\ no.) - (no.\ bonds) - (no.\ unshared\ e^-)$$
$$= +6 - 1 - 6 = 1-$$

The formal charge of the P atom is

$$formal\ charge = +5 - 4 - 0 = 1+$$

The formal charge of each Cl atom is

$$formal\ charge = +7 - 1 - 6 = 0$$

The structure of the molecule is

$$:\overset{..}{\underset{}{O}}:^{\ominus}$$
$$:\overset{..}{\underset{..}{Cl}}-P^{\oplus}-\overset{..}{\underset{..}{Cl}}:$$
$$:\overset{..}{\underset{..}{Cl}}:$$

Notice that the sum of the formal charges in the $POCl_3$ structure is zero. The formal charges of any molecule add up to zero. The sum of the formal charges of the atoms in an ion equals the charge of the ion.

An atom in a Lewis structure that has the number of bonds expected on the basis of its group number has no formal charge. If possible, a Lewis structure should be drawn so that each atom has the number of bonds expected on the basis of its group number. Frequently, however, it is not possible.

Atoms that are bonded to each other in a structure should not have formal charges with the same sign. The repulsion between the charges would tend to break the bond between the atoms. A Lewis structure in which this adjacent charge rule is violated is usually not an accurate representation of the molecule or ion (see Example 6.3 in Section 6.3).

Formal Charges

1. The formal charge of an atom in a Lewis structure can be calculated by use of the formula

$$formal\ charge = +(group\ no.) - [(no.\ bonds) + no.\ unshared\ e^-)]$$

2. In a molecule, the sum of the formal charges is zero. In an ion, the formal charges add up to the charge of the ion.

3. An atom in a Lewis structure that has the number of bonds expected on the basis of its group number has no formal charge. If possible, a Lewis structure should be drawn so that all atoms have these numbers of bonds. Frequently, however, it is not possible.

4. Atoms that are bonded to each other in a structure should not have formal charges with the same sign. A Lewis structure in which this adjacent charge rule is violated is usually not an accurate description of the molecule or ion.

6.3 Lewis Structures

The following examples illustrate how to draw Lewis structures. The steps in the method are described in the first example.

Example 6.1

Diagram the Lewis structure of the chlorate ion ClO_3^-. In the ion, the Cl atom is the central atom to which the three O atoms are bonded.

Solution

1. Find the total number of valence electrons supplied by all the atoms in the structure. The number supplied by each A family element is the same as the group number of the element. For a negative ion, increase the number by the charge of the ion. For a positive ion, decrease the number by the charge of the ion. The total number of valence electrons in ClO_3^- is

$$
\begin{array}{ll}
7 & \text{(from the Cl atom)} \\
18 & \text{(from the three O atoms)} \\
\underline{1} & \text{(from the ionic charge)} \\
26 &
\end{array}
$$

2. Determine the number of electrons that would be required to give 2 electrons to each H atom individually and 8 electrons to each of the other atoms individually. Since there are no H atoms in the ClO_3^- ion,

$$
\begin{aligned}
\textit{no. } e^- \textit{ for individual atoms} &= 2(\textit{no. H atoms}) + 8(\textit{no. other atoms}) \\
&= 2(0) + 8(4) = 32
\end{aligned}
$$

3. The number obtained in step 2 minus the number obtained in step 1 is the number of electrons that must be shared in the final structure:

$$
\begin{aligned}
\textit{no. bonding } e^- &= (\textit{no. } e^- \textit{ for individual atoms}) - (\textit{total no. } e^-) \\
&= 32 - 26 = 6
\end{aligned}
$$

4. One-half the number of bonding electrons (from step 3) is the number of covalent bonds in the final structure:

$$
\begin{aligned}
\textit{no. bonds} &= (\textit{no. bonding } e^-)/2 \\
&= 6/2 = 3
\end{aligned}
$$

5. Write the symbols for the atoms present in the structure, arranging them in the way that they are found in the structure. Indicate covalent bonds by dashes written between the symbols. Indicate one bond between each pair of symbols, and then use any remaining from the number calculated in step 4 to make multiple bonds (note that if the structure contains H atoms, each H atom is limited to one bond):

$$
\begin{array}{c}
O \\
| \\
Cl\!-\!O \\
| \\
O
\end{array}
$$

6. The total number of electrons (step 1) minus the number of bonding electrons (step 3) is the number of unshared electrons. Complete the electron octet of each atom (other than the H atoms) by adding dots to represent unshared electrons:

$$no.\ unshared\ e^- = (total\ no.\ e^-) - (no.\ bonding\ e^-)$$
$$= 26 - 6 = 20$$

$$:\overset{\cdot\cdot}{O}:$$
$$|$$
$$:\overset{\cdot\cdot}{Cl}{-}\overset{\cdot\cdot}{\underset{\cdot\cdot}{O}}:$$
$$|$$
$$:\overset{\cdot\cdot}{\underset{\cdot\cdot}{O}}:$$

7. Indicate the formal charges of the atoms where appropriate. The formal charge of the Cl atom is

$$formal\ charge = +(group\ no.) - (no.\ bonds) - (no.\ unshared\ e^-)$$
$$= +7 - 3 - 2 = 2+$$

The formal charge of each O atom is

$$formal\ charge = +6 - 1 - 6 = 1-$$

The structure is

$$:\overset{\cdot\cdot}{O}:^{\ominus}$$
$$|$$
$$:Cl\overset{2+}{-}\overset{\cdot\cdot}{\underset{\cdot\cdot}{O}}:^{\ominus}$$
$$|$$
$$:\overset{\cdot\cdot}{\underset{\cdot\cdot}{O}}:^{\ominus}$$

Notice that the formal charges add up to the charge of the ion.

Example 6.2

Diagram the Lewis structure of the SO_2 molecule. The molecule is angular and the two O atoms are bonded to a central S atom.

Solution

1. The total number of valence electrons in the molecule is

$$\begin{array}{ll} 6 & \text{(from the S atom)} \\ \underline{12} & \text{(from the two O atoms)} \\ 18 \end{array}$$

2. $no.\ e^-\ for\ individual\ atoms = 2(no.\ H\ atoms) + 8(no.\ other\ atoms)$
$$= 2(0) + 8(3) = 24$$

3. $no.\ bonding\ e^- = (no.\ e^-\ for\ individual\ atoms) - (total\ no.\ e^-)$
$$= 24 - 18 = 6$$

4. $no.\ bonds = (no.\ bonding\ e^-)/2$
$$= 6/2 = 3$$

5.

6. *no. unshared* e^- = (*total no.* e^-) − (*no. bonding* e^-)

= 18 − 6 = 12

7. *formal charge* = +(*group no.*) − (*no. bonds*) − (*no. unshared* e^-)

For the S atom,

formal charge = +6 − 3 − 2 = 1+

For the left-hand O atom,

formal charge = +6 − 2 − 4 = 0

For the right-hand O atom,

formal charge = +6 − 1 − 6 = 1−

The structure is

Notice that an equivalent structure can be drawn, one in which the double bond connects the right-hand O atom to the S atom.

Example 6.3

Diagram the Lewis structure of nitric acid, HNO_3. The N atom is the central atom to which the three O atoms are bonded. The H atom is bonded to one of the O atoms.

Solution

1. The total number of valence electrons in the molecule is

1 (from the H atom)
5 (from the N atom)
18 (from the three O atoms)
——
24

2. *no.* e^- *for individual atoms* = 2(*no. H atoms*) + 8(*no. other atoms*)

= 2(1) + 8(4) = 34

3. *no. bonding* e^- = (*no.* e^- *for individual atoms*) − (*total no.* e^-)

= 34 − 24 = 10

4. *no. bonds* = (*no. bonding* e^-)/2

= 10/2 = 5

How to Diagram Lewis Structures

1. Find the total number of valence electrons supplied by all the atoms in the structure. The number supplied by each A family element is the same as the group number of the element:

    a. For a negative ion, increase the number by the charge of the ion.

    b. For a positive ion, decrease the number by the charge of the ion.

2. Determine the number of electrons that would be required to give 2 electrons to each H atom individually and 8 electrons to each of the other atoms individually:

$$\text{no. } e^- \text{ for individual atoms} = 2(\text{no. H atoms}) + 8(\text{no. other atoms})$$

3. The number obtained in step 2 minus the number obtained in step 1 is the number of electrons that must be shared in the final structure:

$$\text{no. bonding } e^- = (\text{no. } e^- \text{ for individual atoms}) - (\text{total no. } e^-)$$

4. One-half the number of bonding electrons (step 3) is the number of covalent bonds in the final structure:

$$\text{no. bonds} = (\text{no. bonding } e^-)/2$$

5. Write the symbols for the atoms present in the structure, arranging them in the way that they are found in the structure.

6. Indicate covalent bonds by dashes written between the symbols. Indicate one bond between each pair of symbols, and then use any remaining from the total calculated in step 4 to make multiple bonds. Note that each H atom is limited to one bond.

7. The total number of electrons (step 1) minus the number of bonding *electrons* (step 3) is the number of unshared electrons:

$$\text{no. unshared } e^- = (\text{total no. } e^-) - (\text{no. bonding } e^-)$$

Complete the electron octet of each atom (other than the H atoms) by adding dots to represent unshared electrons.

8. Indicate the formal charges of the atoms where appropriate, and evaluate the structure.

---

5. The molecule contains five bonds. If we indicate one bond between each pair of atoms, we use four of the five and have one left over with which to make a double bond. There are three possibilities, which we label a, b, and c:

     (a)   (b)   (c)

Since the H atom is limited to one bond, a structure with a double bond between H and O is not possible.

6. *no. unshared e⁻* = (*total no. e⁻*) − (*no. bonding e⁻*)
   = 24 − 10 = 14

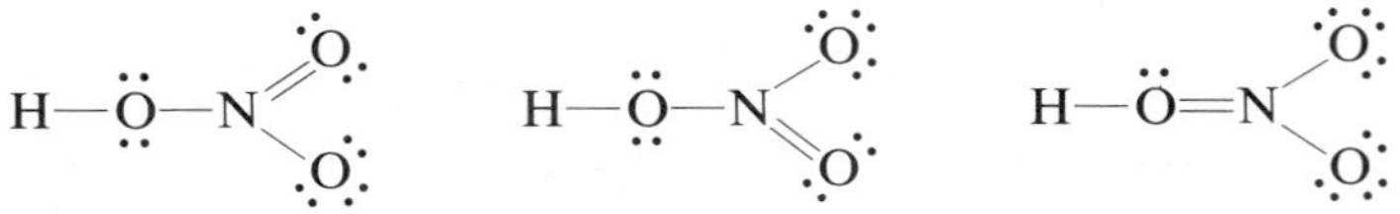

(a) (b) (c)

7. The addition of formal charges to the three structures gives

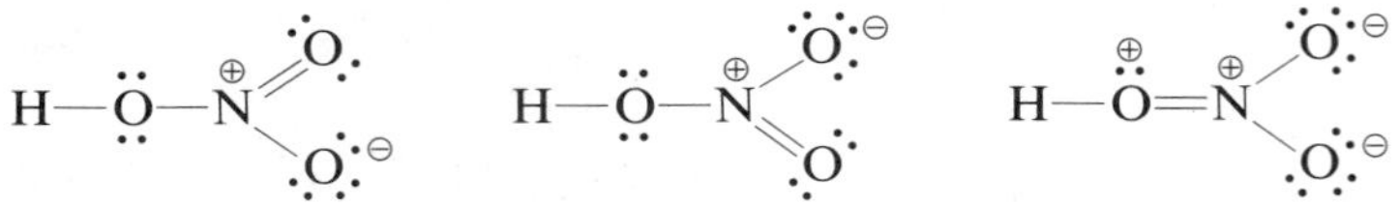

(a) (b) (c)

Structure c must be discarded because adjacent atoms in the molecule carry formal charges with the same sign. Both structures a and b, which are equivalent, are valid Lewis structures. In the next section we will see how the bonding in the nitric acid molecule is based on *both* of these structures.

6.4 Resonance

In some cases the properties of a molecule or ion are not adequately represented by a single Lewis structure. The structure for SO_2 that was derived in Example 6.2, for example,

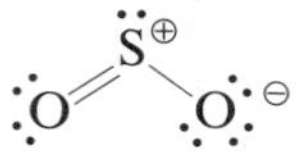

is unsatisfactory in two respects. The structure depicts the S atom to be bonded to one O atom by a double bond and to the other O atom by a single bond. Double bonds are shorter than single bonds, yet experimental evidence shows that both S to O linkages are the same length. The structure also shows one of the O atoms to be more negative than the other. It is known, however, that both of the O atoms are the same in this respect.

In a case of this type, two or more valence-bond structures can be used in combination to depict the molecule. The molecule is said to be a **resonance hybrid** of the structures, which are called **resonance forms.** For SO_2,

The actual structure does not correspond to either resonance form alone. Instead, the true structure is intermediate between the two forms. Each of the S to O bonds is neither the single bond shown in one of the resonance forms nor the double bond shown in the other form. It is intermediate between the two.

Both bonds, therefore, are identical. Each O atom may be considered to have a formal charge of $\frac{1}{2}-$, since in one form it has a formal charge of $1-$ and in the other it has a formal charge of zero. The oxygens, therefore, are equally negative.

There is only one type of SO_2 molecule and only one structure. The electrons of the molecule do *not* flip back and forth so that the molecule is in one form at one moment and in the other form at the next. The molecule always has the same structure. The problem arises because the Lewis theory is limited—not because the SO_2 molecule is unusual.

The two equivalent Lewis structures for the nitric acid molecule that were derived in Example 6.3 are resonance forms for the molecule:

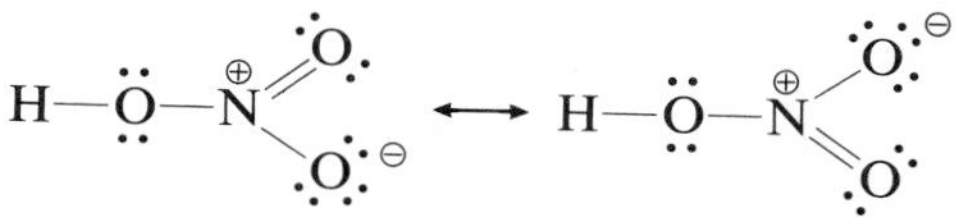

The actual structure of the molecule is intermediate between the two forms. The two right-hand N to O bonds have the same bond distance (121 pm), and each of these bonds is shown as a double bond in one resonance form and as a single bond in the other form. The left-hand N to O bond (shown as a single bond in *both* resonance forms) is longer than the others (141 pm).

The delocalization of charge in an ion is illustrated by the resonance of the carbonate ion, CO_3^{2-}. This ion is planar, all bonds are equivalent (between single and double bonds), and all oxygen atoms are equally negative. The formal charges add up to the charge on the ion:

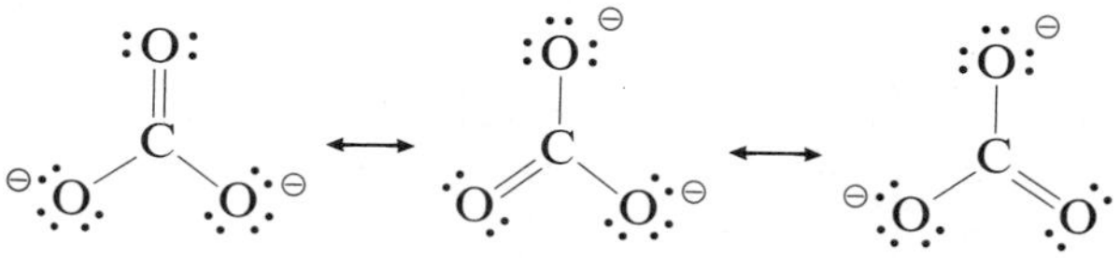

As depicted by resonance, the charge is delocalized; it is impossible to locate the exact position of the "extra" two electrons that give the ion its negative charge.

It is important to note that all resonance forms of a given species *must have the same configuration of nuclei*. The resonance forms of a molecule or an ion differ in the arrangement of electrons and not in the arrangement of nuclei.

Example 6.4

Diagram the resonance forms of the N_2O molecule. Dinitrogen oxide is a linear molecule and the atoms are arranged NNO.

Solution

We follow the rules for drawing Lewis structures given in Example 6.1:

1. The total number of valence electrons in the molecule is

   10 (from the two N atoms)
   <u> 6</u> (from the O atom)
   16

2. *no. e⁻ for individual atoms* = 2(*no. H atoms*) + 8(*no. other atoms*)

$$= 2(0) + 8(3) = 24$$

3. *no. bonding e⁻* = (*no. e⁻ for individual atoms*) − (*total no. e⁻*)

$$= 24 - 16 = 8$$

4. *no. bonds* = (*no. bonding e⁻*)/2

$$= 8/2 = 4$$

5. We can imagine three ways to arrange the four bonds in the molecule:

$$\text{N=N=O} \qquad \text{N≡N—O} \qquad \text{N—N≡O}$$
$$\text{(a)} \qquad\qquad \text{(b)} \qquad\qquad \text{(c)}$$

6. *no. unshared e⁻* = (*total no. e⁻*) − (*no. bonding e⁻*)

$$= 16 - 8 = 8$$

$$:\ddot{\text{N}}\text{=N=}\ddot{\text{O}}: \qquad :\text{N≡N—}\ddot{\text{O}}: \qquad :\ddot{\text{N}}\text{—N≡O}:$$
$$\text{(a)} \qquad\qquad\quad \text{(b)} \qquad\qquad\quad \text{(c)}$$

7. The addition of formal charges gives

$$^{\ominus}:\ddot{\text{N}}\text{=}\overset{\oplus}{\text{N}}\text{=}\ddot{\text{O}}: \qquad :\text{N≡}\overset{\oplus}{\text{N}}\text{—}\ddot{\text{O}}:^{\ominus} \qquad ^{2\ominus}:\ddot{\text{N}}\text{—}\overset{\oplus}{\text{N}}\text{≡O}:^{\oplus}$$
$$\text{(a)} \qquad\qquad\qquad \text{(b)} \qquad\qquad\qquad \text{(c)}$$

Structure c must be discarded since it violates the adjacent charge rule (both the central N atom and the O atom carry positive formal charges). The resonance forms of the N₂O molecule are

$$^{\ominus}:\ddot{\text{N}}\text{=}\overset{\oplus}{\text{N}}\text{=}\ddot{\text{O}}: \longleftrightarrow :\text{N≡}\overset{\oplus}{\text{N}}\text{—}\ddot{\text{O}}:^{\ominus}$$

6.5 Transition between Ionic and Covalent Bonding

The bonding in most compounds is intermediate between purely ionic and purely covalent.

The best examples of ionic bonding are found in compounds formed by a metal with a very low ionization energy (for example, Cs) and a nonmetal that has a strong tendency toward electron gain (for example, F). In a compound such as CsF, the ions exist as separate units in the crystal.

A purely covalent bond is found only in molecules formed from two *identical* atoms, such as Cl_2. One Cl atom attracts electrons to the same extent as any other. The electron cloud of the bond is distributed symmetrically around the two nuclei. In other words, the bonding electrons are shared equally by the two identical atoms.

The bonding in most compounds is somewhere between these two extremes. One approach to the study of these bonds of intermediate character is based on **ion distortion**. The character of the bonding in a compound that contains a metal and a nonmetal can be interpreted in terms of the interactions between the ions. The positively charged ion is believed to attract and deform the electron cloud of the anion. The electron cloud of the negative ion is drawn toward the cation.

In extreme cases, ion deformation can lead to compounds that are more covalent than ionic (see the sequence of drawings in Figure 6.2). The degree of covalent character that a compound has can be considered to correspond to the extent that the anion is distorted.

1. Anions. How easily an anion can be distorted depends upon its size and charge. A *large anion*, in which the outer electrons are far away from the nucleus, is easily deformed. The iodide ion (I^-, ionic radius of 216 pm) is more easily distorted than the fluoride ion (F^-, ionic radius of 136 pm). If an anion has a *high negative charge*, so much the better. In a highly charged anion the electrons outnumber the protons to a greater extent than in an anion with a lower charge. The electron cloud of a highly charged anion is easily distorted. The S^{2-} ion is more easily distorted than the Cl^- ion.

2. Cations. The ability of a cation to distort the electron cloud of a neighboring anion also depends upon size and charge. *A small cation* with a *high positive charge* is the most effective in bringing about anion distortion. A cation of this type has a high concentration of positive charge.

In any group of metals in the periodic table, the member that forms the smallest cation (for example, Li in group I A) has the greatest tendency toward the formation of compounds with a high degree of covalent character. All compounds of beryllium (Be^{2+} is the smallest cation of any group II A element) are significantly covalent. Boron (the smallest member of group III A) forms only covalent compounds. The hypothetical B^{3+} ion would have a comparatively high charge combined with a very small size, which would cause extensive anion distortion leading to covalent bonding.

The first four metals of the fourth period form chlorides in which each of the "cations" is isoelectronic with Ar. In this series, KCl, $CaCl_2$, $ScCl_3$, and $TiCl_4$, the covalent character increases with increasing charge and decreasing size of the "cation." KCl is strongly ionic, and $TiCl_4$, a liquid, is definitely covalent. Truly ionic compounds that contain cations with a charge of $3+$ or higher are rare. Such compounds as $SnCl_4$, $PbCl_4$, $SbCl_5$, and BiF_5 are covalent.

A second approach to bonds of intermediate character considers the **polarization of covalent bonds.** A purely covalent bond results only when two *identical* atoms are bonded. Whenever two *different* atoms are joined by a covalent bond, the electron density of the bond is not symmetrically distributed around the two nuclei. The electrons of the bond are not shared equally. No matter how similar the atoms may be, there will be a difference in their ability to attract electrons.

Chlorine has a greater attraction for electrons than bromine has. In the BrCl molecule the electrons of the covalent bond are more strongly attracted by the chlorine atom than by the bromine atom. The electron cloud of the bond is denser in the vicinity of the chlorine atom. The chlorine end of the bond, therefore, has a partial negative charge. Since the molecule as a whole is electrically neutral, the bromine end is left with a partial positive charge of equal magnitude. Such a bond, with positive and negative poles, is called a **polar covalent bond.** The partial charges of the bond are indicated by the symbols δ^+ and δ^- to distinguish them from full ionic charges.

The greater the difference between the electron-attracting ability of two atoms joined by a covalent bond, the more polar the bond, and the larger the magnitude of the partial charges. If the unequal sharing of electrons were carried to an extreme, one of the bonded atoms would have all of the bonding electrons, and

ionic bond

distorted ions

polarized
covalent bond

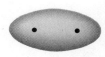

covalent bond

Figure 6.2 Transition between ionic and covalent bonding

separate ions would result (look at the drawings in Figure 6.2 in the reverse order from that shown).

A polar covalent molecule tends to turn in an electric field between the plates of a condenser in such a way that the negative end is toward the positive plate and the positive end is toward the negative plate (Figure 6.3). Polar molecules arranged in this way affect the amount of charge that a pair of electrically charged plates can hold. As a result, measurements can be made that allow the calculation of a value called the dipole moment.

If two equal charges with opposite signs are separated by a known distance, the **dipole moment** is

$$dipole\ moment = (charge)(distance) \tag{6.2}$$

The dipole moments of nonpolar molecules, such as H_2, Cl_2, and Br_2, are zero. The dipole moment of polar diatomic molecules increases as the polarity of the molecule increases.

Linus Pauling has used dipole moment to calculate the **partial ionic character** of a covalent bond. If HCl were completely ionic, the H^+ ion and the Cl^- ion would each have a unit charge (1.60×10^{-19} C). The bond distance in the HCl molecule is 127 pm (which is 1.27×10^{-10} m). The dipole moment of the imaginary H^+Cl^- would be

$$
\begin{aligned}
dipole\ moment &= (charge)(distance) \\
&= (1.60 \times 10^{-19}\ C)(1.27 \times 10^{-10}\ m) \\
&= 2.03 \times 10^{-29}\ C \cdot m = 6.08\ D
\end{aligned}
$$

The debye unit, D, is 3.34×10^{-30} C·m.

The experimentally determined dipole moment of HCl is 1.03 D. The observed dipole moment is

$$1.03\ D/6.08\ D = 0.169$$

times the value for the imaginary ionic compound. The HCl bond appears to be about 17% ionic.

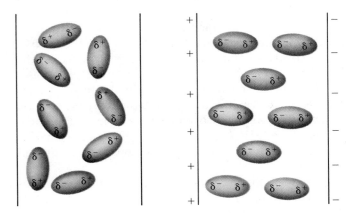

Figure 6.3 Effect of an electrostatic field on the orientation of polar molecules

6.6 Electronegativity

Electronegativity is a measure of the relative ability of an atom in a molecule to attract electrons to itself. The polarity of the HCl bond may be said to arise from the difference between the electronegativity of the Cl atom and that of the H atom. Since the Cl atom is more electronegative than the H atom, the Cl end of the bond bears a partial negative charge, δ^-, and the H end of the bond bears a partial positive charge, δ^+.

The concept of electronegativity is useful but inexact. Electronegativity values are given on an arbitrary scale—they are *relative* values, useful only in making qualitative comparisons between elements. There is no simple, direct way to measure electronegativities, and several methods of calculating them have been proposed.

The relative electronegativity scale of Linus Pauling is based on bond energies. A bond formed between two atoms that have different electronegativities is a polar bond. A polar covalent bond is always stronger than would be expected on the basis of equal electron sharing. The difference represents the amount of energy needed to overcome the δ^+, δ^- partial charges of the polar bond that have been produced by unequal electron sharing. The size of the difference in bond energy, therefore, is related to the size of the difference between the electronegativities of the bonded atoms. The scale was developed by the arbitrary assignment of a value of 4.0 to the F atom (the most electronegative atom).

Some relative electronegativities are given in Table 6.1. In general, electronegativity increases from left to right across any period (with increasing number of valence electrons) and from bottom to top in any group (with decreasing size). The most highly electronegative elements are found in the upper right-hand corner of the periodic table (ignoring the noble gases). The least electronegative elements are found in the lower-left corner of the chart.

Metals are elements that have small attractions for valence electrons (low electronegativities). Nonmetals, except for the noble gases, have large attractions (high electronegativities). Electronegativities, therefore, can be used to rate the reactivities of metals and nonmetals. The positions of the elements in the periodic table are helpful in making predictions concerning chemical reactivity (see Figure 6.4).

Table 6.1   Relative electronegativities

H 2.2							He –
Li 1.0	Be 1.6	B 2.0	C 2.6	N 3.0	O 3.4	F 4.0	Ne –
Na 0.9	Mg 1.3	Al 1.6	Se 1.9	P 2.2	S 2.6	Cl 3.2	Ar –
K 0.8	Ca 1.0	Ga 1.8	Ge 2.0	As 2.2	Se 2.6	Br 3.0	Kr –
Rb 0.8	Sr 0.9	In 1.8	Sn 2.0	Sb 2.1	Te 2.1	I 2.7	Xe –
Cs 0.8	Ba 0.9	Tl 2.0	Pb 2.3	Bi 2.0	Po 2.0	At 2.2	Rn –

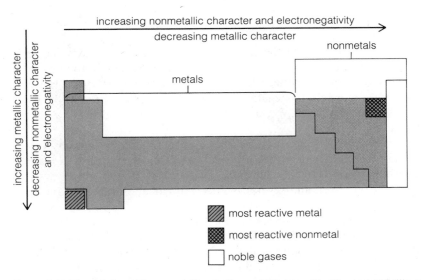

increasing nonmetallic character and electronegativity

decreasing metallic character

increasing metallic character
decreasing nonmetallic character and electronegativity

nonmetals

metals

most reactive metal

most reactive nonmetal

noble gases

Figure 6.4 The relation between position in the periodic classification and metallic or nonmetallic reactivity

Electronegativities can be used to predict the type of bonding that a compound will have. When two elements of widely different electronegativity combine, an ionic compound results. Covalent bonding occurs between nonmetals. The difference between the electronegativities of two nonmetals is not very large. In such cases the electronegativity differences give an indication of the degree of polarity of the covalent bond. If the electronegativity difference is zero or very small, an essentially nonpolar bond can be assumed. The larger the electronegativity difference, the more polar the covalent bond is. The atom with the higher electronegativity is the one that has the negative partial charge of the polar bond. By using electronegativities we can predict that HF is the most polar and has the highest bond energy of any of the hydrogen halides (see Table 6.2).

The concept of electronegativity is inexact. Since this property depends not only upon the structure of the atom under consideration but also upon the number and nature of the atoms to which it is bonded, the electronegativity of an atom is not constant. The electronegativity of phosphorus, for example, is different in PCl_3 than it is in PF_5. Electronegativity values, therefore, are approximations and electronegativity differences cannot be treated in a rigorous manner.

Table 6.2 Some properties of the hydrogen halides				
Hydrogen Halide	Dipole Moment (D)	Bond Energy (kJ/mol)	Electronegativity of Halogen	Electronegativity Difference between Hydrogen and Halogen
HF	1.91	565	F = 4.0	1.8
HCl	1.03	431	Cl = 3.2	1.0
HBr	0.78	364	Br = 3.0	0.8
HI	0.38	297	I = 2.7	0.5

Example 6.5

Which bond is the more polar: (a) N—O or C—O, (b) S—F or O—F?

Solution

(a) The electronegativity differences are

for N—O $3.4 - 3.0 = 0.4$
for C—O $3.4 - 2.6 = 0.8$

The C—O bond is the more polar. In each case the O atom has the δ^- charge.

The same conclusion can be derived from the periodic table alone. Electronegativity increases in a period from left to right. Consequently, the electronegativity values of the atoms fall in the order $C < N < O$. The C—O bond combines the atom with the lowest electronegativity (C) and the atom with the highest electronegativity (O) and therefore is the more polar bond.

(b) The electronegativity differences are

for S—F $4.0 - 2.6 = 1.4$
for O—F $4.0 - 3.4 = 0.6$

The S—F bond is the more polar of the two. In both bonds the F atom has the δ^- charge.

The F atom is the more electronegative atom of each bond since it is the atom closest to the top right. Since S is below O in group VI A, S is *less* electronegative than O. The S—F bond, therefore, is the more polar of the two since it is the combination of the atom with the lowest electronegativity (S) with the atom with the highest electronegativity (F).

Table 6.3 Greek prefixes used in naming covalent binary compounds

Prefix	Value
mono-	1
di-	2
tri-	3
tetra-	4
penta-	5
hexa-	6
hepta-	7
octa-	8
nona-	9
deca-	10

6.7 Nomenclature of Covalent Binary Compounds

A binary compound is one that is formed from only two elements. The nomenclature of ionic binary compounds is discussed in Section 5.8. Covalent binary compounds are formed from two nonmetals. Most of the covalent binary compounds of carbon are classified as organic compounds and their nomenclature is discussed in Chapter 26.

The name of an inorganic covalent binary compound is derived from the names of the two elements that form it with the *less* electronegative element named first. The *more* electronegative element is named second and the ending *-ide* is substituted for the usual ending of the name of that element. Greek prefixes (see Table 6.3) are added to the names of the elements to indicate the numbers of atoms of each type that are found in the molecule. The prefix *mono-* is usually omitted. The oxides of nitrogen serve as examples:

N_2O is dinitrogen oxide
NO is nitrogen oxide
N_2O_3 dinitrogen trioxide

NO_2 is nitrogen dioxide
N_2O_4 is dinitrogen tetroxide
N_2O_5 is dinitrogen pentoxide

Certain binary compounds have acquired nonsystematic names by which they are known exclusively. The list of such substances includes water (H_2O), ammonia (NH_3), hydrazine (N_2H_4), and phosphine (PH_3). Notice that the formulas of the last three compounds are customarily written in inverted form. According to the rules, the symbol H should appear first in each of these formulas since in each case hydrogen is the less electronegative element.

Summary

The topics that have been discussed in this chapter are

1. An introduction to covalent bonding, a type of bonding in which electrons are shared by the bonded atoms.

2. The determination of the formal charge of an atom in a covalent structure and the meaning of this convention.

3. How to draw Lewis structures, which are used to represent covalent molecules and ions.

4. The concept of resonance in which several Lewis structures are used to describe the bonding in a covalent molecule or ion.

5. Bonding that is intermediate in character between purely ionic and purely covalent.

6. Electronegativity and its use in predicting bond type.

7. Nomenclature of covalent binary compounds.

Key Terms

Some of the more important terms introduced in this chapter are listed below. Definitions for terms not included in this list may be located in the text by use of the index.

Adjacent charge rule (Section 6.2)   In a Lewis structure, atoms that are bonded together should not have formal charges with the same sign.

Binary compound (Section 6.7)   A compound formed from two elements.

Covalent bond (Section 6.1)   A bond formed between two atoms by electron sharing.

Dipole moment (Section 6.5)   The distance separating two equal charges with opposite signs times the magnitude of the charge.

Electronegativity (Section 6.6)   A measure of the relative ability of an atom in a molecule to attract electrons to itself.

Formal charge (Section 6.2)   A charge arbitrarily assigned to an atom in a covalent structure by apportioning the bonding electrons equally between the bonded atoms. These charges are useful in interpreting the properties and structures of covalent species, but the concept is merely a convention.

Lewis structure (Sections 6.1 and 6.3)   A representation of a covalent molecule or ion in which only the valence levels of the atoms are shown, a dash is used to represent a covalent bond (a pair of electrons), and dots are used to represent unshared electrons.

Partial ionic character (Section 6.5)   A value (given as a percentage) that relates the polarity of a covalent bond to the polarity that would exist if the atoms were joined by an ionic bond.

Polar covalent bond (Section 6.5)   A covalent bond that has partial charges (δ^+ and δ^-) as a result of the unequal sharing of bonding electrons.

Resonance (Section 6.4)   A concept in which two or more Lewis structures are used to describe the structure of a covalent molecule or ion. The actual structure of the covalent species is said to be a hybrid of the Lewis structures, which are called resonance forms.

Problems

Covalent Compounds, Lewis Structures

6.1 List all the nonmetallic elements that occur as diatomic molecules.

6.2 The formula P_2Br_4 is correct, but it is evident that the formula Ba_2Br_4 is not. Why?

6.3 Complete the following Lewis structures by adding dots for unshared electrons and formal charges. Explain why the first structure given in each pair is better than the second structure:

(a) F—C≡N, F≡C—N

(b)

6.4 Draw Lewis structures for the following molecules (include formal charges): **(a)** HCN, **(b)** H_2S, **(c)** SiH_4, **(d)** HI, **(e)** PH_3. When a subscript is added to a symbol in a formula, the atoms denoted are directly and separately bonded to the atom immediately following or immediately preceding in the formula.

6.5 Draw Lewis structures for the following molecules (include formal charges): **(a)** H_2NOH, **(b)** ClOCl, **(c)** $FClO_3$, **(d)** NCCN, **(e)** HOOH. When a subscript is added to a symbol in a formula, the atoms denoted are directly and separately bonded to the atom immediately following or immediately preceding in the formula.

6.6 Draw Lewis structures for the following molecules (include formal charges): **(a)** HSCl, **(b)** $HCCl_3$, **(c)** H_2CO, **(d)** NSF_3, **(e)** I_2PPI_2. When a subscript is added to a symbol in a formula, the atoms denoted are directly and separately bonded to the atom immediately following or immediately preceding in the formula.

6.7 Draw Lewis structures for the following ions (include formal charges): **(a)** PO_4^{3-}, **(b)** $H_2PO_2^-$, **(c)** ClO_3^-, **(d)** SO_3^{2-}, **(e)** SO_4^{2-}. When a subscript is added to a symbol in a formula, the atoms denoted are directly and separately bonded to the atom immediately following or immediately preceding in the formula.

6.8 Draw Lewis structures for the following (include formal charges): **(a)** H_2NNH_2, **(b)** HNNH, **(c)** F_2NNF_2, **(d)** $ONNO^{2-}$, **(e)** ClSSCl. When a subscript is added to a symbol in a formula, the atoms denoted are directly and separately bonded to the atom immediately following or immediately preceding in the formula.

6.9 The following species are isoelectronic: CO, NO^+, CN^-, and N_2. **(a)** Draw Lewis structures for each and include formal charges. **(b)** Each of the four reacts with metal atoms or metal cations to form complexes. In the formation of a complex, we can assume that an electron pair from one of the four is used to form a covalent bond by occupying an empty orbital of the metal atom or metal cation. For each heteronuclear species, tell which atom is bonded to the metal.

6.10 Draw an acceptable Lewis structure for each of the following (include formal charges): **(a)** O_2NF, **(b)** ONF, **(c)** OSF_2, **(d)** NSF, **(e)** FNNF. Would any of the structures you have drawn constitute a resonance form of a resonance hybrid? When a subscript is added to a symbol in a formula, the atoms denoted are directly and separately bonded to the atom immediately following or immediately preceding in the formula.

6.11 Draw an acceptable Lewis structure for each of the following (include formal charges): **(a)** H_2NCN, **(b)** Cl_2CO, **(c)** HONO, **(d)** NO_3^-, **(e)** SCN^-, **(f)** SCS. Would any of the structures you have drawn constitute a resonance form of a resonance hybrid? When a subscript is added to a symbol in a formula, the atoms denoted are directly and separately bonded to the atom immediately following or immediately preceding in the formula.

Resonance

6.12 When is the concept of resonance applied? What is the difference in meaning between the terms *resonance hybrid* and *resonance form*? Why are H—C≡N: and H—N≡C: not resonance forms of the same molecule? If an atom in a resonance form has a formal charge of $1+$, what restriction is placed on the formal charge of an adjacent atom?

6.13 Diagram the resonance forms of the phospham molecule, NPNH.

6.14 Diagram the resonance forms of the HONS and HNSO molecules.

6.15 Diagram the resonance forms of the OPN molecule.

6.16 In the S_2N_2 molecule, the four atoms are arranged in a ring with alternating S and N atoms. Diagram the resonance forms of the S_2N_2 molecule.

6.17 Diagram the resonance forms of the O_2NCl molecule. The N atom is the central atom to which the three others are directly bonded.

6.18 Diagram the resonance forms of the F_2NNO molecule. The two F atoms are separately bonded to the first N atom.

6.19 Diagram the resonance forms of the HN_3 molecule. The atoms are arranged HNNN.

6.20 Diagram the resonance forms of the OCN^- ion.

6.21 Diagram the resonance forms of the nitrite ion, NO_2^-. The two O atoms are separately bonded to the N atom.

---

* The appendix contains answers to color-keyed problems.

6.22 Diagram the resonance forms of the formate ion, HCO_2^-. The C atom is the central atom to which the three other atoms are separately bonded.

6.23 Diagram the resonance forms of carbon suboxide, C_3O_2, in which the atoms are arranged OCCCO.

Bonds of Intermediate Character, Electronegativity

6.24 On the basis of anion distortion, predict which member of each of the following pairs is the more covalent: **(a)** CuCl or $CuCl_2$, **(b)** MgSe or $MgCl_2$, **(c)** LiI or NaI, **(d)** $PbCl_2$ or $PbCl_4$, **(e)** $CdCl_2$ or CdI_2, **(f)** Al_2O_3 or Al_2S_3, **(g)** SnI_2 or $PbCl_2$.

6.25 On the basis of anion distortion, predict which member of each of the following pairs is the more covalent: **(a)** Li_3N or Li_3P, **(b)** $BeBr_2$ or $MgBr_2$, **(c)** Au_2O or Au_2O_3, **(d)** $SnBr_2$ or $PbBr_2$, **(e)** Ag_2S or AgCl, **(f)** FeS or Fe_2S_3, **(g)** In_2S or Tl_2O.

6.26 The bond distance in the ICl molecule is 232 pm and the dipole moment of ICl is 0.651 D. Calculate the partial ionic character of the ICl bond. The unit charge, e, is 1.60×10^{-19} C and 1 D is 3.34×10^{-30} C·m.

6.27 The bond distance in the BrF molecule is 176 pm and the dipole moment of BrF is 1.29 D. Calculate the partial ionic character of the BrF bond. The unit charge, e, is 1.60×10^{-19} C and 1 D is 3.34×10^{-30} C·m.

6.28 Use electronegativities to arrange the molecules ClF, BrF, IF, BrCl, ICl, and IBr in order of decreasing dipole moment.

6.29 Define and discuss the difference in meaning between the terms *electron affinity* and *electronegativity*.

6.30 Use electronegativities to arrange the following bonds in order of increasing ionic character: **(a)** S—O, **(b)** Si—O, **(c)** In—I, **(d)** Be—I, **(e)** Ca—S, **(f)** C—S, **(g)** Ca—Cl, **(h)** C—Cl, **(i)** O—Cl, **(j)** O—C, **(k)** Al—S, **(l)** Al—O.

6.31 On the basis of electronegativity, state whether you think the bonds formed between the following pairs of elements would be ionic or covalent; if covalent, estimate the degree of polarity of the bond: **(a)** Na, Br; **(b)** N, Br; **(c)** P, Br; **(d)** P, S; **(e)** Pb, I; **(f)** P, H; **(g)** B, H; **(h)** B, Br; **(i)** Ba, Br.

6.32 On the basis of electronegativity, state whether you think the bonds formed between the following pairs of elements would be ionic or covalent; if covalent, estimate the polarity of the bond: **(a)** Ca, O; **(b)** C, O; **(c)** Cl, O; **(d)** C, Cl; **(e)** C, Mg; **(f)** Cs, O; **(g)** C, S; **(h)** C, I; **(i)** C, H.

6.33 For each of the following pairs, use electronegativities to determine which bond is more polar. In each case tell the direction in which the bond is polarized. **(a)** H—C, H—Si; **(b)** O—H, O—F; **(c)** S—H, S—F; **(d)** O—H, S—H; **(e)** O—F, S—F; **(f)** N—H, N—Cl.

Nomenclature of Covalent Binary Compounds

6.34 Give formulas for **(a)** bromine pentafluoride, **(b)** disulfur dichloride, **(c)** tetraphosphorus tetranitride, **(d)** tellurium hexafluoride, **(e)** pentasulfur dinitride.

6.35 Give formulas for **(a)** tetraphosphorus hexoxide, **(b)** boron trifluoride, **(c)** dichlorine trioxide, **(d)** xenon tetrafluoride, **(e)** iodine heptoxide.

6.36 Name the following: **(a)** PCl_5, **(b)** I_2O_5, **(c)** SiF_4, **(d)** SO_3, **(e)** S_4N_4.

6.37 Name the following: **(a)** SF_6, **(b)** P_4S_3, **(c)** Cl_2O_7, **(d)** SiO_2, **(e)** N_2F_4.

MOLECULAR GEOMETRY; MOLECULAR ORBITALS

The simple theory of covalent bonding that was presented in the last chapter has some shortcomings. Lewis structures, based on the octet rule, cannot be drawn for some molecules and for some covalently bonded polyatomic ions. The theory presented thus far fails to account for another important aspect of the subject—the shapes (or molecular geometry) of covalent species. In this chapter, the discussion of covalent bonding is extended and the theory of molecular orbitals (bonding orbitals that extend over whole molecules) is presented.

7.1 Exceptions to the Octet Rule

We have seen that certain ions exist that do not have noble-gas configurations and that some of these ions are relatively stable. Some molecules also exist in which atoms have configurations other than those that the octet principle would lead us to expect.

There are a few molecules (such as NO and NO_2) that have an odd number of valence electrons. The NO molecule has a total of eleven valence electrons (five from the N atom and six from the O atom). The NO_2 molecule contains seventeen valence electrons (five from the N atom and twelve from the two O atoms). It is impossible to divide an odd number of electrons so that each atom of the molecule has a configuration of eight electrons (an even number). There are not many stable odd electron molecules. Odd electron species are usually very reactive and consequently short-lived.

More common are the molecules that have an even number of valence electrons but contain atoms with valence shells of less than, or more than, eight electrons. In the BF_3 molecule, the B atom has six valence electrons around it:

In the PCl_5 molecule, the P atom is bonded to five Cl atoms, and consequently the P atom has ten electrons in its valence shell. The S atom in SF_6 has twelve valence electrons around it since the S atom forms single covalent bonds with six F atoms.

For the elements of the second period, only four bonding orbitals are available ($2s$ and $2p$). The number of covalent bonds on atoms of these elements is therefore limited to a maximum of four. The elements of the third and subsequent periods, however, have more orbitals available in their outer electron shells. Compounds are known in which these elements form four, five, six, or (infrequently) an even higher number of covalent bonds. In valence-bond structures representing compounds containing elements of the third and subsequent periods, therefore, the octet principle is frequently violated. Apparently, the criteria for covalent bond formation should be centered on the electron pair rather than on the attainment of an octet.

7.2 Electron-Pair Repulsions and Molecular Geometry

The geometric arrangement of atoms in molecules and ions may be predicted by means of the **valence-shell electron-pair repulsion (VSEPR) theory**:

1. In the following discussion, we will consider molecules and ions in which a *central atom* is bonded to two or more atoms.

2. Because electron pairs repel one another, the electron pairs in the *valence shell of the central atom* are assumed to take positions as far apart as possible. The shape of the molecule or ion is a consequence of these electron-pair repulsions.

3. All the valence-shell electron pairs of the *central atom* are considered—both the pairs that form covalent bonds (called **bonding pairs**) and the pairs that are unshared (called **nonbonding pairs** or **lone pairs**).

4. The nonbonding pairs help to determine the positions of the atoms in the molecule or ion. The shape of the molecule or ion, however, is described in terms of the positions of the nuclei and not in terms of the positions of the electron pairs.

Some examples will make these points clear. In the diagrams that follow, only the electron pairs in the valence shell of the central atom are shown. Bonding pairs are indicated by dashes, and nonbonding pairs are indicated by dots. The central atom in many of these examples does not obey the octet rule.

1. Two pairs of electrons.   A mercury atom has two electrons in its valence shell $(6s^2)$. Each of these electrons is used to form a covalent bond with an electron of a chlorine atom in the $HgCl_2$ molecule. The molecule is **linear**:

$$Cl—Hg—Cl$$

The $HgCl_2$ molecule adopts a shape in which the two electron-pair bonds are as far apart as possible. Molecules in which the central atom has two bonding electron pairs in its valence shell are always linear. Beryllium, zinc, cadmium, and mercury form molecules of this type.

2. Three pairs of electrons.   A boron atom has three valence electrons (B is a member of group III A). In the BF_3 molecule, each fluorine atom (group VII A) supplies one electron for the formation of a single bond with an electron of the B atom. The boron trifluoride molecule is **triangular** and **planar**:

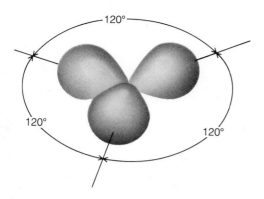

Figure 7.1   Triangular planar arrangement of three pairs of electrons

The angle formed by any two bonds in the molecule (the so-called F—B—F bond angle) is 120°. This arrangement provides the greatest possible separation between the three electron pairs (see Figure 7.1).

Tin (II) chloride vapor, $SnCl_2$, consists of angular molecules:

$$\overset{\displaystyle \overset{..}{Sn}}{\underset{Cl \qquad\qquad Cl}{\diagup \qquad \diagdown}}$$

A tin atom has four electrons in its valence shell (Sn is a member of group IV A), and each chlorine atom supplies one electron to form a bond. These six electrons constitute three electron pairs (two bonding and one nonbonding). The pairs assume a triangular planar configuration because of electron-pair repulsion. The shape of the molecule, however, is described in terms of the positions of its atoms, not its electrons. Tin (II) chloride molecules are therefore described as angular.

The Cl—Sn—Cl bond angle in $SnCl_2$, however, is less than 120° (about 95°). The nonbonding electron pair, which is under the influence of only one positive center, spreads out over a larger volume than a bonding pair, which is under the influence of two nuclei. As a result, the two bonds of the $SnCl_2$ molecule are forced closer together than is normal for the triangular planar arrangement. Nonbonding pairs repel bonding pairs more than bonding pairs repel other bonding pairs.

3. Four pairs of electrons.   In the methane molecule, CH_4, the carbon atom has four bonding pairs of electrons in its valence shell. The Lewis structure for the molecule is

$$\begin{array}{c} H \\ | \\ H\!-\!C\!-\!H \\ | \\ H \end{array}$$

The bond pairs repel each other least when the bonds are directed toward the corners of a regular tetrahedron (see Figures 7.2 and 7.3). All the bonds in this arrangement are equidistant from one another, and all H—C—H bond angles are 109° 28′. The tetrahedral configuration is a common and important one. Many molecules and ions (for example, ClO_4^-, SO_4^{2-}, and PO_4^{3-}) are tetrahedral.

7.2   Electron-Pair Repulsions and Molecular Geometry

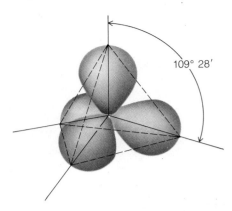

Figure 7.2 Tetrahedral arrangement of four pairs of electrons

The structure of ammonia, NH_3,

$$H \text{—} \overset{\cdot\cdot}{N} \text{—} H$$
$$\underset{H}{|}$$

can also be related to the tetrahedron (Figure 7.3). The N atom has three bonding electron pairs and one nonbonding electron pair in its valence shell. The four pairs assume a tetrahedral configuration which causes the *atoms* of the molecule to have a **trigonal pyramidal** arrangement with the N atom at the apex of the pyramid. The nonbonding electron pair of the molecule squeezes the bonding pairs together so that each H—N—H bond angle is 107° rather than the tetrahedral angle of 109° 28′.

The valence shell of the O atom in the water molecule has two bonding pairs and two nonbonding pairs:

$$:\overset{\cdot\cdot}{O} \text{—} H$$
$$\underset{H}{|}$$

The four electron pairs are arranged in an approximately tetrahedral manner (Figure 7.3) so that the *atoms* of the molecule have a **V-shaped (angular)** configuration. Since there are two nonbonding pairs in the molecule, the bonding pairs are forced together even more than those in the NH_3 molecule. Hence, the H—O—H bond angle in H_2O (105°) is less than the H—N—H bond angle in NH_3 (107°).

4. Five pairs of electrons. In the PCl_5 molecule, the five valence electrons of the phosphorus atom (P is a member of group V A) form five bonding pairs with electrons from five chlorine atoms:

$$\begin{array}{c} Cl \\ | \\ Cl\diagdown \\ Cl\diagup P \text{—} Cl \\ | \\ Cl \end{array}$$

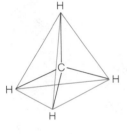

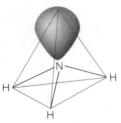

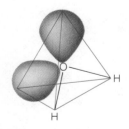

Figure 7.3 Geometries of methane (CH_4), ammonia (NH_3), and water (H_2O) molecules. (Bonding pairs shown as lines.)

The shape that minimizes electron-pair repulsion is the **trigonal bipyramid** (Figure 7.4a). The five bonds in the structure, however, are not equivalent.

The positions that lie around the "equator" (numbers 2, 4, and 5 in the figure) are called **equatorial positions**. The positions at the "north pole" and "south pole" (numbers 1 and 3 in the figure) are called **axial positions**. Equatorial atoms lie in the same plane. Any bond angle formed by two equatorial atoms and the central atom is 120°. The axial atoms are on an axis that is at right angles to the equatorial plane. Any bond angle formed by an axial atom, the central atom, and an equatorial atom is 90°. In addition, the P—Cl axial bond length (219 pm) is slightly longer than the equatorial bond length (204 pm).

When one or more nonbonding electron pairs are present in a molecule of this type, the *nonbonding pairs occupy equatorial positions.* An equatorial position offers the nonbonding pair more room than an axial position. Consider the sulfur tetrafluoride molecule as an example. In SF_4, four of the six valence electrons of the sulfur atom (S is a group VI A element) are used to form bonding pairs, and the remaining two constitute a nonbonding pair:

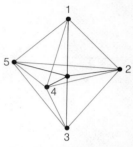

(a) trigonal bipyramid

$$\overset{\displaystyle F}{\underset{\displaystyle F}{\overset{|}{\underset{|}{\ddot{S}{-}F}}}}{\diagdown}F$$

The five electron pairs are arranged in approximately the form of a trigonal bipyramid with the nonbonding pair occupying an equatorial position (Figure 7.4b). The atoms of the molecule form what is described as an **irregular tetrahedron.**

The nonbonding electron pair affects the bond angles in the SF_4 molecule. The bonding pairs appear to be swept back, away from the nonbonding pair, so that the bond angles are less than those observed in the PCl_5 molecule. The two axial bonds form an angle of 173° with each other (instead of 180°). The two equatorial bonds form an angle of 102° with each other (instead of 120°).

In chlorine trifluoride, ClF_3, three of the seven valence electrons of the chlorine atom (a group VII A element) form bonding pairs, and the remaining four electrons make up two nonbonding pairs:

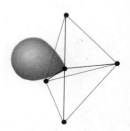

(b) irregular tetrahedron

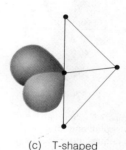

(c) T-shaped

Figure 7.4 Geometries of molecules in which the central atom has five pairs of electrons. (Bonding pairs shown as lines.)

$$\overset{\displaystyle F}{\underset{\displaystyle F}{\overset{|}{\underset{|}{:\ddot{C}l{-}F}}}}$$

Placement of the two nonbonding pairs in equatorial positions produces a **T-shaped** molecule (Figure 7.4c). The distortion introduced by the nonbonding electron pairs cause the F—Cl—F bond angle, formed by an axial bond and the equatorial bond, to be 87° 30′ (rather than 90°).

The xenon atom of the xenon difluoride molecule, XeF_2, has three nonbonding pairs and two bonding pairs in its valence shell since six of the eight valence electrons of Xe (group O) remain unshared after two bonds have formed:

$$\overset{\displaystyle F}{\underset{\displaystyle F}{\overset{|}{\underset{|}{:\ddot{X}e:}}}}$$

All three nonbonding electron pairs occupy equatorial positions. Xenon difluoride is a **linear** molecule (compare Figures 7.4 and 22.11).

(a) octahedron

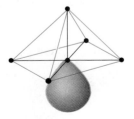

(b) square pyramid

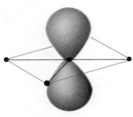

(c) square planar

Figure 7.5 Geometries of
molecules and ions in which
the central atom has six pairs
of electrons. (Bonding pairs
shown as lines.)

5. Six pairs of electrons. In the sulfur hexafluoride molecule, SF_6, the sulfur
atom (group VI A) has six bonding pairs in its valence shell:

$$
\begin{array}{ccc}
 & F & \\
F & | & F \\
 & \diagdown\!S\!\diagup & \\
F & | & F \\
 & F &
\end{array}
$$

The form that minimizes electron-pair repulsion is the regular **octahedron**
(Figure 7.5a). All the positions are equivalent, all the bond distances are equal,
and all the angles formed by any adjacent bonds are 90°.

The bromine atom of the bromine pentafluoride molecule, BrF_5, has five
bonding pairs and one nonbonding pair in its valence shell:

$$
\begin{array}{ccc}
 & F & \\
F & | & F \\
 & \diagdown\!Br\!\diagup & \\
 & \cdot\cdot & \\
F & & F
\end{array}
$$

A Br atom has seven valence electrons (group VII A). In BrF_5, five of these
electrons are engaged in bonding five fluorine atoms, and the other two constitute
a nonbonding pair. The electron pairs are directed to the corners of an octahedron.
The atoms of the molecule form a **square pyramid** (Figure 7.5b).

The nonbonding electron pair causes some distortion in BrF_5. Four of the
bonds in the molecule (the ones that bond the atoms of the base of the pyramid
to the central atom) are swept back—away from the nonbonding pair. As a
result, an angle formed by the F atom at the apex of the pyramid, the Br atom,
and a F atom as the base of the pyramid (see Figure 7.5b) is 85° rather than 90°.

The iodine atom of the IF_4^- ion has four bonding pairs and two nonbonding
pairs in its valence shell:

$$
\begin{array}{ccc}
F & \cdot\cdot & F^- \\
 & \diagdown I \diagup & \\
 & \cdot\cdot & \\
F & & F
\end{array}
$$

To account for the charge on the ion, we can consider that the central I atom
gains an electron and becomes an I^- ion with eight valence electrons. Four of
these electrons are used to form bonding pairs, and the other four electrons
constitute two nonbonding pairs. Since the total number of electron pairs is six,
the electron pairs of this ion assume octahedral positions. The nonbonding pairs
take positions opposite one another which minimizes electron-pair repulsion
(Figure 7.5c). The ion, therefore, has a **square planar** geometry. The relationships
between molecular shape and valence-shell electron pairs are summarized in
Table 7.1.

The concept of electron-pair repulsions can be extended to molecules and ions
that contain multiple bonds. A multiple bond is counted as a unit for the purpose
of applying the concept. Carbon dioxide (CO_2) and hydrogen cyanide (HCN),
for example, are *linear* molecules like species that have two bonding pairs in the
valence shell of the central atom:

$$O\!\!=\!\!C\!\!=\!\!O \qquad H\!\!-\!\!C\!\!\equiv\!\!N$$

Table 7.1   Number of electron pairs in the valence shell of the central atom and molecular shape

| Number of Electron Pairs | | | | |
Total	Bonding	Nonbonding	Shape of Molecule or Ion	Examples
2	2	0	linear	$HgCl_2$, $CuCl_2^-$
3	3	0	triangular planar	BF_3, $HgCl_3^-$
3	2	1	angular	$SnCl_2$, NO_2^-
4	4	0	tetrahedral	CH_4, BF_4^-
4	3	1	trigonal pyramidal	NH_3, PF_3
4	2	2	angular	H_2O, ICl_2^+
5	5	0	trigonal bipyramidal	PCl_5, $SnCl_5^-$
5	4	1	irregular tetrahedral	$TeCl_4$, IF_4^+
5	3	2	T-shaped	ClF_3, BrF_3
5	2	3	linear	XeF_2, ICl_2^-
6	6	0	octahedral	SF_6, PF_6^-
6	5	1	square pyramidal	IF_5, SbF_5^{2-}
6	4	2	square planar	BrF_4^-, XeF_4

The carbonyl chloride molecule, $OCCl_2$, has a *triangular planar* structure similar to that of BF_3:

$$\begin{array}{c} O \\ \parallel \\ C \\ Cl \quad\quad Cl \end{array}$$

A double bond, however, takes up more room than a single bond. In $OCCl_2$, the double bond forces the C—Cl bonds closer together than is normal for the triangular planar arrangement. The Cl—C—Cl bond angle, therefore, is 111° rather than 120°.

The theory can also be applied to resonance structures. Dinitrogen oxide, N_2O, is a *linear* molecule (similar to $HgCl_2$):

$$^{\ominus}N{=}\overset{\oplus}{N}{=}O \longleftrightarrow N{\equiv}\overset{\oplus}{N}{-}O^{\ominus}$$

The nitrite ion, NO_2^-, has an *angular* structure similar to that of $SnCl_2$:

$$\begin{array}{c} \overset{\cdot\cdot}{N} \\ ^{\ominus}O \quad\quad O \end{array} \longleftrightarrow \begin{array}{c} \overset{\cdot\cdot}{N} \\ O \quad\quad O^{\ominus} \end{array}$$

The bond angle of this molecule is 115° (rather than 120°) because of the effect of the nonbonding pair.

The nitrate ion, NO_3^-, is triangular planar (similar to BF_3):

$$\begin{array}{c} O \\ \overset{\oplus}{N} \\ ^{\ominus}O \quad\quad O^{\ominus} \end{array} \longleftrightarrow \begin{array}{c} O^{\ominus} \\ \overset{\oplus}{N} \\ ^{\ominus}O \quad\quad O \end{array} \longleftrightarrow \begin{array}{c} O^{\ominus} \\ \overset{\oplus}{N} \\ O \quad\quad O^{\ominus} \end{array}$$

In this ion, all of the O—N—O bond angles are 120° since neither a nonbonding pair nor a multiple bond introduces distortion. The resonance forms tell us that all of the N—O linkages are identical.

Example 7.1

Use VSEPR theory to predict the shapes of the following ions. All the bonds in these structures are single bonds. Assume that each halogen atom contributes one electron to the valence shell of the central atom for bond formation: (a) $TlCl_2^+$, (b) AsF_2^+, (c) IBr_2^-, (d) $SnCl_3^-$, (e) ClF_4^-.

Solution

The number of valence electrons of the central atom (A), plus one electron for each substituent halogen atom (X), and an adjustment for the charge of the ion (chg), give the total number of electrons in the valence shell of the central atom. One-half of this number is the total number of electron pairs. Since each halogen atom is bonded by a single bond pair, the number of halogen atoms is also the number of bonding electron pairs. The number of nonbonding electron pairs is obtained by subtraction:

	Electrons A + X + chg = total	Electron pairs total	bdg	nonbdg	Shape
(a) $TlCl_2^+$	$3 + 2 - 1 = 4$	2	2	0	linear
(b) AsF_2^+	$5 + 2 - 1 = 6$	3	2	1	angular
(c) IBr_2^-	$7 + 2 + 1 = 10$	5	2	3	linear
(d) $SnCl_3^-$	$4 + 3 + 1 = 8$	4	3	1	trigonal pyramidal
(e) ClF_4^-	$7 + 4 + 1 = 12$	6	4	2	square planar

Example 7.2

Draw Lewis structures for the following and predict their shape: (a) ONF (b) SO_4^{2-}, (c) SO_3^{2-}, (d) CO_3^{2-}.

Solution

angular	tetrahedral	trigonal pyramidal	(a resonance hybrid) triangular planar
(a)	(b)	(c)	(d)

With very few exceptions, the predictions based on VSEPR theory have been shown to be correct. The theory predicts, for example, that the methane molecule (CH_4) is tetrahedral since the central C atom has four bonding pairs of electrons in its valence shell (Figures 7.2 and 7.3). This prediction has been confirmed by a variety of experimental evidence. It is known, with a high degree of certainty, that all four bonds of the CH_4 molecule are equivalent in length and strength and that all H—C—H bond angles are the tetrahedral angle, 109° 28'.

Figure 7.6 Overlap of the 1s orbitals of two hydrogen atoms

According to the valence-bond theory, a covalent bond consists of a pair of electrons (with spins paired) that is shared by two atoms. The formation of a covalent bond (Figure 7.6) may be *imagined* to occur when an orbital of one atom (containing an unpaired electron) overlaps an orbital of another atom (containing an unpaired electron). If we work within the framework of this theory, how can we account for the tetrahedral bonding of CH_4?

The ground-state electron configuration of carbon ($1s^2\ 2s^2\ 2p^1\ 2p^1$) shows only two unpaired electrons. One might incorrectly expect that the C atom would form only two covalent bonds with H atoms. If a $2s$ electron were promoted to the vacant $2p$ orbital, however, the resulting excited state of the C atom ($1s^2\ 2s^1\ 2p^1\ 2p^1\ 2p^1$) would provide four unpaired electrons for bond formation.* How can we construct a model of the bonding in CH_4 based on this excited state of the C atom?

We assume that each bonding orbital of the molecule can be described as the product of the overlap of an atomic orbital of the C atom with a $1s$ orbital of a H atom. Since the four tetrahedral bonding orbitals are equivalent, the four atomic orbitals of the C atom used in making them must be exactly alike with their axes directed at angles of 109° 28' from one another. A $2s$ orbital of C and a $2p$ orbital of C, however, are not equivalent and are not oriented in the manner described (see Figures 4.17 and 4.18). We cannot, therefore, picture the bonds being formed from them in this simple way.

This problem has been traditionally handled by using an alternative description of the excited C atom. Imagine that the *total* electron distribution of the valence shell of the excited C atom—consisting of one unpaired electron in the $2s$ orbital plus three unpaired electrons in the three $2p$ orbitals—is divided into four equal portions that have identical shapes and are arranged in a tetrahedral manner. Since the total electron distribution involves four electrons, the electron density of each of the four equal portions corresponds to one electron. We can say that each portion represents one electron in an atomic orbital of a new type— an sp^3 **hybrid orbital.**

The wave functions that describe the $2s$ and $2p$ orbitals can be mathematically combined to give wave functions for the four equivalent sp^3 hybrid orbitals. The designation sp^3 indicates the number and type of orbitals used in the mathematical combination. The superscripts in notations of this type do not pertain to numbers

---

* Energy is required to unpair and promote one of the $2s$ electrons. This energy, however, is more than recovered in our hypothetical process by the formation of the four covalent bonds of the CH_4 molecule. Energy is required to pull a covalent bond apart (Section 3.7). The reverse process, the formation of a covalent bond, releases energy. Although the formation of CH_2 would not require that energy be used for electron promotion, less energy would be released by the formation of only *two* covalent bonds than would be released by the formation of *four*. On balance, the formation of CH_4 is favored.

sp linear

sp^2 triangular planar

sp^3 tetrahedral

dsp^3 or
sp^3d trigonal bipyramidal

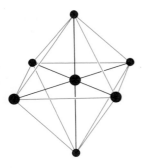

d^2sp^3 or sp^3d^2 octahedral

Figure 7.7 Directional characteristics of hybrid orbitals

of electrons. Each sp^3 hybrid orbital has one-quarter s character and three-quarters p character.

It is equally valid to describe the valence shell of the excited C atom in terms of four electrons separately occupying a 2s orbital and three 2p orbitals or in terms of four electrons separately occupying four equivalent sp^3 hybrid orbitals. The total electron distribution represented by either description is the same. Each solution reflects an equally satisfactory solution of the Schrödinger equation. The bonding in CH_4, therefore, can be described in terms of the overlap of the 1s orbitals of four H atoms with sp^3 hybrid orbitals of C.

Other types of hybrid orbitals are used to describe the bonding in other molecules. These sets need not involve all the atomic orbitals of the valence shell of the central atom. Wave functions for three equivalent sp^2 **hybrid orbitals,** for example, can be obtained by mathematically combining the wave functions for one s and two p orbitals. One of the three p orbitals is not included in this scheme. The axes of the three sp^2 hybrid orbitals lie in a plane and are directed at angles 120° apart. The set is used to account for the bonding of a *triangular planar* molecule in which the central atom has three bonding pairs of electrons (like BF_3).

A set of *sp* **hybrid orbitals** (derived from one s orbital and one p orbital of the central atom) may be used to describe the bonding of *linear* molecules in which the central atom has two bonding pairs of electrons (like $HgCl_2$ or $BeCl_2$). Notice the number of hybrid orbitals of a given type equals the number of simple atomic orbitals used in the mathematical combination that yields that type.

Common types of hybrid orbitals are listed in Table 7.2 and their directional characteristics are illustrated in Figure 7.7. Two of the sets described employ d orbitals of specified types (see Table 7.2 and Figure 4.19). The d orbitals used in a given set may be from the outer shell of the central atom or from the inner shell next to the outer shell. The octahedral hybrid orbitals, for example, may therefore be called d^2sp^3, or sp^3d^2 orbitals. Note that the hybrid orbitals of a dsp^3 (or sp^3d) set are not equivalent (see the discussion of trigonal bipyramidal molecules in Section 7.2).

The concept of hybrid orbitals may also be used to give an approximate description of molecules that contain one or more nonbonding electron pairs in the valence shell of the central atom. In NH_3, for example, we can assume that the central N atom employs sp^3 hybrid orbitals and that one of these orbitals contains a nonbonding electron pair while the others are used to form bonds with H atoms. The H—N—H bond angles in NH_3 (107°) are close to the tetrahedral angle of the sp^3 hybrid orbitals (109° 28′)—the deviation can be ascribed to the influence of the nonbonding electron pair.

Table 7.2 Hybrid orbitals			
Simple Atomic Orbitals	Hybrid Type	Geometry	Example
s, p_x	sp	linear	$HgCl_2$
s, p_x, p_y	sp^2	triangular planar	BF_3
s, p_x, p_y, p_z	sp^3	tetrahedral	CH_4
$d_{x^2}, s, p_x, p_y, p_z$	dsp^3 or sp^3d	trigonal bipyramidal	PF_5
$d_{z^2}, d_{x^2-y^2}, s, p_x, p_y, p_z$	d^2sp^3 or sp^3d^2	octahedral	SF_6

7.4 Molecular Orbitals

The theories of molecular structure that we have discussed so far have described the bonding in molecules in terms of *atomic* orbitals. The method of molecular orbitals is a different approach in which orbitals are associated with the molecule as a whole. The electronic structure of a molecule is derived by adding electrons to these molecular orbitals in an aufbau order. Corresponding to the practice of indicating atomic orbitals by the letters s, p, and so on, molecular orbitals are assigned the Greek letter designation σ (sigma), π (pi), and so on.

If two waves of the same wavelength (λ) and amplitude (a) are combined and are in phase, they reinforce each other (Figure 7.8a). The wavelength of the resultant wave remains the same, but the amplitude of the resultant wave is $a + a = 2a$. Two waves that are completely out of phase, however, cancel each other (Figure 7.8b); the "amplitude" of the resultant is $a + (-a) = 0$. The combination of waves can be *additive* or *subtractive*.

The molecular orbitals of the H_2 molecule can be imagined to result from the overlap of the $1s$ orbitals of two H atoms (Figure 7.6). If the overlap results in wave reinforcement (the additive combination), the electron density in the region between the nuclei is high. The attraction of the nuclei for this extra electronic charge holds the molecule together. The molecular orbital is called a sigma bonding orbital and is given the symbol σ (Figure 7.9).

In the combination, two atomic orbitals are used and therefore two molecular orbitals must be produced. The other molecular orbital is the result of the out-of-phase combination of waves (the subtractive combination). The electron density in the region between the nuclei is low. In this case, the nuclei repel each other since the low charge density between the nuclei does little to counteract this repulsion. This orbital is called a sigma antibonding orbital (given the symbol σ^*) since its net effect is disruptive (Figure 7.9). Sigma orbitals (both σ and σ^*) are

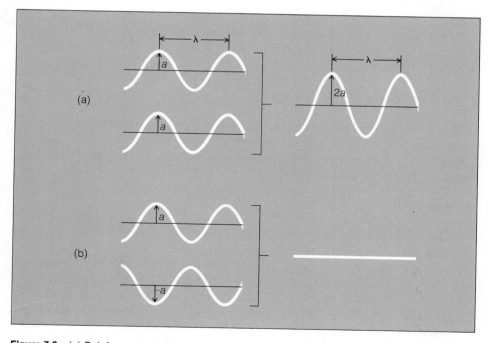

Figure 7.8   (a) Reinforcement of in-phase waves. (b) Cancellation of out-of-phase waves.

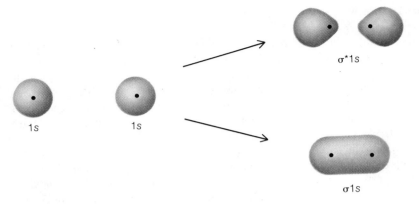

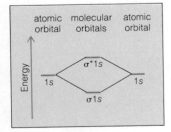

σ^*1s

$1s$ $1s$

$\sigma 1s$

Figure 7.9 Formation of σ and σ^* molecular orbitals from 1s atomic orbitals

cylindrically symmetrical about a line joining the two nuclei. Rotation of the molecule about this axis causes no observable change in orbital shape.

An energy-level diagram for the formation of $\sigma 1s$ and σ^*1s molecular orbitals from the $1s$ atomic orbitals of two atoms is shown in Figure 7.10. The energy of the σ bonding orbital is lower than that of either atomic orbital from which it is derived, whereas the energy of the σ^* antibonding orbitals is higher. When two atomic orbitals are combined, the resulting bonding molecular orbital represents a decrease in energy, and the antibonding molecular orbital represents an increase in energy.

Any orbital (atomic or molecular) can hold two electrons of opposed spin. In the hydrogen molecule, two electrons (with spins paired) occupy the $\sigma 1s$ orbital, the molecular orbital of lowest energy available to them. The σ^*1s orbital is unoccupied. One-half the difference between the number of bonding electrons and the number of antibonding electrons gives the number of bonds in the molecule (the **bond order**):

$$\text{bond order} = \tfrac{1}{2}\left[(\text{no. bonding } e^-) - (\text{no. antibonding } e^-)\right]$$

For H_2,

$$\text{bond order} = \tfrac{1}{2}(2 - 0) = 1$$

If an attempt is made to combine two helium atoms, a total of four electrons must be placed in the two molecular orbitals. Since the $\sigma 1s$ orbital is filled with two electrons, the other two must be placed in the higher σ^*1s orbital. The bond order in He_2 then would be

$$\text{bond order} = \tfrac{1}{2}(2 - 2) = 0$$

He_2 does not exist. The disruptive effect of the antibonding electrons cancels the bonding effect of the bonding electrons.

There is evidence that the hydrogen molecule ion, H_2^+, and the helium molecule ion, He_2^+, exist under proper conditions. The hydrogen molecule ion consists of two protons (H nuclei) and a single electron in the $\sigma 1s$ orbital. The bond order of H_2^+, therefore, is $\tfrac{1}{2}(1 - 0) = \tfrac{1}{2}$. The helium molecule ion consists of two helium nuclei and three electrons. Two of the three electrons, with spins paired, are placed in the $\sigma 1s$ orbital and the other electron is placed in the σ^*1s orbital. The bond order of He_2^+, therefore, is $\tfrac{1}{2}(2 - 1) = \tfrac{1}{2}$.

Figure 7.10 Energy-level diagram for the formation of σ and σ^* molecular orbitals from the 1s orbitals of two atoms

Chapter 7 Molecular Geometry; Molecular Orbitals

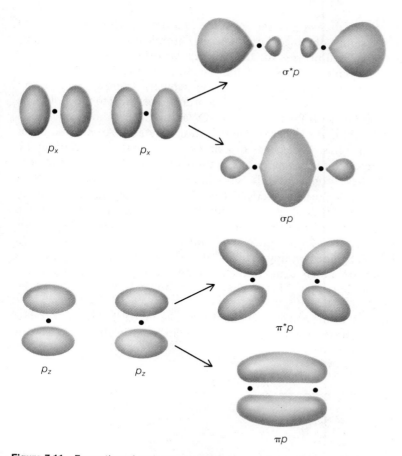

Figure 7.11 Formation of molecular orbitals from *p* atomic orbitals

The combination of two 2*s* orbitals produces σ and σ* molecular orbitals similar to those formed from 1*s* orbitals. The molecular orbitals derived from 2*p* atomic orbitals, however, are slightly more complicated. The three 2*p* orbitals of an atom are directed along the *x*, *y*, and *z* coordinates. If we consider that a diatomic molecule is formed by the atoms approaching each other along the *x* axis, the p_x atomic orbitals approach each other head on and overlap to produce σ2*p* bonding and σ*2*p* antibonding molecular orbitals (Figure 7.11). All sigma orbitals are completely symmetrical about the internuclear axis.

In the formation of a diatomic molecule (Figure 7.11), the p_z atomic orbitals approach each other side to side and produce a **pi bonding molecular orbital** (symbol, π) and a **pi antibonding molecular orbital** (symbol, π*). Pi orbitals are not cylindrically symmetrical about the internuclear axis. Instead, the side-to-side approach of the *p* orbitals leads to a π orbital consisting of two regions of charge density that lie above and below the internuclear axis (see Figure 7.11). The net effect of the π orbital, however, is one that holds the molecule together. The π* orbital has a low electron density in the region between the nuclei (see Figure 7.11). The net effect of the π* orbital is disruptive.

The p_y orbitals, which are not shown in Figure 7.11, also approach each other sideways. They produce another set of π and π* orbitals, which lie at right angles to the first set. The two π2*p* orbitals are degenerate (have equal energies) and the two π*2*p* orbitals are degenerate. Six molecular orbitals, therefore, arise from the two sets of 2*p* atomic orbitals—one σ2*p*, one σ*2*p*, two π2*p*, and two π*2*p*.

These six together with the two derived from the 2s atomic orbitals make a total of eight molecular orbitals obtained from the $n = 2$ atomic orbitals of two atoms.

To illustrate the aufbau process for these molecular orbitals, we will consider the homonuclear diatomic molecules of the second-period elements (molecules formed from two atoms of the same second-period element). There are two aufbau orders for these molecules (Figure 7.12). The first order (a) pertains to the molecules Li_2 to N_2; the second order (b) is the order for O_2 and F_2.

The second order (b) is the easier of the two to understand. An aufbau order is based on orbital energies. The energy of a molecular orbital depends upon the energies of the atomic orbitals used in its derivation and upon the degree and type of overlap between these atomic orbitals. Since 2s orbitals are lower in energy than 2p orbitals, the molecular orbitals derived from 2s orbitals are lower in energy than any molecular orbital derived from 2p orbitals. Because the overlap of the 2p orbitals that form the $\sigma 2p$ orbital is greater than the overlap of 2p orbitals that form $\pi 2p$ orbitals, the $\sigma 2p$ orbital is lower in energy than the degenerate $\pi 2p$ molecular orbitals. The antibonding orbitals of each type represent an increase in energy that is approximately equal to the decrease in energy represented by the bonding orbitals of that type. This aufbau order, however, is believed to be followed only by O_2 and F_2.

In the development of the order shown in (b), it is assumed that the 2s orbitals interact only with each other and that the 2p orbitals used in the derivation of

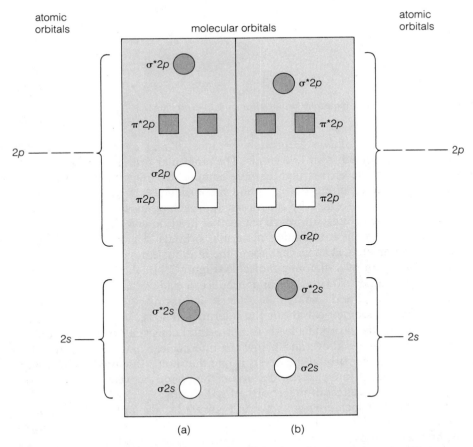

Figure 7.12 Aufbau orders for the homonuclear diatomic molecules of elements of the second period. (a) Li_2 to N_2. (b) O_2 and F_2.

σ and σ^* orbitals interact only with each other. This assumption is approximately valid if the energies of the 2s and 2p orbitals are widely separated (as they are in O and F). If the energies of the 2s and 2p orbitals are close in energy, s-p interaction also occurs. The result of this additional interaction is that the σ and σ^* molecular orbitals derived from the 2s orbitals become more stable (of lower energy) and the σ and σ^* molecular orbitals derived from 2p orbitals become less stable (of higher energy). This effect produces the aufbau order shown in Figure 7.12a. The important difference between Figures 7.12b and 7.12a is that in (a) the $\sigma2p$ orbital has been switched from below the two degenerate $\pi2p$ orbitals to above these two orbitals. The order shown in (a) is followed by the molecules from Li_2 to N_2.

Lithium is a member of group I A, and each Li atom has one valence electron. The Li_2 molecule, therefore, has two electrons with opposed spins in the molecular orbital of lowest energy, the $\sigma2s$ orbital. The bond order of Li_2 is $\frac{1}{2}(2-0)=1$.

If we attempt to make a Be_2 molecule, four electrons must be accommodated since each Be atom has two 2s electrons. The $\sigma2s$ orbital is filled when two electrons are entered. The other two electrons are placed in the σ^*2s orbital. The net effect of two bonding electrons and two antibonding electrons is no bond. The bond order is $\frac{1}{2}(2-2)=0$. Be_2 does not exist.

Molecular orbital energy level diagrams for the molecules B_2, C_2, and N_2 are shown in Figure 7.13. The orbitals are arranged from bottom to top in the pertinent aufbau order (a). Each diagram is obtained by placing the correct number of

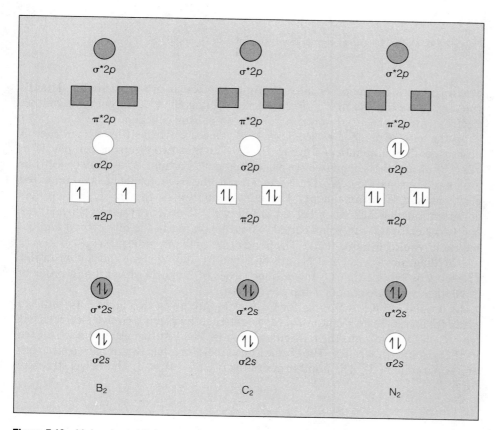

Figure 7.13 Molecular orbital energy-level diagrams for B_2, C_2, and N_2

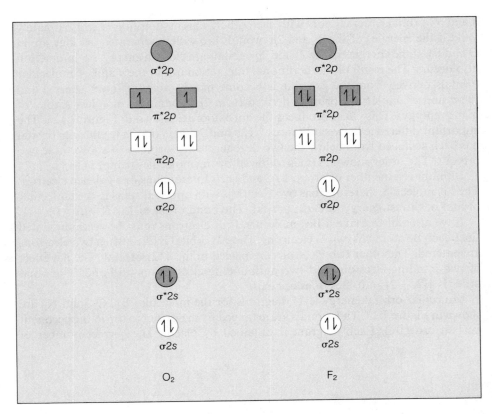

Figure 7.14 Molecular orbital energy-level diagrams for O_2 and F_2

electrons one after the other into the lowest molecular orbital available. Hund's rule is followed in the filling. The two $\pi2p$ orbitals have equal energies, and an electron enters each separately before electron pairing is begun.

In the case of B_2, the molecule has six valence electrons (three from each B atom since B is a member of group III A). The first two electrons are placed in $\sigma2s$ orbital, and the second two are placed in the $\sigma*2s$ orbital. The last two electrons are placed in separate $\pi2p$ orbitals. As a result, the B_2 molecule has two unpaired electrons and is paramagnetic. The paramagnetism of the B_2 molecule offers confirmation that aufbau order (a) is followed. If order (b) were followed, the last two electrons, with spins paired, would be placed in the $\sigma2p$ orbital and the molecule would be diamagnetic. The bond order of the B_2 molecule is $\frac{1}{2}(4-2)=1$.

The diagrams for C_2 and N_2 are obtained by adding eight and ten electrons, respectively. Notice that C_2 has a bond order of 2 and N_2 has a bond order of 3. Neither molecule contains unpaired electrons.

Molecular orbital diagrams for O_2 and F_2 are given in Figure 7.14; aufbau order (b) is used. The diagram for O_2 is obtained by placing twelve electrons (six from each O atom) into the molecular orbitals. The last two electrons are placed in $\pi*2p$ orbitals separately. The O_2 molecule, therefore, has two unpaired electrons and is paramagnetic. The bond order of O_2 is $\frac{1}{2}(8-4)=2$. The Lewis structure for O_2,

$$:\ddot{O}{=}\ddot{O}:$$

Table 7.3 Properties of diatomic molecules of the elements of the second period

Molecule	Number of Electrons in Molecular Orbitals						Bond Order	Bond Length (pm)	Bond Energy (kJ/mol)	Unpaired Electrons
	$\sigma 2s$	$\sigma^* 2s$	$\pi 2p$	$\sigma 2p$	$\pi^* 2p$	$\sigma^* 2p$				
[a]Li_2	2						1	267	106	0
[b]Be_2	2	2					0	—	—	0
[a]B_2	2	2	2				1	159	289	2
[a]C_2	2	2	4				2	131	627	0
N_2	2	2	4	2			3	110	941	0
	$\sigma 2s$	$\sigma^* 2s$	$\sigma 2p$	$\pi 2p$	$\pi^* 2p$	$\sigma^* 2p$				
O_2	2	2	2	4	2		2	121	494	2
F_2	2	2	2	4	4		1	142	155	0
[b]Ne_2	2	2	2	4	4	2	0	—	—	0

[a] Exists only in the vapor state at elevated temperatures.
[b] Does not exist.

is unsatisfactory. It shows the double bond of the O_2 molecule, but it does not show the two unpaired electrons. The diagram for F_2 employs fourteen electrons (seven from each F atom). The F_2 molecule has a bond order of 1.

A summary of the homonuclear diatomic molecules of the second-period elements is given in Table 7.3. As the number of bonds increases, the bond distance shortens and the bonding becomes stronger. The molecule bonded most strongly is N_2, which is held together by a triple bond. Molecules for which the method assigns a bond order of zero (Be_2 and Ne_2) do not exist.

Molecular orbital diagrams can also be drawn for diatomic ions. Diagrams for the N_2^+ and O_2^+ *cations* can be obtained by *removing* one electron from the N_2 and O_2 diagrams, respectively. Diagrams for the O_2^- (superoxide) and O_2^{2-} (peroxide) *anions* can be obtained by *adding* one and two electrons respectively to the O_2 diagram. A diagram for the acetylide ion, C_2^{2-}, results when two electrons are added to the C_2 diagram.

For molecules such as CO and NO, the same types of molecular orbitals, although slightly distorted, are formed. Either aufbau order may be used in most cases with the same qualitative results. The actual order, however, is uncertain.

Since CO is isoelectronic with N_2 (each molecule has ten valence electrons), the molecular orbital energy-level diagram for CO is similar to that for N_2 (Figure 7.13). Carbon monoxide, therefore, has a bond order of 3. The dissociation energy of the CO molecule is about the same as that of the N_2 molecule.

We have said that it is impossible to diagram a Lewis structure for a molecule with an odd number of valence electrons. Nitrogen oxide, NO, is such a molecule. Since five valence electrons are contributed by the N atom and six valence electrons by the O atom, the total number of valence electrons is eleven. An energy-level diagram for the molecular orbitals of NO is given in Figure 7.15. Since there are eight bonding electrons and three antibonding electrons shown in the diagram, a bond order of $\frac{1}{2}(8 - 3)$, or $2\frac{1}{2}$, is indicated. Nitrogen oxide is paramagnetic. The NO molecule has one unpaired electron in a $\pi^* 2p$ orbital.

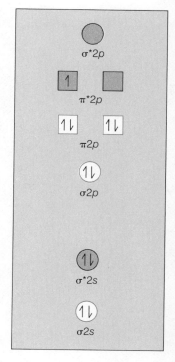

Figure 7.15 Molecular orbital energy-level diagram for NO

7.5 Molecular Orbitals in Polyatomic Species

Molecular orbitals can be derived for molecules that contain more than two atoms, such as H_2O and NH_3. In each case the number of molecular orbitals derived equals the number of atomic orbitals used, and the molecular orbitals encompass the whole molecule. For many purposes, however, it is convenient to think in terms of molecular orbitals that are localized between adjacent atoms. Consider the series

ethane   *ethylene*   *acetylene*

Each C atom in ethane may be considered to use sp^3 hybrid orbitals in the formation of σ bonds with the other C atom and the three H atoms (Figure 7.16). Thus, all bond angles are $109°\,28'$, the tetrahedral angle. Since σ bonding orbitals are symmetrical about the internuclear axis, free rotation about each bond is possible. Rotation about the C—C bond causes a changing atomic configuration (Figure 7.16).

A model for the bonding in a molecule that contains one or more multiple bonds may be derived from a σ bonding skeleton of the molecule (a framework of atoms held together by single bonds). The σ bonding skeleton of ethylene is

The ethylene molecule is planar and the σ bonds around each C atom are arranged in a triangular planar manner (the pattern predicted by VSEPR theory). The H—C—H bond angles are $118°$ and the H—C—C bond angles are $121°$, values that are close to the triangular planar angle of $120°$ (Figure 7.17).

We can account for the geometry of this molecule by assuming that each C atom uses sp^2 hybrid orbitals to form the σ bonding skeleton. One of the three $2p$ orbitals of each carbon is not involved in the formation of the sp^2 hybrid orbitals. These $2p$ orbitals are directed at right angles to the plane of the molecule and overlap to form a π bonding orbital (Figure 7.17). The electron density of the

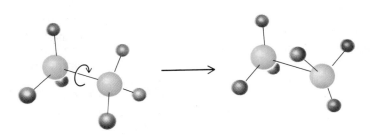

Figure 7.16 Rotation about the C—C bond in ethane

Chapter 7   Molecular Geometry; Molecular Orbitals

π bond is above and below the plane of the molecule. Free rotation about the C—C linkage is impossible without breaking this π bond.

The σ bonding skeleton of acetylene is

$$H—C—C—H$$

The molecule is linear (as would be predicted by VSEPR theory). Each C atom may be assumed to use sp (linear) hybrid orbitals to form two σ bonds. Two $2p$ orbitals of each C atom are not involved in the formation of the sp^2 hybrid orbitals. These $2p$ orbitals overlap to form two π bonding molecular orbitals (Figure 7.18). Note that each π bond has two centers of charge density—on either side of the axis of the σ bonding skeleton.

The multiple bonds of ethylene and acetylene are localized between two nuclei. In some molecules and ions **multicenter** (or **delocalized**) **bonding** exists in which some bonding electrons bond more than two atoms. The description of these species by the valence bond approach requires the use of resonance structures.

An example of delocalized bonding is found in the bonding in the carbonate ion (Figure 7.19). The ion is triangular planar, and each O—C—O bond angle is 120°. The C atom may be assumed to use sp^2 hybrid orbitals to form the σ bonding skeleton. One $2p$ orbital is not used in the sp^2 hybrid set. This $2p$ orbital projects at right angles to the plane of the ion and overlaps similar $2p$ orbitals of the O atoms (Figure 7.19). If we imagine the overlap of these $2p$ orbitals two at a time, we derive the resonance forms of the ion. The $2p$ orbital of C, however, can simultaneously overlap all three of the $2p$ orbitals of the three O atoms. The result is a system of π molecular orbitals that extends over all the atoms in the ion. The structures of sulfur trioxide, SO_3, and the nitrate ion, NO_3^-, are similar.

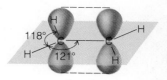

Figure 7.17 Geometric configuration of ethylene. (Shapes of the p orbitals that overlap to form a π bond are simplified; σ bonds are shown as lines.)

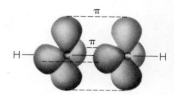

Figure 7.18 Formation of the π bonds of acetylene. (Orbital shapes are simplified; σ bonds are shown as solid lines.)

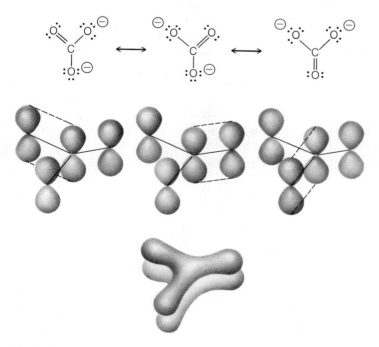

Figure 7.19 Multicenter π bonding system of the carbonate ion and the relationship to resonance structures

7.6 *pπ-dπ* Bonding

The Lewis structure for phosphoric acid (H_3PO_4) is

$$
\begin{array}{c}
\overset{\ominus}{:\ddot{O}:} \\
| \\
H-\overset{..}{\underset{..}{O}}-\overset{\oplus}{P}-\overset{..}{\underset{..}{O}}-H \\
| \\
:\ddot{O}: \\
| \\
H
\end{array}
$$

All the atoms in this structure conform to the octet rule. Many compounds of phosphorus are known, however, in which the P atom forms more than four covalent bonds (for example, PF_5). Since the P atom has $3d$ orbitals available in its valence shell, the octet restriction of a maximum of four covalent bonds does not apply to P.

If we introduce a double bond into the structure

$$
\begin{array}{c}
:O: \\
\| \\
H-\overset{..}{\underset{..}{O}}-P-\overset{..}{\underset{..}{O}}-H \\
| \\
:\ddot{O}: \\
| \\
H
\end{array}
$$

the P atom no longer conforms to the octet rule (the structure shows five bonds on the P atom) and the formal charges are eliminated. In past examples, a π bond has been the result of the overlap of two p orbitals. Here, the π bond is formed by the overlap of a filled $2p$ orbital of the O atom with an empty $3d$ orbital of the P atom. As such, it is an example of what is sometimes called *pπ-dπ* **bonding.**

There is evidence to support the structure that contains the double bond. The P—O bond distance shown in the structure as a double bond is 152 pm, which is shorter than the bond distance for the three remaining P—O bonds (shown as single bonds), 157 pm.

Based on atomic radii, however, the bond distance calculated for a P—O single bond is 176 pm. It appears, therefore, that even those bonds shown in the structure as single bonds are shorter than expected. This shortening may be explained by postulating that *pπ-dπ* interaction, also called **back bonding,** occurs to some extent (less than a bond order of one) in all the P—O bonds. The phosphate ion is sometimes shown as

$$
\left[\begin{array}{c}
O \\
\| \\
O =\!=\! P =\!=\! O \\
\| \\
O
\end{array}\right]^{3-}
$$

with the broken lines indicating *pπ-dπ* interactions. All the bond distances in the structure are 154 pm.

pπ-dπ bonding is particularly important in compounds in which a third-period nonmetal (Si, P, S, or Cl) is bonded to O, N, or F. Since the second-period nonmetals have no d orbitals in their valence shells, *pπ-dπ* bonding does not occur in compounds in which the central atom is a second-period element. Such compounds (for example, $HONO_2$) can contain double bonds but these bonds are formed by the use of s or p orbitals.

The Lewis structure and the double-bond structure for sulfuric acid, H_2SO_4, are

The S—O bond distances in the molecule are 154 pm (for the bonds shown as single bonds) and 142 pm (for the bonds shown as double bonds). The calculated S—O single bond distance (170 pm), however, is longer than either of the measured bond distances, and $p\pi$-$d\pi$ interactions are postulated for all the S—O bonds.

In the sulfate ion

all the S—O bonds are equivalent, and the bond distance of 149 pm indicates some $p\pi$-$d\pi$ character in each.

For perchloric acid, $HClO_4$, the Lewis structure and the double-bond structure are

Three of the Cl—O bonds (those shown as double bonds) have bond distances of 141 pm; the bond distance of the Cl—O bond shown as a single bond is 164 pm. The Cl—O bond distance in the perchlorate ion (ClO_4^-) is 146 pm. Since the calculated Cl—O single bond distance is 165 pm, $p\pi$-$d\pi$ interaction appears to be a characteristic of all the Cl—O bonds, in the ion as well as in the molecule.

Either a Lewis structure or a double-bond representation may be used to describe any one of these acids. In the case of a Lewis structure, the observed shortening of a bond represented in the other structure as a double bond may be ascribed to the effect of the attraction of plus and minus formal charges. No matter which representation is used, $p\pi$-$d\pi$ interactions must be used for even the single bonds to account for the shortness of the measured bond distances.

Summary

The topics that have been discussed in this chapter are

1. Molecules in which the octet rule is violated.

2. The valence-shell electron-pair repulsion theory (which correlates the shapes of covalent species with the repulsions between electron pairs) and predictions based on this theory.

3. The definition of hybridization, a theoretical device that permits the bonding in certain covalent species to be described in terms of orbital overlap.

4. Molecular orbital theory, in which orbitals are defined that are associated with the molecule as a whole.

5. The description, in terms of molecular orbital theory, of the homonuclear diatomic molecules of the first 10 elements and the extension of molecular orbital concepts to polyatomic species.

6. The formation of π bonds from p and d atomic orbitals (called $p\pi$-$d\pi$ bonds) and the existence of $p\pi$-$d\pi$ interactions that cause the shortening of certain single bonds.

Key Terms

Some of the more important terms introduced in this chapter are listed below. Definitions for terms not included in this list may be located in the text by use of the index.

Antibonding molecular orbital (Section 7.4) A molecular orbital in which electron density is low in the internuclear region. The two electrons in an antibonding molecular orbital have higher energies than they would if they were in the atomic orbitals from which the antibonding molecular orbital was derived.

Bond order (Section 7.4) One-half the number of bonding electrons minus one-half the number of antibonding electrons.

Bonding pair of electrons (Section 7.2) A pair of electrons used to form a covalent bond between two atoms.

Bonding molecular orbital (Section 7.4) A molecular orbital in which electron density is high in internuclear regions. The two electrons in a bonding molecular orbital have lower energies than they would if they were in the atomic orbitals from which the bonding molecular orbital was derived.

Hybridization (Section 7.3) A concept used in valence-bond theory in which the wave functions of pertinent atomic orbitals of an atom are mathematically combined to produce the wave functions of a set of equivalent hybrid orbitals. By the use of hybrid orbitals, the bonding in certain covalent species can be described in terms of orbital overlap.

Molecular orbital (Section 7.4) An orbital associated with a molecule rather than an atom.

Nonbonding pair of electrons, lone pair of electrons (Section 7.2) A pair of electrons on an atom of a covalent molecule or ion that is not involved in bonding.

$p\pi$-$d\pi$ bond (Section 7.6) A π bond formed by the overlap of a p orbital with a d orbital.

Pi bond (Section 7.4) A covalent bond in which electron density is concentrated in two regions above and below an axis joining the two nuclei.

Sigma bond (Section 7.4) A covalent bond in which the electron density is high in the region between the two nuclei and which is symmetrical about an axis joining the two nuclei.

Valence-bond theory (Section 7.3) A theory that assumes that a covalent bond is formed by the overlap of two atomic orbitals each of which contains an unpaired electron.

Valence-shell electron-pair repulsion theory (Section 7.2) A theory that permits the prediction of the shape of a covalent molecule or ion on the basis of repulsions between the bonding and nonbonding electron pairs in the valence shell of the central atom.

Problems*

VSEPR Theory, Hybrid Orbitals

7.1 Neither the bonding in NO nor in PCl_5 follows the octet rule. How do they deviate? Why is the type of deviation found in PCl_5 never found in N compounds?

7.2 Let A represent a central atom, B represent an atom bonded by an electron-pair bond to A, and E represent an unshared electron pair on A. What shapes are predicted by VSEPR theory for AB_2, AB_3, AB_2E, AB_4, AB_3E, AB_2E_2, AB_5, AB_4E, AB_3E_2, AB_6, AB_5E, and AB_4E_2?

7.3 Use the VSEPR theory to predict the geometric shape of the following: **(a)** BH_4^-, **(b)** XeF_5^+, **(c)** $BeCl_2$, **(d)** SbF_2^+, **(e)** $SnCl_3^-$, **(f)** AsH_3, **(g)** TeF_4, **(h)** IF_3, **(i)** SiF_5^-.

7.4 What type of hybrid orbitals is employed by the central atom of each of the species listed in Problem 7.3?

7.5 Use the concept of electron-pair repulsions to predict the geometric shape of: **(a)** $TlBr_4^-$, **(b)** XeF_3^+, **(c)** SCl_2, **(d)** AsF_2^+, **(e)** GaI_3, **(f)** ClF_4^-, **(g)** PBr_4^-, **(h)** TeF_5^-, **(i)** SbF_5^{2-}.

---

* The more difficult problems are marked with asterisks. The appendix contains answers to color-keyed problems.

7.6 What type of hybrid orbitals is employed by the central atom of each of the species listed in Problem 7.5?

7.7 Use the VSEPR theory to predict the geometric shape of the following: **(a)** $BiCl_5^{2-}$, **(b)** SeF_5^-, **(c)** ClF_2^+, **(d)** $InCl_2^+$, **(e)** BeF_3^- **(f)** GeF_2, **(g)** AsF_4^-, **(h)** XeF_2, **(i)** AlH_4^-.

7.8 What type of hybrid orbitals is employed by the central atom of each of the species listed in Problem 7.7?

7.9 Use the concept of electron-pair repulsions to predict the geometric shape of **(a)** SbF_4^-, **(b)** $AsCl_4^+$, **(c)** SeF_3^+, **(d)** XeF_4, **(e)** GeF_3^-, **(f)** $CdBr_2$, **(g)** BiI_4^-, **(h)** IBr_2^-, **(i)** SiF_6^{2-}.

7.10 What type of hybrid orbitals is employed by the central atom of each of the species listed in Problem 7.9?

7.11 Draw Lewis structures for the following and predict their shapes: **(a)** Cl_2O, **(b)** AsO_3^{3-}, **(c)** O_2SCl_2, **(d)** $OSCl_2$, **(e)** $OCCl_2$, **(f)** CS_2.

7.12 Draw Lewis structures for the following and predict their shapes: **(a)** $OPCl_3$, **(b)** $OSbCl$, **(c)** OCN^-, **(d)** O_2NF, **(e)** ClO_3^-, **(f)** SCl_2.

7.13 Draw Lewis structures for the following and predict their shapes: **(a)** $H_2PO_2^-$, **(b)** SeO_2, **(c)** ClO_2^-, **(d)** SO_3^{2-}, **(e)** HCN, **(f)** XeO_3.

***7.14** Draw Lewis structures for the following and predict their shapes: **(a)** $O_3ClOClO_3$, **(b)** O_2NONO_2, **(c)** $HONO$, **(d)** $FNNF$, **(e)** $ClSSCl$.

7.15 Draw dot structures for the following molecules in which the central atom does not obey the octet rule (although the other atoms do). Predict the geometric shape of each molecule: **(a)** $OXeF_4$, **(b)** $(HO)_5IO$, **(c)** OSF_4, **(d)** $(HO)_4XeO_2$.

***7.16** The effective volume of a bond pair is assumed to decrease with increasing electronegativity of the atom bonded to the central atom. In the light of this generalization, would you expect the Cl atoms of the trigonal bipyramidal PCl_2F_3 to occupy axial or equatorial positions?

***7.17** In view of the opening statement of Problem 7.16, predict which of the PX_3 molecules (where X is F, Cl, Br, or I) would have the smallest X—P—X bond angles.

Molecular Orbitals, $p\pi$-$d\pi$ Bonding

7.18 By means of molecular-orbital energy-level diagrams, describe the bonding of the homonuclear diatomic molecules of the second-period elements. State the bond order in each molecule and whether the molecule should be paramagnetic or diamagnetic.

7.19 Draw a molecular-orbital energy-level diagram and state the bond order for each of the following: **(a)** H_2, **(b)** H_2^+, **(c)** HHe, **(d)** He_2, **(e)** He_2^+.

7.20 Oxygen forms compounds containing the dioxygenyl ion, O_2^+ (for example, O_2PtF_6), the superoxide ion, O_2^- (for example, KO_2), and the peroxide ion, O_2^{2-} (for example, Na_2O_2). **(a)** Draw molecular-orbital energy-level diagrams for O_2^+, O_2, O_2^-, and O_2^{2-}. **(b)** State the bond order for each species. **(c)** Which of the four are paramagnetic?

7.21 The bond distance in N_2 is 109 pm, in N_2^+ is 112 pm, in O_2 is 121 pm, and in O_2^+ is 112 pm. Draw molecular-orbital energy-level diagrams for these four species and explain why the bond distances vary in the way described.

7.22 The anion of calcium carbide, CaC_2, should properly be called the acetylide ion, C_2^{2-}. **(a)** Draw molecular-orbital energy-level diagrams for C_2 and C_2^{2-}. **(b)** State the bond order in C_2 and in C_2^{2-}. **(c)** With what neutral molecule is C_2^{2-} isoelectronic?

7.23 **(a)** Draw molecular-orbital energy-level diagrams for CO and NO. **(b)** Use your diagrams to determine the bond order of CO, CO^+, CO^-, NO, NO^+, and NO^-. **(c)** Which of these species are paramagnetic?

7.24 Discuss the structure of NO_2^- in terms of resonance and in terms of delocalized π bonding.

7.25 Draw the resonance forms for the ozone molecule, O_3, and the sulfur dioxide molecule, SO_2. The O—O bond distance in O_3 is 127 pm, which is between the O—O single bond distance of 148 pm and the double bond distance of 110 pm. On the other hand, the S—O bond distance in SO_2 is 143 pm, which is shorter than either the S—O single bond distance of 170 pm or the double bond distance of 148 pm. Explain.

7.26 In the SiO_4^{4-} ion, the Si—O bond distance is 163 pm. The Si—O bond distance calculated from atomic radii is 176 pm. Explain the difference.

7.27 The atomic radius of H is 32 pm, of F is 64 pm, and of P is 110 pm. In PH_3, the P—H bond distance is 142 pm and in PF_3 the P—F bond distance is 155 pm. Compare the bond distances in PH_3 and in PF_3 with those expected on the basis of atomic radii. What reason can you give for any discrepancy?

CHAPTER GASES

8

Gases are believed to consist of widely separated molecules in rapid motion. Any two (or more) gases can be mixed in any proportion to prepare a perfectly uniform mixture; no such generalization can be made for liquids. Since the molecules of a gas are separated by comparatively large distances, the molecules of one gas can easily fit between the molecules of another gas. This molecular model can also be used to explain the fact that gases are readily compressed. Compression consists of forcing gas molecules closer together.

A gas expands to fill any container into which it is introduced. When a gas that has an odor is released into a room, it can soon be detected in all parts of the room. Gases diffuse because the gas molecules are in constant, rapid motion. Furthermore, in the course of their random motion, gas molecules strike the walls of the container. These impacts explain the fact that gases exert pressure.

8.1 Pressure

Pressure is defined as force per unit area. The pressure of a gas is equal to the force that the gas exerts on the walls of the container divided by the surface area of the container:

$$\text{pressure} = \frac{\text{force}}{\text{area}}$$

The SI unit of pressure is the pascal (abbreviated Pa), which is defined as the pressure equivalent to a force of one newton ($1 \text{ N} = 1 \text{ kg} \cdot \text{m/s}^2$) acting on one square meter:

$$1 \text{ Pa} = \frac{1 \text{ N}}{1 \text{ m}^2}$$

$$= \frac{1 \text{ kg} \cdot \text{m/s}^2}{1 \text{ m}^2} = 1 \text{ kg/m} \cdot \text{s}^2$$

The chemist, however, usually measures gas pressures by relating them to the pressure of the atmosphere.

A barometer is used to measure the pressure that the atmosphere exerts on the surface of the earth. This instrument was devised in the seventeenth century by Evangelista Torricelli, a pupil of Galileo. A tube, approximately 850 mm in

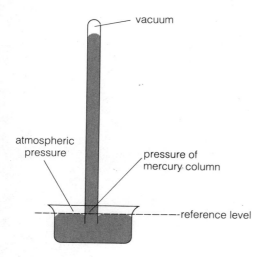

atmospheric pressure

pressure of mercury column

vacuum

reference level

Figure 8.1 Barometer

length and sealed at one end, is filled with mercury and inverted in an open container of mercury (Figure 8.1). The mercury falls in the tube but does not completely run out. The pressure of the atmosphere on the surface of the mercury in the dish supports the column of mercury in the tube.

The space above the mercury inside the tube is a nearly perfect vacuum. Since mercury is not very volatile at room temperature, only a negligible amount of mercury vapor occupies this space. Therefore, practically no pressure is exerted on the upper surface of the mercury in the column. The pressure inside the tube and above the reference level shown in Figure 8.1 results from the weight of the mercury column alone. This pressure is equal to the atmospheric pressure outside the tube and above the reference level.

The height of mercury in the tube serves as a measure of the atmospheric pressure. When the atmospheric pressure rises, it pushes the mercury higher in the tube. Remember that pressure is force *per unit area*. Whether the tube has a relatively large or small cross-sectional area, a given atmospheric pressure will support the mercury in the tube to the same height.

The pressure of the atmosphere changes from day to day and from place to place. The average pressure at sea level supports a column of mercury to a height of 760 mm; this value is called 1 atmosphere (abbreviated atm). The definition of the standard atmosphere, however, is given in terms of the pascal:

1 atm = 101,325 Pa = 101.325 kPa

The pressure equivalent to a height of 1 mm of mercury is called 1 torr (named for Torricelli); therefore,

1 atm = 760 torr

The International Committee of Weights and Measures recommends that pressures not be recorded in torrs. The preceding relationship can be used to convert readings made in terms of the height of a mercury column (in torrs) into atmospheres.

A manometer, a device that is used to measure the pressure of a sample of a gas, is patterned after the barometer. The type of manometer shown in Figure 8.2 consists of a U tube containing mercury. One arm of the U tube is open to the atmosphere; atmospheric pressure is exerted on the mercury in this arm. The

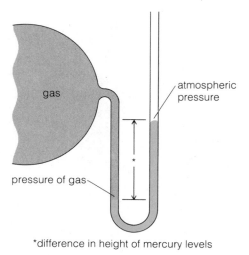

*difference in height of mercury levels

Figure 8.2 A type of manometer

other arm is connected to a container of a gas in such a way that the gas exerts pressure on the mercury in this arm.

If the gas sample were under a pressure *equal* to atmospheric pressure, the mercury would stand at the same level in both arms of the U tube. In the experiment illustrated in Figure 8.2, the gas pressure is *greater* than atmospheric pressure. The difference in height between the two mercury levels (in mm of mercury) must be added to the barometric pressure (in mm of mercury or in torr) to obtain the pressure of the gas (in mm of mercury or in torr). If the pressure of the gas were *less* than atmospheric, the mercury in the left arm of the manometer shown in Figure 8.2 would stand at a higher level than the mercury in the right arm, and the difference in height would have to be subtracted from atmospheric pressure.

8.2 Boyle's Law

The relationship between the volume and the pressure of a sample of a gas was studied by Robert Boyle in 1662. Boyle found that increasing the pressure on a sample of a gas causes the volume of the gas to decrease proportionately. If the pressure is doubled, the volume is cut in half. If the pressure is increased threefold, the volume is decreased to one-third its original value. **Boyle's law** states that at constant temperature the volume of a sample of gas varies inversely with the pressure:

$$V \propto \frac{1}{P}$$

The proportionality can be changed into an equality by the introduction of a constant, k:

$$V = \frac{k}{P} \qquad \text{or} \qquad PV = k \tag{8.1}$$

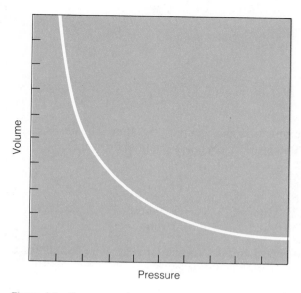

Volume

Pressure

Figure 8.3   Pressure-volume curve for an ideal gas (Boyle's law)

The value of the constant depends upon the size of the sample and the temperature. Pressure–volume data for an ideal gas (see Section 8.5) are plotted in Figure 8.3.

The volume of a gas is customarily measured in liters. The liter is defined as 1 cubic decimeter (1 dm^3 = 1000 cm^3). Since there are 1000 ml in 1 liter, 1 ml is 1 cm^3.

Example 8.1

A sample of a gas occupies 360 ml under a pressure of 0.750 atm. If the temperature is held constant, what volume will the sample occupy under a pressure of 1.000 atm?

Solution

First, we tabulate the data given in the problem:

Initial conditions: V = 360 ml P = 0.750 atm
Final conditions: V = ? ml P = 1.000 atm

The final volume can be obtained by correcting the initial volume for the change in pressure. The units sought are the same as the units of the quantity given:

? ml = 360 ml (pressure correction)

Correction factors are *not* the same as *conversion* factors. A correction factor is not equal to 1. We can derive two pressure-correction factors from the data given:

(1.000 atm/0.750 atm)   (0.750 atm/1.000 atm)

Which one should be used?

Since the pressure *increases* from 0.750 atm to 1.000 atm, the volume must *decrease*. The correction factor must be a fraction less than 1. Hence,

$$? \text{ ml} = 360 \text{ ml} \left(\frac{0.750 \text{ atm}}{1.000 \text{ atm}} \right) = 270 \text{ ml}$$

Example 8.2

At 0°C and 5.00 atm, a given sample of a gas occupies 75.0 liter. The gas is compressed to a final volume of 30.0 liter at 0°C. What is the final pressure?

Solution

Initial conditions:	$V = 75.0$ liter	$P = 5.00$ atm	$t = 0°C$
Final conditions:	$V = 30.0$ liter	$P = ?$ atm	$t = 0°C$

The initial pressure must be corrected for the volume change. The temperature is constant, so no temperature correction is necessary:

$$? \text{ atm} = 5.00 \text{ atm (volume correction)}$$

Volume and pressure are inversely related. Since the volume decreases, the pressure must increase. The volume-correction factor must be a fraction greater than 1. We put the larger volume measurement in the numerator of the factor:

$$? \text{ atm} = 5.00 \text{ atm} \left(\frac{75.0 \text{ liter}}{30.0 \text{ liter}} \right) = 12.5 \text{ atm}$$

8.3 Charles' Law

The relationship between the volume and the temperature of a gas sample was studied by Jacques Charles in 1787. His work was considerably extended by Joseph Gay-Lussac in 1802.

A gas expands when it is heated at constant pressure. Experimental data show that for each Celsius degree rise in temperature, the volume of a gas increases 1/273 of its value at 0°C if the pressure is held constant. A sample of a gas that has a volume of 273 ml at 0°C would expand 1/273 of 273 ml, or 1 ml, for each degree rise in temperature. At 1°C, the volume of the sample would be 274 ml. At 10°C, the volume would have increased by 10 ml to a value of 283 ml. At 273°C, the sample would have expanded 273 ml to a volume of 546 ml, which is double the original volume. These data are recorded in Table 8.1.

Although the volume increases in a regular manner with increase in temperature, the volume is *not* directly proportional to the Celsius temperature. An increase in temperature from 1°C to 10°C, for example, does not increase the volume tenfold but only from 274 ml to 283 ml. An absolute temperature scale, with temperatures measured in kelvins, is defined in such a way that volume *is* directly proportional to kelvin temperature.

A **kelvin reading** (denoted by T) is obtained by adding 273 to the Celsius temperature (denoted by t).

$$T = t + 273$$

Table 8.1 Variation of the volume of a sample of gas with temperature

Volume (ml)	Temperature (°C)	(K)
273	0	273
274	1	274
283	10	283
546	273	546

Note that absolute temperatures are given in kelvins (abbreviated K), not degrees kelvin, and that the degree sign is not used in the abbreviation. Absolute temperatures are listed in the last column in Table 8.1.

The fact that volume is directly proportional to absolute temperature is readily recognized from the data in Table 8.1 since the volume was selected to point up this relationship. For example, when the absolute temperature is doubled (273 K to 546 K), the volume doubles (273 ml to 546 ml). The volume of any sample of a gas varies directly with *absolute* temperature if the pressure is held constant. This generalization is known as Charles' law:

$$V \propto T$$

$$V = kT \tag{8.2}$$

The numerical value of the proportionality constant k depends upon the pressure and the size of the gas sample.

The absolute temperature scale was first proposed by William Thomson, Lord Kelvin, in 1848; the unit is named in his honor. Any absolute measurement scale must be based on a zero point that represents the complete absence of the property being measured. On scales of this type, negative values are impossible. A length given in centimeters is an absolute measurement since 0 cm represents the complete absence of length. One can say that 10 cm is twice 5 cm because these are absolute measurements.

The Celsius temperature scale is not an absolute scale. The zero point, 0°C, is the freezing point of water, not the lowest possible temperature. Negative Celsius temperatures are possible, and doubling the Celsius temperature of a sample of a gas does not double the volume of the gas. On the other hand, the kelvin scale is absolute: 0 K is the lowest possible temperature, and negative kelvin temperatures are as impossible as negative lengths or negative volumes. Doubling the kelvin temperature of a sample of gas doubles the volume.

If volume versus temperature is plotted for a sample of a gas, a straight line results (see Figure 8.4). Since volume is directly proportional to absolute tem-

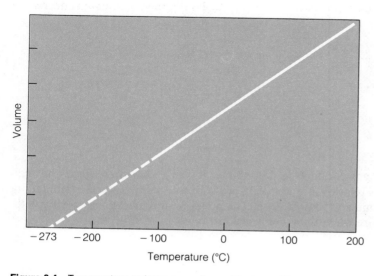

Figure 8.4 Temperature-volume curve for an ideal gas (Charles' law)

perature, the volume of the gas *theoretically* should be zero at absolute zero. Upon cooling, gases liquefy and then solidify before temperatures this low are reached. No substance exists as a gas at a temperature near absolute zero. The straight line temperature-volume curve, however, can be extended to a volume of zero.

The temperature that corresponds to zero volume is $-273.15°C$. The kelvin is the same size as the Celsius degree, but the zero point of the kelvin scale is moved to $-273.15°C$. Exact conversion of a Celsius temperature into kelvins can be accomplished, therefore, by adding 273.15 to the Celsius reading:

$$T = t + 273.15 \tag{8.3}$$

For most problem work, this value can be rounded off to 273 without introducing significant error.

Example 8.3

A sample of a gas has a volume of 79.5 ml at 45°C. What volume will the sample occupy at 0°C when the pressure is held constant?

Solution

We tabulate the data given in the problem. The Celsius temperatures (t) are converted into absolute temperatures (T): $T = t + 273$,

Initial conditions:	$V = 79.5$ ml	$t = 45°C$	$T = 318$ K
Final conditions:	$V = ?$ ml	$t = 0°C$	$T = 273$ K

Therefore,

? ml $= 79.5$ ml (temperature correction)

Since the temperature decreases from 318 K to 273 K, the volume must decrease. A correction factor with a value less than 1 must be used:

$$? \text{ ml} = 79.5 \text{ ml} \left(\frac{273 \text{ K}}{318 \text{ K}}\right) = 68.2 \text{ ml}$$

8.4 Amontons' Law

The pressure of a gas confined in a container increases when the gas is heated. The mathematical relationship between pressure and temperature is similar to that between volume and temperature. The pressure of a gas varies directly with *absolute* temperature when the volume is constant:

$$P \propto T$$

$$P = kT \tag{8.4}$$

In this instance the value of k depends upon the amount of gas considered and its volume.

This generalization is sometimes called **Amontons' law.** In 1703, Guillaume Amontons constructed an air thermometer based on the principle that the pressure of a gas is a measure of the temperature of the gas.

Example 8.4

A 10.0-liter container is filled with a gas to a pressure of 2.00 atm at 0°C. At what temperature will the pressure inside the container be 2.50 atm?

Solution

Initial conditions:	V = 10.0 liter	P = 2.00 atm	T = 273 K
Final conditions:	V = 10.0 liter	P = 2.50 atm	T = ? K

No volume correction is needed since the volume is constant. All temperatures must be expressed in kelvins in any gas problem. Therefore,

$$? \text{ K} = 273 \text{ K (pressure correction)}$$

Pressure varies directly with absolute temperature. The temperature must increase to produce the increase in pressure observed. A factor greater than 1 must be used:

$$? \text{ K} = 273 \text{ K} \left(\frac{2.50 \text{ atm}}{2.00 \text{ atm}} \right) = 341 \text{ K}$$

The answer can then be converted to the Celsius scale:

$$
\begin{aligned}
t &= T - 273 \\
&= 341 \text{ K} - 273 \text{ K} = 68°C
\end{aligned}
$$

8.5 Ideal Gas Law

The volume of a gas, at fixed temperature and pressure, varies directly with the number of moles of gas considered. Obviously, 1 mol of gas (a gram molecular weight) occupies half the volume that 2 mol occupy when the temperature and pressure of both samples are the same. Furthermore, the volume of 1 mol of a given gas is the same as the volume of 1 mol of any other gas if the volumes are measured at the same temperature and pressure (Avogadro's principle, described in Section 8.8). If n is the number of moles of gas,

$$V \propto n$$

or

$$V = kn \tag{8.5}$$

1. Tabulate the data given in the problem. List the *initial conditions* (V, P, and t) and the *final conditions* (V, P, and t).

2. Convert temperature readings that are given in degrees Celsius (t) into kelvins (T). $T = t + 273.15$. For most problem work 273 is acceptable.

3. The solution consists of finding the *final* value of one of the three variables (V, P or T) by correcting the *initial* value of this variable. Multiply the initial value by correction factors to correct for changes in the other two variables.

4. Consider each correction separately. A correction factor consists of a fraction derived from the initial and final values of the same variable (V, P, or T). One of these values is placed in the numerator of the factor, the other in the denominator. Two fractions, therefore, can be derived—one numerically greater than 1, the other, less than 1. Decide whether the change being considered should result in an increase or a decrease in the value being corrected. On this basis, select the fraction to be employed as the correction factor.

5. Since the units in the numerator and denominator of a correction factor are the same, they cancel. The answer has the same units as the value being corrected.

6. If a temperature is sought, it will be obtained in kelvins. The Celsius equivalent may be found at the end of the process: $t = T - 273.15$.

The numerical value of the proportionality constant, k, depends upon the temperature and pressure of the gas.

The relationship can be combined with expressions for Boyle's law and Charles' law to give a general equation relating volume, temperature, pressure, and number of moles. Volume is inversely proportional to pressure and directly proportional to absolute temperature and to number of moles:

$$V \propto \frac{1}{P} \qquad V \propto T \qquad V \propto n$$

Therefore,

$$V \propto \left(\frac{1}{P}\right)(T)(n)$$

The proportionality can be changed into an equality by the use of a constant. In this case, the constant is given the designation R:

$$V = R\left(\frac{1}{P}\right)(T)(n)$$

Rearrangement gives

Table 8.2 Values of the ideal gas constant, R, in various units	
R	Units
8.2056×10^{-2}	liter·atm/(K·mol)
8.3143×10^{3}	liter·Pa/(K·mol)
8.3143	J/(K·mol)[a]

[a] Since the joule is a newton meter (1 J = 1 N·m), this value of R is used when P is expressed in pascals (1 Pa = 1 N/m²), V in cubic meters (m³), n in moles (mol), and T in kelvins (K). This label is the equivalent, therefore, to m³·Pa/(K·mol).

$$PV = nRT \tag{8.6}$$

Under ordinary conditions of temperature and pressure, most gases conform well to the behavior described by this equation. Deviations occur, however, under extreme conditions (low temperature and high pressure; Section 8.13). A hypothetical gas that follows the behavior described by the equation exactly, under all conditions, is called an **ideal gas**. The equation is known, therefore, as the **equation of state for an ideal gas**.

By convention, **standard temperature and pressure (STP)** are defined as 0°C (which is 273.15 K) and exactly 1 atm pressure. The volume of 1 mol of an ideal gas at STP, derived from experimental measurements, is 22.4136 liter. These data can be used to evaluate the ideal gas constant, R. Solution of the equation of state for R yields

$$R = \frac{PV}{nT}$$

Substitution of the data for the STP molar volume of an ideal gas gives

$$R = \frac{(1 \text{ atm})(22.4136 \text{ liter})}{(1 \text{ mol})(273.15 \text{ K})} = 0.082056 \text{ liter·atm/(K·mol)}$$

When this value of R is employed, volume must be expressed in liters, pressure in atmospheres, and temperature in kelvins. Values of R in other units appear in Table 8.2.

The number of moles of gas in a sample, n, is equal to the mass of the sample, g, divided by the molecular weight of the gas, M:

$$n = \frac{g}{M}$$

Substitution of (g/M) for n in $PV = nRT$ gives

$$PV = \left(\frac{g}{M}\right)RT \tag{8.7}$$

Many problems can be solved by use of this form of the equation of state.

Example 8.5

The volume of a sample of a gas is 462 ml at 35°C and 1.15 atm. Calculate the volume of the sample at STP.

Solution

| Initial conditions: | $V = 462$ ml | $T = 308$ K | $P = 1.15$ atm |
| Final conditions: | $V = ?$ ml | $T = 273$ K | $P = 1.00$ atm |

The correction factor approach can be used to solve the problem:

? ml = 462 ml (temperature correction)(pressure correction)

Each of these corrections is considered separately. First, the decrease in temperature, from 308 K to 273 K, causes the volume to decrease by a factor of (273 K/ 308 K). Second, the decrease in pressure, from 1.15 atm to 1.00 atm, causes the volume to increase by a factor of (1.15 atm/1.00 atm) since pressure and volume are inversely related:

$$? \text{ ml} = 462 \text{ ml} \left(\frac{273 \text{ K}}{308 \text{ K}}\right) \left(\frac{1.15 \text{ atm}}{1.00 \text{ atm}}\right) = 471 \text{ ml}$$

Example 8.6

At what pressure will 0.250 mol of $N_2(g)$ occupy 10.0 liter at 100°C?

Solution

$$P = ? \text{ atm} \qquad V = 10.0 \text{ liter} \qquad n = 0.250 \text{ mol} \qquad T = 373 \text{ K}$$

Problems in which only one set of conditions are stated are readily solved by substitution in the equation of state:

$$PV = nRT$$

$$P(10.0 \text{ liter}) = (0.250 \text{ mol})[0.0821 \text{ liter} \cdot \text{atm}/(\text{K} \cdot \text{mol})](373 \text{ K})$$

$$P = 0.766 \text{ atm}$$

Example 8.7

How many moles of CO are present in a 500 ml sample of CO(g) collected at 50°C and 1.50 atm?

Solution

The units of the values substituted into $PV = nRT$ must correspond to the units in which R is expressed. Therefore, we express the volume in liters and the temperature in kelvins:

$$PV = nRT$$

$$(1.50 \text{ atm})(0.500 \text{ liter}) = n[0.0821 \text{ liter} \cdot \text{atm}/(\text{K} \cdot \text{mol})](323 \text{ K})$$

$$n = 0.0283 \text{ mol}$$

Example 8.8

What volume will 10.0 g of $CO_2(g)$ occupy at 27°C and 2.00 atm?

Solution

The problem can be solved in one step if (g/M) is substituted for n in the equation of state. The molecular weight of CO_2 is 44.0:

$$PV = \left(\frac{g}{M}\right) RT$$

$$(2.00 \text{ atm})V = \left(\frac{10.0 \text{ g}}{44.0 \text{ g/mol}}\right)[0.0821 \text{ liter} \cdot \text{atm}/(\text{K} \cdot \text{mol})](300 \text{ K})$$

$$V = 2.80 \text{ liter}$$

Example 8.9

What is the density of $NH_3(g)$ at 100°C and 1.15 atm?

Solution

The density of a gas is the number of grams of gas in 1 liter. We can find the density, therefore, by setting V equal to 1.00 liter and solving for g. The molecular weight of NH_3 is 17.0:

$$PV = \left(\frac{g}{M}\right) RT$$

$$(1.15 \text{ atm})(1.00 \text{ liter}) = \left(\frac{g}{17.0 \text{ g/mol}}\right)[0.0821 \text{ liter} \cdot \text{atm}/(\text{K} \cdot \text{mol})](373 \text{ K})$$

$$g = 0.638 \text{ g}$$

The density is 0.638 g/liter.

Example 8.10

(a) Cyclopropane is a gas that is used as a general anesthetic. The gas has a density of 1.50 g/liter at 50°C and 0.948 atm. What is the molecular weight of cyclopropane? (b) The empirical formula of cyclopropane is CH_2. What is the molecular formula of the compound?

Solution

(a) Since the density is 1.50 g/liter, 1.50 g of the gas will occupy 1.00 liter under the conditions specified:

$$PV = \left(\frac{g}{M}\right) RT$$

$$(0.948 \text{ atm})(1.00 \text{ liter}) = \left(\frac{1.50 \text{ g}}{M}\right)[0.0821 \text{ liter} \cdot \text{atm}/(\text{K} \cdot \text{mol})](323 \text{ K})$$

$$M = 42.0 \text{ g/mol}$$

(b) The formula weight of the empirical formula, CH_2, is 14.0. If we divide this formula weight into the molecular weight, we get $(42.0/14.0) = 3$. There are, therefore, three times as many atoms in the molecule as are indicated by the empirical formula. The molecular formula of cyclopropane is C_3H_6.

8.6 Kinetic Theory of Gases

The kinetic theory of gases provides a model to explain the regularity that is observed in the behavior of all gases. In 1738, Daniel Bernoulli explained Boyle's law by assuming that the pressure of a gas results from the collisions of gas molecules with the walls of the container. Bernoulli's explanation constitutes an early, and simple, expression of key aspects of the kinetic theory. The theory was enlarged and developed in the middle of the nineteenth century by many scientists, notably Krönig, Clausius, Maxwell, and Boltzmann.

The kinetic theory of gases includes the following postulates:

1. Gases consist of molecules widely separated in space. The actual volume of the individual molecules is negligible in comparison to the total volume of the gas as a whole. The word *molecule* is used here to designate the smallest particle of any gas; some gases (for example, the noble gases) consist of uncombined atoms.

2. Gas molecules are in constant, rapid, straight-line motion; they collide with each other and with the walls of the container. Although energy may be transferred from one molecule to another in these collisions, no kinetic energy (which is energy of motion) is lost.

3. The average kinetic energy of the molecules of a gas depends upon the temperature, and it increases as the temperature increases. At a given temperature, the molecules of all gases have the same average kinetic energy.

4. The attractive forces between gas molecules are negligible.

The gas laws can be explained by the kinetic theory. Consider *Boyle's law*. According to the theory, gas pressure is caused by molecular collisions with the walls of the container. If the number of molecules per unit volume (the molecular concentration) is increased, a higher pressure will result because of the larger number of collisions per unit time. Reducing the volume of a gas crowds the molecules into a smaller space, which produces a higher molecular concentration and a proportionally higher pressure.

Charles' law and *Amontons' law* relate the properties of gases to changes in temperature. The average kinetic energy of the molecules of a gas is proportional to the absolute temperature. At absolute zero, the kinetic energy of the molecules is theoretically zero; the molecules are at rest. Since the volume of the molecules of an ideal gas is negligible, the volume of an ideal gas at absolute zero is theoretically zero.

As the temperature is increased, the molecules move at increasing speeds. The collisions of the gas molecules with the walls of the container become increasingly more vigorous and more frequent. As a result, the pressure increases in the manner described by Amontons' law.

The pressure of a gas that is being heated can be held constant if the gas is allowed to expand. The increasing volume keeps the pressure constant by reducing the number of collisions that the molecules make with the container walls in a given time. In this way, the decreasing frequency of the collisions compensates for the increasing intensity of the collisions. Charles' law describes this situation.

8.7 Derivation of the Ideal Gas Law from the Kinetic Theory

The equation of state for an ideal gas may be derived as follows. Consider a sample of gas that contains N molecules that have a mass of m each. If this sample is enclosed in a cube l cm on a side, the total volume of the gas is l^3 cm^3. Although the molecules are moving in every possible direction, the derivation is simplified if we assume that one-third of the molecules ($\frac{1}{3}N$) are moving in the direction of the x axis, one-third in the y direction, and one-third in the z direction. For a very large number of molecules, this is a valid simplification since the velocity of each molecule may be divided into an x component, a y component, and a z component.

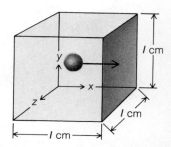

Figure 8.5 Derivation of the ideal gas law

The pressure of the gas on any wall (surface area l^2 cm^2) is due to the impacts of the molecules on that wall. The force of each impact can be calculated from the change in momentum per unit time. Consider the shaded wall in Figure 8.5 and take into account only those molecules moving in the direction of the x axis. A molecule moving in this direction will strike this wall every $2 \times l$ cm of its path, since after an impact it must go to the opposite wall (a distance of l cm) and return (a distance of l cm) before the next impact. If the molecule is moving with a velocity of u cm/s, in one second it will have gone u cm and have made $u/2l$ collisions with the wall under consideration.

Momentum is mass times velocity. Before an impact, the momentum of the molecule is mu; after an impact, the momentum is $-mu$ (the sign is changed because the direction is changed—velocity takes into account speed and direction). Therefore, the *change* in momentum equals $2\,mu$.

In one second a molecule makes $u/2l$ collisions, and the change in momentum per collision is $2mu$. Therefore, in one second the total change in momentum per molecule is

$$\left(\frac{u}{2l}\right)2mu = \frac{mu^2}{l}$$

The total change in momentum (force) for all of the molecules striking the wall in one second is

$$\frac{N}{3} \times \frac{mu^2}{l}$$

In this expression, u^2 is the average of the squares of all the molecular velocities.

Pressure is force per unit area, and the area of the wall is l^2 cm². Therefore, the pressure on the wall under consideration is

$$P = \frac{Nmu^2}{3l} \times \frac{1}{l^2} = \frac{Nmu^2}{3l^3}$$

Since the volume of the cube is l^3 cm³, $V = l^3$, and

$$P = \frac{Nmu^2}{3V} \qquad \text{or} \qquad PV = \tfrac{1}{3}Nmu^2 \tag{8.8}$$

This equation can be written

$$PV = (\tfrac{2}{3}N)(\tfrac{1}{2}mu^2)$$

The kinetic energy of any body is one-half the product of its mass times the square of its speed. The average molecular kinetic energy, therefore, is $\tfrac{1}{2}mu^2$. By substitution:

$$PV = \tfrac{2}{3}N(KE) \tag{8.9}$$

The average molecular kinetic energy, KE, is directly proportional to the absolute temperature, T; the number of molecules, N, is proportional to the number of moles, n. Substitution of these terms requires the inclusion of a constant since these are proportionalities, not equalities. The required constant can be combined with the $\tfrac{2}{3}$ (also a constant); thus,

$$PV = nRT \tag{8.6}$$

8.8 Gay-Lussac's Law of Combining Volumes and Avogadro's Principle

In 1808, Joseph Gay-Lussac reported the results of his experiments with reacting gases. When measured at constant temperature and pressure, the volumes of gases that are used or produced in a chemical reaction can be expressed in ratios of small whole numbers. This statement is called Gay-Lussac's law of combining volumes.

One of the reactions that Gay-Lussac studied is the reaction in which hydrogen chloride *gas* is produced from hydrogen *gas* and chlorine *gas*. If the volumes of all the gases are measured at the same temperature and pressure,

1 volume hydrogen + 1 volume chlorine $\longrightarrow$ 2 volumes hydrogen chloride

For example,

10 liter hydrogen + 10 liter chlorine $\longrightarrow$ 20 liter hydrogen chloride

Gay-Lussac did not know the formulas for these substances and did not write chemical equations. The volume ratio, however, is given by the coefficients of the gases in the chemical equation

$$H_2(g) + Cl_2(g) \longrightarrow 2\,HCl(g)$$

The law is applicable only to gases. The volumes of solids and liquids cannot be treated in this way.

The explanation for the law of combining volumes was proposed in 1811 by Amedeo Avogadro. Avogadro's principle states: equal volumes of all gases at the same temperature and pressure contain the same number of molecules.

Gay-Lussac's data show that equal volumes of $H_2(g)$ and $Cl_2(g)$ react. Since the reaction requires equal numbers of H_2 and Cl_2 molecules, a given volume of either gas must contain the same number of molecules. According to Gay-Lussac, the volume of $HCl(g)$ produced is twice the volume of $H_2(g)$ used. The equation shows that the number of HCl molecules produced is twice the number of H_2 molecules used. We conclude that the number of HCl molecules in a "volume" is the same as the number of H_2 molecules in a "volume." The same comparison can be made between HCl and Cl_2.

The total volume of the reacting gases need not equal the volume of the gases produced. This fact is illustrated by another of Gay-Lussac's examples:

$$2\,CO(g) + O_2(g) \longrightarrow 2\,CO_2(g)$$

The volume ratio for this reaction is

$$2 \text{ volumes } CO(g) + 1 \text{ volume } O_2(g) \longrightarrow 2 \text{ volumes } CO_2(g)$$

The relative numbers of molecules involved in the reaction are given by the chemical equation. If we take $2x$ CO molecules, then

$$2x \text{ molecules } CO(g) + x \text{ molecules } O_2(g) \longrightarrow 2x \text{ molecules } CO_2(g)$$

By comparing these two statements, we see that one "volume" of any of the gases contains the same number of molecules, x.

Example 8.11

(a) What volume of oxygen is required for the complete combustion of 15.0 liter of ethane, $C_2H_6(g)$, if all gases are measured at the same temperature and pressure, according to the following equation:

$$2\,C_2H_6(g) + 7\,O_2(g) \longrightarrow 4\,CO_2(g) + 6\,H_2O(g)$$

(b) What volume of carbon dioxide gas is produced?

Solution

(a) The relationship between the volume of $C_2H_6(g)$ and the volume of $O_2(g)$ is given by the coefficients of the chemical equation

Amedeo Avogadro, 1776–1856. *California Institute of Technology Archives.*

$$2 \text{ liter } C_2H_6 = 7 \text{ liter } O_2$$

This relationship is used to derive a conversion factor:

$$? \text{ liter } O_2 = 15.0 \text{ liter } C_2H_6 \left(\frac{7 \text{ liter } O_2}{2 \text{ liter } C_2H_6} \right) = 52.5 \text{ liter } O_2$$

(b) In this case, the relationship is

$$2 \text{ liter } C_2H_6 = 4 \text{ liter } CO_2$$

Therefore,

$$? \text{ liter } CO_2 = 15.0 \text{ liter } C_2H_6 \left(\frac{4 \text{ liter } CO_2}{2 \text{ liter } C_2H_6} \right) = 30.0 \text{ liter } CO_2$$

According to Avogadro's principle, equal volumes of any two gases at the same temperature and pressure contain the same number of molecules. Conversely, equal numbers of molecules of any two gases under the same conditions of temperature and pressure will occupy equal volumes. A mole of a substance contains 6.022×10^{23} molecules (Avogadro's number, Section 2.4). A mole of a gas, therefore, should occupy the same volume as a mole of any other gas if they are both measured under the same conditions of temperature and pressure. At STP this volume, called the **STP molar volume of a gas**, is 22.414 liters. The molecular weight of a gas, therefore, is the mass, in grams, of 22.4 liter of the gas at STP. For most gases, the deviation from this ideal is less than 1%.

The STP molar volume of a gas can be used to solve some problems. All such calculations, however, can also be made by using the equation of state, $PV = nRT$.

Example 8.12

What is the density of fluorine gas, $F_2(g)$, at STP?

Solution

The molecular weight of F_2 is 38.0:

$$1 \text{ mol } F_2 = 38.0 \text{ g } F_2$$

At STP the volume of a mole of any gas is 22.4 liter:

$$1 \text{ mol } F_2 = 22.4 \text{ liter } F_2$$

Therefore,

$$? \text{ g } F_2 = 1 \text{ liter } F_2 \left(\frac{1 \text{ mol } F_2}{22.4 \text{ liter } F_2} \right) \left(\frac{38.0 \text{ g } F_2}{1 \text{ mol } F_2} \right) = 1.70 \text{ g } F_2$$

The density is 1.70 g/liter.

Example 8.13

What is the molecular weight of a gas that has a density of 1.34 g/liter at STP?

Solution

$$? \, g = 1 \, \text{mol} \left(\frac{22.4 \, \text{liter}}{1 \, \text{mol}} \right) \left(\frac{1.34 \, g}{1 \, \text{liter}} \right) = 30.0 \, g$$

The molecular weight of the gas is 30.0 g/mol.

8.9 Stoichiometry and Gas Volumes

Stoichiometric problems may be based on the volumes of gases that are involved in a chemical reaction. Gay-Lussac's law of combining volumes is used to solve problems that deal with the volumes of two gases (see Section 8.8). Some problems concern the relationship between the *volume* of a gas and the *mass* of another substance. Examples of this type of problem follow. The clue to their solution is, as usual, the mole.

Example 8.14

A 0.400 g sample of sodium azide, $NaN_3(s)$, is heated and decomposes:

$$2 \, NaN_3(s) \longrightarrow 2 \, Na(s) + 3 \, N_2(g)$$

What volume of $N_2(g)$, measured at 25°C and 0.980 atm, is obtained?

Solution

We find the number of moles of NaN_3 in the sample. Since

$$1 \, \text{mol} \, NaN_3 = 65.0 \, g \, NaN_3$$

$$? \, \text{mol} \, NaN_3 = 0.400 \, g \, NaN_3 \left(\frac{1 \, \text{mol} \, NaN_3}{65.0 \, g \, NaN_3} \right) = 0.00615 \, \text{mol} \, NaN_3$$

From the chemical equation, we derive

$$2 \, \text{mol} \, NaN_3 = 3 \, \text{mol} \, N_2$$

Therefore,

$$? \, \text{mol} \, N_2 = 0.00615 \, \text{mol} \, NaN_3 \left(\frac{3 \, \text{mol} \, N_2}{2 \, \text{mol} \, NaN_3} \right) = 0.00923 \, \text{mol} \, N_2$$

We find the volume of $N_2(g)$ by using the equation of state:

$$PV = nRT$$

$$(0.980 \text{ atm})V = (0.00923 \text{ mol})[0.0821 \text{ liter} \cdot \text{atm}/(K \cdot \text{mol})](298 \text{ K})$$

$$V = 0.230 \text{ liter}$$

Example 8.15

How many liters of CO(g), measured at STP, are needed to reduce 1.00 kg of $Fe_2O_3(s)$? The chemical equation is

$$Fe_2O_3(s) + 3\,CO(g) \longrightarrow 2\,Fe(s) + 3\,CO_2(g)$$

Calculations That Are Based on Chemical Equations and That Involve Gas Volumes

The following types of problems may be encountered:

1. A *volume* of gas A is given and the *volume* of gas B is sought.

   a. *The volumes of both gases are measured under the same conditions of temperature and pressure.* Use Gay-Lussac's law of combining volumes. (See Example 8.11.)

   b. *The volumes of the two gases are measured under different conditions.* Use Gay-Lussac's law of combining volumes to find the volume of gas B under the conditions given for gas A. Use correction factors to correct this volume of gas B so that it conforms to the final conditions given in the problem.

2. A *mass* of substance A is given and the *volume* of gas B is sought.

   a. Find the number of moles of A.

   b. Use this number of moles of A to find the number of moles of B. The mole relationship of B to A is given by the chemical equation.

   c. Find the volume of gas B by substituting in $PV = nRT$; n is the number of moles found in step (b); P and T are the conditions under which the volume of B is measured. (See Examples 8.14 and 8.15.)

3. A *volume* of gas A is given and the *mass* of substance B is sought.

   a. Determine the number of moles of gas A by using $PV = nRT$.

   b. Use this number of moles of A to find the number of moles of B. The mole relationship of B to A is given by the chemical equation.

   c. Find the mass of B from the number of moles of B determined in step (b). One mole of B is the molecular weight of B in grams. (See Example 8.16.)

Solutions to problems of types 2 and 3 may be simplified if the gas volumes are measured at STP. The volume of 1 mol of any gas at STP is 22.4 liters. This relationship can be used in steps 2(c) and 3(a) in place of $PV = nRT$. (See Examples 8.15 and 8.16.)

Solution

We first find the number of moles of Fe_2O_3 in 1000 g of Fe_2O_3. Since

$$1 \text{ mol } Fe_2O_3 = 159.6 \text{ g } Fe_2O_3$$

$$? \text{ mol } Fe_2O_3 = 1000 \text{ g } Fe_2O_3 \left(\frac{1 \text{ mol } Fe_2O_3}{159.6 \text{ g } Fe_2O_3} \right) = 6.27 \text{ mol } Fe_2O_3$$

From the equation, we see that

$$1 \text{ mol } Fe_2O_3 = 3 \text{ mol } CO$$

Therefore,

$$? \text{ mol } CO = 6.27 \text{ mol } Fe_2O_3 \left(\frac{3 \text{ mol } CO}{1 \text{ mol } Fe_2O_3} \right) = 18.8 \text{ mol } CO$$

At STP,

$$1 \text{ mol } CO = 22.4 \text{ liter } CO$$

Therefore,

$$? \text{ liter } CO = 18.8 \text{ mol } CO \left(\frac{22.4 \text{ liter } CO}{1 \text{ mol } CO} \right) = 421 \text{ liter } CO$$

The last step could have been solved by using the equation of state:

$$PV = nRT$$

$$(1 \text{ atm})V = (18.8 \text{ mol})[0.0821 \text{ liter} \cdot \text{atm}/(K \cdot \text{mole})](273 \text{ K})$$

$$V = 421 \text{ liter}$$

Example 8.16

How many grams of Fe are needed to produce 100 liter of $H_2(g)$, measured at STP? The equation is

$$3 \, Fe(s) + 4 \, H_2O \longrightarrow Fe_3O_4(s) + 4 \, H_2(g)$$

Solution

First, the number of moles of H_2 is found. At STP,

$$1 \text{ mol } H_2 = 22.4 \text{ liter } H_2$$

$$? \text{ mol } H_2 = 100 \text{ liter } H_2 \left(\frac{1 \text{ mol } H_2}{22.4 \text{ liter } H_2} \right) = 4.46 \text{ mol } H_2$$

(The last step could also be accomplished by substituting in $PV = nRT$.) From the equation,

$$3 \text{ mol } Fe = 4 \text{ mol } H_2$$

we find,

$$? \text{ mol Fe} = 4.46 \text{ mol H}_2 \left(\frac{3 \text{ mol Fe}}{4 \text{ mol H}_2} \right) = 3.35 \text{ mol Fe}$$

The atomic weight of Fe is 55.8, therefore

$$? \text{ g Fe} = 3.35 \text{ mol Fe} \left(\frac{55.8 \text{ g Fe}}{1 \text{ mol Fe}} \right) = 187 \text{ g Fe}$$

8.10 Dalton's Law of Partial Pressures

The behavior of a mixture of gases that do not react with one another is frequently of interest. The pressure that a component of such a mixture would exert if it were the only gas present in the volume under consideration is the **partial pressure** of the component. **Dalton's law of partial pressures** (1801) states that the total pressure of a mixture of gases that do not react is equal to the sum of the partial pressures of all the gases present. If the total pressure is P_{total} and the partial pressures are $p_A, p_B, p_C, \ldots,$

$$P_{total} = p_A + p_B + p_C + \cdots \tag{8.10}$$

Suppose that 1 liter of gas A at 0.2 atm pressure and 1 liter of gas B at 0.4 atm pressure are mixed. If the *final volume is 1 liter* and the temperature is constant, the pressure of the mixture will be 0.6 atm.

According to the kinetic theory, the molecules of gas A have the same average kinetic energy as the molecules of gas B since the two gases are at the same temperature. Furthermore, the kinetic theory assumes that the gas molecules do not attract one another if the gases do not react chemically. Consequently, the act of mixing two or more gases does not change the average kinetic energy of any of the gases. Each gas exerts the same pressure that it would exert if it were the only gas present in the container.

If n_A mol of gas A and n_B mol of gas B are mixed, the total number of moles of gas in the mixture is $(n_A + n_B)$. The ratio of the number of moles of A to the total number of moles present is called the **mole fraction** of A, X_A:

$$X_A = \frac{n_A}{n_A + n_B} = \frac{n_A}{n_{total}} \tag{8.11}$$

The fraction of the total pressure that is due to gas A is given by the mole fraction of A. The partial pressure of A, therefore, is

$$p_A = \left(\frac{n_A}{n_A + n_B} \right) P_{total} = X_A P_{total} \tag{8.12}$$

The partial pressure of B is equal to the mole fraction of B times the total pressure:

$$p_B = \left(\frac{n_B}{n_A + n_B} \right) P_{total} = X_B P_{total} \tag{8.13}$$

Notice that the sum of the mole fraction is one:

$$X_A + X_B = 1$$

$$\frac{n_A}{n_A + n_B} + \frac{n_B}{n_A + n_B} = \frac{n_A + n_B}{n_A + n_B} = 1$$

Suppose that a mixture contains 1 mol of A and 4 mol of B. Then, the total number of moles is five, the mole fraction of A is one-fifth, and the mole fraction of B is four-fifths. The partial pressure of A, therefore, is one-fifth of the total pressure, and the partial pressure of B is four-fifths of the total pressure.

A gas evolved in the course of a laboratory experiment is frequently collected over water if it is not very soluble in water. The gas is conducted into an inverted bottle that has been filled with water. The gas displaces the water, and the collected gas is mixed with water vapor. The total pressure of the mixture is the sum of the partial pressure of the gas and the partial pressure of the water vapor. In Figure 8.6, the total pressure is equal to the barometric pressure since the water stands at the same level inside the bottle as outside it. The pressure of the dry gas is found by subtracting the vapor pressure of water at the temperature of the experiment (Table 8.3) from the barometric pressure.

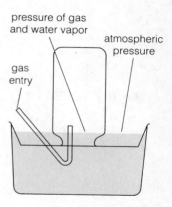

Figure 8.6 Schematic diagram of the collection of gas over water

Table 8.3	Vapor pressure of water					
Temperature (°C)	Pressure (atm)	(torr)	Temperature (°C)	Pressure (atm)	(torr)
0	0.0060	4.6	25	0.0313	23.8
1	0.0065	4.9	26	0.0332	25.2
2	0.0070	5.3	27	0.0352	26.7
3	0.0075	5.7	28	0.0373	28.3
4	0.0080	6.1	29	0.0395	30.0
5	0.0086	6.5	30	0.0419	31.8
6	0.0092	7.0	31	0.0443	33.7
7	0.0099	7.5	32	0.0470	35.7
8	0.0106	8.0	33	0.0496	37.7
9	0.0113	8.6	34	0.0525	39.9
10	0.0121	9.2	35	0.0555	42.2
11	0.0130	9.8	40	0.0728	55.3
12	0.0138	10.5	45	0.0946	71.9
13	0.0148	11.2	50	0.122	92.5
14	0.0158	12.0	55	0.155	118.0
15	0.0168	12.8	60	0.197	149.4
16	0.0179	13.6	65	0.247	187.5
17	0.0191	14.5	70	0.308	233.7
18	0.0204	15.5	75	0.380	289.1
19	0.0217	16.5	80	0.467	355.1
20	0.0231	17.5	85	0.571	433.6
21	0.0245	18.7	90	0.692	525.8
22	0.0261	19.8	95	0.834	633.9
23	0.0277	21.1	100	1.000	760.0
24	0.0294	22.4	105	1.192	906.1

Example 8.17

A 370 ml sample of oxygen is collected over water at 23°C and a barometric pressure of 0.992 atm. What volume would this sample occupy dry and at STP?

Solution

The vapor pressure of water at 23°C is 0.0277 atm. Consequently, the initial pressure of the oxygen is

$$0.992 \text{ atm} - 0.028 \text{ atm} = 0.964 \text{ atm}$$

Therefore,

Initial conditions:	$V = 370$ ml	$P = 0.964$ atm	$T = 296$ K
Final conditions:	$V = ?$ ml	$P = 1.000$ atm	$T = 273$ K

$$? \text{ ml} = 370 \text{ ml} \left(\frac{0.964 \text{ atm}}{1.000 \text{ atm}} \right) \left(\frac{273 \text{ K}}{296 \text{ K}} \right) = 329 \text{ ml}$$

Example 8.18

A mixture of 40.0 g of oxygen and 40.0 g of helium has a total pressure of 0.900 atm. What is the partial pressure of oxygen?

Solution

The molecular weight of O_2 is 32.0. Forty grams of O_2 is 40.0/32.0, or 1.25, mol of O_2. Helium is a monatomic gas. Its atomic weight is 4.00. Therefore, 40.0/4.0, or 10.0, mol of He is present in the mixture. Therefore,

$$X_{O_2} = \frac{n_{O_2}}{n_{O_2} + n_{He}}$$

$$X_{O_2} = \frac{1.25 \text{ mol}}{(1.25 + 10.0) \text{ mol}} = \frac{1.25 \text{ mol}}{11.2 \text{ mol}} = 0.112$$

The partial pressure of O_2 is

$$\begin{aligned} P_{O_2} &= X_{O_2} P_{total} \\ &= (0.112)(0.900 \text{ atm}) \\ &= 0.101 \text{ atm} \end{aligned}$$

8.11 Molecular Speeds

In Section 8.7 we derived the expression

$$PV = \tfrac{1}{3} Nmu^2$$

For one mole of a gas, the number of molecules, N, is Avogadro's number, and N times the mass of a single molecule, m, is the molecular weight, M:

$$PV = \tfrac{1}{3} Mu^2 \tag{8.8}$$

Also for one mole, $PV = RT$; thus,

$$RT = \frac{1}{3}Mu^2$$

Rearranging and solving for the molecular speed, we obtain

$$u = \sqrt{\frac{3RT}{M}} \qquad (8.14)$$

The speed, u, in this equation, as in previous equations, is the root-mean-square speed. The value of the root-mean-square speed is obtained by taking the square root of the average of the squares of all of the molecular speeds; it is the speed of a molecule that possesses average kinetic energy at the temperature under consideration.

In order to solve the equation for the root-mean-square speed, R must be expressed in appropriate units. If u is to be obtained in m/s and M is expressed in g/mol, the appropriate value of R is 8.3143×10^3 g·m^2/(s^2·K·mol). If M is expressed in kg/mol, R is 8.3143 J/(K·mol).

The speed of a hydrogen molecule that has the average kinetic energy at 0°C is 1.84×10^3 m/s. The diffusion of one gas through another gas, however, does not occur this rapidly. Although a given molecule travels at a high speed, its direction is continually being changed through collisions with other molecules. At 1 atm pressure and 0°C a hydrogen molecule, on the average, undergoes about 1.4×10^{10} collisions in one second. The average distance traveled between collisions is only 1.3×10^{-5} cm; this value is called the mean free path of hydrogen.

Not all of the molecules of a gas have the same kinetic energy and travel at the same speed. Since energy can be exchanged in these collisions, the speed as well as the direction of a molecule changes continually. In any sample of a gas, however, there is a large number of molecules, so that the molecular speeds are distributed over a range in a definite manner.

Distributions of the molecular speeds of a gas, called Maxwell-Boltzmann distributions, at two temperatures are shown in Figure 8.7. The fraction of the

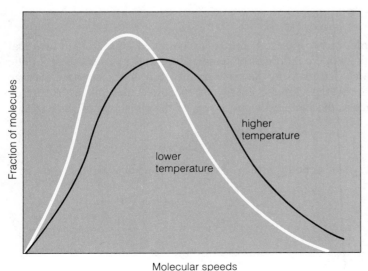

Figure 8.7 Distribution of molecular speeds

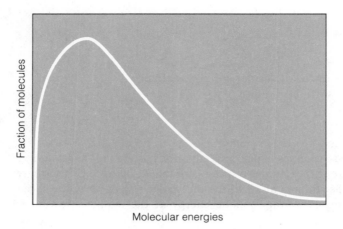

Figure 8.8 Distribution of molecular energies

total number of molecules that has a particular speed is plotted against molecular speed. Each curve has a maximum, and the speed corresponding to the maximum is the most probable speed for that distribution. In other words, more molecules move at this speed than at any other. Comparatively few molecules possess very high or very low speeds.

When the temperature of a gas is increased, the curve broadens and shifts toward higher speeds. Fewer molecules than previously move at the lower speeds and more molecules move at the higher speeds. The addition of heat has caused the molecules, on the average, to move faster. The speed of a hydrogen molecule that has the average kinetic energy at 0°C is 1.84×10^3 m/s; at 100°C, it is 2.15×10^3 m/s.

A typical curve for the distribution of molecular energies is shown in Figure 8.8. Distribution curves of this type can be drawn for liquids and solids as well as gases.

8.12 Graham's Law of Effusion

Suppose that samples of two gases, A and B, are confined separately in identical containers under the same conditions of temperature and pressure. The kinetic theory of gases states that gases at the same temperature have the same average kinetic energy. The average kinetic energy of the molecules of gas A (KE_A), therefore, is the same as the average kinetic energy of the molecules of gas B (KE_B):

$$KE_A = KE_B$$

The kinetic energy of a body with a mass m moving in a straight line at a constant speed u is

$$KE = \tfrac{1}{2}mu^2$$

Therefore,

$$KE_A = \tfrac{1}{2}m_A u_A^2 \qquad \text{and} \qquad KE_B = \tfrac{1}{2}m_B u_B^2$$

The molecules of gas A (or gas B) do not all move at the same speed. The symbol u_A (as well as u_B) stands for the speed of a molecule that has the average kinetic energy. Since

$$KE_A = KE_B$$
$$\tfrac{1}{2}m_A u_A^2 = \tfrac{1}{2}m_B u_B^2$$

or

$$m_A u_A^2 = m_B u_B^2$$

By rearranging the equation, we get

$$\frac{u_A^2}{u_B^2} = \frac{m_B}{m_A}$$

Extracting the square root of both sides of the equation gives us

$$\frac{u_A}{u_B} = \sqrt{\frac{m_B}{m_A}}$$

The ratio of the molecular masses, m_B/m_A, is the same as the ratio of the molecular weights, M_B/M_A. Therefore,

$$\frac{u_A}{u_B} = \sqrt{\frac{M_B}{M_A}}$$

Now let us suppose that each container has an identical, extremely small opening (called an orifice) in it. Gas molecules will escape through these orifices; the process is called **molecular effusion.** The rate of effusion, r, is equal to the rate at which molecules strike the orifice, which in turn is proportional to molecular speed, u. Molecules that move rapidly will effuse at a faster rate than slower moving molecules. The ratio u_A/u_B, therefore, is the same as the ratio of the effusion rates, r_A/r_B:

$$\frac{r_A}{r_B} = \sqrt{\frac{M_B}{M_A}} \tag{8.15}$$

This equation is an expression of **Graham's law of effusion,** which Thomas Graham experimentally derived during the period 1828 to 1833.

The relationship may also be expressed in terms of *gas densities*. Since the density of a gas, d, is proportional to the molecular weight of the gas, M (see Example 8.13), Graham's law may also be written

$$\frac{r_A}{r_B} = \sqrt{\frac{d_B}{d_A}} \tag{8.16}$$

It is not surprising that the lighter of two molecules with the same kinetic energy will effuse more rapidly than the heavier one. (Notice the *inverse* relationship.) The molecular weight of O_2 is 32, and the molecular weight of H_2 is 2.

Thomas Graham, 1805–1869.
*National Portrait Gallery,
Smithsonian Institution.*

Since

$$\frac{r_{H_2}}{r_{O_2}} = \sqrt{\frac{M_{O_2}}{M_{H_2}}}$$

$$\frac{r_{H_2}}{r_{O_2}} = \sqrt{\frac{32}{2}} = \sqrt{16} = 4$$

Hydrogen will effuse four times more rapidly than oxygen.

Example 8.19

What is the molecular weight of gas X if it effuses 0.876 times as rapidly as $N_2(g)$?

Solution

The ratio of the rate of effusion of gas X to the rate of effusion of $N_2(g)$ is

$$\frac{r_X}{r_{N_2}} = 0.876$$

The molecular weight of N_2 is 28.0. Therefore,

$$\sqrt{\frac{M_{N_2}}{M_X}} = \frac{r_X}{r_{N_2}}$$

$$\sqrt{\frac{28.0}{M_X}} = 0.876$$

We square both sides of the equation and solve for M_X:

$$\frac{28.0}{M_X} = 0.767$$

$$M_X = \frac{28.0}{0.767} = 36.5$$

8.13 Real Gases

The gas laws describe the behavior of an ideal or perfect gas—a gas defined by the kinetic theory. Under ordinary conditions of temperature and pressure, real gases follow the ideal gas laws fairly closely. At low temperatures and/or high pressures, however, they do not.

For an ideal gas, $PV = nRT$, and hence

$$\frac{PV}{RT} = n \tag{8.17}$$

If we consider one mole of an ideal gas, $n = 1$, and $PV/RT = 1$. In Figure 8.9, PV/RT (the so called compressibility factor) is plotted against pressure for several gases. The curves for the real gases deviate significantly from that for an ideal gas (a straight line at $PV/RT = 1$). There are two reasons for the deviations.

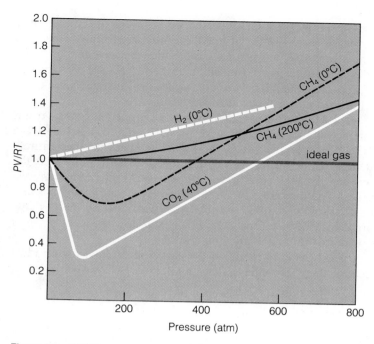

Figure 8.9 *PV/RT* versus pressure for several gases at temperatures indicated

1. Intermolecular forces of attraction. The kinetic theory assumes that there are no attractive forces between gas molecules. Such attractions must exist, however, because all gases can be liquefied. Intermolecular attractions hold the molecules together in the liquid state.

If we assume that P in the expression PV/RT is the applied pressure, a deviation from ideality would be apparent in the measured volume, V. Intermolecular forces of attraction *reduce* the volume by pulling the gas molecules together. In this respect, they augment the applied pressure. Furthermore, the higher the applied pressure, the more the effect of intermolecular attractions will be felt since gas molecules are closer together at higher pressures. This factor tends to cause the value PV/RT to be *less* than 1.

2. Molecular volume. The kinetic theory assumes that gas molecules are points in space and that the actual volume of the molecules is not significant. At absolute zero, the temperature at which molecular motion stops, therefore, the volume of an ideal gas is zero. Real gases, of course, do not have zero molecular volumes. When the applied pressure is increased, the space between the molecules is reduced, but the molecules themselves cannot be compressed. The result is that the measured volume is *larger* than the volume calculated for an ideal gas, where the molecular volume is neglected. Again, the deviation is more pronounced at higher pressures. The molecules are closer together at higher pressures and the molecular volume is a larger fraction of the total volume. This factor tends to cause the value PV/RT to be *greater* than 1.

These two factors operate at the same time and against one another. Which factor predominates depends upon the experimental conditions. In Figure 8.9, those portions of the curves that are *below* the $PV/RT = 1$ line correspond to conditions under which the effect of intermolecular attractive forces is predominant. For those portions that are *above* this line, molecular volume is the predominant effect.

Examine the curves for H_2 (at 0°C), CH_4 (at 0°C), and CO_2 (at 40°C). The curve for CO_2 falls farther below the $PV/RT = 1$ line than the curve for any other gas. We conclude that the attractions between CO_2 molecules are greater than those between the molecules of the other gases. Indeed, since the curve for H_2 lies entirely above this line, the forces of attraction between H_2 molecules must be so weak that at 0°C they cause little deviation from ideality.

Compare the curve for CH_4 at 0°C to that for the same gas at 200°C. As a result of intermolecular attractions, part of the curve for CH_4 at 0°C lies below the $PV/RT = 1$ line. The curve for the gas at the higher temperature lies entirely above this line. At high temperatures, gas molecules move so rapidly that the attractive forces between the molecules have little effect. At low temperatures, however, the molecules move more slowly. The attractive forces pull the molecules together so that the observed volume is less than that predicted by the gas law. The curves of Figure 8.9 show, therefore, that real gases follow the ideal gas laws most closely at low pressures and high temperatures.

Johannes van der Waals in 1873 modified the equation of state for an ideal gas to take into account these two effects. The van der Waals equation is

$$\left(P + \frac{n^2a}{V^2}\right)(V - nb) = nRT \tag{8.18}$$

The numerical values of the constants a and b for each gas are determined by experiment. Typical values are listed in Table 8.4.

The term n^2a/V^2 is added to the measured pressure, P, to correct for the intermolecular attractive forces. Pressure is caused by the collisions of the gas molecules with the walls of the container. The effect of a given collision would be greater if the molecule were not held back by the attraction of other molecules. Consequently, the pressure that is measured is *less* than it would be if these attractive forces did not exist. The term n^2a/V^2 is added to P so that $(P + n^2a/V^2)$ represents the pressure of an ideal gas, one in which there are no molecular forces.

The term (n/V) represents a concentration (mol/liter). If x molecules are confined in a liter, there are $(x - 1)$ ways for a given molecule to collide or interact with another molecule since it cannot collide with itself. This factor applies to all the molecules; therefore, a total of $\frac{1}{2}x(x - 1)$ possible interactions exists for the entire collection of molecules. The fraction $\frac{1}{2}$ is added so that a given interaction is not counted twice—once for each of the molecules entering into it. If a large

Table 8.4 van der Waals constants		
	a (liter^2 atm/mol^2)	b (liter/mol)
H_2	0.244	0.0266
He	0.0341	0.0237
N_2	1.39	0.0391
O_2	1.36	0.0318
Cl_2	6.49	0.0562
NH_3	4.17	0.0371
CO	1.49	0.0399
CO_2	3.59	0.0427

number of molecules are present, $(x - 1)$ is approximately equal to x, and the proportionality is $\frac{1}{2}x^2$ to a good approximation. Hence the number of interactions between gas molecules is proportional to the *square* of the concentration. The van der Waals constant a may be regarded as a proportionality constant (incorporating the $\frac{1}{2}$), and the correction term is n^2a/V^2.

The constant b, multiplied by n, is subtracted from the total volume of the gas to correct for that portion of the volume that is not compressible because of the intrinsic volume of the gas molecules. A given gas molecule cannot move through the entire volume of the container since other molecules are present. The volume through which they can move can be obtained by subtracting an amount called the **excluded volume** from the total volume.

If the molecules are assumed to be spherical and have a radius r, the excluded volume per molecule is not merely the volume of the molecule, $\frac{4}{3}\pi r^3$. Since the closest approach of *two* molecules is $2r$ (Figure 8.10), the excluded volume for *two* molecules is $\frac{4}{3}\pi(2r)^3$ which is $8(\frac{4}{3}\pi r^3)$. For one molecule this volume is $4(\frac{4}{3}\pi r^3)$, which is four times the molecular volume. Hence, for a mole of molecules, N molecules,

$$b = 4N(\tfrac{4}{3}\pi r^3) \tag{8.19}$$

molecular volume, $\frac{4}{3}\pi r^3$

excluded volume for two molecules, $8(\frac{4}{3}\pi r^3)$

Figure 8.10 van der Waals correction, b, for excluded volume

8.14 Liquefaction of Gases

Liquefaction of a gas occurs under conditions that permit the intermolecular attractive forces to bind the gas molecules together in the liquid form. If the pressure is high, the molecules are close together, and the effect of the attractive forces is appreciable. The attractive forces are opposed by the motion of the gas molecules; thus, liquefaction is favored by low temperatures where the average kinetic energy of the molecules is low. The behavior of a gas deviates more and more from ideality as the temperature is lowered and the pressure is raised. At extremes of these conditions, gases liquefy.

The higher the temperature of a gas, the more difficult it is to liquefy and the higher the pressure that must be employed (Table 8.5). For each gas there is a temperature above which it is impossible to liquefy the gas no matter how high the applied pressure. This temperature is called the **critical temperature** of the gas under consideration. The **critical pressure** is the minimum pressure needed to liquefy a gas at its critical temperature. The critical constants of some common gases are listed in Table 8.6.

The critical temperature of a gas gives an indication of the strength of the intermolecular attractive forces of that gas. A substance with weak attractive forces would have a low critical temperature; above this temperature, the molecular motion is too violent to permit the relatively weak forces to hold the molecules in the liquid state. The substances of Table 8.6 are listed in order of increasing critical temperature; the magnitude of the intermolecular attractive forces (related to the a of Table 8.4) increases in this same order. Helium, which has weak attractive forces, can exist as a liquid below 5.3 K only; the strong attractive forces of water permit it to be liquefied up to a temperature of 647.2 K. The critical constants have been used to evaluate the constant of the van der Waals equation.

The data of Table 8.6 show that it is necessary to cool many gases below room temperature (ca. 295 K) before these substances can be liquefied. Commercial

Table 8.5 Pressures needed to liquefy carbon dioxide. (Vapor pressures of liquid CO_2.)

Temperature (°C)	Pressure (atm)
−50	6.7
−30	14.1
−10	26.1
10	44.4
20	56.5
30	71.2
31	72.8

Table 8.6 Critical point data		
Gas	Critical Temperature (K)	Critical Pressure (atm)
He	5.3	2.26
H_2	33.3	12.8
N_2	126.1	33.5
CO	134.0	35.0
O_2	154.4	49.7
CH_4	190.2	45.6
CO_2	304.2	72.8
NH_3	405.6	111.5
H_2O	647.2	217.7

liquefaction procedures make use of the Joule-Thomson effect to cool gases. When a compressed gas is allowed to expand to a lower pressure, the gas cools. In the expansion, work is done against the intermolecular attractive forces. The energy used in performing this work must be taken from the kinetic energy of the gas molecules themselves; hence, the temperature of the gas decreases. This effect was studied by James Joule and William Thomson (Lord Kelvin) during the years 1852–1862. The liquefaction of air is accomplished by first allowing cooled, compressed air to expand. The temperature of the air falls to a lower level. This cooled air is used to precool entering compressed air, and the expansion of this compressed air results in the attainment of even lower temperatures. The cooled, expanded air is recycled through the compression chamber. Eventually the cooling and compression produce liquid air.

Summary

The topics that have been discussed in this chapter are

1. The meaning and measurement of pressure.

2. The simple gas laws: Boyle's law, Charles' law, and Amonton's law.

3. Absolute temperature and its measurement on the Kelvin scale.

4. The equation of state for an ideal gas, which summarizes the simple gas laws.

5. Calculations based on the ideal gas equation including those that involve gas densities and molecular weights.

6. The kinetic theory of gases, a model for understanding the behavior of gases.

7. Gay-Lussac's law of combining volumes, which pertains to the volumes of gases involved in chemical reactions and is the basis for the stoichiometry of this type of reaction.

8. Avogadro's principle, the explanation of Gay-Lussac's law.

9. Dalton's law of partial pressures, used to handle problems involving mixtures of gases.

10. The distribution of molecular speed and kinetic energy among the molecules of a gas.

11. Graham's law of effusion.

12. The ways in which the behavior of real gases deviates from that of an ideal gas and the reasons for the deviations.

13. The van der Waals equation, a modification of the ideal gas equation corrected to take into account nonideal behavior.

14. The liquefaction of gases.

Key Terms

Some of the more important terms introduced in this chapter are listed below. Definitions for terms not included in this list may be located in the text by use of the index.

Amontons' law (Section 8.4)   At constant volume, the pressure of a sample of gas varies directly with the absolute temperature.

Atmosphere, atm (Section 8.1)   A unit of pressure that is defined as 101,325 Pa; 1 atm = 760 torr.

Avogadro's principle (Section 8.8)   Equal volumes of all gases at the same temperature and pressure contain the same number of molecules.

Barometer (Section 8.1)   A device for measuring the pressure that the atmosphere exerts on the surface of the earth.

Boyle's law (Section 8.2)   At constant temperature, the volume of a sample of gas varies inversely with the pressure.

Charles' law (Section 8.3)   At constant pressure, the volume of a sample of gas varies directly with the absolute temperature.

Compressibility factor (Section 8.13) PV/RT where P is the pressure of a gas; V, the volume; R, the ideal gas constant; and T, the absolute temperature. For 1 mol of an ideal gas, the compressibility factor is always equal to 1.

Critical temperature (Section 8.14)   The temperature above which it is impossible to liquefy the gas under study no matter how high the applied pressure.

Critical pressure (Section 8.14)   The pressure required to liquefy a gas at its critical temperature.

Dalton's law of partial pressures (Section 8.10)   The total pressure of a mixture of gases that do not react is equal to the sum of the partial pressures of all the gases present.

Gay-Lussac's law of combining volumes (Section 8.8)   At constant temperature and pressure, the volumes of gases used or produced in a chemical reaction stand in ratios of small, whole numbers.

Graham's law of effusion (Section 8.12)   The rate of effusion of a gas is inversely proportional to the square root of its density or the square root of its molecular weight.

Ideal gas constant, R (Section 8.5)   The proportionality constant in the equation of state for an ideal gas; 0.082056 liter·atm/(K·mol).

Ideal gas law (Section 8.5)   The product of the pressure, P, and the volume, V, of a sample of an ideal gas is proportional to the number of moles of gas, n, times the absolute temperature, T; $PV = nRT$.

Kelvin temperature scale (Section 8.3)   An absolute temperature scale on which a reading can be obtained by adding 273.15 to the value in degrees Celsius.

Kinetic theory of gases (Sections 8.6 and 8.7)   A model on the molecular level that can be used to explain the gas laws and from which the ideal gas equation can be derived.

Maxwell-Boltzmann distribution (Section 8.11)   The way in which kinetic energy or molecular speed is distributed among the molecules of a gas.

Mean free path (Section 8.11)   The average distance that a gas molecule travels between collisions with other gas molecules.

Mole fraction, X (Section 8.10)   The ratio of the number of moles of a component of a mixture to the total number of moles present in the mixture.

Partial pressure (Section 8.10)   The pressure that a component of a mixture of gases would exert if it were the only gas present in the volume under consideration.

Pascal, Pa (Section 8.1)   The SI unit of pressure; equal to a force of one newton (1 kg·m/s^2) acting on a square meter.

Pressure (Section 8.1)   Force per unit area.

Root-mean-square speed, u (Section 8.11)   The square root of the average of the squares of the molecular speeds.

Standard temperature and pressure, STP (Section 8.5)   0°C (which is 273.15 K) and 1 atm pressure.

STP molar volume (Section 8.8)   The volume of one mole of a gas at STP; 22.414 liters.

torr (Section 8.1)   A unit of pressure equivalent to the pressure that will support a column of mercury to a height of 1 mm.

van der Waals equation (Section 8.13)   An equation of state for gases; a modification of the ideal gas equation that takes into account intermolecular attractions and the volumes that gas molecules occupy.

Problems*

Simple Gas Laws

8.1 State **(a)** Boyle's law, **(b)** Charles' law, **(c)** Amontons' law.

8.2 For each of the following pairs of variables, which correspond to measurements made on an ideal gas, draw a rough graph to show how one quantity varies with the

---

* The more difficult problem are marked with asterisks. The appendix contains answers to color-keyed problems.

other: (a) P vs. V, with T constant; (b) T vs. V, with P constant; (c) P vs. T, with V constant; (d) PV vs. V, with T constant.

8.3 The volume of a sample of gas is 500 ml at a pressure of 1.50 atm. If the temperature is held constant, what is the volume of the sample at a pressure of (a) 1.00 atm, (b) 5.00 atm, (c) 0.500 atm?

8.4 The volume of a sample of gas is 12.0 liter at a pressure of 0.250 atm. If the temperature is held constant, what is the pressure of the sample when the volume is (a) 250 ml, (b) 25.0 liter, (c) 0.750 liter?

8.5 A 10.0-liter tank of helium is filled to a pressure of 150 atm. How many 1.50-liter toy balloons can be inflated to a pressure of 1.00 atm from the tank? Assume no change in temperature.

8.6 A McLeod gauge is an instrument used to measure extremely low pressures. Assume that a 250-ml sample of gas from a low pressure system is compressed in a McLeod gauge to a volume of 0.0525 ml, where the pressure of the sample is 0.0355 atm. What is the pressure of the gas in the system?

8.7 The volume of a sample of a gas at 20°C is 2.50 liter. Assume that the pressure is held constant. (a) What is the volume of the gas at 200°C? (b) At what temperature (in °C) will the volume be 3.00 liter? (c) At what temperature (in °C) will the volume be 1.00 liter?

8.8 The volume of a sample of nitrogen gas at 0°C is 125 ml. Assume that the pressure remains constant. (a) What is the volume of the sample at 100°C? (b) At what temperature (in °C) will the volume be 100 ml? (c) At what temperature (in °C) will the volume be 175 ml?

8.9 A gas thermometer contains 250.00 ml of gas at 0°C and 1.00 atm pressure. If the pressure remains at 1.00 atm, how many milliliters will the volume increase for every one Celsius degree that the temperature rises?

8.10 A container is filled with a gas to a pressure of 2.00 atm at 25°C. (a) What pressure will develop within the sealed container if it is warmed to 75°C? (b) At what temperature (in °C) will the pressure be 10.0 atm? (c) At what temperature (in °C) will the pressure be 1.50 atm?

8.11 A container is filled with a gas to a pressure of 5.00 atm at 50°C. (a) What pressure will develop within the sealed container if it is warmed to 150°C? (b) What would the pressure inside the container be at −50°C? (c) At what temperature (in °C) would the pressure be 7.50 atm?

Ideal Gas Law

8.12 A sample of a gas occupies 400 ml at STP. What volume will the sample occupy at 77°C and 2.50 atm?

8.13 A sample of a gas occupies 15.0 liter at 50°C and 0.750 atm. What volume will the sample occupy at STP?

8.14 The volume of a sample of gas is 750 ml at 75°C and 0.750 atm. At what temperature (in °C) will the sample occupy 1.000 liter under a pressure of 1.000 atm?

8.15 A 1.00 liter sample of a gas is collected at 25°C and 1.25 atm. What is the pressure of the gas at 200°C if the volume is 4.00 liter?

8.16 Complete the following table for an ideal gas.

Pressure P	Volume V	Moles n	Temperature T
1.00 atm	_____	1.00 mol	273 K
0.500 atm	_____	1.00 mol	0°C
5.00 atm	10.0 liter	_____	100°C
0.452 atm	5.00 ml	1.00×10^{-3} mol	_____
_____	5.00 liter	1.25 mol	300°C

8.17 What volume will 3.00 g of CO_2 gas occupy at 100°C and 0.350 atm?

8.18 What is the density of N_2O gas at 25°C and 0.750 atm?

8.19 If the temperature is held constant at 50°C, at what pressure will the density of N_2 gas be 0.500 g/liter?

8.20 If the pressure is held constant at 1.00 atm, at what temperature (in °C) will the density of O_2 gas be 1.00 g/liter?

8.21 A gas has a density of 0.572 g/liter at 90°C and a pressure of 0.500 atm. What is the molecular weight of the gas?

8.22 A gas has a density of 0.991 g/liter at 75°C and a pressure of 0.350 atm. What is the molecular weight of the gas?

8.23 A 0.300 g sample of a liquid was vaporized at 150°C. The vapor occupied 180 ml under a pressure of 0.998 atm. What is the molecular weight of the liquid?

Gay-Lussac's Law of Combining Volumes and Avogadro's Principle

8.24 Hydrogen cyanide, a highly poisonous compound, is commercially prepared by the following reaction run at a high temperature in the presence of a catalyst:

$$2\,CH_4(g) + 3\,O_2(g) + 2\,NH_3(g) \longrightarrow 2\,HCN(g) + 6\,H_2O(g)$$

How many liters of $CH_4(g)$, $O_2(g)$ and $NH_3(g)$ are required and how many liters of $H_2O(g)$ are produced in the preparation of 30.0 liters of $HCN(g)$? Assume that all gas volumes are measured under the same conditions of temperature and pressure.

8.25 How many liters of $HCN(g)$ can be prepared from 10.0 liters of $CH_4(g)$, 20.0 liters of $O_2(g)$, and 30.0 liters of $NH_3(g)$ in a reaction such as the one described by the equation given in Problem 8.24? Assume that the volumes are measured at constant conditions of temperature and pressure.

8.26 Ammonia, $NH_3(g)$, reacts with oxygen at 850°C in the presence of a Pt catalyst to yield $NO(g)$ and $H_2O(g)$. Write an equation for the reaction. What volume of $NO(g)$ can be obtained from 50.0 liter of $NH_3(g)$ and 50.0 liter of $O_2(g)$? The volumes of all gases are measured under the same conditions.

8.27 In the absence of a catalyst, ammonia, $NH_3(g)$, reacts with oxygen gas to yield nitrogen gas and water vapor. Write an equation for the reaction. What volume of $N_2(g)$ can be obtained from 25.0 liter of $NH_3(g)$ and

20.0 liter of $O_2(g)$? The volumes of all gases are measured under the same conditions.

8.28 A mixture is prepared from 3.00 liter of ammonia and 5.00 liter of chlorine. These substances react according to the equation

$$2 NH_3(g) + 3 Cl_2(g) \longrightarrow N_2(g) + 6 HCl(g)$$

If the volumes of all of the gases are measured at the same temperature and pressure, list the volumes of all of the substances present at the conclusion of the reaction.

8.29 At temperatures above 50°C, nitrogen oxide, $NO(g)$, decomposes to yield dinitrogen oxide, $N_2O(g)$, and nitrogen dioxide, $NO_2(g)$. **(a)** Write an equation for the reaction. **(b)** What total volume of $N_2O(g)$ and $NO_2(g)$ together would result from the decomposition of 125 ml of $NO(g)$ at 250°C and 1.00 atm? Assume that all gas volumes are measured under the same conditions. **(c)** What are the partial pressures of $N_2O(g)$ and $NO_2(g)$ in this gas mixture?

8.30 The reaction of $NH_3(g)$ and $F_2(g)$ in the presence of a copper catalyst yields $NF_3(g)$ and $NH_4F(s)$. **(a)** Write the chemical equation for the reaction. **(b)** How many milliliters of $NH_3(g)$ and of $F_2(g)$ are required to make 150 ml of $NF_3(g)$ if a 60.0% yield is obtained? Assume that all gases are measured under the same conditions of temperature and pressure.

8.31 What is the density of hydrogen iodide gas, HI, at STP?

8.32 What is the molecular weight of a gas with a density of 2.59 g/liter at STP?

8.33 About 1.50×10^8 metric tonnes (1 metric tonne is 1000 kg) of $CO(g)$ are released into the atmosphere each year. **(a)** What is the volume of this quantity of $CO(g)$ at STP? **(b)** How many CO molecules are contained in this volume?

8.34 Federal standards limit the amount of $SO_2(g)$ in the air to a maximum of 80 $\mu g/m^3$. One cubic meter is 1.00×10^3 liter. **(a)** How many grams of SO_2 is this? **(b)** How many moles of SO_2? **(c)** What is the partial pressure of SO_2 in air that meets this standard at STP? **(d)** What percentage of the total number of molecules are SO_2 molecules?

8.35 Federal standards limit the amount of $CO(g)$ in the air to a maximum of 10,000 $\mu g/m^3$. **(a)** What is the maximum partial pressure of CO in air at STP that meets this standard? **(b)** How many molecules of CO would be present in 1.00 liter of this air? **(c)** What percentage of the total number of molecules are CO molecules?

Stoichiometry and Gas Volumes

8.36 In liquid ammonia, Na(s) reacts with $N_2O(g)$ and $NH_3(l)$ to yield $NaN_3(s)$, NaOH(s), and $N_2(g)$. **(a)** Write the chemical equation for the reaction. **(b)** How many milliliters of $N_2O(g)$, measure at STP, would be required to make 12.0 g of NaN_3? **(c)** How many milliliters of $N_2(g)$, also at STP, would be produced by the reaction?

8.37 Cyanogen, C_2N_2, is a flammable, highly poisonous gas. It can be prepared by a catalyzed gas-phase reaction between $HCN(g)$ and $NO_2(g)$. The products of the reaction are $C_2N_2(g)$, $NO(g)$, and $H_2O(g)$. **(a)** Write the chemical equation for the reaction. **(b)** How many milliliters of $HCN(g)$ and of $NO_2(g)$, both at STP, are required to make 6.50 g of C_2N_2? **(c)** How many milliliters of $NO(g)$, also at STP, are produced by the reaction?

8.38 Calcium hydride, $CaH_2(s)$, reacts with water to yield $H_2(g)$ and $Ca(OH)_2(s)$. **(a)** Write the chemical equation for this reaction. **(b)** How many grams of CaH_2 are needed to prepare 5.00 liter of $H_2(g)$ at STP?

8.39 Calcium metal, Ca(s), reacts with water to yield $H_2(g)$ and $Ca(OH)_2(s)$. **(a)** Write the chemical equation for this reaction. **(b)** How many grams of Ca(s) are needed to prepare 5.00 liter of $H_2(g)$ at STP? Compare your answer to that of Problem 8.38.

8.40 Aluminum carbide, $Al_4C_3(s)$ reacts with water to yield methane gas, $CH_4(g)$ and $Al(OH)_3(s)$. **(a)** Write the chemical equation for this reaction. **(b)** What volume of CH_4 (measured at 20°C and 0.750 atm) would be obtained by the reaction of 1.50 g of Al_4C_3?

8.41 Lanthanum carbide, $La_2(C_2)_3(s)$, reacts with water to yield acetylene gas, $C_2H_2(g)$ and $La(OH)_3(s)$. **(a)** Write the chemical equation for this reaction. **(b)** What volume of C_2H_2 (measured at 35°C and 0.300 atm) would be obtained by the reaction of 0.500 g of $La_2(C_2)_3$?

8.42 The complete combustion of octane yields carbon dioxide and water:

$$2 C_8H_{18}(g) + 25 O_2(g) \longrightarrow 16 CO_2(g) + 18 H_2O(g)$$

What volume of gas is produced from the complete combustion of 0.536 g octane if the temperature is 500°C and the pressure is 10.0 atm?

8.43 Magnesium and aluminum react with aqueous acids to give hydrogen gas:

$$Mg(s) + 2 H^+(aq) \longrightarrow H_2(g) + Mg^{2+}(aq)$$

$$2 Al(s) + 6 H^+(aq) \longrightarrow 3 H_2(g) + 2 Al^{3+}(aq)$$

What volume of $H_2(g)$ measured at 25°C and a pressure of 0.985 atm can be obtained from 1.50 g of an alloy that is 70.0% Mg and 30.0% Al?

*8.44 A 10.0 g sample of an alloy of Mg and Al produces 10.5 liter of H_2 gas (measured at STP) when reacted with an acid. The equations for the reactions are given in Problem 8.43. What percentage of the alloy is Al?

8.45 The combustion of 2.80 liter of a gaseous compound that contains only C and H requires 18.2 liter of $O_2(g)$ and produces 11.2 liter of $CO_2(g)$ and 11.25 g of $H_2O(l)$. All gas measurements are made at STP. **(a)** Calculate the number of moles of each substance involved in the reaction. **(b)** Use your answers from part (a) to derive whole-number coefficients for the chemical equation for the reaction. **(c)** Determine the formula for the hydrocarbon and write the equation for the reaction.

8.46 The complete combustion of 0.250 g of a compound that contains only C and H produced 392 ml of $CO_2(g)$, measured at STP, and 0.360 g of H_2O. **(a)** How many moles of C and of H did the sample contain? **(b)** The compound is a gas at 100°C and 1.000 atm. Under these conditions, the sample occupied 76.6 ml. How many moles of compound did the sample represent? **(c)** What is the molecular formula of the compound?

Dalton's Law of Partial Pressures

8.47 A mixture of 0.770 g of $N_2O(g)$ and 0.770 g of $N_2(g)$ exerts a pressure of 0.500 atm. What is the partial pressure of each gas?

8.48 A mixture of 0.300 g of He(g) and 0.505 g of Ne(g) exerts a pressure of 0.250 atm. What is the partial pressure of each gas?

8.49 The partial pressure of CO(g) is 0.200 atm and of $CO_2(g)$ is 0.600 atm in a mixture of the two gases. **(a)** What is the mole fraction of each gas in the mixture? **(b)** If the mixture occupies 11.6 liter at 50°C, what is the total number of moles present in the mixture? **(c)** How many grams of each gas are present in the mixture?

8.50 A sample of a gas, collected over water at 20°C and a total pressure of 1.03 atm, occupies 200 ml. What volume would the gas occupy if dry and at STP?

8.51 A 75.0 ml sample of a gas is collected over water at 22°C and a barometric pressure of 0.987 atm. What volume would the gas occupy if dry and at 100°C and a pressure of 1.000 atm?

8.52 A 1.00 liter sample of neon is collected over water at 27°C and 0.983 atm pressure. If the gas is dried and placed in a 1.50 liter container at 72°C, what pressure will it exert?

*8.53 A sample of a gas, collected over water at 32°C, occupies a volume of 1.000 liter. The wet gas exerts a pressure of 1.000 atm. When dried, the sample occupies 1.000 liter and exerts a pressure of 1.000 atm at 47°C. What is the vapor pressure of water at 32°C?

*8.54 One mole of $N_2O_4(g)$ was placed in a container and allowed to dissociate:

$$N_2O_4(g) \longrightarrow 2\, NO_2(g)$$

The mixture resulting from the dissociation (N_2O_4 and NO_2) occupied 36.0 liter at a total pressure of 1.000 atm and 45°C. **(a)** Use the equation of state to find the total number of moles of mixed gases present. **(b)** Allow x to equal the number of moles of $N_2O_4(g)$ that dissociate. In terms of x, how many moles of $N_2O_4(g)$ remain undissociated and how many moles of $NO_2(g)$ are produced by the dissociation? Use your answer to (a) to find these values. **(c)** What are the mole fractions of $N_2O_4(g)$ and $NO_2(g)$ in the mixture? **(d)** What are the partial pressures of $N_2O_4(g)$ and $NO_2(g)$?

Kinetic Theory of Gases, Graham's Law

8.55 List the postulates of the kinetic theory of gases.

8.56 How does the kinetic theory explain Boyle's law, Charles' law, and Amontons' law?

8.57 Draw rough graphs to show for a collection of molecules **(a)** the distribution of molecular speeds for two different temperatures, **(b)** the distribution of molecular energies.

8.58 What is the root-mean-square speed of the O_2 molecule at 100 K and at 500 K?

8.59 At what temperature would the root-mean-square speed of the O_2 molecule equal the root-mean-square speed of the H_2 molecule at 100 K?

8.60 Compare the rate of effusion of CO(g) with that of $CO_2(g)$ under the same conditions.

8.61 A gas, X, effuses 1.66 times faster than $N_2O(g)$. What is the molecular weight of gas X?

8.62 At 25°C and 0.500 atm, the density of $N_2(g)$ is 0.572 g/liter. The rate of effusion of $N_2(g)$ through an apparatus is 9.50 ml/s. **(a)** What is the density of an unknown gas if it effuses at a rate of 6.60 ml/s through the same apparatus under the same conditions? **(b)** What is the molecular weight of the gas?

8.63 Under certain conditions of temperature and pressure, the density of gas X is 1.25 g/liter. A volume of 15.0 ml of gas X effuses through an appartatus in 1.00 s. The rate of effusion of gas Y through the same apparatus and under the same conditions is 20.4 ml/s. Calculate the density of gas Y under the experimental conditions.

8.64 Calculate the density of a gas at STP if a given volume of the gas effuses through an apparatus in 5.00 min and the same volume of oxygen, at the same temperature and pressure, effuses through this apparatus in 6.30 min.

8.65 Use Graham's law to calculate the molecular weight of a gas if a given volume of the gas effuses through an apparatus in 300 s and the same volume of $CH_4(g)$, under the same conditions of temperature and pressure, effuses through the same apparatus in 219 s.

Real Gases

8.66 What are the reasons that real gases deviate from ideal behavior?

8.67 Which member of each of the following pairs would follow the ideal gas law more closely: **(a)** $H_2(g)$, molecular weight 2.0, or HI(g), molecular weight 127.9, **(b)** a gas at 100°C or the same gas at 100 K, **(c)** a gas under a pressure of 1.0 atm or the same gas under a pressure of 10.0 atm, **(d)** a gas with a critical temperature of 100 K or one with a critical temperature of 300 K? Give the reasons for your predictions.

8.68 Consider 1.00 mol of $CO_2(g)$ at 40°C. At 50.0 atm, the volume of the sample is 0.314 liter. At 800 atm, the volume of the sample is 0.0421 liter. **(a)** What should these volumes be according to the ideal gas law? **(b)** Account for the deviations. **(c)** What are the values of PV/RT for the two pressures?

8.69 Which gas of those listed in Table 8.4 would you expect to have the highest critical temperature? Why?

8.70 Calculate the pressure exerted by 1.000 mol of $N_2(g)$ confined to a volume of 1.000 liter at 0°C by **(a)**

the ideal gas law, and **(b)** the van der Waals equation. **(c)** Compare the results.

8.71 Calculate the pressure exerted by 1.000 mol of $N_2(g)$ confined to a volume of 10.00 liter at 0°C by **(a)** the ideal gas law and **(b)** the van der Waals equation. **(c)** Compare the results to each other and to those of Problem 8.70.

8.72 Calculate the pressure exerted by 1.000 mol of $N_2(g)$ confined to a volume of 1.000 liter at 100°C by **(a)** the ideal gas law and **(b)** the van der Waals equation. **(c)** Compare the results to each other and to those of Problem 8.70.

8.73 Use the values of a and b that are given in Table 8.4 to compare the strength of the intermolecular attractive forces in $H_2(g)$ and $He(g)$. Note that the normal boiling point of H_2 is 20.4 K and that of He is 4.2 K (the lowest of any known substance).

8.74 (a) Use the value of the van der Waals constant b for CO_2 (0.0427 liter/mol) to calculate the volume of a molecule of CO_2. **(b)** At STP, what percentage of the total volume of CO_2 gas is molecular volume?

8.75 The value of the van der Waals constant b for $Kr(g)$ is 0.0398 liter/mol. Use this value to calculate the radius of a krypton atom.

CHAPTER 9 LIQUIDS AND SOLIDS

The kinetic energies of gas molecules decrease when the temperature is lowered. Consequently, the intermolecular attractive forces cause the gas molecules to condense into a liquid when the gas has been cooled sufficiently. The molecules are closer together and the attractive forces exert a greater influence in a liquid than in a gas. Molecular motion, therefore, is more restricted in the liquid state than in the gaseous state.

Additional cooling causes the kinetic energies of the molecules to decrease further and ultimately produces a solid. In a crystalline solid the molecules assume positions in a crystal lattice, and the motion of the molecules is restricted to vibration about these fixed points.

The comparatively *high* kinetic energies of gas molecules cause the intermolecular attractive forces to assume a role that can be minimized in the development of a satisfactory theory of gases. The comparatively *low* kinetic energies of molecules (or ions) in crystals are easily overcome by the attractive forces to produce highly ordered, crystalline structures that have been well characterized by diffraction techniques. Our understanding of the intermediate state, the liquid state, however, is not so complete as that of the other two states.

9.1 Intermolecular Forces of Attraction

Atoms are held together in molecules by covalent bonds, but what forces attract molecules to each other in the liquid and solid states? There are several types of attractive forces that hold molecules together which, taken as a group, are called intermolecular attractive forces. Two types are discussed in this section, and a third is presented in the next section.

Dipole-dipole forces occur between polar molecules. Molecules of this type have dipoles and line up in an electric field (see Section 6.5). Dipole-dipole forces are caused by the attractions of the positive and negative poles of the molecules for one another. In a crystal of a polar molecular substance, the molecules are lined up in a way that reflects the dipole-dipole forces (Figure 9.1).

Electronegativity differences between atoms can be used to predict the degree of polarity of a diatomic molecule as well as the positions of the positive and negative poles. A prediction concerning the polarity of a molecule that contains more than two atoms, however, must be based upon a knowledge of the geometry of the molecule, the polarities of the bonds, *and the arrangement of nonbonding electron pairs.*

Figure 9.1 Orientation of polar molecules in a crystal

Consider the three molecules (CH_4, NH_3, and H_2O) represented in Figure 9.2. The dipole moment of a molecule is the result of the individual bond dipoles and the nonbonding electron pairs of the molecule. In each of the molecules under consideration, the central atom is more electronegative than the H atoms bonded to it. The negative end of each *bond* dipole, therefore, points toward the central atom. In CH_4, the tetrahedral arrangement of the four polar C—H bonds produces a molecule that is not polar; CH_4 has no dipole moment. The center of positive charge of the *molecule* (derived by considering all four bonds) falls in the center of the C atom and coincides with the center of negative charge for the molecule.

On the other hand, the trigonal pyramidal NH_3 molecule is polar (its dipole moment is 1.49 D). The three polar bonds and nonbonding electron pair are arranged so that the molecule has a dipole with the negative end directed toward the top of the trigonal pyramid and the positive end toward the bottom. Similarly, the angular H_2O molecule is polar (its dipole moment is 1.85 D). The polar bonds and the nonbonding electron pairs contribute to a dipole with the negative end directed toward the O atom and the positive end directed toward a point halfway between the two H atoms.

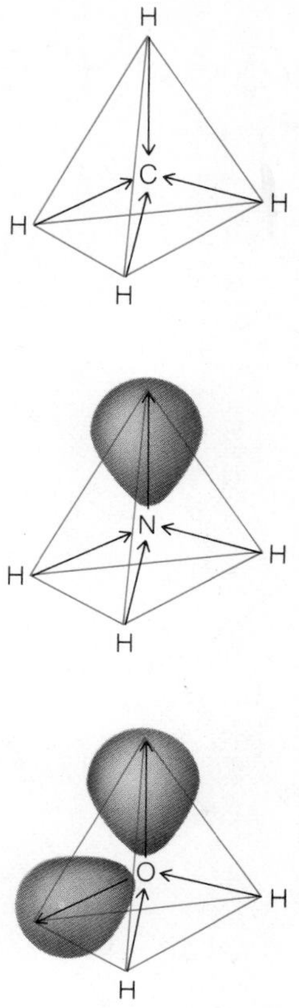

The influence that a nonbonding pair of electrons has on the dipole moment of a molecule is seen in the case of NF_3. The NF_3 molecule has a structure similar to that of NH_3 (Figure 9.2), but the direction of the polarity of the bonds is the reverse of that in NH_3 since F is more electronegative than N. Nitrogen trifluoride has a dipole moment of 0.24 D, a surprisingly low value in view of the highly polar nature of the N—F bonds. The N—F bond dipoles combine to give the molecule a dipole with the negative end in the direction of the base of the pyramid, but the contribution of the nonbonding electron pair works in the opposite direction and reduces the total polarity of the molecule.

What intermolecular forces attract *nonpolar* molecules to each other in the liquid and solid state? Such molecules do not have permanent dipoles, but nevertheless, they can be liquefied. Some type of intermolecular force, therefore, must exist in addition to the dipole-dipole force.

The existence of **London forces (dispersion forces)** is postulated.* These forces are thought to arise from the motion of electrons. At one instant of time, the electron cloud of a molecule may be distorted so that a dipole is produced in which one part of the molecule is slightly more negative than the rest. At the next instant, the positions of the negative and positive poles of the dipole will be different because the electrons have moved. Over a period of time (a very short period of time—electrons move rapidly), the effects of these **instantaneous dipoles** cancel so that a nonpolar molecule has no permanent dipole moment.

The instantaneous, fluctuating dipole of a molecule, however, induces matching dipoles in neighboring molecules (lined up in the same way that permanent dipoles are aligned). The motion of the electrons of neighboring molecules is synchronized (see Figure 9.3). The force of attraction between these instantaneous dipoles constitutes the London force. *The strongest London forces occur between large, complex molecules, which have large electron clouds that are easily distorted, or polarized.*

Figure 9.2 Analysis of the polarities of the methane (CH_4), ammonia (NH_3), and water (H_2O) molecules. (Arrows point toward the negative end of the individual dipoles that compose the dipole moment of the molecule.)

* Johannes van der Waals postulated the existence of intermolecular attractive forces between gas molecules in 1873 (Section 8.13). The explanation of the origin of the type of intermolecular force discussed here was proposed by Fritz London in 1930. Although there is a lack of uniformity in the use of terms, current usage appears to favor calling these specific forces *London forces* and intermolecular forces in general *van der Waals forces.*

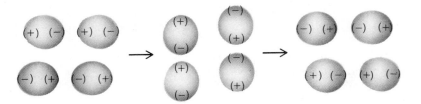

Figure 9.3 Instantaneous dipoles

		Attractive Energies (kJ/mol)			
Molecule	Dipole Moment (D)	Dipole-Dipole	London	Melting Point (K)	Boiling Point (K)
CO	0.12	0.0004	8.74	74	82
HI	0.38	0.025	27.9	222	238
HBr	0.78	0.69	21.9	185	206
HCl	1.03	3.31*	16.8	158	188
NH_3	1.49	13.3*	14.7	195	240
H_2O	1.84	36.4*	9.0	273	373

Table 9.1 Intermolecular attractive energies in some simple molecular crystals

* Caused by hydrogen bonding (see Section 9.2).

Since all molecules contain electrons, London forces also exist between polar molecules; in the case of nonpolar molecular substances, London forces are the *only* intermolecular forces that exist. The values listed in Table 9.1 show that London forces are the principal intermolecular forces for most molecular substances. The hydrogen bond, a special type of dipole-dipole interaction that is discussed in the next section, is responsible for the magnitude of the dipole-dipole energy listed for H_2O, NH_3, and (to a lesser extent) HCl.

The dipole moments of the molecules listed in Table 9.1 increase in the order given, and the dipole-dipole energies increase in the same order. The London energies, however, depend upon the sizes of the molecules. The largest molecule listed is HI, and it has the strongest London forces. HCl is a more polar molecule than HI; the electronegativity of Cl is 3.2, and the electronegativity of I is 2.7. The dipole-dipole energy of HCl is higher than that of HI. The London energy of HI, however, is so much higher than the London energy of HCl, that the *total* effect causes the molecules of HI to be more strongly attracted to one another than the molecules of HCl are. The boiling point of HI is 238 K, which is higher than the boiling point of HCl (188 K).

9.2 The Hydrogen Bond

The intermolecular attractions of certain hydrogen-containing compounds are unusually strong. These attractions occur in compounds in which hydrogen is covalently bonded to highly electronegative elements of small atomic size. In these compounds the atom of the electronegative element exerts such a strong attraction

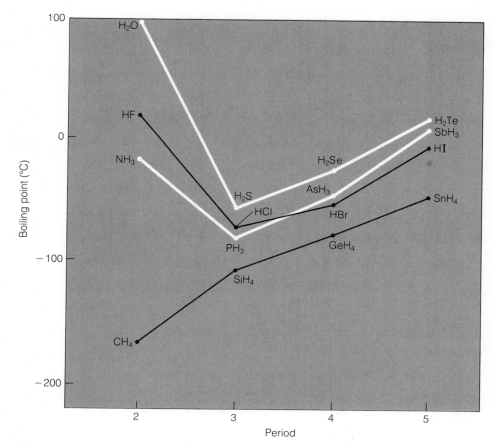

Figure 9.4 Boiling points of the hydrogen compounds of the elements of groups IV A, V A, VI A, and VII A

on the bonding electrons that the hydrogen atom is left with a significant δ^+ charge. In fact, the hydrogen atom is almost an exposed proton since it has no screening electrons.

The hydrogen atom of one molecule and a pair of unshared electrons on the electronegative atom of another molecule are mutually attracted and form what is called a **hydrogen bond.** Because of its small size, each hydrogen atom is capable of forming only one hydrogen bond. The association of HF, H_2O, and NH_3 by hydrogen bonds (indicated by dotted lines) can be roughly diagramed as follows:

$$H—F \cdots H—F \cdots \qquad H—\overset{\displaystyle ..}{\underset{\displaystyle H}{O}} \cdots H—\overset{\displaystyle ..}{\underset{\displaystyle H}{O}} \cdots \qquad H—\overset{\displaystyle H}{\underset{\displaystyle H}{N}} \cdots H—\overset{\displaystyle H}{\underset{\displaystyle H}{N}} \cdots$$

Unusual properties are characteristic of compounds in which hydrogen bonding occurs. In Figure 9.4 the normal boiling points of the hydrogen compounds of the elements of groups IV A, V A, VI A, and VII A are plotted. The series CH_4, SiH_4, GeH_4, and SnH_4 illustrates the expected trend in boiling point for compounds in which the only intermolecular forces are London forces; the boiling point increases as the molecular size increases. The hydrogen compounds of the IV A elements are nonpolar molecules; the central atom of each molecule has no unshared electron pair.

In each compound of the other three series of Figure 9.4, however, London forces are aided by dipole-dipole forces in holding the molecules together. Nevertheless, the boiling point of the first member of each series (HF, H_2O, and NH_3) is unusually high in comparison to those of the other members of the series. In each of these three compounds, hydrogen bonding increases the difficulty of separating the molecules from the liquid state. Significant hydrogen bonding is not found in any of the other compounds for which a boiling point is plotted in Figure 9.4. In addition to high boiling points, compounds that are associated by hydrogen bonding have abnormally high melting points, heats of vaporization, heats of fusion, and viscosities.

Hydrogen bonding occurs not only between the identical molecules of some pure compounds but also between the different molecules that make up certain solutions. There are two requirements for strong hydrogen bonding:

1. The molecule that supplies the proton for the formation of the hydrogen bond (the *proton donor*) must be highly polar so that the hydrogen atom will have a relatively high positive charge. The increasing strength of the hydrogen bonds $N—H \cdots N < O—H \cdots O < F—H \cdots F$ parallels the increasing electronegativity of the atom bonded to hydrogen, $N < O < F$. The high positive charge on the hydrogen atom attracts the electron pair from another molecule strongly, and the small size of the hydrogen atom permits the second molecule to approach closely.

2. The atom (of the *proton acceptor*) that supplies the electron pair for the hydrogen bond must be relatively small. Really effective hydrogen bonds are formed only by fluorine, oxygen, and nitrogen compounds. Chlorine compounds form weak hydrogen bonds, as evidenced by the slight displacement of the boiling point of HCl (Figure 9.4). Chlorine has approximately the same electronegativity as nitrogen. A chlorine atom, however, is larger than a nitrogen atom, and the electron cloud of the chlorine atom is, therefore, more diffuse than that of the nitrogen atom.

An examination of Figure 9.4 will show that hydrogen bonding has a greater effect on the boiling point of water than on the boiling point of hydrogen fluoride. This effect is observed even though the $O—H \cdots O$ bond is only about two-thirds as strong as the $F—H \cdots F$ bond. There are, *on the average*, twice as many hydrogen bonds *per molecule* in H_2O as there are *per molecule* in HF. The oxygen atom of each water molecule has two hydrogen atoms and two unshared electron pairs. The fluorine atom of the hydrogen fluoride molecule has three electron pairs that can bond with hydrogen atoms but only one hydrogen atom with which it can form a hydrogen bond.

Other properties of water are affected to an unusual degree by hydrogen bonding. The tetrahedral arrangement of the hydrogen atoms and the unshared electron pairs of oxygen in water cause the hydrogen bonds of the ice crystal to be arranged in this manner and lead to the open structure of the ice crystal. Ice, therefore, has a comparatively low density. In water at the freezing point, the molecules are arranged more closely together; water, therefore, has a higher density than ice—an unusual situation. It should be noted that H_2O molecules are associated by hydrogen bonds in the liquid state but not to the same extent, nor in the rigid manner, as they are associated in ice.

Hydrogen bonding also accounts for the unexpectedly high solubilities of some compounds containing oxygen, nitrogen, and fluorine in certain hydrogen-containing solvents, notably water. Thus, ammonia and methanol dissolve in water through the formation of hydrogen bonds:

$$
\begin{array}{ccc}
& H & & & & H \\
& | & & & & | \\
H-N\cdots H-O & & & H-C-O\cdots H-O \\
& | & | & & | \ \ | & | \\
& H & H & & H \ \ H & H
\end{array}
$$

In addition, certain oxygen-containing anions (e.g., the sulfate ion, SO_4^{2-}) dissolve in water through hydrogen-bond formation.

9.3 The Liquid State

In gases, the molecules move rapidly in a completely random manner. In solids, the molecules are held together in the orderly arrangements typical of crystals. The liquid state is intermediate between the gaseous state and the solid state.

In liquids, the molecules move more slowly than they do in gases. The intermolecular attractive forces are able, therefore, to hold them together into a definite volume. The molecular motion, however, is too rapid for the attractive forces to fix the molecules into definite positions in a crystal lattice. A liquid, consequently, retains its volume but not its shape. Liquids flow and assume the shapes of their containers.

A change in pressure has almost no effect on the volume of a liquid since there is little free space between the molecules. An increase in temperature, however, increases the volume of most liquids slightly and, consequently, decreases the liquid density. When the temperature of a liquid is increased, the average kinetic energy of the molecules increases, and this increased molecular motion works against the attractive forces. The expansion, however, is much less than that observed for gases, in which the effect of attractive forces is negligible.

Two liquids that are soluble in one another will diffuse into each other when placed together. If one liquid is carefully poured on top of another more dense liquid, the boundary between the two liquids will be sharp and easily seen. This boundary will gradually become less distinct and, in time, will disappear as the molecules of the two liquids diffuse into each other.

The diffusion of liquids is a much slower process than the diffusion of gases. Since the molecules of liquids are relatively close together, a molecule undergoes a tremendous number of collisions in a given time period. The average distance it travels between collisions, the mean free path, is much shorter for a molecule of a liquid than it is for a molecule of a gas. Gases, therefore, diffuse much more rapidly than liquids.

Any liquid exhibits resistance to flow, a property known as viscosity. One way of determining the viscosity of a liquid is to measure the time that it takes for a definite amount of the liquid to pass through a tube of small diameter under a given pressure. Resistance to flow is largely due to the attractions between molecules, and the measurement of the viscosity of a liquid gives a simple estimate of the strength of these attractions. In general, as the temperature of a liquid is increased, the cohesive forces are less able to cope with increasing molecular motion, and the viscosity decreases. On the other hand, increasing the pressure generally increases the viscosity of a given liquid.

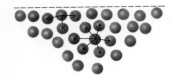

Figure 9.5 Schematic diagram indicating the unbalanced intermolecular forces on the surface molecules of a liquid as compared to the balanced intermolecular forces on the interior molecules

Surface tension is another property of liquids caused by the intermolecular attractive forces. A molecule in the center of a liquid is attracted equally in all directions by surrounding molecules. Molecules on the surface of a liquid, however, are attracted only toward the interior of the liquid (Figure 9.5). The surface molecules, therefore, are pulled inward, and the surface area of a liquid tends to be minimized. This behavior accounts for the spherical shape of liquid drops. Surface tension is a measure of this inward force on the surface of a liquid, the force which must be overcome to expand the surface area. The surface tension of a liquid decreases with increasing temperature since the increased molecular motion tends to decrease the effect of the intermolecular attractive forces.

9.4 Evaporation

The kinetic energies of the molecules of a liquid follow a Maxwell-Boltzmann distribution similar to the distribution of kinetic energy among gas molecules (Figure 9.6). The kinetic energy of a given molecule of a liquid is continually changing as the molecule collides with other molecules. At any given instant, however, some of the molecules of the total collection have relatively high energies and some have relatively low energies. The molecules with kinetic energies high enough to overcome the attractive forces of surrounding molecules can escape from the liquid and enter the gas phase if they are close to the surface and are moving in the right direction. They use part of their energy to work against the attractive forces when they escape.

In time, the loss of a number of high-energy molecules causes the average kinetic energy of the molecules remaining in the liquid to decrease, and the temperature of the liquid falls. When liquids evaporate from an open container, heat flows into the liquid from the surroundings to maintain the temperature of the liquid. In this way the supply of high-energy molecules is replenished, and the process continues until all of the liquid has evaporated. The total quantity

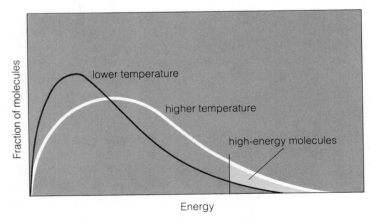

Figure 9.6 Distribution of kinetic energy among molecules of a liquid

of heat required to vaporize a mole of liquid at a given temperature is called the **molar enthalpy of vaporization** of that liquid. At 25°C, for example,

$$H_2O(l) \longrightarrow H_2O(g) \qquad \Delta H_v = +43.8 \text{ kJ}$$

The transfer of heat from the surroundings explains why swimmers emerging from the water become chilled as the water evaporates from their skin. Likewise, the regulation of body temperature is, in part, accomplished by the evaporation of perspiration from the skin. Various cooling devices have made use of this principle. A water cooler of the Middle East consists of a jar of unglazed pottery filled with water. The water saturates the clay of the pottery and evaporates from the outer surface of the jar, thus cooling the water remaining in the jar.

The rate of evaporation increases as the temperature of a liquid is raised. When the temperature is increased, the average kinetic energy of the molecules increases, and the number of molecules with energies high enough for them to escape into the vapor phase increases (Figure 9.6).

9.5 Vapor Pressure

When a liquid in a closed container evaporates, the vapor molecules cannot escape from the vicinity of the liquid. In the course of their random motion, some of the vapor molecules return to the liquid. We can represent the process for water by using a double arrow:

$$H_2O(l) \rightleftharpoons H_2O(g)$$

The rate of return of the vapor molecules to the liquid depends upon the concentration of the molecules in the vapor. The more molecules that there are in a given volume of vapor, the greater the chance that some of them will strike the liquid and be recaptured.

At the start, the rate of return of molecules from the vapor to the liquid is low since there are few molecules in the vapor. The continued vaporization, however, causes the concentration of the molecules in the vapor to increase. The rate of condensation, therefore, also increases. Eventually the system reaches a point in which the rate of condensation equals the rate of vaporization.

This condition, in which the rates of two opposite tendencies are equal, is called a state of **equilibrium.** At equilibrium, the concentration of molecules in the vapor state is constant because molecules leave the vapor through condensation at the same rate that molecules add to the vapor through vaporization. Similarly, the quantity of liquid is a constant because molecules are returning to the liquid at the same rate that they are leaving it.

It is important to note that a condition of equilibrium does not imply that nothing is going on. In any system, the numbers of molecules present in the liquid and in the vapor are constant because the two opposing changes are taking place at the same rates and *not* because vaporization and condensation have stopped.

Since the concentration of the molecules in the vapor is a constant at equilibrium, the pressure that the vapor exerts is a constant too. The pressure of vapor in equilibrium with a liquid at a given temperature is called the **equilibrium vapor**

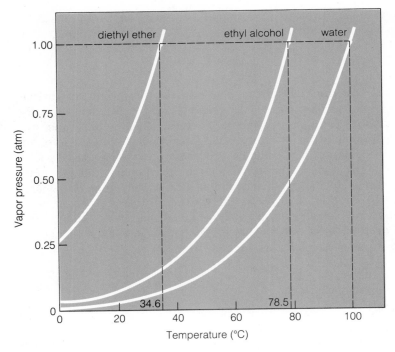

Figure 9.7 Vapor pressure curves for diethyl ether, ethyl alcohol, and water

pressure of the liquid. The vapor pressure of a given liquid is determined by the temperature and increases when the temperature is increased.

Figure 9.7 shows the temperature-vapor pressure curves for diethyl ether, ethyl alcohol, and water. The curves show the increase in vapor pressure that accompanies an increase in temperature. The curve for each substance could be extended to the critical temperature of that substance. At the critical temperature, the vapor pressure equals the critical pressure, and the curves end at this point. Above the critical temperature only one phase can exist—the gas phase.

The magnitude of the vapor pressure of a liquid gives an indication of the strength of the intermolecular attractive forces of that liquid. Liquids that have strong attractive forces have low vapor pressures. At 20°C, the vapor pressure of water is 0.023 atm, of ethyl alcohol is 0.058 atm, and of diethyl ether is 0.582 atm. The forces of attraction are strongest in water and weakest in diethyl ether. A list of the vapor pressures of water at various temperatures is given in Table 8.3.

9.6 Boiling Point

The temperature at which the vapor pressure of a liquid equals the external pressure is called the boiling point of the liquid. At this temperature, vapor produced in the interior of a liquid results in the bubble formation and turbulence that is characteristic of boiling. Bubble formation is impossible at temperatures below the boiling point. The atmospheric pressure on the surface of the liquid prevents the formation of bubbles with internal pressures that are less than the pressure of the atmosphere.

The temperature of a boiling liquid remains constant until all the liquid has been vaporized. In an open container the maximum vapor pressure that can be attained by any liquid is the atmospheric pressure. This vapor pressure corresponds to the boiling point. Heat must be added to a boiling liquid to maintain the temperature because in the boiling process, the high-energy molecules are lost by the liquid. The higher the rate at which heat is added to a boiling liquid, the faster it boils. The temperature of the liquid, however, does not rise.

The boiling point of a liquid changes with changes in external pressure. Water, for example, will boil at 98.6°C at a pressure of 0.950 atm and at 101.4°C at a pressure of 1.05 atm. Only at a pressure of 1.00 atm will water boil at 100°C. The **normal boiling point** of a liquid is defined as the temperature at which the vapor pressure of the liquid equals 1 atm. Boiling points given in reference books are understood to be normal boiling points.

The normal boiling points of diethyl ether (34.6°C), ethyl alcohol (78.5°C), and water are indicated on the vapor pressure curves of Figure 9.7. The boiling point of a liquid can be read from its vapor pressure curve by finding the temperature at which the vapor pressure of the liquid equals the prevailing pressure.

The changes in atmospheric pressure at any one geographic location cause a maximum variation of about 2°C in the boiling point of water. The variations from place to place, however, can be greater than this. The average barometric pressure at sea level is 1 atm. At higher elevations, average barometric pressures are less. At an elevation of 1524 m above sea level, for example, the average barometric pressure is 0.836 atm; at this pressure, water boils at 95.1°C. Water boils at 90.1°C at 0.695 atm, which is the average atmospheric pressure at 3048 m above sea level.

If a liquid has a high normal boiling point or decomposes when heated, it can be made to boil at lower temperatures by reducing the pressure. This procedure is followed in vacuum distillation. Water can be made to boil at 10°C, which is considerably below room temperature, by adjusting the pressure to 0.0121 atm (see Table 8.3). Unwanted water is removed from many food products by boiling it away under reduced pressure. In these procedures, the product is not subjected to temperatures that bring about decomposition or discoloration.

9.7 Enthalpy of Vaporization

The quantity of heat that must be supplied to vaporize a mole of a liquid at a specified temperature is called the **molar enthalpy of vaporization, ΔH_v**. Enthalpies of vaporization are usually recorded at the normal boiling point in kilojoules per mole (see Table 9.2).

Table 9.2 Enthalpies of vaporization of liquids at their normal boiling points			
Liquid	Formula	t_b Normal Boiling Point (°C)	ΔH_v Enthalpy of Vaporization (kJ/mol)
water	H_2O	100.0	40.7
benzene	C_6H_6	80.1	30.8
ethyl alcohol	C_2H_5OH	78.5	38.6
carbon tetrachloride	CCl_4	76.7	30.0
chloroform	$CHCl_3$	61.3	29.4
diethyl ether	C_2H_5OH	34.6	26.0

The magnitude of the molar enthalpy of vaporization gives an indication of the strength of the intermolecular attractive forces. A high enthalpy of vaporization indicates that these forces are strong. The enthalpy of vaporization of a liquid, however, includes both the energy required to overcome the intermolecular attractive forces and the energy needed to expand the vapor. The volume of a gas is considerably larger than the volume of the liquid from which it is derived. A volume of about 1700 ml of steam, for example, is produced by the vaporization of 1 ml of water at 100°C. Energy must be supplied to do the work of pushing back the atmosphere to make room for the vapor.

When a mole of vapor is condensed into a liquid, energy is released, not absorbed. This enthalpy change is called the **molar enthalpy of condensation.** It has a negative sign, but it is numerically equal to the molar enthalpy of vaporization at the same temperature.

The enthalpy of vaporization of a liquid decreases as the temperature increases, and it equals zero at the critical temperature of the substance. This trend parallels an increase in the fraction of high-energy molecules. At the critical temperature, all the molecules have sufficient energy to vaporize.

9.8 The Freezing Point

When a liquid is cooled, the molecules move more and more slowly. Eventually a temperature is reached at which some of the molecules have kinetic energies that are low enough to allow the intermolecular attractions to hold them in crystal lattice. The substance then starts to freeze. Gradually the low-energy molecules assume positions in the crystal lattice. The molecules remaining in the liquid have a higher temperature because of the loss of these low-energy molecules. Heat must be removed from the liquid to maintain the temperature.

The **normal freezing point** of a liquid is the temperature at which solid and liquid are in equilibrium under a total pressure of 1 atm. At the freezing point the temperature of the solid-liquid system remains constant until all of the liquid is frozen. The quantity of heat that must be removed to freeze a mole of a substance at the freezing point is called the **molar enthalpy of crystallization.** This quantity represents the difference between the enthalpies of the liquid and the solid.

At times the molecules of a liquid, as they are cooled, continue the random motion characteristic of the liquid state at temperatures below the freezing point. Such liquids are referred to as **undercooled** or **supercooled liquids.** These systems can usually be caused to revert to the freezing temperature and the stable solid-liquid equilibrium by scratching the interior walls of the container with a stirring rod or by adding a seed crystal around which crystallization can occur. The crystallization process supplies heat, and the temperature is brought back to the freezing point until normal crystallization is complete.

Some supercooled liquids can exist for long periods, or even permanently, in this state. When these liquids are cooled, molecules solidify in a random arrangement typical of the liquid state rather than in an orderly geometric pattern of a crystal. Substances of this type have complex molecular forms for which crystallization is difficult. They are frequently called **amorphous solids, vitreous materials,** or **glasses;** examples include glass, tar, and certain plastics. Amorphous solids have no definite freezing or melting points. These transitions take place over a temperature range. They break into fragments that have curved, shell-like surfaces. Crystalline materials break into fragments that resemble the parent crystals.

Table 9.3 Enthalpies of fusion of solids at their melting points

Solid	Formula	t_f Melting Point (°C)	ΔH_f Enthalpy of Fusion (kJ/mol)
water	H_2O	0.0	6.02
benzene	C_6H_6	5.5	9.83
ethyl alcohol	C_2H_5OH	−117.2	4.60
carbon tetrachloride	CCl_4	−22.9	2.51
chloroform	$CHCl_3$	−63.5	9.20
diethyl ether	$C_2H_5OC_2H_5$	−116.3	7.26

When a crystalline substance is heated, the temperature at which solid-liquid equilibrium is attained under air at 1 atm pressure is called the **melting point.** It is, of course, the same temperature as the freezing point of the substance. The quantity of heat that must be *added* to melt a mole of the material at the melting point is called the **molar enthalpy of fusion,** ΔH_f, and is numerically equal to the enthalpy of crystallization but opposite in sign (Table 9.3).

9.9 Vapor Pressure of a Solid

Molecules in crystal vibrate about their lattice positions. A distribution of kinetic energies exists among these molecules similar to the distribution for liquids and gases but on a lower level. Energy is transmitted from molecule to molecule within a crystal; the energy of any one molecule, therefore, is not constant. High-energy molecules on the surface of the crystal can overcome the attractive forces of the crystal and escape into the vapor phase. If the crystal is in a closed container, an equilibrium is eventually reached in which the rate of the molecules leaving the solid equals the rate at which the vapor molecules return to the crystal. The vapor pressure of a solid at a given temperature is a measure of the number of molecules in a given volume of the vapor at equilibrium.

Every solid has a vapor pressure, although some pressures are very low. The size of the vapor pressure is inversely proportional to the strength of the attractive forces. Ionic crystals, therefore, have very low vapor pressures.

Since the ability of molecules to overcome the intermolecular forces of attraction depends upon their kinetic energies, the vapor pressure of a solid increases as the temperature increases. The temperature-vapor pressure curve for ice is illustrated in Figure 9.8. This curve intersects the vapor pressure curve for water at the freezing point. At the freezing point, the vapor pressures of solid and liquid are equal.

In the absence of air, the normal freezing point of water (1 atm total pressure) is 0.0025°C. *In air*, however, and under a total pressure of 1 atm, the freezing point of water is 0.0000°C, which is the commonly reported value. The difference in freezing point is caused by the presence of dissolved air in the water (Section 10.8). The vapor pressures plotted in Figure 9.8 are the partial pressures of H_2O in air with the total pressure equal to 1 atm. Freezing points are usually determined in air; however, in any event, any change in freezing point of a given substance caused by the presence of air is generally very small.

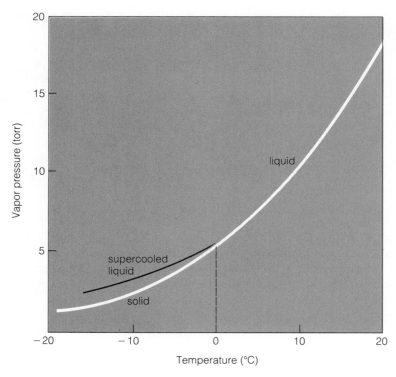

Figure 9.8 Vapor pressure curves for ice and water near the freezing point. (Vapor pressures are partial pressures of H_2O under air at a total pressure of 1.00 atm.)

9.10 Phase Diagrams

The temperature-pressure **phase diagram** for water conveniently illustrates the conditions under which water can exist as solid, liquid, or vapor, as well as the conditions that bring about changes in the state of water. Figure 9.9 is a schematic representation of the water system. It is not drawn to scale, and some of its features are exaggerated so that important details can be easily seen. Every substance has its own phase diagram, which describes only systems in equilibrium. These diagrams are derived from experimental observations.

The diagram of Figure 9.9 relates to what is called a **one-component system**; that is, it pertains to the behavior of water in the absence of any other substance. No part of the total pressure of any system described by the diagram is due to the pressure of a gas other than water vapor. The vapor pressure curves plotted in Figure 9.8 (measured in air under a constant total pressure of 1 atm) therefore deviate slightly, but only very slightly, from the vapor pressure curves of Figure 9.9 (for which the vapor pressure of water is the *total* pressure). The easiest way to interpret the phase diagram for water is to visualize the total pressure acting on a system in mechanical terms, for example, as a piston acting on the material of the system contained in cylinder.

In Figure 9.9 curve OC is a vapor pressure curve for liquid and terminates at the critical point, C. Any point on this line describes a set of temperature and pressure conditions under which liquid and vapor can exist in equilibrium. The extension DO is the curve for supercooled liquid; systems between liquid and vapor described by points on this line are **metastable.** (The term *metastable* is

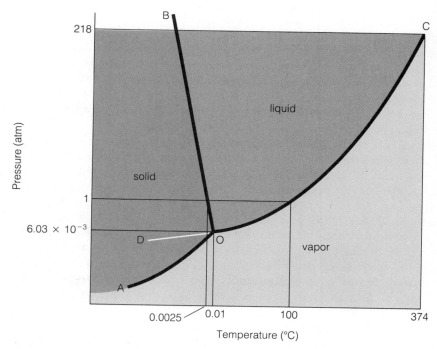

Figure 9.9 Phase diagram for water. (Not drawn to scale.)

applied to systems that are not in the most stable state possible at the temperature in question.) Curve AO is a vapor pressure curve for solid and represents a set of points that describe the possible temperature and pressure conditions for solid-vapor equilibria. The line BO, the melting point curve, represents conditions for equilibria between solid and liquid.

These three curves intersect at point O, a **triple point**. Solid, liquid, and vapor can exist together in equilibrium under the conditions represented by this point: 0.01°C (which is 273.16 K) and a pressure of 0.00603 atm (or 4.58 torr).

The phases (solid, liquid, and vapor) that exist in equilibrium under a set of temperature and pressure conditions can be read from the phase diagram. The temperature and pressure define a point on the diagram. The phases can be read from the position of the point. If the point falls

1. *In a region* labeled solid, liquid, or vapor, only *one phase* exists—the phase noted on the diagram

2. *On a line*, *two phases* exist; the phases are those that are marked in the regions on both sides of the line

3. *On a point*, all *three phases* exist; there is only one such point in the phase diagram given for water—the triple point

The slope of the melting point (or freezing point) curve, BO, shows that the freezing point decreases as the pressure is increased. A slope of this type is observed for only a few substances such as gallium, bismuth, and water. It indicates an unusual situation in which the liquid expands upon freezing. At 0°C a mole of water occupies 18.00 cm³, and a mole of ice occupies 19.63 cm³. The system expands, therefore, when one mole of liquid water freezes into ice. An increase in pressure on the system would oppose this expansion and the freezing process.

Hence, the freezing point of water is lowered when the total pressure is increased. In Figure 9.9 the slope of the line BO is exaggerated.

Phase changes brought about by temperature changes at constant pressure may be read from a phase diagram by interpreting a horizontal line drawn at the reference pressure (like the line drawn at 1.00 atm in Figure 9.9). The point where this line intersects curve BO indicates the normal melting point (or freezing point), and the point where the 1.00 atm line intersects curve CO represents the normal boiling point. Beyond this point, only vapor exists.

Phase changes brought about by pressure changes at constant temperature may be read from a vertical line drawn at the reference temperature. If the pressure is increased, for example, at 0.0025°C (Figure 9.9), the point where the vertical line crosses AO is the pressure where vapor changes to solid, and the point where the vertical line crosses BO represents the pressure where solid changes to liquid. Above this point, only liquid exists.

For materials that contract upon freezing (that is, the solid phase is more dense than the liquid phase), the freezing-point curve slants in the opposite direction, and the freezing point increases as the pressure is increased. This behavior is characteristic of most substances. The freezing-point curves of most phase diagrams slant to the right as is seen in the phase diagram for carbon dioxide in Figure 9.10.

The process in which a solid goes directly into a vapor without going through the liquid state is known as **sublimation**; this process is reversible. The phase diagram for carbon dioxide is typical for substances that sublime at ordinary pressures rather than melt and then boil. The triple point of the carbon dioxide system is −56.6°C at a pressure of 5.11 atm. Liquid carbon dioxide exists only at pressures greater than 5.11 atm. When solid carbon dioxide (dry ice) is heated

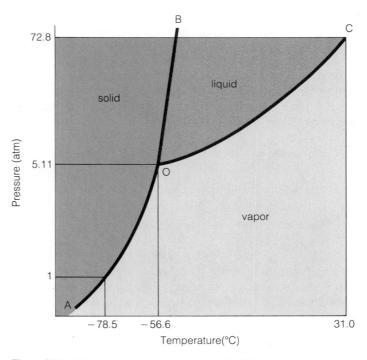

Figure 9.10   Phase diagram for carbon dioxide. (Not drawn to scale.)

Chapter 9   Liquids and Solids

at 1 atm pressure, it is converted directly into gas at $-78.5°C$. This relationship is shown in Figure 9.10. The **molar enthalpy of sublimation** is the heat that must be added to a mole of solid to convert it directly into a gas.

9.11 Types of Crystalline Solids

Crystals are formed by atoms, ions, and molecules. We can classify crystals into four types according to the kind of particles that make up the crystal and the forces that hold them together:

1. Ionic crystals. Positive and negative ions are held in the crystal lattice arrangement by electrostatic attraction. Because these forces are strong, ionic substances have high melting points. Ionic crystals are hard and brittle. Figure 9.11 shows what happens if an attempt is made to deform an ionic crystal. Because of the movement of one plane of ions over another, ions with the same charge are brought next to one another. The crystal breaks into fragments. Ionic compounds are good conductors of electricity when molten or in solution but not in the crystalline state where the ions are not free to move.

2. Molecular crystals. Molecules occupy lattice positions in crystals of covalent compounds. The intermolecular forces that hold the molecules in the crystal structure are not nearly so strong as the electrostatic forces that hold ionic crystals together. Molecular crystals, therefore, are soft and have low melting points.

London forces hold nonpolar molecules in the lattice. In crystals of polar molecules, dipole-dipole forces as well as London forces occur. Polar compounds, therefore, generally melt at slightly higher temperatures than nonpolar compounds of *comparable molecular size and shape.*

In general, molecular substances do not conduct electricity in the solid or liquid states. A few molecular compounds, such as water, dissociate to a very slight extent and produce low concentrations of ions; these liquids are poor electrical conductors.

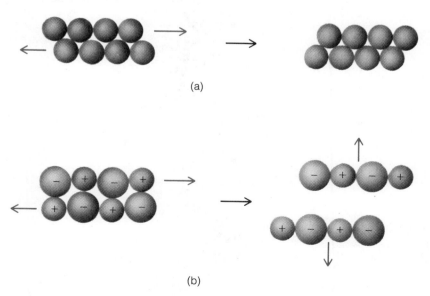

(a)

(b)

Figure 9.11 Effect of deformation on (a) a metallic crystal and (b) an ionic crystal

Figure 9.12 Arrangement of atoms in a diamond crystal

3. Network crystals. In these crystals, atoms occupy lattice positions and they are joined by a network of covalent bonds. The entire crystal can be looked at as one giant molecule. In diamond, an example of this type of crystal, carbon atoms are bonded by covalent bonds into a three-dimensional structure (see Figure 9.12). Materials of this type have high melting points and are extremely hard because of the large number of covalent bonds that would have to be broken to destroy the crystal structure. Network crystals do not conduct electricity.

4. Metallic crystals. The outer electrons of metal atoms are loosely held and move freely throughout a metallic crystal. The remainder of the metal atoms, positive ions, occupy fixed positions in the crystal. The negative cloud of the freely moving electrons, sometimes called an electron gas or a sea of electrons, binds the crystal together. This binding force, called a **metallic bond,** is described more fully in Section 23.1.

The metallic bond is strong. Most metals have high melting points, high densities, and structures in which the positive ions are packed together closely (called **close-packed** arrangements). Unlike ionic crystals, the positions of the positive ions can be altered without destroying the crystal because of the uniform cloud of negative charge provided by the freely moving electrons (see Figure 9.11). Most metallic crystals, therefore, are easily deformed, and most metals are malleable and ductile. The freely moving electrons are also responsible for the fact that most metals are good conductors of electricity.

The properties of the four types of crystals are summarized in Table 9.4.

Table 9.4	Types of crystalline solids			
Crystal	Particles	Attractive Forces	Properties	Examples
ionic	positive and negative ions	electrostatic attractions	high m.p. hard, brittle good electrical conductor in fused state	NaCl, BaO, KNO_3
molecular	polar molecules	London and dipole-dipole	low m.p. soft nonconductor or extremely poor conductor of electricity in liquid state	H_2O, NH_3, SO_2
	nonpolar molecules	London		H_2, Cl_2, CH_4
network	atoms	covalent bonds	very high m.p. very hard nonconductor of electricity	C (diamond) SiC, AlN, SiO_2
metallic	positive ions and mobile electrons	metallic bonds	fairly high m.p. hard or soft malleable and ductile good electrical conductor	Ag, Cu, Na, Fe, K

9.12 Crystals

A crystal is a symmetrical array of atoms, ions, or molecules arranged in a repeating three-dimensional pattern. If the centers of the material units are replaced with points, the resulting system of points is called a **space lattice** or **crystal lattice**. By using a network of lines joining lattice points, a crystal lattice can be divided into identical parts called **unit cells** (Figure 9.13). A crystal lattice can be reproduced, in theory, by stacking its unit cells in three dimensions.

The simplest types of unit cells are the cubic unit cells (Figure 9.14). Notice that it is possible to have points at positions other than the corners of the unit cells. In the body-centered cubic unit cell, a point occurs in the center of the cell. In the face-centered cubic unit cell, a point occurs in the center of each face of the cell.

In crystals of metals, atoms occupy the lattice positions. In counting the number of atoms per unit cell, one must keep in mind that atoms on corners or faces are shared with adjoining cells. Eight unit cells share each corner atom, and two unit cells share each face-centered atom (Figure 9.15).

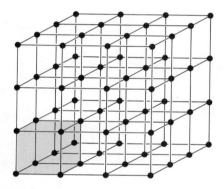

Figure 9.13 Simple cubic space lattice. (A unit cell is shown in color.)

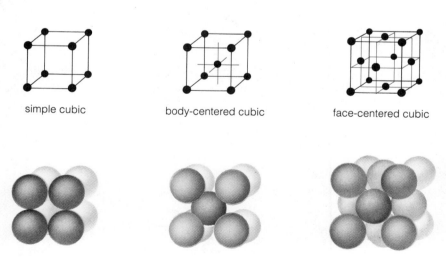

simple cubic   body-centered cubic   face-centered cubic

Figure 9.14 Cubic structures

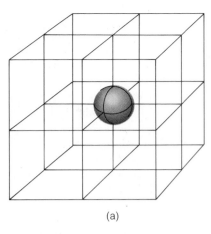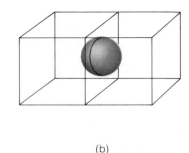

(a)   (b)

Figure 9.15 In cubic crystals, (a) a corner atom is shared by eight unit cells and (b) a face-centered atom is shared by two unit cells

1. The simple cubic unit cell contains the equivalent of only one atom (8 corners at $\frac{1}{8}$ atom each).

2. The body-centered unit cell contains two atoms (8 corners at $\frac{1}{8}$ atom each and one unshared atom in the center).

3. The face-centered unit cell contains the equivalent of four atoms (8 corners at $\frac{1}{8}$ atom each and 6 face-centered atoms at $\frac{1}{2}$ atom each).

Example 9.1

Nickel crystallizes in a face-centered cubic crystal. The edge of a unit cell is 352 pm. The atomic weight of nickel is 58.7, and its density is 8.94 g/cm^3. Calculate Avogadro's number from these data.

Solution

Since 1 pm $= 10^{-10}$ cm,

$$352 \text{ pm} = 3.52 \times 10^{-8} \text{ cm}$$

The volume of one unit cell is $(3.52 \times 10^{-8} \text{ cm})^3$, or 4.36×10^{-23} cm^3. Since the unit cell is face-centered, it contains 4 atoms. Therefore,

$$4 \text{ atoms} = 4.36 \times 10^{-23} \text{ cm}^3$$

We derive from the density,

$$1 \text{ cm}^3 = 8.94 \text{ g Ni}$$

The number of atoms in 58.7 g of Ni is Avogadro's number:

$$? \text{ atoms} = 58.7 \text{ g Ni} \left(\frac{1 \text{ cm}^3}{8.94 \text{ g Ni}} \right) \left(\frac{4 \text{ atoms}}{4.36 \times 10^{-23} \text{ cm}^3} \right) = 6.02 \times 10^{23} \text{ atoms}$$

Example 9.2

Sodium crystallizes in a cubic lattice, and the edge of a unit cell is 430 pm. The density of sodium is 0.963 g/cm³, and the atomic weight of sodium is 23.0. How many atoms of sodium are contained in one unit cell? What type of cubic unit cell does sodium form?

Solution

The edge of the unit cell is 4.30×10^{-8} cm. The volume of the unit cell, therefore, is $(4.30 \times 10^{-8} \text{ cm})^3$, or 7.95×10^{-23} cm³. We must find the number of Na atoms in this volume.

We derive our conversion factors from the density of Na:

$$0.963 \text{ g Na} = 1 \text{ cm}^3$$

and the fact that 1 mol of Na (which is 23.0 g of Na) contains Avogadro's number of Na atoms:

$$6.02 \times 10^{23} \text{ atoms Na} = 23.0 \text{ g Na}$$

The solution is

$$? \text{ atoms Na} = 7.95 \times 10^{-23} \text{ cm}^3 \left(\frac{0.963 \text{ g Na}}{1 \text{ cm}^3} \right) \left(\frac{6.02 \times 10^{23} \text{ atoms Na}}{23.0 \text{ g Na}} \right)$$

$$= 2.00 \text{ atoms Na}$$

Sodium crystallizes in a body-centered cell since the body-centered cubic unit cell contains 2 atoms.

Crystal data can be used to calculate atomic radii:

1. In the case of a simple cubic unit cell, the atomic radius, r, is one-half the length of the edge of the cell, a (see Figure 9.14):

$$r = a/2 \tag{9.1}$$

2. In a face-centered cubic unit cell, the atoms that lie along an edge do not touch. We must calculate the length of the face diagonal (see Figure 9.16a). From the Pythagorean theorem for right triangles,

$$\text{hypotenuse}^2 = \text{side}^2 + \text{side}^2$$
$$(\text{face diagonal})^2 = a^2 + a^2$$
$$= 2a^2$$
$$\text{face diagonal} = a\sqrt{2} \tag{9.2}$$

This diagonal is equal to four radii:

$$4r = a\sqrt{2}$$
$$r = a/\sqrt{8} \tag{9.3}$$

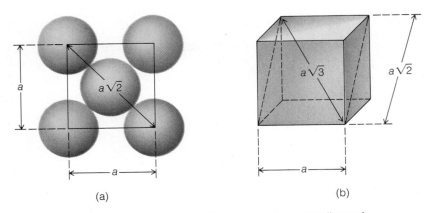

(a)   (b)

Figure 9.16 Determination of (a) a face diagonal and (b) a cube diagonal

3. We must determine the length of a cube diagonal to find the atomic radius of an atom that forms a body-centered cubic unit cell (see Figure 9.16b). From the figure, we see that the diagonal of a cube is the diagonal of a rectangle formed by the edge of the cube, a, and the diagonal of a face, $a\sqrt{2}$. Therefore,

$$(\text{cube diagonal})^2 = a^2 + (a\sqrt{2})^2$$
$$= 3a^2$$
$$\text{cube diagonal} = a\sqrt{3} \tag{9.4}$$

This diagonal is equal to four atomic radii:

$$4r = a\sqrt{3}$$
$$r = a\sqrt{3}/4 \tag{9.5}$$

Example 9.3

Sodium crystallizes in a body-centered cubic unit cell with the length of the edge equal to 430 pm. What is the atomic radius of Na?

Solution

The cube diagonal of the unit cell is

$$\text{cube diagonal} = a\sqrt{3}$$
$$= (430 \text{ pm})\sqrt{3}$$
$$= 745 \text{ pm}$$

This length is four atomic radii:

$$4r = 745 \text{ pm}$$
$$r = 186 \text{ pm}$$

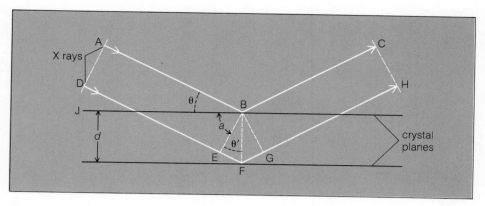

Figure 9.17 Derivation of the Bragg equation

9.13 X-ray Diffraction of Crystals

Much of what is known about the internal structure of crystals has been learned from X-ray diffraction experiments. When two X rays that have the same wavelength are in phase, they reinforce each other and produce a wave that is stronger than either of the original waves. Two waves that are completely out of phase cancel each other (see Figure 7.8).

Figure 9.17 illustrates the way that crystal spacings can be determined by use of X rays of a single wavelength, λ. The rays strike the parallel planes of the crystal at an angle θ. Some of the rays are reflected from the upper plane, some from the second plane, and some from the lower planes. A strong reflected beam will result only if all the reflected rays are in phase.

In Figure 9.17 the lower ray travels farther than the upper ray by an amount equal to EF + FG. The rays will be in phase at BG only if the difference is equal to a whole number of wavelengths:

$$EF + FG = n\lambda$$

where n is a simple integer.

Since angle ABE is a right angle,

$$\theta + a = 90°$$

Angle JBF is also a right angle, and

$$\theta' + a = 90°$$

The angle θ' therefore is equal to θ. The sine of angle θ' is equal to EF/BF (the ratio of the side opposite the angle to the hypotenuse). Since line BF is equal to d,

$$\sin \theta = \frac{EF}{d}$$

or

$$EF = d \sin \theta$$

The expression

$$FG = d \sin \theta$$

can be derived in the same way. Therefore,

$$EF + FG = 2d \sin \theta$$

Since EF + FG is equal to $n\lambda$,

$$n\lambda = 2d \sin \theta \tag{9.6}$$

This equation, derived by William Henry Bragg and his son William Lawrence Bragg in 1913, is called the **Bragg equation.**

With X rays of a definite wavelength, reflections at various angles will be observed for a given set of planes separated by a given distance, d. These reflections correspond to $n = 1, 2, 3$, and so on, and are spoken of as first order, second order, third order, and so on. With each successive order, the angle θ increases, and the intensity of the reflected beam weakens.

Figure 9.18 is a schematic representation of an X-ray spectrometer. An X-ray beam defined by a slit system impinges upon a crystal that is mounted on a turn-table. A detector (photographic plate, ionization chamber, or Geiger counter) is positioned as shown in the figure. As the crystal is rotated, strong signals flash out as angles are passed that satisfy the Bragg equation. Any set of regularly positioned planes that contain atoms can give rise to reflections—not only those that form the faces of the unit cells. Thus, the value of d is not necessarily the edge of the unit cell, although the two are always mathematically related.

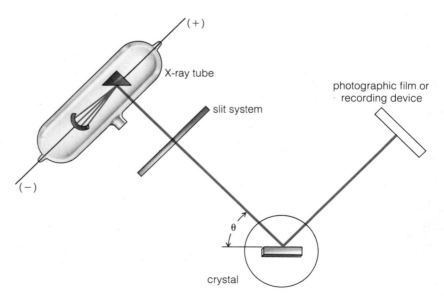

Figure 9.18   X-ray diffraction of crystals. (Schematic.)

Example 9.4

The diffraction of a crystal of barium with X radiation of wavelength 229 pm gives a first-order reflection at 27° 8′. What is the distance between the diffracted planes?

Solution

Substitution into the Bragg equation gives

$$n\lambda = 2d \sin \theta$$

$$1(229 \text{ pm}) = 2d(0.456)$$

$$d = 251 \text{ pm}$$

9.14 Crystal Structure of Metals

In the large majority of cases, metal crystals belong to one of three classifications: body-centered cubic (Figure 9.14), face-centered cubic (Figure 9.14), and hexagonal close-packed (Figure 9.19). The geometric arrangement of atoms in the face-centered cubic and the hexagonal close-packed crystals is such that each atom has a coordination number of 12 (each is surrounded by 12 other atoms at equal distances). If the atoms are viewed as spheres, there is a minimum of empty space in these two types of crystals (about 26%), and both of the crystal lattices are called **close-packed structures.** The body-centered cubic arrangement is slightly more open than either of the close-packed arrangements (about 32% empty space); each atom of a body-centered cubic crystal has a coordination number of 8.

The difference between the two close-packed structures may be derived from a consideration of Figure 9.20. The shaded circles of the diagram represent the first layer of spheres, which are placed as close together as possible. The second layer of spheres (open circles of Figure 9.20) are placed in the hollows formed by adjacent spheres of the first layer. The first two layers of both the face-centered cubic and the hexagonal close-packed arrangements are the same; the difference arises in the third and subsequent layers.

Figure 9.19 Hexagonal close-packed structure

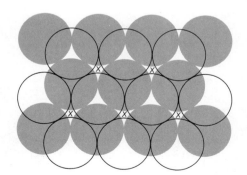

Figure 9.20 Schematic representation of the first two layers of the close-packed arrangements

Table 9.5 Crystal structures of metals

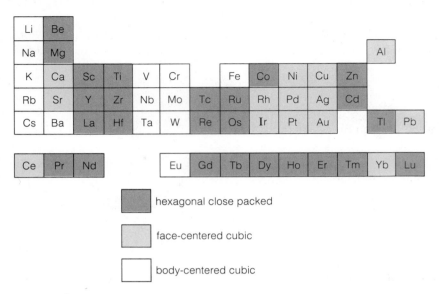

| hexagonal close packed |
| face-centered cubic |
| body-centered cubic |

In the hexagonal close-packed arrangement, the spheres of the third layer are placed so that they are directly over those of the first layer; the sequence of layers may be represented as *ababab* . . . . In the face-centered cubic structure, however, the spheres of the third layer are placed over the holes (marked *x* in Figure 9.20) formed by the arrangements of the first two layers. The spheres of the fourth layer of the face-centered cubic structure are placed so that they are directly over those of the first layer, and the sequence of layers is *abcabc* . . . .

The crystal structures of metals are summarized in Table 9.5. The division between the three structures (hexagonal close-packed, face-centered cubic, and body-centered cubic) is about even. The close packing of most metallic crystals helps to explain the relatively high densities of metals. The structures of a few metals (for example, manganese and mercury) do not fall into any of the three categories, and the symbols for these metals do not appear in the table. Some metals exhibit **crystal allotropy**; that is, different crystal forms of the same metal are stable under different conditions. For example, a modification of calcium exists in each of the three structures. The form of each metal that is indicated in Table 9.5 is the one that is stable under ordinary conditions. In addition to many metals, the noble gases also crystallize in the face-centered cubic lattice (except for He).

9.15 Ionic Crystals

The structures of ionic crystals are more complicated than those of metallic crystals. An ionic crystal must accommodate ions of opposite charge and different size in the proper stoichiometric ratio and in such a way that electrostatic attractions outweigh electrostatic repulsions.

The potential energy of interaction (*PE*) between two ions is directly proportional to the product of the charges between the two ions (q_1 and q_2) and inversely proportional to the distance between the centers of the two ions (*d*):

$$PE = \frac{q_1 q_2}{d} \tag{9.7}$$

cesium chloride
CsCl

sodium chloride
NaCl

zinc blende
ZnS

Figure 9.21 Crystal structures of ionic compounds of the MX type. (Colored spheres represent cations.)

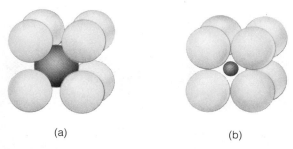

(a)

(b)

Figure 9.22 The cesium chloride structure (a) with cation-anion contact and (b) without cation-anion contact. (The colored spheres represent the cation.)

If the charges have the same sign (both positive or both negative), they will repel each other and the potential energy will be a *positive* value (energy is *required* to push the ions together). On the other hand, if the charges have unlike signs, they will attract each other and the potential energy will be a *negative* value (energy is *released* when the ions come together). The most stable structure for a given compound, therefore, is one in which the largest possible number of cation-anion attractions exist and one in which the positive and negative ions are as close together as possible (small value of d).

The three most common crystal types for ionic compounds of formula MX are shown in Figure 9.21. In each diagram, there are as many cations per unit cell as there are anions (one each in CsCl and four each in NaCl and ZnS, taking into account sharing by adjacent cells). A 1:1 stoichiometric ratio is represented, therefore, in each case.

In the CsCl structure, the Cs^+ ion (the central ion in the CsCl structure shown in Figure 9.21) has eight Cl^- ions as nearest neighbors (the Cs^+ ion is said to have a coordination number of 8). In the NaCl crystal, a Na^+ ion has six Cl^- ions as nearest neighbors (a coordination number of 6). In the ZnS structure, each Zn^{2+} ion has four S^{2-} ions as nearest neighbors (a coordination number of 4). In terms of crystal stability brought about by plus-minus attractions, the CsCl crystal, in which each Cs^+ ion has a coordination number of 8, would appear to be best.

There is, however, another factor to take into consideration: the distance between the cation and the anion, d, which becomes a matter of the comparison of the size of the cation to the size of the anion. In the CsCl structure, the Cs^+ ion in the center is assumed to touch each of the Cl^- ions at the corners (see Figure 9.22a). Consider a compound in which the cation is much smaller

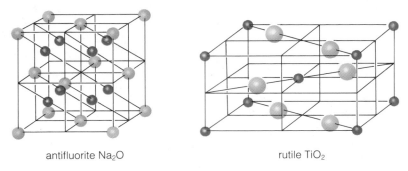

antifluorite Na$_2$O   rutile TiO$_2$

Figure 9.23 Crystal structures of ionic compounds of the MX$_2$ type. (Colored spheres represent cations.)

Table 9.6 Crystal structures of some ionic compounds	
Structure	**Examples**
cesium chloride	CsCl, CsBr, CsI, TlCl, TlBr, TlI, NH$_4$Cl, NH$_4$Br
sodium chloride	halides of Li$^+$, Na$^+$, K$^+$, Rb$^+$ oxides and sulfides of Mg^{2+}, Ca^{2+}, Sr^{2+}, Ba^{2+}, Mn^{2+}, Ni^{2+} AgF, AgCl, AgBr, NH$_4$I
zinc blende	sulfides of Be^{2+}, Zn^{2+}, Cd^{2+}, Hg^{2+} CuCl, CuBr, CuI, AgI
fluorite	fluorides of Ca^{2+}, Sr^{2+}, Ba^{2+}, Cd^{2+}, Pb^{2+} BaCl$_2$, SrCl$_2$, ZrO$_2$, ThO$_2$, UO$_2$
antifluorite	oxides and sulfides of Li$^+$, Na$^+$, K$^+$, Rb$^+$
rutile	fluorides of Mg^{2+}, Ni^{2+}, Mn^{2+}, Zn^{2+}, Fe^{2+} oxides of Ti^{4+}, Mn^{4+}, Sn^{4+}, Te^{4+}

than the anion. A small cation could not touch the anions that surround it (see Figure 9.22b) since electrostatic repulsions between the anions would prevent the anions from being squeezed together. As a result, the size of the plus-minus attraction would be comparatively low (d would be larger than necessary). In such an event, the compound would crystallize in a pattern with a lower coordination number for the cation but one that would permit closer cation-anion contact.

Two structures for ionic compounds of formula MX$_2$ or M$_2$X are shown in Figure 9.23. A third structure, the fluorite structure (named for the mineral fluorite, CaF$_2$), is similar to the antifluorite structure shown in Figure 9.23 except that cation and anion positions are exchanged. Examples of ion compounds that crystallize in the types of crystal structures we have mentioned are listed in Table 9.6.

9.16 Defect Structures

Few crystals are perfect; many have some type of lattice defect. **Dislocations** are crystal imperfections that occur where planes of atoms are misaligned. For example, one type of dislocation is caused by the insertion (perpendicular to a face of the crystal) of an extra plane of atoms part way through a crystal. Atoms within the part of the crystal containing the extra plane are compressed.

Point defects are caused by missing or misplaced ions. One type of defect consists of a cation that has been moved from its proper position (thus creating a vacancy) to a place between regular lattice sites (an interstitial position). Another type of point defect consists of a pair of vacancies—one cation and one anion; the ions for these lattice positions are missing from the structure completely. These two point defects do not alter the stoichiometry of the crystal.

Certain crystals are imperfect because their compositions are not stoichiometric. Thus, samples of iron(II) oxide, FeO, usually contain more oxygen atoms than iron atoms, whereas zinc oxide, ZnO, usually has more zinc atoms than oxygen atoms. These departures from stoichiometry are usually quite small, on the order of 0.1%.

There are several causes for nonstoichiometry, and in each the electrical neutrality of the crystal is preserved. Extra metal *atoms* or nonmetal *atoms* may be included in interstitial positions between the *ions* of the crystal (such as extra Zn atoms in ZnO). In addition, metal *atoms* or nonmetal *atoms* may assume regular lattice positions in place of *ions;* in these cases, ionic vacancies exist in the crystal lattice so that the whole lattice is electrically neutral. In FeO, for example, oxygen atoms assume positions normally occupied by oxide ions, and there are missing Fe^{2+} ions so that a 1 to 1 cation-anion ratio is maintained. In some crystals of this type (for example, KCl), electrons from the "extra" metal atoms occupy the holes created by the anion vacancies.

Nonstoichiometric substances are sometimes referred to as berthollides, named for Claude Louis Berthollet, who believed that the compositions of compounds varied continuously within limits. Joseph Proust, who upheld the law of definite proportions, engaged Berthollet in an argument of eight years' duration (1799–1807) concerning the composition of compounds.

The presence of impurities frequently accounts for crystal defects. For example, a Mg^{2+} ion may occur in a lattice position in a NaCl crystal in place of a Na^+ ion; electrical neutrality requires that another lattice position be vacant, since the charge of the Mg^{2+} ion is twice that of the Na^+ ion. The addition of impurities to certain crystals produces materials that are called semiconductors (see Section 23.2).

Summary

The topics that have been discussed in this chapter are

1. Intermolecular forces of attraction, which hold molecules together in condensed states (liquids and solids); dipole-dipole forces and London forces.

2. The hydrogen bond, an unusually strong type of intermolecular attraction.

3. Some properties of liquids, the process of evaporation, vapor pressure, and the boiling point.

4. The enthalpy changes involved in condensation and vaporization.

5. The freezing point, enthalpy of crystallization, enthalpy of fusion, and vapor pressure of solids.

6. Phase diagrams, which summarize the phases of a substance that exist in equilibrium under a given set of temperature and pressure conditions.

7. The types of crystalline solids; ionic, molecular, network, and metallic crystals.

8. Cubic unit cells and calculations involving these types of crystal lattice.

9. The determination of crystal spacing by X-ray diffraction; the Bragg equation.

10. The crystal structure of metals.

11. Types of ionic crystals.

12. Crystal imperfections.

Key Terms

Some of the more important terms introduced in this chapter are listed below. Definitions for terms not included in this list may be located in the text by use of the index.

Berthollide (Section 9.16)   A nonstoichiometric substance.

Boiling point (Section 9.6)   The temperature at which the vapor pressure of a liquid equals the external pressure. The **normal boiling point** is the temperature at which the vapor pressure of a liquid equals 1 atm.

Bragg equation (Section 9.13)   An equation that relates the angles at which an X ray with a known wavelength is reflected from a crystal to the distance between the crystal planes.

Close-packed crystal (Sections 9.11 and 9.14)   A crystal structure in which the atoms are so efficiently packed that a maximum number are included in a given volume; a face-centered cubic or hexagonal close-packed crystal.

Crystal (Sections 9.11 and 9.12)   A solid composed of a symmetrical array of atoms, ions, or molecules arranged in a repeating three-dimensional pattern.

Crystal defect (Section 9.16)   A crystal imperfection caused by a dislocation, missing or misplaced ions, or the presence of impurities.

Crystal lattice (Section 9.12)   A three-dimensional, symmetrical pattern of points that defines a crystal.

Dipole-dipole force (Section 9.1)   An intermolecular force caused by the mutual attraction of oppositely charged poles of neighboring polar molecules.

Enthalpy of condensation (Section 9.7)   The enthalpy change associated with the condensation of a given quantity of vapor (usually one mole or one gram) into a liquid at a specified temperature.

Enthalpy of crystallization (Section 9.8)   The enthalpy change associated with the conversion of a given quantity of a liquid (usually one mole or one gram) into a solid at a specified temperature.

Enthalpy of fusion (Section 9.8)   The energy required to melt a given quantity of a solid (usually one mole or one gram) at a specified temperature.

Enthalpy of vaporization (Sections 9.4 and 9.7)   The energy required to vaporize a given quantity of a liquid (usually one mole or one gram) at a specified temperature.

Equilibrium (Section 9.5)   A condition in which the rates of two opposite tendencies are equal.

Evaporation; vaporization (Section 9.4)   The process in which a liquid is converted into a gas.

Freezing point (Section 9.8)   The temperature at which solid and liquid phases are in equilibrium. If the total pressure is 1 atm, the value is called a **normal freezing point**.

Hydrogen bond (Section 9.2)   An intermolecular attraction that occurs between molecules in which hydrogen is bonded to a small, highly electronegative atom (principally N, O, and F).

Instantaneous dipole (Section 9.1)   A fluctuating, temporary dipole induced in molecules by the motion of electrons.

Intermolecular forces (Section 9.1)   Forces of attraction between molecules; principally evident in liquids and solids.

London forces (Section 9.1)   Intermolecular forces brought about by attractions between instantaneous dipoles.

Melting point (Section 9.8)   See freezing point.

Phase diagram (Section 9.10)   A diagram that presents, graphically, the number and type of phases in which a chemical system exists under a given set of conditions of temperature and pressure.

Sublimation (Section 9.10)   The process in which a solid goes directly into a vapor without going through the liquid state.

Surface tension (Section 9.3)   A measure of the inward force on the surface of a liquid caused by intermolecular attractive forces.

Triple point (Section 9.10)   The temperature and pressure at which a substance can simultaneously exist as a solid, a liquid, and a gas in equilibrium with each other.

Unit cell (Section 9.12)   The smallest part of a crystal that will reproduce the crystal when repeated in three dimensions.

Vapor pressure (Section 9.5)   The pressure of vapor in equilibrium with a pure liquid or a pure solid at a given temperature.

Viscosity (Section 9.3)   A property of liquids; resistance to flow.

Problems*

Intermolecular Attractive Forces

9.1 Describe the difference between London forces and dipole-dipole forces. In what types of molecular substances do each exist? Which type is stronger in *most* molecular substances where both exist?

9.2 Give an explanation for the following. **(a)** The dipole moment of OF_2 is 0.30 D but the dipole moment of BeF_2 is zero. **(b)** The dipole moment of PF_3 is 1.03 D but the dipole moment of BF_3 is zero. **(c)** The dipole moment of SF_4 is 0.63 D but the dipole moment of SnF_4 is zero.

9.3 Which molecules (not ions) given as examples in Table 7.1 would you expect to have a dipole moment of zero?

9.4 How could measurement of the dipole moment of the trigonal bipyramidal molecule PCl_2F_3 help to determine whether the Cl atoms occupy axial or equatorial positions?

9.5 The dipole moment of PF_3 is 1.03 D, whereas PF_5 has no dipole moment. Explain.

9.6 The dipole moment of NH_3 (1.49 D) is greater than that of NF_3 (0.24 D). On the other hand, the dipole moment of PH_3 (0.55 D) is less than that of PF_3 (1.03 D). Explain these results.

9.7 Explain why the dipole moment of SCO is 0.72 D, whereas the dipole moment of CO_2 is zero. Would CS_2 have a dipole moment?

9.8 In Table 6.1 the electronegativity of C is given as 2.6 and that of O is given as 3.4. On the other hand, in Table 9.1 the dipole moment of CO is listed as only 0.12 D (the dipole-dipole forces of CO are negligible). Draw the Lewis structure of CO and offer an explanation for the low dipole moment of CO.

9.9 Explain why the following melting points fall in the order given: F_2 ($-233°C$), Cl_2 ($-103°C$), Br_2 ($-7°C$), and I_2 (113.5°C).

9.10 Consider the following molecules, each of which is tetrahedral with the C atom as the central atom: CH_4, CH_3Cl, CH_2Cl_2, $CHCl_3$, CCl_4. In which of these compounds would dipole-dipole forces exist in the liquid state? In what order would you expect the boiling points of the compounds to fall?

The Hydrogen Bond

9.11 The effect of hydrogen bonding on the properties of the following hydrides falls in the order given: $H_2O >$ $HF > NH_3$. Explain this observation.

9.12 The compound KHF_2 can be prepared from the reaction of KF and HF in water solution. Explain the structure of the HF_2^- ion.

9.13 Diagram the structures of the following molecules and explain how the water solubility of each compound is enhanced by hydrogen bonding: **(a)** NH_3, **(b)** H_2N—OH, **(c)** H_3COH, **(d)** H_2CO.

9.14 Although there are exceptions, most acid salts (such as $NaHSO_4$) are more soluble in water than the corresponding normal salts (Na_2SO_4). Offer an explanation for this generalization.

9.15 Offer an explanation as to why chloroform, $HCCl_3$, and acetone,

$$CH_3 - \overset{\overset{\displaystyle O}{\|}}{C} - CH_3$$

mixtures have higher boiling points than either pure component.

9.16 The normal boiling point of the compound ethylene diamine, $H_2NCH_2CH_2NH_2$, is 117°C and that of propyl amine, $CH_3CH_2CH_2NH_2$, is 49°C. The molecules, however, are similar in size and molecular weight. What reason can you give for the difference in boiling point?

The Liquid State

9.17 Briefly explain how and why each of the following gives an indication of the strength of the intermolecular forces of attraction of a substance: **(a)** critical temperature, **(b)** surface tension, **(c)** viscosity, **(d)** vapor pressure, **(e)** enthalpy of vaporization, **(f)** normal boiling point.

9.18 Explain, using a Maxwell-Boltzmann distribution curve, why an evaporating liquid becomes cool.

9.19 What is an equilibrium state? Describe the condition that exists when an evaporating liquid is placed in a container.

9.20 Why does the boiling point of a liquid vary with pressure? What is the normal boiling point? Use the curves of Figure 9.7 to estimate the boiling point of diethyl ether, ethyl alcohol, and water at a pressure of 0.50 atm.

9.21 Use the data of Table 8.3 to estimate the boiling point of water at **(a)** 0.010 atm and **(b)** 0.025 atm.

Phase Diagrams

9.22 Use the following data to draw a rough phase diagram for hydrogen: Normal melting point, 14.01 K; normal boiling point, 20.38 K; triple point, 13.95 K, 7×10^{-2} atm; critical point, 33.3 K, 12.8 atm; vapor pressure of solid at 10 K, 1×10^{-3} atm.

9.23 Use the following data to draw a rough phase diagram for krypton: Normal boiling point, $-152°C$;

---

* The more difficult problems are marked with asterisks. The appendix contains answers to color-keyed problems.

normal melting point, $-157°C$; triple point, $-169°C$, 0.175 atm; critical point, $-63°C$, 54.2 atm; vapor pressure of solid at $-199°C$, 1.3×10^{-3} atm. Which has the higher density at a pressure of 1 atm: solid Kr or liquid Kr?

9.24 Figure 9.9 is the phase diagram for water. Describe the phase changes that occur, and the approximate pressures at which they occur, when the pressure on a H_2O system is gradually increased **(a)** at a constant temperature of $-1°C$, **(b)** at a constant temperature of $50°C$, **(c)** at a constant temperature of $-50°C$.

9.25 Refer to Figure 9.9 and describe the phase changes that occur, and the approximate temperatures at which they occur, when water is heated from $-10°C$ to $110°C$ **(a)** under a pressure of 1×10^{-3} atm, **(b)** under a pressure of 0.5 atm, **(c)** under a pressure of 1.1 atm.

9.26 Refer to Figure 9.10 and describe the phase changes that occur and the approximate pressures at which they occur when the pressure on a CO_2 system is gradually increased **(a)** at a constant temperature of $-60°C$, **(b)** at a constant temperature of $0°C$.

9.27 Refer to Figure 9.10 and describe the phase changes that occur, and the approximate temperatures at which they occur, when carbon dioxide is heated **(a)** at a constant pressure of 2.0 atm, **(b)** at a constant pressure of 6.0 atm.

9.28 Briefly explain how and why the slope of the melting-point curve of a phase diagram for a substance depends upon the relative densities of the solid and liquid forms of the substance.

9.29 Sublimation is sometimes used to purify solids. The impure material is heated, and the pure crystalline product condenses on a cold surface. Is it possible to purify ice by sublimation? What conditions would have to be employed?

Types of Crystalline Solids

9.30 List the types of crystals; tell the particles of which each is composed and the kinds of forces that hold them together.

9.31 What type of forces must be overcome in order to melt crystals of the following: **(a)** Si, **(b)** Ba, **(c)** F_2, **(d)** BaF_2, **(e)** BF_3, **(f)** PF_3.

9.32 What type of forces must be overcome in order to melt crystals of the following: **(a)** O_2, **(b)** Br_2, **(c)** Br_2O, **(d)** Ba, **(e)** $BaBr_2$, **(f)** BaO?

9.33 Which substance of each of the following pairs would you expect to have the higher melting point: **(a)** ClF or BrF, **(b)** BrCl or Cl_2, **(c)** CsBr or BrCl, **(d)** Cs or Br_2, **(e)** C (diamond) or Cl_2? Why?

9.34 Which substance of each of the following pairs would you expect to have the higher melting point: **(a)** Sr or Cl_2, **(b)** $SrCl_2$ or $SiCl_4$, **(c)** $SiCl_4$ or $SiBr_4$, **(d)** $SiCl_4$ or SCl_4, **(e)** SiC (carborundum) or $SiCl_4$? Why?

9.35 Which substance of each of the following pairs would you expect to have the higher melting point:

(a) Li or H_2, **(b)** LiH or H_2, **(c)** Li or LiH, **(d)** H_2 or Cl_2, **(e)** H_2 or HCl ?

Crystals

9.36 Neon crystallizes in a face-centered cubic lattice, and the edge of the unit cell is 452 pm. The atomic weight of neon is 20.2. What is the density of crystalline neon?

9.37 Barium crystallizes in a body-centered cubic lattice, and the edge of the unit cell is 502 pm. The atomic weight of barium is 137. What is the density of crystalline barium?

9.38 Polonium crystallizes in a cubic system, and the edge of the unit cell is 336 pm. The density of Po is 9.20 g/cm^3, and the atomic weight of Po is 210. What type of cubic unit cell does Po form?

9.39 Gold crystallizes in a cubic system, and the edge of the unit cell is 407 pm. The density of Au is 19.4 g/cm^3, and the atomic weight of Au is 197. What type of cubic unit cell does Au form?

9.40 An element crystallizes in a body-centered cubic lattice, and the edge of the unit cell is 286 pm. The density of the element is 7.92 g/cm^3. Find the atomic weight of the element.

9.41 An element crystallizes in a face-centered cubic lattice, and the edge of the unit cell is 556 pm. The density of the element is 1.55 g/cm^3. Find the atomic weight of the element.

9.42 Tungsten crystallizes in a body-centered cubic lattice. The density of W is 19.4 g/cm^3, and the atomic weight of W is 184. What is the length of an edge of the unit cell?

9.43 Palladium crystallizes in a face-centered cubic lattice. The density of Pd is 12.0 g/cm^3, and the atomic weight of Pd is 106. What is the length of an edge of a unit cell?

9.44 Potassium crystallizes in a body-centered cubic lattice, and the length of an edge of the unit cell is 533 pm. The atomic weight of K is 39.1, and the density of K is 0.858 g/cm^3. Use these data to calculate Avogadro's number.

9.45 Silver crystallizes in a face-centered cubic lattice, and the edge of a unit cell is 407.7 pm. Calculate the dimensions of a cube that would contain one mole of Ag (107.7 g).

9.46 Copper crystallizes in a face-centered cubic lattice, and the edge of a unit cell is 361 pm. Calculate the atomic radius of Cu.

9.47 Chromium crystallizes in a body-centered cubic lattice, and the edge of a unit cell is 287.5 pm. Calculate the atomic radius of Cr.

9.48 Molybdenum has an atomic radius of 136 pm and crystallizes in a body-centered cubic lattice. What is the length of an edge of a unit cell?

X-Ray Diffraction of Crystals

9.49 In the diffraction of a crystal using X rays with a wavelength of 135 pm, a first-order reflection was obtained at an angle of 11.2°. What is the distance between the diffracted planes? The sine of 11.2° is 0.1942.

9.50 In the diffraction of a crystal using X rays with a wavelength of 194 pm, a first-order reflection was obtained at an angle of 25.9°. What is the distance between the diffracted planes? The sine of 25.9° is 0.4368.

9.51 In the X-ray diffraction of a set of crystal planes for which d is 248 pm, a first-order reflection was obtained at an angle of 8.21°. What is the wavelength of the X rays that were employed? The sine of 8.21° is 0.1428.

9.52 In the diffraction of a crystal using X rays with a wavelength of 179 pm, a first-order reflection is found at an angle of 21.0°. What is the wavelength of X rays that show the same reflection at an angle of 25.0°? The sine of 21.0° is 0.3584, and the sine of 25.0° is 0.4226.

9.53 At what angle would a first-order reflection be observed in the X-ray diffraction of a set of crystal planes for which d is 300 pm if the X rays used have a wavelength of 154 pm? At what angle would a second-order reflection be observed?

9.54 At what angle would a first-order reflection be observed in the X-ray diffraction of a set of crystal planes for which d is 349 pm if the X rays used have a wavelength of 194 pm? At what angle would a second-order reflection be observed?

Ionic Crystals

9.55 (a) How many ions of each type are shown in the unit cell of the cesium chloride crystal diagrammed in Figure 9.21? **(b)** The density of cesium chloride is 3.99 g/cm³. What is the length of the edge of the unit cell shown in the figure? **(c)** Use the equations given in Section 9.12 to determine the shortest distance between a Cs^+ ion and a Cl^- ion.

9.56 (a) How many ions of each type are shown in the unit cell of the sodium chloride crystal depicted in Figure 9.21? **(b)** The density of silver chloride, which crystallizes in a sodium chloride lattice, is 5.57 g/cm³. What is the length of the edge of the AgCl unit cell of the type shown in the figure? **(c)** Use the equations given in Section 9.12 to determine the shortest distance between a Ag^+ ion and a Cl^- ion.

9.57 (a) How many ions of each type are shown in the unit cell of the zinc sulfide crystal depicted in Figure 9.21? **(b)** The density of copper (I) chloride, which crystallizes in a zinc blende lattice, is 4.14 g/cm³. What is the length of the edge of the CuCl unit cell of the type shown in the figure? **(c)** Use the equations given in Section 9.12 to determine the shortest distance between a Cu^+ ion and a Cl^- ion.

*****9.58 (a)** Thallium(I) chloride crystallizes in the cesium chloride lattice. The shortest distance between a Tl^+ ion and a Cl^- ion is 333 pm. What is the length of the edge of a unit cell of TlCl similar to that shown in Figure 9.21? **(b)** What is the density, in g/cm³, of TlCl?

*****9.59 (a)** Lead(II) sulfide crystallizes in the sodium chloride lattice. The shortest distance between a Pb^{2+} ion and a S^{2-} ion is 297 pm. What is the length of the edge of a unit cell of PbS similar to that shown in Figure 9.21? **(b)** What is the density, in g/cm³, of PbS?

*****9.60 (a)** Cadmium sulfide crystallizes in the zinc blende lattice. The shortest distance between a Cd^{2+} ion and a S^{2-} ion is 253 pm. What is the length of the edge of a unit cell similar to that shown in Figure 9.21? **(b)** What is the density of CdS?

9.61 In a given crystal, the distance between the centers of a cation and a neighboring anion is approximately equal to the sum of the ionic radii of the two ions. Some ionic radii are: Na^+, 95 pm; K^+, 133 pm; Ca^{2+}, 99 pm; Ba^{2+}, 135 pm; Ni^{2+}, 69 pm; Ag^+, 126 pm; Cl^-, 181 pm; Br^-, 195 pm; O^{2-}, 140 pm; S^{2-}, 184 pm. Refer to Equation 9.7 in Section 9.15 and arrange the following (each of which crystallizes in a sodium-chloride type lattice) in decreasing order of lattice energy (most negative value first): AgCl, BaO, CaS, KCl, NaBr, and NiS.

Defect Structures

9.62 List and describe the types of defects found in crystals.

*****9.63** Cadmium oxide, CdO, crystallizes in the sodium chloride structure, with four Cd^{2+} and four O^{2-} ions per unit cell (see Figure 9.21). The compound, however, is usually nonstoichiometric, with a formula that approximates $CdO_{0.995}$. The defect occurs because some cation positions in the crystal are occupied by Cd atoms instead of Cd^{2+} ions and an equivalent number of anion positions are vacant. **(a)** What percent of the anion sites are vacant? **(b)** If the edge of a unit cell is 469.5 pm, what would be the density of a perfect crystal? **(c)** What is the density of the nonstoichiometric crystal? The atomic weight of Cd is 112.40 and of O is 16.00.

*****9.64** Iron(II) oxide crystallizes in the sodium chloride structure, with four Fe^{2+} and four O^{2-} ions per unit cell (see Figure 9.21). The crystals, however, are always deficient in iron. Some cation sites are vacant and some cation sites contain Fe^{3+} ions instead of Fe^{2+} ions, but the combination is such that the structure is electrically neutral. The formula usually approximates $Fe_{0.93}O$. **(a)** What is the ratio of Fe^{2+} ions to Fe^{3+} ions in the crystal? **(b)** What percentage of the cation sites are vacant? Hint: consider a crystal that contains 100 O^{2-} ions.

*****9.65 (a)** The edge of the NaCl unit cell shown in Figure 9.21 is 563.8 pm and the density of NaCl is 2.165 g/cm³. Use these data to calculate the apparent molecular weight of NaCl to four significant figures. **(b)** The difference between the value calculated from crystal data and the actual molecular weight of NaCl (58.44) is ascribed to a type of lattice defect in which Na atoms replace a number of Na^+ ions in the crystal and an equal number of Cl^- ions are missing from lattice positions. On the basis of your answer to part (a), calculate the percentage of anion sites that are vacant.

CHAPTER SOLUTIONS
10

Solutions are homogeneous mixtures. They are usually classified according to their physical state; gaseous, liquid, and solid solutions can be prepared. Dalton's law of partial pressures describes the behavior of gaseous solutions, of which air is the most common example. Certain alloys are solid solutions; coinage silver is copper dissolved in silver, and brass is a solid solution of zinc in copper. Not all alloys are solid solutions, however. Some are heterogeneous mixtures, and some are intermetallic compounds. Liquid solutions are the most common and are probably the most important to the chemist.

10.1 Nature of Solutions

The component of a solution that is present in greatest quantity is usually called the **solvent,** and all other components are called **solutes.** This terminology is loose and arbitrary. It is sometimes convenient to designate a component as the solvent even though it is present in only small amounts. At other times, the assignment of the terms solute and solvent has little significance (for example, in describing gaseous solutions).

Certain pairs of substances will dissolve in each other in all proportions. Complete miscibility is characteristic of the components of all gaseous solutions and some pairs of components of liquid and solid solutions. For most materials, however, there is a limit on the amount of the substance that will dissolve in a given solvent. The **solubility** of a substance in a particular solvent at a specified temperature is the maximum amount of the solute that will dissolve in a definite amount of the solvent and produce a stable system.

For a given solution, the amount of solute dissolved in an amount of solvent given is the **concentration** of the solute. Solutions containing a relatively low concentration of solute are called **dilute solutions;** those of relatively high concentrations are called **concentrated solutions.**

If an excess of solute (more than will normally dissolve) is added to a quantity of a liquid solvent, an equilibrium is established between the pure solute and the dissolved solute:

$$\text{solute}_{\text{pure}} \rightleftharpoons \text{solute}_{\text{dissolved}}$$

The pure solute may be a solid, liquid, or gas. At equilibrium in such a system, the rate at which the pure solute dissolves equals the rate at which the dissolved solute comes out of solution. The concentration of the dissolved solute, therefore, is a constant. A solution of this type is called a **saturated solution,** and its concentration is the solubility of the solute in question.

That such dynamic equilibria exist has been shown experimentally. If small crystals of a solid solute are placed in contact with a saturated solution of the solute, the crystals are observed to change in size and shape. Throughout this experiment, however, the concentration of the saturated solution does not change, nor does the quantity of excess solute decrease or increase.

An **unsaturated solution** has a lower concentration of solute than a saturated solution. On the other hand, it is sometimes possible to prepare a **supersaturated solution,** one in which the concentration of solute is higher than that of a saturated solution. A supersaturated solution, however, is metastable, and if a very small amount of pure solute is added to it, the solute that is in excess of that needed to saturate the solution will precipitate.

10.2 The Solution Process

London forces are the only intermolecular forces between nonpolar covalent molecules. On the other hand, the intermolecular attractions between polar covalent molecules are due to dipole–dipole forces as well as to London forces. In substances in which there is hydrogen bonding the intermolecular forces are unusually strong.

Nonpolar substances and polar substances are generally insoluble in one another. Carbon tetrachloride (a nonpolar substance) is insoluble in water (a polar substance). The attraction of one water molecule for another water molecule is much greater than an attraction between a carbon tetrachloride molecule and a water molecule. Hence, carbon tetrachloride molecules are "squeezed out," and these two substances form a two-liquid-layer system.

Iodine, a nonpolar material, is soluble in carbon tetrachloride. The attractions between I_2 molecules in solid iodine are approximately of the same type and magnitude as those between CCl_4 molecules in pure carbon tetrachloride. Hence, significant iodine–carbon tetrachloride attractions are possible, and iodine molecules can mix with carbon tetrachloride molecules. The resulting solution is a random molecular mixture.

Methyl alcohol, CH_3OH, like water, consists of polar molecules that are highly associated. In both pure liquids the molecules are attracted to one another through hydrogen bonding:

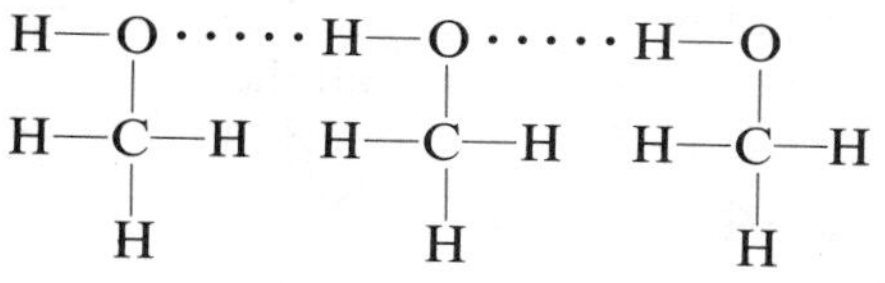

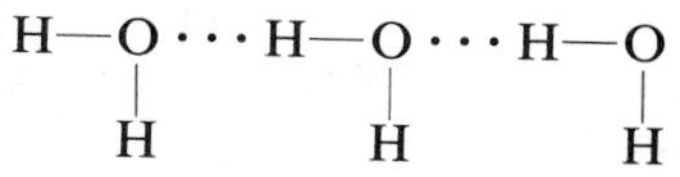

Methyl alcohol and water are miscible in all proportions. In solutions of methyl alcohol in water, CH_3OH and H_2O molecules are associated through hydrogen bonding:

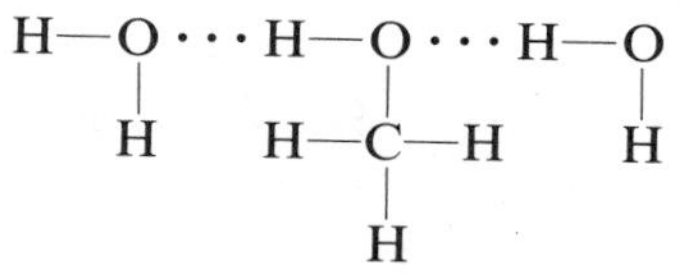

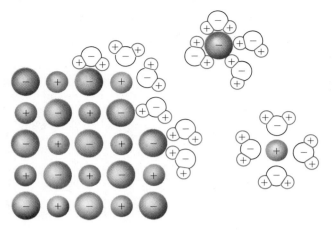

Figure 10.1   Solution of an ionic crystal in water

Methyl alcohol does not dissolve in nonpolar solvents. The strong intermolecular attractions of pure methyl alcohol are not overcome unless the solvent molecules can form attractions of equal, or almost equal, strength with the methyl alcohol molecules.

In general, polar materials dissolve only in polar solvents, and nonpolar substances are soluble in nonpolar solvents. This is the first rule of solubility: "like dissolves like." Network crystals (the diamond, for example), in which the atoms that make up the crystal are held together by covalent bonds, are insoluble in all liquids. This crystalline structure is far too stable to be broken down by a solution process. Any potential solute-solvent attractions cannot approach the strength of the covalent bonding of the crystal.

Polar liquids (water, in particular) can function as solvents for many ionic compounds. The ions of the solute are electrostatically attracted by the polar solvent molecules—negative ions by the positive poles of the solvent molecules, positive ions by the negative poles of the solvent molecules. These ion–dipole attractions can be relatively strong.

Figure 10.1 diagrams the solution of an ionic crystal in water. The ions in the center of the crystal are attracted equally in all directions by the oppositely charged ions of the crystal. The electrostatic attractions on the ions of the surface of the crystal, however, are unbalanced. Water molecules are attracted to these surface ions, the positive ends of the water molecules to the anions and the negative ends of the water molecules to the cations. The ion–dipole attractions formed in this way allow the ions to escape from the crystal and drift into the liquid phase. The dissolved ions are **hydrated** and move through the solution surrounded by a sheath of water molecules. All ions are hydrated in water solution.

10.3 Hydrated Ions

Negative ions are hydrated in water solution by means of attractions between the ion and the hydrogen atoms of the water molecule. In some cases (the sulfate ion, for example) these attractions may be one or more hydrogen bonds:

$$\left[\begin{array}{c} \overset{\displaystyle H}{\underset{\displaystyle}{O}}-H\cdots O \qquad O\cdots H-\overset{\displaystyle H}{O} \\[2pt] \qquad\qquad S \\[2pt] \underset{\displaystyle H}{O}-H\cdots O \qquad O\cdots H-\underset{\displaystyle H}{O} \end{array}\right]^{2-}$$

Positive ions are hydrated by means of attractions between the ion and the unshared electron pairs of the oxygen atom of the water molecule. These attractions are strong. In many cases, each cation is hydrated by a definite number of H_2O molecules:

$$\left[\begin{array}{c} \overset{\displaystyle H}{H-O} \qquad \overset{\displaystyle H}{O-H} \\[2pt] \qquad Be \\[2pt] \underset{\displaystyle H}{H-O} \qquad \underset{\displaystyle H}{O-H} \end{array}\right]^{2+}$$

Additional water molecules form hydrogen bonds with those molecules that are bonded to the cation or anion. These outer layers of water molecules, however, are more loosely held.

What factors lead to the formation of strong interactions between the ion and water molecules?

1. Ions with *high charges* strongly attract the H or O atoms of the H_2O molecule.

2. *Small ions* are more effective than large ones because the charge is more highly concentrated in small ions.

A number of *covalent* compounds of metals produce hydrated *ions* in water solution. The compounds of beryllium, for example, are covalent when pure. The same factor that is largely responsible for the covalent character of the compounds of beryllium (a high ratio of ionic charge to ion size) also leads to the formation of very stable hydrated ions:

$$BeCl_2(s) + 4\,H_2O \longrightarrow Be(H_2O)_4^{2+}(aq) + 2\,Cl^-(aq)$$

The formation of a bond always liberates energy, and the breaking of a bond always requires energy. The energy *released* by a hypothetical process in which hydrated ions are formed from gaseous ions is called the **enthalpy of hydration** of the ions. For example,

$$K^+(g) + Cl^-(g) \longrightarrow K^+(aq) + Cl^-(aq) \qquad \Delta H = -684.1 \text{ kJ}$$

The size of the enthalpy of hydration depends upon the concentration of the final solution. If no distinction is made (as in the example previously given), it is assumed that the enthalpy change pertains to a process in which the ions are hydrated to the greatest possible extent. This high degree of hydration would occur only if the solution were *very dilute*. The corresponding ΔH values are called enthalpies of hydration at **infinite dilution.**

The value of an enthalpy of hydration gives an indication of the strength of the attractions between the ions and the water molecules that hydrate them. A large negative value (which indicates a large quantity of energy evolved) shows that the ions are strongly hydrated.

Frequently, hydrated ions remain in the crystalline solids that are obtained by the evaporation of aqueous solutions of salts. Thus,

$FeCl_3 \cdot 6\,H_2O$ consists of $Fe(H_2O)_6^{3+}$ and Cl^- ions
$BeCl_2 \cdot 4\,H_2O$ consists of $Be(H_2O)_4^{2+}$ and Cl^- ions
$ZnSO_4 \cdot 7\,H_2O$ consists of $Zn(H_2O)_6^{2+}$ and $SO_4(H_2O)^{2-}$ ions
$CuSO_4 \cdot 5\,H_2O$ consists of $Cu(H_2O)_4^{2+}$ and $SO_4(H_2O)^{2-}$ ions

Water molecules can also occur in crystalline hydrates by taking lattice positions in the crystal structure without associating with any specific ion ($BaCl_2 \cdot 2\,H_2O$ is an example) or by occupying interstices (holes) of a crystal structure (the hydrous silicates called zeolites are examples).

10.4 Enthalpy of Solution

The enthalpy change associated with the process in which a solute dissolves in a solvent is called the **enthalpy of solution**. The value of an enthalpy of solution (given in kJ/mol of solute) depends upon the concentration of the final solution in the same way that an enthalpy of hydration does. Unless otherwise noted, an enthalpy of solution is considered to apply to the preparation of a solution that is infinitely dilute. The enthalpy of solution is virtually constant for all the dilute solutions of a given solute–solvent pair.

The enthalpy change observed when a solution is prepared is the net result of the energy *required* to break apart certain chemical bonds or attractions (solute–solute and solvent–solvent) and the energy *released* by the formation of new ones (solute–solvent). The enthalpy of solution for the preparation of a solution of KCl in water, for example, can be considered to be the sum of two enthalpy changes:

1. The energy required to break apart the KCl crystal lattice and form gaseous ions:

$$KCl(s) \longrightarrow K^+(g) + Cl^-(g) \qquad \Delta H = +701.2 \text{ kJ}$$

2. The *enthalpy of hydration* of KCl, which is the energy released when the gaseous ions are hydrated:

$$K^+(g) + Cl^-(g) \longrightarrow K^+(aq) + Cl^-(aq) \qquad \Delta H = -684.1 \text{ kJ}$$

This enthalpy change is actually the sum of two enthalpy changes: the energy required to break the hydrogen bonds between some of the water molecules and the energy released when these water molecules hydrate the ions. It is, however, difficult to investigate these two effects separately.

In this example, the overall process is endothermic. The enthalpy of solution is positive because more energy is required in step 1 than is released in step 2:

$$KCl(s) \longrightarrow K^+(aq) + Cl^-(aq) \qquad \Delta H = +17.1 \text{ kJ}$$

Some enthalpies of solution are negative because more energy is liberated by the hydration of the ions of the solute than is required to break apart the crystal lattice:

$$
\begin{array}{ll}
AgF(s) \longrightarrow Ag^+(g) + F^-(g) & \Delta H = +910.9 \text{ kJ} \\
\underline{Ag^+(g) + F^-(g) \longrightarrow Ag^+(aq) + F^-(aq)} & \underline{\Delta H = -931.4 \text{ kJ}} \\
AgF(s) \longrightarrow Ag^+(aq) + F^-(aq) & \Delta H = -20.5 \text{ kJ}
\end{array}
$$

The factor that leads to a high lattice energy (high charge to size ratio for the ions) also produces high enthalpies of hydration. Consequently, the values for the steps are usually numerically close and the enthalpy of solution itself is a much lower value than either of the values that go into it. A relatively small error in the lattice energy or the enthalpy of hydration can, therefore, lead to a proportionately large error in the enthalpy of solution.

If a solvent other than water is employed in the preparation of a solution, the same type of analysis can be made. The values for the second step are called enthalpies of solvation.

Similar considerations apply to the solution of nonionic materials. The lattice energies of molecular crystals are not so large as those of ionic crystals since the forces holding molecular crystals together are not so strong as those holding ionic crystals together. Solvation energies for these nonionic materials, however, are also low. For molecular substances that dissolve in nonpolar solvents without ionization and without appreciable solute–solvent interaction, the enthalpy of solution is endothermic and has about the same magnitude as the enthalpy of fusion of the solute.

Gases generally dissolve in liquids with the evolution of heat. Since no energy is required to separate the molecules of a gas, the predominant enthalpy change of such a solution process is the solvation of the gas molecules; the process, therefore, is exothermic. Calculation of the enthalpy of solution of hydrogen chloride gas in water, however, requires that the heat evolved by the formation of hydrated ions be reduced by the energy required for the ionization of the HCl molecules.

10.5 Effect of Temperature and Pressure on Solubility

The effect of a temperature change on the solubility of a substance depends upon whether heat is absorbed or evolved *when a saturated solution is prepared.* Suppose that a small quantity of solute dissolves in a nearly saturated solution with the absorption of heat. We can represent the equilibrium between excess solid solute and the saturated solution as

$$\text{energy} + \text{solute} + H_2O \rightleftharpoons \text{saturated solution}$$

The effect of a temperature change on this system can be predicted by means of a principle proposed by Henri Le Chatelier in 1884. Le Chatelier's principle states that a system in equilibrium reacts to a stress in a way that counteracts the stress and establishes a new equilibrium state.

Suppose that we have a beaker containing a saturated solution of the type previously described together with some excess solid solute in equilibrium with it. What will be the effect if we raise the temperature? According to Le Chatelier's principle, the system will react in a way that lowers the temperature. It can do that if it shifts to the direction in which heat is absorbed (to the right in the equation given previously). The shift means that more solute will dissolve. We conclude that increasing the temperature increases the solubility of this particular solute.

What will happen if we lower the temperature? Le Chatelier's principle predicts that the system will react in a way that raises the temperature—a way that liberates energy. The reaction will shift to the left; solute will precipitate out of solution. We conclude that lowering the temperature decreases the solubility of this particular solute. Our two conclusions are actually the same.

The solubilities of substances that *absorb heat* when they dissolve in nearly saturated solutions *increase* with *increasing temperature*. Most ionic solutes behave in this way. The enthalpy of solution of many ionic compounds for solutions that are *infinitely dilute* are *exothermic*. The same compounds, however, usually dissolve in *nearly saturated solutions* with the *absorption* of energy. When a solute dissolves in a nearly saturated solution, the enthalpy of hydration is less than when it dissolves in a very dilute solution. An ion has a lower degree of hydration (is hydrated by fewer water molecules) in a concentrated solution than in a dilute solution. For this reason, enthalpies of solution for nearly saturated solutions are more positive than enthalpies of solution for dilute solutions.

When a substance dissolves in a nearly saturated solution with the evolution of heat,

$$\text{solute} + H_2O \rightleftharpoons \text{saturated solution} + \text{energy}$$

Le Chatelier's principle indicates that the solubility of the solute increases when the temperature is lowered, and the solubility decreases when the temperature is raised. A few ionic compounds (such as Li_2CO_3 and Na_2SO_4) behave in this fashion. In addition, the solubility of all gases decreases as the temperature is raised. Warming a soft drink causes bubbles of carbon dioxide gas to come out of solution.

How much the solubility changes when the temperature changes depends upon the magnitude of the enthalpy of solution. The solubilities of substances with small enthalpies of solution do not change much with changes in temperature.

Changes in pressure ordinarily have little effect upon the solubility of solid and liquid solutes. However, increasing or decreasing the pressure on a solution that contains a dissolved gas has a definite effect. William Henry in 1803 discovered that the amount of a gas that dissolves in a given quantity of a liquid at constant temperature is directly proportional to the partial pressure of the gas above the solution. **Henry's law** is valid only for dilute solutions and relatively low pressures. Gases that are extremely soluble generally react chemically with the solvent (for example, hydrogen chloride gas in water reacts to produce hydrochloric acid); these solutions do not follow Henry's law.

10.6 Concentrations of Solutions

The concentration of a solute in a solution can be expressed in several different ways. We have discussed some of these in previous sections, and we will now review the common ways to express solution concentrations:

1. The **percentage by mass** of a solute in a solution is 100 times the mass of the solute divided by the *total mass* of the solution. A 10% aqueous solution of sodium chloride contains 10 g of NaCl and 90 g of H_2O. Volume percentages are seldom used. Figures recorded as percentages should be understood to be based on mass unless a specific notation to the contrary is made.

2. The **mole fraction,** X, of a component of a solution is the ratio of the number of moles of that component to the *total number* of moles of all the substances present in the solution (see Section 8.10):

$$X_A = \frac{n_A}{n_A + n_B + n_C + \cdots} \tag{10.1}$$

where X_A is the mole fraction of A and $n_A, n_B, n_C, \ldots,$ are the number of moles of A, B, C, ... The sum of the mole fractions of all the components present in the solution must equal one.

Example 10.1

A gaseous solution contains 2.00 g of He and 4.00 g of O_2. What are the mole fractions of He and O_2 in the solution?

Solution

We first find the number of moles of each component present in the solution:

$$? \text{ mol He} = 2.00 \text{ g He} \left(\frac{1 \text{ mol He}}{4.00 \text{ g He}} \right) = 0.500 \text{ mol He}$$

$$? \text{ mol } O_2 = 4.00 \text{ g } O_2 \left(\frac{1 \text{ mol } O_2}{32.0 \text{ g } O_2} \right) = 0.125 \text{ mol } O_2$$

We use these values to find the mole fractions:

$$X_{He} = \frac{n_{He}}{n_{He} + n_{O_2}}$$

$$= \frac{0.500 \text{ mol}}{0.500 \text{ mol} + 0.125 \text{ mol}} = \frac{0.500 \text{ mol}}{0.625 \text{ mol}} = 0.800$$

$$X_{O_2} = \frac{n_{O_2}}{n_{He} + n_{O_2}}$$

$$= \frac{0.125 \text{ mol}}{0.625 \text{ mol}} = 0.200$$

Notice that the sum of the mole fraction is equal to one.

---

3. The **molarity,** M, of a solution is the number of moles of solute per liter of *solution* (see Section 2.9). Thus, 1 liter of a 6 M solution of sodium chloride is prepared by taking 6 mol of NaCl and adding sufficient water to make exactly 1 liter of solution.

Figure 10.2 Volumetric flask

Notice that the definition is based on the *total volume of the solution*. When a liquid solution is prepared, the volume of the solution rarely equals the sum of the volumes of the pure components. Usually the final volume of the solution is larger or smaller than the total of the volumes of the materials used to prepare it. It is not practical, therefore, to attempt to predict the amount of solvent that should be employed to prepare a given solution. Molar solutions (as well as others that are based on total volume) are generally prepared by the use of the volumetric flasks (Figure 10.2). In the preparation of a solution, the correct amount of solute is placed in the flask, and water is then added, with careful and constant mixing, until the solution fills the flask to the calibration mark on the neck.

It is a simple matter to calculate the quantity of solute present in a given volume of a solution when the concentration of the solution is expressed in terms of molarity. Concentrations defined on the basis of the total volume of solution have this decided advantage. Thus, 1 liter of a 3 M solution contains 3 mol of solute, 500 ml contains 1.5 mol, 250 ml contains 0.75 mol, and 100 ml contains 0.3 mol. A disadvantage of basing concentrations on volume of solution, however, is that such concentrations change slightly with temperature changes because of expansion or contraction of liquid solutions. For exact work, therefore, a solution should be prepared at the temperature at which it is to be used, and a volumetric flask calibrated for this temperature should be employed.

Simple problems involving the preparation of molar solutions are found in Examples 2.18 and 2.19 in Section 2.9.

Example 10.2

(a) How many grams of concentrated nitric acid solution should be used to prepare 250 ml of 2.00 M HNO$_3$? The concentrated acid is 70.0% HNO$_3$.
(b) If the density of the concentrated nitric acid solution is 1.42 g/ml, what volume should be used?

Solution

(a) The factors used to solve the problem are derived from the following facts (in order):

1. Since the desired solution is 2.00 M, there must be 2.00 mol HNO$_3$ in 1000 ml.

2. The molecular weight of HNO$_3$ is 63.0.

3. In 100.0 g of concentrated nitric acid (70.0% HNO$_3$) there are 70.0 g of HNO$_3$:

$$? \text{ g conc HNO}_3 = 250 \text{ ml sol'n} \left(\frac{2 \text{ mol HNO}_3}{1000 \text{ ml sol'n}}\right)\left(\frac{63.0 \text{ g HNO}_3}{1 \text{ mol HNO}_3}\right)\left(\frac{100.0 \text{ g conc HNO}_3}{70.0 \text{ g HNO}_3}\right)$$

$$= 45.0 \text{ g conc HNO}_3$$

(b) The density of the concentrated acid is used to convert the answer to part (a) into ml of concentrated HNO$_3$:

$$? \text{ ml conc HNO}_3 = 45.0 \text{ g conc HNO}_3 \left(\frac{1.00 \text{ ml conc HNO}_3}{1.42 \text{ g conc HNO}_3} \right) = 31.7 \text{ ml conc HNO}_3$$

Example 10.3

What is the molarity of concentrated HCl if the solution contains 37.0% HCl by mass and if the density of the solution is 1.18 g/ml?

Solution

In order to find the molarity of the solution, we must determine the number of moles of HCl in 1 liter of solution. The problem can be solved by using factors derived in the following way:

1. The mass of 1 liter of the solution is found by means of the density.

2. The mass of pure HCl in this quantity of solution is obtained by use of the percent composition.

3. The molecular weight of HCl (36.5) is used to convert the mass of HCl into moles of HCl:

$$? \text{ mol HCl} = 1000 \text{ ml sol'n} \left(\frac{1.18 \text{ g sol'n}}{1 \text{ ml sol'n}} \right) \left(\frac{37.0 \text{ g HCl}}{100.0 \text{ g sol'n}} \right) \left(\frac{1 \text{ mol HCl}}{36.5 \text{ g HCl}} \right)$$

$$= 12.0 \text{ mol HCl}$$

Since 1 liter of the solution contains 12.0 mol HCl, the solution is 12.0 M.

Often solutions must be prepared by diluting concentrated reagents. The molarities of common concentrated reagents are listed in Table 10.1. These values can be used to determine the proportions of reagent and water required to prepare a solution of a desired concentration.

The number of moles of solute in a sample of a solution may be obtained by multiplying the volume of the solution (V_1, in liters) by the molarity of the solution (M_1, the number of moles of solute in 1 liter):

Table 10.1 Composition of some common concentrated reagents

Reagent	Formula	Molecular Weight	Percent by Mass	Density (g/ml)	Molarity
acetic acid	$HC_2H_3O_2$	60.05	100	1.05	17.5
hydrochloric acid	HCl	36.46	37	1.18	12.0
nitric acid	HNO_3	63.01	70	1.42	15.8
phosphoric acid	H_3PO_4	98.00	85	1.70	14.7
sulfuric acid	H_2SO_4	98.07	96	1.84	18.0
ammonia	NH_3	17.03	28	0.90	14.8

$$\text{number of moles of solute} = V_1 M_1$$

When the solution is diluted to a new volume, V_2, *it still contains the same number of moles of solute*. The concentration has decreased to M_2 but the product $V_2 M_2$ equals the same number of moles. Therefore,

$$V_1 M_1 = V_2 M_2 \tag{10.2}$$

Since a volume term is found on both sides of this equation, any volume unit can be used to express V_1 and V_2 provided the same unit is used for each. Note that this equation is used for dilution problems only.

Example 10.4

What volume of concentrated HCl should be used to prepare 500 ml of a 3.00 M HCl solution?

Solution

From Table 10.1 (as well as from Example 10.3) we note that concentrated HCl is 12.0 M:

$$V_1 M_1 = V_2 M_2$$
$$V_1(12.0 \ M) = (500 \text{ ml})(3.00 \ M)$$
$$V_1 = 125 \text{ ml}$$

---

4. The molality of a solution, m, is defined as the number of moles of solute dissolved in 1 kg of *solvent*. A 1 m solution of urea, $CO(NH_2)_2$, is made by dissolving 1 mol of urea (60.06 g) in 1000 g of water. Notice that the definition is *not* based on the total volume of solution. The final volume of a molal solution is of no importance. One molal solutions of different solutes, each containing 1000 g of water, will have different volumes. All these solutions, however, will have the same mole fractions of solute and solvent (see Example 10.5).

Example 10.5

What are the mole fractions of solute and solvent in a 1.00 m aqueous solution?

Solution

The molecular weight of H_2O is 18.0. We find the number of moles of water in 1000 g of H_2O:

$$? \text{ mol } H_2O = 1000 \text{ g } H_2O \left(\frac{1 \text{ mol } H_2O}{18.0 \text{ g } H_2O} \right) = 55.6 \text{ mol } H_2O$$

A 1.00 *m* aqueous solution contains

$$n_{\text{solute}} = 1.0 \text{ mol}$$
$$\underline{n_{\text{H}_2\text{O}} = 55.6 \text{ mol}}$$
$$n_{\text{total}} = 56.6 \text{ mol}$$

The mole fractions are

$$X_{\text{solute}} = \frac{n_{\text{solute}}}{n_{\text{total}}} = \frac{1.0 \text{ mol}}{56.6 \text{ mol}} = 0.018$$

$$X_{\text{H}_2\text{O}} = \frac{n_{\text{H}_2\text{O}}}{n_{\text{total}}} = \frac{55.6 \text{ mol}}{56.6 \text{ mol}} = 0.982$$

These mole fractions pertain to all 1.00 *m* aqueous solutions.

Molal solutions that employ solvents other than water are often prepared. In each case, all the 1 *m* solutions that employ the same solvent have the same mole fractions of solute and solvent. The mole fraction of solute in any 1 *m* carbon tetrachloride solution is 0.133; the mole fraction of solvent is 0.867. These numbers differ from those in Example 10.5 because the molecular weight of carbon tetrachloride is different from that of water. A kilogram of CCl_4, therefore, is a different number of moles than a kilogram of H_2O.

The molality of a given solution does not vary with temperature since the solution is prepared on the basis of the masses of the components; mass does not vary with temperature changes. The *molality* of a *very dilute* aqueous solution is approximately the same as the *molarity* of the solution since 1000 g of water occupies approximately 1000 ml.

5. The **normality** of a solution, *N*, is the number of *equivalent weights* per liter of solution. Equivalent weights and normal solutions are discussed in Section 11.8. We note here, however, that normal concentrations, like molar concentrations, are based on the *total volume* of solution. Consequently, volumetric flasks are used in the preparation of normal solutions, and the normality of a solution, like the molarity, varies slightly with temperature.

10.7 Vapor Pressure of Solutions

The vapor pressure of any solution (P_{total}) is the sum of the partial pressures of the components of the solution (p_A, p_B, . . .). For a solution of two components, A and B,

$$P_{\text{total}} = p_A + p_B \tag{10.3}$$

The partial pressures in this equation are given by a relationship known as **Raoult's law,** which describes what is called an **ideal solution.** The partial pressure of A, p_A, for example, is given by the equation

$$p_A = X_A P_A^\circ \tag{10.4}$$

where X_A is the mole fraction of A in the solution and P_A° is the vapor pressure of *pure* A at the temperature of the experiment.

Suppose that we prepare a solution from 4 mol of A and 1 mol of B. Since the total number of moles in the solution is 5, the mole fraction of A is 4/5. Only 4 out of every 5 molecules on the surface of the solution are A molecules. The partial pressure of A, therefore, is 4/5 the vapor pressure of pure A (where 5 out of 5 molecules on the surface are A molecules).

The partial pressure of B can be found by use of a similar equation:

$$p_B = X_B P_B^\circ \tag{10.5}$$

where X_B is the mole fraction of B in the solution and P_B° is the vapor pressure of *pure* B at the temperature of the experiment. In our hypothetical solution, the mole fraction of B is 1/5. The partial pressure of B, therefore, is 1/5 the vapor pressure of pure B.

The vapor pressure of the solution is equal to the sum of the two partial pressures, according to Equation 10.3:

$$P_{\text{total}} = X_A P_A^\circ + X_B P_B^\circ \tag{10.6}$$

The vapor pressure of an *ideal solution*, therefore, can be derived from the vapor pressures of the pure components by taking into account the proportion of the components (by moles) present in the solution.

In Figure 10.3, the partial pressures of A and B as well as the total vapor pressure of solutions of A and B are plotted against solution concentration.

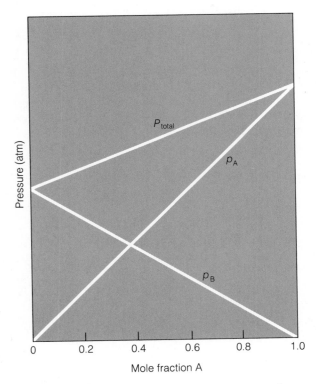

Figure 10.3 Typical total and partial pressure curves for solutions that follow Raoult's law

Chapter 10   Solutions

The total vapor pressure curve is the sum of the two partial pressure curves. In the figure, pressures are plotted against mole fraction of A. Since the mole fractions of A and of B must add up to 1, the mole fraction of B for a given point is easily derived from the scale of the x axis. When X_A is 0.2, for example, X_B is 0.8.

Actually, a solution containing A and B will be ideal only if the intermolecular attractive forces between A and B molecules are similar to those between molecules of the pure components (A and A, B and B). In such a situation, the partial pressure of a component is merely a matter of the proportion of molecules of that component to the total number of molecules on the surface of the solution.

Few solutions are ideal. Two types of deviations from Raoult's law are observed:

1. Positive deviations. The partial pressures of A and B and the total vapor pressure are *higher* than predicted (Figure 10.4). This type of deviation is observed when the attractive forces between A and B molecules are *weaker* than those between two A molecules or two B molecules. In such a situation, A molecules find it easier to escape from the liquid and the partial pressure of A is higher than predicted. The behavior of B molecules is similar.

2. Negative deviations. The partial pressures of A and B and the total vapor pressure are *lower* than predicted (Figure 10.5). The A—B attractions are *stronger* than A—A or B—B attractions. The A molecules find it more difficult to leave the liquid, and the partial pressure of A is lower than predicted. The behavior of B molecules is similar.

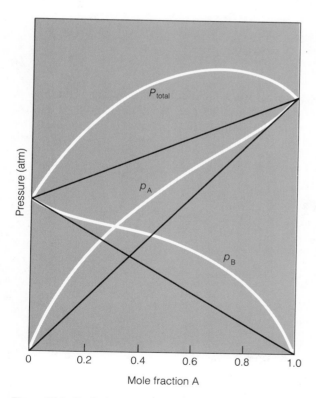

Figure 10.4 Typical total and partial vapor pressure curves for solutions that show positive deviations from Raoult's law. (Colored lines are based on Raoult's law.)

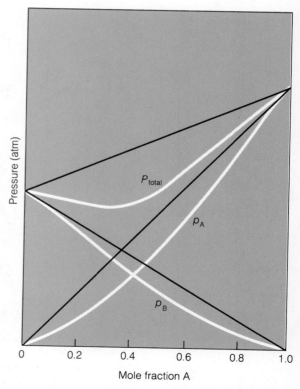

Figure 10.5 Typical total and partial vapor pressure curves for solutions that show negative deviations from Raoult's law. (Colored lines are based on Raoult's law.)

Consider a dilute solution prepared from a solute (which we will designate as B) that is *nonvolatile* ($P_B^\circ = 0$, for all practical purposes) and that does not dissociate in solution. The vapor pressure of the solution is caused by solvent molecules alone (A molecules). Such solutions usually follow Raoult's law:

$$P_{total} = X_A P_A^\circ \tag{10.7}$$

Since $X_A + X_B = 1$, $X_A = 1 - X_B$. Therefore,

$$P_{total} = (1 - X_B)P_A^\circ \tag{10.8}$$

or

$$P_{total} = P_A^\circ - X_B P_A^\circ \tag{10.9}$$

which means that the vapor pressure of pure A, P_A°, is lowered by an amount equal to $X_B P_A^\circ$.

The vapor pressure of a solution prepared from 1 mol of a nonvolatile, nondissociating solute and 99 mol of solvent is 99% of the vapor pressure of the pure solvent at the same temperature. The escape of the solvent molecules to the vapor is reduced because only 99% of the surface molecules in the solution are solvent molecules. The other 1% of the surface molecules are not volatile. The vapor pressure of the solvent is reduced by an amount that is proportional to the mole fraction of nonvolatile solute present.

Example 10.6

Heptane (C_7H_{16}) and octane (C_8H_{18}) form ideal solutions. What is the vapor pressure at 40°C of a solution that contains 3.00 mol of heptane and 5.00 mol of octane? At 40°C, the vapor pressure of heptane is 0.121 atm and the vapor pressure of octane is 0.041 atm.

Solution

The total number of moles is 8.00. Therefore,

$$X_{heptane} = \frac{3.00 \text{ mol}}{8.00 \text{ mol}} = 0.375$$

$$X_{octane} = \frac{5.00 \text{ mol}}{8.00 \text{ mol}} = 0.625$$

The vapor pressure is

$$\begin{aligned}
P_{total} &= X_{heptane}P_{heptane}^\circ + X_{octane}P_{octane}^\circ \\
&= 0.375(0.121 \text{ atm}) + 0.625(0.041 \text{ atm}) \\
&= 0.045 \text{ atm} + 0.026 \text{ atm} \\
&= 0.071 \text{ atm}
\end{aligned}$$

Example 10.7

Assuming ideality, calculate the vapor pressure of a 1.00 m solution of a non-volatile, nondissociating solute in water at 50°C. The vapor pressure of water at 50°C is 0.122 atm.

Solution

The mole fraction of water in a 1.00 m solution is 0.982 (from Example 10.5). The vapor pressure of a 1.00 m solution of this type at 50°C is

$$P_{total} = X_{H_2O}P^{\circ}_{H_2O}$$
$$= (0.982)(0.122 \text{ atm})$$
$$= 0.120 \text{ atm}$$

10.8 Boiling Point and Freezing Point of Solutions

The lowering of the vapor pressure in solutions of nonvolatile solutes affects the boiling points and freezing points of these solutions.

The boiling point of a liquid is defined as the temperature at which the vapor pressure of the liquid is equal to the prevailing atmospheric pressure. Boiling points measured under 1 atmosphere pressure are called normal boiling points. Since the addition of a nonvolatile solute decreases the vapor pressure of a liquid, a solution will not boil at the normal boiling point of the solvent. It is necessary to increase the temperature above this point in order to attain a vapor pressure over the solution of 1 atmosphere. The boiling point of a solution containing a nonvolatile molecular solute, therefore, is higher than that of the pure solvent. The elevation is proportional to the concentration of solute in the solution.

This effect is illustrated by the vapor pressure curves plotted in Figure 10.6. The extent to which the vapor pressure curve of the solution lies below the vapor pressure curve of the solvent is proportional to the mole fraction of solute in the solution. The elevation of the boiling point, Δt_b, reflects this displacement of the vapor pressure curve. For a given solvent the elevation is a constant for all solutions of the same concentration.

Concentrations are customarily expressed in molalities, rather than mole fractions, for problems involving boiling-point elevations. The boiling point of a 1 m aqueous solution, for example, is 0.512°C higher than the boiling point of water. Table 10.2 lists molal boiling-point elevation constants for several solvents.

The boiling point of a 0.5 m solution would be expected to be elevated by an amount equal to $\frac{1}{2}$ the molal constant. Thus, the boiling-point elevation, Δt_b, of a solution can be calculated by multiplying the molal boiling-point elevation constant of the solvent, k_b, by the molality of the solution, m:

$$\Delta t_b = mk_b \tag{10.10}$$

In reality, this relationship is only approximate. A more exact statement would require that the concentration be expressed in mole fraction of solute, not in

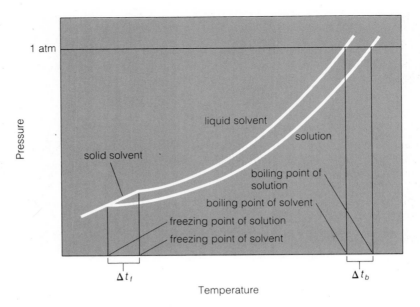

Figure 10.6 Vapor pressure curves of a pure solvent and a solution of a nonvolatile solute. (Not drawn to scale.)

	Boiling Point (°C)	k_b (°C/m)	Freezing Point (°C)	k_f (°C/m)
Table 10.2 Molal boiling-point elevation and freezing-point depression constants				
Solvent				
acetic acid	118.1	+3.07	16.6	−3.90
benzene	80.1	+2.53	5.5	−5.12
camphor	—	—	179.	−39.7
carbon tetrachloride	76.8	+5.02	−22.8	−29.8
chloroform	61.2	+3.63	−63.5	−4.68
ethyl alcohol	78.4	+1.22	−114.6	−1.99
naphthalene	—	—	80.2	−6.80
water	100.0	+0.512	0.0	−1.86

molality. However, molalities of *dilute* solutions are proportional (at least with sufficient accuracy) to mole fractions of solute. And since Raoult's law only describes the behavior of most real solutions satisfactorily if they are dilute, the use of molalities in these calculations is justified.

At the freezing point the vapor pressure of solid and liquid are equal. In Figure 10.6 the vapor pressure curves of the liquid solvent and the solid solvent intersect at the freezing point of the solvent. At this temperature, however, the vapor pressure of the solution is lower than the equilibrium vapor pressure of the pure solvent. The vapor pressure curve of the solution intersects the vapor pressure curve of the solid solvent at a lower temperature. The freezing point of the solution, therefore, is lower than that of the pure solvent. As in the case of boiling-point elevations, freezing-point depressions depend upon the concentration of the solution and the solvent employed. Molal freezing-point depression constants for some solvents are listed in Table 10.2. The freezing-point depression, Δt_f, of a solution can be calculated from the molality of the solution and the constant for the solvent k_f:

$$\Delta t_f = mk_f \tag{10.11}$$

This statement assumes that the solute does not form a solid solution with the solvent. If this is not the case, the relationship is not valid.

Example 10.8

What are the boiling point and freezing point of a solution prepared by dissolving 2.40 g of biphenyl ($C_{12}H_{10}$) in 75.0 g of benzene? The molecular weight of biphenyl is 154.

Solution

The molality of the solution is the number of moles of biphenyl dissolved in 1000 g of benzene:

$$? \text{ mol } C_{12}H_{10} = 1000 \text{ g benzene} \left(\frac{2.40 \text{ g } C_{12}H_{10}}{75.0 \text{ g benzene}}\right)\left(\frac{1 \text{ mol } C_{12}H_{10}}{154 \text{ g } C_{12}H_{10}}\right)$$

$$= 0.208 \text{ mol } C_{12}H_{10}$$

The molal boiling-point elevation constant for benzene solutions is $+2.53°C/m$ (Table 10.2):

$$\Delta t_b = mk_b \tag{10.10}$$
$$= (0.208 \text{ } m)(+2.53°C/m)$$
$$= +0.526°C$$

The normal boiling point of benzene is 80.1°C (Table 10.2). The boiling point of the solution, therefore, is

$$80.1°C + 0.5°C = 80.6°C$$

The molal freezing-point depression constant for benzene solutions is $-5.12°C/m$ (Table 10.2):

$$\Delta t_f = mk_f \tag{10.11}$$
$$= (0.208 \text{ } m)(-5.12°C/m)$$
$$= -1.06°C$$

The normal freezing point of benzene is 5.5°C (Table 10.2). The freezing point of the solution, therefore, is

$$5.5°C - 1.1°C = 4.4°C$$

Example 10.9

A solution prepared by dissolving 0.300 g of an unknown nonvolatile solute in 30.0 g of carbon tetrachloride has a boiling point that is 0.392°C higher than that of pure CCl_4. What is the molecular weight of the solute?

Solution

For CCl_4 solutions, k_b is $+5.02°C/m$ (Table 10.2). We find the molality of the solution from the boiling-point elevation:

$$\Delta t_b = mk_b \qquad (10.10)$$
$$+0.392°C = m(+5.02°C/m)$$
$$m = 0.0781 \ m$$

Next, we find the number of grams of solute dissolved in 1000 g of CCl_4:

$$? \text{ g solute} = 1000 \text{ g } CCl_4 \left(\frac{0.300 \text{ g solute}}{30.0 \text{ g } CCl_4}\right) = 10.0 \text{ g solute}$$

Since the solution is 0.0781 m, 10.0 g of solute is 0.0781 mol of solute:

$$? \text{ g solute} = 1 \text{ mol solute}\left(\frac{10.0 \text{ g solute}}{0.0781 \text{ mol solute}}\right) = 128 \text{ g solute}$$

10.9 Osmosis

The properties of solutions that depend principally upon the concentration of dissolved particles, rather than upon the nature of these particles, are called **colligative properties.** For solutions of nonvolatile solutes these properties include: vapor-pressure lowering, freezing-point depression, boiling-point elevation, and osmotic pressure. The last of these, osmotic pressure, is the topic of this section.

A membrane, such as cellophane or parchment, that permits some molecules or ions, but not all, to pass through it is called a **semipermeable membrane.** Figure 10.7 shows a membrane that is permeable to water but not to sucrose (cane sugar) placed between pure water and a sugar solution. Water molecules, but not sugar molecules, can go through the membrane in either direction. However, there are more water molecules per unit volume on the left side (the side containing pure water) than on the right. The rate of passage through the membrane from left to right, therefore, exceeds the rate in the opposite direction.

As a result, the number of water molecules on the right increases, the sugar solution becomes more dilute, and the height of the solution in the right arm of the U tube increases. This process is called **osmosis.** The difference in height between the levels of the liquids in the two arms of the U tube is a measure of the **osmotic pressure.**

The increased hydrostatic pressure on the right tends to force water molecules through the membrane from right to left, so that ultimately the rate of passage to the left equals the rate of passage to the right. The final condition, therefore, is an equilibrium state with equal rates of passage of water molecules through

Purification of tap water for laboratory use by reverse osmosis. *Millipore Corporation.*

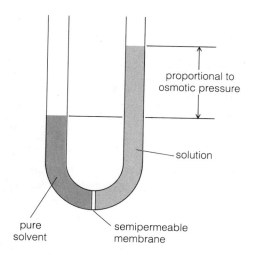

proportional to osmotic pressure

solution

pure solvent

semipermeable membrane

Figure 10.7 Osmosis

the membrane in both directions. If a pressure that is higher than the equilibrium pressure is applied to the solution in the right arm, water is forced in a direction contrary to that normally observed. This process, called **reverse osmosis,** is used to secure pure water from salt water.

Similarities exist between the behavior of water molecules in osmosis and the behavior of gas molecules in diffusion. In both processes molecules diffuse from regions of high concentrations to regions of low concentration. In 1887 Jacobus van't Hoff discovered the relation

$$\pi V = nRT \tag{10.12}$$

where π is the osmotic pressure (in atm), n is the number of moles of solute dissolved in volume V (in liters), T is absolute temperature, and R is the gas constant (0.08206 liter atm/K mol). The similarity between this equation and the equation of state for an ideal gas is striking. The equation may be written in the form

$$\pi = \left(\frac{n}{V}\right)RT \tag{10.13}$$

$$\pi = MRT \tag{10.14}$$

where M is the molarity of the solution. Osmosis plays an important role in plant and animal physiological processes; the passage of substances through the semipermeable walls of a living cell, the action of the kidneys, and the rising of sap in trees are prime examples.

Example 10.10

Find the osmotic pressure of blood at body temperature (37°C) if the blood behaves as if it were a 0.296 M solution of a nonionizing solute.

Solution

$$\pi = MRT$$
$$= (0.296 \text{ mol/liter})(0.0821 \text{ liter atm/K mol})(310 \text{ K})$$
$$= 7.53 \text{ atm}$$

Example 10.11

One liter of an aqueous solution contains 30.0 g of a protein. The osmotic pressure of the solution is 0.0167 atm at 25°C. What is the approximate molecular weight of the protein?

Solution

We use the van't Hoff equation to find the number of moles of protein present in the solution:

$$\pi = \left(\frac{n}{V}\right) RT \tag{10.13}$$

$$0.0167 \text{ atm} = \left(\frac{n}{1 \text{ liter}}\right)[0.0821 \text{ liter} \cdot \text{atm/(K} \cdot \text{mol)}] (298 \text{ K})$$

$$n = 6.83 \times 10^{-4} \text{ mol}$$

Since there are 30.0 g of protein in the solution,

$$? \text{ g protein} = 1 \text{ mol protein} \left(\frac{30.0 \text{ g protein}}{6.83 \times 10^{-4} \text{ mol protein}}\right)$$

$$= 4.39 \times 10^4 \text{ g protein}$$

The molecular weight of the protein is approximately 43,900.

Since proteins have high molecular weights, the molal and molar concentrations of saturated solutions of proteins are very low. For the solution described in this example, the following effects can be calculated:

vapor pressure lowering	0.000000385 atm
boiling-point elevation	0.000350°C
freezing-point depression	−0.00127°C
osmotic pressure	0.0167 atm

Clearly, the first three are too small for accurate measurement. The osmotic pressure of this solution, however, would make the difference in the height of the two columns shown in Figure 10.7 approximately 17 cm, an amount which is easily measured.

10.10 Distillation

A solution of a nonvolatile solute can be separated into its components by **simple distillation**. This procedure consists of boiling away the volatile solvent from the solute. The solvent is collected by condensing the vapor. The solute is the residue that remains after the distillation.

A solution of two volatile components that follows Raoult's law (Figure 10.3) can be separated into its components by a process known as **fractional distillation**. According to Raoult's law, each component contributes to the vapor pressure of the solution in proportion to its mole fraction times its vapor pressure in the pure state (P_A° and P_B°):

$$P_{total} = X_A P_A^\circ + X_B P_B^\circ \qquad (10.6)$$

Let us consider a two-component solution in which the mole fraction of A is 0.75 and the mole fraction of B is 0.25. Assume that, at the temperature of the experiment, the vapor pressure of pure A is 1.20 atm and the vapor pressure of pure B is 0.40 atm. Then,

$$
\begin{aligned}
P_{total} &= X_A P_A^\circ + X_B P_B^\circ \qquad (10.6)\\
&= 0.75(1.20 \text{ atm}) + 0.25(0.40 \text{ atm})\\
&= 0.90 \text{ atm} + 0.10 \text{ atm}\\
&= 1.00 \text{ atm}
\end{aligned}
$$

Since the total pressure is 1.00 atm, the temperature of the experiment is the normal boiling point of the solution. The partial pressures of A and B are 0.90 atm and 0.10 atm.

The composition of the *vapor* in equilibrium with this solution can be calculated by comparing the partial pressure of each component with the total vapor pressure of the solution (Section 8.10). In the vapor, therefore,

$$X_{A, vapor} = \frac{0.90 \text{ atm}}{1.00 \text{ atm}} = 0.90 \qquad X_{B, vapor} = \frac{0.10 \text{ atm}}{1.00 \text{ atm}} = 0.10$$

The solution, in which $X_A = 0.75$, is in equilibrium with vapor in which $X_{A, vapor} = 0.90$. For ideal solutions, the vapor is always richer than the liquid in the more volatile component (which in this instance is A—it has the higher vapor pressure).

In a distillation of the solution of A and B, the vapor that comes off and condenses is richer in A than the liquid remaining behind. The actual compositions of vapor and liquid change as the distillation proceeds, but at any given time this generalization is true. By collecting the condensed vapor in several fractions and subjecting these fractions to repeated distillation, the components of the original mixture can eventually be obtained in substantially pure form.

For systems that deviate from Raoult's law, the situation is somewhat different. A positive deviation may lead to a maximum in the total vapor pressure curve (Figure 10.4). The maximum corresponds to a solution, of definite composition, that has a vapor pressure higher than either pure component. A solution of this type, called a **minimum boiling azeotrope**, will boil at a *lower* temperature than either of the two pure components. Ethyl alcohol and water form a minimum boiling azeotrope that contains 4.0% water and has a normal boiling point of 78.17°C. Ethyl alcohol and water boil at 78.3°C and 100°C, respectively.

If a system shows a negative deviation from Raoult's law (Figure 10.5), there may be a minimum in the P_{total} curve. The solution that has a concentration corresponding to this minimum will have a vapor pressure, at any given temperature, lower than either pure component. A solution of this type boils at a temperature *higher* than either pure component and is called a **maximum boiling**

azeotrope. Hydrochloric acid and water form such an azeotrope containing 20.22% HCl and boiling at 108.6°C. Pure HCl has a boiling point of −80°C.

The vapor in equilibrium with a maximum or minimum boiling azeotrope has the same concentration as the liquid. Azeotropes, therefore, like pure substances, distill without change. Fractional distillation of a solution containing two components that form an azeotrope will eventually produce one pure component and the azeotrope but not both pure components.

10.11 Solutions of Electrolytes

If an aqueous solution contains ions, it will conduct electricity. Pure water itself is slightly ionized and is a poor conductor:

$$2\,H_2O \rightleftharpoons H_3O^+(aq) + OH^-(aq)$$

The solute of an aqueous solution that is a better electrical conductor than pure water alone is called an **electrolyte.** An electrolyte is wholly or partly ionized in water solution. Covalent solutes that are exclusively molecular in solution and therefore do nothing to enhance the conductivity of the solvent are called **non-electrolytes**; sucrose (cane sugar) is an example.

Electrolytes can be further divided into two groups:

1. **Strong electrolytes** are virtually completely ionic in water solution.

2. **Weak electrolytes** are polar covalent substances that are incompletely dissociated in water solution. The conductivity of a 1 m solution of a weak electrolyte is lower than the conductivity of a 1 m solution of a strong electrolyte.

Svante Arrhenius, 1859–1927. *Smithsonian Institution.*

The boiling-point elevations and freezing-point depressions of dilute solutions of electrolytes are different from those of solutions of nonelectrolytes with the same concentrations. Since 1 mol of NaCl contains 2 mol of ions (1 mol of Na$^+$ and 1 mol of Cl$^-$) and since colligative properties depend upon the number of dissolved particles and not their nature, we would expect the freezing-point depression of a 1 m solution of NaCl to be twice that of a 1 m solution of a non-electrolyte. We would also expect the freezing-point depression of a solution of K$_2$SO$_4$ (which contains 3 mol of ions per mole of K$_2$SO$_4$) to be three times that of a solution that has the same concentration and that contains a solute that does note dissociate into ions.

The observed freezing-point depressions listed in Table 10.3 agree approximately with these predictions. The observed values agree best with the calculated values when the solutions are most dilute. Data such as these, together with data from electrical conductivity experiments led Svante Arrhenius to propose his "chemical theory of electrolytes" in 1887. The boiling-point elevations of solutions of electrolytes are proportionately higher than the boiling-point elevations of solutions of nonelectrolytes with the same concentrations.

10.12 Interionic Attractions in Solution

The **van't Hoff factor,** i, is defined as the ratio of the observed freezing-point depression of a solution, Δt_f, to the depression calculated on the basis that the solute does not dissociate:

Table 10.3 Observed freezing-point depressions for some equeous solutions compared to calculated depressions[a]

Solute	Concentration of Solution		
	0.001 m	0.01 m	0.1 m
nonelectrolyte sucrose	0.00186°C 0.00186	0.0186°C 0.0186	0.186°C 0.188
2 ions/formula NaCl	0.00372 0.00366	0.0372 0.0360	0.372 0.348
3 ions/formula K_2SO_4	0.00558 0.00528	0.0558 0.0501	0.558 0.432
4 ions/formula $K_3[Fe(CN)_6]$	0.00744 0.00710	0.0744 0.0626	0.744 0.530

[a] Calculated on the assumptions that k_f for water is $-1.86°C$/molal over the entire range of concentrations, the salts are 100% ionic in solution, and the ions act independently of one another in their effect on the freezing point of the solution.

Table 10.4 van't Hoff factor, i, for various strong electrolytes in solution[a]

Electrolyte	Concentration of Solution		
	0.001 m	0.01 m	0.1 m
NaCl	1.97	1.94	1.87
$MgSO_4$	1.82	1.53	1.21
K_2SO_4	2.84	2.69	2.32
$K_3[Fe(CN)_6]$	3.82	3.36	2.85

[a] From freezing-point determinations.

$$i = \frac{\Delta t_f}{mk_f}$$

(10.15)

The preceding equation can be rearranged to

$$\Delta t_f = imk_f$$

(10.16)

When dissociation of a solute occurs, it is necessary to correct the molality of the solution for calculations involving colligative properties. The i factor will do just that. Since 1 mol of NaCl dissociates into 2 mol of ions in solution (1 mol of Na^+ and 1 mol of Cl^-), the i factor for a solution of NaCl is theoretically 2. If we assume that each ion acts independently, the effective concentration of a 0.001 m solution of NaCl would be 0.002 m. In other words, $i = 2, m = 0.001\ m$, and

$$\Delta t_f = 2(0.001\ m)k_f$$

Inspection of the i values recorded in Table 10.4 reveals that the i factors do not exactly equal the number of ions per formula unit for each of the strong electrolytes listed. Thus, for 0.001 m solutions, the i factor for NaCl is 1.97 (not 2), for K_2SO_4 is 2.84 (not 3), and for $K_3[Fe(CN)_6]$ is 3.28 (not 4). Furthermore, the i value changes with concentration of the solution and approaches the value expected for complete dissociation as the solution becomes more dilute.

Interionic attractions occur in solutions of electrolytes and the ions are not completely independent of one another in the way that uncharged solute molecules are. The electrical forces that operate between oppositely charged ions reduce the effectiveness of these ions. As a solution is diluted, the ions spread farther and farther apart, their influence on one another diminishes, and the i factor approaches its limiting value. Notice that the interionic attractions in solutions of $MgSO_4$ produce a stronger effect than those in solutions of NaCl, even though both solutes contain two moles of ions per mole of compound. For 0.001 m $MgSO_4$, $i = 1.82$, whereas for 0.001 m NaCl, $i = 1.97$. Both the ions of $MgSO_4$ are doubly charged (Mg^{2+}, SO_4^{2-}), whereas the ions of NaCl are singly charged (Na^+, Cl^-). The interionic attractions are stronger in solutions of magnesium sulfate.

Some characteristics of solutes are summarized in Table 10.5. The van't Hoff factor, i, is used in calculations of freezing-point depression, boiling-point elevation, and osmotic pressure when the solute is an electrolyte.

Table 10.5   Characteristics of solutes

Solute	Form of Solute in 1 m Solution	van't Hoff Factor for Δt_f; Dilute Solutions of Molality, m $\Delta t_f = ik_f m$	Examples
nonelectrolytes	molecules	$i = 1$	$C_{12}H_{22}O_{11}$ (sucrose) CON_2H_4 (urea) $C_3H_5(OH)_3$ (glycerol)
strong electrolytes	ions	$i \approx n^a$	NaCl KOH
weak electrolytes	molecules and ions	$1 < i < n^a$	$HC_2H_3O_2$ NH_3 $HgCl_2$

^{a} n = number of moles of ions per mole of solute.

Summary

The topics that have been discussed in this chapter are

1. The nature of solutions; saturated, unsaturated, and supersaturated solutions.

2. The solution process.

3. The hydration of ions; enthalpy of hydration.

4. The energy changes that occur when solutions are prepared; enthalpy of solution.

5. The effect of temperature and pressure on solubility; Henry's law.

6. Ways to express the concentration of a solution: percentage by mass, mole fraction, molarity, and molality.

7. Vapor pressure of solutions; Raoult's law.

8. Boiling-point elevations and freezing-point depressions of solutions of nonvolatile solutes.

9. Osmotic pressure.

10. Distillation.

11. Solutions of electrolytes and the difference between the properties of these solutions and solutions of nonelectrolytes.

Key Terms

Some of the more important terms introduced in this chapter are listed below. Definitions for terms not included in this list may be located in the text by use of the index.

Azeotrope (Section 10.10)   A solution that has a higher or lower vapor pressure than any of the pure components of which it is composed. If the vapor pressure is higher, the solution is a **minimum-boiling azeotrope**; if lower, a **maximum-boiling azeotrope**.

Colligative property (Section 10.9)   A property of a solution that depends principally upon the concentration of dissolved particles rather than the nature of these particles; vapor-pressure lowering, freezing-point depression, boiling-point elevation, and osmotic pressure.

Distillation (Section 10.10)   The separation of a liquid solution into its components by vaporization and condensation.

Electrolyte (Section 10.11)   A solute that dissolves in water to produce a solution that is a better conductor of electricity than pure water alone.

Enthalpy of hydration (Sections 10.3 and 10.4)   The enthalpy change associated with the process in which gaseous ions of a given quantity of solute (usually one mole) are hydrated.

Enthalpy of solution (Section 10.4)   The enthalpy change associated with the process in which a given quantity of a solute (usually one mole) dissolves in a solvent. The value depends upon the concentration of the final solution and the temperature.

Henry's law (Section 10.5)   The amount of gas that dissolves without reaction in a given quantity of a liquid is directly proportional to the partial pressure of the gas above the solution.

Hydration (Sections 10.3 and 10.4)   The process in which water molecules are attracted to and surround solute particles.

Le Chatelier's principle (Section 10.5)   A system in equilibrium reacts when the conditions are changed in a way that tends to counteract the change.

Molal boiling-point elevation constant, k_b (Section 10.8)   The elevation of the boiling point of a solvent brought about by dissolving one mole of a nonvolatile, nondissociating solute in 1000 g of the solvent (a 1 m solution); the value of k_b is specific to the solvent considered.

Molal freezing-point depression constant, k_f (Section 10.8)   The depression of the freezing point of a solvent brought about by dissolving one mole of a nonvolatile, nondissociating solute in 1000 g of the solvent (a 1 m solution); the value of k_f is specific to the solvent considered.

Molality, m (Section 10.6)   A solution concentration; the number of moles of solute per 1000 g of solvent.

Osmosis (Section 10.9)   The process in which there is a net movement of solvent molecules through a semipermeable membrane separating two solutions in the direction of the more concentrated solution.

Raoult's law (Section 10.7)   The partial pressure of vapor of a component of an ideal solution is equal to the mole fraction of the component in the solution times the vapor pressure of the pure component.

van't Hoff factor, i (Section 10.12)   The ratio of the observed freezing-point depression (or boiling-point elevation) to the value calculated on the assumption that the solute is a nonelectrolyte.

Problems*

The Solution Process

10.1 Why is Br_2 more soluble in CCl_4 than I_2?

10.2 Which of each of the following pairs is the more soluble in water **(a)** CH_3OH or CH_3CH_3, **(b)** CCl_4 or NaCl, **(c)** CH_3F or CH_3Cl.

10.3 Why is the enthalpy of solution of I_2 in CCl_4 about the same as the enthalpy of fusion of pure I_2? Why cannot a similar statement be made concerning the dissolution of ionic substances in water?

10.4 What factors cause a solute to have a high enthalpy of hydration?

10.5 Which ion of each of the following pairs would be expected to be the more strongly hydrated: **(a)** Fe^{2+} or Fe^{3+}, **(b)** Li^+ or Na^+, **(c)** F^- or Cl^-, **(d)** Sn^{2+} or Pb^{2+}, **(e)** Mg^{2+} or Al^{3+}, **(f)** Mg^{2+} or Ba^{2+}?

10.6 Describe three ways in which water occurs in crystalline hydrates.

---

* The more difficult problems are marked with asterisks. The appendix contains answers to color-keyed problems.

10.7 How does the sign of an enthalpy of solution determine the effect of a temperature change on the solubility of a solute?

10.8 The lattice energy of $SrCl_2$ is -2114 kJ/mol. The enthalpy of hydration of $SrCl_2$ (infinite dilution) at 298 K is -2161 kJ/mol. What is the enthalpy of solution of $SrCl_2$ for the preparation of very dilute solutions?

10.9 The lattice energy of KF is -804 kJ/mol. The enthalpy of hydration of KF (infinite dilution) at 298 K is -819 kJ/mol. What is the enthalpy of solution of KF for the preparation of very dilute solutions?

10.10 The enthalpy of solution of NaCl at 298 K for the preparation of very dilute aqueous solutions is $+3.9$ kJ/mol. The lattice energy of NaCl is -789 kJ/mol. What is the enthalpy of hydration of the ions of this compound? Describe the processes to which this value pertains.

10.11 The enthalpy of solution of RbF at 298 K for the preparation of very dilute aqueous solutions is -24 kJ/mol. The lattice energy of RbF is -768 kJ/mol. What is the enthalpy of hydration of the ions of this compound? Describe the processes to which this value pertains.

10.12 Henry's law may be stated as $p = KX$, where p is the partial pressure of a gas over a saturated solution, X is the mole fraction of the dissolved gas in the solution, and K is a constant. For aqueous solutions of $CO_2(g)$ at 10°C, K is 1.04×10^3 atm. How many moles of CO_2 will dissolve in 900 g of H_2O if the partial pressure of CO_2 over the solution is 1.00 atm? How many grams?

10.13 Henry's law may be stated as $p = KX$, where p is the partial pressure of a gas over a saturated solution, X is the mole fraction of the dissolved gas in the solution, and K is a constant. For aqueous solutions of $N_2O(g)$ at 20°C, K is 2.04×10^3 atm. How many moles of N_2O will dissolve in 100 g of H_2O if the partial pressure over the solution is 1.50 atm? How many grams?

Concentrations of Solutions

10.14 The volume of an aqueous solution changes as the temperature changes. Thus, a concentration of a given solution determined at one temperature may not be correct for that same solution at another temperature. Which methods of expressing concentration yield results that are temperature-independent, and which yield results that are temperature-dependent?

*__10.15__ A given aqueous solution has a density of x g/ml and is $y\%$ solute by mass. Derive a mathematical equation relating the molarity of the solution, M, to the molality of the solution, m.

10.16 What are the mole fractions of methanol (CH_3OH) and water in a solution that contains 40.0 g CH_3OH and 54.0 g H_2O?

10.17 What are the mole fractions of phenol (C_6H_5OH) and ethyl alcohol (C_2H_5OH) in a solution that contains 23.5 g of phenol and 41.4 g of ethyl alcohol?

10.18 What is the mole fraction of urea (CON_2H_4) in an aqueous solution that is 10.0% urea by mass?

10.19 In an aqueous solution of glucose ($C_6H_{12}O_6$), the mole fraction of glucose is 0.200. What is the percent by mass of glucose in the solution?

10.20 A solution of NaOH is 4.00 M and has a density of 1.20 g/ml. What is the concentration of the solution in terms of percent NaOH by mass?

10.21 How many grams of KOH should be used to prepare 250 ml of 0.500 M KOH solution?

10.22 How many grams of $KMnO_4$ should be used to prepare 750 ml of a 0.125 M $KMnO_4$ solution?

10.23 Concentrated HBr is 48.0% HBr by mass and has a density of 1.50 g/ml. **(a)** How many grams of concentrated HBr should be used to prepare 500 ml of a 0.600 M HBr solution? **(b)** How many milliliters of concentrated HBr should be used to prepare this solution?

10.24 Concentrated H_3PO_4 is 85.0% H_3PO_4 by mass and has a density of 1.70 g/ml. **(a)** How many grams of concentrated H_3PO_4 should be used to prepare 250 ml of a 2.00 M H_3PO_4 solution? **(b)** How many milliliters of concentrated H_3PO_4 should be used to prepare this solution?

10.25 **(a)** What is the molarity of concentrated HI if the solution is 47.0% HI by mass and has a density of 1.50 g/ml? **(b)** What is the molality of the solution?

10.26 A solution is 30.0% KBr by mass and has a density of 1.26 g/ml. What is **(a)** the molarity and **(b)** the molality of the solution?

10.27 A 35.0 ml sample of concentrated HCl is diluted to a final volume of 125 ml. The concentrated acid is 37.0% HCl by mass and has a density of 1.18 g/ml. What is the molarity of the final solution?

10.28 A solution that is 40.0% NaBr by mass has a density of 1.42 g/ml. If 50.0 ml of this solution is diluted to a final volume of 350 ml, what is the molarity of the final solution?

10.29 What volume of concentrated H_3PO_4 should be used to prepare 125 ml of 3.00 M H_3PO_4 solution? See Table 10.1.

10.30 What volume of concentrated NH_3 should be used to prepare 200 ml of 0.600 M NH_3 solution? See Table 10.1.

10.31 A 10.0 ml sample of concentrated HNO_3 is diluted to a final volume of 500 ml. What is the molarity of the final solution? See Table 10.1.

10.32 A 25.0 ml sample of concentrated HCl is diluted to a final volume of 750 ml. What is the molarity of the final solution? See Table 10.1.

10.33 What are the mole fractions of solute and solvent in a 1.00 m solution in which toluene ($C_6H_5CH_3$) is the solvent?

*__10.34__ A quantity of sodium peroxide, Na_2O_2, is added to water and 61.6 ml of oxygen gas (measured dry and at STP) together with 150 ml of a solution of NaOH are produced. **(a)** Write the chemical equation for the reaction. **(b)** What is the molarity of the NaOH solution?

Vapor Pressure, Freezing Point, Boiling Point of Solutions

10.35 Benzene, C_6H_6, and toluene, C_7H_8, form ideal solutions. At 60°C the vapor pressure of pure benzene is 0.507 atm and the vapor pressure of pure toluene is 0.184 atm. What is the vapor pressure at 60°C of a solution containing 6.5 g of benzene and 23.0 g of toluene?

10.36 Use the data given in Problem 10.35 to calculate the mole fraction of benzene in a solution that has a vapor pressure of 0.350 atm at 60°C.

10.37 Benzene, C_6H_6, and octane, C_8H_{18}, form ideal solutions. At 60°C, the vapor pressure of pure benzene is 0.507 atm, and the vapor pressure of octane is 0.103 atm. What is the vapor pressure at 60°C of a solution that contains 9.75 g of benzene and 57.0 g of octane?

10.38 Use the data of Problem 10.37 to calculate the mole fraction of benzene in a solution that has a vapor pressure of 0.461 atm at 60°C.

10.39 A solution prepared from 30.0 g of a nonvolatile, nondissociating solute in 10.0 mol of H_2O has a vapor pressure of 0.117 atm at 50°C. What is the approximate molecular weight of the solute? The vapor pressure of water at 50°C is 0.122 atm.

10.40 A solution prepared from 20.0 g of a nonvolatile, nondissociating solute in 4.00 mol of CCl_4 has a vapor pressure of 0.663 atm at 65°C. What is the approximate molecular weight of the solute? The vapor pressure of carbon tetrachloride at 65°C is 0.699 atm.

10.41 Liquids A and B form ideal solutions. The vapor pressure of pure B is 0.650 atm at the boiling point of a solution prepared from 0.200 mol of B and 0.600 mol of A. **(a)** What is the vapor pressure of pure A at this temperature? **(b)** What is the mole fraction of A in the vapor that is in equilibrium with this solution when the solution first begins to boil?

10.42 A solution containing 1 mol of chloroform and 4 mol of acetone has a vapor pressure of 0.400 atm at 35°C. At this temperature the vapor pressure of pure chloroform is 0.359 atm and that of acetone is 0.453 atm. **(a)** What would the vapor pressure of the solution be if chloroform and acetone formed ideal solutions? **(b)** Does the vapor pressure of the solution show a positive or negative deviation from that predicted by Raoult's law? **(c)** Is heat evolved or absorbed when this solution is prepared? **(d)** Do chloroform and acetone form a minimum- or maximum-boiling azeotrope?

10.43 A solution containing 1 mol of carbon disulfide and 4 mol of acetone has a vapor pressure of 0.750 atm at 35°C. At this temperature the vapor pressure of pure carbon disulfide is 0.674 atm and that of acetone is 0.453 atm. **(a)** What would the vapor pressure of the solution be if it were an ideal solution? **(b)** Does the vapor pressure of the solution show a positive or negative deviation from that predicted by Raoult's law? **(c)** Is heat evolved or absorbed when this solution is prepared? **(d)** Do carbon disulfide and acetone form a minimum- or maximum-boiling azeotrope?

*10.44 A solution containing 20.0 g of a nonvolatile solute in exactly 1.00 mol of a volatile solvent has a vapor pressure of 0.500 atm at 20°C. A second mole of solvent is added to the mixture, and the resulting solution has a vapor pressure of 0.550 atm at 20°C. **(a)** What is the molecular weight of the solute? **(b)** What is the vapor pressure of the pure solvent at 20°C?

10.45 The antifreeze commonly used in car radiators is ethylene glycol, $C_2H_4(OH)_2$. How many grams of ethylene glycol should be added to 1.00 kg of water to produce a solution that freezes at −10.0°C?

10.46 How many grams of naphthalene, $C_{10}H_8$, should be dissolved in 300 g of nitrobenzene to produce a solution that freezes at 3.00°C? The normal freezing point of nitrobenzene is 5.70°C and k_f for nitrobenzene is −7.00°C/m.

10.47 What is the freezing point of a solution that contains 64.3 g of sucrose, $C_{12}H_{22}O_{11}$, in 200 g of water?

10.48 A solution that contains 4.32 g of naphthalene, $C_{10}H_8$, in 150 g of ethylene dibromide freezes at 7.13°C. The normal freezing point of ethylene dibromide is 9.79°C. What is the freezing point constant, k_f, for ethylene dibromide?

10.49 A solution that contains 6.79 g of a solute in 50.0 g of camphor freezes at 135.0°C. What is the molecular weight of the solute?

10.50 A solution that contains 13.2 g of a solute in 250 g of CCl_4 freezes at −33.0°C. What is the molecular weight of the solute?

10.51 A solution that contains 22.0 g of ascorbic acid (vitamin C) in 100 g of water freezes at −2.33°C. What is the molecular weight of ascorbic acid?

*10.52 (a) Air is approximately 80.0% N_2 and 20.0% O_2 by volume. What are the partial pressures of N_2 and O_2 in air when the total pressure is 1.00 atm? **(b)** Henry's law may be stated as $p = KX$, where p is the partial pressure of the gas over a saturated solution, X is the mole fraction of the dissolved gas, and K is a constant. At 0°C the values of K for N_2 and O_2 are 5.38×10^4 atm and 2.51×10^4 atm, respectively. What are the mole fractions of N_2 and O_2 in a saturated aqueous solution at 0°C under air at a pressure of 1.00 atm? **(c)** What is the freezing point of the solution?

10.53 A solution that contains a certain quantity of X dissolved in 75.0 g of acetic acid freezes at 14.40°C. A solution that contains this same quantity of X dissolved in 75.0 g of cyclohexane boils at 82.32°C. The normal freezing point of acetic acid is 16.60°C and k_f for acetic acid is −3.90°C/m. The normal boiling point of cyclohexane is 80.74°C. What is the boiling point constant, k_b, for cyclohexane?

10.54 How many grams of glycerol, $C_3H_5(OH)_3$, should be added to 100 g of water to produce a solution that boils at 101.5°C?

10.55 What is the boiling point of a solution that contains 12.5 g of biphenyl, $C_{12}H_{10}$, in 100 g of bromobenzene? The normal boiling point of bromobenzene is 156.0°C and k_b for bromobenzene is +6.26°C/m.

10.56 A solution that contains 3.86 g of X in 150 g of ethyl acetate boils at 78.21°C. What is the molecular weight of X? The normal boiling point of ethyl acetate is 77.06°C and k_b for ethyl acetate is +2.77°C/m.

10.57 A solution that contains 4.20 g of X in 200 g of chloroform boils at 62.29°C. What is the molecular weight of X?

10.58 Lauryl alcohol is obtained from coconut oil and is used to make detergents. A solution of 5.00 g of lauryl alcohol in 100.0 g of benzene boils at 80.78°C. What is the molecular weight of lauryl alcohol?

Osmotic Pressure

10.59 What is the osmotic pressure of an aqueous solution containing 2.50 g of urea, CON_2H_4, in 350 ml of solution at 25°C?

10.60 A protein has a molecular weight of 3000. The osmotic pressure of a saturated solution of this protein in water is 0.274 atm at 25°C. How many grams of the protein are dissolved in one liter of the saturated solution?

10.61 A solution that contains 9.30 g of hemoglobin per 200 ml of solution has an osmotic pressure of 0.0171 atm at 27°C. What is the molecular weight of hemoglobin?

10.62 An aqueous solution that contains 0.157 g of penicillin G in 100 ml of solution has an osmotic pressure of 0.115 atm at 25°C. What is the molecular weight of penicillin G?

10.63 (a) Insulin has an approximate molecular weight of 5700, but in water solution, it forms a dimer with a molecular weight of 11,400. What is the osmotic pressure of a solution that contains 0.250 g of insulin in 50.0 ml of solution at 20°C? **(b)** Assuming that the density of the insulin solution is 1.00 g/cm³, what is the difference in height between the two arms of an apparatus similar to that shown in Figure 11.7? Since mercury has a density of 13.6 g/cm³ and water has a density of 1.00 g/cm³,

in the measurement of pressure, the height of a column of water is 13.6 times the height of a column of mercury.

10.64 An aqueous solution that contains 0.500 g of a sugar in 250 ml of solution has an osmotic pressure of 0.272 atm at 25°C. What is the molecular weight of the sugar?

10.65 One of the problems of producing fresh water from salt water by reverse osmosis is developing semipermeable membranes strong enough to withstand the high pressure required. Sea water contains dissolved salts (3.50% by mass). Assume that the solute is entirely NaCl (over 90% is), that the density of sea water is 1.03 g/ml, and that the i factor for a NaCl solution of this concentration is 1.83. For solutions of electrolytes, $\pi = iMRT$. What is the osmotic pressure of sea water at 20°C? What pressure would have to be applied to effect reverse osmosis?

Solutions of Electrolytes

10.66 A solution prepared by adding a given weight of a solute to 20.0 g of benzene freezes at a temperature 0.384°C below the freezing point of pure benzene. The freezing point of a solution prepared from the same weight of the solute and 20.00 g of water freezes at −0.4185°C. Assume that the solute is undissociated in benzene solution but is completely dissociated into ions in water solution. How many ions result from the dissociation of one molecule in water solution?

10.67 A solution containing 3.81 g of $MgCl_2$ in 400 g of water freezes at −0.497°C. What is the van't Hoff factor, i, for the freezing point of this solution?

10.68 A solution is prepared from 3.00 g of NaOH and 75.0 g of water. If the van't Hoff factor, i, for this solution is 1.83, at what temperature will the solution freeze?

10.69 What is the freezing point of a 0.125 m aqueous solution of a weak acid, HX, if the acid is 4.00% ionized? Ignore interionic attractions.

10.70 A 0.0200 m solution of a weak acid, HY, in water freezes at −0.0385°C. Ignore interionic attractions and calculate the percentage ionization of the weak acid.

REACTIONS IN AQUEOUS SOLUTION

CHAPTER 11

Reactions that are run in aqueous solution are usually rapid for several reasons. The reactants are in a small state of subdivision (solutes exist as molecules, atoms, or ions). The attractions between the particles of pure solute (molecules, atoms, or ions) are at least partially overcome when the solution is prepared. These particles are free to move about through the solution, and ensuing collisions between solute particles result in reaction. In this chapter, we will discuss several types of reactions that take place in aqueous solution.

11.1 Metathesis Reactions

A metathesis reaction has the general form

$$AB + CD \longrightarrow AD + CB$$

In the reaction, the cations and anions of the substances involved exchange partners. This type of reaction is a common one, particularly in aqueous solution.

A specific example of a metathesis reaction is

$$AgNO_3(aq) + NaCl(aq) \longrightarrow AgCl(s) + NaNO_3(aq)$$

Since we are talking about reactions in aqueous solution, the symbol (aq) will be omitted from future examples. The symbols (g) for gaseous and (s) for solid, however, will be used.

The preceding equation is written in *molecular form*. These substances, however, are ionic. A solution of a soluble ionic substance in water contains hydrated ions. The *ionic form* of the equation is

$$Ag^+ + NO_3^- + Na^+ + Cl^- \longrightarrow AgCl(s) + Na^+ + NO_3^-$$

Why did the reaction occur? Collisions of the Ag^+ and Cl^- ions result in the formation of AgCl, which is insoluble and falls out of solution. An insoluble substance formed in this way is called a **precipitate,** and the process is called a **precipitation.** Sodium nitrate ($NaNO_3$), the other product of the reaction, is soluble in water and remains in solution in the form of hydrated ions.

If we eliminate the spectator ions (ions that appear on both sides of the equation and hence do not enter into the reaction), we get the *net ionic form* of the equation:

$$Ag^+ + Cl^- \longrightarrow AgCl(s)$$

This is the most general form of the equation. It tells us that a solution of any soluble Ag^+ salt and a solution of any soluble Cl^- salt will produce insoluble AgCl when mixed.

Suppose that we mix solutions of NaCl and NH_4NO_3. The ionic equation for the "reaction" is

$$Na^+ + Cl^- + NH_4^+ + NO_3^- \longrightarrow Na^+ + NO_3^- + NH_4^+ + Cl^-$$

All of the compounds involved in the "reaction" are soluble in water. If we eliminate spectator ions from the equation, nothing is left. There is, therefore, no reaction, which is usually indicated in the following way:

$$Na^+ + Cl^- + NH_4^+ + NO_3^- \longrightarrow N.R.$$

There are other reasons for a metathesis reaction to occur in addition to the formation of an insoluble solid. Consider the reaction that occurs when a hydrochloric acid solution and a solution of sodium sulfide are mixed:

molecular equation: $\qquad 2\,HCl + Na_2S \longrightarrow H_2S(g) + 2\,NaCl$

ionic equation: $\qquad 2\,H^+ + 2\,Cl^- + 2\,Na^+ + S^{2-} \longrightarrow H_2S(g) + 2\,Na^+ + 2\,Cl^-$

net ionic equation: $\qquad 2\,H^+ + S^{2-} \longrightarrow H_2S(g)$

In this reaction, collisions of the H^+ and S^{2-} ions have formed a gas, H_2S, which is only slightly soluble and which escapes from the solution.

In a third type of metathesis reaction, a weak electrolyte is produced. Soluble weak electrolytes do not dissociate completely into ions in aqueous solution. They exist principally in the form of molecules. An acid-base neutralization reaction (see Section 11.4) is a metathesis reaction of this type:

molecular equation: $\qquad HCl + NaOH \longrightarrow H_2O + NaCl$

ionic equation: $\qquad H^+ + Cl^- + Na^+ + OH^- \longrightarrow H_2O + Na^+ + Cl^-$

net ionic equation: $\qquad H^+ + OH^- \longrightarrow H_2O$

The reaction occurs because the H^+ and OH^- ions form H_2O molecules, and water is a weak electrolyte.

Metathesis reactions, then, occur when a precipitate, an insoluble gas, or a weak electrolyte is formed. In ionic equations for these reactions,

1. A *soluble salt* is indicated by the formulas of the ions that make up the compound.

2. An *insoluble* or *slightly soluble compound* is indicated by the molecular formula of the compound followed by the symbol (s).

3. An *insoluble* or *slightly soluble gas* is indicated by the molecular formula of the gas followed by the symbol (g).

4. A *weak electrolyte* is indicated by the molecular formula of the compound. Weak electrolytes are partly dissociated into ions in aqueous solution, but they exist principally in molecular form.

In order to write equations for metathesis reactions, we need to identify these types of compounds. The following rules are designed for this purpose:

1. Solubility rules. The classification of ionic substances according to their solubility in water is difficult. Nothing is completely "insoluble" in water. The degree of solubility varies greatly from one "soluble" substance to another. Nevertheless, a solubility classification scheme is useful even though it must be regarded as approximate. The rules given in Table 11.1 apply to compounds of the following cations:

a. *1+ cations.* Li^+, Na^+, K^+, Rb^+, Cs^+, NH_4^+, Ag^+

b. *2+ cations.* Mg^{2+}, Ca^{2+}, Sr^{2+}, Ba^{2+}, Mn^{2+}, Fe^{2+}, Co^{2+}, Ni^{2+}, Cu^{2+}, Zn^{2+}, Cd^{2+}, Hg^{2+}, Hg_2^{2+}, Sn^{2+}, Pb^{2+}

c. *3+ cations.* Fe^{3+}, Al^{3+}, Cr^{3+}

Compounds that dissolve to the extent of at least 10 g/liter are listed as *soluble*. Those that fail to dissolve to the extent of 1 g/liter are classed as *insoluble*. Those compounds with solubilities that are intermediate between these limits are listed as *slightly soluble* (and marked with a star in the table). These standards are common but arbitrary. Most acids are water soluble.

2. Gases. The common insoluble gases formed in metathesis reactions are listed in Table 11.2. Reactions in which three of the gases (CO_2, SO_2, and NH_3) are produced may be viewed as involving the initial formation of a substance that then breaks down to give the gas and H_2O. The reaction of Na_2SO_3 and HCl, for example, produces H_2SO_3:

$$2\,Na^+ + SO_3^{2-} + 2\,H^+ + 2\,Cl^- \longrightarrow H_2SO_3 + 2\,Na^+ + 2\,Cl^-$$

Table 11.1   Solubilities of some ionic compounds in water[a]

	Mainly Water-Soluble
NO_3^-	All nitrates are soluble.
$C_2H_3O_2^-$	All acetates are soluble.
ClO_3^-	All chlorates are soluble.
Cl^-	All chlorides are soluble except $AgCl$, Hg_2Cl_2, and $PbCl_2$*.
Br^-	All bromides are soluble except $AgBr$, Hg_2Br_2, $PbBr_2$*, and $HgBr_2$*.
I^-	All iodides are soluble except AgI, Hg_2I_2, PbI_2, and HgI_2.
SO_4^{2-}	All sulfates are soluble except $CaSO_4$*, $SrSO_4$, $BaSO_4$, $PbSO_4$, Hg_2SO_4, and Ag_2SO_4*.

	Mainly Water-Insoluble
S^{2-}	All sulfides are insoluble except those of the I A and II A elements and $(NH_4)_2S$.
CO_3^{2-}	All carbonates are insoluble except those of the I A elements and $(NH_4)_2CO_3$.
SO_3^{2-}	All sulfites are insoluble except those of the I A elements and $(NH_4)_2SO_3$.
PO_4^{3-}	All phosphates are insoluble except those of the I A elements and $(NH_4)_3PO_4$.
OH^-	All hydroxides are insoluble except those of the I A elements, $Ba(OH)_2$, $Sr(OH)_2$*, and $Ca(OH)_2$*.

[a] The following cations are considered: those of the I A and II A families, NH_4^+, Ag^+, Al^{3+}, Cd^{2+}, Co^{2+}, Cr^{3+}, Cu^{2+}, Fe^{2+}, Fe^{3+}, Hg_2^{2+}, Hg^{2+}, Mn^{2+}, Ni^{2+}, Pb^{2+}, Sn^{2+}, and Zn^{2+}.

* Soluble compounds dissolve to the extent of at least 10 g/liter. Slightly soluble compounds (marked with an *) dissolve to the extent of from 1 g/liter to 10 g/liter.

Gas	
H_2S	Any sulfide (salt of S^{2-}) and any acid form $H_2S(g)$ and a salt.
CO_2	Any carbonate (salt of CO_3^{2-}) and any acid form $CO_2(g)$, H_2O, and a salt.
SO_2	Any sulfite (salt of SO_3^{2-}) and any acid form $SO_2(g)$, H_2O, and a salt.
NH_3	Any ammonium salt (salt of NH_4^+) and any soluble strong hydroxide form $NH_3(g)$, H_2O, and a salt.

The H_2SO_3 is unstable and decomposes to give SO_2 and H_2O:

$$H_2SO_3 \longrightarrow H_2O + SO_2(g)$$

The ionic equation for the complete reaction, therefore, is

$$2\,Na^+ + SO_3^{2-} + 2\,H^+ + 2\,Cl^- \longrightarrow H_2O + SO_2(g) + 2\,Na^+ + 2\,Cl^-$$

A typical reaction of a carbonate and an acid is

$$2\,Na^+ + CO_3^{2-} + 2\,H^+ + 2\,Cl^- \longrightarrow H_2O + CO_2(g) + 2\,Na^+ + 2\,Cl^-$$

Ammonium salts and soluble strong bases react in the following way:

$$NH_4^+ + Cl^- + Na^+ + OH^- \longrightarrow H_2O + NH_3(g) + Na^+ + Cl^-$$

3. **Weak electrolytes.** A simplified list of rules follows (see Section 11.4):

 a. **Acids.** Most common acids are weak electrolytes except $HClO_4$, $HClO_3$, HCl, HBr, HI, HNO_3, and H_2SO_4 (first ionization only).

 b. **Bases.** The hydroxides of the I A elements and of Ca^{2+}, Sr^{2+} and Ba^{2+} are soluble and strong. Most others are insoluble and weak.

 c. **Salts.** Most common salts are strong electrolytes.

 d. **Water.** Water is a weak electrolyte.

Example 11.1

Write balanced chemical equations for the reactions that occur when aqueous solutions of the following pairs of compounds are mixed. Show all substances (both reactants and products) in proper form:

 (a) $FeCl_3$ and $(NH_4)_3PO_4$
 (b) Na_2SO_4 and $CuCl_2$
 (c) $ZnSO_4$ and $Ba(OH)_2$
 (d) $CaCO_3$ and HNO_3

Solution

 (a) $Fe^{3+} + 3\,Cl^- + 3\,NH_4^+ + PO_4^{3-} \longrightarrow FePO_4(s) + 3\,NH_4^+ + 3\,Cl^-$
 (b) $2\,Na^+ + SO_4^{2-} + Cu^{2+} + 2\,Cl^- \longrightarrow$ N. R.

Settling ponds in which magnesium hydroxide is precipitated from sea water. The $Mg(OH)_2$ is produced by a metathesis reaction between the Mg^{2+} ions of sea water and OH^- ions from $Ca(OH)_2$, which is added. The $Mg(OH)_2$ is used in a process for the production of Mg metal (see Section 23.5). *Dow Chemical U.S.A.*

(c) $Zn^{2+} + SO_4^{2-} + Ba^{2+} + 2\,OH^- \longrightarrow Zn(OH)_2(s) + BaSO_4(s)$

(d) $CaCO_3(s) + 2\,H^+ + 2\,NO_3^- \longrightarrow H_2O + CO_2(g) + Ca^{2+} + 2\,NO_3^-$

Metathesis reactions are reversible to some extent. Aqueous equilibrium systems of this type are discussed in Chapters 15 and 16.

Metathesis reactions also occur in the absence of water. Examples are

$CaF_2(s) + H_2SO_4(l) \longrightarrow CaSO_4(s) + 2\,HF(g)$

$2\,NaNO_3(s) + H_2SO_4(l) \longrightarrow Na_2SO_4(s) + 2\,HNO_3(g)$

The reactants in these reactions are heated, and the acids (HF and HNO_3, which are soluble in water) are driven off as gases.

11.2 Oxidation Numbers

An important type of reaction that occurs in aqueous solution, the oxidation-reduction reaction, is discussed in the next section. Before this type of reaction is considered, however, we must discuss the concept of oxidation numbers, an arbitrary but useful convention.

Oxidation numbers are charges (fictitious charges in the case of covalent species) assigned to the atoms of a compound according to some arbitrary rules. The oxidation numbers of monatomic ions in binary ionic compounds are the same as the charges on the ions. In NaCl, for example, the oxidation number of sodium in Na^+ is $1+$ and the oxidation number of chlorine in Cl^- is $1-$.

The oxidation numbers of the atoms in a covalent molecule can be derived by assigning the electrons of each bond to the more electronegative of the bonded atoms. For the molecule

$$H\!:\!\overset{..}{\underset{..}{Cl}}\!:$$

both electrons of the covalent bond are assigned to the chlorine atom since chlorine is more electronegative than hydrogen. The chlorine atom is said to have an oxidation number of $1-$ since the assignment has given it one more electron than it has as a neutral atom. The hydrogen atom is said to have an oxidation number of $1+$ because its only valence electron has been assigned to the chlorine atom.

In the case of a nonpolar bond between identical atoms, in which there is no electronegativity difference, the bonding electrons are divided equally between the bonded atoms in deriving oxidation numbers. Thus, the oxidation numbers of both chlorine atoms are zero in the molecule

$$:\overset{..}{\underset{..}{Cl}}:\overset{..}{\underset{..}{Cl}}:$$

The following rules, based on these ideas, can be used to assign oxidation numbers:

1. Any uncombined atom or any atom in a molecule of an element is assigned an oxidation number of zero.

2. The sum of the oxidation numbers of the atoms in a compound is zero, since compounds are electrically neutral.

3. The oxidation number of a monatomic ion is the same as the charge on the ion. In their compounds, group I A elements (Li, Na, K, Rb, and Cs) always have oxidation numbers of $1+$, group II A elements (Be, Mg, Ca, Sr, and Ba) always have oxidation numbers of $2+$.

4. The sum of the oxidation numbers of the atoms that constitute a polyatomic ion equals the charge on the ion.

5. The oxidation number of fluorine, the most electronegative element, is $1-$ in all fluorine-containing compounds.

6. In most oxygen-containing compounds, the oxidation number of oxygen is $2-$. There are, however, a few exceptions:

  a. In peroxides each oxygen has an oxidation number of $1-$. The two O atoms of the peroxide ion, O_2^{2-}, are equivalent. Each must be assigned an oxidation number of $1-$ so that the sum equals the charge on the ion.

  b. In the superoxide ion, O_2^-, each oxygen has an oxidation number of $1/2-$.

  c. In OF_2 the oxygen has an oxidation number of $2+$ (see rule 5).

7. The oxidation number of hydrogen is $1+$ in all its compounds except the metallic hydrides (CaH_2 and NaH are examples) in which hydrogen is in the $1-$ oxidation state.

8. In a combination of two nonmetals (either a molecule or a polyatomic ion) the oxidation number of the more electronegative element is negative and equal to the charge on the common monatomic ion of that element. In PCl_3, for example, the oxidation number of Cl is $1-$, and of P is $3+$. In CS_2, the oxidation number of S is $2-$, and of C is $4+$.

Example 11.2

What is the oxidation number of the P atom in H_3PO_4?

Solution

The oxidation numbers of the molecule must add up to zero. Therefore,

$$3(ox.\ no.\ H) + (ox.\ no.\ P) + 4(ox.\ no.\ O) = 0$$

Each H is assigned an oxidation number of $1+$ (see rule 7), and each O is assigned an oxidation number of $2-$ (see rule 6):

$$3(1+) + x + 4(2-) = 0$$
$$x = 5+$$

Example 11.3

What is the oxidation number of Cr in the dichromate ion, $Cr_2O_7^{2-}$?

Solution

The sum of the oxidation numbers must equal the charge of the ion, $2-$. The oxidation number of O is $2-$ (see rule 6):

$$2(ox.\ no.\ Cr) + 7(ox.\ no.\ O) = 2-$$
$$2x + 7(2-) = 2-$$
$$2x = 12+$$
$$x = 6+$$

Notice that the oxidation state of an element is reported on the basis of a single atom. It would be misleading to say that oxygen has an oxidation number of $2-$ in H_2O and $4-$ in SO_2. In both compounds, the O atom has an oxidation number of $2-$.

Example 11.4

What is the oxidation number of Cl in calcium perchlorate, $Ca(ClO_4)_2$?

Solution

$$(ox.\ no.\ Ca) + 2(ox.\ no.\ Cl) + 8(ox.\ no.\ O) = 0$$

Since Ca is a member of group II A, its oxidation number is $2+$, which is the charge on a Ca^{2+} ion. The oxidation number of O is $2-$. Therefore,

$$(2+) + 2x + 8(2-) = 0$$
$$2x = 14+$$
$$x = 7+$$

There is another way to solve problems of this type. Since the charge on the Ca^{2+} ion is $2+$, and since there are two perchlorate ions for every one Ca^{2+} ion in the compound, the charge on the perchlorate ion must be $1-$. The ion has the formula ClO_4^-. Therefore,

$$(ox.\ no.\ Cl) + 4(ox.\ no.\ O) = 1-$$
$$x + 4(2-) = 1-$$
$$x = 7+$$

These values should be interpreted with care. The ionic charges have physical significance; the oxidation numbers of O and Cl are merely conventions.

Frequently an element exhibits a range of oxidation numbers in its compounds. Nitrogen, for example, exhibits oxidation numbers from $3-$ (as in NH_3) to $5+$ (as in HNO_3).

1. The highest oxidation number of an A family element is the same as its group number. The number of valence electrons that an A family element has is equal to its group number. It is not logical to expect an atom to lose more valence electrons than it has. The highest possible positive charge—even a hypothetical one, therefore—is the same as the group number.

2. The lowest oxidation number of an A family element in a compound containing the element is the same as the charge of a monatomic ion of the element.

The highest oxidation number of sulfur (a member of group VI A) is $6+$ (in H_2SO_4, for example). The lowest oxidation number of sulfur is $2-$ (as in Na_2S and H_2S). In its compounds, the highest oxidation number of sodium (found in group I A) is the same as the lowest oxidation number, $1+$. The oxidation number of uncombined sodium is, of course, zero. There are, however, exceptions to these generalizations (fluorine and oxygen, for example).

Oxidation numbers are not the same as formal charges. In the assignment of formal charges to the atoms of a covalent molecule, the bonding electrons are divided equally between the bonded atoms, and any bond polarity caused by the unequal sharing of electrons is ignored. In the assignment of oxidation numbers, the bonding electrons are assigned to the more electronegative atom. Both concepts are merely conventions. Formal charges are useful in interpreting the structure and some of the properties of covalent molecules. Oxidation numbers are useful in describing oxidation-reduction reactions, the topic of the next section.

11.3 Oxidation-Reduction Reactions

The term oxidation was originally applied to reactions in which substances combined with oxygen, and reduction was defined as the removal of oxygen from an oxygen-containing compound. The meanings of the terms have gradually been broadened. Today oxidation and reduction are defined on the basis of change in oxidation number.

Oxidation is the process in which an atom undergoes an algebraic increase in oxidation number, and **reduction** is the process in which an atom undergoes an algebraic decrease in oxidation number. On this basis, oxidation-reduction is involved in the reaction

$$\overset{0}{S} + \overset{0}{O_2} \longrightarrow \overset{4+}{S}\overset{2-}{O_2}$$

The oxidation number of each type of atom is written above its symbol. Since the oxidation number of the S atom *increases* from 0 to 4+, sulfur is said to be *oxidized*. The oxidation number of the O atom *decreases* from 0 to 2−, and oxygen is said to be *reduced*. Oxidation-reduction is not involved in the reaction

$$\overset{4+}{S}\overset{2-}{O_2} + \overset{1+}{H_2}\overset{2-}{O} \longrightarrow \overset{1+}{H_2}\overset{4+}{S}\overset{2-}{O_3}$$

since no atom undergoes a change in oxidation number.

It is apparent, from the way in which oxidation numbers are assigned, that *neither oxidation nor reduction can occur by itself.* Furthermore, *the total increase in oxidation number must equal the total decrease in oxidation number.* In the reaction of sulfur and oxygen, the sulfur undergoes an increase of 4. Each oxygen atom undergoes a decrease of 2. Since two O atoms appear in the equation, the total decrease is 4.

Since one substance cannot be reduced unless another is simultaneously oxidized, the substance that is reduced is responsible for the oxidation. This substance is called, therefore, the **oxidizing agent** or **oxidant**. Because of the interdependence of the two processes, the opposite is also true. The material that is itself oxidized is called the **reducing agent** or **reductant**. Therefore,

$$\underset{\substack{\text{oxidized} \\ \text{reducing agent}}}{\overset{0}{S}} + \underset{\substack{\text{reduced} \\ \text{oxidizing agent}}}{\overset{0}{O_2}} \longrightarrow \overset{4+}{S}\overset{2-}{O_2}$$

Equations for oxidation-reduction reactions are usually more difficult to balance than those for reactions that do not entail oxidation and reduction. Two methods are commonly used to balance oxidation-reduction equations: the oxidation-number method and the ion-electron method. Both of these methods will be discussed. For clarity, the physical state of the reactants and products will not be indicated in the examples that follow. In addition, the symbol H^+, instead of H_3O^+ or $H^+(aq)$, will be used.

There are three steps in the **oxidation-number method** of balancing oxidation-reduction equations. We will use the equation for the reaction of nitric acid and hydrogen sulfide to illustrate this method. The unbalanced expression for the reaction is

$$HNO_3 + H_2S \longrightarrow NO + S + H_2O$$

1. The oxidation numbers of the atoms in the equation are determined in order to identify those undergoing oxidation or reduction. Thus,

$$H\overset{5+}{N}O_3 + H_2\overset{2-}{S} \longrightarrow \overset{2+}{N}O + \overset{0}{S} + H_2O$$

Nitrogen is reduced (from 5+ to 2+, a decrease of 3), and sulfur is oxidized (from 2− to 0, an increase of 2).

2. Coefficients are added so that the total decrease and the total increase in oxidation number will be equal. We have a decrease of 3 and an increase of 2 indicated in the unbalanced expression. The lowest common multiple of 3 and 2 is 6. We therefore indicate $2\,HNO_3$ and $2\,NO$ (for a total decrease of 6) and $3\,H_2S$ and $3\,S$ (for a total increase of 6):

$$2\,HNO_3 + 3\,H_2S \longrightarrow 2\,NO + 3\,S + H_2O$$

3. Balancing is completed by inspection. This method takes care of only those substances that are directly involved in oxidation-number change. In this example, the method does not assign a coefficient to H_2O. We note, however, that there are now 8 H atoms on the left of the equation. We can indicate the same number of H atoms on the right by showing $4\,H_2O$:

$$2\,HNO_3 + 3\,H_2S \longrightarrow 2\,NO + 3\,S + 4\,H_2O$$

The final, balanced equation should be checked to ensure that there are as many atoms of each element on the right as there are on the left.

The oxidation-number method can be used to balance net ionic equations, in which only those ions and molecules that take part in the reaction are shown. Consider the reaction between $KClO_3$ and I_2:

$$H_2O + I_2 + ClO_3^- \longrightarrow IO_3^- + Cl^- + H^+$$

The K^+ ion does not take part in the reaction and is not shown in the equation. The steps in balancing the equation follow:

1. $H_2O + \overset{0}{I_2} + \overset{5+}{Cl}O_3^- \longrightarrow \overset{5+}{I}O_3^- + \overset{1-}{Cl}^- + H^+$

2. *Each* iodine atom undergoes an increase of 5 (from 0 to 5+), but there are *two* iodine atoms in I_2. The increase in oxidation number is therefore 10. Chlorine undergoes a decrease of 6 (from 5+ to 1−). The lowest common multiple of 6 and 10 is 30. Therefore, $3\,I_2$ molecules must be indicated (a total increase of 30) and $5\,ClO_3^-$ ions are needed (a total decrease of 30). The coefficients of the products, IO_3^- and Cl^-, follow from this assignment:

$$H_2O + 3\,I_2 + 5\,ClO_3^- \longrightarrow 6\,IO_3^- + 5\,Cl^- + H^+$$

3. If H_2O is ignored, there are now 15 oxygen atoms on the left and 18 oxygen atoms on the right. To make up 3 oxygen atoms on the left, we must indicate $3\,H_2O$ molecules. It then follows that the coefficient of H^+ must be 6 to balance the hydrogens of the H_2O molecules:

$$3\,H_2O + 3\,I_2 + 5\,ClO_3^- \longrightarrow 6\,IO_3^- + 5\,Cl^- + 6\,H^+$$

An ionic equation must indicate charge balance as well as mass balance. Since the algebraic sum of the charges on the left (5−) equals that on the right (5−), the equation is balanced.

Reactions in which electrons are transferred are clearly examples of oxidation-reduction reactions. In the reaction of sodium and chlorine, a sodium atom loses its valence electron to a chlorine atom:

$$2\,\overset{0}{Na} + \overset{0}{Cl_2} \longrightarrow 2\,\overset{1+}{Na}^+ + 2\,\overset{1-}{Cl}^-$$

For simple ions the oxidation number is the same as the charge on the ion. It follows, then, that electron loss represents a type of oxidation, and electron gain represents a type of reduction. This equation can be divided into two **partial equations** that represent **half reactions**:

Oxidation: $$2\,Na \longrightarrow 2\,Na^+ + 2e^-$$

Reduction: $$2e^- + Cl_2 \longrightarrow 2\,Cl^-$$

The **ion-electron method** of balancing oxidation-reduction equations employs partial equations. One partial equation is used for the oxidation (in which electrons are lost), and another partial equation is used for the reduction (in which electrons are gained). The final equation is obtained by combining the partial equations in such a way that the number of electrons lost equals the number of electrons gained.

Two slightly different procedures are employed to balance equations by the ion-electron method. One is used for reactions that take place in acid solution; the other, for reactions that occur in alkaline solution. We will give examples of both procedures.

An example of a reaction that occurs in *acid solution* is

$$Cr_2O_7^{2-} + Cl^- \longrightarrow Cr^{3+} + Cl_2$$

In this unbalanced expression, H_2O and H^+ are not shown. The proper numbers of H_2O molecules and H^+ ions, as well as their positions in the final equation (whether on the left or the right), are determined in the course of the balancing:

1. Two skeleton partial equations for the half reactions are written. The central elements of the partial equations (Cr and Cl) are balanced:

$$Cr_2O_7^{2-} \longrightarrow 2\ Cr^{3+}$$
$$2\ Cl^- \longrightarrow Cl_2$$

2. Now the H and O atoms are balanced. Since the reaction occurs in acid solution, H_2O and H^+ can be added where needed. For each O atom that is needed, one H_2O is added to the side that is deficient. The hydrogen is then brought into balance by the addition of H^+.

Seven O atoms are needed on the right side of the first partial equation, and, therefore, 7 H_2O are added to this side. The H atoms of the first partial equation are then balanced by the addition of 14 H^+ to the left side. The second partial equation is already in material balance:

$$14\ H^+ + Cr_2O_7^{2-} \longrightarrow 2\ Cr^{3+} + 7\ H_2O$$
$$2\ Cl^- \longrightarrow Cl_2$$

3. The next step is to balance the partial equations electrically. In the first partial equation, the net charge is 12+ on the left side of the equation (14+ and 2−) and 6+ on the right. Six electrons must be added to the left. In this way, the net charge on both sides of the equation will be 6+. The second partial equation is balanced electrically by the addition of 2 electrons to the right:

$$6e^- + 14\ H^+ + Cr_2O_7^{2-} \longrightarrow 2\ Cr^{3+} + 7\ H_2O$$
$$2\ Cl^- \longrightarrow Cl_2 + 2e^-$$

4. The number of electrons gained must equal the number of electrons lost. The second partial equation, therefore, is multiplied through by 3:

$$6e^- + 14\,H^+ + Cr_2O_7^{2-} \longrightarrow 2\,Cr^{3+} + 7\,H_2O$$
$$6\,Cl^- \longrightarrow 3\,Cl_2 + 6e^-$$

5. Addition of the two partial equations gives the final equation. In the addition, the electrons cancel:

$$14\,H^+ + Cr_2O_7^{2-} + 6\,Cl^- \longrightarrow 2\,Cr^{3+} + 3\,Cl_2 + 7\,H_2O$$

As a second example, consider the reaction

$$MnO_4^- + As_4O_6 \longrightarrow Mn^{2+} + H_3AsO_4$$

which also occurs in *acid solution*. The same steps are followed:

1. The equation is divided into two skeleton partial equations. The As atoms in the second partial equation are balanced:

$$MnO_4^- \longrightarrow Mn^{2+}$$
$$As_4O_6 \longrightarrow 4\,H_3AsO_4$$

2. The first partial equation can be brought into material balance by the addition of $4\,H_2O$ to the right side and $8\,H^+$ to the left side. In the second partial equation, $10\,H_2O$ must be added to the left side to make up the needed 10 oxygens. If we stopped at this point, there would be 20 hydrogen atoms on the left and 12 on the right. Therefore, $8\,H^+$ must be added to the right:

$$8\,H^+ + MnO_4^- \longrightarrow Mn^{2+} + 4\,H_2O$$
$$10\,H_2O + As_4O_6 \longrightarrow 4\,H_3AsO_4 + 8\,H^+$$

3. To balance the net charges, electrons are added:

$$5e^- + 8\,H^+ + MnO_4^- \longrightarrow Mn^{2+} + 4\,H_2O$$
$$10\,H_2O + As_4O_6 \longrightarrow 4\,H_3AsO_4 + 8\,H^+ + 8e^-$$

4. The first partial equation must be multiplied through by 8 and the second by 5 so that the same number of electrons are lost in the oxidation partial equation as are gained in the reduction partial equation:

$$40e^- + 64\,H^+ + 8\,MnO_4^- \longrightarrow 8\,Mn^{2+} + 32\,H_2O$$
$$50\,H_2O + 5\,As_4O_6 \longrightarrow 20\,H_3AsO_4 + 40\,H^+ + 40e^-$$

5. When these two partial equations are added, water molecules and hydrogen ions must be canceled as well as electrons. It is poor form to leave an equation with $64\,H^+$ on the left and $40\,H^+$ on the right:

$$24\,H^+ + 18\,H_2O + 5\,As_4O_6 + 8\,MnO_4^- \longrightarrow 20\,H_3AsO_4 + 8\,Mn^{2+}$$

Equations for reactions that take place in *alkaline solution* are balanced in a manner slightly different from those that occur in acidic solution. All the steps are the same except the second one; H^+ cannot be used to balance equations for reactions that occur in alkaline solution. As an example, consider the reaction

$$MnO_4^- + N_2H_4 \longrightarrow MnO_2 + N_2$$

that takes place in alkaline solution:

1. The equation is divided into two partial equations:

$$MnO_4^- \longrightarrow MnO_2$$
$$N_2H_4 \longrightarrow N_2$$

2. For reactions occurring in alkaline solution, OH^- and H_2O are used to balance oxygen and hydrogen. For each oxygen that is needed, $2\,OH^-$ ions are added to the side of the partial equation that is deficient, and one H_2O molecule is added to the opposite side. For each hydrogen that is needed, one H_2O molecule is added to the side that is deficient, and one OH^- ion is added to the opposite side.

In the first partial equation, the right side is deficient by 2 oxygen atoms. We add, therefore, $4\,OH^-$ to the right side and $2\,H_2O$ to the left:

$$2\,H_2O + MnO_4^- \longrightarrow MnO_2 + 4\,OH^-$$

In order to bring the second partial equation into material balance, we must add four hydrogen atoms to the right side. For *each* hydrogen atom needed, we add one H_2O to the side deficient in hydrogen and one OH^- to the opposite side. In the present case we add $4\,H_2O$ to the right side and $4\,OH^-$ to the left to make up the four hydrogen atoms needed on the right:

$$4\,OH^- + N_2H_4 \longrightarrow N_2 + 4\,H_2O$$

3. Electrons are added to effect charge balances:

$$3e^- + 2\,H_2O + MnO_4^- \longrightarrow MnO_2 + 4\,OH^-$$
$$4\,OH^- + N_2H_4 \longrightarrow N_2 + 4\,H_2O + 4e^-$$

4. The lowest common multiple of 3 and 4 is 12. Therefore, the first partial equation is multiplied through by 4 and the second by 3 so that the number of electrons gained equals the number lost:

$$12e^- + 8\,H_2O + 4\,MnO_4^- \longrightarrow 4\,MnO_2 + 16\,OH^-$$
$$12\,OH^- + 3\,N_2H_4 \longrightarrow 3\,N_2 + 12\,H_2O + 12e^-$$

5. Addition of these partial equations, with cancellation of OH^- ions and H_2O molecules as well as electrons, gives the final equation:

$$4\,MnO_4^- + 3\,N_2H_4 \longrightarrow 4\,MnO_2 + 3\,N_2 + 4\,H_2O + 4\,OH^-$$

As a final example, consider the following skeleton equation for a reaction in *alkaline solution:*

$$Br_2 \longrightarrow BrO_3^- + Br^-$$

In this reaction the same substance, Br_2, is both oxidized and reduced. Such reactions are called **disproportionations** or **auto-oxidation-reduction reactions**:

1.
$$Br_2 \longrightarrow 2\,BrO_3^-$$
$$Br_2 \longrightarrow 2\,Br^-$$

2.
$$12\,OH^- + Br_2 \longrightarrow 2\,BrO_3^- + 6\,H_2O$$
$$Br_2 \longrightarrow 2\,Br^-$$

3.
$$12\,OH^- + Br_2 \longrightarrow 2\,BrO_3^- + 6\,H_2O + 10e^-$$
$$2e^- + Br_2 \longrightarrow 2\,Br^-$$

4.
$$12\,OH^- + Br_2 \longrightarrow 2\,BrO_3^- + 6\,H_2O + 10e^-$$
$$10e^- + 5\,Br_2 \longrightarrow 10\,Br^-$$

5. $12\,OH^- + 6\,Br_2 \longrightarrow 2\,BrO_3^- + 10\,Br^- + 6\,H_2O$

When the ion-electron method is applied to a disproportionation reaction, the coefficients of the resulting equation usually are divisible by some common number since one reactant was used in both partial equations. The coefficients of this equation are all divisible by 2 and should be reduced to the lowest possible terms:

$$6\,OH^- + 3\,Br_2 \longrightarrow BrO_3^- + 5\,Br^- + 3\,H_2O$$

The Ion-Electron Method for Balancing Oxidation-Reduction Equations

1. Divide the equation into two skeleton partial equations.

2. Balance the atoms that change their oxidation numbers in each partial equation.

3. Balance the O and H atoms in each partial equation.

   a. For reaction in *acid solution:*
   i. For each O atom that is needed, add one H_2O to the side of the partial equation that is deficient in oxygen.
   ii. Add H^+ where needed to bring the hydrogen into balance.

   b. For reactions in *alkaline solution:*
   i. For each O atom that is needed, add *two* OH^- ions to the side of the partial equation that is deficient in O, and add *one* H_2O to the opposite side.
   ii. For each H atom that is needed, add *one* H_2O to the side of the partial equation that is deficient in H, and add *one* OH^- ion to the opposite side.

4. To each partial equation, add electrons in such a way that the net charge on the left side of the equation equals the net charge on the right side.

5. If necessary, multiply one or both partial equations by numbers that will make the number of electrons lost in one partial equation equal the number of electrons gained in the other partial equation.

6. Add the partial equations. In the addition, cancel terms common to both sides of the final equation.

Most oxidation-reduction equations may be balanced by the ion-electron method, which is especially convenient for electrochemical reactions and reactions of ions in water solution. However, several misconceptions that can arise must be pointed out. Half reactions cannot occur alone, and partial equations do not represent complete chemical changes. Even in electrochemical cells, where the two half reactions take place at different electrodes, the two half reactions always occur simultaneously.

Whereas the partial equations probably represent an overall, if not detailed, view of the way an oxidation-reduction reaction occurs in an electrochemical cell, the same reaction in a beaker may not take place in this way at all. The method should *not* be interpreted as necessarily giving the correct mechanism by which a reaction occurs. It is, at times, difficult to recognize whether a given reaction is a legitimate example of an electron-exchange reaction. The reaction

$$\overset{4+}{SO_3^{2-}} + \overset{5+}{ClO_3^{-}} \longrightarrow \overset{6+}{SO_4^{2-}} + \overset{3+}{ClO_2^{-}}$$

looks like an electron-exchange reaction, can be made to take place in an electrochemical cell, and can be balanced by the ion-electron method. However, this reaction has been shown to proceed by direct oxygen exchange (from ClO_3^- to SO_3^{2-}) and *not* by electron exchange.

11.4 Arrhenius Acids and Bases

The several concepts of acids and bases that are in current use are the topic of Chapter 14. The **Arrhenius concept of acids and bases,** the oldest of these, is presented in this section.

An **acid** is defined as a substance that dissociates in water to produce H_3O^+ ions, which are sometimes shown as $H^+(aq)$ ions. For example,

$$H-\ddot{O}: + H-\ddot{Cl}:(g) \longrightarrow \left[H-\underset{H}{\overset{H}{\ddot{O}}}-H \right]^+ (aq) + :\ddot{Cl}:^- (aq)$$

Pure HCl gas consists of covalent molecules. In water, the H^+ (which is nothing more than a proton) of the HCl molecule is strongly attracted by an electron pair of the O atom of a H_2O molecule. The transfer of the proton to the H_2O molecule produces what is called a **hydronium ion** (H_3O^+) and leaves behind a Cl^- ion.

Every ion is hydrated in water solution, which is indicated by the symbol (aq) placed behind the formula of the ion. This symbolism does not give the number of water molecules associated with each ion. The number in most cases is not known and in many cases is variable. The H^+ ion, however, is a special case. The positive charge of the H^+ ion, the proton, is not shielded by any electrons and in comparison to other ions is extremely small. The H^+ ion, therefore, is strongly attracted to an electron pair of a H_2O molecule.

There is, however, evidence that the H_3O^+ ion has three additional water molecules associated with it in an ion that has the formula $H_9O_4^+$. Other evidence supports the idea that several types of hydrated ions exist simultaneously in water solution. Some chemists, therefore, prefer to represent the hydrated proton as $H^+(aq)$. The process in which HCl molecules dissolve in water would be indicated:

$$HCl(g) \longrightarrow H^+(aq) + Cl^-(aq)$$

In the Arrhenius system, a **base** is a substance that contains hydroxide ions, OH^-, or dissolves in water to produce hydrated hydroxide ions, $OH^-(aq)$:

$$NaOH(s) \longrightarrow Na^+(aq) + OH^-(aq)$$

$$Ca(OH)_2(s) \longrightarrow Ca^{2+}(aq) + 2OH^-(aq)$$

The only soluble metal hydroxides are those of the group I A elements and $Ba(OH)_2$, $Sr(OH)_2$, and $Ca(OH)_2$ of group II A. Insoluble hydroxides, however, react as bases with acids.

The reaction of an acid and a base is called a **neutralization,** which is a metathesis reaction in which water is produced. The *net* ionic equation is

$$H_3O^+(aq) + OH^-(aq) \longrightarrow 2H_2O$$

which may also be written

$$H^+(aq) + OH^-(aq) \longrightarrow H_2O$$

Ionic equations for two neutralization reactions are

$$Ba^{2+}(aq) + 2OH^-(aq) + 2H^+(aq) + 2Cl^-(aq) \longrightarrow$$
$$Ba^{2+}(aq) + 2Cl^-(aq) + 2H_2O$$

$$Fe(OH)_3(s) + 3H^+(aq) + 3NO_3^-(aq) \longrightarrow Fe^{3+}(aq) + 3NO_3^-(aq) + 3H_2O$$

The barium chloride ($BaCl_2$) and iron(III) nitrate [$Fe(NO_3)_3$] produced by these reactions are called **salts,** which are ionic compounds with cations derived from bases and anions derived from acids.

Acids are classified as strong or weak depending upon the extent of their dissociation in water (see Table 11.3). A **strong acid** is 100% dissociated in water solution. The common strong acids are HCl, HBr, HI, HNO_3, H_2SO_4 (first H^+ dissociation only), $HClO_4$, and $HClO_3$. Other common acids are **weak acids,** which are less than 100% dissociated in water solution. Acetic acid ($HC_2H_3O_2$), for example, is a weak acid:

$$H_2O + HC_2H_3O_2(l) \rightleftharpoons H_3O^+(aq) + C_2H_3O_2^-(aq)$$

The reversible arrow ($\rightleftharpoons$) is used in this equation to show that reactions occur in both directions. In a 1 *M* solution of acetic acid, a balance is achieved with 0.4% of the $HC_2H_3O_2$ dissociated into ions.

All soluble metal hydroxides are strong bases. After all, these substances are 100% ionic when pure. There are a few molecular weak bases. The most common one is an aqueous solution of ammonia, NH_3:

$$NH_3(g) + H_2O \rightleftharpoons NH_4^+(aq) + OH^-(aq)$$

In this reaction, the ammonia molecule accepts a proton from the water molecule to form the ammonium ion and the hydroxide ion. The reaction, however, is not complete.

Acids that can lose only one proton per molecule (such as HCl, $HC_2H_3O_2$, and HNO_3) are called **monoprotic acids.** Some acids can lose more than one proton per molecule; they are called **polyprotic acids.** A molecule of sulfuric acid, for example, can lose two protons:

$$H_2SO_4(l) + H_2O \longrightarrow H_3O^+(aq) + HSO_4^-(aq)$$

$$HSO_4^-(aq) + H_2O \rightleftharpoons H_3O^+(aq) + SO_4^{2-}(aq)$$

Only one proton is neutralized if 1 mol of H_2SO_4 is reacted with 1 mol of NaOH:

$$H_2SO_4(aq) + NaOH(aq) \longrightarrow NaHSO_4(aq) + H_2O$$

The salt obtained, $NaHSO_4$, is called an **acid salt** because it contains an acid hydrogen. If 1 mol of H_2SO_4 is reacted with 2 mol of NaOH, both acid hydrogens are neutralized. The **normal salt,** Na_2SO_4, is obtained:

$$H_2SO_4(aq) + 2\,NaOH(aq) \longrightarrow Na_2SO_4(aq) + 2\,H_2O$$

The acid salt can be reacted with NaOH to produce the normal salt:

$$NaHSO_4(aq) + NaOH(aq) \longrightarrow Na_2SO_4(aq) + H_2O$$

The products of the neutralization of a polyprotic acid, therefore, depend upon the quantities of acid and base employed.

Phosphoric acid, H_3PO_4, has three acid hydrogens, and three salts can be obtained from it:

$$NaH_2PO_4 \qquad Na_2HPO_4 \qquad Na_3PO_4$$

11.5 Acidic and Basic Oxides

The oxides of metals are called **basic oxides.** The oxides of the group I A metals and those of Ca, Sr, and Ba dissolve in water to produce hydroxides. All these oxides are ionic. When one of them dissolves in water, it is the oxide ion that reacts with the water:

$$O^{2-}(aq) + H_2O \longrightarrow 2\,OH^-(aq)$$

The oxides (as well as the hydroxides) of other metals are insoluble in water.

Nevertheless, metal oxides and hydroxides are chemically related. When heated, most hydroxides are converted into oxides:

$$Mg(OH)_2(s) \longrightarrow MgO(s) + H_2O(g)$$

Metal oxides, as well as hydroxides, can be neutralized by acids:

$$MgO(s) + 2\,H^+(aq) \longrightarrow Mg^{2+}(aq) + H_2O$$

$$Mg(OH)_2(s) + 2\,H^+(aq) \longrightarrow Mg^{2+}(aq) + 2\,H_2O$$

The insoluble Fe_2O_3 will react with acids, even though it will not react with water to produce a hydroxide:

$$Fe_2O_3(s) + 6\,H^+(aq) \longrightarrow 2\,Fe^{3+}(aq) + 3\,H_2O$$

Most of the oxides of the nonmetals are **acidic oxides.** Many of them react with water to produce oxyacids:

$$Cl_2O + H_2O \longrightarrow 2\,HOCl$$

$$Cl_2O_7 + H_2O \longrightarrow 2\,HClO_4$$

$$N_2O_5 + H_2O \longrightarrow 2\,HNO_3$$

$$P_4O_{10} + 6\,H_2O \longrightarrow 4\,H_3PO_4$$

$$SO_3 + H_2O \longrightarrow H_2SO_4$$

$$SO_2 + H_2O \rightleftharpoons H_2SO_3$$

$$CO_2 + H_2O \rightleftharpoons H_2CO_3$$

In the last two cases, both the oxides $[SO_2(g)$ and $CO_2(g)]$ and the acids $[H_2SO_3(aq)$ and $H_2CO_3(aq)]$ exist in the solutions. For some nonmetal oxides, there are no corresponding acids.

Oxides of nonmetals will neutralize bases. The same products are obtained from the reaction of an acidic oxide as are obtained from the reaction of the corresponding acid:

$$H_2SO_3(aq) + 2\,OH^-(aq) \longrightarrow SO_3^{2-}(aq) + 2\,H_2O$$

$$SO_2(g) + 2\,OH^-(aq) \longrightarrow SO_3^{2-}(aq) + H_2O$$

Mortar consists of lime $[Ca(OH)_2]$, sand $[SiO_2]$, and water. The initial hardening of mortar occurs when the mortar dries out. Over a long period of time, however, the mortar sets by absorbing $CO_2(g)$ from the air and forming insoluble $CaCO_3$:

$$Ca(OH)_2(s) + CO_2(g) \longrightarrow CaCO_3(s) + H_2O$$

Some oxides have both acidic and basic properties (for example, BeO and Al_2O_3). They are called **amphoteric oxides** and are formed principally by the elements in the center of the periodic table, near to the borderline between the metals and the nonmetals:

$$Al_2O_3(s) + 6\,H^+(aq) \longrightarrow 2\,Al^{3+}(aq) + 3\,H_2O$$

$$Al_2O_3(s) + 2\,OH^-(aq) + 3\,H_2O \longrightarrow 2\,Al(OH)_4^-(aq)$$
aluminate ion

$$ZnO(s) + 2\,H^+(aq) \longrightarrow Zn^{2+}(aq) + H_2O$$

$$ZnO(s) + 2\,OH^-(aq) + H_2O \longrightarrow Zn(OH)_4^{2-}(aq)$$
zincate ion

Acidic and basic oxides react directly, and many of these reactions are of industrial significance. In the manufacture of pig iron (see Section 23.6), $CaCO_3$ (limestone) is used as a flux. The $CaCO_3$ decomposes into CaO and CO_2 at the high temperatures of the blast furnace. The CaO, a basic oxide, reacts with SiO_2, an acidic oxide present in the iron ore, to form slag ($CaSiO_3$) and thereby remove the unwanted SiO_2:

$$CaCO_3(s) + SiO_2(s) \longrightarrow CaSiO_3(l) + CO_2(g)$$

In the open hearth process for the manufacture of steel from pig iron, the hearth is often lined with CaO or MgO (basic oxides) to remove the oxides of silicon, phosphorus, and sulfur (acidic oxides), which are present in the pig iron as impurities.

Glass is made from various combinations of acidic and basic oxides. Ordinary soft glass is made from lime ($CaCO_3$), soda (Na_2CO_3), and silica (SiO_2). The corresponding basic oxides are CaO and Na_2O, the acidic oxide is SiO_2, and the product is a mixture of sodium and calcium silicates.

Substitutions are sometimes made for these oxides. If boric oxide, an acidic oxide, is substituted for a part of the SiO_2, a borosilicate (Pyrex) glass is produced. The use of PbO as a part of the basic-oxide component produces flint glass (used for lenses). The use of some basic oxides produces colored glasses; for example, FeO (light green), Cr_2O_3 (dark green), and CoO (blue).

11.6 Nomenclature of Acids and Salts

Some common acids are listed in Table 11.3. Rules for naming these compounds and the salts derived from them follow.

1. Aqueous solutions of *binary compounds that function as acids* are named by modifying the root of the name of the element that is combined with hydrogen. The prefix *hydro-* and the suffix *-ic* are used followed by the word *acid*:

Table 11.3 Some common acids

ACIDS
Binary Compounds

Monoprotic Acids		Polyprotic Acids	
HF*	hydrofluoric acid	H_2S*	hydrosulfuric acid
HCl	hydrochloric acid		
HBr	hydrobromic acid		
HI	hydroiodic acid		

Ternary Compounds

Monoprotic Acids		Polyprotic Acids	
HNO_3	nitric acid	H_2SO_4**	sulfuric acid
HNO_2*	nitrous acid	H_2SO_3*	sulfurous acid
$HClO_4$	perchloric acid	H_3PO_4*	phosphoric acid
$HClO_3$	chloric acid		
$HClO_2$*	chlorous acid	H_2CO_3*	carbonic acid
HOCl*	hypochlorous acid	H_3BO_3*	boric acid
$HC_2H_3O_2$*	acetic acid		

* Weak acid.
** The second dissociation is weak.

HCl forms hydrochloric acid
H$_2$S forms hydrosulfuric acid

2. *Metal hydroxides* are named in the manner described in Section 5.8:

Mg(OH)$_2$ is magnesium hydroxide
Fe(OH)$_2$ is iron(II) hydroxide or ferrous hydroxide

3. *Salts of binary acids* are themselves binary compounds, and their names have the customary *-ide* ending. They are named according to the rules given in Section 5.8.

4. *Ternary acids* are composed of three elements. When oxygen is the third element, the compound is called an *oxyacid.*

a. If an element forms only one oxyacid, the acid is named by changing the ending of the name of the element to *-ic* and adding the word *acid:*

H$_3$BO$_3$ is boric acid

b. If there are two common oxyacids of the same element, the ending *-ous* is used in naming the oxyacid of the element in its lower oxidation state; the ending *-ic* is used to denote the higher oxidation state (see Table 11.3):

HNO$_2$ is nitrous acid
HNO$_3$ is nitric acid

c. There are a few series of oxyacids for which two names are not enough. See the names of the oxyacids of chlorine in Table 11.3. The prefix *hypo-* is added to the name of an *-ous* acid to indicate an oxidation state of the central element lower than that of the *-ous* acid:

HClO$_2$ is chlorous acid (Cl has an oxidation number of 3+)
HOCl is hypochlorous acid (Cl has an oxidation number of 1+)

The prefix *per-* is added to the name of an *-ic* acid to indicate an oxidation state of the central atom that is higher than that of the *-ic* acid:

HClO$_3$ is chloric acid (Cl has an oxidation number of 5+)
HClO$_4$ is perchloric acid (Cl has an oxidation number of 7+)

5. The names of the *anions of normal salts* are derived from the names of the acids from which the salts are obtained. The *-ic* ending is changed to *-ate.* The *-ous* ending is changed to *-ite.* Prefixes, if any, are retained:

SO$_4^{2-}$ (from sulfuric acid) is the sulfate ion
OCl$^-$ (from hypochlorous acid) is the hypochlorite ion

The name of the salt itself is obtained by combining the name of the cation with the name of the anion:

NaNO$_2$ is sodium nitrite
Fe(ClO$_4$)$_3$ is iron(III) perchlorate or ferric perchlorate

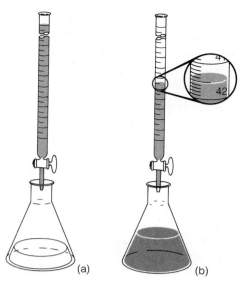

Figure 11.1 An acid-base titration. (a) The buret contains the standard solution. The unknown solution and indicator are placed in the flask. (b) Solution is added from the buret to the flask. Equivalence point is reached when the indicator changes color.

6. In naming the *anion of an acid salt*, the number of acid hydrogens retained by the anion must be indicated. The prefix *mono-* is usually omitted:

$H_2PO_4^-$ is the dihydrogen phosphate ion
HPO_4^{2-} is the hydrogen phosphate ion
PO_4^{3-} (the anion of the normal salt) is the phosphate ion

The prefix *bi-* may be used in place of the word *hydrogen* in the name of the anion of an acid salt derived from a diprotic acid:

HCO_3^- is the hydrogen carbonate ion or the bicarbonate ion
HSO_3^- is the hydrogen sulfite ion or the bisulfite ion

11.7 Volumetric Analysis

A **volumetric analysis** is one that relies on the measurement of the volume of a solution that has an exactly known concentration. A procedure called a **titration** is employed (see Figure 11.1). In a titration, a solution of known concentration, called a **standard solution,** is added to a measured volume of an unknown solution until the reaction is complete. The standard solution is placed in a graduated tube called a buret. The buret is fitted with a stopcock at the lower end to permit the solution to be withdrawn in controlled amounts. A measured volume of the unknown solution, or a weighed mass of a solid unknown dissolved in water, is placed in a flask together with a few drops of a substance known as an **indicator.** The standard solution from the buret is slowly added to the flask until the indicator changes color. Throughout the addition, the contents of the flask are kept well mixed by swirling. At the equivalence point, as shown by the color change of the indicator, equivalent amounts of the two reactants have been used. The volume of standard solution employed is read from the buret.

Three types of volumetric analyses are in use. They are based on precipitation reactions, acid-base neutralizations, and oxidation-reduction reactions. These three types are illustrated in the examples that follow.

Example 11.5

An effluent from a manufacturing process is analyzed for Cl^- content. A 10.00 g sample of the waste water requires 30.20 ml of a 0.1050 M $AgNO_3$ solution for reaction. What is the percent Cl^- in the waste water? The equation for the reaction is

$$Cl^-(aq) + AgNO_3(aq) \longrightarrow AgCl(s) + NO_3^-(aq)$$

Solution

First we find the number of moles of $AgNO_3$ that have been used:

$$? \text{ mol } AgNO_3 = 30.20 \text{ ml sol'n} \left(\frac{0.1050 \text{ mol } AgNO_3}{1000 \text{ ml sol'n}} \right)$$

$$= 3.171 \times 10^{-3} \text{ mol } AgNO_3$$

From the chemical equation, we see that

$$1 \text{ mol } Cl^- = 1 \text{ mol } AgNO_3$$

and since the atomic weight of Cl is 35.45 g/mol, we can find the mass of Cl^- in the sample in the following way:

$$? \text{ g } Cl^- = 3.171 \times 10^{-3} \text{ mol } AgNO_3 \left(\frac{1 \text{ mol } Cl^-}{1 \text{ mol } AgNO_3} \right) \left(\frac{35.45 \text{ g } Cl^-}{1 \text{ mol } Cl^-} \right)$$

$$= 0.1124 \text{ g } Cl^-$$

The mass percent of Cl^- in the sample is

$$\left(\frac{0.1124 \text{ g } Cl^-}{10.00 \text{ g sample}} \right) 100\% = 1.124\% \ Cl^-$$

Example 11.6

A 25.00 g sample of vinegar, which contains acetic acid ($HC_2H_3O_2$), requires 37.50 ml of 0.4600 M NaOH solution for neutralization. What is the mass percent of acetic acid in the vinegar? The equation is

$$NaOH(aq) + HC_2H_3O_2(aq) \longrightarrow NaC_2H_3O_2(aq) + H_2O$$

Solution

The number of moles of NaOH employed can be found in the following way:

$$? \text{ mol NaOH} = 37.50 \text{ ml sol'n} \left(\frac{0.4600 \text{ mol NaOH}}{1000 \text{ ml sol'n}} \right)$$

$$= 1.725 \times 10^{-2} \text{ mol NaOH}$$

Since the equation shows

$$1 \text{ mol HC}_2\text{H}_3\text{O}_2 = 1 \text{ mol NaOH}$$

and since the molecular weight of $HC_2H_3O_2$ is 60.05 g/mol,

$$? \text{ g HC}_2\text{H}_3\text{O}_2 = 1.725 \times 10^{-2} \text{ mol NaOH} \left(\frac{1 \text{ mol HC}_2\text{H}_3\text{O}_2}{1 \text{ mol NaOH}} \right) \left(\frac{60.05 \text{ g HC}_2\text{H}_3\text{O}_2}{1 \text{ mol HC}_2\text{H}_3\text{O}_2} \right)$$

$$= 1.036 \text{ g HC}_2\text{H}_3\text{O}_2$$

The mass percent of $HC_2H_3O_2$ in the vinegar sample is

$$\left(\frac{1.036 \text{ g HC}_2\text{H}_3\text{O}_2}{25.00 \text{ g vinegar}} \right) 100\% = 4.144\% \text{ HC}_2\text{H}_3\text{O}_2 \text{ in vinegar}$$

Example 11.7

A 0.4308 g sample of iron ore is dissolved in acid and the iron converted into the Fe^{2+} state. This solution is reacted with a solution of potassium permanganate. The reaction requires 27.35 ml of 0.02496 M $KMnO_4$ solution. What is the mass percent of iron in the ore? The chemical equation is

$$8 \text{ H}^+ + 5 \text{ Fe}^{2+} + \text{MnO}_4^- \longrightarrow 5 \text{ Fe}^{3+} + \text{Mn}^{2+} + 4 \text{ H}_2\text{O}$$

Solution

First we find the number of moles of $KMnO_4$ consumed in the reaction:

$$? \text{ mol KMnO}_4 = 27.35 \text{ ml sol'n} \left(\frac{0.02496 \text{ mol KMnO}_4}{1000 \text{ ml sol'n}} \right)$$

$$= 6.827 \times 10^{-4} \text{ mol KMnO}_4$$

Since the equation shows

$$5 \text{ mol Fe}^{2+} = 1 \text{ mol KMnO}_4$$

and since the atomic weight of iron is 55.85 g/mol,

$$? \text{ g Fe} = 6.827 \times 10^{-4} \text{ mol KMnO}_4 \left(\frac{5 \text{ mol Fe}}{1 \text{ mol KMnO}_4} \right) \left(\frac{55.85 \text{ g Fe}}{1 \text{ mol Fe}} \right)$$

$$= 0.1906 \text{ g Fe}$$

This is the quantity of iron found in the 0.4308 g sample of ore. Therefore,

$$\left(\frac{0.1906 \text{ g Fe}}{0.4308 \text{ g ore}}\right) 100\% = 44.24\% \text{ Fe in ore}$$

11.8 Equivalent Weights and Normal Solutions

All volumetric analysis problems can be solved in the way that was used in the last section, on the basis of the mole and molar solutions. There is, however, another method, one based on what are called *equivalents* and *normal solutions*. The definition of an equivalent depends upon the type of reaction being considered. The definition, however, is always framed so that one equivalent of a given reactant will react with exactly one equivalent of another.

Two types of reactions for which equivalents are defined are neutralization reactions and oxidation-reduction reactions. The mass of one equivalent of a compound is called an **equivalent weight.** In general,

$$equivalent\ weight = \frac{molecular\ weight}{a} \tag{11.1}$$

where the value of a depends upon the type of reaction considered.

1. For *neutralization reactions*, equivalent weights are based on the fact that one $H^+(aq)$ ion reacts with one $OH^-(aq)$ ion. One equivalent weight of an acid is the amount of the acid that supplies one mole of $H^+(aq)$ ions, and one equivalent weight of a base is the amount of the base that supplies one mole of $OH^-(aq)$ ions. The value of a in Equation 11.1, therefore, is the number of moles of $H^+(aq)$ supplied by one mole of the acid or the number of moles of $OH^-(aq)$ supplied by one mole of the base in the reaction considered.

2. For *oxidation-reduction reactions*, equivalent weights are based on oxidation number changes. In an oxidation-reduction reaction, the increase in oxidation number of one element must equal the decrease in oxidation number of another. An equivalent is defined in terms of an oxidation number change of one unit, and a in Equation 11.1 is the total change in oxidation number (either up or down) that the atoms in a formula unit of the compound undergo in the reaction under consideration. The equivalent weight of $KMnO_4$ for a reaction in which the MnO_4^- ion is reduced to the Mn^{2+} ion (a change in oxidation number of 5 units) is the formula weight of $KMnO_4$ divided by 5. The equivalent weight of $KMnO_4$ for a reaction in which the MnO_4^- ion is reduced to MnO_2 (a change in oxidation number of 3 units) is the formula weight of $KMnO_4$ divided by 3.

The **normality,** N, of a solution is the number of gram equivalent weights of solute dissolved in one liter of solution. The normality of a solution and its molarity are related:

$$N = aM \tag{11.2}$$

The number of equivalents of A in a sample of a solution of A, e_A, can be obtained by multiplying the volume of the sample, V_A (in liters), by the normality of the solution, N_A (which is the number of equivalents of A in one liter of solution):

$$e_A = V_A N_A \qquad (V_A \text{ in liters}) \tag{11.3}$$

By design, $e_A = e_B$, and therefore

$$V_A N_A = V_B N_B \tag{11.4}$$

Since a volume term appears on both sides of Equation 11.4, any volume unit can be used to express V_A and V_B provided that both volumes are expressed in the same unit.

Example 11.8

(a) What is the normality of a solution of H_2SO_4 if 50.00 ml of the solution requires 37.52 ml of 0.1492 N NaOH for complete neutralization?
(b) What is the molarity of the solution?

Solution

(a) $$V_A N_A = V_B N_B$$

$$(50.00 \text{ ml})N_A = (37.52 \text{ ml})(0.1492 \text{ N})$$

$$N_A = 0.1120 \text{ N}$$

(b) Since 1 mol of H_2SO_4 is 2 equivalents, $a = 2$. Therefore

$$N = aM$$

$$0.1120 \text{ equiv/liter} = (2 \text{ equiv/mol})M$$

$$M = 0.05600 \text{ mol/liter}$$

Example 11.9

A 0.4308 g sample of iron ore is dissolved in acid and the iron converted to the Fe^{2+} state. The solution is reacted with a solution of potassium permanganate. The reaction, in which Fe^{2+} is oxidized to Fe^{3+}, requires 27.35 ml of 0.1248 N $KMnO_4$. What is the mass percent of iron in the ore?

Solution

This problem is the same as that given in Example 11.7. Here, however, we will solve it by using equivalents and normality. We find the number of equivalents of $KMnO_4$ used:

$$e_A = V_A N_A$$
$$= (0.02735 \text{ liter})(0.1248 \text{ equiv/liter})$$
$$= 3.413 \times 10^{-3} \text{ equiv}$$

The number of equivalents of $KMnO_4$ used is the same as the number of equivalents of iron in the ore sample.

In the reaction, the oxidation number of the iron increases by one unit (from $2+$ to $3+$). The equivalent weight of iron, therefore, is the same as the atomic weight, 55.85:

$$? \text{ g Fe} = 3.413 \times 10^{-3} \text{ equiv Fe} \left(\frac{55.85 \text{ g Fe}}{1 \text{ equiv Fe}} \right) = 0.1906 \text{ g Fe}$$

The mass percent of Fe in the ore sample, therefore, is

$$\left(\frac{0.1906 \text{ g Fe}}{0.4308 \text{ g ore}} \right) 100\% = 44.24\% \text{ Fe in ore}$$

Summary

The topics that have been discussed in this chapter are

1. Aqueous metathesis reactions, which occur because of the formation of a precipitate, gas, or a weak electrolyte.

2. The assignment of oxidation numbers to the atoms in compounds.

3. Oxidation and reduction.

4. How to balance oxidation-reduction equations by the oxidation-number method and by the ion-electron method.

5. The Arrhenius concept of acids and bases.

6. Acidic and basic oxides.

7. Nomenclature of acids and salts.

8. Volumetric analyses based on precipitation, neutralization and oxidation-reduction reactions, and the stoichiometric calculations involved in these procedures.

9. The use of equivalent weights and normality in stoichiometric calculations.

Key Terms

Some of the more important terms introduced in this chapter are listed below. Definitions for terms not included in this list may be located in the text by use of the index.

Acid (Section 11.4)   A covalent compound of hydrogen that dissociates in water to produce $H^+(aq)$ ions (or H_3O^+ ions).

Acidic oxide (Section 11.5)   An oxide of a nonmetal that reacts with water to form an acid.

Acid salt (Section 11.4)   A salt formed by the incomplete neutralization of a polyprotic acid. The anions of these salts retain one or more hydrogen atoms of the parent acid.

Amphoteric oxide (Section 11.5)   An oxide that has both acidic and basic properties and that will react with both bases and acids to form salts.

Base (Section 11.4)   In the Arrhenius system, a compound that dissociates in water to produce $OH^-(aq)$ ions.

Basic oxide (Section 11.5)   An oxide of a metal that reacts with water to form a base.

Disproportionation (Section 11.3)   A reaction in which a substance is both oxidized and reduced; an auto-oxidation-reduction reaction.

Equivalent weight (Section 11.8)   A quantity defined on the basis of the reaction being considered in such a way that one equivalent weight of one reactant will react with exactly one equivalent weight of another. For an acid-base neutralization, the mass of acid or base that will supply one mole of $H^+(aq)$ or one mole of $OH^-(aq)$. For an oxidation-reduction reaction, the molecular weight of the oxidant or reductant divided by the total change in oxidation number for that reactant.

Half reaction (Section 11.3) Half of an oxidation-reduction reaction; an oxidation *or* a reduction.

Hydronium ion (Section 11.4) An ion formed from a proton and a water molecule: H_3O^+.

Metathesis reaction (Section 11.1) A reaction between two compounds in which the cations and anions exchange partners.

Monoprotic acid (Sections 11.4 and 11.6) An acid that can lose only one proton per molecule.

Net-ionic equation (Section 11.1) A chemical equation that shows only those ions that are involved in the reaction.

Neutralization (Section 11.4) A reaction between an acid and a base.

Normality (Section 11.8) A solution concentration; the number of equivalents of solute per liter of solution.

Normal salt (Section 11.4) A salt of a polyprotic acid formed by the loss of all possible protons by the acid.

Oxidation (Section 11.3) That part of an oxidation-reduction reaction characterized by an algebraic increase in oxidation number or by electron loss.

Oxidation number (Section 11.2) A charge that is assigned to an atom in a compound according to arbitrary rules that take into account bond polarity. The concept is merely a convention.

Oxidizing agent (Section 11.3) A substance that is reduced in a chemical reaction and thereby causes the oxidation of another substance.

Oxyacid (Section 11.6) An acid composed of three elements with oxygen as one of the three.

Partial equation (Section 11.3) A chemical equation for a half reaction written to show electron loss or electron gain.

Polyprotic acid (Sections 11.4 and 11.6) An acid that can lose more than one proton per molecule.

Precipitation (Section 11.1) The formation of an insoluble substance (called a **precipitate**) in an aqueous reaction.

Reducing agent (Section 11.3) A substance that is oxidized in a chemical reaction and thereby causes the reduction of another substance.

Reduction (Section 11.3) That part of an oxidation-reduction reaction characterized by an algebraic decrease in oxidation number or by electron gain.

Salt (Section 11.4) A compound derived from the reaction of an acid and a base; it contains the cation of the base and the anion of the acid.

Spectator ion (Section 11.1) An ion that is present during the course of an aqueous reaction but one that does not take part in the reaction.

Standard solution (Section 11.7) A solution that has a known concentration of solute.

Strong acids and bases (Section 11.4) Acids and bases that are completely ionized in water solution.

Titration (Section 11.7) A process in which a standard solution is reacted with a solution of unknown concentration in order to determine the unknown concentration.

Volumetric analysis (Section 11.7) A chemical analysis that is based on the measurement of the volume of a standard solution that is required for reaction with a solution of unknown concentration.

Weak acids and bases (Section 11.4) Acids and bases that are only partly dissociated in water solution.

Problems*

Metathesis Reactions

11.1 Write balanced ionic equations for the reactions that occur between **(a)** ZnS and HCl, **(b)** Na_2CO_3 and $Sr(C_2H_3O_2)_2$, **(c)** $SnCl_2$ and $(NH_4)_2SO_4$, **(d)** $Mg(NO_3)_2$ and $Ba(OH)_2$, **(e)** Na_3PO_4 and HBr.

11.2 Write balanced ionic equations for the reactions that occur between **(a)** $AgNO_3$ and CdI_2, **(b)** $ZnSO_3$ and HCl, **(c)** $AlCl_3$ and $LiClO_3$, **(d)** $Sr(OH)_2$ and $NiSO_4$, **(e)** $Mg(OH)_2$ and HNO_3.

11.3 Write balanced ionic equations for the reactions that occur between **(a)** $Pb(NO_3)_2$ and H_2S, **(b)** BaS and $ZnSO_4$, **(c)** Na_3PO_4 and $NH_4C_2H_3O_2$, **(d)** Hg_2CO_3 and HCl, **(e)** $Fe(OH)_3$ and H_3PO_4.

11.4 Write balanced ionic equations for the reactions that occur between **(a)** $(NH_4)_2SO_4$ and $Ca(OH)_2$, **(b)** $MnCl_2$ and $CoSO_4$, **(c)** $Cd(ClO_3)_2$ and K_2S, **(d)** $Fe_2(CO_3)_3$ and HNO_3, **(e)** $Pb(NO_3)_2$ and $MgSO_4$.

Oxidation Numbers

11.5 State the oxidation number of the atom other than O in each of the oxy-anions found in Table 5.5.

11.6 State the oxidation number of **(a)** S in $S_2O_5Cl_2$, **(b)** U in Mg_3UO_6, **(c)** P in $Na_3P_3O_9$, **(d)** N in CaN_2O_2, **(e)** V in Ca_2VO_4, **(f)** N in N_2H_4, **(g)** W in $W_2Cl_9^{3-}$, **(h)** N in NO_2^+.

---

* The more difficult problems are marked with asterisks. The appendix contains answers to color-keyed problems.

11.7 State the oxidation number of **(a)** B in B_2Cl_4, **(b)** N in NH_2OH, **(c)** Mo in $K_2Mo_4O_{13}$, **(d)** Xe in XeO_6^{4-}, **(e)** Sb in $Ca_2Sb_2O_7$, **(f)** U in UO_2^{2+}, **(g)** Ta in $TaOCl_3$, **(h)** Xe in Cs_2XeF_8.

11.8 State the oxidation number of **(a)** Mo in $Na_3Mo_2Br_9$, **(b)** B in $Mg(BF_4)_2$, **(c)** Ti in $K_2Ti_2O_5$, **(d)** Ta in $Ta_6O_{19}^{8-}$, **(e)** Sn in K_2SnO_3, **(f)** V in $Na_6V_{10}O_{28}$, **(g)** Bi in BiO^+, **(h)** U in U_2Cl_{10}.

11.9 State the oxidation number of **(a)** Ta in Na_3TaF_8, **(b)** W in $W_4O_{13}^{2-}$, **(c)** Zr in $K_2Zr_2O_5$, **(d)** Mo in $Mo_8O_{26}^{4-}$, **(e)** U in $Li_2U_2O_7$, **(f)** W in $K_2W_4O_{13}$, **(g)** Nb in K_3NbOF_6, **(h)** Te in Cs_2TeF_8, **(i)** U in $U(OH)_3^{3+}$.

Oxidation-Reduction Reactions

11.10 For each of the following reactions, identify the substance oxidized, the substance reduced, the oxidizing agent, and the reducing agent:

(a) $Zn + Cl_2 \longrightarrow ZnCl_2$

(b) $2\,ReCl_5 + SbCl_3 \longrightarrow 2\,ReCl_4 + SbCl_5$

(c) $Mg + CuCl_2 \longrightarrow MgCl_2 + Cu$

(d) $2\,NO + O_2 \longrightarrow 2\,NO_2$

(e) $WO_3 + 3\,H_2 \longrightarrow W + 3\,H_2O$

11.11 For each of the following reactions, identify the substance oxidized, the substance reduced, the oxidizing agent and the reducing agent:

(a) $Cl_2 + 2\,NaBr \longrightarrow 2\,NaCl + Br_2$

(b) $Zn + 2\,HCl \longrightarrow ZnCl_2 + H_2$

(c) $Fe_2O_3 + 2\,Al \longrightarrow Al_2O_3 + 2\,Fe$

(d) $OF_2 + H_2O \longrightarrow O_2 + 2\,HF$

(e) $2\,HgO \longrightarrow 2\,Hg + O_2$

11.12 Balance the following by the change-in-oxidation-number method:

(a) $H_2O + MnO_4^- + ClO_2^- \longrightarrow$
$$MnO_2 + ClO_4^- + OH^-$$

(b) $H^+ + Cr_2O_7^{2-} + H_2S \longrightarrow Cr^{3+} + S + H_2O$

(c) $H_2O + P_4 + HOCl \longrightarrow H_3PO_4 + Cl^- + H^+$

(d) $Cu + H^+ + NO_3^- \longrightarrow Cu^{2+} + NO + H_2O$

(e) $PbO_2 + HI \longrightarrow PbI_2 + I_2 + H_2O$

11.13 Balance the following by the change-in-oxidation-number method:

(a) $Sb + H^+ + NO_3^- \longrightarrow Sb_4O_6 + NO + H_2O$

(b) $NaI + H_2SO_4 \longrightarrow H_2S + I_2 + Na_2SO_4 + H_2O$

(c) $IO_3^- + H_2O + SO_2 \longrightarrow I_2 + SO_4^{2-} + H^+$

(d) $NF_3 + AlCl_3 \longrightarrow N_2 + Cl_2 + AlF_3$

(e) $As_4O_6 + Cl_2 + H_2O \longrightarrow H_3AsO_4 + HCl$

11.14 Balance the following by the change-in-oxidation-number method:

(a) $Fe^{2+} + H^+ + ClO_3^- \longrightarrow Fe^{3+} + Cl^- + H_2O$

(b) $Pt + H^+ + NO_3^- + Cl^- \longrightarrow$
$$PtCl_6^{2-} + NO + H_2O$$

(c) $Cu + H^+ + SO_4^{2-} \longrightarrow Cu^{2+} + SO_2 + H_2O$

(d) $Pb + PbO_2 + H^+ + SO_4^{2-} \longrightarrow PbSO_4 + H_2O$

(e) $MnO_2 + HI \longrightarrow MnI_2 + I_2 + H_2O$

11.15 Complete and balance the following equations by the ion-electron method. All reactions occur in acid solution.

(a) $AsH_3 + Ag^+ \longrightarrow As_4O_6 + Ag$

(b) $Mn^{2+} + BiO_3^- \longrightarrow MnO_4^- + Bi^{3+}$

(c) $NO + NO_3^- \longrightarrow N_2O_4$

(d) $MnO_4^- + HCN + I^- \longrightarrow Mn^{2+} + ICN$

(e) $Zn + H_2MoO_4 \longrightarrow Zn^{2+} + Mo^{3+}$

11.16 Complete and balance the following equations by the ion-electron method. All reactions occur in acid solution.

(a) $S_2O_3^{2-} + IO_3^- + Cl^- \longrightarrow SO_4^{2-} + ICl_2^-$

(b) $Se + BrO_3^- \longrightarrow H_2SeO_3 + Br^-$

(c) $H_3AsO_3 + MnO_4^- \longrightarrow H_3AsO_4 + Mn^{2+}$

(d) $H_5IO_6 + I^- \longrightarrow I_2$

(e) $Pb_3O_4 \longrightarrow Pb^{2+} + PbO_2$

11.17 Complete and balance the following equations by the ion-electron method. All reactions occur in acid solution.

(a) $ClO_3^- + I^- \longrightarrow Cl^- + I_2$

(b) $Zn + NO_3^- \longrightarrow Zn^{2+} + NH_4^+$

(c) $H_3AsO_3 + BrO_3^- \longrightarrow H_3AsO_4 + Br^-$

(d) $H_2SeO_3 + H_2S \longrightarrow Se + HSO_4^-$

(e) $ReO_2 + Cl_2 \longrightarrow HReO_4 + Cl^-$

11.18 Complete and balance the following equations by the ion-electron method. All reactions occur in acid solution.

(a) $Fe^{2+} + Cr_2O_7^{2-} \longrightarrow Fe^{3+} + Cr^{3+}$

(b) $HNO_2 + MnO_4^- \longrightarrow NO_3^- + Mn^{2+}$

(c) $As_2S_3 + ClO_3^- \longrightarrow H_3AsO_4 + S + Cl^-$

(d) $IO_3^- + N_2H_4 \longrightarrow I^- + N_2$

(e) $Cu + NO_3^- \longrightarrow Cu^{2+} + NO$

11.19 Complete and balance the following equations by the ion-electron method. All reactions occur in acid solution.

(a) $P_4 + HOCl \longrightarrow H_3PO_4 + Cl^-$

(b) $XeO_3 + I^- \longrightarrow Xe + I_3^-$

(c) $UO^{2+} + Cr_2O_7^{2-} \longrightarrow UO_2^{2+} + Cr^{3+}$

(d) $H_2C_2O_4 + BrO_3^- \longrightarrow CO_2 + Br^-$

(e) $Te + NO_3^- \longrightarrow TeO_2 + NO$

*11.20 Complete and balance the following equations by the ion-electron method. All reactions occur in acid solution.

(a) $Hg_5(IO_6)_2 + I^- \longrightarrow HgI_4^{2-} + I_2$

(b) $MnO_4^- + Mn^{2+} + H_2P_2O_7^{2-} \longrightarrow Mn(H_2P_2O_7)_3^{3-}$

(c) $CS(NH_2)_2 + BrO_3^- \longrightarrow CO(NH_2)_2 + SO_4^{2-} + Br^-$

(d) $Co(NO_2)_6^{3-} + MnO_4^- \longrightarrow Co^{2+} + NO_3^- + Mn^{2+}$

(e) $CNS^- + IO_3^- + Cl^- \longrightarrow CN^- + SO_4^{2-} + ICl_2^-$

(f) $CrI_3 + Cl_2 \longrightarrow CrO_4^{2-} + IO_3^- + Cl^-$

11.21 Complete and balance the following equations by the ion-electron method. All reactions occur in alkaline solution.

(a) $S^{2-} + I_2 \longrightarrow SO_4^{2-} + I^-$

(b) $CN^- + MnO_4^- \longrightarrow CNO^- + MnO_2$

(c) $Au + CN^- + O_2 \longrightarrow Au(CN)_2^- + OH^-$

(d) $Si + OH^- \longrightarrow SiO_3^{2-} + H_2$

(e) $Cr(OH)_3 + BrO^- \longrightarrow CrO_4^{2-} + Br^-$

11.22 Complete and balance the following equations by the ion-electron method. All reactions occur in alkaline solution.

(a) $Al + H_2O \longrightarrow Al(OH)_4^- + H_2$

(b) $S_2O_3^{2-} + OCl^- \longrightarrow SO_4^{2-} + Cl^-$

(c) $I_2 + Cl_2 \longrightarrow H_3IO_6^{2-} + Cl^-$

(d) $Bi(OH)_3 + Sn(OH)_4^{2-} \longrightarrow Bi + Sn(OH)_6^{2-}$

(e) $NiO_2 + Fe \longrightarrow Ni(OH)_2 + Fe(OH)_3$

11.23 Complete and balance the following equations by the ion-electron method. All reactions occur in alkaline solution.

(a) $HClO_2 \longrightarrow ClO_2 + Cl^-$

(b) $MnO_4^- + I^- \longrightarrow MnO_4^{2-} + IO_4^-$

(c) $P_4 \longrightarrow HPO_3^{2-} + PH_3$

(d) $SbH_3 + H_2O \longrightarrow Sb(OH)_4^- + H_2$

(e) $CO(NH_2)_2 + OBr^- \longrightarrow CO_2 + N_2 + Br^-$

11.24 Complete and balance the following equations by the ion-electron method. All reactions occur in alkaline solution.

(a) $Mn(OH)_2 + O_2 \longrightarrow Mn(OH)_3$

(b) $Cl_2 \longrightarrow ClO_3^- + Cl^-$

(c) $HXeO_4^- \longrightarrow XeO_6^{4-} + Xe + O_2$

(d) $As + OH^- \longrightarrow AsO_3^{3-} + H_2$

(e) $S_2O_4^{2-} + O_2 \longrightarrow SO_3^{2-} + OH^-$

11.25 Complete and balance the following equations by the ion-electron method. All reactions occur in alkaline solution.

(a) $Al + NO_3^- \longrightarrow Al(OH)_4^- + NH_3$

(b) $Ni^{2+} + Br_2 \longrightarrow NiO(OH) + Br^-$

(c) $S \longrightarrow SO_3^{2-} + S^{2-}$

(d) $S_2O_3^{2-} + I_2 \longrightarrow SO_4^{2-} + I^-$

(e) $S^{2-} + HO_2^- \longrightarrow SO_4^{2-} + OH^-$

11.26 Because the oxygen of H_2O_2 can be either oxidized (to O_2) or reduced (to H_2O), hydrogen peroxide can function as a reducing agent or as an oxidizing agent. Write and balance equations (by the ion-electron method) to show the following reactions of H_2O_2: (a) the oxidization of PbS to $PbSO_4$ in acid solution, (b) the oxidation of $Cr(OH)_3$ to CrO_4^{2-} in alkaline solution, (c) the reduction of MnO_4^- to Mn^{2+} in acid solution, (d) the reduction of Ag_2O to Ag in alkaline solution.

Acids and Bases; Acidic and Basic Oxides

11.27 What is the hydronium ion? What are the ways in which the H^+ ion is different from other cations?

11.28 Give examples of monoprotic and polyprotic acids, normal salts, and acid salts.

11.29 Write chemical equations for the reactions of HNO_3 with (a) NaOH, (b) $Mg(OH)_2$, (c) $Al(OH)_3$. Assume complete neutralization.

11.30 Write chemical equations for the reactions of KOH with (a) HBr, (b) H_2SO_4, (c) H_3PO_4. Assume complete neutralization.

11.31 Show by means of chemical equations how to prepare (a) Na_3PO_4, (b) Na_2HPO_4, (c) NaH_2PO_4 from H_3PO_4.

11.32 Give names for (a) $HBrO_3$, (b) $KBrO_3$, (c) HBr(aq), (d) $NaNO_2$, (e) $KHSO_4$, (f) K_2SO_3, (g) $NaHCO_3$, (h) NaH_2PO_4, (i) $Cu(NO_3)_2$.

11.33 Give formulas for (a) perchloric acid, (b) hydroiodic acid, (c) hypoiodous acid, (d) iodic acid, (e) sodium hydrogen sulfite, (f) calcium hydrogen carbonate, (g) copper(II) nitrite.

11.34 Write chemical equations for the reactions of the following with water: (a) Cl_2O, (b) N_2O_5, (c) SO_3, (d) CO_2, (e) CaO, (f) Na_2O, (g) P_4O_{10}.

11.35 Give the formulas for the anhydrides of the following: (a) $HClO_4$, (b) HNO_2, (c) H_2SO_3, (d) H_3BO_3, (e) $Al(OH)_3$, (f) $Zn(OH)_2$, (g) KOH.

11.36 Write chemical equations for the reactions of the following (assume complete neutralization): (a) $Ca(OH)_2$ and CO_2, (b) CO_2 and OH^-(aq), (c) ZnO and H^+(aq), (d) BaO and SO_2, (e) FeO and HCl.

11.37 What is an amphoteric oxide? Give the formulas of the ions formed by ZnO in acidic solution and in alkaline solution.

Volumetric Analysis

11.38 What is the molarity of a solution of H_2SO_4 if 25.00 ml of this solution requires 32.15 ml of a 0.6000 M solution of NaOH for complete neutralization?

11.39 What is the molarity of a solution of $Ba(OH)_2$ if 25.00 ml of this solution requires 15.27 ml of a 0.1000 M solution of HCl for complete neutralization?

11.40 A 1.250 g sample of impure $Mg(OH)_2$ requires 29.50 ml of 0.6000 M HCl for neutralization. If the impurity is $MgCl_2$, what is the mass percent of $Mg(OH)_2$ in the impure sample?

11.41 A 0.300 g sample of impure oxalic acid ($H_2C_2O_4$) is completely neutralized by 27.0 ml of a 0.179 M solution of NaOH. What is the mass percent of $H_2C_2O_4$ in the sample?

11.42 Potassium hydrogen phthalate, $KHC_8H_4O_4$, functions as a monoprotic acid. If a 1.46 g sample of impure $KHC_8H_4O_4$ requires 34.3 ml of 0.145 M NaOH for neutralization, what percentage of $KHC_8H_4O_4$ is in the material?

11.43 Potassium hydrogen phthalate, $KHC_8H_4O_4$, functions as a monoprotic acid. A 0.625 g sample of pure $KHC_8H_4O_4$ requires 27.8 ml of a solution of NaOH for neutralization. What is the molarity of the NaOH solution?

11.44 A 5.00 g sample of $NaNO_3$ contains NaCl as an impurity. The sample requires 15.3 ml of a 0.0500 M solution of $AgNO_3$ to precipitate all of the chloride as AgCl. (a) What mass of NaCl was present in the sample? (b) What is the mass percent of NaCl in the material?

11.45 A 1.00 g sample that contains Fe^{2+} is dissolved in 30.0 ml of water and the solution titrated against 0.0200 M $KMnO_4$. In the reaction, Fe^{2+} is oxidized to Fe^{3+} and MnO_4^- is reduced to Mn^{2+}. It takes 35.8 ml of the $KMnO_4$ solution to reach the equivalence point. **(a)** Write the chemical equation for the reaction. **(b)** What is the mass percent of Fe in the sample?

11.46 Hydrazine, N_2H_4, reacts with BrO_3^- in acid solution to produce N_2 and Br^-. **(a)** Write the chemical equation for the reaction. **(b)** A 0.132 g sample of impure hydrazine requires 38.3 ml of 0.0172 M $KBrO_3$ for complete reaction. What is the mass percent of hydrazine in the sample?

11.47 A 5.00 g sample of hemoglobin is treated in such a way as to produce small water-soluble molecules and ions by destroying the hemoglobin molecule. The iron in the aqueous solution that results from this procedure is reduced to Fe^{2+} and titrated against standard $KMnO_4$. In the titration, Fe^{2+} is oxidized to Fe^{3+} and MnO_4^- is reduced to Mn^{2+}. The sample requires 30.5 ml of 0.00200 M $KMnO_4$. What is the mass percent of iron in hemoglobin?

11.48 Iodine, I_2, reacts with the thiosulfate ion, $S_2O_3^{2-}$, to form the iodide ion, I^-, and the tetrathionate ion, $S_4O_6^{2-}$. **(a)** Write the chemical equation for this reaction. **(b)** How many grams of I_2 will react with 25.00 ml of a 0.0500 M solution of $Na_2S_2O_3$?

Equivalent Weights and Normal Solutions

11.49 What fraction of a mole is one equivalent weight of each of the following: **(a)** N_2H_4 for a reaction in which N_2 is produced, **(b)** $KBrO_3$ for a reaction in which Br^- is produced, **(c)** $KBrO_3$ for a reaction in which Br_2 is produced, **(d)** $K_2Cr_2O_7$ for a reaction in which Cr^{3+} is produced, **(e)** H_3PO_4 for a reaction in which HPO_4^{2-} is produced, **(f)** $Ca(OCl)_2$ for a reaction in which Cl^- is produced?

11.50 What are the normalities of 6.00 M HCl, 6.00 M H_2SO_4, and 6.00 M H_3PO_4? Assume complete neutralization of the acids.

11.51 What are the molarities of 6.00 N HCl, 6.00 N H_2SO_4, and 6.00 N H_3PO_4? Assume complete neutralization of the acids.

11.52 How many milliliters of 0.300 N H_2SO_4 would be required to neutralize 38.0 ml of 0.450 N NaOH?

11.53 A 25.0 ml sample of a solution of an acid requires 43.5 ml of 0.235 N NaOH for neutralization. What is the normality of the acid?

11.54 A 10.0 ml sample of a solution of a base requires 37.2 ml of 0.125 N H_2SO_4 for neutralization. What is the normality of the base?

11.55 Lactic acid, the acid in sour milk, has the molecular formula $C_3H_6O_3$. A 0.612 g sample of pure lactic acid requires 39.3 ml of 0.173 N NaOH for complete neutralization. **(a)** What is the equivalent weight of lactic acid? **(b)** How many acidic hydrogens per molecule does lactic acid have?

11.56 Citric acid, which may be obtained from lemon juice, has the molecular formula $C_6H_8O_7$. A 0.571 g sample of citric acid requires 42.5 ml of 0.210 N NaOH for complete neutralization. **(a)** What is the equivalent weight of citric acid? **(b)** How many acidic hydrogens per molecule does citric acid have?

11.57 A sample of a Fe^{2+} solution requires 26.0 ml of a 0.0200 M $K_2Cr_2O_7$ solution for a reaction in which Fe^{2+} is oxidized to Fe^{3+} and $Cr_2O_7^{2-}$ is reduced to Cr^{3+}. An identical sample of the same Fe^{2+} solution requires 41.6 ml of $KMnO_4$ solution for a reaction in which Fe^{2+} is oxidized to Fe^{3+} and MnO_4^- is reduced to Mn^{2+}. **(a)** What is the normality of the $K_2Cr_2O_7$ solution? **(b)** What is the normality of the $KMnO_4$ solution? **(c)** What is the molarity of the $KMnO_4$ solution?

11.58 A 0.6324 g sample of an iron ore is dissolved in an acid solution and the iron converted into the Fe^{2+} state. The resulting solution requires 32.37 ml of a 0.2024 N solution of $K_2Cr_2O_7$ for reaction. In the reaction, the Fe^{2+} is oxidized to Fe^{3+} and the $Cr_2O_7^{2-}$ is reduced to Cr^{3+}. What is the mass percent of iron in the ore?

CHEMICAL KINETICS

CHAPTER 12

Chemical kinetics is the study of the speeds, or rates, of chemical reactions. A small number of factors control how fast a reaction will occur. Investigation of these factors provides clues to the ways in which reactants are transformed into products in chemical reactions. The detailed description of the way a reaction occurs, based on the behavior of atoms, molecules, and ions, is called a **reaction mechanism.** Most chemical changes take place by mechanisms that consist of several steps. We can never be sure that a proposed mechanism represents reality—mechanisms are only educated guesses based on kinetic studies.

12.1 Reaction Rates

Consider a hypothetical reaction:

$$A_2(g) + B_2(g) \longrightarrow 2\ AB(g)$$

Throughout the time that the reaction is occurring, A_2 and B_2 are being gradually used up. The concentrations of these two substances, which are usually expressed in moles per liter, are decreasing. Since AB is being produced at the same time, the concentration of AB is increasing. The rate of the reaction is a measure of how fast these changes are taking place.

The symbol for the concentration of a substance consists of the formula of the substance enclosed in brackets. The symbol [AB], for example, stands for the concentration of AB. The symbol $\Delta[AB]$, therefore, stands for a change in the concentration of AB.

The rate of the reaction between A_2 and B_2 could be expressed in terms of the increase in the concentration of AB, $\Delta[AB]$, that occurs in a given time interval, Δt:

$$\text{rate of appearance of AB} = \frac{\Delta[AB]}{\Delta t}$$

If the concentration of AB is expressed in mol/liter and time in seconds, the rate would have the units

$$\frac{\text{mol/liter}}{\text{s}} = \text{mol/(liter} \cdot \text{s)}$$

The rate of the reaction could also be expressed in terms of the decrease in the concentration of A_2 or B_2 that takes place in a time interval. The rate based on the concentration of A_2, for example, would be

$$\text{rate of disappearance of } A_2 = \frac{-\Delta[A_2]}{\Delta t}$$

Since the concentration of A_2 is becoming smaller, $\Delta[A_2]$ is a negative value. The minus sign is included in this expression so that the rate will be a positive value.

The rate based on the concentration of AB is not numerically the same as the rate based on the concentration of A_2 or B_2. Suppose that at a given instant the concentration of A_2 is decreasing by 0.02 mol/liter in one second. The rate of decrease in the concentration of A_2, therefore, is 0.02 mol/(liter·s).

The chemical equation shows two moles of AB produced for every one mole of A_2 used. In the same time interval that the concentration of A_2 decreases by 0.02 mol/liter, the concentration of AB must increase by 0.04 mol/liter. The rate of increase in the concentration of AB, therefore, is 0.04 mol/(liter·s). These two values—the rate of disappearance of A_2, 0.02 mol/(liter·s), and the rate of appearance of AB, 0.04 mol/(liter·s)—describe the rate of the same reaction at the same instant. The rate of a reaction may be described in terms of the rate of appearance of a product or the rate of disappearance of a reactant, but *the basis of the rate measurement must be specified.*

The rate of a reaction usually changes as the reaction proceeds. In Figure 12.1, the concentrations of AB and A_2 are plotted against time. If the initial concentration of B_2 is the same as the initial concentration of A_2, then a curve for $[B_2]$ against time is the same as the one shown for $[A_2]$ against time.

In the figure, the concentration of the product, AB, starts at zero and rises rapidly at the beginning of the reaction. During this period, the concentration of the reactant, A_2, drops rapidly. The curves show, however, that the concentrations

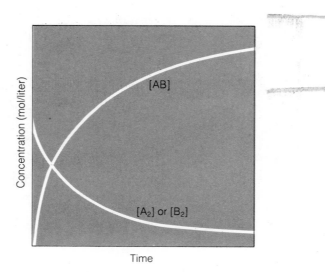

Figure 12.1 Curves showing changes in concentrations of substances with time for the reaction $A_2 + B_2 \rightarrow 2\,AB$

change more slowly as the reaction goes on. The rates of most chemical reactions depend upon the concentrations of the reactants. As these substances are used up, the reactions slow down. The rate at the start of the reaction is called the **initial rate**.

The rate of decrease in the concentration of A_2 at a given time can be obtained from the slope of a tangent drawn to the $[A_2]$ curve at the point that corresponds to the time of interest. In Figure 12.2, a tangent is drawn to the curve at $t = 0$ s. The tangent is extended so that it may be clearly seen that $[A_2]$ is changing by -0.05 mol/liter ($\Delta[A_2]$) for a 10 s time interval (Δt):

$$\text{rate of disappearance of } A_2 = \frac{-\Delta[A_2]}{\Delta t}$$

$$= \frac{-(-0.05 \text{ mol/liter})}{10 \text{ s}} = 0.005 \text{ mol/(liter} \cdot \text{s)}$$

This value is the initial rate of the reaction in terms of the disappearance of A_2.

At $t = 20$ s, the rate has decreased. Notice that the tangent to the curve at $t = 20$ s declines by -0.006 mol/liter for a 10 s interval:

$$\text{rate of disappearance of } A_2 = \frac{-(-0.006 \text{ mol/liter})}{10 \text{ s}} = 0.0006 \text{ mol/(liter} \cdot \text{s)}$$

Obtaining the data for a concentration curve is often a difficult thing to do. The concentrations must be determined at definite times throughout the course

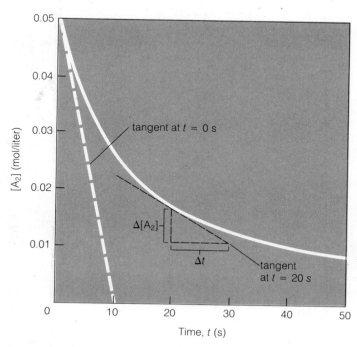

Figure 12.2 Determination of reaction rate by drawing tangents to the curve for $[A_2]$ against time

of the reaction without disturbing the reaction. The best methods for such determinations are based on the continuous measurement of a property that changes as the reaction occurs. Changes in pressure, color (the appearance or disappearance of a colored substance), acidity, conductivity, volume, and viscosity have been used.

12.2 Concentrations and Reaction Rates

Reaction rates usually depend upon the concentrations of the reacting substances. For most reactions, the rates are highest when the concentrations of reactants are high. This effect can be explained on the basis of the collision theory of reaction rates (see Section 12.3). The high concentrations mean that a relatively large number of molecules are crowded into a given volume. Under these conditions, the collisions between reacting molecules that convert them into molecules of product are relatively frequent, and consequently the reaction is more rapid.

For each chemical reaction there is a mathematical expression, called a **rate equation** or a **rate law**, that relates the concentrations of the reactants to the reaction rate. For the reaction

$$2\,N_2O_5(g) \longrightarrow 4\,NO_2(g) + O_2(g)$$

the rate equation is

$$\text{rate} = k[N_2O_5]$$

The expression tells us that the rate is directly proportional to the concentration of N_2O_5. If the concentration is doubled, the rate is doubled. If the concentration is tripled, the rate is tripled. The proportionality constant, k, is called the **rate constant**. *The form of the rate equation and the value of k must be determined by experiment.* The numerical value of k depends upon the temperature and the terms in which the rate is expressed.

The rate of the reaction

$$NO_2(g) + 2\,HCl(g) \longrightarrow NO(g) + H_2O(g) + Cl_2(g)$$

is proportional to the concentration of NO_2 times the concentration of HCl:

$$\text{rate} = k[NO_2][HCl]$$

Doubling the concentration of NO_2 would double the reaction rate. Doubling the concentration of HCl would also double the reaction rate. If the concentrations of both reactants were doubled at the same time, the reaction rate would increase fourfold.

For the reaction

$$2\,NO(g) + 2\,H_2(g) \longrightarrow N_2(g) + 2\,H_2O(g)$$

the rate equation is

$$\text{rate} = k[NO]^2[H_2]$$

The rate of the reaction is directly proportional to the *square* of the concentration of NO times the concentration of H_2. When the concentration of NO is increased by a factor of *two*, the rate increases by a factor of *four* (since 2^2 is 4). When the concentration of H_2 is increased by a factor of *two*, the rate increases by the factor of *two*. If the concentrations of both NO and H_2 are doubled, the rate would increase eightfold (since $2^2 \times 2 = 8$).

The **order** of a reaction is given by the *sum of the exponents* of the concentration terms in the rate equation. The decomposition of N_2O_5 is said to be first order since the exponent of $[N_2O_5]$ in the rate equation is one:

$$\text{rate} = k[N_2O_5]$$

The reaction between NO_2 and HCl is said to be first order in NO_2, first order in HCl, and second order overall:

$$\text{rate} = k[NO_2][HCl]$$

The reaction between NO and H_2 is second order in NO, first order in H_2, and third order overall:

$$\text{rate} = k[NO]^2[H_2]$$

The rate equation for a reaction, and consequently the order of a reaction, *must* be determined experimentally. They cannot be derived from the chemical equation for the reaction. The order of a reaction need not be a whole number. Reactions of a fractional order, as well as zero order, are known. For the decomposition of acetaldehyde (CH_3CHO),

$$CH_3CHO(g) \longrightarrow CH_4(g) + CO(g)$$

at 450°C, the rate equation is

$$\text{rate} = k[CH_3CHO]^{3/2}$$

The reaction, therefore, is three-halves order.

The decomposition of $N_2O(g)$ on gold surfaces at relatively high pressures of N_2O is zero order:

$$2\,N_2O(g) \xrightarrow{\text{Au}} 2\,N_2(g) + O_2(g) \qquad \text{rate} = k$$

When the pressure of N_2O is high, the decomposition proceeds at a steady rate that does not depend upon the concentration of N_2O.

Chemically similar reactions do not necessarily have the same type of rate equation. Consider the following two reactions:

$$H_2(g) + I_2(g) \longrightarrow 2\,HI(g) \qquad \text{rate} = k[H_2][I_2]$$

$$H_2(g) + Br_2(g) \longrightarrow 2\,HBr(g) \qquad \text{rate} = \frac{k[H_2][Br_2]^{1/2}}{k' + [HBr]/[Br_2]}$$

The last example shows that some reactions do not correspond to any simple order. Note also that this rate equation includes a term for the concentration of a product (HBr).

Example 12.1

The data given in the following table pertain to the reaction

$$2\,NO(g) + O_2(g) \longrightarrow 2\,NO_2(g)$$

run at 25°C. Determine the form of the rate equation and the value of the rate constant, k.

	Initial concentration		Initial rate
Experiment	NO mol/liter	O_2 mol/liter	Appearance of NO_2 mol/(liter·s)
A	1×10^{-3}	1×10^{-3}	7×10^{-6}
B	1×10^{-3}	2×10^{-3}	14×10^{-6}
C	1×10^{-3}	3×10^{-3}	21×10^{-6}
D	2×10^{-3}	3×10^{-3}	84×10^{-6}
E	3×10^{-3}	3×10^{-3}	189×10^{-6}

Solution

The rate equation is in the form

$$\text{rate of appearance of } NO_2 = k[NO]^x[O_2]^y$$

Data from the table are used to find the values of the exponents x and y.

In the first three experiments (A, B, and C), the concentration of NO is held constant and the concentration of O_2 is changed. Any change in the rate observed in this series of experiments, therefore, is caused by the change in the concentration of O_2. The concentration of O_2 in experiment B is double that in experiment A, and the rate observed in experiment B is twice the rate in experiment A. Comparison of the data from experiment C with the data from experiment A shows that when the concentration of O_2 is increased threefold, the rate increases threefold. The value of y, therefore, is 1. The rate is directly proportional to the first power of $[O_2]$.

In the last three experiments (C, D, and E), the concentration of O_2 is held constant (at 3×10^{-3} mol/liter) and the concentration of NO is changed. The increase in rate that is observed in this series of experiments is caused by the increase in the concentration of NO. The concentration of NO in experiment D is *two* times the concentration of NO in experiment C. The rate observed in experiment D, however, is *four* times the rate observed in experiment C. It appears that the square of $[NO]$ must appear in the rate equation since 2^2 is 4.

We can check this conclusion by comparing the data from experiment E to the data from experiment C. The concentration of NO increases by a factor of 3:

$$\frac{3 \times 10^{-3} \text{ mol/liter}}{1 \times 10^{-3} \text{ mol/liter}} = 3$$

The rate increases by a factor of 9:

$$\frac{189 \times 10^{-6} \text{ mol}/(\text{liter} \cdot \text{s})}{21 \times 10^{-6} \text{ mol}/(\text{liter} \cdot \text{s})} = 9$$

Since 3^2 is 9, the exponent x must be 2; the term for the concentration of NO is squared. The rate equation is

rate of the appearance of $NO_2 = k[NO]^2[O_2]$

The value of k can be obtained by using the data from any of the experiments. The same value should be obtained in each case. The data from experiment A are used in the following way:

rate of appearance of $NO_2 = k[NO]^2[O_2]$

$$7 \times 10^{-6} \text{ mol}/(\text{liter} \cdot \text{s}) = k(1 \times 10^{-3} \text{ mol}/\text{liter})^2 (1 \times 10^{-3} \text{ mol}/\text{liter})$$

$$7 \times 10^{-6} \text{ mol}/(\text{liter} \cdot \text{s}) = k(1 \times 10^{-9} \text{ mol}^3/\text{liter}^3)$$

$$k = \frac{7 \times 10^{-6} \text{ mol}/(\text{liter} \cdot \text{s})}{1 \times 10^{-9} \text{ mol}^3/\text{liter}^3}$$

$$k = 7 \times 10^3 \text{ liter}^2/(\text{mol}^2 \cdot \text{s})$$

The decay of all radioactive substances follows a rate expression like that for a first-order chemical reaction (see Section 25.3). Equation 25.6 (found in Section 25.3) relates concentration of reactant to time elapsed for a first-order transformation and Equation 25.7 shows how to find the half-life for a first-order reaction (the time required for one-half of the reactant to disappear).

12.3 Single-Step Reactions

The chemical equation for a reaction gives the stoichiometric relationships between the initial reactants and the final products. Usually, however, a reaction occurs by way of a mechanism that consists of several steps. A product of one step may be a reactant in the next step.

Consider the formation of nitrosyl fluoride (ONF) as an example:

$$2\,NO(g) + F_2(g) \longrightarrow 2\,ONF(g)$$

This reaction is believed to follow a two-step mechanism:

1. $NO(g) + F_2(g) \longrightarrow ONF(g) + F(g)$
2. $NO(g) + F(g) \longrightarrow ONF(g)$

Notice that the steps of the mechanism add up to the chemical equation for the overall reaction. The F atoms that are produced in the first step are used in the second step and therefore cancel in the addition. The F atoms are reaction intermediates—substances that are produced and used in the course of a reaction and are therefore neither reactants nor products of the reaction.

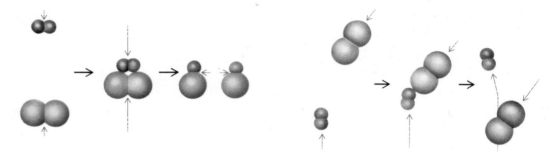

Figure 12.3 Collision between an A_2 molecule and a B_2 molecule resulting in a reaction

Figure 12.4 Collision between an A_2 molecule and a B_2 molecule producing no reaction

There are some reactions, however, that are believed to occur by a single step. The reaction between methyl bromide (CH_3Br) and sodium hydroxide in aqueous ethyl alcohol as a solvent,

$$CH_3Br + OH^- \longrightarrow CH_3OH + Br^-$$

and the gas-phase reaction

$$CO(g) + NO_2(g) \longrightarrow CO_2(g) + NO(g)$$

are examples. In this section we will consider how reactions of this type occur. Our discussion also applies to the manner in which a single step of a multistep mechanism takes place.

The **collision theory of reaction rates** describes reactions in terms of collisions between reacting molecules. Assume that the hypothetical gas-phase reaction

$$A_2(g) + B_2(g) \longrightarrow 2\,AB(g)$$

takes place through collisions between A_2 and B_2 molecules as shown in Figure 12.3. An A_2 and a B_2 molecule strike one another. The old A—A and B—B bonds break while simultaneously two new A—B bonds form, and two AB leave the scene of the collision. The rate of the reaction is proportional to the number of these collisions that occur in a given time interval.

Calculations show, however, that the number of molecular collisions per unit time in a situation such as this is extremely large. At room temperature and 1 atm pressure, about 10^{31} collisions/liter occur in 1 s. If each collision between an A_2 and B_2 molecule resulted in a reaction, the process would be over in a fraction of a second. Most reactions are not this rapid.

Not every A_2–B_2 collision, therefore, can lead to a reaction. Usually the number of collisions that produce reactions, called **effective collisions**, is a very small fraction of the total number of collisions between A_2 and B_2 molecules.

There are two reasons why a collision may not be effective. First, the molecules may be improperly aligned (see Figure 12.4). Second, the collision may be so gentle that the molecules rebound unchanged. The electron cloud of a molecule is, of course, negatively charged. When two slow-moving molecules approach one another closely, they rebound because of the repulsion due to the charges of their electron clouds. Faster-moving molecules, however, are not deterred by this repulsion; the impact of the collision causes the reaction to occur. For an

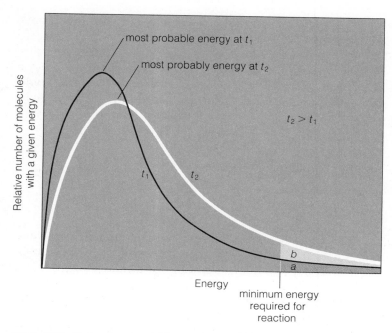

Figure 12.5 Molecular energy distributions at temperatures t_1 and t_2

effective collision, the sum of the energies of the colliding molecules must equal, or exceed, some minimum value.

The effect of temperature on reaction rates reinforces this view. The rates of almost all chemical reactions increase when the temperature is raised. The effect is observed for endothermic as well as exothermic reactions. An increase of 10°C in the temperature, at temperatures near room temperature, is often found to increase the reaction rate from 100% to 300%. The more rapid molecular motion that results from an increase in temperature brings about a larger number of molecular collisions per unit time. This factor alone, however, cannot account for the increase in speed. Raising the temperature from 25°C to 35°C causes an increase in the total number of molecular collisions of only about 2%. Obviously, increasing the temperature must increase the fraction of molecular collisions that are effective, and this factor must be the more important of the two.

By an examination of Figure 12.5 we can understand why proportionately more molecular collisions result in reactions at a higher temperature than at a lower temperature. Two molecular energy distribution curves are shown—one for a temperature, t_1, and another for a higher temperature, t_2. The minimum energy required for reaction is indicated on Figure 12.5. The number of molecules at t_1 with energies equal to or greater than this minimum energy is proportional to the area, a, under the curve for t_1.

The curve for temperature t_2 is shifted only slightly in the direction of higher energy. At t_2, however, the number of molecules possessing sufficient energy to react successfully upon collision is greatly increased and is proportional to the area $a + b$. An increase in temperature, therefore, produces an increase in reaction rate principally because the proportion of effective collisions is increased. The increase in the total number of collisions per unit time is only a minor factor. The influence of temperature on reaction rates is analyzed mathematically in Section 12.6.

Flashbulb before and after firing. The bulb contains magnesium wire in an atmosphere of pure oxygen. Passing a small electric current through the Mg wire heats it and provides the activation energy for the reaction, which rapidly produces magnesium oxide, heat, and light. *Fundamental Photographs.*

The energy requirements for a successful collision are described in a slightly different way by the **transition state theory.** Let us again consider the reaction between A_2 and B_2. In a gentle collision, the A_2 and B_2 molecules are repelled by the charges of their electron clouds and never get close enough for A—B bonds to form. In a successful collision, however, high-energy A_2 and B_2 molecules are assumed to form a short-lived **activated complex,** A_2B_2. The A_2B_2 complex may split to form two AB molecules or may split to re-form A_2 and B_2 molecules:

$$A_2 + B_2 \rightleftharpoons \begin{bmatrix} A\text{---}A \\ | \quad | \\ B\text{---}B \end{bmatrix} \longrightarrow 2\,AB$$

An activated complex, usually shown between brackets, is not a molecule that can be isolated or detected. Instead, it is an unstable arrangement of atoms that exists only for a moment. It is sometimes called a **transition state.** In the activated complex, the A—A and B—B bonds are weakened and partly broken and the A—B bonds are only partly formed. The activated complex is a state of relatively high potential energy.

A potential energy diagram for the reaction between A_2 and B_2 is shown in Figure 12.6. The figure illustrates how the potential energy of the substances involved in the reaction changes in the course of the reaction. Distances along the reaction coordinate indicate how far the formation of products from reactants has progressed.

The difference between the potential energy of the reactants, A_2 and B_2, and the potential energy of the activated complex, A_2B_2, is called the **energy of activation** and is given the symbol $E_{a,f}$. In any collision between an A_2 and a B_2 molecule, the total energy of the molecules remains constant, but kinetic energy (energy of motion) and potential energy may be converted from one form into the other. In a successful collision, part of the kinetic energy of fast-moving A_2 and B_2 molecules is used to provide the energy of activation and produce the high-energy activated complex.

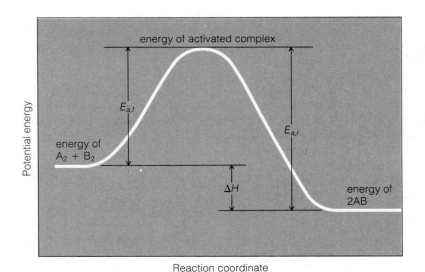

Figure 12.6 Potential energy diagram for the hypothetical reaction $A_2 + B_2 \rightleftharpoons A_2B_2 \rightleftharpoons 2\,AB$

The activated complex can split in two ways. If it re-forms the reactants, A_2 and B_2, the energy of activation, $E_{a,f}$, is released in the form of kinetic energy to the A_2 and B_2 molecules. In this case, there is no net reaction. If the complex splits into the product, two AB molecules, the energy indicated as $E_{a,r}$ on Figure 12.6 is released as kinetic energy to the AB molecules. The difference between the energy absorbed, $E_{a,f}$, and the energy evolved, $E_{a,r}$, is the enthalpy change, ΔH, for the reaction:

$$\Delta H = E_{a,f} - E_{a,r}$$

Since $E_{a,r}$ is larger than $E_{a,f}$, ΔH is negative and the reaction is exothermic.

The energy of activation is a potential energy barrier between the reactants and products. Even though the energy of the reactant molecules is higher than the energy of the product molecules, the system must climb a potential energy hill before it can coast down to a state of lower energy. When A_2 and B_2 molecules that have relatively low kinetic energies approach each other, they do not have enough energy between them to produce the activated complex. The repulsion between their electron clouds prevents them from approaching each other closely enough to form the complex. In this case the molecules have only enough energy to get part way up the hill. Then, repelling each other, they coast back down the hill and fly apart unchanged.

Suppose that the reaction diagrammed in Figure 12.6 were reversible. The reverse reaction can be interpreted by reading the figure from right to left. The energy of activation for the reverse reaction is $E_{a,r}$, and the energy released by the formation of the products (in this case A_2 and B_2) from the activated complex is $E_{a,f}$. The enthalpy change for the reverse reaction is

$$\Delta H = E_{a,r} - E_{a,f}$$

The enthalpy change is positive since $E_{a,r}$ is larger than $E_{a,f}$. The reverse reaction is endothermic.

Two potential energy diagrams are shown in Figure 12.7—one for an exothermic single-step reaction and one for an endothermic single-step reaction.

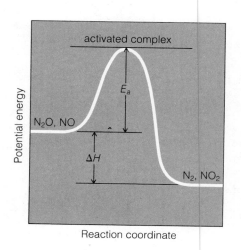

(a)

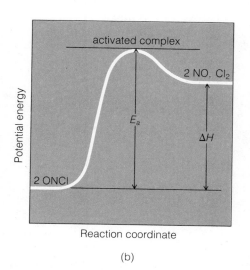

(b)

Figure 12.7 Potential energy diagrams for single-step reactions: (a) an exothermic reaction, and (b) an endothermic reaction

The exothermic reaction, between N_2O and NO, can be represented as follows:

$$N\equiv N-O + N=O \rightleftharpoons \left[N\equiv N\cdots O \cdots N\overset{\displaystyle O}{\diagup\diagup} \right] \longrightarrow N\equiv N + O-N\overset{\displaystyle O}{\diagup\diagup}$$

The diagram of the activated complex shows the N—O bond of the N_2O molecule stretched and weakened and a new bond, between the O atom and the N atom of the NO molecule, partially formed. The energy of activation for the reaction is 209 kJ/mol; ΔH is -138 kJ/mol.

The endothermic reaction diagrammed in Figure 12.7 is

$$O=N\diagdown_{Cl} + \diagup_{Cl}N=O \rightleftharpoons \left[O=N\diagdown_{Cl\cdots Cl}N=O \right] \longrightarrow$$

$$O=N + Cl-Cl + N=O$$

In the activated complex the N—Cl bonds of both ONCl molecules are in the process of breaking and a new Cl—Cl bond is starting to form. The energy of activation for the reaction is 98 kJ/mol; ΔH is $+76$ kJ/mol.

12.4 Rate Equations for Single-Step Reactions

The **molecularity** of a single-step reaction refers to the number of molecules that participate in an individual reaction. Thus, a step of a mechanism may be called **unimolecular, bimolecular,** or **termolecular** depending upon whether one, two, or three molecules react in the step. Most reactions do not occur in a single step, but rather proceed through a sequence of steps. Each step may be described in terms of its molecularity, but *such a description is not applied to a reaction as a whole when it consists of more than one step*.

The molecularity of a single-step reaction determines its reaction order. The coefficients in the chemical equation for the step appear as exponents in the rate equation. For example,

$$2A + B \longrightarrow \text{products} \qquad \text{rate} = k[A]^2[B]$$

the coefficient (2) of A in the chemical equation is the exponent of $[A]$ in the rate equation; the coefficient of B is the exponent of $[B]$. Since the reaction involves three molecules, it is termolecular and the rate equation is third order overall. This method for deriving rate equations cannot be applied routinely to all chemical equations; *it is used only if the chemical equation pertains to a reaction that occurs in one step*. The following types of steps are encountered.

1. Unimolecular steps. A unimolecular reaction is first order:

$$A \longrightarrow \text{products} \qquad \text{rate} = k[A]$$

A reaction of this type occurs when a high-energy A molecule breaks into smaller molecules or rearranges into a new molecular structure. The rate of the reaction is proportional to the concentration of A molecules present.

2. Bimolecular steps. There are two types of bimolecular steps. The first one is

$$A + B \longrightarrow products \qquad rate = k[A][B]$$

The reaction occurs by collisions between A and B molecules. The rate of the reaction is proportional to the number of A–B collisions per second. If we double the concentration of A, the rate would double since there would be twice as many A molecules in a given volume and twice as many A–B collisions per second. If we tripled the concentration of A, the number of collisions per second would increase by a factor of 3. Whatever we do to the concentration of A is reflected in the rate. The rate, therefore, is proportional to the first power of $[A]$.

In like manner, changes in the concentration of B produce similar changes in the number of A–B collisions per second. The rate is also proportional to the first power of $[B]$. The reaction, therefore, is first order in A, first order in B, and second order overall as shown by the rate equation given previously.

The second type of bimolecular step is

$$2\,A \longrightarrow products \qquad rate = k[A]^2$$

The step occurs by collisions between two A molecules. Suppose that there are n molecules of A in a container. The number of collisions per second for a *single* A molecule is proportional to the number of other A molecules present, $n - 1$. The *total* number of collisions per second for all n molecules might incorrectly be expected to be proportional to n times$(n - 1)$. It is, however, proportional to $\frac{1}{2}n(n - 1)$. The factor $\frac{1}{2}$ is included so that a given collision is not counted twice—once for a collision in which molecule 1 hits molecule 2 and again for a collision in which molecule 2 hits molecule 1.

Since n is a very large number, $(n - 1)$ is equal to n for all practical purposes. We can say that the total number of collisions per second is proportional to $\frac{1}{2}n^2$. Since the rate of the reaction is proportional to the total number of collisions per second,

$$rate \propto \tfrac{1}{2}n^2$$

The number of molecules in the container, n, determines the concentration of A; n^2, therefore, is proportional to $[A]^2$. The constant $\frac{1}{2}$ can be incorporated into the proportionality constant, k. Thus,

$$rate = k[A]^2$$

The collision theory, therefore, can be used to justify the fact that the molecularity of a step determines the reaction order.

3. Termolecular steps. There are, in theory, three types of termolecular reactions:

$$
\begin{array}{ll}
A + B + C \longrightarrow products & rate = k[A][B][C] \\
2\,A + B \longrightarrow products & rate = k[A]^2[B] \\
3\,A \longrightarrow products & rate = k[A]^3
\end{array}
$$

Termolecular steps are encountered in reaction mechanisms. They are, however, not common since they involve three-body collisions. Such collisions in which three bodies must come together simultaneously are rare.

The steps in the preceding list are the only types that are thought to occur in reaction mechanisms. Mechanism steps that involve a molecularity higher than three are never postulated. The chance that an effective four-body collision will occur with any regularity is so slight that such collisions are never proposed as a part of a reaction mechanism.

12.5 Reaction Mechanisms

The rate equation for a reaction must be determined by experimentation. A mechanism for the reaction is proposed on the basis of this rate equation and any other available evidence—such as the detection of a reaction intermediate. A mechanism, therefore, is only a hypothesis.

The following rate equation for the formation of nitrosyl fluoride was obtained experimentally:

$$2\,NO + F_2 \longrightarrow 2\,ONF \qquad rate = k[NO][F_2]$$

The suggested mechanism and the corresponding bimolecular rate equations are

1. $NO + F_2 \longrightarrow ONF + F \qquad rate_1 = k_1[NO][F_2]$
2. $NO + F \longrightarrow ONF \qquad rate_2 = k_2[NO][F]$

The two steps add up to the overall reaction with the F atoms as a reaction intermediate. The first step is assumed to be much slower than the second step. Step 1 produces F atoms slowly. As soon as they are produced, they are rapidly used by step 2. Step 1, therefore, is the bottleneck of the reaction; the overall reaction cannot be faster than it is. Since it controls the overall rate, it is called the **rate-determining step.** For this reason, the rate equation for step 1 is the rate equation for the overall change with $k = k_1$.

There is no way to tell from a chemical equation whether the reaction it describes proceeds by one step or several. Consider two chemically similar reactions. The reaction of methyl bromide, CH_3Br, and OH^- ion is second order:

$$OH^- + CH_3Br \longrightarrow CH_3OH + Br^- \qquad rate = k[CH_3Br][OH^-]$$

A single-step mechanism is consistent with this rate equation. The reaction is believed to proceed through a transition state in which the OH^- ion approaches the C atom on the opposite side from the Br atom:

$$H-O^- + H-\overset{H}{\underset{H}{C}}-Br \rightleftharpoons \left[H-O\cdots\overset{H}{\underset{H}{C}}\cdots Br \right] \longrightarrow H-O-\overset{H}{\underset{H}{C}}-H + Br^-$$

The reaction between tertiary butyl bromide, $(CH_3)_3CBr$, and OH^- ion is chemically similar, but it is first order:

$$OH^- + CH_3-\overset{CH_3}{\underset{CH_3}{C}}-Br \longrightarrow CH_3-\overset{CH_3}{\underset{CH_3}{C}}-OH + Br^- \qquad rate = k[(CH_3)_3CBr]$$

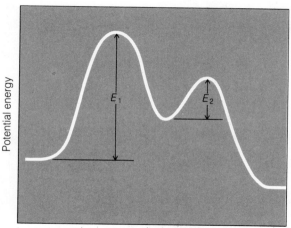

Figure 12.8 Potential energy diagram for a two-step mechanism in which the first step is rate-determining

The approach of the OH^- ion to the central C atom is blocked by the CH_3-groups and this reaction follows a different mechanism from that of the reaction between CH_3Br and OH^-. This reaction is thought to occur by two steps:

1. $$(CH_3)_3CBr \longrightarrow (CH_3)_3C^+ + Br^- \qquad rate_1 = k_1[(CH_3)_3CBr]$$
2. $$(CH_3)_3C^+ + OH^- \longrightarrow (CH_3)_3COH \qquad rate_2 = k_2[(CH_3)_3C^+][OH^-]$$

The first step, a unimolecular step in which the $(CH_3)_3CBr$ molecule ionizes, is thought to be the rate-determining step. The overall rate equation, therefore, corresponds to the unimolecular rate equation of step 1.

Each step of a multistep mechanism has a transition state and an activation energy. A two-step mechanism in which the first step is rate-determining, such as the last one or the one for the reaction of NO and F_2, would have a reaction profile similar to that shown in Figure 12.8. The energy of activation for the first step, E_1 on the diagram, is higher than the energy of activation for the second step, E_2. The overall rate, therefore, depends upon the rate at which reacting molecules get over the first potential energy barrier.

What about a multistep mechanism in which the first step is not rate-determining? Consider the reaction

$$CH_3OH + H^+ + Br^- \longrightarrow CH_3Br + H_2O \qquad rate = k[CH_3OH][H^+][Br^-]$$

The exponents of the experimentally derived rate equation are the same as the coefficients in the chemical equation. A termolecular single-step mechanism would be consistent with this rate equation. The reaction, however, is believed to occur by a series of steps, none of which is a three-body collision. The third step is thought to be the slowest:

1. $$CH_3OH + H^+ \longrightarrow CH_3OH_2^+ \qquad rate_1 = k_1[CH_3OH][H^+]$$
2. $$CH_3OH_2^+ \longrightarrow CH_3OH + H^+ \qquad rate_2 = k_2[CH_3OH_2^+]$$
3. $$Br^- + CH_3OH_2^+ \longrightarrow CH_3Br + H_2O \qquad rate_3 = k_3[CH_3OH_2^+][Br^-]$$

In the first step, the reaction intermediate $CH_3OH_2^+$ is formed. This intermediate can decompose back into CH_3OH and H^+ (step 2) or react with Br^- to form the products (step 3).

Since the third step is the rate-determining step, the overall rate depends upon it:

$$\text{rate} = \text{rate}_3 = k_3[CH_3OH_2^+][Br^-]$$

This expression, however, contains a term for the concentration of the reaction intermediate, $[CH_3OH_2^+]$. To eliminate this term, we assume that the concentration of the reaction intermediate $CH_3OH_2^+$ becomes constant after the reaction has been going on for a while. That is to say that the intermediate is used as fast as it is produced. The intermediate is produced in step 1:

$$\text{rate of appearance of } CH_3OH_2^+ = k_1[CH_3OH][H^+]$$

It is used in steps 2 and 3:

$$\text{rate of disappearance of } CH_3OH_2^+ = k_2[CH_3OH_2^+] + k_3[CH_3OH_2^+][Br^-]$$

Since the third step is *much* slower than the second, k_3 is *much* smaller than k_2. We can, therefore, neglect the term $k_3[CH_3OH_2^+][Br^-]$ since it is much smaller than the term $k_2[CH_3OH_2^+]$:

$$\text{rate of appearance of } CH_3OH_2^+ = \text{rate of disappearance of } CH_3OH_2^+$$

$$k_1[CH_3OH][H^+] = k_2[CH_3OH_2^+]$$

Therefore,

$$[CH_3OH_2^+] = \frac{k_1[CH_3OH][H^+]}{k_2}$$

If we substitute this value into the rate equation for the third step, we get

$$\text{rate} = \text{rate}_3 = k_3[CH_3OH_2^+][Br^-]$$

$$\text{rate} = k_3\left(\frac{k_1[CH_3OH][H^+]}{k_2}\right)[Br^-]$$

$$\text{rate} = \frac{k_1 k_3}{k_2}[CH_3OH][H^+][Br^-]$$

The constants can be combined into a single constant to give

$$\text{rate} = k[CH_3OH][H^+][Br^-]$$

which is the experimentally determined rate equation. Note that

$$k = \frac{k_1 k_3}{k_2}$$

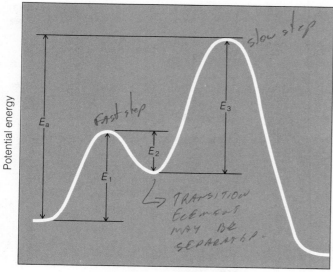

Handwritten annotations on figure:
slow step
Fast step
TRANSITION
SCHEMES
MAY BE
SEPARATE.

Figure 12.9 Potential energy diagram for a three-step mechanism in which the third step is rate-determining

A reaction profile for a three-step mechanism such as the preceding one might look something like that shown in Figure 12.9. The activation energies for the steps (E_1, E_2, and E_3) are indicated on the diagram. Remember that step 2 is the reverse of step 1.

The reaction between H_2 gas and Br_2 vapor at temperatures around 200°C provides an example of an important type of reaction mechanism called a **chain mechanism**:

$$H_2 + Br_2 \longrightarrow 2\,HBr$$

The mechanism for the reaction can be described in four parts:

1. Chain initiation. Some Br_2 molecules dissociate into atoms:

$$Br_2 \longrightarrow 2\,Br$$

2. Chain propagation. The Br atoms are reactive intermediates called **chain carriers.** A Br atom reacts with a H_2 molecule:

$$Br + H_2 \longrightarrow HBr + H$$

The reaction produces a molecule of product, HBr, and a second chain carrier, an H atom. The H atom reacts with a Br_2 molecule:

$$H + Br_2 \longrightarrow HBr + Br$$

producing another molecule of HBr and the original chain carrier, a Br atom. The Br atom reacts with a H_2 molecule, and the cycle starts again. These two steps are repeated over and over again.

3. Chain inhibition. If a H atom collides with a HBr molecule, the reaction that results is said to inhibit the overall reaction:

$$H + HBr \longrightarrow H_2 + Br$$

Since a molecule of product is used (HBr) and a molecule of reactant is produced (H_2), this step slows down the overall reaction. It does not, however, break the chain or stop the reaction since a chain carrier (Br) is also produced.

4. Chain termination. Two chains are terminated when two chain carriers come together:

$$2\,Br \longrightarrow Br_2$$
$$2\,H \longrightarrow H_2$$
$$H + Br \longrightarrow HBr$$

The reaction of $H_2(g)$ and $Cl_2(g)$ is believed to follow a similar mechanism. A mixture of these two gases at room temperature can be kept for a long period of time *in the dark* without reacting. If the mixture is exposed to light, a violently rapid reaction occurs. Light is believed to start the chain reaction by dissociating some Cl_2 molecules into atoms. The reaction of H_2 and Br_2 is also light sensitive, but the reaction at room temperature is slower.

Many reactions are thought to occur by a chain mechanism. Chain reactions are usually very rapid; some of them are explosive.

12.6 Rate Equations and Temperature

The rate constant, k, varies with temperature in a manner described by the following equation:

$$k = Ae^{-E_a/RT} \tag{12.1}$$

where A is a constant that is characteristic of the reaction being studied, e is the base of natural logarithms (2.718. . .), E_a is the energy of activation for the reaction (in J/mol), R is the molar gas constant [8.3143 J/(K·mol)], and T is the absolute temperature. The equation was first proposed by Svanté Arrhenius in 1889 and is known as the **Arrhenius equation.**

For a single-step reaction, the factor $e^{-E_a/RT}$ represents the fraction of molecules that acquires the energy of activation needed for a successful reaction (see Figure 12.5 in Section 12.3). The constant A, called the **frequency factor,** incorporates other factors that influence reaction rate, such as the frequency of molecular collisions and the geometric requirements for the alignment of colliding molecules that react. The Arrhenius equation is only approximate, but in most cases the approximation is a good one.

The Arrhenius equation also applies to multistep reactions. For a reaction that follows a mechanism such as that to which Figure 12.8 applies, the Arrhenius parameters A and E_a are those for the first step (A_1 and E_1) since the first step is the rate-determining step. In most multistep reactions, however, A and E_a are composites of the values for the individual steps.

In the three-step reaction that was discussed in the last section and in which the third step was rate-determining (see Figure 12.9),

$$k = \frac{k_1 k_3}{k_2}$$

The rate constant for each step may be expressed in terms of the Arrhenius equation (see Equation 12.1). Therefore,

$$k = \frac{A_1 e^{-E_1/RT} A_3 e^{-E_3/RT}}{A_2 e^{-E_2/RT}}$$

or

$$k = \frac{A_1 A_3}{A_2} e^{-(E_1 + E_3 - E_2)/RT}$$

Hence the Arrhenius parameters for the overall rate constant are

$$A = \frac{A_1 A_3}{A_2}$$

and

$$E_a = E_1 + E_3 - E_2$$

If we trace these energy terms in Figure 12.9—E_1 minus E_2 plus E_3—we can see that the overall energy of activation in this case (indicated as E_a in the figure) is equal to the height of the potential energy barrier of the third step above the potential energy of the initial reactants.

If we take the natural logarithm of the Arrhenius equation, we get

$$\ln k = \ln A - \frac{E_a}{RT} \tag{12.2}$$

which may be transformed into

$$2.303 \log k = 2.303 \log A - \frac{E_a}{RT} \tag{12.3}$$

or

$$\log k = \log A - \frac{E_a}{2.303RT} \tag{12.4}$$

There are two variables in this equation, k and T. If we rearrange it into

$$\log k = -\frac{E_a}{2.303R} \left(\frac{1}{T}\right) + \log A \tag{12.5}$$

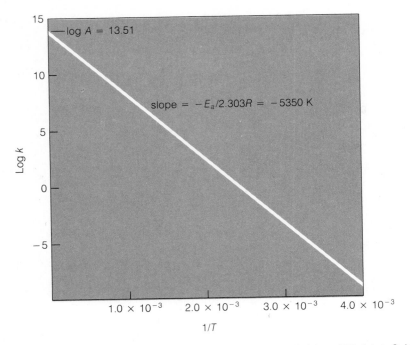

Figure 12.10   Plot of log k versus $1/T$ for the reaction $2\,N_2O_5(g) \rightarrow 4\,NO_2(g) + O_2(g)$

we can see that the equation is in the form of an equation for a straight line $(y = mx + b)$. A plot of $\log k$ against $(1/T)$ is a straight line with a slope of $-E_a/2.303R$ and a y-intercept of $\log A$ (Figure 12.10). If values of k are determined at several temperatures and the data plotted in this manner, E_a for the reaction can be calculated from the slope of the curve and A can be obtained by taking the antilogarithm of the y-intercept.

The curve shown in Figure 12.10 pertains to the first-order reaction:

$$2\,N_2O_5(g) \longrightarrow 4\,NO_2(g) + O_2(g)$$

The slope of the curve is -5350 K, from which we obtain

$$-\frac{E_a}{2.30R} = -5350 \text{ K}$$

$$E_a = (5350 \text{ K})(2.30)[8.31 \text{ J/(K} \cdot \text{mol})]$$
$$= 102,000 \text{ J/mol} = 102 \text{ kJ/mol}$$

The y-intercept is 13.51, and therefore

$$\log A = 13.51$$
$$A = 3.2 \times 10^{13}/\text{s}$$

The values of E_a and A for a reaction can also be found from the rate constants for two different temperatures. If the rate constant at T_1 is k_1 and at T_2 is k_2, then

$$\log k_2 = \log A - \frac{E_a}{2.303RT_2} \qquad (12.6)$$

and

$$\log k_1 = \log A - \frac{E_a}{2.303RT_1} \qquad (12.7)$$

Subtraction of Equation 12.7 from Equation 12.6 gives

$$\log k_2 - \log k_1 = -\frac{E_a}{2.303RT_2} + \frac{E_a}{2.303RT_1} \qquad (12.8)$$

Since $\log x - \log y$ is $\log(x/y)$,

$$\log\left(\frac{k_2}{k_1}\right) = \frac{E_a}{2.303R}\left(\frac{1}{T_1} - \frac{1}{T_2}\right) \qquad (12.9)$$

or

$$\log\left(\frac{k_2}{k_1}\right) = \frac{E_a}{2.303R}\left(\frac{T_2 - T_1}{T_1 T_2}\right) \qquad (12.10)$$

If this equation is solved for the energy of activation, E_a, the following relationship is obtained:

$$E_a = 2.303R\left(\frac{T_1 T_2}{T_2 - T_1}\right)\log\left(\frac{k_2}{k_1}\right) \qquad (12.11)$$

The uses of the last two equations are illustrated in the following examples.

Example 12.2

For the reaction

$$2\,NOCl(g) \longrightarrow 2\,NO(g) + Cl_2(g)$$

the rate equation is

rate of production of $Cl_2 = k[NOCl]^2$

The rate constant, k, is 2.6×10^{-8} liter/(mol·s) at 300 K and 4.9×10^{-4} liter/(mol·s) at 400 K. What is the energy of activation, E_a, for the reaction?

Solution

Let

$$T_1 = 300\ K$$
$$T_2 = 400\ K$$

$$k_1 = 2.6 \times 10^{-8} \text{ liter/(mol·s)}$$

$$k_2 = 4.9 \times 10^{-4} \text{ liter/(mol·s)}$$

R is 8.31 J/(K · mol), and substitution into Equation 12.11 gives

$$E_a = 2.30R\left(\frac{T_1 T_2}{T_2 - T_1}\right)\log\left(\frac{k_2}{k_1}\right)$$

$$= 2.30[8.31 \text{ J/(K·mol)}]\left(\frac{(300 \text{ K})(400 \text{ K})}{400 \text{ K} - 300 \text{ K}}\right)\log\left(\frac{4.9 \times 10^{-4} \text{ liter/(mol·s)}}{2.6 \times 10^{-8} \text{ liter/(mol·s)}}\right)$$

$$= [19.1 \text{ J/(K·mol)}](1200 \text{ K})\log(1.88 \times 10^4)$$
$$= (22{,}900 \text{ J/mol})(4.28)$$
$$= 98{,}000 \text{ J/mol} = 98 \text{ kJ/mol}$$

Example 12.3

Given the data found in Example 12.2, find the value of k at 500 K.

Solution

Let

$$T_1 = 400 \text{ K}$$

$$T_2 = 500 \text{ K}$$

$$k_1 = 4.9 \times 10^{-4} \text{ liter/(mol·s)}$$

$$k_2 = \text{unknown}$$

$$E_a = 9.8 \times 10^4 \text{ J/mol}$$

We use Equation 12.10:

$$\log\left(\frac{k_2}{k_1}\right) = \frac{E_a}{2.30R}\left(\frac{T_2 - T_1}{T_1 T_2}\right)$$

$$= \frac{9.8 \times 10^4 \text{ J/mol}}{2.30 \, [8.31 \text{ J/(K·mol)}]}\left(\frac{500 \text{ K} - 400 \text{ K}}{(400 \text{ K})(500 \text{ K})}\right)$$

$$= (5.13 \times 10^3 \text{ K})(5.00 \times 10^{-4}/\text{K})$$
$$= 2.57$$

Therefore,

$$\frac{k_2}{k_1} = \text{antilog} \, 2.57 = 3.7 \times 10^2$$

or

$$k_2 = (3.7 \times 10^2)k_1$$
$$= (3.7 \times 10^2)[4.9 \times 10^{-4} \text{ liter/(mol·s)}]$$
$$= 0.18 \text{ liter/(mol·s)}$$

Since the relation between k and T is exponential, a small change in T causes a relatively large change in k. Any change in the rate constant is, of course, reflected in the rate of the reaction. For the reaction considered in Examples 12.2 and 12.3, a 100° increase in the temperature causes the following effects:

300 K to 400 K	rate increases 18,800 times
400 K to 500 K	rate increases 367 times

The marked effect of a temperature increase is obvious. Notice, however, that the reaction rate is affected more at low temperatures than at high temperatures.

The energies of activation of many reactions range from 60 kJ/mol to 250 kJ/mol, values which are on the same scale as bond energies. For a 10° rise in temperature, from 300 K to 310 K, the rate of reaction varies with the energy of activation in the following way:

$E_a = 60$ kJ/mol	rate increases about 2 times
$E_a = 250$ kJ/mol	rate increases about 25 times

12.7 Catalysts

A **catalyst** is a substance that increases the rate of a chemical reaction without being used up in the reaction. A catalyst may be recovered unchanged at the end of the reaction. Oxygen can be prepared by heating potassium chlorate ($KClO_3$) by itself. Or, a small amount of manganese dioxide (MnO_2) can be used as a catalyst for this reaction. When MnO_2 is present, the reaction is much more rapid and the decomposition of $KClO_3$ takes place at a satisfactory rate at a lower temperature:

$$2\,KClO_3(s) \xrightarrow{MnO_2} 2\,KCl(l) + 3\,O_2(g)$$

The catalyst is written over the arrow in the chemical equation since a catalyst does not affect the overall stoichiometry of the reaction. The MnO_2 may be recovered unchanged at the conclusion of the reaction.

The mere *presence* of a catalyst does not cause the effect on the reaction rate. A catalyzed reaction takes place by a pathway, or mechanism, that is different from the one that the uncatalyzed reaction follows. Suppose, for example, that an uncatalyzed reaction occurs by collisions between X and Y molecules:

$$X + Y \longrightarrow XY$$

The catalyzed reaction might follow a two-step mechanism consisting of

1. $X + C \longrightarrow XC$
2. $XC + Y \longrightarrow XY + C$

where C is the catalyst. Notice that the catalyst is used in the first step and regenerated in the second. It is, therefore, used over and over again. Consequently, only a small amount of a catalyst is needed to do the job.

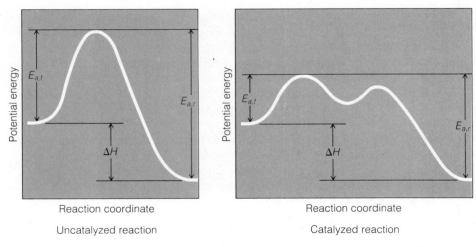

Figure 12.11 Potential energy diagrams for a reaction in the absence and presence of a catalyst

A catalyst, therefore, works by opening a new path by which the reaction can take place. The catalyzed path has a lower overall energy of activation than the uncatalyzed path does (see Figure 12.11), which accounts for the more rapid reaction rate. Two additional points can be derived from Figure 12.11:

1. The enthalpy change, ΔH, for the catalyzed reaction is the same as the ΔH for the uncatalyzed reaction.

2. For reversible reactions, the catalyst has the same effect on the reverse reaction that it has on the forward reaction. The energy of activation for the reverse reaction, $E_{a,r}$ is lowered by the catalyst to the same extent that energy of activation for the forward reaction, $E_{a,f}$, is lowered.

A **homogeneous catalyst** is present in the same phase as the reactants. An example of homogeneous catalysis in the gas phase is the effect of chlorine gas on the decomposition of dinitrogen oxide gas. Dinitrogen oxide, N_2O, is relatively unreactive at room temperature, but at temperatures near $600°C$ it decomposes according to the equation

$$2\,N_2O(g) \longrightarrow 2\,N_2(g) + O_2(g)$$

The uncatalyzed reaction is thought to occur by means of a complex mechanism that includes the following steps:

1. Through collisions between N_2O molecules, some N_2O molecules gain enough energy to split apart:

$$N_2O(g) \longrightarrow N_2(g) + O(g)$$

2. The oxygen atoms are very reactive. They readily react with other N_2O molecules:

$$O(g) + N_2O(g) \longrightarrow N_2(g) + O_2(g)$$

The final products of the reaction are N_2 and O_2. The O atom is a reaction intermediate and not a final product. The energy of activation for the uncatalyzed reaction is about 240 kJ/mol.

The reaction is catalyzed by a trace of chlorine gas. The path that has been proposed for the catalyzed reaction consists of the following steps:

1. At the temperature of the decomposition, and particularly in the presence of light, some chlorine molecules dissociate into chlorine atoms:

$$Cl_2(g) \longrightarrow 2\,Cl(g)$$

2. The chlorine atoms readily react with N_2O molecules:

$$N_2O(g) + Cl(g) \longrightarrow N_2(g) + ClO(g)$$

3. The decomposition of the unstable ClO molecules follows:

$$2\,ClO(g) \longrightarrow Cl_2(g) + O_2(g)$$

Notice that the catalyst (Cl_2) is returned to its original state in the last step. The final products of the catalyzed reaction ($2\,N_2$ and O_2) are the same as those of the uncatalyzed reaction. Cl and ClO are not products because they are used in steps that follow the ones in which they are produced. The energy of activation for the reaction catalyzed by chlorine is about 140 kJ/mol, which is considerably lower than E_a for the uncatalyzed reaction (240 kJ/mol).

In heterogeneous catalysis the reactants and catalyst are present in different phases. Reactant molecules are *adsorbed* on the surface of the catalyst in these processes, and the reaction takes place on that surface. Adsorption is a process in which molecules adhere to the surface of a solid. Charcoal, for example, is used in gas masks as an adsorbent for noxious gases. In ordinary physical adsorption, the molecules are held to the surface by London forces.

Heterogeneous catalysis, however, usually takes place through chemical adsorption (or chemisorption) in which the adsorbed molecules are held to the surface by bonds that are similar in strength to those in chemical compounds. When these bonds form, the chemisorbed molecules undergo changes in the arrangement of their electrons. Some bonds of the molecules may be stretched and weakened, and in some cases, even broken. Hydrogen molecules, for example, are believed to be adsorbed as hydrogen atoms on the surface of platinum, palladium, nickel, and other metals. The chemisorbed layer of molecules or atoms, therefore, functions as a reaction intermediate in a surface catalyzed reaction.

The decomposition of N_2O is catalyzed by gold. A proposed mechanism for the gold-catalyzed decomposition is diagrammed in Figure 12.12. The steps are

1. Molecules of $N_2O(g)$ are chemisorbed on the surface of the gold:

$$N_2O(g) \longrightarrow N_2O(\text{on Au})$$

2. The bond between the O atom and the adjacent N atom of a N_2O molecule is weakened when the O atom bonds to the gold. This N—O bond breaks and an N_2 molecule leaves:

$$N_2O(\text{on Au}) \longrightarrow N_2(g) + O(\text{on Au})$$

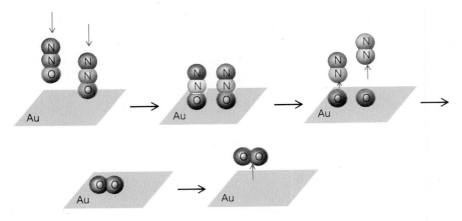

Figure 12.12 Proposed mode of decomposition of N_2O on Au

3. Two O atoms on the surface of the gold combine to form an O_2 molecule which enters the gas phase:

$$O(\text{on Au}) + O(\text{on Au}) \longrightarrow O_2(g)$$

The energy of activation for the gold-catalyzed decomposition is about 120 kJ/mol, which is lower than E_a for either the uncatalyzed decomposition (240 kJ/mol) or the chlorine-catalyzed decomposition (140 kJ/mol).

The second step of the mechanism of the gold-catalyzed decomposition is believed to be the rate-determining step. The rate of this step is proportional to the fraction of the gold surface that holds chemisorbed N_2O molecules. If one-half the surface is covered, step 2 is faster than if only one-quarter of the surface is occupied. This fraction, however, is directly proportional to the pressure of $N_2O(g)$. If the pressure is low, the fraction of surface covered will be low. The rate of the reaction, therefore, is proportional to the concentration of $N_2O(g)$, and the decomposition is first order:

$$\text{rate} = k[N_2O]$$

At high pressures of N_2O, the surface of the gold becomes completely covered; the fraction is equal to 1. Under these conditions, the reaction becomes zero order, that is, the rate is unaffected by changes in the concentration of $N_2O(g)$:

$$\text{rate} = k$$

The gold surface is holding all the N_2O molecules that it can, and the pressure of $N_2O(g)$ is high enough to keep the surface saturated. Small changes in the pressure of $N_2O(g)$ do not cause the chemisorbed N_2O molecules to decompose any more slowly or rapidly.

The electronic structure and arrangement of atoms on the surface of a catalyst determine its activity. Lattice defects and irregularities are thought to be active sites for catalysis. The surface of some catalysts can be changed by the addition of substances called **promoters,** which enhance the catalytic activity. In the synthesis of ammonia,

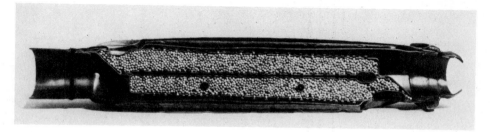

A cross-sectional view of a catalytic converter used in automobiles. Engine exhaust, which enters on the right, is conducted to the top of the converter and forced to pass through dual beds of catalytic beads before exiting at the bottom and to the left. Air is inducted into the chamber between the catalytic beds. The beads contain Pt, Pd, and Rh and are designed to catalyze the oxidation of CO and hydrocarbons to CO_2 and the transformation of the oxides of nitrogen into N_2 and O_2. *General Motors Corporation.*

$$N_2(g) + 3H_2(g) \xrightarrow{\text{Fe}} 2NH_3(g)$$

an iron catalyst is made more effective when traces of potassium or vanadium are added to it.

Catalytic **poisons** are substances that inhibit the activity of catalysts. For example, small amounts of arsenic destroy the power of platinum to catalyze the preparation of sulfur trioxide from sulfur dioxide:

$$2SO_2(g) + O_2(g) \xrightarrow{\text{Pt}} 2SO_3(g)$$

Presumably, platinum arsenide forms on the surface of the platinum and destroys its catalytic activity.

Catalysts are generally highly specific in their activity. In some cases a given substance will catalyze the synthesis of one set of products from certain reagents, whereas another substance will catalyze the synthesis of completely different products from the same reactants. In these cases, both reactions are possible and the products obtained are those that are produced most rapidly. Carbon monoxide and hydrogen can be made to yield a wide variety of products depending upon the catalyst employed and the conditions of the reaction.

If a cobalt or nickel catalyst is used, CO and H_2 produce mixtures of hydrocarbons. One hydrocarbon produced, for example, is methane, CH_4:

$$CO(g) + 3H_2(g) \xrightarrow{\text{Ni}} CH_4(g) + H_2O(g)$$

On the other hand, methanol is the product of the reaction of CO and H_2 when a mixture of zinc and chromic oxides is employed as a catalyst:

$$CO(g) + 2H_2(g) \xrightarrow{\text{ZnO/Cr}_2\text{O}_3} CH_3OH(g)$$

The catalytic converter installed on car mufflers is a recent application of surface catalysis. Carbon monoxide and hydrocarbons from unburned fuel are present in automobile engine exhaust and are serious air pollutants. In the converter, the exhaust gases and additional air are passed over a catalyst that consists of metal oxides. The CO and hydrocarbons are converted into CO_2 and H_2O, which are relatively harmless and are released to the atmosphere. Since the catalyst is poisoned by lead, unleaded gasoline must be used in automobiles equipped with catalytic converters.

Many industrial processes depend upon catalytic procedures, but the natural catalysts known as **enzymes** are even more important to man. These extremely complicated substances catalyze life processes such as digestion, respiration, and cell synthesis. The large number of complex chemical reactions that occur in the body, and are necessary for life, can occur at the relatively low temperature of the body because of the action of enzymes. Thousands of enzymes are known, and each serves a specific function. Research into the structure and action of enzymes may lead to a better understanding of the causes of disease and the mechanism of growth.

Summary

The topics that have been discussed in this chapter are

1. How reaction rates are expressed.

2. Rate equations, which give the quantitative relationship between reaction rate and reactant concentrations.

3. The collision theory, which accounts for the rate of a chemical reaction on the basis of effective collisions between the reacting molecules, and the transition state theory, which describes a step of a chemical reaction on the basis of the attainment of a transition state arrangement of the reacting molecules.

4. The molecularity of the steps of reaction mechanisms and the corresponding rate equations.

5. Reaction mechanisms, which may consist of one step or of several and which describe the ways in which reactions occur on an atomic, molecular, or ionic level.

6. The effect of a temperature change on reaction rate; the Arrhenius equation.

7. Catalysts and how they function.

Key Terms

Some of the more important terms introduced in this chapter are listed below. Definitions for terms not included in this list may be located in the text by use of the index.

Activated complex (Section 12.3) An unstable arrangement of atoms that exists only for a moment in the course of a chemical reaction, also called a **transition state.**

Arrhenius equation (Section 12.6) An equation that describes how the rate constant for a chemical reaction varies with temperature and energy of activation.

Catalyst (Section 12.7) A substance that increases the rate of a chemical reaction without being used up in the reaction.

Chain mechanism (Section 12.5) A multistep mechanism for a reaction in which, after an initiation step, two steps are repeated over and over again. A product of the first of these steps is a reactant in the second, and a product of the second of these steps is a reactant in the first.

Chemical adsorption (Section 12.7) A process through which solid, heterogeneous catalysts work. The adsorbed reactant molecules are held to the surface of the catalyst by bonds that are similar in strength to chemical bonds, and in the process the adsorbed molecules become activated.

Chemical kinetics (Introduction) The study of the rates and mechanisms of chemical reactions.

Collision theory (Section 12.3) A theory that describes reactions in terms of collisions between reacting particles (atoms, molecules, or ions).

Effective collision (Section 12.3) A collision between two particles (atoms, molecules, or ions) that results in a reaction.

Energy of activation (Section 12.3) The difference in energy between the potential energy of the reactants of a reaction and the potential energy of the activated complex.

Enzyme (Section 12.7)   A natural catalyst that is effective in a biochemical process (such as digestion, respiration, and cell synthesis).

Heterogeneous catalyst (Section 12.7)   A catalyst that is present in a different phase from that of the reactants.

Homogeneous catalyst (Section 12.7)   A catalyst that is present in the same phase as the reactants.

Molecularity (Section 12.4)   The number of reacting particles (atoms, molecules, or ions) that participate in a single step of a reaction mechanism. A step may be unimolecular, bimolecular, or termolecular depending upon whether *one*, *two*, or *three* particles react in it.

Order of a chemical reaction (Section 12.2)   The sum of the exponents of the concentration terms in the rate equation for the reaction.

Rate constant (Section 12.2)   The proportionality constant in a rate equation.

Rate-determining step (Section 12.5)   The slowest step in a multistep reaction mechanism; the one that determines how fast the overall reaction proceeds.

Rate equation (Section 12.2)   A mathematical equation that relates the rate of a chemical reaction to the concentrations of reactants.

Reaction intermediate (Section 12.3)   A substance that is produced and used in the course of a chemical reaction and is, therefore, neither a reactant nor a product of the reaction.

Reaction mechanism (Section 12.5)   The detailed description of the way a reaction occurs based on the behavior of atoms, molecules, or ions; may include more than one step.

Reaction rate (Section 12.1)   The rate at which a reaction proceeds, expressed in terms of the increase in the concentration of a product per unit time or the decrease in the concentration of a reactant per unit time; value changes during the course of the reaction.

Transition state theory (Section 12.3)   A theory that assumes that reacting particles must assume a specific arrangement, called a transition state, before they can form the products of the reaction.

Problems*

Reaction Rates, Rate Equations

12.1 The rate equation for the reaction

$$2\ NO(g) + 2\ H_2(g) \longrightarrow N_2(g) + 2\ H_2O(g)$$

is second order in $NO(g)$ and first order in $H_2(g)$. **(a)** Write an equation for the rate of appearance of $N_2(g)$. **(b)** If concentrations are expressed in mol/liter, what units would the rate constant, k, have? **(c)** Write an equation for the rate of disappearance of $NO(g)$. Would k in this equation have the same numerical value as k in the equation of part **(a)**?

12.2 For a reaction in which A and B form C, the following data were obtained from three experiments:

[A] (mol/liter)	[B] (mol/liter)	Rate of formation of C (mol/liter·s)
0.30	0.15	7.0×10^{-4}
0.60	0.30	2.8×10^{-3}
0.30	0.30	1.4×10^{-3}

(a) What is the rate equation for the reaction?
(b) What is the numerical value of the rate constant, k?

12.3 For a reaction in which A and B form C, the following data were obtained from three experiments:

[A] (mol/liter)	[B] (mol/liter)	Rate of formation of C (mol/liter·s)
0.03	0.03	0.3×10^{-4}
0.06	0.06	1.2×10^{-4}
0.06	0.09	2.7×10^{-4}

(a) What is the rate equation for the reaction?
(b) What is the numerical value of the rate constant, k?

12.4 The rate equation for the reaction $A \rightarrow B + C$ is expressed in the form

$$\text{rate of disappearance of } A = k[A]^x$$

The value of the rate constant, k, is 0.100 (units unspecified) and $[A] = 0.050$ mol/liter. What are the units of k and the rate of the reaction (in mol/liter·s) if the reaction is: **(a)** zero order in A, **(b)** first order in A, **(c)** second order in A?

12.5 The rate equation for the reaction $A \rightarrow B + C$ is expressed in terms of the concentration of A only. The rate of disappearance of A is 0.0080 mol/liter·s when $[A] = 0.20$ mol/liter. Calculate the value of k if the reaction is: **(a)** zero order in A, **(b)** first order in A, **(c)** second order in A.

---

* The more difficult problems are marked with asterisks. The appendix contains answers to color-keyed problems.

12.6 For the reaction

$$2\,NO(g) + Cl_2(g) \longrightarrow 2\,ONCl(g)$$

the rate equation is second order in $NO(g)$, first order in $Cl_2(g)$, and third order overall. Compare the initial rate of reaction of a mixture of 0.02 mol $NO(g)$ and 0.02 mol $Cl_2(g)$ in a 1 liter container with **(a)** the rate of the reaction when half the $NO(g)$ has been consumed, **(b)** the rate of the reaction when half the $Cl_2(g)$ has been consumed, **(c)** the rate of the reaction when two-thirds of $NO(g)$ has been consumed, **(d)** the initial rate of a mixture of 0.04 mol $NO(g)$ and 0.02 mol $Cl_2(g)$ in a 1 liter container, **(e)** the initial rate of a mixture of 0.02 mol $NO(g)$ and 0.02 mol $Cl_2(g)$ in a 0.5 liter container.

12.7 The single-step reaction

$$NO_2Cl(g) + NO(g) \rightleftharpoons NO_2(g) + ONCl(g)$$

is reversible; $E_{a,f}$ is 28.9 kJ and $E_{a,r}$ is 41.8 kJ. Draw a potential energy diagram for the reaction. Indicate $E_{a,f}$, $E_{a,r}$, and ΔH on the diagram.

12.8 A common and serious mistake is to assume that the rate equation for reaction can be derived from the balanced chemical equation for the reaction by using the coefficients of the chemical equation as exponents in the rate expression. Why cannot rate equations be derived in this way?

12.9 Why are some collisions between molecules of reactants not effective?

12.10 The synthesis of perbromates has only recently been accomplished. The best preparation involves the oxidation of bromates in alkaline solution by fluorine. What reason can you give to account for the difficulty encountered in the synthesis of perbromates? How well would you expect perbromates to function as oxidizing agents?

Reaction Mechanisms, Catalysis

12.11 Define the following: **(a)** activated complex, **(b)** energy of activation, **(c)** reaction order, **(d)** catalyst, **(e)** chemisorption, **(f)** rate-determining step, **(g)** reaction intermediate, **(h)** zero-order reaction.

12.12 The reaction

$$2\,ICl(g) + H_2(g) \longrightarrow I_2(g) + 2\,HCl(g)$$

at temperatures above 200°C is first order in H_2 and first order in ICl. Suggest a mechanism of two steps with the first step rate-determining to account for the rate equation.

*12.13 For the reaction

$$N_2O_5(g) + NO(g) \longrightarrow 3\,NO_2(g)$$

The rate equation is

$$\text{rate of disappearance of } N_2O_5 = \frac{k_1 k_3 [N_2O_5][NO]}{k_2[NO_2] + k_3[NO]}$$

The suggested mechanism is

$$N_2O_5 \xrightarrow{k_1} NO_2 + NO_3$$
$$NO_2 + NO_3 \xrightarrow{k_2} N_2O_5$$
$$NO + NO_3 \xrightarrow{k_3} 2\,NO_2$$

Assume that $[NO_3]$ becomes constant after the reaction has been going on for a while (that is, NO_3 is used as fast as it is produced). Show that the mechanism leads to the observed rate equation.

*12.14 According to the collision theory, a first-order decomposition (A → products) proceeds by the following steps:

$$A + A \xrightarrow{k_1} A^* + A$$
$$A^* + A \xrightarrow{k_2} A + A$$
$$A^* \xrightarrow{k_3} \text{products}$$

In step 1, two A molecules collide, energy is transferred, and one molecule, marked A*, attains a high-energy state. A subsequent collision of A* can cause the process to reverse (step 2). Some A* molecules, however, decompose to form the products in the rate-determining step. Derive a rate equation from this mechanism by assuming that $[A^*]$ becomes constant after the reaction has been going on for a while (that is, A* is used as fast as it is produced). Remember that since step 3 is slow, k_3 is much smaller than k_2.

*12.15 The rate equation for the reaction

$$2\,O_3 \longrightarrow 3\,O_2$$

has been determined experimentally as

$$\text{rate of disappearance of } O_3 = k\,\frac{[O_3]^2}{[O_2]}$$

The following mechanism has been postulated for the decomposition of ozone.

$$O_3 \xrightarrow{k_1} O_2 + O$$
$$O_2 + O \xrightarrow{k_2} O_3$$
$$O + O_3 \xrightarrow{k_3} 2\,O_2$$

The third step of the mechanism is the rate-determining step. Assume that the concentration of O becomes constant after the reaction has been going on for a while (that is, O is used as fast as it is produced) and find an expression for [O]. Remember that k_3 is much smaller than either k_1 or k_2. Show how the suggested mechanism leads to the observed rate expression.

12.16 For the mechanism outlined in problem 12.15, the proposed activation energies for the three steps are 100 kJ, 4 kJ, and 24 kJ. The value of ΔH for the overall reaction is -285 kJ. Draw a potential energy diagram for the reaction.

*12.17 The rate expression for the reaction

$$2\,NO(g) + O_2(g) \longrightarrow 2\,NO_2(g)$$

is second order in $NO(g)$ and first order in $O_2(g)$. The following mechanism has been suggested:

$$NO(g) + O_2(g) \xrightarrow{k_1} NO_3(g)$$
$$NO_3(g) \xrightarrow{k_2} NO(g) + O_2(g)$$
$$NO_3(g) + NO(g) \xrightarrow{k_3} 2\,NO_2(g)$$

The third step of the mechanism is the rate-determining step. Assume that the concentration of NO_3 becomes constant after the reaction has been going on for a while (that is, NO_3 is used as fast as it is produced). Remember that k_3 is much smaller than either k_1 or k_2. Show that this mechanism leads to the observed rate equation.

12.18 The reaction

$$CH_4(g) + Cl_2(g) \longrightarrow CH_3Cl(g) + HCl(g)$$

proceeds by a chain mechanism. The chain propagators are Cl atoms and CH_3 radicals, and it is believed that free H atoms are not involved. Write a series of equations showing the mechanism and identify the chain-initiating, chain-sustaining, and chain-terminating steps.

12.19 Use potential energy diagrams to explain how a catalyst functions. Does a catalyst affect the value of ΔH for the reaction? Can a catalyst slow a reaction down? Can a catalyst be found that affects only the forward reaction of a reversible reaction? Explain each of your answers.

12.20 What is the difference between a heterogeneous and a homogeneous catalyst? Describe the way in which each type of catalyst functions.

Rate Equations and Temperature

12.21 The reaction: $2\,NO(g) \rightarrow N_2(g) + O_2(g)$ is second order in NO. The rate constant is 0.143 liter/mol·s at 1400 K and 0.659 liter/mol·s at 1500 K. What is the energy of activation for the reaction?

12.22 The reaction: $HI(g) + CH_3I(g) \rightarrow CH_4(g) + I_2(g)$ is first order in each of the reactants and second order overall. The rate constant is 1.7×10^{-5} liter/mol·s at 430 K and 9.6×10^{-5} liter/mol·s at 450 K. What is the energy of activation for the reaction?

12.23 The reaction: $C_2H_5Cl(g) \rightarrow C_2H_4(g) + HCl(g)$ is first order in C_2H_5Cl. The rate constant is 3.5×10^{-8}/s at 600 K and 1.6×10^{-6}/s at 650 K. What is the energy of activation for the reaction?

12.24 For the reaction: $NO_2Cl(g) + NO(g) \rightarrow NO_2(g) + ONCl(g)$, A is 8.3×10^8 and E_a is 28.9 kJ/mol. The rate equation is first order in NO_2Cl and first order in NO. What is the rate constant, k, at 500 K?

12.25 For the reaction: $NO(g) + N_2O(g) \rightarrow NO_2(g) + N_2(g)$, A is 2.5×10^{11} and E_a is 209 kJ/mol. The rate equation is first order in NO and first order in N_2O. What is the rate constant, k, at 1000 K?

12.26 The reaction: $C_2H_4(g) + H_2(g) \rightarrow C_2H_6(g)$ is first order in C_2H_4, first order in H_2, and second order overall. The energy of activation for the reaction is 181 kJ/mol and k is 1.3×10^{-3} liter/mol·s at 700 K. What is the value of k for the reaction conducted at 730 K?

12.27 The reaction: $NO(g) + N_2O(g) \rightarrow NO_2(g) + N_2(g)$ is first order in each of the reactants and second order overall. The energy of activation of the reaction is 209 kJ/mol, and k is 0.77 liter/mol·s at 950 K. What is the value of k for a reaction conducted at 1000 K?

12.28 The reaction: $C_2H_5Br(g) \rightarrow C_2H_4(g) + HBr(g)$ is first order in C_2H_5Br. The rate constant is 2.0×10^{-5}/s at 650 K, and the energy of activation is 226 kJ/mol. At what temperature is the rate constant 6.0×10^{-5}/s?

12.29 What is the energy of activation of a reaction that increases ten-fold in rate when the temperature is increased from 300 K to 310 K?

12.30 At 400 K a certain reaction is 50.0% complete in 1.50 min. At 430 K the same reaction is 50.0% complete in 0.50 min. Calculate the activation energy for the reaction.

CHAPTER 13 CHEMICAL EQUILIBRIUM

Under suitable conditions, nitrogen and hydrogen react to form ammonia:

$$N_2(g) + 3H_2(g) \longrightarrow 2NH_3(g)$$

On the other hand, ammonia decomposes at high temperatures to yield nitrogen and hydrogen:

$$2NH_3(g) \longrightarrow N_2(g) + 3H_2(g)$$

The reaction is reversible, and the equation for the reaction can be written

$$N_2(g) + 3H_2(g) \rightleftharpoons 2NH_3(g)$$

The double arrow ($\rightleftharpoons$) indicates that the equation can be read in either direction.

All reversible processes tend to attain a state of equilibrium. For a reversible chemical reaction, an equilibrium state is attained when the rate at which a chemical reaction is proceeding equals the rate at which the reverse reaction is proceeding. Equilibrium systems that involve reversible chemical reactions are the topic of this chapter.

13.1 Reversible Reactions and Chemical Equilibrium

Consider a hypothetical reversible reaction

$$A_2(g) + B_2(g) \rightleftharpoons 2AB(g)$$

This equation may be *read either forward or backward*. If A_2 and B_2 are mixed, they will react to produce AB. A sample of pure AB, on the other hand, will decompose to form A_2 and B_2.

Suppose that we place a mixture of A_2 and B_2 in a container. They will react to produce AB, and their concentrations will gradually decrease as the forward reaction occurs (see Figure 13.1). Since the concentrations of A_2 and B_2 decrease, the rate of the forward reaction decreases.

At the start of the experiment, the reverse reaction cannot occur since there is no AB present. As soon as the forward reaction produces some AB, however, the reverse reaction begins. The reverse reaction starts slowly (since the concentration of AB is low) and gradually picks up speed.

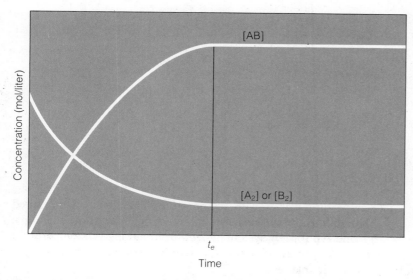

Figure 13.1 Curves showing changes in concentrations of materials with time for the reaction $A_2 + B_2 \rightleftharpoons 2\,AB$. Equilibrium is attained at time t_e.

As time passes, the rate of the forward reaction decreases and the rate of the reverse reaction increases until the two rates are equal. At this point, **chemical equilibrium** is established. An equilibrium condition is one in which the rates of two opposing tendencies are equal.

At equilibrium, the concentrations of all the substances are constant. The concentration of AB is constant since AB is produced by the forward reaction at the same rate that it is used by the reverse reaction. In like manner, A_2 and B_2 are being made (by the reverse reaction) at the same rate that they are being used (by the forward reaction). It is important to note that the concentrations are constant because the rates of the opposing reactions are equal and *not* because all activity has stopped. Data for the experiment are plotted in Figure 13.1. Equilibrium is attained at time t_e.

If we assume that the forward and reverse reactions occur by simple one-step mechanisms, the rate of the forward reaction is

$$\text{rate}_f = k_f[A_2][B_2]$$

and the rate of the reverse reaction is

$$\text{rate}_r = k_r[AB]^2$$

At equilibrium, these two rates are equal, and therefore

$$\text{rate}_f = \text{rate}_r$$
$$k_f[A_2][B_2] = k_r[AB]^2$$

This equation can be rearranged to

$$\frac{k_f}{k_r} = \frac{[AB]^2}{[A_2][B_2]}$$

The rate constant of the forward reaction, k_f, divided by the rate constant of the reverse reaction, k_r, is equal to a third constant, which is called the **equilibrium constant**, K:

$$\frac{k_f}{k_r} = K \tag{13.1}$$

Therefore,

$$K = \frac{[AB]^2}{[A_2][B_2]}$$

The numerical value of K varies with temperature. There are an unlimited number of possible equilibrium systems for this reaction. The concentrations, however, of A_2, B_2, and AB for *any* system in equilibrium at a given temperature will, when expressed in the preceding manner, equal the same value of K.

In general, for any reversible reaction

$$wW + xX \rightleftharpoons yY + zZ \tag{13.2}$$

at equilibrium,

$$K = \frac{[Y]^y[Z]^z}{[W]^w[X]^x} \tag{13.3}$$

By convention, the concentration terms for the materials on the *right* of the chemical equation are written in the *numerator* of the expression for the equilibrium constant.

If the equation is written in the reverse form,

$$yY + zZ \rightleftharpoons wW + xX$$

the expression for the equilibrium constant (which we will indicate as K') is

$$K' = \frac{[W]^w[X]^x}{[Y]^y[Z]^z}$$

Notice that K' is the reciprocal of K:

$$K' = \frac{1}{K}$$

In our derivation, we assumed that the forward and reverse reactions occurred by mechanisms each consisting of a single step. Does the law of chemical equilibrium hold for reactions that occur by mechanisms of more than one step? The answer to the question is that it does.

Consider the reaction

$$2\,NO_2Cl(g) \rightleftharpoons 2\,NO_2(g) + Cl_2(g)$$

for which the expression for the equilibrium constant is

$$K = \frac{[NO_2]^2[Cl_2]}{[NO_2Cl]^2}$$

The reaction is believed to occur by means of a mechanism consisting of two steps:

1. $\quad NO_2Cl \underset{k_1'}{\overset{k_1}{\rightleftharpoons}} NO_2 + Cl$

2. $NO_2Cl + Cl \underset{k_2'}{\overset{k_2}{\rightleftharpoons}} NO_2 + Cl_2$

Symbols for the rate constants appear above and below the arrows in these equations. The symbol k is used for the rate constant of the forward reaction, and k' is used for the reverse reaction. The subscripts of these symbols designate the steps.

Since the overall reaction is reversible, each step of the mechanism must also be reversible. When equilibrium is established for the overall reaction, each step of the mechanism must be in equilibrium. Therefore,

$$K_1 = \frac{k_1}{k_1'} = \frac{[NO_2][Cl]}{[NO_2Cl]}$$

and

$$K_2 = \frac{k_2}{k_2'} = \frac{[NO_2][Cl_2]}{[NO_2Cl][Cl]}$$

The product of these expressions is

$$K_1K_2 = \frac{k_1k_2}{k_1'k_2'} = \frac{[NO_2][Cl]}{[NO_2Cl]} \frac{[NO_2][Cl_2]}{[NO_2Cl][Cl]} = \frac{[NO_2]^2[Cl_2]}{[NO_2Cl]^2}$$

which is the same as the expression for the equilibrium constant that we derived directly from the equation for the overall change. In this case, the equilibrium constant for the overall change is the product of the equilibrium constants of each of the steps:

$$K = K_1K_2$$

13.2 Equilibrium Constants $\quad$ $L5P1$

For the reaction

$$H_2(g) + I_2(g) \rightleftharpoons 2\,HI(g)$$

the equilibrium constant at 425°C is

$$K = \frac{[HI]^2}{[H_2][I_2]} = 54.8$$

The numerical value of K must be determined experimentally. If the concentrations of the materials (in mol/liter) present in any equilibrium mixture at 425°C are substituted into the expression for the equilibrium constant, the result will equal 54.8. If any value other than 54.8 is obtained, the mixture is not in equilibrium.

The equilibrium condition may be approached from either direction. That is, an equilibrium mixture can be obtained by mixing hydrogen and iodine, by allowing pure hydrogen iodide to dissociate, or by mixing all three materials.

The magnitude of the value of the equilibrium constant gives an indication of the position of equilibrium. Recall that the concentration terms of substances on the right of the equation are written in the numerator of the expression for the equilibrium constant. For the reaction

$$CO(g) + Cl_2(g) \rightleftharpoons COCl_2(g)$$

at 100°C,

$$K = \frac{[COCl_2]}{[CO][Cl_2]} = 4.57 \times 10^9 \text{ liter/mol}$$

From this relatively large value of K, we conclude that equilibrium concentrations of CO and Cl_2 are small and that the synthesis of $COCl_2$ is virtually complete. In other words, the reaction to the right is fairly complete at equilibrium.

For the reaction

$$N_2(g) + O_2(g) \rightleftharpoons 2\,NO(g)$$

at 2000°C,

$$K = \frac{[NO]^2}{[N_2][O_2]} = 4.08 \times 10^{-4}$$

We conclude from this small value of K that NO is largely dissociated into N_2 and O_2 at equilibrium. The reaction to the left is fairly complete.

Equilibria between substances in two or more phases are called **heterogeneous equilibria.** The concentration of a pure solid or a pure liquid is constant when the temperature and pressure are constant. For any heterogeneous equilibrium, therefore, the values of the concentrations of solids or liquids involved are included in the value of K, and concentration terms for these substances do not appear in the expression for the equilibrium constant.

For example, for the reaction

$$CaCO_3(s) \rightleftharpoons CaO(s) + CO_2(g)$$

the values for the concentrations of CaO and $CaCO_3$ are included in the value of K, and the expression for the equilibrium constant is

$$K = [CO_2]$$

Hence, at any fixed temperature the equilibrium concentration of CO_2 over a mixture of the solids is a definite value. The equilibrium constant for the reaction

$$3\,Fe(s) + 4\,H_2O(g) \rightleftharpoons Fe_3O_4(s) + 4\,H_2(g)$$

is expressed in the following terms:

$$K = \frac{[H_2]^4}{[H_2O]^4}$$

Some facts dealing with expressions for equilibrium constants may be summarized as follows:

1. The concentration terms for substances that appear on the *right side* of the chemical equation are written in the *numerator* of the expression for K. The concentration terms for substances that appear on the *left side* are written in the *denominator*.

2. No concentration terms are included for pure solids or pure liquids. The value of K includes these terms.

3. The value of K for a given equilibrium is constant if the equilibrium temperature does not change. At a different temperature, the value of K is different.

4. The magnitude of K for a given equilibrium indicates the position of equilibrium. A *large* value of K indicates that the reaction to the *right* is fairly complete. A *small* value of K indicates that the reaction to the *left* is fairly complete. A value of K that is neither very large nor very small indicates an intermediate situation.*

The following examples illustrate how some simple problems involving equilibrium constants may be solved.

Example 13.1

For the reaction

$$N_2O_4(g) \rightleftharpoons 2\,NO_2(g)$$

the concentrations of the substances present in an equilibrium mixture at 25°C are

$$[N_2O_4] = 4.27 \times 10^{-2} \text{ mol/liter}$$

$$[NO_2] = 1.41 \times 10^{-2} \text{ mol/liter}$$

What is the value of K for this temperature?

---

* To be strictly accurate, equations for equilibrium constants should be written in terms of activities rather than concentrations or pressures (see Section 17.7). At low concentrations and pressures up to a few atmospheres, however, concentrations may be used with reasonable accuracy.

Solution

$$K = \frac{[NO_2]^2}{[N_2O_4]}$$

$$= \frac{(1.41 \times 10^{-2} \text{ mol/liter})^2}{(4.27 \times 10^{-2} \text{ mol/liter})}$$

$$= 4.66 \times 10^{-3} \text{ mol/liter}$$

Example 13.2

At 500 K, 1.00 mol of ONCl(g) is introduced into a 1 liter container. At equilibrium, the ONCl(g) is 9.0% dissociated:

$$2\,ONCl(g) \rightleftharpoons 2\,NO(g) + Cl_2(g)$$

Calculate the value of K for the equilibrium at 500 K.

Solution

By considering exactly 1 liter, we simplify the problem since the number of moles of a gas is the same as the concentration of the gas (in mol/liter).

Since the ONCl is 9.0% dissociated,

number of moles dissociated = 0.090(1.00 mol) = 0.090 mol ONCl

We must subtract this quantity from the number of moles of ONCl present initially. The concentration of ONCl at equilibrium, therefore, is

[ONCl] = 1.00 mol/liter − 0.090 mol/liter = 0.91 mol/liter

The ONCl that dissociates produces NO and Cl_2. We can derive the amounts of these substances produced from the coefficients of the chemical equation:

$$2\,ONCl \rightleftharpoons 2\,NO + Cl_2$$
$$ 0.090 \text{ mol} \quad 0.045 \text{ mol}$$

Since 2 mol of ONCl produces 2 mol of NO, 0.090 mol of ONCl will produce 0.090 mol of NO. The equation shows that 2 mol of ONCl produces only 1 mol of Cl_2, therefore, 0.090 mol of ONCl will produce 0.045 mol of Cl_2. The equilibrium concentrations are

[ONCl] = 0.91 mol/liter

[NO] = 0.090 mol/liter

$[Cl_2]$ = 0.045 mol/liter

Therefore,

$$K = \frac{[NO]^2[Cl_2]}{[ONCl]^2}$$

$$= \frac{(0.090 \text{ mol/liter})^2(0.045 \text{ mol/liter})}{(0.91 \text{ mol/liter})^2}$$

$$= 4.4 \times 10^{-4} \text{ mol/liter}$$

Example 13.3

K for the HI equilibrium at 425°C is 54.8:

$$H_2(g) + I_2(g) \rightleftharpoons 2\,HI(g)$$

A quantity of HI(g) is placed in a 1.00 liter container and allowed to come to equilibrium at 425°C. What are the concentrations of $H_2(g)$ and $I_2(g)$ in equilibrium with 0.50 mol/liter of HI(g)?

Solution

The concentrations of $H_2(g)$ and $I_2(g)$ must be equal since they are produced in equal amounts by the decomposition of HI(g). Therefore, let

$$[H_2] = [I_2] = x$$

We are told that the equilibrium concentration of HI is

$$[HI] = 0.50 \text{ mol/liter}$$

We substitute these values into the equation for the equilibrium constant and solve for x:

$$K = \frac{[HI]^2}{[H_2][I_2]} = 54.8$$

$$\frac{(0.50 \text{ mol/liter})^2}{x^2} = 54.8$$

$$54.8x^2 = 0.25 \text{ mol}^2/\text{liter}^2$$

$$x^2 = 0.00456 \text{ mol}^2/\text{liter}^2$$

$$x = 0.068 \text{ mol/liter}$$

The equilibrium concentrations are

$$[HI] = 0.50 \text{ mol/liter}$$

$$[H_2] = [I_2] = 0.068 \text{ mol/liter}$$

Example 13.4

For the reaction

$$H_2(g) + CO_2(g) \rightleftharpoons H_2O(g) + CO(g)$$

K is 0.771 at 750°C. If 0.0100 mol of H_2 and 0.0100 mol of CO_2 are mixed in a 1 liter container at 750°C, what are the concentrations of all substances present at equilibrium?

Solution

If x mol of H_2 reacts with x mol of CO_2 out of the total amounts supplied, x mol of H_2O and x mol of CO will be produced. Since the container has a volume of 1 liter, at equilibrium the concentrations are (in mol/liter)

$$H_2(g) \quad + \quad CO_2(g) \quad \rightleftharpoons H_2O(g) + CO(g)$$
$$(0.0100 - x) \quad (0.0100 - x) \qquad x \qquad x$$

$$K = \frac{[H_2O][CO]}{[H_2][CO_2]} = 0.771$$

$$= \frac{x^2}{(0.0100 - x)(0.0100 - x)} = 0.771$$

If we find the square root of both sides of this equation, we get

$$\frac{x}{(0.0100 - x)} = 0.878$$

$$x = 0.00468$$

Therefore, at equilibrium

$$[H_2] = [CO_2] = (0.0100 - x)$$
$$= 0.0053 \text{ mol/liter}$$
$$= 5.3 \times 10^{-3} \text{ mol/liter}$$

$$[H_2O] = [CO] = x = 0.00468 \text{ mol/liter}$$
$$= 4.68 \times 10^{-3} \text{ mol/liter}$$

13.3 Equilibrium Constants Expressed in Pressures

The partial pressure of a gas is a measure of its concentration. Equilibrium constants for reactions involving gases, therefore, may be written in terms of the partial pressures of the reacting gases. An equilibrium constant of this type is given the designation K_p.

For the calcium carbonate equilibrium

$$CaCO_3(s) \rightleftharpoons CaO(s) + CO_2(g)$$

the equilibrium constant in terms of partial pressures is

$$K_p = p_{CO_2}$$

For the equilibrium

$$N_2(g) + 3\,H_2(g) \rightleftharpoons 2\,NH_3(g)$$

The K_p is

$$K_p = \frac{(p_{NH_3})^2}{(p_{N_2})(p_{H_2})^3}$$

There is a simple relation between the K_p for a reaction and the equilibrium constant derived from concentrations. Consider the reaction

$$wW + xX \rightleftharpoons yY + zZ \tag{13.4}$$

If *all* these materials are *gases*,

$$K_p = \frac{(p_Y)^y(p_Z)^z}{(p_W)^w(p_X)^x} \tag{13.5}$$

Assume that each of the gases follows the ideal gas law

$$PV = nRT$$

Then the partial pressure of any gas, p, is

$$p = \frac{n}{V}RT$$

The concentration of a gas in mol/liter is equal to n/V. Therefore, for gas W

$$p_W = [W]RT \tag{13.6}$$

$$(p_W)^w = [W]^w(RT)^w \tag{13.7}$$

If we substitute expressions such as Equation 13.7 for the partial pressure terms in the expression for K_p (Equation 13.5), we get

$$K_p = \frac{[Y]^y(RT)^y[Z]^z(RT)^z}{[W]^w(RT)^w[X]^x(RT)^x}$$

$$= \frac{[Y]^y[Z]^z}{[W]^w[X]^x}(RT)^{+y+z-w-x} \tag{13.8}$$

The fractional term in the last equation is equal to K:

$$K_p = K(RT)^{+y+z-w-x} \tag{13.9}$$

If we read the chemical equation for the reaction

$$wW + xX \rightleftharpoons yY + zZ \tag{13.4}$$

in molar quantities,

$y + z$ = number of moles of gases on the right

$w + x$ = number of moles of gases on the left

We let Δn equal the change in the number of moles of gases when the equation is read from left to right:

$$\Delta n = (y + z) - (w + x) = +y + z - w - x \tag{13.10}$$

Therefore,

$$K_p = K(RT)^{\Delta n} \tag{13.11}$$

Partial pressures are expressed in atmospheres, concentrations are expressed in moles per liter, R is 0.08206 liter atm/(K · mol), T is the absolute temperature in K.

For the reaction

$$PCl_5(g) \rightleftharpoons PCl_3(g) + Cl_2(g)$$

Δn is $+1$. Therefore,

$$K_p = K(RT)^{+1}$$

For the reaction

$$CO(g) + Cl_2(g) \rightleftharpoons COCl_2(g)$$

$\Delta n = -1$. Therefore,

$$K_p = K(RT)^{-1} \qquad \text{or} \qquad K_p = \frac{K}{(RT)}$$

For the reaction

$$H_2(g) + I_2(g) \rightleftharpoons 2\,HI(g)$$

$\Delta n = 0$. Therefore,

$$K_p = K(RT)^0 \qquad \text{or} \qquad K_p = K$$

Example 13.5

At 1100 K, the equilibrium constant for the reaction

$$2\,SO_3(g) \rightleftharpoons 2\,SO_2(g) + O_2(g)$$

is 0.0271 mol/liter. What is K_p at this temperature?

Solution

Two moles of $SO_3(g)$ produce a total of 3 mol of gases. Therefore,

$$\Delta n = +1$$

$$\begin{aligned}
K_p &= K(RT)^{+1} \\
&= (0.0271 \text{ mol/liter}) \{[0.0821 \text{ liter} \cdot \text{atm}/(\text{K} \cdot \text{mol})](1100 \text{ K})\} \\
&= 2.45 \text{ atm}
\end{aligned}$$

Example 13.6

What is K for the reaction

$$N_2(g) + 3 H_2(g) \rightleftharpoons 2 NH_3(g)$$

at 500°C if K_p is 1.50×10^{-5}/atm^2 at this temperature?

Solution

There are 4 mol of gases indicated on the left of the chemical equation and 2 mol of gases indicated on the right. Therefore,

$$\Delta n = -2$$

The temperature is 500°C, which is

$$T = 773 \text{ K}$$

Therefore,

$$K_p = K(RT)^{\Delta n}$$

$$(1.50 \times 10^{-5}/\text{atm}^2) = K\{[0.0821 \text{ liter} \cdot \text{atm}/(\text{K} \cdot \text{mol})](773 \text{ K})\}^{-2}$$

$$(1.50 \times 10^{-5}/\text{atm}^2) = \frac{K}{(63.5 \text{ liter} \cdot \text{atm/mol})^2}$$

$$K = (1.50 \times 10^{-5}/\text{atm}^2)(4.03 \times 10^3 \text{ liter}^2 \cdot \text{atm}^2/\text{mol}^2)$$
$$= 6.04 \times 10^{-2} \text{ liter}^2/\text{mol}^2$$

Example 13.7

For the reaction

$$C(s) + CO_2(g) \rightleftharpoons 2 CO(g)$$

K_p is 167.5 atm at 1000°C. What is the partial pressure of $CO(g)$ in an equilibrium system in which the partial pressure of $CO_2(g)$ is 0.100 atm?

Solution

$$K_p = \frac{(p_{CO})^2}{p_{CO_2}} = 167.5 \text{ atm}$$

$$\frac{(p_{CO})^2}{0.100 \text{ atm}} = 167.5 \text{ atm}$$

$$(p_{CO})^2 = 16.8 \text{ atm}^2$$

$$p_{CO} = 4.10 \text{ atm}$$

Example 13.8

K_p for the equilibrium

$$FeO(s) + CO(g) \rightleftharpoons Fe(s) + CO_2(g)$$

at 1000°C is 0.403. If CO(g), at a pressure of 1.000 atm, and excess FeO(s) are placed in a container at 1000°C, what are the pressures of CO(g) and CO_2(g) when equilibrium is attained?

Solution

Let x equal the partial pressure of CO_2 when equilibrium is attained. Since 1 mol of CO_2 is produced for every 1 mol of CO used, the partial pressure of CO_2 is equal to the decrease in the pressure of CO:

$$FeO(s) \;+\; CO(g) \rightleftharpoons Fe(s) + CO_2(g)$$
$$(1.000 - x)\,\text{atm} \qquad\qquad x\,\text{atm}$$

$$K_p = \frac{p_{CO_2}}{p_{CO}} = 0.403$$

$$\frac{x\,\text{atm}}{(1.000 - x)\,\text{atm}} = 0.403$$

$$x = p_{CO_2} = 0.287\,\text{atm}$$

$$1.000 - x = p_{CO} = 0.713\,\text{atm}$$

13.4 Le Chatelier's Principle

What happens to an equilibrium system if an experimental condition (such as temperature or pressure) is changed? The effects of such changes were summarized in 1884 by Henri Le Chatelier. Le Chatelier's principle states that a system in equilibrium reacts to a stress in a way that counteracts the stress and establishes a new equilibrium state. This important generalization is very simple to apply:

1. **Concentration changes.** If the concentration of a substance is *increased*, the equilibrium will shift in a way that will *decrease* the concentration of the substance that was added. Suppose that we have a system in equilibrium

$$H_2(g) + I_2(g) \rightleftharpoons 2\,HI(g)$$

and we *increase* the concentration of H_2 by adding more H_2 to the system. The equilibrium is upset, and the system will react in a way that will *decrease* the concentration of H_2. It can do that by using up some of the H_2 (and some I_2 as well) to form more HI. When equilibrium is established again, the concentration of HI will be higher than it was initially. The position of equilibrium is said to have shifted to the right.

If the concentration of HI is increased by adding HI to the system, the position of equilibrium will shift to the left. In this way some HI will be used up. When equilibrium is established again, the concentrations of H_2 and I_2 will be higher than they were initially.

Removal of one of the substances from an equilibrium system will also cause the position of equilibrium to shift. If, for example, HI could be removed, the

Henri Le Chatelier, 1850–1936.
Smithsonian Institution.

position of equilibrium would shift to the right. Additional HI would be produced and the concentrations of H_2 and I_2 would decrease.

By the continuous removal of a product it is possible to drive some reversible reactions "to completion," Complete conversion of $CaCO_3(s)$ into $CaO(s)$:

$$CaCO_3(s) \rightleftharpoons CaO(s) + CO_2(g)$$

can be accomplished by removing the CO_2 gas as fast as it is produced.

2. Pressure changes. Le Chatelier's principle may also be used to make qualitative predictions of the effect of pressure changes on systems in equilibrium. Consider the effect of a pressure increase on an equilibrium mixture of SO_2, O_2, and SO_3:

$$2\,SO_2(g) + O_2(g) \rightleftharpoons 2\,SO_3(g)$$

In the forward reaction two gas molecules ($2\,SO_3$) are produced by the disappearance of three gas molecules ($2\,SO_2 + O_2$). Two gas molecules do not exert as high a pressure as three gas molecules. When the pressure on an equilibrium mixture is increased (or the volume of the system decreased), the position of equilibrium shifts to the right. In this way the system counteracts the change. Alternatively, decreasing the pressure (or increasing the volume) causes the position of this equilibrium to shift to the left.

For reactions in which $\Delta n = 0$, pressure changes have no effect on the position of equilibria. Equilibria involving the systems

$$H_2(g) + I_2(g) \rightleftharpoons 2\,HI(g)$$

$$N_2(g) + O_2(g) \rightleftharpoons 2\,NO(g)$$

$$H_2(g) + CO_2(g) \rightleftharpoons H_2O(g) + CO(g)$$

are not influenced by changing the pressure since there is no difference in the total volume in either the forward or reverse direction for any one of these reactions.

For a system that involves only liquids and solids, the effect of pressure on the position of equilibrium is slight and may usually be ignored for ordinary changes in pressure. Large pressure changes, however, can significantly alter such equilibria; and at times, even slight changes in such equilibria are of interest. For example, the position of equilibrium

$$H_2O(s) \rightleftharpoons H_2O(l)$$

is forced to the right by an increase in pressure because a given quantity of water occupies a smaller volume in the liquid state than in the solid state (its density is higher in the liquid state).

Pressure changes affect equilibria involving gases to a much greater degree. For example, a high pressure would favor the production of a high yield of ammonia from the equilibrium

$$N_2(g) + 3\,H_2(g) \rightleftharpoons 2\,NH_3(g)$$

Hence, Le Chatelier's principle is of practical importance as an aid in determining favorable reaction conditions for the production of a desired substance.

For heterogeneous equilibria the effect of pressure is predicted by counting the number of moles of *gas* indicated on each side of the equation. For example, the position of equilibrium

$$3 \, Fe(s) + 4 \, H_2O(g) \rightleftharpoons Fe_3O_4(s) + 4 \, H_2(g)$$

is virtually unaffected by pressure because there are four moles of gas indicated on each side of the equation.

3. Temperature changes. In order to predict the effect of a temperature change on a system in equilibrium, the nature of the heat effect that accompanies the reaction must be known. At 25°C, the thermochemical equation for the synthesis of ammonia is

$$N_2(g) + 3 \, H_2(g) \rightleftharpoons 2 \, NH_3(g) \qquad \Delta H = -92.4 \text{ kJ}$$

Since ΔH is negative, the reaction to the right evolves heat. We can write the equation to indicate heat as a product:

$$N_2(g) + 3 \, H_2(g) \rightleftharpoons 2 \, NH_3(g) + 92.4 \text{ kJ}$$

The forward reaction is *exothermic*, and the reverse reaction is *endothermic*. In other words, the forward reaction produces heat, and the reverse reaction uses heat. If heat is added (the temperature of the system is raised), the position of equilibrium will shift to the left—the direction in which heat is absorbed. If the mixture is cooled, the position of equilibrium will shift to the right—the direction in which heat is evolved. The highest yields of NH_3 will be obtained at the lowest temperatures. Unfortunately, if the temperature is too low, the reaction will be extremely slow. High pressures and temperatures around 500°C are employed in the commercial process.

Consider the reaction

$$CO_2(g) + H_2(g) \rightleftharpoons CO(g) + H_2O(g) \qquad \Delta H = +41.1 \text{ kJ}$$

Since ΔH is positive, the forward reaction is endothermic. We write the equation

$$41.1 \text{ kJ} + CO_2(g) + H_2(g) \rightleftharpoons CO(g) + H_2O(g)$$

Increasing the temperature *always* favors the endothermic change, and decreasing the temperature *always* favors the exothermic change. In this case, the reaction is forced to the right by an increase in temperature. A decrease in temperature causes the position of equilibrium to shift to the left.

The numerical value of the equilibrium constant changes when the temperature is changed. The reaction of CO_2 and H_2 is shifted to the right by an increase in temperature. The concentrations of the substances on the right are increased by this shift. The terms for the concentrations of these substances appear in the numerator of the expression for K. As the temperature is increased, therefore, the value of K increases. For this reaction, at 700°C, $K = 0.63$, and at 1000°C, $K = 1.66$.

4. Addition of a catalyst. The presence of a catalyst has no effect on the position of a chemical equilibrium since a catalyst affects the rates of the forward and reverse reactions equally (see Section 12.7). A catalyst will, however, cause a system to attain equilibrium more rapidly than it otherwise would.

Summary

The topics that have been discussed in this chapter are

1. Reversible reactions and chemical equilibrium; the derivation of equilibrium constants.

2. The meaning of an equilibrium constant and its use in problem solving.

3. Equilibrium constants expressed in partial pressures, K_p's, and their relation to constants expressed in molar concentrations, K's or K_c's

4. Le Chatelier's principle applied to chemical equilibria; the effects of a change in conditions on a system in equilibrium.

Key Terms

Some of the more important terms introduced in this chapter are listed below. Definitions for terms not included in this list may be located in the text by use of the index.

Chemical equilibrium (Section 13.1) A state in which the rate of a reversible reaction in one direction equals the rate in the other direction.

Equilibrium constant (Section 13.1) A constant for an equilibrium system equal to a fraction in which the numerator is the product of the concentrations of the substances on the right of the chemical equation, each raised to a power equal to its coefficient in the balanced chemical equation, and the denominator is the product of the concentrations of the substances on the left of the chemical

equation, each raised to a power equal to its coefficient in the balanced chemical equation. A value obtained by the use of molar concentrations is given the symbol K or K_c; one written in terms of the partial pressures of reacting gases is given the symbol K_p.

Heterogeneous equilibrium (Section 13.2) An equilibrium state between substances in more than one phase.

Homogeneous equilibrium (Section 13.2) An equilibrium state between substances in the same phase.

Le Chatelier's principle (Section 13.4) A system in equilibrium reacts to a change in conditions in a way that tends to offset the change and establish a new equilibrium state.

Problems*

Chemical Equilibrium, Le Chatelier's Principle

13.1 For each of the following reactions, write an expression for the equilibrium constant, K:

(a) $2\,CO_2(g) \rightleftharpoons 2\,CO(g) + O_2(g)$
(b) $2\,Pb_3O_4(s) \rightleftharpoons 6\,PbO(s) + O_2(g)$
(c) $2\,Ag_2O(s) \rightleftharpoons 4\,Ag(s) + O_2(g)$
(d) $CH_4(g) + 2\,H_2S(g) \rightleftharpoons CS_2(g) + 4\,H_2(g)$
(e) $C(s) + CO_2(g) \rightleftharpoons 2\,CO(g)$

13.2 Indicate in which direction each of the equilibria given in problem 13.1 will shift with an increase in pressure.

13.3 For the equilibrium

$$CO_2(g) + H_2(g) \rightleftharpoons CO(g) + H_2O(g)$$

K is 0.08 at 400°C and 0.41 at 600°C. Is the reaction as written exothermic or endothermic?

13.4 For the equilibrium

$$2\,SO_2(g) + O_2(g) \rightleftharpoons 2\,SO_3(g)$$

K_p is 1.0×10^5/atm at 700 K and 1.1×10^3/atm at 800 K. Is the reaction as written exothermic or endothermic?

13.5 The reaction $A(g) + B(g) \rightleftharpoons C(g)$ is exothermic as written. (a) What effect would an increase in temperature have on the numerical value of K? (b) How would the numerical values of K and K_p vary with an increase in total pressure?

---

* The more difficult problems are marked with asterisks. The appendix contains answers to color-keyed problems.

13.6 The reaction $A(g) + B(g) \rightleftharpoons C(g)$ is exothermic as written. Assume that an equilibrium system is established. How would the equilibrium concentration of $C(g)$ change with **(a)** an increase in temperature, **(b)** an increase in pressure, **(c)** the addition of $A(g)$, **(d)** the addition of a catalyst, **(e)** the removal of $B(g)$? How would the numerical value of K change with **(f)** an increase in temperature, **(g)** an increase in pressure, **(h)** the addition of a catalyst, **(i)** the addition of $A(g)$?

13.7 State the direction in which each of the following equilibrium systems would be shifted upon the application of the stress listed after the equation:

(a) $2 SO_2(g) + O_2(g) \rightleftharpoons 2 SO_3(g)$ (exothermic)
decrease temperature

(b) $C(s) + CO_2(g) \rightleftharpoons 2 CO(g)$ (endothermic)
increase temperature

(c) $N_2O_4(g) \rightleftharpoons 2 NO_2(g)$
increase total pressure

(d) $CO(g) + H_2O(g) \rightleftharpoons CO_2(g) + H_2(g)$
decrease total pressure

(e) $2 NOBr(g) \rightleftharpoons 2 NO(g) + Br_2(g)$
decrease total pressure

(f) $3 Fe(s) + 4 H_2O(g) \rightleftharpoons Fe_3O_4(s) + 4 H_2(g)$
add $Fe(s)$

(g) $2 SO_2(g) + O_2(g) \rightleftharpoons 2 SO_3(g)$
add a catalyst

(h) $CaCO_3(s) \rightleftharpoons CaO(s) + CO_2(g)$
remove $CO_2(g)$

(i) $N_2(g) + 3 H_2(g) \rightleftharpoons 2 NH_3(g)$
increase the concentration of $H_2(g)$

Problems Based on Equilibrium Constants

13.8 What is the value of the equilibrium constant, K, for the following system at 395°C:

$$H_2(g) + I_2(g) \rightleftharpoons 2 HI(g)$$

The equilibrium concentrations for such a system are: $[H_2] = 0.0064$ mol/liter, $[I_2] = 0.0016$ mol/liter, $[HI] = 0.0250$ mol/liter.

13.9 Solid NH_4HS was introduced into an evacuated container at 24°C. When equilibrium was established,

$$NH_4HS(s) \rightleftharpoons NH_3(g) + H_2S(g)$$

the total pressure (of NH_3 and H_2S taken together) was 0.614 atm. What is K_p for the equilibrium at 24°C?

13.10 At 250°C, 0.110 mol of $PCl_5(g)$ was introduced into a one-liter container. Equilibrium was established:

$$PCl_5(g) \rightleftharpoons PCl_3(g) + Cl_2(g)$$

At equilibrium, the concentration of $PCl_3(g)$ was 0.050 mol/liter. **(a)** What were the equilibrium concentrations of $Cl_2(g)$ and $PCl_5(g)$? **(b)** What is the value of K at 250°C?

13.11 A mixture of 0.0080 mol of $SO_2(g)$ and 0.0056 mol of $O_2(g)$ is placed in a one-liter container at 1000 K. When

equilibrium is established, 0.0040 mol of $SO_3(g)$ is present:

$$2 SO_2(g) + O_2(g) \rightleftharpoons 2 SO_3(g)$$

(a) What are the equilibrium concentrations of $SO_2(g)$ and $O_2(g)$? **(b)** What is the value of K for the equilibrium at 1000 K?

13.12 If 0.025 mol of $COCl_2(g)$ is placed in a one-liter container at 400°C, 16.0% of the $COCl_2$ is dissociated when equilibrium is established:

$$COCl_2(g) \rightleftharpoons CO(g) + Cl_2(g)$$

Calculate the value of K for the equilibrium at 400°C.

13.13 For the equilibrium

$$CO(g) + 2 H_2(g) \rightleftharpoons CH_3OH(g)$$

at 225°C, K is 10.2 liter^2/mol^2. What concentration of $CH_3OH(g)$ is in equilibrium with $CO(g)$ at a concentration of 0.020 mol/liter and $H_2(g)$ at a concentration of 0.020 mol/liter?

13.14 For the equilibrium

$$2 NO(g) \rightleftharpoons N_2(g) + O_2(g)$$

at 1800 K, K is 8.36×10^3. What concentration of $NO(g)$ is in equilibrium with $N_2(g)$ at a concentration of 0.0500 mol/liter and $O_2(g)$ at a concentration of 0.0500 mol/liter?

13.15 For the equilibrium

$$Br_2(g) + Cl_2(g) \rightleftharpoons 2 BrCl(g)$$

at 400 K, K is 7.0. If 0.060 mol of $Br_2(g)$ and 0.060 mol of $Cl_2(g)$ are introduced into a one-liter container at 400 K, what is the concentration of $BrCl(g)$ when equilibrium is established?

13.16 At 425°C, K is 54.8 for the equilibrium

$$H_2(g) + I_2(g) \rightleftharpoons 2 HI(g)$$

If 1.000 mol of $H_2(g)$, 1.000 mol of $I_2(g)$, and 1.000 mol of $HI(g)$ are placed in one-liter container at 425°C, what are the concentrations of all gases present at equilibrium?

13.17 At 425°C, K is 1.82×10^{-2} for the equilibrium

$$2 HI(g) \rightleftharpoons H_2(g) + I_2(g)$$

Assume that an equilibrium is established at 425°C by adding only $HI(g)$ to the reaction flask. **(a)** What are the concentrations of $H_2(g)$ and $I_2(g)$ in equilibrium with 0.0100 mol/liter of $HI(g)$? **(b)** What was the initial concentration of $HI(g)$ before equilibrium was established? **(c)** What percent of the $HI(g)$ added is dissociated at equilibrium?

13.18 For the equilibrium

$$2 IBr(g) \rightleftharpoons I_2(g) + Br_2(g)$$

K is 8.5×10^{-3} at 150°C. If 0.0300 mol of IBr(g) is introduced into a one-liter container, what is the concentration of this substance after equilibrium is established?

13.19 For the equilibrium

$$3 \, Fe(s) + 4 \, H_2O(g) \rightleftharpoons Fe_3O_4(s) + 4 \, H_2(g)$$

at 900°C, K is 5.1. If 0.050 mol of $H_2O(g)$ and excess solid Fe are placed in a one-liter container, what is the concentration of $H_2(g)$ when equilibrium is established at 900°C?

K_p's

13.20 A mixture consisting of 1.000 mol of CO(g) and 1.000 mol of $H_2O(g)$ is placed in a 10.00 liter container at 800 K. At equilibrium, 0.665 mol of $CO_2(g)$ and 0.665 mol of $H_2(g)$ are present:

$$CO(g) + H_2O(g) \rightleftharpoons CO_2(g) + H_2(g)$$

(a) What are the equilibrium concentrations of all four gases? **(b)** What is the value of K at 800 K? **(c)** What is the value of K_p at 800 K?

13.21 At 585 K and a total pressure of 1.00 atm, NOCl(g) is 56.4% dissociated:

$$2 \, ONCl(g) \rightleftharpoons 2 \, NO(g) + Cl_2(g)$$

Assume that 1.00 mol of ONCl(g) was present before dissociation. **(a)** How many moles of ONCl(g), NO(g), and $Cl_2(g)$ are present at equilibrium? **(b)** What is the total number of moles of gas present at equilibrium? **(c)** What are the equilibrium partial pressures of the three gases? **(d)** What is the numerical value of K_p at 585 K?

13.22 At 100°C, K is 4.57×10^9 liter/mol for the equilibrium

$$CO(g) + Cl_2(g) \rightleftharpoons COCl_2(g)$$

What are **(a)** K and **(b)** K_p for the equilibrium

$$COCl_2(g) \rightleftharpoons CO(g) + Cl_2(g)$$

at this temperature?

13.23 For the equilibrium

$$4 \, HCl(g) + O_2(g) \rightleftharpoons 2 \, Cl_2(g) + 2 \, H_2O(g)$$

K is 889 liter/mol at 480°C. **(a)** What is the value of K_p at this temperature? **(b)** Calculate the values of K and K_p at 480°C for the equilibrium

$$2 \, Cl_2(g) + 2 \, H_2O(g) \rightleftharpoons 4 \, HCl(g) + O_2(g)$$

13.24 For the equilibrium

$$CH_3OH(g) \rightleftharpoons CO(g) + 2 \, H_2(g)$$

at 275°C, K_p is 1.14×10^3 atm². **(a)** What is the value of K? **(b)** What are K and K_p for the following equilibrium at 275°C:

$$CO(g) + 2H_2(g) \rightleftharpoons CH_3OH(g)$$

13.25 At 800 K, K_p is 0.220 atm for the equilibrium

$$CaCO_3(s) \rightleftharpoons CaO(s) + CO_2(g)$$

What is the concentration of $CO_2(g)$, in mol/liter, that is in equilibrium with solid $CaCO_3$ and CaO at this temperature?

13.26 For the equilibrium

$$N_2(g) + O_2(g) \rightleftharpoons 2 \, NO(g)$$

K is 2.5×10^{-3} at 2400 K. The partial pressure of $N_2(g)$ is 0.50 atm, and the partial pressure of $O_2(g)$ is 0.50 atm in a mixture of these two gases. What is the partial pressure of NO(g) when equilibrium is established at 2400 K?

13.27 For the equilibrium

$$N_2(g) + 3 \, H_2(g) \rightleftharpoons 2 \, NH_3(g)$$

at 350°C, K_p is $7.73 \times 10^{-4}/atm^2$. If the partial pressure of $N_2(g)$ is 9.4 atm and the partial pressure of $H_2(g)$ is 28.0 atm in an equilibrium mixture at 350°C, what is the partial pressure of $NH_3(g)$? What is the total pressure? What is the mole fraction of $NH_3(g)$ present?

13.28 For the equilibrium

$$PCl_5(g) \rightleftharpoons PCl_3(g) + Cl_2(g)$$

at a given temperature, K_p is 2.25 atm. A quantity of $PCl_5(g)$ is introduced into an evacuated flask at the reference temperature. When equilibrium is established, the partial pressure of $PCl_5(g)$ is 0.25 atm. What are the equilibrium partial pressures of $PCl_3(g)$ and $Cl_2(g)$? What was the initial pressure of $PCl_5(g)$ before any of it had dissociated into $PCl_3(g)$ and $Cl_2(g)$? What mole percent of $PCl_5(g)$ has dissociated in this system at equilibrium?

***13.29** At a given temperature and a total pressure of 1.00 atm, the partial pressures of an equilibrium mixture

$$N_2O_4(g) \rightleftharpoons 2 \, NO_2(g)$$

are $p_{N_2O_4} = 0.50$ atm and $p_{NO_2} = 0.50$ atm. **(a)** What is K_p at this temperature? **(b)** If the total pressure is increased to 2.00 atm and the temperature is constant, what are the partial pressures of the components of an equilibrium mixture? Note that the quadratic formula must be used to solve this problem.

CHAPTER 14 THEORIES OF ACIDS AND BASES

Throughout the history of chemistry various acid-base concepts have been proposed and used. In this chapter four concepts in current use are reviewed. Each of the definitions can be applied with advantage in appropriate circumstances. In a given situation the chemist uses the concept that best suits the purpose.

The earliest criteria for the characterization of acids and bases were the experimentally observed properties of aqueous solutions. An acid was defined as a substance that in water solution tastes sour, turns litmus red, neutralizes bases, and so on. A substance was a base if its aqueous solution tastes bitter, turns litmus blue, neutralizes acids, and so on. Concurrent with the development of generalizations concerning the structure of matter, scientists searched for a correlation between acidic and basic properties and the structure of compounds that exhibit these properties.

14.1 The Arrhenius Concept

When Svanté Arrhenius published a "chemical theory of electrolytes" in 1887, he proposed that an electrolyte dissociates into ions in water solution. On this basis, an *acid* came to be defined as a compound that produces $H^+(aq)$ ions in water solution and a *base* a compound that produces $OH^-(aq)$ ions in water solution. The strength of an acid or a base is determined by the extent that the compound dissociates in water. A strong acid or base is one that dissociates completely. Note that the Arrhenius concept is based on the ions of water. The net ionic equation for a *neutralization* is

$$H^+(aq) + OH^-(aq) \longrightarrow H_2O$$

The Arrhenius concept is presented in Section 11.4.

Oxides may be incorporated into the Arrhenius scheme (see Section 11.5.). The oxides of many nonmetals react with water to form acids and are therefore called *acidic oxides* or *acid anhydrides*. For example,

$$N_2O_5(s) + H_2O \longrightarrow 2H^+(aq) + 2NO_3^-(aq)$$

Many oxides of metals dissolve in water to form hydroxides. Metal oxides are called *basic oxides* or *base anhydrides*:

$$Na_2O(s) + H_2O \longrightarrow 2Na^+(aq) + 2OH^-(aq)$$

$L6P1$

Acidic oxides and basic oxides react in the absence of water to produce salts. It must be noted, however, that not all acids and bases can be derived from oxides (HCl and NH_3 are examples of compounds that cannot be).

The Arrhenius concept is severely limited by its emphasis on water and reactions in aqueous solution. Later definitions are more general, serve to correlate more reactions, and are applicable to reactions in nonaqueous media.

14.2 The Brønsted-Lowry Concept

In 1923 Johannes Brønsted and Thomas Lowry independently proposed a broader concept of acids and bases. According to the Brønsted-Lowry definitions, an acid is a substance that can donate a proton and a base is a substance that can accept a proton. In these terms, the reaction of an acid with a base is the transfer of a proton from the acid to the base; this is the only type of reaction formally treated by this theory. Acids and bases may be either molecules or ions.

In the reaction

$$HC_2H_3O_2(aq) + H_2O(aq) \rightleftharpoons H_3O^+(aq) + C_2H_3O_2^-(aq)$$

the acetic acid molecule, $HC_2H_3O_2$, functions as an *acid* and releases a proton to the H_2O molecule, which functions as a *base*. This reaction is reversible (as shown by the double arrow) and the system exists in equilibrium.

Now, consider the reverse reaction (from right to left). In this reaction, the H_3O^+ ion donates a proton to the acetate ion, $C_2H_3O_2^-$. The H_3O^+ ion, therefore, functions as an *acid* and the $C_2H_3O_2^-$ ion, because it accepts a proton from this acid, functions as a *base*. It follows, then, that two acids ($HC_2H_3O_2$ and H_3O^+) and two bases (H_2O and $C_2H_3O_2^-$) are involved in this reversible reaction, which is actually a competition between two bases for a proton.

In the forward direction, the base H_2O gains a proton and becomes the acid H_3O^+ and in the reverse direction, the acid H_3O^+ loses a proton and becomes the base H_2O. Such an acid-base pair, which is related through the loss or gain of a proton, is called a conjugate pair:

H_2O is the conjugate base of H_3O^+
H_3O^+ is the conjugate acid of H_2O

In like manner, $HC_2H_3O_2$ and $C_2H_3O_2^-$ are related and form a second conjugate acid-base pair in this reversible system. We can indicate conjugate relationships by the use of subscripts in the following manner:

$$\overset{Acid_1}{HC_2H_3O_2(aq)} + \overset{Base_2}{H_2O} \rightleftharpoons \overset{Acid_2}{H_3O^+(aq)} + \overset{Base_1}{C_2H_3O_2^-(aq)}$$

There are many molecules and ions that can function as acids in certain reactions and as bases in other reactions. Water, for example, acts as a base in the preceding reaction; but when it reacts with ammonia, NH_3, it functions as an acid:

$$\overset{Acid_1}{H_2O} + \overset{Base_2}{NH_3(aq)} \rightleftharpoons \overset{Acid_2}{NH_4^+(aq)} + \overset{Base_1}{OH^-(aq)}$$

| | Conjugate Pair | |
Amphiprotic Substance	Acid	Base
H_2O	H_2O H_3O^+	OH^- H_2O
NH_3	NH_3 NH_4^+	NH_2^- NH_3
HSO_4^-	HSO_4^- H_2SO_4	SO_4^{2-} HSO_4^-
HPO_4^{2-}	HPO_4^{2-} $H_2PO_4^-$	PO_4^{3-} HPO_4^{2-}

Table 14.1 Some amphiprotic substances

In this reaction, the conjugate *base* of H_2O is the OH^- ion. In like manner, NH_3 functions as a base in the preceding reaction with H_2O. In the reaction with the hydride ion, H^-, in liquid ammonia, NH_3 acts as an acid:

$$\underset{Acid_1}{NH_3} + \underset{Base_2}{H^-} \rightleftharpoons \underset{Acid_2}{H_2} + \underset{Base_1}{NH_2^-}$$

Substances that can function as acids or bases are called *amphiprotic;* several amphiprotic substances are listed in Table 14.1.

The neutralization reaction of the Arrhenius system, therefore, may be interpreted in terms of the Brønsted definitions. Such a reaction is merely the acid-base reaction between the conjugate acid and the conjugate base of the amphiprotic solvent, water:

$$\underset{Acid_1}{H_3O^+(aq)} + \underset{Base_2}{OH^-(aq)} \rightleftharpoons \underset{Acid_2}{H_2O} + \underset{Base_1}{H_2O}$$

14.3 Strength of Brønsted Acids and Bases

In Brønsted terms the strength of an acid is determined by its tendency to donate protons, and the strength of a base is dependent upon its tendency to receive protons. The reaction

$$\underset{Acid_1}{HCl(aq)} + \underset{Base_2}{H_2O} \rightleftharpoons \underset{Acid_2}{H_3O^+(aq)} + \underset{Base_1}{Cl^-(aq)}$$

proceeds virtually to completion (from left to right). We must conclude, therefore, that HCl is a stronger acid than H_3O^+ since it has the stronger tendency to lose protons and the equilibrium is displaced far to the right. In addition, it is apparent that H_2O is a stronger base than Cl^- since, in the competition for protons, water molecules succeed in holding practically all of them. The strong acid, HCl, has a weak conjugate base, Cl^-.

A strong acid, which has a great tendency to lose protons, is necessarily conjugate to a weak base, which has a small tendency to gain and hold protons.

Table 14.2 Relative strengths of some conjugate acid-base pairs

	Acid		Base	
	$HClO_4$	$\xrightarrow{\text{100\% in } H_2O}$	ClO_4^-	
	HCl	$\xrightarrow{\text{100\% in } H_2O}$	Cl^-	
↑ Increasing Acid Strength	HNO_3	$\xrightarrow{\text{100\% in } H_2O}$	NO_3^-	Increasing Base Strength ↓
	H_3O^+		H_2O	
	H_3PO_4		$H_2PO_4^-$	
	$HC_2H_3O_2$		$C_2H_3O_2^-$	
	H_2CO_3		HCO_3^-	
	H_2S		HS^-	
	NH_4^+		NH_3	
	HCN		CN^-	
	HCO_3^-		CO_3^{2-}	
	HS^-		S^{2-}	
	H_2O		OH^-	
	NH_3	$\xleftarrow{\text{100\% in } H_2O}$	NH_2^-	
	H_2	$\xleftarrow{\text{100\% in } H_2O}$	H^-	

Hence, *the stronger the acid, the weaker its conjugate base*. In like manner, a strong base attracts protons strongly and is necessarily conjugate to a weak acid, one that does not readily lose protons. *The stronger the base, the weaker its conjugate acid.*

Acetic acid in 1.0 M solution is 0.42% ionized at 25°C (see Section 15.1). The equilibrium

$$\underset{Acid_1}{HC_2H_3O_2(aq)} + \underset{Base_2}{H_2O} \rightleftharpoons \underset{Acid_2}{H_3O^+(aq)} + \underset{Base_1}{C_2H_3O_2^-(aq)}$$

is displaced to the left. This equation may be said to represent a competition between bases, acetate ions, and water molecules, for protons. The position of the equilibrium shows that the $C_2H_3O^-$ ion is a stronger base than H_2O; at equilibrium more protons form $HC_2H_3O_2$ molecules than form H_3O^+ ions. We may also conclude that H_3O^+ is a stronger acid than $HC_2H_3O_2$; at equilibrium more H_3O^+ ions than $HC_2H_3O_2$ molecules have lost protons. In the preceding example we note again that the stronger acid, H_3O^+, is conjugate to the weaker base, H_2O, and the stronger base, $C_2H_3O_2^-$, is conjugate to the weaker acid, $HC_2H_3O_2$.

One further conclusion should be stated. In a given reaction, *the position of equilibrium favors the formation of the weaker acid and the weaker base*. Thus, in the reaction of HCl and H_2O, the equilibrium concentrations of H_3O^+ and Cl^- (the *weaker* acid and base, respectively) are *high*, whereas in the solution of acetic acid, the equilibrium concentrations of H_3O^+ and $C_2H_3O_2^-$ (the *stronger* acid and base, respectively) are *low*.

In Table 14.2, some acids are listed in decreasing order of acid strength (the ability to donate protons). A second column in the table lists the conjugate bases

14.3 Strength of Brønsted Acids and Bases

357

of these acids. Perchloric acid, $HClO_4$, is the strongest acid listed, and its conjugate base, the perchlorate ion, ClO_4^-, is the weakest base. Hydride ion, H^- is the strongest base in the table, and its conjugate acid, H_2, is the weakest acid. The inverse relation between the acid and base strengths of a conjugate pair is evident.

The strength of an acid is determined by the ability of the acid to donate a proton. The reactions of water and the first three acids listed in Table 14.2 go virtually to completion:

$$HClO_4(aq) + H_2O \longrightarrow H_3O^+(aq) + ClO_4^-(aq)$$

$$HCl(aq) + H_2O \longrightarrow H_3O^+(aq) + Cl^-(aq)$$

$$HNO_3(aq) + H_2O \longrightarrow H_3O^+(aq) + NO_3^-(aq)$$

Each of these acids is a stronger acid than H_3O^+ and in a Brønsted acid-base reaction, the weaker acid is formed predominantly.

Aqueous solutions of $HClO_4$, HCl, and HNO_3 of the same concentration appear to be of the same acid strength. The acid properties of the solutions are due to the H_3O^+ ion, which the compounds produce to an equivalent extent in their reactions with water. Water is said to have a **leveling effect** on acids stronger than H_3O^+. The strongest acid that can exist in water solution is the conjugate acid of water, H_3O^+. Acids that are weaker than H_3O^+ are not leveled by water. Thus, $HC_2H_3O_2$, H_3PO_4, H_2S, and other weak acids show a wide variation in their degree of ionization—the extent to which they form H_3O^+ in their reactions with water (see Section 15.1).

Water also levels bases. The strongest base capable of existing in water solution is the conjugate base of water, OH^-. Many substances, such as NH_2^- and H^-, are stronger bases than OH^-. In water solution, however, these strongly basic substances accept protons to form OH^- ions; these reactions are essentially complete. The apparent basicity of strongly basic materials in water is reduced to the level of the OH^- ion:

$$H_2O + NH_2^-(aq) \longrightarrow NH_3(aq) + OH^-(aq)$$

$$H_2O + H^-(aq) \longrightarrow H_2(g) + OH^-(aq)$$

Substances, such as ammonia, that are less basic than OH^- are not leveled by water and show varying degrees of ionization in aqueous solution (see Section 15.1).

The leveling effect is observed for solvents other than water. The strongest acid in liquid ammonia solutions is the conjugate acid of ammonia, the ammonium ion, NH_4^+. Acetic acid (which is incompletely ionized in water) is completely ionized in liquid ammonia since $HC_2H_3O_2$ is a stronger acid than NH_4^+:

$$HC_2H_3O_2 + NH_3 \longrightarrow NH_4^+ + C_2H_3O_2^-$$

The strongest base that can exist in liquid ammonia is the conjugate base of ammonia, the amide ion, NH_2^-. In liquid ammonia, the hydride ion, H^-, reacts completely and rapidly to form hydrogen and the amide ion:

$$NH_3 + H^- \longrightarrow H_2 + NH_2^-$$

The hydride ion, therefore, is a stronger base than the amide ion. Liquid ammonia reduces H^- to the level of the conjugate base of ammonia, NH_2^-. Notice that this reaction establishes the order of base strength for H^- and NH_2^- (see Table 14.2).

14.4 Acid Strength and Molecular Structure

In order to analyze the relationships between molecular structure and acid strength, we will divide acids into two types: covalent hydrides and oxyacids. Each of these types will be considered in turn.

1. Hydrides. Some covalent binary compounds of hydrogen (such as H_2S and HCl) are acidic. Two factors influence the acid strength of the hydride of an element: the electronegativity of the element and the atomic size of the element. The first of these factors is best understood by comparing the hydrides of the elements of a period. The second is important when group comparisons are made.

a. Hydrides of the elements of a period. The acid strengths of the hydrides of the elements of a period increase from left to right across the period in the same order that the electronegativities of the elements increase. We would expect a highly electronegative element to withdraw electrons from the hydrogen and facilitate its release as a proton.

Consider the hydrides of nitrogen, oxygen, and fluorine of the second period. The electronegativity of these elements increases in the order

$$N < O < F$$

and acid strength of the hydrides increases in the same order:

$$NH_3 < H_2O < HF$$

A water solution of ammonia (NH_3) is basic:

$$NH_3(g) + H_2O \rightleftharpoons NH_4^+(aq) + OH^-(aq)$$

Water dissociates to a very small extent and forms extremely low concentrations of both $H_3O^+(aq)$ and $OH^-(aq)$:

$$H_2O + H_2O \rightleftharpoons H_3O^+(aq) + OH^-(aq)$$

A water solution of hydrogen fluoride is acidic:

$$HF(g) + H_2O \rightleftharpoons H_3O^+(aq) + F^-(aq)$$

The electronegativities of the following third-period elements fall in the order

$$P < S < Cl$$

The acid strengths of the hydrides of these elements increase in the same order; PH_3 does not react with water, H_2S is a weak acid, and HCl is a strong acid:

$$PH_3 < H_2S < HCl$$

b. Hydrides of the elements of a group. The acidity of the hydrides of the elements of a group increases with increasing size of the central atom. Consider the hydrides of the group VI A and group VII A elements:

$$H_2O < H_2S < H_2Se < H_2Te$$
$$HF < HCl < HBr < HI*$$

This order is the reverse of that expected on the basis of electronegativity. The first hydride of each series (H_2O and HF) is the weakest acid of the series and is formed by the element with the highest electronegativity.

The two factors that influence acid strength are the electronegativity of the central atom and the size of the central atom. When these factors work against each other, the effect of atomic size outweighs the electronegativity effect. A proton is more easily removed from a hydride in which the central atom is large and its electron cloud therefore diffuse than from one in which the central atom is small.

When the hydrogen compounds of the elements of a period are compared, the small differences in the sizes of the central atom are unimportant. The size of the central atom, however, becomes very important when the hydrides of the elements of a group are compared. The atomic radii of the members of a group increase markedly from the lightest to the heaviest members. Fluorine has an atomic radius of 71 pm, and iodine has an atomic radius of 133 pm; HF is a weak acid, and HI is a strong acid.

Consider the hydrides of carbon, sulfur, and iodine. The electronegativites of C, S, and I, which belong to different groups, are about the same (2.5). The atomic radius of C is 77 pm, of S is 103 pm, and of I is 133 pm. There is a marked increase in the acidity of the hydride with the increase in the size of the central atom. Methane, CH_4, does not dissociate in water, H_2S is a weak acid, and HI is a strong acid.

2. Oxyacids. The oxyacids are compounds that are derived from the structure

$$\overset{a \quad b}{H{-}O{-}Z}$$

In each of these compounds, the acidic hydrogen is bonded to an O atom, and the variation in the size of this atom is very small. The key to the acidity of these oxyacids, therefore, lies with the electronegativity of the atom Z.

If Z is an atom of a metal with a low electronegativity, the electron pair that is marked *b* will belong completely to the O atom, which has a high electronegativity. The compound will be an ionic hydroxide—a base. Sodium hydroxide (HO^-Na^+, usually written Na^+OH^-) falls into this category.

If Z is an atom of a nonmetal with a high electronegativity, the situation is different. The bond marked *b* will be a strong covalent bond, not an ionic bond. Instead of adding to the electron density around the O atom, Z will tend to reduce

---

* In water solution, HCl, HBr, and HI are virtually completely dissociated.

the electron density, even though oxygen is itself highly electronegative. The effect will be felt in bond *a*. The O atom will draw the electron density of this H—O bond away from the H atom, which will allow the proton to dissociate and make the compound acidic. Hypochlorous acid, HOCl, is an acid of this type.

The higher the electronegativity of Z, the more the electrons of the H—O bond are drawn away from the H atom and the more readily the proton is lost. In the series

$$HOI < HOBr < HOCl$$

the electronegativity of Z increases (I < Br < Cl), and the acid strength increases in the same order.

In some molecules, additional O atoms are bonded to Z. For example,

$$H—O—\overset{\overset{\displaystyle O}{|}}{Z}—O$$

These O atoms draw electrons away from the Z atom and make it more positive. The Z atom, therefore, becomes more effective in withdrawing electron density away from the O atom that is bonded to H. In turn, the electrons of the H—O bond are drawn more strongly away from the H atom. The net effect makes it easier for the proton to dissociate and increases the acidity of the compound.

The more O atoms bonded to Z, the stronger the acid is. This effect is illustrated by the following series of acids, which are arranged by increasing order of acid strength:

$$H—\overset{..}{\underset{..}{O}}—\overset{..}{\underset{..}{Cl}}: \; < \; H—\overset{..}{\underset{..}{O}}—\overset{\oplus}{\underset{..}{Cl}}—\overset{..}{\underset{..}{O}}:^{\ominus} \; < \; H—\overset{..}{\underset{..}{O}}—Cl^{2+} \cdots \; < \; H—\overset{..}{\underset{..}{O}}—Cl^{3+}$$

Notice that the formal charge on the central atom increases in this series. As the formal charge on the Cl increases, the electron density of the H—O bond shifts away from the H atom. As a result, the acidity increases.

Chemists frequently correlate the acid strength of a series of oxyacids such as this with the oxidation number of the central atom rather than with the formal charge of the central atom as we have done. In the series of the oxyacids of chlorine, formal charge and oxidation number increase in the same order so that it may appear that either could be used:

	HOCl	HOClO	HOClO$_2$	HOClO$_3$
formal charge of Cl	0	1+	2+	3+
oxidation number of Cl	1+	3+	5+	7+

In some cases, however, oxidation number is not a reliable indicator—formal charge must be used. The oxyacids of phosphorus, for example, are all weak acids—about of equal strength:

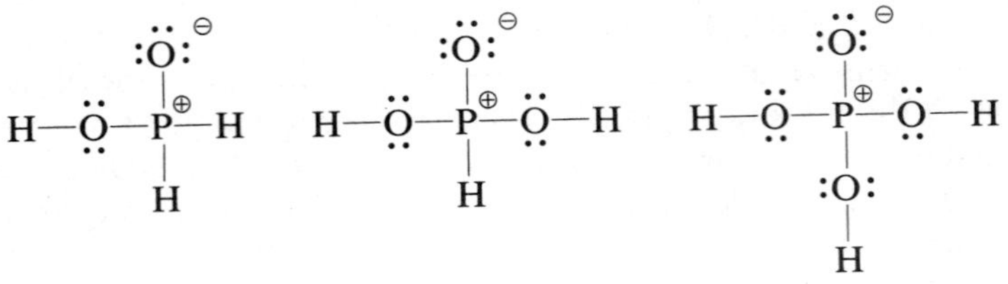

formal charge of P	1+	1+	1+
oxidation number of P	1+	3+	5+

Any prediction based on oxidation number is incorrect; there is practically no difference in acid strength between any of these compounds. This conclusion, however, could be reached by noting the formal charge of P in the structures.

We can rate the acid strength of compounds of this type by counting the number of O atoms bonded to Z but *not bonded to H atoms*:

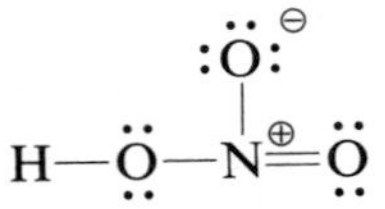

H—Ö—N⁺=Ö   is a stronger acid than   H—Ö—N=Ö

nitric acid
(a resonance hybrid)

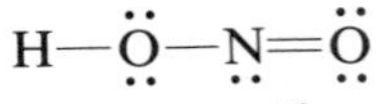

nitrous acid

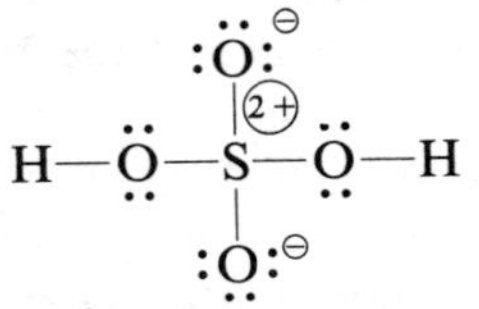

H—Ö—S—Ö—H   is a stronger acid than   H—Ö—S⁺—Ö—H

sulfuric acid

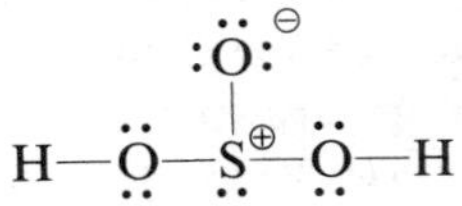

sulfurous acid

In general, the strengths of acids that have the general formula

$$(HO)_m ZO_n$$

can be related to the value of n:

a. If $n = 0$, the acid is *very weak*: $HOCl$, $(HO)_3B$

b. If $n = 1$, the acid is *weak*: $HOClO$, $HONO$, $(HO)_2SO$, $(HO)_3PO$

c. If $n = 2$, the acid is *strong*: $HOClO_2$, $HONO_2$, $(HO)_2SO_2$

d. If $n = 3$, the acid is *very strong*: $HOClO_3$, $HOIO_3$

The first proton of each acid that belongs to the last two categories ($n = 2$ and $n = 3$) is virtually completely dissociated in water. The distinction between the two groups applies to dissociations in solvents other than water.

The effect of electron-withdrawing groups is also seen in organic acids. None of the H atoms of ethanol:

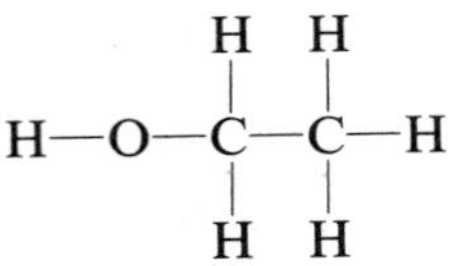

dissociates as a proton in water solution. The introduction of an O atom into the molecule:

$$\text{H}-\text{O}-\overset{\displaystyle\text{O}}{\underset{}{\overset{\|}{\text{C}}}}-\overset{\displaystyle\text{H}}{\underset{\displaystyle\text{H}}{\text{C}}}-\text{H}$$

produces the compound called acetic acid (which we have previously written $HC_2H_3O_2$). Acetic acid is a weak monoprotic acid; only the H of the H—O group is acidic. There are a large number of organic acids that contain the

$$\text{H}-\text{O}-\overset{\displaystyle\text{O}}{\overset{\|}{\text{C}}}-$$

group, which is called the carboxyl group. Most carboxylic acids are weak acids and may be considered to belong to the (b) category of our previous classification ($n = 1$).

The carboxylic acids may be assigned the general formula

$$\text{H}-\text{O}-\overset{\displaystyle\text{O}}{\overset{\|}{\text{C}}}-\text{R}$$

Modifications in the R group can bring about enhanced acidity. If one or more of the H atoms that are bonded to C in acetic acid are replaced by highly electronegative atoms (such as Cl), the acidity is increased. Trichloroacetic acid, for example,

$$\text{H}-\text{O}-\overset{\displaystyle\text{O}}{\overset{\|}{\text{C}}}-\overset{\displaystyle\text{Cl}}{\underset{\displaystyle\text{Cl}}{\text{C}}}-\text{Cl}$$

is a much stronger acid than acetic acid.

Trends in base strength are readily derived from conjugate relationships. A strong acid is conjugate to a weak base. We can, for example, predict that S^{2-} is a weaker base than O^{2-} since H_2S is a stronger acid than H_2O.

14.5 The Lewis Concept

L6PA

In reality, the Brønsted concept enlarges the definition of a base much more than it does that of an acid. In the Brønsted system a base is a molecule or ion that has an unshared electron pair with which it can attract and hold a proton, and an acid is a substance that can supply a proton to a base. If a molecule or ion can share an electron pair with a proton, it can do the same thing with other substances as well.

Gilbert N. Lewis proposed a broader concept of acids and bases which liberated acid-base phenomena from the proton. Although Lewis first proposed his system in 1923, he did little to develop it until 1938. Lewis defined a **base** as a substance that has an unshared electron pair with which it can form a covalent bond with an atom, molecule, or ion. An **acid** is a substance that can form a covalent bond by accepting an electron pair from a base. The emphasis has been shifted by the Lewis concept from the proton to the electron pair and covalent-bond formation.

An example of an acid-base reaction that is not treated as such by any other acid-base concept is

$$
\begin{array}{ccc}
\underset{\text{acid}}{
\begin{array}{c}
\ddot{\text{:F:}} \quad \text{H} \\
| \qquad | \\
\ddot{\text{:F}}\!-\!\text{B} \;+\; \ddot{\text{:N}}\!-\!\text{H} \\
| \qquad | \\
\ddot{\text{:F:}} \quad \text{H}
\end{array}}
\quad
\underset{\text{base}}{}
& \longrightarrow &
\begin{array}{c}
\ddot{\text{:F:}} \quad \text{H} \\
| \qquad | \\
\ddot{\text{:F}}\!-\!\text{B}\!-\!\text{N}\!-\!\text{H} \\
| \qquad | \\
\ddot{\text{:F:}} \quad \text{H}
\end{array}
\end{array}
$$

Many Lewis acids and bases of this type can be titrated against one another by the use of suitable indicators in the same way that traditional acids and bases can be titrated.

Substances that are bases in the Brønsted system are also bases according to the Lewis concept. However, the Lewis definition of an acid considerably expands the number of substances that are classified as acids. A Lewis acid must have an empty orbital capable of receiving the electron pair of the base; the proton is but a single example of a Lewis acid.

Chemical species that can function as Lewis acids include the following:

1. Molecules or atoms that have incomplete octets:

$$
\begin{array}{c}
\ddot{\text{:F:}} \\
| \\
\ddot{\text{:F}}\!-\!\text{B} \;+\; \ddot{\text{:F:}}^{-} \\
| \\
\ddot{\text{:F:}}
\end{array}
\longrightarrow
\left[
\begin{array}{c}
\ddot{\text{:F:}} \\
| \\
\ddot{\text{:F}}\!-\!\text{B}\!-\!\ddot{\text{F:}} \\
| \\
\ddot{\text{:F:}}
\end{array}
\right]^{-}
$$

$$
\ddot{\text{:S}} \;+\;
\left[
\begin{array}{c}
\ddot{\text{:O:}} \\
| \\
\text{:S}\!-\!\ddot{\text{O:}} \\
| \\
\ddot{\text{:O:}}
\end{array}
\right]^{2-}
\longrightarrow
\left[
\begin{array}{c}
\ddot{\text{:O:}} \\
| \\
\ddot{\text{:S}}\!-\!\text{S}\!-\!\ddot{\text{O:}} \\
| \\
\text{:O:}
\end{array}
\right]^{2-}
$$

$$
\begin{array}{c}
\ddot{\text{:Cl:}} \\
| \\
\ddot{\text{:Cl}}\!-\!\text{Al} \;+\; \ddot{\text{:Cl:}}^{-} \\
| \\
\ddot{\text{:Cl:}}
\end{array}
\longrightarrow
\left[
\begin{array}{c}
\ddot{\text{:Cl:}} \\
| \\
\ddot{\text{:Cl}}\!-\!\text{Al}\!-\!\ddot{\text{Cl:}} \\
| \\
\ddot{\text{:Cl:}}
\end{array}
\right]^{-}
$$

Aluminum chloride, although it reacts as $AlCl_3$, is actually a dimer—Al_2Cl_6. The formation of the dimer from the monomer may be regarded as a Lewis acid-base reaction in itself since a chlorine atom in each $AlCl_3$ unit supplies an electron pair to the aluminum atom of the other $AlCl_3$ unit to complete the octet of the aluminum atom; these bonds are indicated by $\cdots$ in the diagram

$$
\begin{array}{ccccc}
\text{Cl} & & \text{Cl} & & \text{Cl} \\
\diagdown & & \diagup \;\cdot\cdot & \cdot & \diagup \\
& \text{Al} \cdot & & \cdot \; \text{Al} & \\
\diagup & & \cdot & & \diagdown \\
\text{Cl} & & \text{Cl} & & \text{Cl}
\end{array}
$$

2. Many simple cations can function as Lewis acids—for example,

$$
Cu^{2+} + 4:NH_3 \longrightarrow Cu(:NH_3)_4^{2+}
$$

$$
Fe^{3+} + 6:C\!\equiv\!N:^{-} \longrightarrow Fe(:C\!\equiv\!N:)_6^{3-}
$$

3. Some metals atoms can function as acids in the formation of compounds such as the carbonyls, which are produced by the reaction of the metal with carbon monoxide:

$$Ni + 4\,:C\!\equiv\!O: \longrightarrow Ni(:C\!\equiv\!O:)_4$$

4. Compounds that have central atoms capable of expanding their valence shells are Lewis acids in reactions in which this expansion occurs. Examples are

$$SnCl_4 + 2\,Cl^- \longrightarrow SnCl_6^{2-}$$

$$SiF_4 + 2\,F^- \longrightarrow SiF_6^{2-}$$

$$PF_5 + F^- \longrightarrow PF_6^-$$

In each of the first two reactions, the valence shell of the central atom (Sn and Si) is expanded from eight to twelve electrons, and in the third reaction the valence shell of P goes from ten to twelve electrons.

5. Some compounds have an acidic site because of one or more multiple bonds in the molecule. Examples are

The reactions of silica, SiO_2, with metal oxides are analogous to the reaction of carbon dioxide with the oxide ion, although both silica and the silicate products (compounds of SiO_3^{2-}) are polymeric. This reaction is important in high-temperature metallurgical processes in which a basic oxide is added to an ore to remove silica in the form of silicates (slag). Many of the processes used in the manufacture of glass, cement, and ceramics involve the reaction of the base O^{2-} (from metal oxides, carbonates, and so forth) with acid oxides (such as SiO_2, Al_2O_3, and B_2O_3).

Arrhenius and Brønsted acid-base reactions may be interpreted in Lewis terms by focusing attention on the proton, as a Lewis acid:

$$H^+(aq) + OH^-(aq) \longrightarrow H_2O$$

in which case the Brønsted acid is termed a **secondary** Lewis acid since it serves to provide the **primary** Lewis acid, the proton. Probably a better interpretation is that which classifies Brønsted acid-base reactions as Lewis base displacements. The Brønsted acid is interpreted as a complex in which the Lewis acid (proton) is already combined with a base; the reaction is viewed as a displacement of this base by another, stronger base:

$$H_3O^+ + OH^- \longrightarrow H_2O + H_2O$$

In this reaction the base OH^- displaces the weaker base H_2O from its combination with the acid, the proton.

All Brønsted acid-base reactions are Lewis base displacements. In the reaction

$$HCl + H_2O \longrightarrow H_3O^+ + Cl^-$$

the base H_2O displaces the weaker base, Cl^-. A base supplies an electron pair to a nucleus and is therefore called **nucleophilic** (from Greek, meaning "nucleus loving"). Base displacements are nucleophilic displacements.

Nucleophilic displacements may be identified among reactions that are not Brønsted acid-base reactions. The formation of $Cu(NH_3)_4^{2+}$ has been previously used as an illustration of a Lewis acid-base reaction. Since the reaction occurs in water, the formation of this complex is more accurately interpreted as the displacement of the base H_2O from the complex $Cu(H_2O)_4^{2+}$ by the stronger base NH_3:

$$Cu(H_2O)_4^{2+} + 4\,NH_3 \longrightarrow Cu(NH_3)_4^{2+} + 4\,H_2O$$

Lewis acids accept an electron pair in a reaction with a base; they are **electrophilic** (from Greek, meaning "electron loving"). Acid displacements, or electrophilic displacements, are not so common as base displacements, but this type of reaction is known. For example, if $COCl_2$ is viewed as a combination of $COCl^+$ (an acid) with Cl^- (a base), the reaction

$$COCl_2 + AlCl_3 \longrightarrow COCl^+ + AlCl_4^-$$

is an electrophilic displacement in which the acid $AlCl_3$ displaces the weaker acid $COCl^+$ from its complex with the base Cl^- (Table 14.3). The reaction

$$SeOCl_2 + BCl_3 \longrightarrow SeOCl^+ + BCl_4^-$$

may be similarly interpreted.

The Lewis theory is frequently used to interpret reaction mechanisms. Examples of such interpretations are found in Sections 26.5 and 26.6.

14.6 Solvent Systems

The principles of the Arrhenius water concept can be used to devise acid-base schemes for many solvents. In a **solvent system** an **acid** is a substance that gives the cation characteristic of the solvent, and a **base** is a substance that yields the anion characteristic of the solvent. Thus, the reaction of an acid and a base, a **neutralization,** yields the solvent as one of its products. Many solvent systems of acids and bases have been developed (Table 14.3); the water concept is but a single example of a solvent system.

The ammonia system has been investigated more extensively than any other with the exception of the water system. The properties of liquid ammonia (boiling point, $-33.4°C$) are strikingly similar to those of water. Liquid ammonia is

Table 14.3 Some solvent systems

Solvent	Acid Ion	Base Ion	Typical Acid	Typical Base
H_2O	H_3O^+ $(H^+ \cdot H_2O)$	OH^-	HCl	NaOH
NH_3	NH_4^+ $(H^+ \cdot NH_3)$	NH_2^-	NH_4Cl	$NaNH_2$
NH_2OH	NH_3OH^+ $(H^+ \cdot NH_2OH)$	$NHOH^-$	$NH_2OH \cdot HCl$ (NH_3OH^+, Cl^-)	$K(NHOH)$
$HC_2H_3O_2$	$H_2C_2H_3O_2^+$ $(H^+ \cdot HC_2H_3O_2)$	$C_2H_3O_2^-$	HCl	$NaC_2H_3O_2$
SO_2	SO^{2+}	SO_3^{2-}	$SOCl_2$	Cs_2SO_3
N_2O_4	NO^+	NO_3^-	NOCl	$AgNO_3$
$COCl_2$	$COCl^+$	Cl^-	$(COCl)AlCl_4$	$CaCl_2$

associated through hydrogen bonding, and the NH_3 molecule is polar. Hence, liquid ammonia is an excellent solvent for ionic and polar compounds, and it functions as an ionizing solvent for electrolytes. Many compounds form ammoniates, which are analogous to hydrates ($BaBr_2 \cdot 8NH_3$ and $CaCl_2 \cdot 6NH_3$ are examples), and ions are solvated in liquid ammonia solutions [$Ag(NH_3)_2^+$ and $Cr(NH_3)_6^{3+}$ are examples]. Whereas solutions of electrolytes in ammonia are good conductors of electricity, pure liquid ammonia, like water, has a relatively low conductance.

The autoionization of ammonia:

$$2\,NH_3 \rightleftharpoons NH_4^+ + NH_2^-$$

which occurs only to a low degree, is responsible for the electrical conductivity of pure solvent, just as the autoionization of water:

$$2\,H_2O \rightleftharpoons H_3O^+ + OH^-$$

is responsible for the electrical properties of this compound.

Any compound that produces ammonium ion, NH_4^+, in liquid ammonia solution is an acid, and any compound that yields amide ion, NH_2^-, is a base. Thus, the neutralization reaction is the reverse of the autoionization reaction:

$$NH_4^+ + NH_2^- \longrightarrow 2\,NH_3$$

Indicators may be used to follow an acid-base reaction in liquid ammonia. For example, phenolphthalein is red in a liquid ammonia solution of potassium amide, KNH_2, and becomes colorless after a stoichiometrically equivalent amount of ammonium chloride has been added.

In addition to the neutralization reaction, the ammonium ion in liquid ammonia undergoes other reactions analogous to the reactions of the hydronium ion in water. For example, metals such as sodium react with the ammonium ion to liberate hydrogen:

$$2\,Na(s) + 2\,NH_4^+ \longrightarrow 2\,Na^+ + H_2(g) + 2\,NH_3(l)$$

The reactions of the amide ion are analogous to those of the hydroxide ion:

$$Zn(OH)_2(s) + 2\,OH^- \longrightarrow Zn(OH)_4^{2-}$$

$$Zn(NH_2)_2(s) + 2\,NH_2^- \longrightarrow Zn(NH_2)_4^{2-}$$

$$Hg^{2+} + 2\,OH^- \longrightarrow HgO(s) + H_2O$$

$$3\,Hg^{2+} + 6\,NH_2^- \longrightarrow Hg_3N_2(s) + 4\,NH_3$$

Many properties and reactions of compounds belonging to the ammonia system have been predicted and correlated by comparison to the better known chemistry of the compounds of the water system. In fact, the study of various solvent systems has been responsible for greatly increasing our knowledge of the reactions that occur in solvents other than water.

Summary

The topics that have been discussed in this chapter are

1. A review of the Arrhenius concept of acids and bases.

2. The Brønsted-Lowry theory of acids and bases: the transfer of a proton as the criterion for acid-base reactions.

3. The strengths of Brønsted acids and bases and the leveling effects of solvents.

4. The interpretation of acid strengths on the basis of molecular structure.

5. The Lewis concept of acids and bases: the formation of a covalent bond as the criterion of acid-base reactions; displacement reactions.

6. Solvent systems of acids and bases: the use of the principles of the Arrhenius concept to devise schemes for solvents other than water.

Key Terms

Some of the more important terms introduced in this chapter are listed below. Definitions for terms not included in this list may be located in the text by use of the index.

Amphiprotic substance (Section 14.2) A substance that can function as a Brønsted acid (through proton loss) or as a Brønsted base (through proton gain).

Arrhenius acid (Sections 11.4 and 14.1) A compound that dissociates in water to produce H^+ (aq) ions (or H_3O^+ ions).

Arrhenius base (Sections 11.4 and 14.1) A compound that dissolves in water to produce OH^-(aq) ions.

Arrhenius neutralization (Sections 11.4 and 14.1) A reaction in which H^+(aq) from an acid and OH^-(aq) from a base react to form H_2O.

Brønsted acid (Section 14.2) A substance that can donate protons.

Brønsted base (Section 14.2) A substance that can accept protons.

Conjugate pair (Section 14.2) A Brønsted acid-base pair that is related through the loss or gain of a proton; for example, NH_4^+ (a Brønsted acid) and NH_3 (a Brønsted base) form a conjugate pair.

Electrophilic displacement (Section 14.5) A reaction in which a Lewis acid displaces a second, weaker, Lewis acid from an acid-base complex; a Lewis **acid displacement.**

Leveling effect (Section 14.3) An effect of a solvent on the strength of a Brønsted acid or a Brønsted base. An acid in solution can be no stronger than the conjugate acid of the solvent; a base in solution can be no stronger than the conjugate base of the solvent.

Lewis acid (Section 14.5) A substance that can form a covalent bond by sharing an electron pair that is donated by a Lewis base; an **electrophilic substance.**

Lewis base (Section 14.5) A substance that can form a covalent bond with a Lewis acid by donating an electron pair toward the formation of the bond; a **nucleophilic substance.**

Nucleophilic displacement (Section 14.5) A reaction in which a Lewis base displaces a second, weaker, Lewis base from an acid-base complex; a Lewis base displacement.

Solvent-system acid (Section 14.6) A substance that yields the cation characteristic of the solvent.

Solvent-system base (Section 14.6) A substance that yields the anion characteristic of the solvent.

Solvent-system neutralization (Section 14.6) A reaction between an acid and a base that yields the solvent as one of the products.

Problems*

Acid-Base Concepts

14.1 Give one-sentence descriptions of an *acid*, a *base*, and a *neutralization reaction* based on **(a)** the Arrhenius concept, **(b)** the solvent system concept, **(c)** the Brønsted-Lowry concept, and **(d)** the Lewis concept.

14.2 Briefly discuss, using chemical equations, how the compound H_2O is classified according to **(a)** the Arrhenius concept, **(b)** the Brønsted-Lowry concept, and **(c)** the Lewis concept.

14.3 Do acid-base reactions ever involve oxidation and reduction? Consider all of the acid-base theories presented in this chapter.

14.4 Ammonium chloride (NH_4Cl) reacts with sodium amide ($NaNH_2$) in liquid ammonia to produce sodium chloride and ammonia. Interpret this reaction in terms of the solvent system, Brønsted, and Lewis theories of acids and bases. State clearly what acid(s) and base(s) are involved in each case.

14.5 Write equations analogous to the following but employing compounds of the water system instead of compounds of the ammonia system:

(a) $KNH_2 + NH_4NO_3 \longrightarrow KNO_3 + 2\,NH_3$
(b) $Ca_3N_2 + 4\,NH_3 \longrightarrow 3\,Ca(NH_2)_2$
(c) $Cl_2 + 2\,NH_3 \longrightarrow NH_4Cl + H_2NCl$
(d) $2\,Na + 2\,NH_3 \longrightarrow 2\,NaNH_2 + H_2$
(e) $Zn(NH_2)_2 + 2\,NaNH_2 \longrightarrow Na_2[Zn(NH_2)_4]$

Brønsted-Lowry Concept

14.6 What is the conjugate base of **(a)** H_3AsO_3, **(b)** $H_2AsO_4^-$, **(c)** HNO_2, **(d)** HS^-? :$H_2A_sO_3^-$, $HAsO_4^{2-}$, NO_2^-, S^{2-}

14.7 What is the conjugate base of **(a)** NH_3, **(b)** $HC_2O_4^-$, **(c)** $HOBr$, **(d)** H_3PO_4?

14.8 What is the conjugate acid of **(a)** $H_2AsO_4^-$, **(b)** PH_3, **(c)** $C_2H_3O_2^-$, **(d)** S^{2-}, **(e)** PO_4^{3-}? :H_3AsO_4, PH_4^+, $HC_2H_3O_2$, HS^-, HPO_4^{2-}

14.9 What is the conjugate acid of **(a)** NH_3, **(b)** $HC_2O_4^-$, **(c)** OCl^-, **(d)** NH_2^-?

14.10 Identify all the Brønsted acids and bases in the following equations:

(a) $HF + HF \rightleftharpoons H_2F^+ + F^-$ (in liquid HF)
(b) $HNO_3 + HF \rightleftharpoons H_2NO_3^+ + F^-$ (in liquid HF)
(c) $HF + CN^- \rightleftharpoons HCN + F^-$ (in liquid HF)
(d) $H_2PO_4^- + CO_3^{2-} \rightleftharpoons HPO_4^{2-} + HCO_3^-$ (in water)
(e) $N_2H_4 + HSO_4^- \rightleftharpoons N_2H_5^+ + SO_4^{2-}$ (in water)
(f) $HC_2O_4^- + HS^- \rightleftharpoons C_2O_4^{2-} + H_2S$ (in water)

14.11 Write chemical equations to illustrate the behavior of the following as Brønsted acids: **(a)** $HOCl$, **(b)** H_2O, **(c)** HCO_3^-, **(d)** NH_3.

14.12 Write chemical equations to illustrate the behavior of the following as Brønsted bases: **(a)** H_2NOH, **(b)** N^{3-}, **(c)** H^-, **(d)** HSO_3^-.

14.13 **(a)** What is an amphiprotic substance? **(b)** Give four examples of such substances. Select both molecules and ions as your examples. **(c)** Write chemical equations to illustrate the characteristic behavior of the substances that you listed in part **(b)**.

14.14 What is the leveling effect of a solvent? Use chemical equations in your answer, and describe the leveling of both acids and bases.

14.15 What is the difference in meaning between the terms amphiprotic and amphoteric?

14.16 Each of the following reactions is displaced to the right. **(a)** Arrange all the Brønsted acids that appear in these equations according to decreasing acid strength. **(b)** Make a similar list for the Brønsted bases.

(a) $H_3O^+ + H_2PO_4^- \rightleftharpoons H_3PO_4 + H_2O$
(b) $HCN + OH^- \rightleftharpoons H_2O + CN^-$
(c) $H_3PO_4 + CN^- \rightleftharpoons HCN + H_2PO_4^-$
(d) $H_2O + NH_2^- \rightleftharpoons NH_3 + OH^-$

---

* The more difficult problems are marked with asterisks. The appendix contains answers to color-keyed problems.

14.17 On the basis of your lists from problem 14.16 would you expect an appreciable reaction (over 50%) between the species listed in each of the following?

(a) $H_3O^+ + CN^- \longrightarrow$

(b) $NH_3 + CN^- \longrightarrow$

(c) $HCN + H_2PO_4^- \longrightarrow$

(d) $H_3PO_4 + NH_2^- \longrightarrow$

14.18 Each of the following reactions is displaced to the right. (a) Arrange all the Brønsted acids that appear in equations according to decreasing acid strength. (b) Make a similar list for the Brønsted bases.

(a) $HCO_3^- + OH^- \rightleftharpoons H_2O + CO_3^{2-}$

(b) $HC_2H_3O_2 + HS^- \rightleftharpoons H_2S + C_2H_3O_2^-$

(c) $H_2S + CO_3^{2-} \rightleftharpoons HCO_3^- + HS^-$

(d) $HSO_4^- + C_2H_3O_2^- \rightleftharpoons HC_2H_3O_2 + SO_4^{2-}$

14.19 On the basis of your lists from problem 14.18 would you expect an appreciable reaction (over 50%) between the species listed in each of the following?

(a) $HCO_3^- + C_2H_3O_2^- \longrightarrow$

(b) $HSO_4^- + HS^- \longrightarrow$

(c) $HC_2H_3O_2 + CO_3^{2-} \longrightarrow$

(d) $H_2S + C_2H_3O_2^- \longrightarrow$

Acid Strength and Molecular Structure

14.20 Compare the acidity of AsH_3, H_2Se, and HBr. How does acidity vary among the hydrides of the elements of a period?

14.21 Compare the acidity of H_2S, H_2Se, and H_2Te. How does acidity vary among the hydrides of elements of a group?

14.22 Which compound of each of the following pairs is the stronger acid:

(a) H_3PO_4 or H_3AsO_4

(b) H_3AsO_3 or H_3AsO_4

(c) H_2SO_4 or H_2SO_3

(d) H_3BO_3 or H_2CO_3

(e) H_2Se or HBr

14.23 Which compound of each of the following pairs is the stronger base?

(a) P^{3-} or S^{2-}

(b) PH_3 or NH_3

(c) SiO_3^{2-} or SO_3^{2-}

(d) NO_2^- or NO_3^-

(e) Br^- or F^-

14.24 Perchloric acid, $HClO_4$, is a stronger acid than chloric acid, $HClO_3$. On the other hand, periodic acid, H_5IO_6, is a weaker acid than iodic acid, HIO_3. (a) Draw

Lewis structures for the four compounds, and assign formal charges to the atoms of the molecules. (b) Explain the differences in acid strength on the basis of your structures of part (a).

14.25 (a) Draw Lewis structures (include formal charges) for selenic acid, H_2SeO_4, and telluric acid, H_6TeO_6. (b) Which is the stronger acid?

The Lewis Concept

14.26 Interpret the following reactions in terms of the Lewis theory:

(a) $S{=}C{=}S + SH^- \longrightarrow S_2CSH^-$

(b) $Fe(CO)_5 + H^+ \longrightarrow FeH(CO)_5^+$

(c) $Ag^+ + 2NH_3 \longrightarrow Ag(NH_3)_2^+$

(d) $Mn(CO)_5^- + H^+ \longrightarrow HMn(CO)_5$

(e) $HF + F^- \longrightarrow HF_2^-$

14.27 Interpret the following reactions in terms of the Lewis theory:

(a) $BeF_2 + 2F^- \longrightarrow BeF_4^{2-}$

(b) $H_2O + BF_3 \longrightarrow H_2OBF_3$

(c) $S + S^{2-} \longrightarrow S_2^{2-}$

(d) $H^- + H_2C{=}O \longrightarrow H_3CO^-$

(e) $AuCN + CN^- \longrightarrow Au(CN)_2^-$

*14.28 Interpret the following as Lewis displacement reactions. For each, state the type of displacement, what is displaced, and the agent for the displacement.

(a) $NH_3 + H^- \longrightarrow H_2 + NH_2^-$

(b) $HONO_2 + H^+ \longrightarrow NO_2^+ + H_2O$

(c) $[Co(NH_3)_5(H_2O)]^{3+} + Cl^- \longrightarrow$
$[Co(NH_3)_5Cl]^{2+} + H_2O$

(d) $Ge + GeS_2 \longrightarrow 2GeS$

(e) $O^{2-} + H_2O \longrightarrow 2OH^-$

(f) $Br_2 + FeBr_3 \longrightarrow Br^+ + FeBr_4^-$

(g) $CH_3Cl + AlCl_3 \longrightarrow CH_3^+ + AlCl_4^-$

(h) $CH_3I + OH^- \longrightarrow CH_3OH + I^-$

14.29 Sulfur dioxide is a significant contributor to atmospheric pollution. The sulfurous acid produced by the reaction of SO_2 and atmospheric moisture adversely affects the eyes, skin, and respiratory systems of humans and causes the corrosion of metals and the deterioration of other materials. The reaction of SO_2 and water can be roughly indicated as

$$H_2O + SO_2 \longrightarrow H_2OSO_2 \longrightarrow (HO)_2SO$$

(a) Draw Lewis structures (complete with formal charges) for the compounds given. (b) Interpret the two steps of the sequence in terms of the Lewis acid-base theory. (c) Why does the proton migration of the second step occur?

IONIC EQUILIBRIUM, PART I

CHAPTER

15

The principles of chemical equilibrium can be applied to equilibrium systems that involve molecules and ions in water solution. In pure water, H_3O^+ and OH^- ions exist in equilibrium with H_2O molecules from which they are derived. Other molecular substances (weak electrolytes) are partially ionized in water solution and exist in equilibrium with their ions. An understanding of these systems is important to the study of analytical chemistry.

15.1 Weak Electrolytes

Strong electrolytes are completely ionized in water solution. A 0.01 M solution of $CaCl_2$, for example, contains Ca^{2+} ions (at a concentration of 0.01 M), Cl^- ions (at a concentration of 0.02 M), and no $CaCl_2$ molecules at all. Weak electrolytes, however, are *incompletely* ionized in water solution. Dissolved molecules exist in equilibrium with their ions in such solutions. The following chemical equation represents the dissociation of acetic acid in water:

$$HC_2H_3O_2 + H_2O \rightleftharpoons H_3O^+ + C_2H_3O_2^-$$

The equilibrium constant for this reaction as written is

$$\frac{[H_3O^+][C_2H_3O_2^-]}{[HC_2H_3O_2][H_2O]} = K_a'$$

In dilute solutions, the concentration of water may be considered to be a constant. The number of moles of water used in forming H_3O^+ (which is about 0.001 mol in 1 liter of a 0.1 M solution of acetic acid) is very small in comparison to the large number of moles of water present (which is about 55.5 mol in 1 liter of the solution). If we combine $[H_2O]$ with K_a', we get

$$\frac{[H_3O^+][C_2H_3O_2^-]}{[HC_2H_3O_2]} = K_a'[H_2O] = K_a$$

We shall follow the practice of representing the concentration of hydronium ion, $[H_3O^+]$, by the symbol $[H^+]$. The simplified expression for the equilibrium constant K_a, sometimes called an acid dissociation constant, and the corresponding simplified chemical equation are

$$HC_2H_3O_2 \rightleftharpoons H^+ + C_2H_3O_2^-$$

$$\frac{[H^+][C_2H_3O_2^-]}{[HC_2H_3O_2]} = K_a$$

By convention, the *ions* are written on the *right* of the chemical equation for a reversible ionization. Consequently, the terms for the concentrations of the *ions* appear in the *numerator* of the expression for K_a

The degree of dissociation, α, of a weak electrolyte in water solution is the fraction of the total concentration of the electrolyte that is in ionic form at equilibrium. These values are frequently given in terms of percent ionized, which is 100α.

Example 15.1

At 25°C, a 0.1000 M solution of acetic acid ($HC_2H_3O_2$) is 1.34% ionized. What is the ionization constant, K_a, for acetic acid?

Solution

Let us assume that we have 1 liter of solution. Since the acid is 1.34% ionized, the number of moles of $HC_2H_3O_2$ in *ionic form* is

$$(0.0134)(0.1000 \text{ mol } HC_2H_3O_2) = 0.00134 \text{ mol } HC_2H_3O_2$$

We subtract this number from the total number of moles of $HC_2H_3O_2$ to get the the number of moles of acid in *molecular form*:

$$(0.1000 \text{ mol } HC_2H_3O_2) - (0.00134 \text{ mol } HC_2H_3O_2) = 0.0987 \text{ mol } HC_2H_3O_2$$

According to the chemical equation for the ionization, 1 mol of H^+ and 1 mol of $C_2H_3O_2^-$ are produced for every 1 mol of $HC_2H_3O_2$ that ionizes. Therefore, the concentrations at equilibrium are

$$HC_2H_3O_2 \rightleftharpoons H^+ + C_2H_3O_2^-$$
$$0.0987 \ M \qquad 0.00134 \ M \qquad 0.00134 \ M$$

These concentrations may be used to find the numerical value of the equilibrium constant:

$$K_a = \frac{[H^+][C_2H_3O_2^-]}{[HC_2H_3O_2]}$$

$$= \frac{(0.00134)(0.00134)}{(0.0987)}$$

$$= 1.82 \times 10^{-5}$$

In future problem work, we shall express equilibrium constants to two significant figures. Higher accuracy is usually not warranted.*

---

* Equilibrium concentrations should be written in terms of activities, not concentrations. The activity of an ion is a theoretical concentration that takes into account interionic attractions (see Section 10.12). The concentrations of ions in dilute solutions of weak electrolytes are so small, however, that interionic attractions are negligible. Under these conditions molar concentrations may be used rather than activities. Reasonably accurate results are obtained from this practice.

Example 15.2

What are the concentrations of all species present in 1.00 M acetic acid at 25°C? For $HC_2H_3O_2$, K_a is 1.8×10^{-5}.

Solution

If we let x equal the number of moles of $HC_2H_3O_2$ in ionic form in 1 liter of solution, the equilibrium concentrations are

$$HC_2H_3O_2 \rightleftharpoons H^+ + C_2H_3O_2^-$$
$$(1.00 - x)\ M \qquad x\ M \qquad x\ M$$

We substitute these values into the expression for K_a:

$$K_a = \frac{[H^+][C_2H_3O_2^-]}{[HC_2H_3O_2]}$$

$$1.8 \times 10^{-5} = \frac{x^2}{(1.00 - x)}$$

The result can be rearranged to

$$x^2 + (1.8 \times 10^{-5})x - 1.8 \times 10^{-5} = 0$$

An equation in the form

$$ax^2 + bx + c = 0$$

is called a **quadratic equation** (see the appendix). Solutions to a quadratic equation can be found by substitution in the **quadratic formula**:

$$x = \frac{-b \pm \sqrt{b^2 - 4ac}}{2a}$$

In the present problem, $a = 1$, $b = 1.8 \times 10^{-5}$, and $c = -1.8 \times 10^{-5}$. If we substitute these values into the quadratic formula, we get

$$x = 4.2 \times 10^{-3}\ M*$$

Therefore,

$$[H^+] = [C_2H_3O_2^-] = x = 4.2 \times 10^{-3}\ M$$
$$[HC_2H_3O_2] = (1.00 - x)\ M$$
$$= (1.00 - 0.0042)\ M$$
$$= 0.9958\ M$$

---

* There are always two solutions to a quadratic equation. One of them must be discarded because it represents a physical impossibility. The two solutions obtained in the present case are

$$x = 4.2 \times 10^{-3}\ M \qquad \text{and} \qquad x = -4.2 \times 10^{-3}\ M$$

The second solution is impossible. The concentration of $HC_2H_3O_2$ is $(1.00 - x)$. Substitution of the second value (which is a negative number) for x gives a final $HC_2H_3O_2$ concentration that would require more $HC_2H_3O_2$ than was supplied.

Notice that the value of K_a that we employed has two significant figures. Our answer, therefore, cannot be given to more than two significant figures. The concentration of $HC_2H_3O_2$ must be rounded off:

$$[HC_2H_3O_2] = 1.0\ M$$

This value is the same as the initial concentration of $HC_2H_3O_2$.

The use of the quadratic formula in problem solving may be avoided by making an approximation that is frequently employed in calculations involving aqueous equilibria. The subtraction of a very small number from a large number does not significantly change the value of the large number and may be neglected.

In the preceding example such a small amount of acetic acid is ionized (x) that the quantity $(1.00 - x)$, which is the concentration of acetic acid molecules, is for all practical purposes equal to 1.00 (as previously noted). By using 1.00 instead of $(1.00 - x)$ for the concentration of $HC_2H_3O_2$, we find

$$1.8 \times 10^{-5} = \frac{[H^+][C_2H_3O_2^-]}{[HC_2H_3O_2]}$$

$$= \frac{x^2}{1.00}$$

$$x = 4.2 \times 10^{-3}\ M$$

which is the same as the result obtained through the use of the quadratic formula.

The subtraction of a *small* number from another *small* number may *not* be neglected. Consequently, the quadratic formula *must* be used to solve some problems. A good procedure to determine when the simplified method is justified is the following:

1. Set up the problem in the same way as that used in Example 15.2. That is,

$$[HC_2H_3O_2] = (1.00 - x)\ M$$
$$[H^+] = x\ M$$
$$[C_2H_3O_2^-] = x\ M$$

2. Solve the problem using the simplified method. In the example, substitute 1.00 M for $[HC_2H_3O_2]$ instead of $(1.00 - x)\ M$, on the assumption that x is small in comparison to 1.00 M.

3. Check the answer obtained by the simplified method against the assumption made. In the case at hand, since $x = 4.2 \times 10^{-3}\ M$,

$$(1.00 - x) = (1.00 - 0.0042) = 0.9958\ M$$

which to two significant figures is 1.0 M. Our assumption was justified.

4. The values of ionization constants are given to two significant figures, and we will solve problems to this limit of accuracy. If the subtraction in step 3 changes the value (when it is recorded to two significant figures), then the assumption was not justified. The problem must be solved again, and this time the quadratic formula must be used.

Example 15.3

What are the concentrations of all species present in a 0.10 M solution of HNO_2 at 25°C? For HNO_2, K_a is 4.5×10^{-4}.

Solution

If we let x equal the number of moles of HNO_2 in ionic form in one liter of solution, the equilibrium concentrations are

$$HNO_2 \quad \rightleftharpoons \quad H^+ + NO_2^-$$
$$(0.10 - x)\,M \qquad x\,M \qquad x\,M$$

We attempt the simplified method and substitute 0.10 M for $[HNO_2]$ instead of $(0.10 - x)\,M$:

$$4.5 \times 10^{-4} = \frac{[H^+][NO_2^-]}{[HNO_2]}$$

$$= \frac{x^2}{0.10}$$

$$x = 6.7 \times 10^{-3}\,M$$

On this basis,

$$[HNO_2] = (0.10 - x)\,M$$
$$= (0.10 - 0.0067)\,M$$
$$= 0.093\,M$$

We can see that the use of the simplified method was not justified; x is not negligible in comparison to 0.10 M since the subtraction changes the value (when it is recorded to two significant figures) from 0.10 M to 0.093 M. We must solve the problem again and this time use the quadratic formula:

$$4.5 \times 10^{-4} = \frac{x^2}{(0.10 - x)}$$

$$x^2 + (4.5 \times 10^{-4})x - 4.5 \times 10^{-6} = 0$$

By use of the quadratic formula:

$$x = [H^+] = [NO_2^-] = 6.5 \times 10^{-3}\,M$$
$$(0.10 - x) = [HNO_2] = 0.094\,M$$

The percent ionization of a weak electrolyte is found by dividing the number of moles of solute in the ionic form by the total number of moles of solute and

Table 15.1   Ion concentrations and percent ionization of solutions of acetic acid at 25°C

Concentration of Solution (M)	$[H^+]$ or $[C_2H_3O_2^-]$ (M)	Percent Ionization
1.00	0.00426	0.426
0.100	0.00134	1.34
0.0100	0.000418	4.18
0.00100	0.000126	12.6

multiplying the fraction obtained by 100. The percent ionization of the acetic acid solution described in Example 15.2 is

$$\frac{4.2 \times 10^{-3}\ M}{1.00\ M} \times 100 = 0.42\%$$

The ion concentration and percent ionization of solutions of acetic acid of various concentrations are listed in Table 15.1. The percent ionization increases with dilution, which is consistent with Le Chatelier's principle. Addition of water to an equilibrium system of a weak electrolyte:

$$HC_2H_3O_2 + H_2O \rightleftharpoons H_3O^+ + C_2H_3O_2^-$$

shifts the equilibrium to the right. Proportionately more of the solute is found in the ionic form.

An equilibrium system of a weak electrolyte can be prepared from compounds that supply the ions of the weak electrolyte. The following example is an illustration of this procedure.

Example 15.4

What are the concentrations of all species present in a solution made by diluting 0.10 mol of HCl and 0.50 mol of $NaC_2H_3O_2$ to 1.0 liter?

Solution

Hydrochloric acid and sodium acetate are strong electrolytes. We may assume, therefore, that before equilibrium is attained, the concentration of H^+ is 0.10 M and the concentration of $C_2H_3O_2^-$ is 0.50 M. If we let x equal the number of moles per liter of acetic acid at equilibrium, the equilibrium concentrations are

$$HC_2H_3O_2 \rightleftharpoons \quad H^+ \quad + \quad C_2H_3O_2^-$$
$$(x)\ M \qquad (0.10 - x)\ M \qquad (0.50 - x)\ M$$

Substitution of these terms into the expression for K_a gives

$$1.8 \times 10^{-5} = \frac{(0.10 - x)(0.50 - x)}{x}$$

We note that x must be a comparatively large value since x appears in the denominator of the expression and K_a (which is 1.8×10^{-5}) is a relatively small number.

This equation, therefore, must be solved by means of the quadratic formula since x is not negligible in comparison with 0.10 or 0.50.

A simpler way to solve the problem is to assume that the reaction goes as far to the left as is possible and that then a portion of the acetic acid that has been formed dissociates into ions. In the reverse reaction, 0.10 mol of H^+ will react with 0.10 mol of $C_2H_3O_2^-$ to form 0.10 mol of $HC_2H_3O_2$. Therefore, all the H^+ will be used, 0.40 mol of $C_2H_3O_2^-$ will be left of the original 0.50 mol supplied, and 0.10 mol of $HC_2H_3O_2$ will be formed. We then assume that y mol of $HC_2H_3O_2$ dissociates into ions:

$$HC_2H_3O_2 \rightleftharpoons H^+ + C_2H_3O_2^-$$

	$HC_2H_3O_2$	H^+	$C_2H_3O_2^-$
after mixing:	—	0.10 M	0.50 M
reaction to left:	0.10 M	—	0.40 M
equilibrium:	$(0.10 - y)\ M$	$(y)\ M$	$(0.40 + y)\ M$

Now, y is indeed negligible in comparison with both 0.10 and 0.40, and we may simplify the problem by neglecting y in the terms for the concentrations of acetic acid and acetate ion. Thus,

$$1.8 \times 10^{-5} = \frac{[H^+][C_2H_3O_2^-]}{[HC_2H_3O_2]}$$

$$= \frac{y(0.40)}{(0.10)}$$

$$y = 4.5 \times 10^{-6}\ M$$

We can see from the value obtained for y that the approximations applied to $[C_2H_3O_2^-]$ and $[HC_2H_3O_2]$ are justified. To two significant figures, the equilibrium concentrations are

$$[H^+] = 4.5 \times 10^{-6}\ M$$

$$[C_2H_3O_2^-] = 0.40\ M$$

$$[HC_2H_3O_2] = 0.10\ M$$

The hydroxides of most metals are either strong electrolytes or slightly soluble compounds. There are some water-soluble compounds, however, that produce alkaline solutions in which equilibria exist between molecules and ions. The most important such compound is ammonia, NH_3, and the chemical equation for the reversible reaction is

$$H_2O + NH_3 \rightleftharpoons NH_4^+ + OH^-$$

The expression for the equilibrium constant (a base dissociation constant) for this reaction,

$$K_b' = \frac{[NH_4^+][OH^-]}{[NH_3][H_2O]}$$

simplifies to the following if the concentration of water is assumed to be a constant:

Table 15.2 Ionization constants at 25°C

	Weak Acids		
acetic	$HC_2H_3O_2 \rightleftharpoons H^+ + C_2H_3O_2^-$		1.8×10^{-5}
benzoic	$HC_7H_5O_2 \rightleftharpoons H^+ + C_7H_5O_2^-$		6.0×10^{-5}
chlorous	$HClO_2 \rightleftharpoons H^+ + ClO_2^-$		1.1×10^{-2}
cyanic	$HOCN \rightleftharpoons H^+ + OCN^-$		1.2×10^{-4}
formic	$HCHO_2 \rightleftharpoons H^+ + CHO_2^-$		1.8×10^{-4}
hydrazoic	$HN_3 \rightleftharpoons H^+ + N_3^-$		1.9×10^{-5}
hydrocyanic	$HCN \rightleftharpoons H^+ + CN^-$		4.0×10^{-10}
hydrofluoric	$HF \rightleftharpoons H^+ + F^-$		6.7×10^{-4}
hypobromous	$HOBr \rightleftharpoons H^+ + BrO^-$		2.1×10^{-9}
hypochlorous	$HOCl \rightleftharpoons H^+ + ClO^-$		3.2×10^{-8}
nitrous	$HNO_2 \rightleftharpoons H^+ + NO_2^-$		4.5×10^{-4}
	Weak Bases		
ammonia	$NH_3 + H_2O \rightleftharpoons NH_4^+ + OH^-$		1.8×10^{-5}
aniline	$C_6H_5NH_2 + H_2O \rightleftharpoons C_6H_5NH_3^+ + OH^-$		4.6×10^{-10}
dimethylamine	$(CH_3)_2NH + H_2O \rightleftharpoons (CH_3)_2NH_2^+ + OH^-$		7.4×10^{-4}
hydrazine	$N_2H_4 + H_2O \rightleftharpoons N_2H_5^+ + OH^-$		9.8×10^{-7}
methylamine	$CH_3NH_2 + H_2O \rightleftharpoons CH_3NH_3^+ + OH^-$		5.0×10^{-4}
pyridine	$C_5H_5N + H_2O \rightleftharpoons C_5H_5NH^+ + OH^-$		1.5×10^{-9}
trimethylamine	$(CH_3)_3N + H_2O \rightleftharpoons (CH_3)_3NH^+ + OH^-$		7.4×10^{-5}

$$K_b = K_b'[H_2O] = \frac{[NH_4^+][OH^-]}{[NH_3]}$$

Table 15.2 lists the ionization constants for some compounds that function as weak acids and some that function as weak bases. Certain ions also can function as weak acids and others as weak bases. These systems are discussed in later sections of this chapter.

The values of ionization constants change with temperature. At 25°C, K_a for acetic acid is 1.8×10^{-5}; at 100°C, K_a for this weak acid is 1.1×10^{-5}. Ionization constants are usually reported for equilibrium systems at 25°C.

15.2 The Ionization of Water

Pure water is itself a very weak electrolyte and ionizes according to the equation

$$H_2O + H_2O \rightleftharpoons H_3O^+ + OH^-$$

In simplified form this equation is

$$H_2O \rightleftharpoons H^+ + OH^-$$

The expression for the ionization constant derived from the simplified equation is

$$K = \frac{[H^+][OH^-]}{[H_2O]}$$

In dilute solutions the concentration of water is virtually a constant, and we may combine $[H_2O]$ with the constant K. Thus,

$$K[H_2O] = [H^+][OH^-]$$

This constant, $K[H_2O]$, is called the ion product of water, or the **water dissociation constant,** and is given the symbol K_w. At 25°C,

$$K_w = 1.0 \times 10^{-14} = [H^+][OH^-] \tag{15.1}$$

In pure water,

$$[H^+] = [OH^-] = x$$
$$[H^+][OH^-] = 1.0 \times 10^{-14}$$
$$x^2 = 1.0 \times 10^{-14}$$
$$x = 1.0 \times 10^{-7} \, M$$

The concentrations of both of the ions of water are equal to $1.0 \times 10^{-7} \, M$ in pure water or in any neutral solution at 25°C. In 1 liter, only 10^{-7} mol of water is in ionic form out of a total of approximately 55.5 mol.

Both $H^+(aq)$ and $OH^-(aq)$ exist in any aqueous solution. In an acid solution the concentration of $H^+(aq)$ is larger than $1.0 \times 10^{-7} \, M$ and larger than the $OH^-(aq)$ concentration. In an alkaline solution the $OH^-(aq)$ concentration is larger than $1.0 \times 10^{-7} \, M$ and larger than the $H^+(aq)$ concentration.

Example 15.5

What are $[H^+]$ and $[OH^-]$ in a 0.020 M solution of HCl?

Solution

The quantity of $H^+(aq)$ ion obtained from the ionization of water is negligible compared to that derived from the hydrochloric acid. Since HCl is a strong electrolyte, $[H^+] = 0.020 \, M$:

$$[H^+][OH^-] = 1.0 \times 10^{-14}$$
$$(2.0 \times 10^{-2})[OH^-] = 1.0 \times 10^{-14}$$
$$[OH^-] = \frac{1.0 \times 10^{-14}}{2.0 \times 10^{-2}}$$
$$[OH^-] = 5.0 \times 10^{-13} \, M$$

Notice that $[OH^-]$ is extremely small. In this solution there would be one OH^- ion for every 40 billion H^+ ions.

Example 15.6

What are $[H^+]$ and $[OH^-]$ in a 0.0050 M solution of NaOH?

Solution

Sodium hydroxide is a strong electrolyte, and therefore $[OH^-] = 5.0 \times 10^{-3}\ M$:

$$[H^+][OH^-] = 1.0 \times 10^{-14}$$

$$[H^+](5.0 \times 10^{-3}) = 1.0 \times 10^{-14}$$

$$[H^+] = \frac{1.0 \times 10^{-14}}{5.0 \times 10^{-3}}$$

$$[H^+] = 2.0 \times 10^{-12}\ M$$

15.3 pH

The concentration of H^+(aq) in a solution may be expressed in terms of the *pH* scale. The *pH* of a solution is defined as

$$pH = \log \frac{1}{[H^+]} = -\log[H^+] \tag{15.2}$$

The common logarithm of a number is the power to which 10 must be raised in order to get the number (see the appendix). If

$$a = 10^n$$

then

$$\log a = n$$

Therefore, if

$$a = 10^{-3}$$

then

$$\log a = -3$$

The *pH* is the *negative* logarithm of the hydrogen ion concentration. Therefore, if

$$[H^+] = 10^{-3}\ M$$

$$\log[H^+] = -3$$

$$pH = -\log[H^+] = 3$$

For a neutral solution, therefore,

A modern digital *pH* meter.
Corning Medical and Scientific Instruments.

$$[H^+] = 1.0 \times 10^{-7} \, M$$
$$pH = -\log(10^{-7}) = -(-7)$$
$$pH = 7$$

The pOH of a solution is defined in the same terms:

$$pOH = -\log[OH^-] \tag{15.3}$$

The relationship between pH and pOH can be derived from the water constant:

$$[H^+][OH^-] = 10^{-14}$$

We take the logarithm of each term:

$$\log[H^+] + \log[OH^-] = \log(10^{-14})$$

and multiply through by -1

$$-\log[H^+] - \log[OH^-] = -\log(10^{-14})$$

or

$$pH + pOH = 14 \tag{15.4}$$

For a 0.01 M solution of NaOH, which is a strong base,

$$[OH^-] = 10^{-2} \, M$$
$$pOH = -\log[OH^-] = 2$$

Since the sum of the pH and pOH of a solution at 25°C equals 14,

$$pH = 12$$

Example 15.7

What is the pH of a solution that is 0.050 M in H^+?

Solution

$$[H^+] = 5.0 \times 10^{-2} \, M$$
$$\log[H^+] = \log 5.0 + \log 10^{-2}$$

The logarithm of 5.0 can be found in the logarithm table that appears in the appendix:

$$\log[H^+] = 0.70 - 2.00 = -1.30$$
$$pH = 1.30$$

Example 15.8

What is the pH of a solution for which $[OH^-] = 0.030\ M$?

Solution

$$[OH^-] = 3.0 \times 10^{-2}$$

$$\log[OH^-] = \log 3.0 + \log 10^{-2}$$

$$= 0.48 - 2.00 = -1.52$$

$$pOH = 1.52$$

$$pH = 14.00 - pOH$$

$$= 14.00 - 1.52 = 12.48$$

An alternative solution is

$$[H^+][OH^-] = 1.0 \times 10^{-14}$$

$$[H^+] = \frac{1.0 \times 10^{-14}}{3.0 \times 10^{-2}} = 3.3 \times 10^{-13}$$

$$\log[H^+] = \log 3.3 + \log 10^{-13}$$
$$= 0.52 - 13.00 = -12.48$$

$$pH = 12.48$$

Example 15.9

What is the $[H^+]$ of a solution with a pH of 10.60?

Solution

Since

$$pH = \log[H^+]$$

$$\log[H^+] = -pH$$

In the present example, $pH = -10.60$, and therefore

$$\log[H^+] = -10.60$$

$$[H^+] = 10^{-10.60} = 2.5 \times 10^{-11}\ M$$

If a logarithm table is used, the value of $\log[H^+]$ must be divided into two parts: a decimal portion (which is called a mantissa and which *must* be positive) and a negative whole number (which is called a characteristic). Thus,

$$\log[H^+] = -10.60 = 0.40 - 11.00$$

$$[H^+] = \text{antilog}\,0.40 \times \text{antilog}(-11)$$

The antilogarithm of 0.40 is obtained by finding the number that corresponds to this logarithm in the logarithm table (it is 2.5). The antilogarithm of -11 is 10^{-11}. Therefore,

$$[H^+] = 2.5 \times 10^{-11}\ M$$

The preceding example illustrates an important attribute of logarithms. The only significant figures in a logarithmic term, such as a pH, are found in the decimal portion. In Example 15.9, the mantissa (0.40) has two significant figures, and its antilogarithm (2.5) has two significant figures. The characteristic (-11) serves only to locate the decimal point in the final antilogarithm (10^{-11}). Hence, the pH values 1.30, 10.60, and 14.00 all have only *two* significant figures.

It should be kept in mind that pH relates to a power of 10. Hence, a solution of $pH = 1$ has a H^+(aq) concentration 100 times that of a solution of $pH = 3$ (not three times). Furthermore, since the pH is related to a *negative* exponent, the lower the pH value, the larger the concentration of H^+(aq). At $pH = 7$, a solution is neutral. Solutions with pH's below 7 are acidic. Those with pH's above 7 are alkaline. These relationships are summarized in Table 15.3.

The ionization constant of a weak acid or a weak base can be determined by measuring the pH of a solution of known concentration of the weak electrolyte.

Example 15.10

The pH of a 0.10 M solution of a weak acid HX is 3.30. What is the ionization constant of HX?

Table 15.3	The pH scale		
pH	$[H^+]$	$[OH^-]$	
14	10^{-14}	10^0	
13	10^{-13}	10^{-1}	
12	10^{-12}	10^{-2}	
11	10^{-11}	10^{-3}	increasing alkalinity
10	10^{-10}	10^{-4}	
9	10^{-9}	10^{-5}	
8	10^{-8}	10^{-6}	
7	10^{-7}	10^{-7}	neutrality
6	10^{-6}	10^{-8}	
5	10^{-5}	10^{-9}	
4	10^{-4}	10^{-10}	
3	10^{-3}	10^{-11}	increasing acidity
2	10^{-2}	10^{-12}	
1	10^{-1}	10^{-13}	
0	10^0	10^{-14}	

Solution

$$HX \rightleftharpoons H^+ + X^-$$

$$pH = 3.30$$

$$\log[H^+] = -3.30$$

$$[H^+] = 10^{-3.30} = 5.0 \times 10^{-4} \, M$$

Since the solution was prepared from HX alone,

$$[H^+] = [X^-] = 5.0 \times 10^{-4} \, M$$

The concentration of HX in the solution is for all practical purposes equal to 0.10 *M*. The small quantity of HX that dissociated need not be considered. The equilibrium concentrations, therefore, are

$$HX \rightleftharpoons H^+ + X^-$$
$$0.10 \, M \quad 5.0 \times 10^{-4} \, M \quad 5.0 \times 10^{-4} \, M$$

The value of the ionization constant is

$$K = \frac{[H^+][X^-]}{[HX]}$$

$$= \frac{(5.0 \times 10^{-4})^2}{1.0 \times 10^{-1}} = \frac{25 \times 10^{-8}}{1.0 \times 10^{-1}}$$

$$K = 2.5 \times 10^{-6}$$

15.4 Indicators

Indicators are organic compounds of complex structure that change color in solution as the *p*H changes. Methyl orange, for example, is red in solutions of *p*H below 3.1 and yellow in solutions of *p*H above 4.5. The color of this indicator is a varying mixture of yellow and red in the *p*H range between 3.1 and 4.5. Many indicators have been described and used. A few are listed in Table 15.4.

The *p*H of a solution is usually determined by use of a *p*H meter, but indicators may also be used for this purpose. If thymol blue is yellow in a test solution and methyl orange is red in another sample of the same solution, the *p*H of the solution is between 2.8 and 3.1. Reference to Table 15.4 shows that thymol blue is yellow only in solutions of *p*H greater than 2.8, and methyl orange is red only in solutions of *p*H less than 3.1. If enough indicators are employed, it is possible to determine *p*H values that are accurate to the first decimal place.

Indicators are weak acids or weak bases. Since they are intensely colored, only a few drops of a dilute solution of an indicator need be employed in any determination. Hence, the acidity of the solution in question is not significantly altered by the addition of the indicator.

If we let the symbol HIn stand for the litmus molecule (which is red) and the symbol In⁻ stand for the anion (which is blue) derived from the weak acid, the equation for the litmus equilibrium may be written

Table 15.4 Some indicators

Indicator	Acid Color	pH Range of Color Change	Alkaline Color
thymol blue	red	1.2–2.8	yellow
methyl orange	red	3.1–4.5	yellow
bromcresol green	yellow	3.8–5.5	blue
methyl red	red	4.2–6.3	yellow
litmus	red	5.0–8.0	blue
bromthymol blue	yellow	6.0–7.6	blue
thymol blue	yellow	8.0–9.6	blue
phenolphthalein	colorless	8.3–10.0	red
alizarin yellow	yellow	10.0–12.1	lavender

$$\underset{red}{HIn} \rightleftharpoons H^+ + \underset{blue}{In^-}$$

According to the principle of Le Chatelier, increasing the concentration of H^+ shifts the equilibrium to the left, and the red (or acid) color of HIn is observed. On the other hand, addition of OH^- decreases the concentration of H^+. The equilibrium shifts to the right, and the blue (or alkaline) color of In^- is observed.

The ionization constant for litmus is approximately equal to 10^{-7}:

$$10^{-7} = \frac{[H^+][In^-]}{[HIn]}$$

We may rearrange this expression in the following manner:

$$\frac{10^{-7}}{[H^+]} = \frac{[In^-]}{[HIn]}$$

At a pH of 5 or below, the red color of litmus is observed. If we substitute $[H^+] = 10^{-5}$, which corresponds to $pH = 5$, into the preceding expression, we get

$$\frac{10^{-7}}{10^{-5}} = \frac{1}{100} = \frac{[In^-] \longleftarrow blue}{[HIn] \longleftarrow red}$$

Thus, the mixture appears red to the eye when the concentration of the red HIn is 100 times (or more) that of the blue In^-.

The blue color of litmus is observed in solutions of $pH = 8$ or higher. If $[H^+] = 10^{-8}$,

$$\frac{10^{-7}}{10^{-8}} = \frac{10}{1} = \frac{[In^-] \longleftarrow blue}{[HIn] \longleftarrow red}$$

When the concentration of the blue In^- ion is 10 times that of the red HIn, or approximately 91% of the indicator is in ionic form, the blue color of the In^- ions completely masks the red color of the HIn molecules, and the mixture appears blue.

Thus, the blue color predominates when the concentration of the blue species is 10 times that of the red, whereas the red color predominates only when the concentration of the red form is 100 times that of the blue. This difference is not surprising since the blue color of litmus is a much stronger color than the red. Only when $[H^+] = K = 10^{-7}$ will the factor $K/[H^+] = 1$ and $[In^-] = [HIn]$. Hence, at $pH = 7$, litmus exhibits its "neutral" color (purple).

The pH range over which a given indicator changes color depends, therefore, upon the ionization constant of the indicator. For indicators that are weak acids, the smaller the value of K, the higher is the pH range of the color change.

15.5 The Common-Ion Effect

In a 0.1 M solution of acetic acid, methyl orange assumes its acid color, red. If sodium acetate is added to this solution, the color changes to yellow, showing that the addition causes the acidity of the solution to decrease. This experimental observation is readily explained on the basis of Le Chatelier's principle. The equilibrium

$$HC_2H_3O_2 \rightleftharpoons H^+ + C_2H_3O_2^-$$

is shifted to the left by the addition of the acetate ion from the sodium acetate, and the concentration of $H^+(aq)$ correspondingly decreases. Since acetic acid and sodium acetate have the acetate ion in common, this phenomenon is called the **common-ion effect**.

The concentration of $H^+(aq)$ in a solution of acetic acid to which acetate ion has been added can be calculated by means of the ionization constant of acetic acid.

Example 15.11

What is the concentration of H^+ in a 0.10 M solution of acetic acid that has been made 0.15 M in sodium acetate ($NaC_2H_3O_2$)?

Solution

Sodium acetate is a strong electrolyte. Therefore, the concentration of $C_2H_3O_2^-$ from this source is 0.15 M. The addition of this excess $C_2H_3O_2^-$ shifts the acetic acid equilibrium to the left:

$$HC_2H_3O_2 \rightleftharpoons H^+ + C_2H_3O_2^-$$

Consequently, the concentration of $HC_2H_3O_2$ may be taken as equal to the total $HC_2H_3O_2$ concentration (0.10 M) since so little is ionized. The concentration of $C_2H_3O_2^-$ may be taken as equal to the concentration of $C_2H_3O_2^-$ from the $NaC_2H_3O_2$ (0.15 M) since very little is supplied by the ionization of $HC_2H_3O_2$:

$$HC_2H_3O_2 \rightleftharpoons H^+ + C_2H_3O_2^-$$

0.10 M (x) M 0.15 M

$$1.8 \times 10^{-5} = \frac{[H^+][C_2H_3O_2^-]}{[HC_2H_3O_2]}$$

$$= \frac{[H^+](0.15)}{(0.10)}$$

$$[H^+] = 1.2 \times 10^{-5} \ M$$

The concentration of H^+ in a 0.10 M solution of pure $HC_2H_3O_2$ is $1.3 \times 10^{-3} \ M$ (see Table 15.1).

Example 15.12

What is the concentration of hydroxide ions in a solution made by dissolving 0.020 mol of ammonium chloride (NH_4Cl) in 100 ml of 0.15 M ammonia? Assume that the condition of the solid does not cause the volume of the solution to change.

Solution

Ammonium chloride is a strong electrolyte. The concentration of NH_4^+ from the NH_4Cl added is

$$? \ \text{mol} \ NH_4^+ = 1000 \ \text{ml soln} \left(\frac{0.020 \ \text{mol} \ NH_4^+}{100 \ \text{ml soln}}\right) = 0.20 \ \text{mol} \ NH_4^+$$

The concentration of NH_4^+, therefore, is 0.20 M:

$$NH_3 + H_2O \rightleftharpoons NH_4^+ + OH^-$$

0.15 M 0.20 M (x) M

$$1.8 \times 10^{-5} = \frac{[NH_4^+][OH^-]}{[NH_3]}$$

$$= \frac{(0.20)[OH^-]}{(0.15)}$$

$$[OH^-] = 1.4 \times 10^{-5} \ M$$

15.6 Buffers

It is sometimes necessary that a solution with a definite pH be prepared and stored. The preservation of such a solution is even more difficult than its preparation. If the solution comes in contact with the air, it will absorb carbon dioxide (an acid anhydride) and become more acidic. If the solution is stored in a glass bottle, alkaline impurities leached from the glass may alter the pH. **Buffer solutions** are capable of maintaining their pH at some fairly constant value even when small amounts of acid or base are added.

A buffer solution based on a weak acid contains relatively high concentrations of both the weak acid and a salt of the acid. Consider an acetic acid-acetate buffer in which the concentration of acetic acid and the concentration of acetate ion (from sodium acetate) are both 1.00 M:

$$HC_2H_3O_2 \rightleftharpoons H^+ + C_2H_3O_2^-$$
$$ 1.00\ M \qquad ? \qquad 1.00\ M$$

$$\frac{[H^+][C_2H_3O_2^-]}{[HC_2H_3O_2]} = 1.8 \times 10^{-5}$$

Since $[HC_2H_3O_2] = [C_2H_3O_2^-]$:

$$[H^+] = 1.8 \times 10^{-5}\ M$$
$$pH = -\log(1.8 \times 10^{-5}) = 4.74$$

The pK of a weak electrolyte may be defined in a manner analogous to pH:

$$pK = -\log K \tag{15.5}$$

A solution of a weak acid in which the concentration of the anion is the same as the concentration of the undissociated acid has a pH equal to the pK_a of the acid.

In a sample of the buffer previously described, the quantities of $HC_2H_3O_2$ and $C_2H_3O_2^-(aq)$ are much larger than the quantity of $H^+(aq)$ present—approximately 50,000 times larger:

$$HC_2H_3O_2 \rightleftharpoons H^+ + C_2H_3O_2^-$$
$$ 1.0\ M \qquad 1.8 \times 10^{-5}\ M \qquad 1.0\ M$$

If a small quantity of $H^+(aq)$ ion is added, the large reserve of acetate ion will quickly convert it to acetic acid. If a small amount of $OH^-(aq)$ ion is added to the buffer, it will neutralize $H^+(aq)$ ion, but the large quantity of acetic acid present will by dissociation replace any $H^+(aq)$ ion removed and maintain the pH at a fairly constant value.

Buffers cannot withstand the addition of large amounts of acids or alkalies. The addition of 0.01 mol per liter of $H^+(aq)$ or $OH^-(aq)$ is about the maximum that any buffer can be expected to withstand. How successfully a buffer withstands such an addition is illustrated in the following example.

Example 15.13

The ionization constant of acetic acid to three significant figures is 1.82×10^{-5}. A buffer containing 1.00 M concentrations of acetic acid and sodium acetate has a pH of 4.742. (a) What is the pH of the solution after 0.01 mol of HCl has been added to 1 liter of the buffer? (b) What is the pH of the solution after the addition of 0.01 mol of NaOH?

Solution

(a) The 0.01 mol of H^+ from HCl converts an equivalent amount of acetate ion into acetic acid. Thus, the concentrations are

$$HC_2H_3O_2 \rightleftharpoons H^+ + C_2H_3O_2^-$$

	$HC_2H_3O_2$	H^+	$C_2H_3O_2^-$
buffer:	1.00 M	1.82×10^{-5} M	1.00 M
after H^+ addition:	1.01 M	?	0.99 M

$$\frac{[H^+][C_2H_3O_2^-]}{[HC_2H_3O_2]} = 1.82 \times 10^{-5}$$

$$\frac{[H^+](0.99)}{(1.01)} = 1.82 \times 10^{-5}$$

$$[H^+] = 1.82 \times 10^{-5} \frac{(1.01)}{(0.99)}$$

$$= 1.86 \times 10^{-5} \ M$$

$$pH = 4.731$$

Thus, the addition causes the pH to change 0.009 pH units. A similar addition to pure water would change the pH from 7.0 to 2.0—a change of 5.0 pH units.

(b) The addition of 0.0100 mol of OH^- would change the concentration of acetate ion to 1.01 M and the concentration of acetic acid to 0.99 M. Thus, the concentrations are

$$HC_2H_3O_2 \rightleftharpoons H^+ + C_2H_3O_2^-$$

	$HC_2H_3O_2$	H^+	$C_2H_3O_2^-$
buffer:	1.00 M	1.81×10^{-5} M	1.00 M
after OH^- addition:	0.99 M	?	1.01 M

$$\frac{[H^+][C_2H_3O_2^-]}{[HC_2H_3O_2]} = 1.82 \times 10^{-5}$$

$$\frac{[H^+](1.01)}{(0.99)} = 1.82 \times 10^{-5}$$

$$[H^+] = 1.82 \times 10^{-5} \frac{(0.99)}{(1.01)}$$

$$= 1.78 \times 10^{-5} \ M$$

$$pH = 4.749$$

A similar addition to water would cause the pH to rise from 7.0 to 12.0.

Alkaline buffers may also be prepared. If the base and its derived ion are present in equal concentrations,

$$pOH = pK_b$$

$$pH = 14.00 - pK_b$$

A solution with $[NH_3] = 0.10 \ M$ and $[NH_4^+] = 0.10 \ M$ is an example of this type of buffer:

$$NH_3 + H_2O \rightleftharpoons NH_4^+ + OH^-$$

0.10 M 0.10 M ?

$$\frac{[NH_4^+][OH^-]}{[NH_3]} = 1.8 \times 10^{-5}$$

$$[OH^-] = 1.8 \times 10^{-5} \, M$$

$$pOH = 4.74$$

$$pH = 9.26$$

Buffers may also be prepared in which the ratio of the concentration of weak electrolyte to the concentration of common ion is not 1:1; this technique may be used to obtain a buffer that has a pH (or pOH) different from the pK_a of the weak acid (or pK_b of the weak base). Let us assume that a buffer is to be prepared from a hypothetical weak acid, HA:

$$HA \rightleftharpoons H^+ + A^-$$

for which

$$\frac{[H^+][A^-]}{[HA]} = K_a$$

The logarithm of this expression is

$$\log \frac{[H^+][A^-]}{[HA]} = \log K_a$$

which can be also written

$$\log[H^+] + \log \frac{[A^-]}{[HA]} = \log K_a$$

If we multiply both sides of the equation by -1, we get

$$-\log[H^+] - \log \frac{[A^-]}{[HA]} = -\log K_a$$

or

$$pH - \log \frac{[A^-]}{[HA]} = pK_a$$

Upon rearrangement, we get what is called the **Henderson-Haselbalch equation**:

$$pH = pK_a + \log \frac{[A^-]}{[HA]} \tag{15.6}$$

In general, the ratio of ionic species to molecular species for an effective buffer should be between 1/10 and 10/1. This concentration range is equivalent to the following pH range:

$$pH = pK_a + \log \frac{1}{10} = pK_a + \log 10^{-1}$$

$$= pK_a - 1$$

$$pH = pK_a + \log \frac{10}{1}$$

$$= pK_a + 1$$

A buffer, therefore, can be prepared with a pH of any value between $(pK_a + 1)$ and $(pK_a - 1)$. An acetic acid-acetate buffer, for example, can be prepared that has any desired pH between 3.74 and 5.74 since the pK_a of acetic acid is 4.74.

Example 15.14

What concentration should be used to prepare a cyanic acid-cyanate buffer with a pH of 3.50?

Solution

$$pH = 3.50$$

$$\log[H^+] = -3.50 = 0.50 - 4.00$$

$$[H^+] = 3.2 \times 10^{-4} \, M$$

$$HOCN \rightleftharpoons H^+ + OCN^-$$

$$\frac{[H^+][OCN^-]}{[HOCN]} = 1.2 \times 10^{-4}$$

$$\frac{(3.2 \times 10^{-4})[OCN^-]}{[HOCN]} = 1.2 \times 10^{-4}$$

$$\frac{[OCN^-]}{[HOCN]} = \frac{1.2 \times 10^{-4}}{3.2 \times 10^{-4}}$$

$$\frac{[OCN^-]}{[HOCN]} = 0.38$$

Any solution in which $[OCN^-]/[HOCN]$ is 0.38 will have a pH of 3.50. For example, if $[OCN^-] = 0.38 \, M$, then $[HOCN]$ must be $1.00 \, M$; or if $[OCN^-] = 0.76 \, M$, $[HOCN] = 2.00 \, M$.

An alternative solution utilizes the Henderson-Haselbalch equation:

$$pH = pK_a + \log \frac{[OCN^-]}{[HOCN]}$$

$$3.50 = 3.92 + \log \frac{[OCN^-]}{[HOCN]}$$

$$\log \frac{[OCN^-]}{[HOCN]} = -0.42$$

$$\frac{[OCN^-]}{[HOCN]} = 0.38$$

Example 15.15

What is the *pH* of a solution made by mixing 100 ml of 0.15 *M* HCl and 200 ml of 0.20 *M* aniline ($C_6H_5NH_2$)? Assume that the volume of the final solution is 300 ml.

Solution

The number of moles of HCl used and the number of moles of aniline used are

$$? \text{ mol HCl} = 100 \text{ ml sol'n} \left(\frac{0.15 \text{ mol HCl}}{1000 \text{ ml sol'n}} \right)$$

$$= 0.015 \text{ mol HCl}$$

$$? \text{ mol aniline} = 200 \text{ ml sol'n} \left(\frac{0.20 \text{ mol aniline}}{1000 \text{ ml sol'n}} \right)$$

$$= 0.040 \text{ mol aniline}$$

One mole of aniline reacts with one mole of H^+. Thus,

$$C_6H_5NH_2 \quad + \quad H^+ \quad \longrightarrow C_6H_5NH_3^+$$

before reaction:	0.040 mol	0.015 mol	–
after reaction:	0.025 mol	–	0.015 mol

The molar concentrations in the final solution are

$$? \text{ mol } C_6H_5NH_3^+ = 1000 \text{ ml sol'n} \left(\frac{0.015 \text{ mol } C_6H_5NH_3^+}{300 \text{ ml sol'n}} \right)$$

$$= 0.050 \text{ mol } C_6H_5NH_3^+$$

$$? \text{ mol } C_6H_5NH_2 = 1000 \text{ ml sol'n} \left(\frac{0.025 \text{ mol } C_6H_5NH_2}{300 \text{ ml sol'n}} \right)$$

$$= 0.083 \text{ mol } C_6H_5NH_2$$

Therefore, the concentrations in the solution are

$$C_6H_5NH_2 + H_2O \rightleftharpoons C_6H_5NH_3^+ + OH^-$$
$$0.083 \; M \qquad\qquad\qquad 0.050 \; M$$

$$\frac{[C_6H_5NH_3^+][OH^-]}{[C_6H_5NH_2]} = 4.6 \times 10^{-10}$$

$$\frac{(0.050)[OH^-]}{(0.083)} = 4.6 \times 10^{-10}$$

$$[OH^-] = 7.6 \times 10^{-10} \; M$$

$$[H^+] = 1.3 \times 10^{-5} \; M$$

$$pH = 4.89$$

The use of buffers is an important part of many industrial processes. Examples are electroplating and the manufacture of leather, photographic materials, and dyes. In bacteriological research, culture media are generally buffered to maintain the pH required for the growth of the bacteria being studied. Buffers are used extensively in analytical chemistry and are used to calibrate pH meters. Human blood is buffered to a pH of 7.4 by means of bicarbonate, phosphate, and complex protein systems.

15.7 Polyprotic Acids

Polyprotic acids are acids that contain more than one acid hydrogen per molecule. Examples include sulfuric acid (H_2SO_4), oxalic acid ($H_2C_2O_4$), phosphoric acid (H_3PO_4), and arsenic acid (H_3AsO_4). Polyprotic acids ionize in a stepwise manner, and there is an ionization constant for each step. Numbers are added to the subscripts of the symbol K_a in order to specify the step to which the constant applies.

Phosphoric acid is triprotic and ionizes in three steps:

$$H_3PO_4 \rightleftharpoons H^+ + H_2PO_4^- \qquad \frac{[H^+][H_2PO_4^-]}{[H_3PO_4]} = K_{a1} = 7.5 \times 10^{-3}$$

$$H_2PO_4^- \rightleftharpoons H^+ + HPO_4^{2-} \qquad \frac{[H^+][HPO_4^{2-}]}{[H_2PO_4^-]} = K_{a2} = 6.2 \times 10^{-8}$$

$$HPO_4^{2-} \rightleftharpoons H^+ + PO_4^{3-} \qquad \frac{[H^+][PO_4^{3-}]}{[HPO_4^{2-}]} = K_{a3} = 1 \times 10^{-12}$$

Thus, in a solution of phosphoric acid three equilibria occur together with the water equilibrium.

The ionization of phosphoric acid is typical of all polyprotic acids in that the primary ionization is stronger than the secondary, and the secondary ionization is stronger than the tertiary. This trend in the value of the ionization constant is consistent with the nature of the particle that releases a proton in each step. One would predict that a proton would be released more readily by an uncharged molecule than by an uninegative ion and more readily by a uninegative ion than by a binegative ion.

No polyprotic acid is known for which all ionizations are complete. The primary ionization of sulfuric acid is essentially complete:

$$H_2SO_4 \longrightarrow H^+ + HSO_4^-$$

but the secondary ionization is not:

$$HSO_4^- \rightleftharpoons H^- + SO_4^{2-} \qquad \frac{[H^+][SO_4^{2-}]}{[HSO_4^-]} = K_{a2} = 1.3 \times 10^{-2}$$

Solutions of carbon dioxide are acidic. Carbon dioxide reacts with water to form carbonic acid, H_2CO_3. The reaction, however, is not complete—most of the carbon dioxide exists in solution as CO_2 molecules. Therefore, we shall indicate the primary ionizations as

$$CO_2 + H_2O \rightleftharpoons H^+ + HCO_3^- \qquad \frac{[H^+][HCO_3^-]}{[CO_2]} = K_{a1} = 4.2 \times 10^{-7}$$

Table 15.5   Ionization constants of some polyprotic acids

acid	equilibria	constants
arsenic	$H_3AsO_4 \rightleftharpoons H^+ + H_2AsO_4^-$ $H_2AsO_4^- \rightleftharpoons H^+ + HAsO_4^{2-}$ $HAsO_4^{2-} \rightleftharpoons H^+ + AsO_4^{3-}$	$K_{a1} = 2.5 \times 10^{-4}$ $K_{a2} = 5.6 \times 10^{-8}$ $K_{a3} = 3 \times 10^{-13}$
carbonic	$CO_2 + H_2O \rightleftharpoons H^+ + HCO_3^-$ $HCO_3^- \rightleftharpoons H^+ + CO_3^{2-}$	$K_{a1} = 4.2 \times 10^{-7}$ $K_{a2} = 4.8 \times 10^{-11}$
hydrosulfuric	$H_2S \rightleftharpoons H^+ + HS^-$ $HS^- \rightleftharpoons H^+ + S^{2-}$	$K_{a1} = 1.1 \times 10^{-7}$ $K_{a2} = 1.0 \times 10^{-14}$
oxalic	$H_2C_2O_4 \rightleftharpoons H^+ + HC_2O_4^-$ $HC_2O_4^- \rightleftharpoons H^+ + C_2O_4^{2-}$	$K_{a1} = 5.9 \times 10^{-2}$ $K_{a2} = 6.4 \times 10^{-5}$
phosphoric	$H_3PO_4 \rightleftharpoons H^+ + H_2PO_4^-$ $H_2PO_4^- \rightleftharpoons H^+ + HPO_4^{2-}$ $HPO_4^{2-} \rightleftharpoons H^+ + PO_4^{3-}$	$K_{a1} = 7.5 \times 10^{-3}$ $K_{a2} = 6.2 \times 10^{-8}$ $K_{a3} = 1 \times 10^{-12}$
phosphorus (diprotic)	$H_3PO_3 \rightleftharpoons H^+ + H_2PO_3^-$ $H_2PO_3^- \rightleftharpoons H^+ + HPO_3^{2-}$	$K_{a1} = 1.6 \times 10^{-2}$ $K_{a2} = 7 \times 10^{-7}$
sulfuric	$H_2SO_4 \rightleftharpoons H^+ + HSO_4^-$ $HSO_4^- \rightleftharpoons H^+ + SO_4^{2-}$	strong $K_{a2} = 1.3 \times 10^{-2}$
sulfurous	$SO_2 + H_2O \rightleftharpoons H^+ + HSO_3^-$ $HSO_3^- \rightleftharpoons H^+ + SO_3^{2-}$	$K_{a1} = 1.3 \times 10^{-2}$ $K_{a2} = 5.6 \times 10^{-8}$

where the symbol $[CO_2]$ is used to represent the total concentration of $CO_2(aq)$ and H_2CO_3. The second ionization step is

$$HCO_3^- \rightleftharpoons H^+ + CO_3^{2-} \qquad \frac{[H^+][CO_3^{2-}]}{[HCO_3^-]} = K_{a2} = 4.8 \times 10^{-11}$$

An analogous situation exists for solutions of sulfur dioxide in water. The acidity of aqueous SO_2 has been attributed to the ionization of sulfurous acid, H_2SO_3. However, H_2SO_3 has never been isolated in pure form. In solution it apparently exists in equilibrium with $SO_2(aq)$:

$$SO_2(aq) + H_2O \rightleftharpoons H_2SO_3(aq)$$

We shall represent the primary ionization of sulfurous acid as

$$SO_2 + H_2O \rightleftharpoons H^+ + HSO_3^-$$

The ionization constants of some polyprotic acids are listed in Table 15.5.

Polyprotic acids form more than one salt. Depending upon the stoichiometric ratio of reactants, the reaction of NaOH and H_2SO_4 yields either the normal salt Na_2SO_4 (sodium sulfate) or the acid salt $NaHSO_4$ (sodium hydrogen sulfate or sodium bisulfate). Three salts may be derived from phosphoric acid: NaH_2PO_4 (sodium dihydrogen phosphate), Na_2HPO_4 (sodium hydrogen phosphate), and Na_3PO_4 (sodium phosphate).

Example 15.16

Calculate $[H^+]$, $[H_2PO_4^-]$, $[HPO_4^{2-}]$, $[PO_4^{3-}]$, and $[H_3PO_4]$ in a 0.10 M solution of phosphoric acid.

Solution

The principal source of H^+ is the primary ionization. The H^+ produced by the other ionizations, as well as that from the ionization of water, is negligible in comparison. Furthermore, the concentration of $H_2PO_4^-$ derived from the primary ionization is not significantly diminished by the secondary ionization. Thus, we write

$$H_3PO_4 \rightleftharpoons H^+ + H_2PO_4^-$$
$$(0.10 - x)\,M \qquad (x)\,M \qquad (x)\,M$$

The problem must be solved by means of the quadratic formula since x is not negligible in comparison to $0.10\,M$:

$$\frac{[H^+][H_2PO_4^-]}{[H_3PO_4]} = 7.5 \times 10^{-3}$$

$$\frac{x^2}{(0.10 - x)} = 7.5 \times 10^{-3}$$

$$x = [H^+] = [H_2PO_4^-] = 2.4 \times 10^{-2}\,M$$

$$(0.10 - x) = [H_3PO_4] = 7.6 \times 10^{-2}\,M$$

The $[H^+]$ and $[H_2PO_4^-]$ apply to the secondary ionization. Therefore,

$$H_2PO_4^- \rightleftharpoons H^+ + HPO_4^{2-}$$
$$2.4 \times 10^{-2}\,M \qquad 2.4 \times 10^{-2}\,M$$

$$\frac{[H^+][HPO_4^{2-}]}{[H_2PO_4^-]} = 6.2 \times 10^{-8}$$

$$\frac{(2.4 \times 10^{-2})[HPO_4^{2-}]}{(2.4 \times 10^{-2})} = 6.2 \times 10^{-8}$$

$$[HPO_4^{2-}] = 6.2 \times 10^{-8}\,M$$

In any solution of H_3PO_4 that does not contain ions derived from another electrolyte, the concentration of the secondary ion is equal to K_{a2}.

For the tertiary ionization,

$$HPO_4^{2-} \rightleftharpoons H^+ + PO_4^{3-}$$
$$6.2 \times 10^{-8}\,M \qquad 2.4 \times 10^{-2}\,M$$

$$\frac{[H^+][PO_4^{3-}]}{[HPO_4^{2-}]} = 1 \times 10^{-12}$$

$$\frac{(2.4 \times 10^{-2})[PO_4^{3-}]}{(6.2 \times 10^{-8})} = 1 \times 10^{-12}$$

$$[PO_4^{3-}] = 3 \times 10^{-18}\,M$$

Example 15.17

What are $[H^+]$, $[HS^-]$, $[S^{2-}]$, and $[H_2S]$ in a $0.10\,M$ solution of H_2S?

Solution

K_{a1} for H_2S is 1.1×10^{-7}. Therefore, the small amount of H_2S that ionizes is negligible in comparison with the original concentration of H_2S. In addition, the concentrations of H^+ and HS^- are not significantly altered by the secondary ionization ($K_{a2} = 1.0 \times 10^{-14}$). Thus:

$$H_2S \rightleftharpoons H^+ + HS^-$$
$$0.10\,M \qquad (x)\,M \quad (x)\,M$$

$$\frac{[H^+][HS^-]}{[H_2S]} = 1.1 \times 10^{-7}$$

$$\frac{x^2}{0.10} = 1.1 \times 10^{-7}$$

$$x = [H^+] = [HS^-] = 1.0 \times 10^{-4}\,M$$

These concentrations also apply to the secondary ionization:

$$HS^- \rightleftharpoons H^+ + S^{2-}$$
$$1.0 \times 10^{-4}\,M \qquad 1.0 \times 10^{-4}\,M \qquad ?$$

$$\frac{[H^+][S^{2-}]}{[HS^-]} = 1.0 \times 10^{-14}$$

$$\frac{(1.0 \times 10^{-4})[S^{2-}]}{(1.0 \times 10^{-4})} = 1.0 \times 10^{-4}\,M$$

$$[S^{2-}] = 1.0 \times 10^{-14}\,M$$

The concentration of the secondary ion is equal to K_{a2} in any solution of H_2S that does not contain ions derived from another electrolyte.

The product of the expressions for the two ionizations of H_2S is

$$\left(\frac{[H^+][HS^-]}{[H_2S]}\right)\left(\frac{[H^+][S^{2-}]}{[HS^-]}\right) = K_{a1}K_{a2}$$

$$\frac{[H^+]^2\,[S^{2-}]}{[H_2S]} = (1.1 \times 10^{-7})(1.0 \times 10^{-14}) = 1.1 \times 10^{-21}$$

This very convenient relationship can be misleading. Superficially it looks as though it applies to a process in which one sulfide ion is produced for every two H^+ ions. However, the ionization of H_2S does not proceed in this manner. In any solution of H_2S, the concentration of $H^+(aq)$ is much larger than the concentration of sulfide ion (see Example 15.17). The majority of the H_2S molecules that ionize do so only to the HS^- stage, and S^{2-} ions result only from the small ionization of the secondary ion.

At $25°C$ a saturated solution of H_2S is $0.10\,M$. For a *saturated solution*, therefore,

$$\frac{[H^+]^2[S^{2-}]}{(0.10)} = 1.1 \times 10^{-21}$$

$$[H^+]^2[S^{2-}] = 1.1 \times 10^{-22}$$

This relation can be used to calculate the sulfide ion concentration of a solution of known pH that has been saturated with H_2S.

Example 15.18

What is the sulfide ion concentration of a dilute HCl solution that has been saturated with H_2S if the pH of the solution is 3.00?

Solution

Since the $pH = 3.00$,

$$[H^+] = 1.0 \times 10^{-3} \, M$$

Therefore,

$$[H^+]^2 [S^{2-}] = 1.1 \times 10^{-22}$$
$$(1.0 \times 10^{-3})^2 [S^{2-}] = 1.1 \times 10^{-22}$$
$$[S^{2-}] = 1.1 \times 10^{-16} \, M$$

In a saturated solution of pure H_2S (see Example 15.17), $[S^{2-}] = 1.0 \times 10^{-14}$ M. In the H_2S solution described in this problem, the common ion, H^+, has repressed the ionization of H_2S. In addition, since the solution contains H^+ ions from a source other than H_2S, $[H^+]$ does not equal $[HS^-]$, and consequently $[S^{2-}]$ does not equal K_{a2}.

15.8 Ions That Function as Acids and Bases

That the anions of polyprotic acids (such as $H_2PO_4^-$ and HS^-) have acidic properties is not surprising. What is perhaps unexpected, however, is that certain ions derived from normal salts (such as $C_2H_3O_2^-$, NO_2^-, NH_4^+, and Fe^{3+}) form acidic or basic solutions:

1. Anions derived from weak acids (such as $C_2H_3O_2^-$ and NO_2^-) form basic solutions.

2. Cations derived from weak bases (such as NH_4^+ and Fe^{3+}) form acidic solutions.

We will consider the anions of weak acids first. In water solution, the acetate ion reacts with water to increase the concentration of OH^- ions:

$$C_2H_3O_2^- + H_2O \rightleftharpoons HC_2H_3O_2 + OH^-$$

This reaction of the acetate ion is similar to that of any other weak base, such as NH_3, with water:

$$NH_3 + H_2O \rightleftharpoons NH_4^+ + OH^-$$

The fact that the acetate ion has a charge and the ammonia molecule does not is unimportant. Both species are bases, and both equilibria have corresponding K_b

values. The reaction of an ion with water, however, is sometimes called a **hydrolysis reaction.**

According to the Brønsted theory (see Section 14.3), the acetate ion is the conjugate base of acetic acid. The chemical equations for the two equilibria and corresponding expressions for the equilibrium constants are

$$HC_2H_3O_2 \rightleftharpoons H^+ + C_2H_3O_2^- \qquad K_a = \frac{[H^+][C_2H_3O_2^-]}{[HC_2H_3O_2]}$$

$$C_2H_3O_2^- + H_2O \rightleftharpoons HC_2H_3O_2 + OH^- \qquad K_b = \frac{[HC_2H_3O_2][OH^-]}{[C_2H_3O_2^-]}$$

The two constants are related in the following way. The product of K_a and K_b is

$$K_aK_b = \left(\frac{[H^+][C_2H_3O_2^-]}{[HC_2H_3O_2]}\right)\left(\frac{[HC_2H_3O_2][OH^-]}{[C_2H_3O_2^-]}\right)$$

$$K_aK_b = [H^+][OH^-]$$

Since $[H^+][OH^-]$ is equal to the water dissociation constant, K_w,

$$K_aK_b = K_w \qquad\qquad\qquad (15.7)$$

This relation provides a convenient way to obtain the value of K_b for the anion derived from a weak acid:

$$K_b = K_w/K_a \qquad\qquad\qquad (15.8)$$

In the case of the K_b for the hydrolysis of the acetate ion,

$$K_b = (1.0 \times 10^{-14})/(1.8 \times 10^{-5}) = 5.6 \times 10^{-10}$$

Anions derived from *strong* acids (such as Cl^- from HCl) and cations derived from *strong* bases (such as Na^+ from $NaOH$) do not react with water to affect the *p*H. An equilibrium of this type (a hydrolysis equilibrium) results only when the ion can form a molecule or ion that is a *weak electrolyte* in the reaction with water. Strong acids and bases do not exist as molecules in water solution.

Example 15.19

What is the *p*H of a 0.10 *M* solution of $NaC_2H_3O_2$?

Solution

The salt $NaC_2H_3O_2$ completely dissociates into Na^+ and $C_2H_3O_2^-$ ions in water solution. The Na^+ ion does react with H_2O. Let *x* equal the equilibrium concentration of $HC_2H_3O_2$ that results from the hydrolysis of $C_2H_3O_2^-$ ions:

$$C_2H_3O_2^- + H_2O \rightleftharpoons HC_2H_3O_2 + OH^-$$

| 0.10 *M* | | (*x*) *M* | (*x*) *M* |

Since the value of K_b is very small (5.6×10^{-10}), x is small and $[C_2H_3O_2^-]$ may be assumed to be equal to 0.10 M rather than $(0.10 - x)$ M:

$$\frac{[HC_2H_3O_2][OH^-]}{[C_2H_3O_2^-]} = K_b$$

$$\frac{x^2}{0.10} = 5.6 \times 10^{-10}$$

$$x^2 = 5.6 \times 10^{-11}$$

$$x = [HC_2H_3O_2] = [OH^-] = 7.5 \times 10^{-6} \ M$$

$$pOH = -\log(7.5 \times 10^{-6}) = 5.12$$

$$pH = 14.00 - 5.12 = 8.88$$

The weaker the electrolyte from which an ion is derived, the more extensive is its reaction with water. A 0.1 M solution of sodium acetate has a pH of 8.9, and the pH of a 0.1 M solution of sodium cyanide is 11.2. In each case it is the hydrolysis of the anion that causes the solution to be alkaline. The sodium ion does not undergo hydrolysis:

$$C_2H_3O_2^- + H_2O \rightleftharpoons HC_2H_3O_2 + OH^-$$

$$CN^- + H_2O \rightleftharpoons HCN + OH^-$$

Since HCN ($K_a = 4.0 \times 10^{-10}$) is a weaker electrolyte than $HC_2H_3O_2$ ($K_a = 1.8 \times 10^{-5}$), HCN does a better job of tying up protons than $HC_2H_3O_2$ does. Therefore, the hydrolysis of CN^- is more complete than that of $C_2H_3O_2^-$, and the concentration of OH^- is higher in the NaCN solution than in the $NaC_2H_3O_2$ solution. Note, however, that in both of these systems the position of equilibrium is such that the reverse reaction, which may be regarded as a neutralization, proceeds to a greater extent than the forward reaction since H_2O is a weaker electrolyte than either HCN or $HC_2H_3O_2$.

A cation derived from a weak base reacts with water to form an acidic solution. Consider the reaction of the ammonium ion:

$$NH_4^+ + H_2O \rightleftharpoons NH_3 + H_3O^+$$

This reaction of NH_4^+ is similar to the ionization of any other weak acid such as $HC_2H_3O_2$. If we represent the acid dissociation of NH_4^+ ion our customary way, we get

$$NH_4^+ \rightleftharpoons H^+ + NH_3$$

The equilibrium constant for this system, a K_a, can be derived by using the K_b for the base dissociation of NH_3, which is the conjugate base of NH_4^+:

$$NH_3 + H_2O \rightleftharpoons NH_4^+ + OH^-$$

According to the relation

$$K_aK_b = K_w$$

the acid dissociation constant may be found by substitution into

$$K_a = K_w/K_b \qquad (15.9)$$

For the NH_4^+/NH_3 system,

$$K_a = (1.0 \times 10^{-14})/(1.8 \times 10^{-5}) = 5.6 \times 10^{-10}$$

Example 15.20

What is the pH of a 0.30 M solution of NH_4Cl?

Solution

In water solution, NH_4Cl completely dissociates into NH_4^+ and Cl^- ions. The Cl^- ion, since it is derived from a strong acid, does not react with water. Let x equal the equilibrium concentration of NH_3 from the hydrolysis of NH_4^+:

$$NH_4^+ \rightleftharpoons H^+ + NH_3$$
$$0.30\ M \qquad (x)\ M \qquad (x)\ M$$

$$\frac{[H^+][NH_3]}{[NH_4^+]} = 5.6 \times 10^{-10}$$

$$\frac{x^2}{0.30} = 5.6 \times 10^{-10}$$

$$x^2 = 1.7 \times 10^{-10}$$

$$x = [H^+] = [NH_3] = 1.3 \times 10^{-5}\ M$$

$$pH = -\log(1.3 \times 10^{-5}) = 4.89$$

The cations of the I A metals, as well as Ca^{2+}, Sr^{2+}, and Ba^{2+}, do not react with H_2O since they are derived from strong bases. Most other metal cations, however, do hydrolyze. In the hydrolysis of a metal cation a coordinated water molecule of the hydrated cation donates a proton to a free water molecule:

$$Fe(H_2O)_6^{3+}(aq) + H_2O \rightleftharpoons Fe(OH)(H_2O)_5^{2+}(aq) + H_3O^+(aq)$$

Such equations are sometimes written without indicating the coordinated water:

$$Fe^{3+}(aq) + H_2O \rightleftharpoons Fe(OH)^{2+}(aq) + H^+(aq)$$

Additional steps in the hydrolysis of the ion produce $Fe(OH)_2^+$, $[Fe_2(OH)_2(H_2O)_8]^{4+}$, polynuclear ions of a higher order, and ultimately a precipitate of hydrous Fe_2O_3.

Mathematical analysis of the hydrolysis of metal cations is complicated by several factors. As in the hydrolysis of the Fe^{3+} ion, more than one hydrolytic product usually exists, and many of these are polynuclear. For many systems reliable values for the equilibrium constants are not available, and in many instances not all of the equilibria have been identified. Occasionally, reaction proceeds to the point where the metal hydroxide or the hydrous metal oxide precipitates.

The hydrolysis of an anion derived from a weak polyprotic acid proceeds in several steps. The acid dissociation constants for H_2S are

$$H_2S \rightleftharpoons H^+ + HS^- \qquad K_{a1} = 1.1 \times 10^{-7}$$

$$HS^- \rightleftharpoons H^+ + S^{2-} \qquad K_{a2} = 1.0 \times 10^{-14}$$

Therefore, the base dissociation constants for the ions are

$$S^{2-} + H_2O \rightleftharpoons HS^- + OH^- \qquad K_{b1} = \frac{K_w}{K_{a2}} = 1.0$$

$$HS^- + H_2O \rightleftharpoons H_2S + OH^- \qquad K_{b2} = \frac{K_w}{K_{a1}} = 9.1 \times 10^{-8}$$

Notice that the *first* base dissociation constant is obtained by dividing the water constant by the *second* acid dissociation constant of H_2S.

In solutions of a soluble sulfide the first step of the hydrolysis of the sulfide ion is so nearly complete that it far overshadows the second, and the acidity of the solution may be calculated by neglecting the hydrolysis of the HS^- ion.

Example 15.21

What is the pH of a 0.10 M solution of Na_2S?

Solution

$$S^{2-} + H_2O \rightleftharpoons HS^- + OH^-$$

$$(0.10 - x) \, M \qquad\qquad (x) \, M \quad (x) \, M$$

$$\frac{[HS^-][OH^-]}{[S^{2-}]} = 1.0$$

$$\frac{x^2}{(0.10 - x)} = 1.0$$

$$x^2 + x - 0.10 = 0$$

This equation is solved by using the quadratic formula:

$$x = \frac{-1.00 + \sqrt{1.00 + 0.40}}{2.00}$$

$$x = [OH^-] = 9.2 \times 10^{-2} \, M$$

$$pOH = 1.04$$

$$pH = 12.96$$

The pH of a solution of a normal salt can be predicted on the basis of the strengths of the acid and base from which the salt is derived:

1. **Salt of a strong base and a strong acid.** Examples are: NaCl, KNO_3, and $Ba(ClO_3)_2$. Neither cation nor anion hydrolyzes. The solution has a pH of 7.

2. **Salt of a strong base and a weak acid.** Examples are: KNO_2, $Ca(C_2H_3O_2)_2$, and NaCN. The anion hydrolyzes to produce OH^- ions. The solution has a pH that is higher than 7.

3. **Salt of a weak base and a strong acid.** Examples are: NH_4NO_3, $FeBr_2$, and $AlCl_3$. The cation hydrolyzes to produce H_3O^+ ions. The pH of the solution is below 7.

4. **Salt of a weak base and a weak acid.** Examples are: $NH_4C_2H_3O_2$, NH_4CN, and $Cu(NO_2)_2$. Both cation and anion hydrolyze. The pH of the solution depends upon the extent to which each ion hydrolyzes. The pH of a solution of $NH_4C_2H_3O_2$ is 7 since NH_3 ($K_b = 1.8 \times 10^{-5}$) and $HC_2H_3O_2$ ($K_a = 1.8 \times 10^{-5}$) are equally weak. The pH of a solution of NH_4CN, on the other hand, is above 7 because HCN ($K_a = 4.0 \times 10^{-10}$) is a weaker acid than NH_3 ($K_b = 1.8 \times 10^{-5}$) is a base. As a consequence, the CN^- hydrolyzes to a greater extent (producing OH^-) than the NH_4^+ does (producing H_3O^+).

The pH of a solution of an acid salt (such as NaHS, NaH_2PO_4, Na_2HPO_4, and $NaHCO_3$) is affected not only by the hydrolysis of the anion but also by the acid dissociation of the anion. Solutions of these salts may be acidic or alkaline. The two important equilibria in solutions of NaH_2PO_4, for example, are the acid dissociation of the $H_2PO_4^-$ ion:

$$H_2PO_4^- \rightleftharpoons H^+ + HPO_4^{2-} \qquad K_{a2} = 6.2 \times 10^{-8}$$

and the hydrolysis of the $H_2PO_4^-$ ion:

$$H_2PO_4^- + H_2O \rightleftharpoons H_3PO_4 + OH^-$$

$$K_{b3} = \frac{K_w}{K_{a1}} = \frac{1.0 \times 10^{-14}}{7.5 \times 10^{-3}} = 1.3 \times 10^{-12}$$

Since the equilibrium constant for the dissociation (which produces H^+) is larger than the equilibrium constant for the hydrolysis (which produces OH^-), a solution of NaH_2PO_4 is acidic.

On the other hand, a solution of Na_2HPO_4 is alkaline. The two pertinent equilibria are

$$HPO_4^{2-} \rightleftharpoons H^+ + PO_4^{3-} \qquad\qquad K_{a3} = 1 \times 10^{-12}$$

$$HPO_4^{2-} + H_2O \rightleftharpoons H_2PO_4^- + OH^- \qquad K_{b2} = \frac{K_w}{K_{a2}} = \frac{1.0 \times 10^{-14}}{6.2 \times 10^{-8}} = 1.6 \times 10^{-7}$$

The base dissociation occurs to a greater extent than the acid dissociation, and the pH of the solution is higher than 7.

15.9 Acid-Base Titrations

We are now in a position to study acid-base titrations in some detail. Let us consider the titration of a 50.0 ml sample of 0.100 M HCl with a 0.100 M NaOH solution. Since both HCl and NaOH are strong electrolytes, the only equilibrium to be considered is the water equilibrium.

The concentration of $H^+(aq)$ in the original 50.0 ml sample of acid in the titration flask is $0.100\ M$ (or $10^{-1}\ M$). The pH, therefore, is 1.00.

After the addition of 10.0 ml of $0.100\ M$ NaOH from the buret, the equivalent of 40.0 ml of $0.100\ M$ HCl remains unneutralized. The number of moles of $H^+(aq)$ in the titration flask, therefore, is

$$? \text{ mol } H^+ = 40.0 \text{ ml} \left(\frac{0.100 \text{ mol } H^+}{1000 \text{ ml}} \right) = 4.00 \times 10^{-3} \text{ mol } H^+$$

The total volume of the solution after the addition is 60.0 ml (which is 6.00×10^{-2} liter), and, therefore,

$$[H^+] = \frac{4.00 \times 10^{-3} \text{ mol } H^+}{6.00 \times 10^{-2} \text{ liter}} = 6.67 \times 10^{-2}\ M$$

$$pH = 1.18$$

When 50.0 ml of $0.100\ M$ NaOH has been added from the buret, the equivalence point is reached. All the acid is neutralized, and the pH is 7.00.

As NaOH solution is added beyond the equivalence point, the solution in the titration flask becomes increasingly alkaline. When 60.0 ml of $0.100\ M$ NaOH has been added, for example, the solution contains the equivalent of 10.0 ml of $0.100\ M$ NaOH:

$$? \text{ mol } OH^- = 10.0 \text{ ml} \left(\frac{0.100 \text{ mol } OH^-}{1000 \text{ ml}} \right) = 1.00 \times 10^{-3} \text{ mol } OH^-$$

The total volume of the solution at this point is 110.0 ml (which is 1.10×10^{-1} liter), and, therefore,

$$[OH^-] = \frac{1.00 \times 10^{-3} \text{ mol } OH^-}{1.10 \times 10^{-1} \text{ liter}} = 9.09 \times 10^{-3}\ M$$

$$pOH = 2.04$$

$$pH = 11.96$$

The values in Table 15.6 were obtained from calculations such as these. The data of Table 15.6 are plotted in Figure 15.1. Notice that the curve rises sharply in the section around the equivalence point. Whereas the first 49.9 ml of NaOH solution added causes the pH to change by three units, the next 0.2 ml added causes a change of *six* units in the pH.

The color-change ranges of three indicators are shown in Figure 15.1. Each indicator exhibits its acid color at pH's below its range and its alkaline color at pH's above its range. In the course of the titration, the pH of the solution changes along the curve from left to right. The end point of the titration is signaled when the indicator changes to its alkaline color.

Any of the three indicators would be satisfactory to use for this titration. The color-change ranges of all three indicators fall in the straight portion of the pH curve where one drop of NaOH solution causes a sharp increase in the pH.

Let us now consider the titration of 50.0 ml of $0.100\ M$ acetic acid—a weak acid—with $0.100\ M$ sodium hydroxide. The concentration of H^+ in the original

Table 15.6   Titration of 50.0 ml of 0.100 *M* HCl with 0.100 *M* NaOH

Volume of 0.100 *M* NaOH Added (ml)	pH
0.0	1.00
10.0	1.18
20.0	1.37
30.0	1.60
40.0	1.96
49.0	3.00
49.9	4.00
50.0	7.00
50.1	10.00
51.0	11.00
60.0	11.96
70.0	12.22
80.0	12.36
90.0	12.46
100.0	12.52

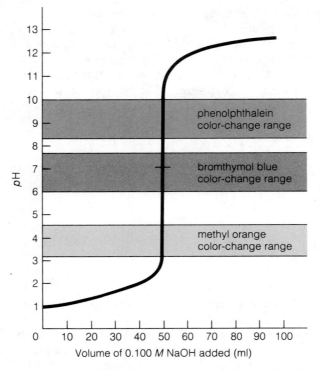

Figure 15.1   Titration of 50.0 ml of 0.100 M HCl with 0.100 M NaOH

50.0 ml sample of acid may be calculated by using the equilibrium constant of acetic acid (see Example 15.2). From Table 15.1 we see that $[H^+] = 1.34 \times 10^{-3}\ M$. The pH of the solution is therefore 2.87.

The solution resulting from the addition of 10.0 ml of 0.100 M NaOH to the 50.0 ml sample of 0.100 M $HC_2H_3O_2$ is, in effect, a buffer since it contains a mixture of $C_2H_3O_2^-$ ions, produced by the neutralization, together with un-neutralized $HC_2H_3O_2$. The pH of a buffer is conveniently calculated by the use of the relation

$$pH = pK_a + \log\left(\frac{[A^-]}{[HA]}\right) \tag{15.6}$$

which was derived in Section 15.6. For acetic acid the pK_a is 4.74 (the negative logarithm of 1.8×10^{-5}). The ratio $[C_2H_3O_2^-]/[HC_2H_3O_2]$ is easily calculated. After 10.0 ml of NaOH is added, 10/50 of the acid of the original 50.0 ml sample has been neutralized; it has, in effect, been converted into the salt sodium acetate. Since 40/50 of the acid remains unneutralized, the ratio is 1 to 4. Thus,

$$pH = pK_a + \log\left(\frac{[C_2H_3O_2^-]}{[HC_2H_3O_2]}\right)$$
$$= 4.74 + \log(1/4)$$
$$= 4.14$$

At the equivalence point all of the acid has been neutralized, and the solution (100 ml) is effectively 0.0500 M in sodium acetate. The calculation of the pH of

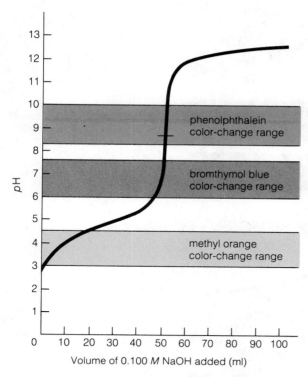

Figure 15.2 Titration of 50.0 ml of 0.100 M $HC_2H_3O_2$ with 0.100 M NaOH

this solution must take into account the hydrolysis of the $C_2H_3O_2^-$ ion (see Example 15.19):

$$C_2H_3O_2^- + H_2O \rightleftharpoons HC_2H_3O_2 + OH^-$$
$$0.0500\ M \qquad\qquad (x)\ M \qquad (x)\ M$$

$$\frac{[HC_2H_3O_2][OH^-]}{[C_2H_3O_2^-]} = 5.56 \times 10^{-10}$$

$$\frac{x^2}{5.00 \times 10^{-2}} = 5.56 \times 10^{-10}$$

$$x = [OH^-] = 5.27 \times 10^{-6}$$

$$pOH = 5.28$$

$$pH = 8.72$$

Notice that the equivalence point in this titration does not occur at a pH of 7.

After the equivalence point the addition of NaOH causes the solution to become increasingly alkaline. The added OH^- shifts the hydrolysis equilibrium to the left. The effect of the hydrolysis on the pH is negligible. Thus, the calculations from this point on are identical to those for the HCl–NaOH titration.

Data for a $HC_2H_3O_2$–NaOH titration are summarized in Table 15.7 and plotted in Figure 15.2. The equivalence point of this titration occurs at a higher pH than that of the preceding titration, and all pH values on the acid side of the equivalence point are higher. Consequently, the rapidly ascending portion of

Table 15.7 Titration of 50.0 ml of 0.100 M $HC_2H_3O_2$ with 0.100 M NaOH

Volume of 0.100 M NaOH Added (ml)	pH
0.0	2.87
10.0	4.14
20.0	4.56
30.0	4.92
40.0	5.34
49.0	6.44
49.9	7.45
50.0	8.72
50.1	10.00
51.0	11.00
60.0	11.96
70.0	12.22
80.0	12.36
90.0	12.46
100.0	12.52

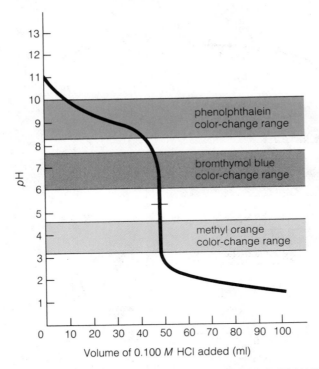

Figure 15.3 Titration of 50.0 ml of 0.100 M NH_3 with 0.100 M HCl

the curve around the equivalence point is reduced in length. From the curve of Figure 15.2 we can see that methyl orange is not a suitable indicator for this titration. Neither is bromthymol blue suitable, since its color change would start after 47.34 ml of NaOH has been added and continue until 49.97 ml has been added, which would hardly constitute a sharp end point for a titration. Phenolphthalein, however, would be a satisfactory indicator to employ.

Other titration curves may be drawn following the general line of approach outlined here. Figure 15.3 represents the titration curve for the titration of 50.0 ml of 0.100 M NH_3 with 0.100 M HCl. In this instance methyl orange could be used as the indicator. Figure 15.4 is the titration curve for the titration of 50.0 ml of 0.100 M $HC_2H_3O_2$ with 0.100 M NH_3; both solutes are weak electrolytes. No indicator can be found that would function satisfactorily for this titration, and such titrations—between weak electrolytes—are not usually run.

Titrations may be conducted potentiometrically. For example, a titration may be performed with the electrodes of a pH meter immersed in the solution being analyzed. The pH of the solution is determined after successive additions of reagent. The equivalence point of the titration is indicated by an abrupt change in the pH, but additions and readings are continued beyond this point. The equivalence point may be determined by graphing the data and estimating the volume corresponding to the midpoint of the steeply rising portion of the titration curve.

The potentiometric titration of a weak electrolyte serves as an important method of determining the dissociation constant of the electrolyte. Since

$$pH = pK_a + \log([A^-]/[HA]) \tag{15.6}$$

the pK_a of the electrolyte is equal to the pH of the solution at half neutralization

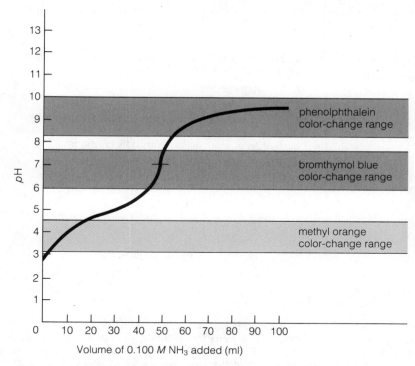

Figure 15.4 Titration of 50.0 ml of 0.100 M $HC_2H_3O_2$ with 0.100 M NH_3

(where $[A^-] = [HA]$), but the pK_a, and hence the K_a itself, may be determined from any point on the titration curve.

Example 15.22

The equivalence point in the titration of 40.00 ml of a solution of a weak mono-protic acid occurs when 35.00 ml of a 0.100 M NaOH solution has been added. The pH of the solution is 5.75 after 20.00 ml of the NaOH solution has been added. What is the dissociation constant of the acid?

Solution

Since 35.00 ml of NaOH solution is required for complete neutralization, the acid is 20/35 neutralized after 20.00 ml of NaOH has been added. In other words, 15/35 of the acid is in the form HA, and 20/35 is in the form A^-. The ratio $[A^-]/[HA]$ is therefore 20 to 15, or 1.33:

$$pH = pK_a + \log\left(\frac{[A^-]}{[HA]}\right)$$

$$5.75 = pK_a + \log(1.33)$$

$$pK_a = 5.63$$

$$K_a = 2.4 \times 10^{-6}$$

Summary

The topics that have been discussed in this chapter are

Equilibria involving the aqueous dissociations of weak electrolytes; ionization constants and their use in problem solving.

2. The ionization of water and the water constant, K_w.

3. The pH and pOH of aqueous solutions.

4. How indicators function.

5. The common-ion effect, a method of shifting an ionic equilibrium by the addition of an ion involved in the equilibrium.

6. The theory of buffer solutions, their preparation, and calculations involving buffer systems.

7. The stepwise ionization of polyprotic acids; the H_2S system.

8. Acidic and basic solutions produced by the reactions of water with cations derived from weak bases or anions derived from weak acids (hydrolysis reactions).

9. The graphic analysis of acid-base titrations.

Key Terms

Some of the more important terms introduced in this chapter are listed below. Definitions for terms not included in this list may be located in the text by the use of the index.

Acid-base indicator (Section 15.4) A compound that changes color in solution as the pH of the solution changes.

Acid dissociation constant, K_a (Section 15.1) An equilibrium constant that pertains to an equilibrium involving a weak acid and the ions derived from it in water solution.

Base dissociation constant, K_b (Section 15.1) An equilibrium constant that pertains to an equilibrium involving a weak base and the ions derived from it in water solution.

Buffer (Section 15.6) A solution that is capable of maintaining its pH at a fairly constant value even when small amounts of acids or bases are added.

Common-ion effect (Section 15.5) The effect on an equilibrium system caused by the addition of a compound that has an ion in common with one present in the system.

Degree of dissociation, α (Section 15.1) The fraction of the total concentration of a weak electrolyte that is in ionic form in water solution at equilibrium.

End point (Section 15.9) That point in a titration when the indicator changes color.

Equivalence point (Section 15.9) That point in a titration when an equivalent amount of base or acid has been added to the sample of acid or base being titrated.

Henderson-Haselbalch equation (Section 15.6) An equation that permits the calculation of the pH of a buffer; $p\text{H} = pK_a + \log([A^-]/[HA])$, where pK_a is the negative logarithm of the acid dissociation constant of the weak acid used to prepare the buffer, $[A^-]$ is the concentration of anions, and $[HA]$ is the concentration of molecules of the weak acid.

Hydrolysis (Section 15.8) A reaction of a cation or an anion with water that affects the pH.

pH (Section 15.3) The negative logarithm (base 10) of the concentration of $H^+(aq)$; $p\text{H} = -\log[H^+]$.

pK (Section 15.6) The negative logarithm of an ionization constant; $pK = -\log K$.

pOH (Section 15.3) The negative logarithm of the concentration of $OH^-(aq)$ ions in an aqueous solution; $p\text{OH} = -\log[OH^-]$.

Titration curve (Section 15.9) A graph that shows how the pH of the solution changes during the course of a titration; the pH is plotted against the volume of base or acid added.

Water dissociation constant, K_w (Section 15.2) The product of the concentration of $H^+(aq)$ and the concentration of $OH^-(aq)$ in any aqueous system; at 25°C, $K_w = [H^+][OH^-] = 1.0 \times 10^{-14}$.

Problems*

Weak Electrolytes

15.1 Lactic acid, $HC_3H_5O_3$, is 2.4% ionized in 0.25 M solution. What is the ionization constant of this acid?

15.2 A 0.200 M solution of dichloroacetic acid, $H(O_2CCHCl_2)$, a weak monobasic acid, is 33% ionized. What is the ionization constant of this acid?

15.3 A solution is prepared by dissolving 0.300 mol of cyanoacetic acid, $H(O_2CCH_2CN)$, in sufficient water to make exactly one liter. The concentration of $H^+(aq)$ in the resulting solution is 0.032 M. What is the ionization constant of cyanoacetic acid?

15.4 In a 0.25 M solution of benzyl amine, $C_7H_7NH_2$, the concentration of $OH^-(aq)$ is 2.4×10^{-3} M. What is the ionization constant for this weak base? The reaction is

$$C_7H_7NH_2 + H_2O \rightleftharpoons C_7H_7NH_3^+ + OH^-$$

15.5 Morphine is a weak base, and the aqueous ionization can be indicated as

$$M(aq) + H_2O \rightleftharpoons MH^+(aq) + OH^-(aq)$$

In a 0.0050 M solution of morphine, the ratio of $OH^-(aq)$ ions to morphine molecules is 1.0/83. What is the value of K_b for the aqueous ionization of morphine?

15.6 (a) What concentration of $HClO_2$ molecules is in equilibrium with 0.030 M $H^+(aq)$ in a solution prepared from pure chlorous acid? (b) How many moles of $HClO_2$ should be used to prepare one liter of this solution?

15.7 The ionization constant for cacodylic acid, a monobasic, arsenic-containing, organic acid, is 6.4×10^{-7}. What is the concentration of $H^+(aq)$ in a 0.30 M solution of cacodylic acid?

15.8 (a) What is the concentration of $H^+(aq)$ in a 0.20 M solution of benzoic acid? (b) What is the percent ionization of $HC_7H_5O_2$ in this solution?

15.9 The ionization constant of propanoic acid, $HC_3H_5O_2$, is 1.3×10^{-5}. (a) What is the concentration of $H^+(aq)$ in a 0.25 M solution of this acid? (b) What is the percent ionization of $HC_3H_5O_2$ in this solution?

15.10 (a) What are the concentrations of $H^+(aq)$, $OCl^-(aq)$ and $HOCl(aq)$ in a 0.20 M solution of $HOCl$? (b) What is the percent ionization of $HOCl$ in this solution?

15.11 What are the concentrations of $C_6H_5NH_3^+(aq)$, $OH^-(aq)$, and $C_6H_5NH_2(aq)$ in a 0.30 M solution of aniline?

15.12 In an aqueous solution of NH_3, the concentration of $OH^-(aq)$ is 1.8×10^{-3} M. What is the concentration of $NH_3(aq)$?

15.13 What is the concentration of $OH^-(aq)$ in a 0.15 M aqueous solution of hydrazine (N_2H_4)?

15.14 A weak acid, HX, is 0.80% ionized in 0.015 M solution. What percent of HX is ionized in a 0.10 M solution?

15.15 A weak acid, HX, is 0.10% ionized in 0.10 M solution. At what concentration is the acid 1.0% ionized?

15.16 What are the concentrations of $H^+(aq)$, $N_3^-(aq)$, and $HN_3(aq)$ in a solution prepared from 0.23 mol of NaN_3 and 0.10 mol of HCl in a total volume of 1.0 liter?

15.17 What is the concentration of $H^+(aq)$ in a solution prepared from 0.22 mol of sodium formate ($NaCHO_2$) and 0.15 mol of HCl in a total volume of 1.0 liter?

15.18 What are the concentrations of $H^+(aq)$, $C_7H_5O_2^-(aq)$, and $HC_7H_5O_2(aq)$ in a solution prepared by adding 10 ml of 0.30 M sodium benzoate, $NaC_7H_5O_2$, to 10 ml of 0.20 M HCl? Assume that the total volume of the solution is 20 ml.

15.19 What are the concentrations of $NH_3(aq)$, $NH_4^+(aq)$, and $OH^-(aq)$ in a solution prepared from 0.15 mol of NH_4Cl and 0.25 mol of NaOH in a total volume of 1.0 liter?

15.20 What are the concentrations of $OH^-(aq)$, $NH_4^+(aq)$, and $NH_3(aq)$ in a solution prepared by adding 150 ml of 0.45 M NH_4Cl to 300 ml of 0.30 M NaOH? Assume that the total volume of the solution is 450 ml.

Ionization of Water, pH

15.21 What are the concentrations of $H^+(aq)$ and $OH^-(aq)$ in (a) a 0.020 M solution of HNO_3 and (b) a 0.020 M solution of $Ba(OH)_2$?

15.22 What pH corresponds to each of the following: (a) $[H^+] = 0.35 M$, (b) $[OH^-] = 0.15 M$, (c) $[OH^-] = 6.0 \times 10^{-6} M$, (d) $[H^+] = 2.5 \times 10^{-8} M$?

15.23 What pH corresponds to each of the following: (a) $[H^+] = 7.3 \times 10^{-5} M$, (b) $[H^+] = 0.084 M$, (c) $[OH^-] = 3.3 \times 10^{-4} M$, (d) $[OH^-] = 0.042 M$?

15.24 What concentration of $H^+(aq)$ corresponds to each of the following: (a) $pH = 3.33$, (b) $pOH = 3.33$, (c) $pH = 6.78$, (d) $pOH = 11.11$?

15.25 What concentration of $OH^-(aq)$ corresponds to each of the following: (a) $pH = 0.55$, (b) $pOH = 4.32$, (c) $pH = 12.13$, (d) $pOH = 12.34$?

15.26 A 0.37 M solution of a weak acid, HX, has a pH of 3.70. What is the ionization constant of the acid?

---

* The more difficult problems are marked with asterisks. The appendix contains answers to color-keyed problems.

15.27 How many moles of chlorous acid, $HClO_2$, must be used to prepare 1.00 liter of solution that has a pH of 2.60?

15.28 How many moles of benzoic acid, $HC_7H_5O_2$, must be used to prepare 500 ml of solution that has a pH of 2.44?

15.29 A 0.23 M solution of a weak acid, HX, has a pH of 2.89. What is the ionization constant of the acid?

15.30 What is the pH of a 0.30 M NH_3 solution?

15.31 Cocaine is a weak base. The aqueous ionization can can be indicated as

$$C(aq) + H_2O \rightleftharpoons CH^+(aq) + OH^-(aq)$$

A 5.0×10^{-3} M solution of cocaine has a pH of 10.04. What is the ionization constant for this substance?

15.32 An indicator, HIn, has an ionization constant of 9.0×10^{-9}. The acid color of the indicator is yellow, and the alkaline color is red. The yellow color is visible when the ratio of yellow form to red form is 30 to 1, and the red color is predominant when the ratio of red form to yellow form is 2 to 1. What is the pH range of color change for the indicator?

15.33 In each of the following parts, a given solution affects the indicators listed in the way shown. Refer to Table 15.4, and describe the pH of each solution as completely as possible: **(a)** Thymol blue turns blue and alizarin yellow turns yellow. **(b)** Bromcresol green turns blue and bromthymol blue turns yellow. **(c)** Methyl orange turns yellow and litmus turns red.

Common-Ion Effect, Buffers

15.34 A solution is prepared by adding 0.010 mol of sodium formate, $Na(CHO_2)$, to 100 ml of 0.025 M formic acid, $HCHO_2$. Assume that no volume change occurs. Calculate **(a)** the pH of the solution and **(b)** the percent ionization of $HCHO_2$.

15.35 A solution is prepared by adding 0.010 mol of sodium nitrite, $NaNO_2$, to 100 ml of 0.035 M nitrous acid, HNO_2. Assume that the volume of the final solution is 100 ml. Calculate **(a)** the pH of the solution and **(b)** the percent ionization of HNO_2.

15.36 A solution prepared from 0.028 mol of a weak acid, HX, and 0.0070 mol of NaX diluted to 200 ml has a pH of 3.66. What is the ionization constant of HX?

15.37 A solution is prepared from 3.0×10^{-3} mol of a weak acid, HX, and 6.0×10^{-4} mol of NaX diluted to 200 ml. The solution has a pH of 4.80. What is the ionization constant of HX?

15.38 A 0.10 M solution of hydrazine, N_2H_4, containing an unknown concentration of hydrazine hydrochloride, $N_2H_5^+Cl^-$, has a pH of 7.15. What is the concentration of hydrazine hydrochloride in the solution?

15.39 A 0.24 M solution of dimethylamine, $(CH_3)_2NH$, containing an unknown concentration of dimethylamine hydrochloride, $(CH_3)_2 NH_2^+Cl^-$, has a pH of 10.40. What is the concentration of dimethylamine hydrochloride?

15.40 A solution prepared from 0.060 mol of a weak acid, HX, diluted to 250 ml has a pH of 2.89. What is the pH of the solution after 0.030 mol of solid NaX is dissolved in it? Assume that no significant volume change occurs when the NaX is dissolved in the solution.

15.41 A 0.050 M solution of formic acid, $HCHO_2$, containing an unknown concentration of formate ion, CHO_2^-, derived from sodium formate has a pH of 3.70. What is the concentration of formate ion in the solution?

15.42 A 1.00 liter solution prepared from 0.049 mol of a weak acid, HX, has a pH of 3.55. What is the pH of the solution after 0.020 mol of solid NaX is dissolved in it? Assume that the volume of the solution does not change when the NaX is dissolved in it.

15.43 What concentrations should be used to prepare a benzoic acid–benzoate ion buffer with a pH of 5.00?

15.44 What concentrations should be used to prepare an ammonia–ammonium ion buffer with a pH of 9.50?

15.45 How many moles of sodium hypochlorite, NaOCl, should be added to 200 ml of 0.22 M hypochlorous acid, HOCl, to produce a buffer with a pH of 6.75? Assume that no volume change occurs when the NaOCl is added to the solution.

Polyprotic Acids

15.46 **(a)** What are the concentrations of $H^+(aq)$, $HCO_3^-(aq)$, $CO_3^{2-}(aq)$, and $CO_2(aq)$ in a saturated solution of carbonic acid (0.034 M in CO_2)? **(b)** What is the pH of the solution?

15.47 Calculate the concentrations of $H^+(aq)$, $H_2AsO_4^-(aq)$, $HAsO_4^{2-}(aq)$, $AsO_4^{3-}(aq)$, and $H_3AsO_4(aq)$ in a 0.30 M solution of arsenic acid.

15.48 A 0.15 M solution of HCl is saturated with H_2S. **(a)** What is the concentration of $S^{2-}(aq)$? **(b)** What is the concentration of $HS^-(aq)$?

15.49 A solution with a pH of 3.30 is saturated with H_2S. What is the concentration of $S^{2-}(aq)$?

15.50 What should the concentration of $H^+(aq)$ be in order to have a sulfide ion concentration of 3.0×10^{-17} M when the solution is saturated with H_2S?

Ions that Function as Acids and Bases

15.51 What is the pH of a 0.10 M solution of sodium nitrite ($NaNO_2$)?

15.52 What is the pH of a 0.10 M solution of aniline hydrochloride ($C_6H_5NH_3^+Cl^-$)?

15.53 What is the pH of a 0.10 M solution of sodium benzoate $NaC_7H_5O_2$)?

15.54 What is the pH of a 0.10 M solution of hydrazine hydrochloride ($N_2H_5^+Cl^-$)?

15.55 A solution of sodium benzoate ($NaC_7H_5O_2$) has a pH of 9.00. What is the concentration of sodium benzoate?

15.56 What concentration of ammonium chloride, NH_4Cl, will produce a solution with a pH of 5.20?

15.57 The pH of a 0.15 M solution of NaX is 9.77. What is the value of K_a of the weak acid HX?

15.58 The pH of a 0.30 M solution of NaY is 9.50. What is the value of K_a of the weak acid HY?

*15.59** The $Zn(OH)^+$ ion is believed to be the only zinc-containing ion resulting from the reaction of $Zn^{2+}(aq)$ with water. What is the pH of a 0.010 M solution of $Zn(NO_3)_2$? The K_{inst} of $Zn(OH)^+$ is 4.1×10^{-5}.

*15.60** The $Cd(OH)^+$ ion is believed to be the only cadmium-containing ion resulting from the reaction of $Cd^{2+}(aq)$ with water. What is the concentration of $Cd^{2+}(aq)$ in a solution of $Cd(ClO_4)_2$ that has a pH of 5.48? The K_{inst} of $Cd(OH)^+$ is 6.9×10^{-5}.

Acid-Base Titrations

15.61 Determine pH values for the titration curve pertaining to the titration of 30.00 ml of 0.100 M benzoic acid, $HC_7H_5O_2$, with 0.100 M NaOH **(a)** after 10.00 ml of NaOH solution has been added, **(b)** after 30.00 ml of NaOH solution has been added, **(c)** after 40.00 ml of NaOH solution has been added.

15.62 Determine pH values for the titration curve pertaining to the titration of 25.00 ml of 0.100 M NH_3 solution with a 0.100 M solution of HCl **(a)** after 10.00 ml of the HCl solution has been added, **(b)** after 25.00 ml of HCl solution has been added, **(c)** after 35.00 ml of HCl solution has been added.

15.63 In the titration of 25.00 ml of a solution of a weak acid HX with 0.250 M NaOH, the pH of the solution is 4.50 after 5.00 ml of the NaOH solution have been added. The equivalence point in the titration occurs when 30.40 ml of the NaOH solution have been added. What is the equilibrium constant for the ionization of HX?

15.64 The equivalence point in the titration of 50.00 ml of a solution of acetic acid occurs when 33 ml of 0.200 M NaOH have been added. How many ml of the NaOH solution should be added to 50.00 ml of the acetic acid solution to produce a solution with a pH of 5.04?

CHAPTER 16 IONIC EQUILIBRIUM, PART II

Homogeneous, aqueous equilibrium systems involving weak acids and bases were discussed in Chapter 15. This chapter begins with a study of heterogeneous systems in which slightly soluble solids are in equilibrium with their constituent ions in solution. Equilibrium principles apply to the precipitation of such solids in the qualitative or quantitative determinations of dissolved ions. Several additional types of homogeneous systems are also considered, including the equilibria that occur in the formation of complex ions in solution and equilibria associated with amphoteric substances.

16.1 The Solubility Product

Most substances are soluble in water to at least some slight extent. If an "insoluble" or "slightly soluble" material is placed in water, an equilibrium is established when the rate of dissolution of ions from the solid equals the rate of precipitation of ions from the *saturated solution*. Thus, an equilibrium exists between solid silver chloride and a saturated solution of silver chloride:

$$AgCl(s) \rightleftharpoons Ag^+(aq) + Cl^-(aq)$$

The equilibrium constant is

$$K' = \frac{[Ag^+][Cl^-]}{[AgCl]}$$

Since the concentration of a pure solid is a constant, $[AgCl]$ may be combined with K' to give

$$K_{SP} = K'[AgCl] = [Ag^+][Cl^-]$$

The constant K_{SP} is called a **solubility product.** The ionic concentrations of the expression are those for a saturated solution at the reference temperature. Since the solubility of a salt usually varies widely with temperature, the numerical value for K_{SP} for a salt changes with temperature. A table of solubility products at 25°C is given in the appendix.

The numerical value of K_{SP} for a salt may be found from the molar solubility of the salt.

Example 16.1

At 25°C, 0.00188 g of AgCl dissolves in 1 liter of water. What is the K_{SP} of AgCl?

Solution

The molar solubility of AgCl (molecular weight, 143) is

$$? \text{ mol AgCl} = 0.00188 \text{ g AgCl} \left(\frac{1 \text{ mol AgCl}}{143 \text{ g AgCl}} \right)$$

$$= 1.31 \times 10^{-5} \text{ mol AgCl}$$

For each mole of AgCl dissolving, 1 mol of Ag^+ and 1 mol of Cl^- are formed:

$$\text{AgCl(s)} \rightleftharpoons \underset{1.31 \times 10^{-5} M}{Ag^+} + \underset{1.31 \times 10^{-5} M}{Cl^-}$$

$$K_{SP} = [Ag^+][Cl^-]$$
$$= (1.31 \times 10^{-5})^2$$
$$= 1.7 \times 10^{-10}$$

For salts that have more than two ions per formula unit, the ion concentrations must be raised to the powers indicated by the coefficients of the balanced chemical equation:

$$\text{Mg(OH)}_2(s) \rightleftharpoons Mg^{2+} + 2OH^- \qquad K_{SP} = [Mg^{2+}][OH^-]^2$$
$$\text{Bi}_2S_3(s) \rightleftharpoons 2Bi^{3+} + 3S^{2-} \qquad K_{SP} = [Bi^{3+}]^2[S^{2-}]^3$$
$$\text{Hg}_2Cl_2(s) \rightleftharpoons Hg_2^{2+} + 2Cl^- \qquad K_{SP} = [Hg_2^{2+}][Cl^-]^2$$

For a salt of this type the calculation of the K_{SP} from the molar solubility requires that the chemical equation representing the dissociation process be carefully interpreted.

Example 16.2

At 25°C, 7.8×10^{-5} mol of silver chromate dissolves in 1 liter of water. What is the K_{SP} of Ag_2CrO_4?

Solution

For each mole of Ag_2CrO_4 that dissolves, 2 mol of Ag^+ and 1 mol of CrO_4^{2-} are formed. Therefore,

$$Ag_2CrO_4(s) \rightleftharpoons 2\,Ag^+ + CrO_4^{2-}$$
$$\qquad\qquad 2(7.8 \times 10^{-5})\,M \quad 7.8 \times 10^{-5}\,M$$

$$K_{SP} = [Ag^+]^2\,[CrO_4^{2-}]$$
$$= (1.56 \times 10^{-4})^2\,(7.8 \times 10^{-5})$$
$$= 1.9 \times 10^{-12}$$

Example 16.3

The K_{SP} of CaF_2 is 3.9×10^{-11} at 25°C. What is the concentration of Ca^{2+} and of F^- in the saturated solution? How many grams of calcium fluoride will dissolve in 100 ml of water at 25°C?

Solution

Let x equal the molar solubility of CaF_2:

$$CaF_2(s) \rightleftharpoons Ca^{2+} + 2\,F^-$$
$$\qquad\quad x \qquad\quad 2x$$

$$K_{SP} = [Ca^{2+}][F^-]^2 = 3.9 \times 10^{-11}$$
$$x(2x)^2 = 3.9 \times 10^{-11}$$
$$4x^3 = 3.9 \times 10^{-11}$$
$$x = 2.1 \times 10^{-4}\,M$$

Therefore,

$$[Ca^{2+}] = x = 2.1 \times 10^{-4}\,M$$
$$[F^-] = 2x = 4.2 \times 10^{-4}\,M$$

$$?\,g\,CaF_2 = 100\,ml\,H_2O \left(\frac{2.1 \times 10^{-4}\,mol\,CaF_2}{1000\,ml\,H_2O}\right)\left(\frac{78\,g\,CaF_2}{1\,mol\,CaF_2}\right)$$
$$= 1.6 \times 10^{-3}\,g\,CaF_2$$

The solubilities of some salts in water are slightly higher than those predicted by calculations based on the K_{SP} values. Consider the barium carbonate system

$$BaCO_3(s) \rightleftharpoons Ba^{2+}(aq) + CO_3^{2-}(aq)$$

The carbonate ion in this solution undergoes hydrolysis since it is a weak base (see Section 15.8):

$$H_2O + CO_3^{2-}(aq) \rightleftharpoons HCO_3^-(aq) + OH^-(aq)$$

The concentration of the CO_3^{2-} ion is reduced by this reaction with water, the $BaCO_3$ equilibrium is forced to the right, and more $BaCO_3$ dissolves than would otherwise be the case. A calculation of molar solubility based on the K_{SP} for $BaCO_3$ would give the concentration of CO_3^{2-} that must be present in the solution at equilibrium. More than the corresponding amount of $BaCO_3$ must dissolve, however, in order to provide not only this concentration of CO_3^{2-} but also the CO_3^{2-} that is converted into HCO_3^- by hydrolysis. In the solutions of some salts (PbS, for example) both cation and anion hydrolyze.

The **salt effect** is another factor that may cause a calculated solubility to be in error. The solubility of a salt is increased by the addition of another electrolyte to the solution. Silver chloride, for example, is about 20% more soluble in 0.02 M KNO_3 solutions than it is in pure water:

$$AgCl(s) \rightleftharpoons Ag^+(aq) + Cl^-(aq)$$

The K^+ and NO_3^- ions help create an ionic atmosphere in the solution so that Ag^+ and Cl^- ions are surrounded by ions of opposite charges. As a result, the Ag^+ and Cl^- ions are held in solution more firmly and are less apt to recombine to form AgCl(s). The net effect is that the solubility equilibrium shifts to the right.

Other types of ionic equilibria are influenced by the salt effect. A 0.10 M solution of acetic acid in pure water is 1.3% ionized:

$$HC_2H_3O_2 \rightleftharpoons H^+ + C_2H_3O_2^-$$

A 0.10 M solution of acetic acid in 0.10 M NaCl is 1.7% ionized.

Interionic attractions, therefore, make a solution behave as though its ion concentrations were less than they actually are. The activity of an ion, its "effective concentration," is obtained by correcting the actual ion concentration to take interionic attractions into account. Equilibrium constants are properly expressed in terms of activities, rather than concentrations.

16.2 Precipitation and the Solubility Product

The numerical value of the K_{SP} of a salt is a quantitative statement of the limit of solubility of the salt. When the values for the concentrations of the ions of a salt solution are substituted into an expression similar to that for the K_{SP} of the salt, the result is called the **ion product** of the solution. The K_{SP} is the ion product of a saturated solution.

We can calculate an ion product for a test solution and compare the result to the K_{SP} for the salt under consideration. Three types of comparisons are possible:

1. The ion product is less than the K_{SP}. This solution is unsaturated. Additional solid can be dissolved in it—up to the limit described by the K_{SP}.

2. The ion product is greater than the K_{SP}. The solution is momentarily supersaturated. Precipitation will occur until the ion product equals the K_{SP}.

3. The ion product equals the K_{SP}. This solution is saturated.

The following examples illustrate the application of this method.

Example 16.4

Will a precipitate form if 10 ml of 0.010 M AgNO$_3$ and 10 ml of 0.00010 M NaCl are mixed? Assume that the final volume of the solution is 20 ml. For AgCl, $K_{SP} = 1.7 \times 10^{-10}$.

Solution

Diluting a solution to twice its original volume reduces the concentrations of ions in the solution to half their original value. If there were no reaction, the ion concentrations would be

$$[Ag^+] = 5.0 \times 10^{-3} \ M$$
$$[Cl^-] = 5.0 \times 10^{-5} \ M$$

The ion product is

$$[Ag^+][Cl^-] = ?$$
$$(5.0 \times 10^{-3})(5.0 \times 10^{-5}) = 2.5 \times 10^{-7}$$

Since the ion product is larger than the K_{SP} (1.7×10^{-10}), precipitation of AgCl should occur.

Example 16.5

Will a precipitate of $Mg(OH)_2$ form in a 0.0010 M solution of $Mg(NO_3)_2$ if the pH of the solution is adjusted to 9.0? The K_{SP} of $Mg(OH)_2$ is 8.9×10^{-12}.

Solution

If the pH = 9.0, the pOH = 5.0:

$$[OH^-] = 1.0 \times 10^{-5} \ M$$

Since $[Mg^{2+}] = 1.0 \times 10^{-3} \ M$, the ion product is

$$[Mg^{2+}][OH^-]^2 = ?$$
$$(1.0 \times 10^{-3})(1.0 \times 10^{-5})^2 = 1.0 \times 10^{-13}$$

Since the ion product is less than 8.9×10^{-12}, no precipitate will form.

The common-ion effect pertains to solubility equilibria. As an example, consider the system

$$BaSO_4(s) \rightleftharpoons Ba^{2+}(aq) + SO_4^{2-}(aq)$$

The addition of sulfate ion, from sodium sulfate, to a saturated solution of barium sulfate will cause the equilibrium to shift to the left; the concentration of Ba^{2+} will decrease, and $BaSO_4$ will precipitate. Since the product $[Ba^{2+}][SO_4^{2-}]$ is a constant, increasing $[SO_4^{2-}]$ will cause $[Ba^{2+}]$ to decrease.

The amount of barium ion in a solution may be determined by precipitating the Ba^{2+} as $BaSO_4$. The precipitate is then removed by filtration and is dried and weighed. The concentration of Ba^{2+} left in solution after the precipitation may be reduced to a very low value if excess sulfate ion is employed in the precipitation. As a general rule, however, too large an excess of the common ion should be avoided. At high ionic concentrations the salt effect increases the solubility of a salt, and for certain precipitates the formation of a complex ion may lead to enhanced solubility.

Example 16.6

At 25°C a saturated solution of $BaSO_4$ is 3.9×10^{-5} M. The K_{SP} of $BaSO_4$ is 1.5×10^{-9}. What is the solubility of $BaSO_4$ in 0.050 M Na_2SO_4?

Solution

The SO_4^{2-} derived from the $BaSO_4$ is negligible in comparison to the $[SO_4^{2-}]$ already present in the solution (5.0×10^{-2} M):

$$[Ba^{2+}][SO_4^{2-}] = K_{SP}$$
$$[Ba^{2+}][SO_4^{2-}] = 1.5 \times 10^{-9}$$
$$[Ba^{2+}](5.0 \times 10^{-2}) = 1.5 \times 10^{-9}$$
$$[Ba^{2+}] = 3.0 \times 10^{-8}\ M$$

The solubility of $BaSO_4$ has been reduced from 3.9×10^{-5} M to 3.0×10^{-8} M by the common-ion effect.

At times, the common-ion effect is used to *prevent* the formation of a precipitate. Consider the precipitation of magnesium hydroxide from a solution that contains Mg^{2+} ions:

$$Mg(OH)_2(s) \rightleftharpoons Mg^{2+}(aq) + 2OH^-(aq)$$

The precipitation can be prevented by holding the concentration of OH^- to a low value. If the hydroxide ion is supplied by the weak base ammonia

$$NH_3 + H_2O \rightleftharpoons NH_4^+ + OH^-$$

the concentration of OH^- can be controlled by the addition of NH_4^+. When the common ion NH_4^+ is added, the ammonia equilibrium shifts to the left, which reduces the concentration of OH^-. In this way, the concentration of OH^- can be held to a level that will not cause $Mg(OH)_2$ to precipitate.

Example 16.7

What concentration of NH_4^+, derived from NH_4Cl, is necessary to prevent the formation of an $Mg(OH)_2$ precipitate in a solution that is 0.050 M in Mg^{2+} and 0.050 M in NH_3? The K_{SP} of $Mg(OH)_2$ is 8.9×10^{-12}.

Solution

We first calculate the maximum concentration of hydroxide ion that can be present in the solution without causing $Mg(OH)_2$ to precipitate:

$$[Mg^{2+}][OH^-]^2 = 8.9 \times 10^{-12}$$
$$(5.0 \times 10^{-2})[OH^-]^2 = 8.9 \times 10^{-12}$$
$$[OH^-]^2 = 1.8 \times 10^{-10}$$
$$[OH^-] = 1.3 \times 10^{-5} \ M$$

The concentration of OH^-, therefore, can be no higher than $1.3 \times 10^{-5} \ M$. From the expression for the ionization constant for NH_3, we can derive the concentration of NH_4^+ that will maintain the concentration of OH^- at this level:

$$\frac{[NH_4^+][OH^-]}{[NH_3]} = 1.8 \times 10^{-5}$$

$$\frac{[NH_4^+](1.3 \times 10^{-5})}{(5.0 \times 10^{-2})} = 1.8 \times 10^{-5}$$

$$[NH_4^+] = 6.9 \times 10^{-2} \ M$$

Thus, the *minimum* concentration of NH_4^+ that must be present is $0.069 \ M$.

Frequently, a solution contains more than one ion capable of forming a precipitate with another ion which is to be added to the solution. For example, a solution might contain both Cl^- and CrO_4^{2-} ions, both of which form insoluble salts with Ag^+. When Ag^+ is added to the solution, the less soluble silver salt will precipitate first. If the addition is continued, eventually a point will be reached where the more soluble salt will begin to precipitate along with the less soluble.

Example 16.8

A solution is $0.10 \ M$ in Cl^- and $0.10 \ M$ in CrO_4^{2-}. If solid $AgNO_3$ is gradually added to this solution, which will precipitate first, $AgCl$ or Ag_2CrO_4? Assume that the addition causes no change in volume. For $AgCl$, $K_{SP} = 1.7 \times 10^{-10}$, for Ag_2CrO_4, $K_{SP} = 1.9 \times 10^{-12}$.

Solution

When a precipitate *begins* to form, the pertinent ion product *just* exceeds the K_{SP} of the solid. Therefore, we calculate the concentrations of Ag^+ needed to precipitate $AgCl$ and Ag_2CrO_4:

$$AgCl(s) \rightleftharpoons Ag^+ + Cl^- \qquad\qquad Ag_2CrO_4(s) \rightleftharpoons 2\,Ag^+ + CrO_4^{2-}$$

?	0.10 M	?

$$[Ag^+][Cl^-] = 1.7 \times 10^{-10} \qquad\qquad [Ag^+]^2[CrO_4^{2-}] = 1.9 \times 10^{-12}$$

$$[Ag^+](0.10) = 1.7 \times 10^{-10} \qquad\qquad [Ag^+]^2(0.10) = 1.9 \times 10^{-12}$$

$$[Ag^+] = 1.7 \times 10^{-9} \ M \qquad\qquad [Ag^+]^2 = 1.9 \times 10^{-11}$$

$$[Ag^+] = 4.4 \times 10^{-6} \ M$$

Therefore, $AgCl$ will precipitate first.

Example 16.9

(a) In the experiment described in Example 16.8, what will be the concentration of the Cl^- ion when Ag_2CrO_4 begins to precipitate? (b) At this point, what percent of the chloride ion originally present remains in solution?

Solution

(a) From the preceding example, we see that $[Ag^+] = 4.4 \times 10^{-6}\ M$ when Ag_2CrO_4 starts to precipitate. At this point, the concentration of chloride ion will be

$$[Ag^+][Cl^-] = 1.7 \times 10^{-10}$$

$$(4.4 \times 10^{-6})[Cl^-] = 1.7 \times 10^{-10}$$

$$[Cl^-] = \frac{1.7 \times 10^{-10}}{4.4 \times 10^{-6}} = 3.9 \times 10^{-5}\ M$$

Thus, until the $[Cl^-]$ is reduced to $3.9 \times 10^{-5}\ M$, no Ag_2CrO_4 will form. (b) Since the original concentration of chloride was 0.10 M, the percent of Cl^- remaining in solution when Ag_2CrO_4 starts to precipitate is

$$\frac{3.9 \times 10^{-5}}{1.0 \times 10^{-1}} \times 100 = 0.039\%$$

Chromate ion is used as a **precipitation indicator.** The concentration of chloride ion in a solution can be determined by titrating a sample of the solution against a standard $AgNO_3$ solution using a few drops of K_2CrO_4 as an indicator. In the course of the titration white $AgCl$ precipitates. The appearance of red Ag_2CrO_4 indicates that the precipitation of chloride ion is essentially complete. In a quantitative procedure of this type, the concentration of CrO_4^{2-} used is much less than that employed in the preceding problem. Hence, a higher concentration of Ag^+ is required to start the precipitation of Ag_2CrO_4, and a lower concentration of Cl^- will be present in the solution when the Ag_2CrO_4 begins to precipitate.

16.3 Precipitation of Sulfides

The sulfide ion concentration in an acid solution that has been saturated with H_2S is extremely low. In 10 ml of a saturated H_2S solution that has been made 0.3 M in H^+, there are approximately seven S^{2-} ions. Nevertheless, when Pb^{2+} ions are added to such a solution, PbS precipitates immediately. It seems unlikely that the precipitate forms as a result of the reaction of Pb^{2+} ions with S^{2-} ions.

There is evidence that in this and in similar sulfide precipitations, a hydrosulfide forms initially and then decomposes to give the normal sulfide:

$$Pb(HS)_2(s) \rightleftharpoons PbS(s) + H_2S(aq)$$

The hydroxides of many metals are known to produce oxides in a parallel manner. In fact, what is known as lead hydroxide, $Pb(OH)_2$, is in reality hydrous lead oxide, $PbO \cdot xH_2O$. The hydrosulfide of lead could form as a result of the reaction of Pb^{2+} ions with either H_2S molecules or HS^- ions. The concentrations of H_2S and HS^- are much higher than the concentration of S^{2-}.

An equilibrium constant does not depend upon the reaction mechanism by which the equilibrium is attained (see Section 13.1). Provided the system is in equilibrium, the relationship expressed by the solubility product principle is valid no matter what series of reactions produces the precipitate. We may, therefore, use K_{SP} to calculate favorable reaction conditions for the formation of a desired precipitate or reaction conditions that will prevent the formation of a precipitate.

Example 16.10

A solution that is 0.30 M in H^+, 0.050 M in Pb^{2+}, and 0.050 M in Fe^{2+} is saturated with H_2S. Should PbS and/or FeS precipitate? The K_{SP} of PbS is 7×10^{-29}, and the K_{SP} of FeS is 4×10^{-19}.

Solution

For any saturated solution of H_2S,

$$[H^+]^2[S^{2-}] = 1.1 \times 10^{-22}$$

Since this solution is 0.30 M in H^+,

$$(3.0 \times 10^{-1})^2[S^{2-}] = 1.1 \times 10^{-22}$$
$$[S^{2-}] = 1.2 \times 10^{-21} \ M$$

Both Pb^{2+} and Fe^{2+} are $2+$ ions, and the form of the ion product is

$$[M^{2+}][S^{2-}]$$

where M^{2+} stands for either metal ion. Since both are present in concentrations of 0.050 M,

$$[M^{2+}][S^{2-}]$$
$$(5.0 \times 10^{-2})(1.2 \times 10^{-21}) = 6.0 \times 10^{-23}$$

This ion product is greater than the K_{SP} of PbS; therefore, PbS will precipitate. The ion product, however, is less than the K_{SP} of FeS. The solubility of FeS has not been exceeded. No FeS will form.

Example 16.11

What must be the H^+ concentration of a solution that is 0.050 M in Ni^{2+} to prevent the precipitation of NiS when the solution is saturated with H_2S? The K_{SP} of NiS is 3×10^{-21}.

Solution

$$[Ni^{2+}][S^{2-}] = 3 \times 10^{-21}$$

$$(0.050)[S^{2-}] = 3 \times 10^{-21}$$

$$[S^{2-}] = 6 \times 10^{-20} \, M$$

Therefore, the $[S^{2-}]$ must be less than $6 \times 10^{-20} \, M$ if NiS is not to precipitate. For a solution saturated with H_2S,

$$[H^+]^2[S^{2-}] = 1.1 \times 10^{-22}$$

$$[H^+]^2(6 \times 10^{-20}) = 1.1 \times 10^{-22}$$

$$[H^+] = 0.04 \, M$$

The $[H^+]$ must be greater than $0.04 \, M$ to prevent the precipitation of NiS.

The preceding examples illustrate an important analytical technique. In the usual qualitative analysis scheme, certain cations are separated into groups on the basis of whether their sulfides precipitate in acidic solution. In a solution that has an H^+ concentration of $0.3 \, M$, the sulfides of Hg^{2+}, Pb^{2+}, Cu^{2+}, Bi^{3+}, Cd^{2+}, and Sn^{2+} are insoluble, whereas the sulfides of Fe^{2+}, Co^{2+}, Ni^{2+}, Mn^{2+}, and Zn^{2+} are soluble.

In considering systems that involve sulfide precipitation, one must note that the H^+ concentration of a solution increases when a sulfide precipitates from the solution.

Example 16.12

A solution that is $0.050 \, M$ in Cd^{2+} and $0.10 \, M$ in H^+ is saturated with H_2S. What concentration of Cd^{2+} remains in solution after CdS has precipitated? The K_{SP} of CdS is 1.0×10^{-28}.

Solution

For each Cd^{2+} ion precipitated, two H^+ ions are added to the solution:

$$Cd^{2+}(aq) + H_2S(aq) \rightleftharpoons CdS(s) + 2H^+(aq)$$

We shall assume that virtually all the Cd^{2+} precipitates as CdS. Hence, the precipitation introduces 0.10 mol of H^+ per liter of solution, and the final $[H^+]$ is $0.20 \, M$. Therefore,

$$[H^+]^2[S^{2-}] = 1.1 \times 10^{-22}$$

$$(0.20)^2[S^{2-}] = 1.1 \times 10^{-22}$$

$$[S^{2-}] = 2.8 \times 10^{-21}$$

The concentration of Cd^{2+} after the CdS has precipitated may be derived from the K_{SP} of CdS:

$$[Cd^{2+}][S^{2-}] = 1.0 \times 10^{-28}$$

$$[Cd^{2+}](2.8 \times 10^{-21}) = 1.0 \times 10^{-28}$$

$$[Cd^{2+}] = 3.6 \times 10^{-8} \, M$$

From our answer we can see that our assumption was justified. The value of $[H^+]$ that we used was well within our limits of accuracy.

16.4 Equilibria Involving Complex Ions

Complex ions are discussed in Chapter 24. These ions, however, take part in some aqueous equilibria that should be discussed in this chapter. A **complex ion** is an aggregate consisting of a central metal cation (usually a transition-metal ion) surrounded by a number of ligands. The **ligands** of a complex may be anions, molecules, or a combination of the two.

A ligand must have an unshared pair of electrons with which it can bond to the central ion. Thus, the ammonia molecule, NH_3, functions as a ligand in the formation of complex ions—the ammonium ion, NH_4^+, does not:

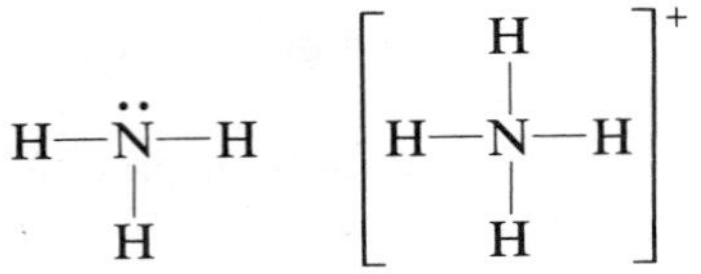

In addition to ammonia, the requirement is met by many molecules and anions:

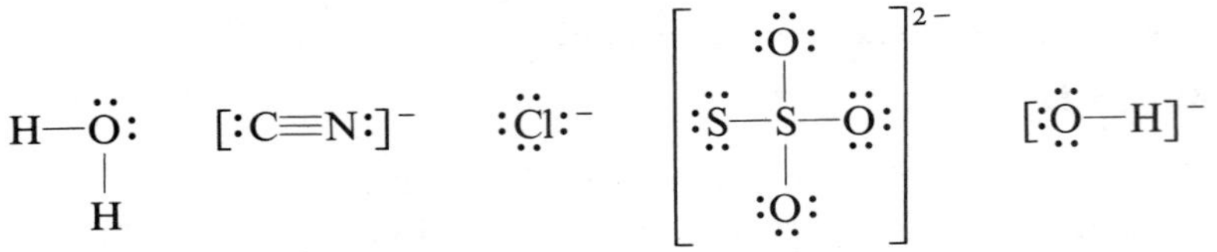

The charge of a complex ion can be found by adding the charges of all the particles of which it is composed. The charge of the $Fe(CN)_6^{4-}$ ion, for example, equals the sum of the charges of one Fe^{2+} ion and six CN^- ions:

$$Fe^{2+} + 6\,CN^- \longrightarrow Fe(CN)_6^{4-}$$

Examples of complex ions are: $Ag(NH_3)_2^+$, $Cd(NH_3)_4^{2+}$, $Cu(NH_3)_4^{2+}$, $Fe(CN)_6^{3-}$, $Fe(CN)_6^{4-}$, $CdCl_4^{2-}$, $Ag(S_2O_3)_3^{5-}$, $Cu(H_2O)_4^{2+}$, and $Zn(OH)_4^{2-}$.

All ions are hydrated in water solution. Most hydrated metal ions should be regarded as complex ions since each metal ion has a definite number of water molecules tightly bonded to it. This number is difficult to determine and is not known with certainty for some species. Nevertheless, in water solution, complexes probably form by the replacement of water ligands by other ligands. For example,

$$Cu(H_2O)_6^{2+} + NH_3 \rightleftharpoons Cu(H_2O)_5(NH_3)^{2+} + H_2O$$

For simplicity, however, the coordinated water molecules are usually not shown:

$$Cu^{2+} + NH_3 \rightleftharpoons Cu(NH_3)^{2+}$$

The dissociation, as well as the formation, of a complex ion occurs in steps. Thus, the $Ag(NH_3)_2^+$ ion dissociates as follows:

$$Ag(NH_3)_2^+ \rightleftharpoons Ag(NH_3)^+ + NH_3 \qquad K_{d1} = \frac{[Ag(NH_3)^+][NH_3]}{[Ag(NH_3)_2^+]}$$

$$= 1.4 \times 10^{-4}$$

$$Ag(NH_3)^+ \rightleftharpoons Ag^+ + NH_3 \qquad K_{d2} = \frac{[Ag^+][NH_3]}{[Ag(NH_3)^+]}$$

$$= 4.3 \times 10^{-4}$$

The product of the two dissociation constants is called the **instability constant** of the $Ag(NH_3)_2^+$ ion. This instability constant corresponds to the overall dissociation of the ion:

$$Ag(NH_3)_2^+ \rightleftharpoons Ag^+ + 2\,NH_3$$

$$\left(\frac{[Ag(NH_3)^+][NH_3]}{[Ag(NH_3)_2^+]}\right)\left(\frac{[Ag^+][NH_3]}{[Ag(NH_3)^+]}\right) = \frac{[Ag^+][NH_3]^2}{[Ag(NH_3)_2^+]}$$

$$K_{inst} = K_{d1}K_{d2} = (1.4 \times 10^{-4})(4.3 \times 10^{-4}) = 6.0 \times 10^{-8}$$

The reciprocals of dissociation constants, which pertain to equilibrium reactions written in reverse form, are called **formation constants** or **stability constants.**

For many complexes the values of the equilibrium constants for the individual steps of the dissociation are not known, although the value for the overall dissociation may have been determined. Note, however, that an overall instability constant for a complex ion must be cautiously interpreted since the chemical equation to which it applies does not take into consideration any intermediate species. Some instability constants are given in the appendix.

Example 16.13

A 0.010 M solution of $AgNO_3$ is made 0.50 M in NH_3 and the $Ag(NH_3)_2^+$ complex forms. (a) What is the concentration of $Ag^+(aq)$ in the solution? (b) What percentage of the total concentration of silver is in the form $Ag^+(aq)$?

Solution

(a) Since a large excess of NH_3 is employed, we assume that virtually all of the Ag^+ is converted into the form of the higher complex, $Ag(NH_3)_2^+$. To two significant figures, therefore,

$$[Ag(NH_3)_2^+] = 0.010 \ M$$

The formation of this complex would reduce the concentration of NH_3 by 0.020 M (each $Ag(NH_3)_2^+$ ion contains two NH_3 molecules). Therefore,

$$[NH_3] = 0.50\ M - 0.020\ M = 0.48\ M$$

We neglect the very small decrease in the concentration of NH_3 brought about by the reaction of NH_3 with water:

$$Ag(NH_3)_2^+ \rightleftharpoons Ag^+ + 2\,NH_3$$

$$\frac{[Ag^+][NH_3]^2}{[Ag(NH_3)_2^+]} = 6.0 \times 10^{-8}$$

$$\frac{[Ag^+](0.48)^2}{(0.010)} = 6.0 \times 10^{-8}$$

$$[Ag^+] = 2.6 \times 10^{-9}\ M$$

(b) $\left(\dfrac{2.6 \times 10^{-9}\ M}{1.0 \times 10^{-2}\ M}\right)100\% = 2.6 \times 10^{-5}\%$

$2.6 \times 10^{-5}\%$ of the silver is in the form $Ag^+(aq)$.

Slightly soluble substances can often be dissolved through the formation of complex ions. The solubility of AgCl, for example, is increased if the solution contains NH_3. The equilibrium

$$AgCl(s) \rightleftharpoons Ag^+(aq) + Cl^-(aq)$$

is forced to the right by the removal of Ag^+ through the formation of $Ag(NH_3)_2^+$.

Example 16.14

What is the solubility of AgCl in 0.10 M NH_3?

Solution

We write the equation for the complete transformation and let x equal the molar solubility of AgCl:

$$\begin{array}{ccccc} AgCl(s) + & 2NH_3 & \rightleftharpoons & Ag(NH_3)_2^+ & + & Cl^- \\ & 0.10 - 2x & & x & & x \end{array}$$

Notice that since two molecules of NH_3 are required for every $Ag(NH_3)_2^+$ formed, the concentration of NH_3 is $(0.10 - 2x)$. An equilibrium constant corresponding to this equation may be derived by dividing the K_{SP} of AgCl by the K_{inst} of $Ag(NH_3)_2^+$:

$$K_{SP}\left(\frac{1}{K_{inst}}\right) = [Ag^+][Cl^-]\left(\frac{[Ag(NH_3)_2^+]}{[Ag^+][NH_3]^2}\right)$$

$$(1.7 \times 10^{-10})\left(\frac{1}{6.0 \times 10^{-8}}\right) = 2.8 \times 10^{-3}$$

Therefore,

$$\frac{[Ag(NH_3)_2^+][Cl^-]}{[NH_3]^2} = 2.8 \times 10^{-3}$$

$$\frac{x^2}{(0.10 - 2x)^2} = 2.8 \times 10^{-3}$$

If we extract the square root of both sides of this equation, we get

$$\frac{x}{(0.10 - 2x)} = 5.3 \times 10^{-2}$$

$$x = 4.8 \times 10^{-3} \ M$$

The solubility of AgCl in pure water is $1.3 \times 10^{-5} \ M$.

A few metals form complex ions with S^{2-} (for example, HgS_2^{2-}, AsS_3^{3-}, SbS_2^{3-}, SnS_3^{2-}). The true structure of these anions, which are called thio complex ions, may involve the HS^- ion. The formation of a thio complex ion may be used to separate sulfide precipitates. For example, a mixture of CuS (which does not form a thio complex) and As_2S_3 may be separated by dissolving the As_2S_3 in an alkaline solution containing S^{2-} ion:

$$CuS(s) + S^{2-}(aq) \longrightarrow \text{no reaction}$$
$$As_2S_3(s) + 3S^{2-}(aq) \rightleftharpoons 2AsS_3^{3-}(aq)$$

If a solution containing Al^{3+} and Zn^{2+} is treated with a buffer of NH_3 at a controlled alkaline pH, $Al(OH)_3$ will precipitate, but Zn^{2+} will stay in solution as $Zn(NH_3)_4^{2+}$. The precipitation of $Zn(OH)_2$ is prevented by the formation of the complex ion; Al^{3+} does not form an ammonia complex:

$$Al^{3+}(aq) + 3OH^-(aq) \rightleftharpoons Al(OH)_3(s)$$
$$Zn^{2+}(aq) + 4NH_3(aq) \rightleftharpoons Zn(NH_3)_4^{2+}(aq)$$

Complex ions are frequently highly colored; $Fe(SCN)^{2+}$, for example, is deep red. Thus, the production of a deep red color when the colorless SCN^- ion is added to a solution serves as a test for the Fe^{3+} ion.

16.5 Amphoterism

The hydroxides of certain metals, called **amphoteric hydroxides,** can function as acids or bases. These water-insoluble compounds dissolve in solutions of low pH or high pH. Zinc hydroxide, for example, dissolves in hydrochloric acid to produce solutions of zinc chloride, $ZnCl_2$:

$$Zn(OH)_2(s) + 2H^+(aq) \longrightarrow Zn^{2+}(aq) + 2H_2O$$

Zinc hydroxide will also dissolve in sodium hydroxide solutions:

$$Zn(OH)_2(s) + 2\,OH^-(aq) \longrightarrow Zn(OH)_4^{2-}(aq)$$

The $Zn(OH)_4^{2-}$ complex ion is called the zincate ion. The oxide corresponding to an amphoteric hydroxide reacts in the same way. For example,

$$ZnO(s) + 2\,H^+(aq) \longrightarrow Zn^{2+}(aq) + H_2O$$

$$ZnO(s) + 2\,OH^-(aq) + H_2O \longrightarrow Zn(OH)_4^{2-}(aq)$$

Other examples of amphoteric compounds are: $Al(OH)_3$, $Sn(OH)_2$, $Cr(OH)_3$, $Be(OH)_2$, Sb_2O_3, and As_2O_3. Many other compounds, including $Cu(OH)_2$ and Ag_2O, exhibit this property to a lesser degree.

The following series of equations can be used to describe the reactions that occur when the OH^- concentration of a $Zn^{2+}(aq)$ solution is gradually increased:

$$Zn^{2+}(aq) + OH^-(aq) \rightleftharpoons Zn(OH)^+(aq)$$

$$Zn(OH)^+(aq) + OH^-(aq) \rightleftharpoons Zn(OH)_2(s)$$

$$Zn(OH)_2(s) + OH^-(aq) \rightleftharpoons Zn(OH)_3^-(aq)$$

$$Zn(OH)_3^-(aq) + OH^-(aq) \rightleftharpoons Zn(OH)_4^{2-}(aq)$$

The precipitation of $Zn(OH)_2$ is observed when the proper pH is reached, and upon further increase in pH, the precipitate dissolves. The formulas of the amphoterate anions used here are based upon the fact that $NaZn(OH)_3$ and $Na_2Zn(OH)_4$ have been isolated from such solutions.

The process can be reversed. Acidification of solutions of the zincate ion results in the stepwise removal of OH^-. For example,

$$Zn(OH)_4^{2-}(aq) + H^+(aq) \rightleftharpoons Zn(OH)_3^-(aq) + H_2O$$

The cation of an amphoteric hydroxide has the ability to form complex ions with H_2O and OH^-. Equations for the preceding reactions can be written showing coordinated water molecules. For example,

$$Zn(H_2O)_6^{2+}(aq) + OH^-(aq) \rightleftharpoons Zn(OH)(H_2O)_5^+(aq) + H_2O$$

The reactions can be interpreted, therefore, as replacements of the ligands of the zinc complex ion. In the Brønsted interpretation of this reaction, the $Zn(H_2O)_6^{2+}$ ion is an acid that releases a proton to the OH^- ion, a base.

In general, amphoteric systems are very complicated and much remains to be learned about them. The equilibrium constants are of doubtful validity. Indeed, there is evidence that not all the ions and equilibria in systems of this type have been identified. Probably the reactions of most systems are much more complicated than we have depicted them.

We have considered only mononuclear complexes (complexes with but a single coordinated metal ion). Polynuclear complexes, however, appear to be common; $Sn_2(OH)_2^{2+}$ and $Sn_3(OH)_4^{2+}$ as well as $SnOH^+$ have been identified in Sn^{2+} solutions. The formula of the aluminate ion is usually written as $Al(OH)_4^-$, and such a tetrahedral ion has been identified in 0.1 M to 1.5 M aluminate solutions in which the pH is 13 or higher. Under other conditions, however, polynuclear complexes of aluminum are thought to occur.

Advantage is taken of the amphoteric nature of some hydroxides in analytical chemistry and in some commercial processes. For example, Mg^{2+} and Zn^{2+} may be separated from a solution containing the two ions by making the solution alkaline:

$$Mg^{2+} + 2OH^- \rightleftharpoons Mg(OH)_2(s)$$

$$Zn^{2+} + 4OH^- \rightleftharpoons Zn(OH)_4^{2-}$$

The insoluble magnesium hydroxide, which is not amphoteric, may be removed by filtration from the solution containing the zincate ion.

In the production of aluminum metal from bauxite (impure hydrated Al_2O_3), the ore is purified prior to its reduction to aluminum metal. This purification is accomplished by dissolving the aluminum oxide in a solution of sodium hydroxide and removing the insoluble impurities by filtration:

$$Al_2O_3(s) + 2OH^- + 3H_2O \longrightarrow 2Al(OH)_4^-$$

When the filtered solution is acidified, pure aluminum hydroxide precipitates and is recovered.

Summary

The topics that have been discussed in this chapter are

1. Solubility products, equilibrium constants for equilibria between solid substances and saturated solutions of these substances.

2. The use of solubility products to determine whether precipitation will occur when solutions with given ion concentrations are mixed.

3. The common-ion effect on solubility equilibria.

4. Precipitation from a solution by the addition of an ion capable of forming two different precipitates with two ions present in the solution.

5. The precipitation of sulfides; the effect of pH on these reactions.

6. Equilibria involving complex ions.

7. Amphoteric hydroxides and equilibria involving them.

Key Terms

Some of the more important terms introduced in this chapter are listed below. Definitions for terms not included in this list may be located in the text by use of the index.

Amphoterism (Section 16.5) A property of the hydroxides of certain metals that permits them to function as acids or bases; these water-insoluble substances dissolve in solutions of low pH or of high pH.

Complex ion (Section 16.4) An aggregate consisting of a central metal cation surrounded by a number of ligands.

Instability constant (Section 16.4) An equilibrium constant for the overall dissociation of a complex ion into a metal cation and ligands; the reciprocal is called a **formation constant**.

Ion product (Section 16.2) A value calculated and compared to the pertinent K_{SP} in order to determine whether a precipitate will form when the ion concentrations of a solution are adjusted to given values. The ion product is the value obtained by substituting the proposed concentrations into the concentration terms in an expression similar to that of the solubility product.

Ligand (Section 16.4) An anion or a molecule that has an unshared pair of electrons with which it can bond to a metal cation in the formation of a complex ion.

Salt effect (Section 16.1) The increase in solubility of a slightly soluble substance observed following the addition of another electrolyte to the solution.

Solubility product, K_{SP} (Section 16.1) The equilibrium constant for a system involving a slightly soluble substance in equilibrium with a saturated solution of its ions. The constant is the product of the ion concentrations with the concentrations raised to the powers indicated by the coefficients of the balanced chemical equation.

Problems*

The Solubility Product

16.1 At $25°C$, 1.7×10^{-5} mol of $Cd(OH)_2$ dissolves in one liter of water. Calculate the K_{SP} of $Cd(OH)_2$.

16.2 The K_{SP} of $NiCO_3$ is 1.4×10^{-7}. What is the molar solubility of $NiCO_3$? Neglect the hydrolysis of ions.

16.3 At $25°C$, 5.2×10^{-6} mol of $Ce(OH)_3$ dissolves in one liter of water. Calculate the K_{SP} of $Ce(OH)_3$.

16.4 The K_{SP} of CaF_2 is 3.9×10^{-11}. What is the molar solubility of CaF_2? Neglect the hydrolysis of ions.

16.5 Which carbonate has the lower molar solubility, Ag_2CO_3 $(K_{SP}, 8.2 \times 10^{-12})$ or $CuCO_3$ $(K_{SP}, 2.5 \times 10^{-10})$? Neglect the hydrolysis of ions.

16.6 Which sulfide has the lower molar solubility, Ag_2S $(K_{SP}, 5.5 \times 10^{-51})$ or CuS $(K_{SP}, 8.0 \times 10^{-37})$? Neglect the hydrolysis of ions.

16.7 A saturated solution of a slightly soluble hydroxide, $M(OH)_2$, has a pH of 9.53. What is the K_{SP} of $M(OH)_2$?

16.8 How many moles of $Ni(OH)_2$ will dissolve in one liter of a solution of $NaOH$ with a pH of 12.34? Neglect hydrolysis of ions.

16.9 How many moles of Ag_2CO_3 will dissolve in 150 ml of $0.15\ M$ Na_2CO_3 solution? Neglect hydrolysis of ions.

Precipitation and K_{SP}

16.10 Calculate the final concentrations of $Na^+(aq)$, $C_2O_4^{2-}(aq)$, $Ba^{2+}(aq)$, and $Cl^-(aq)$ in a solution prepared by adding 100 ml of $0.20\ M$ $Na_2C_2O_4$ to 150 ml of $0.25\ M$ $BaCl_2$. Neglect the hydrolysis of ions.

16.11 Calculate the final concentrations of $Sr^{2+}(aq)$, $NO_3^-(aq)$, $Na^+(aq)$, and $F^-(aq)$ in a solution prepared by adding 50 ml of $0.30\ M$ $Sr(NO_3)_2$ to 150 ml of $0.12\ M$ NaF. Neglect the hydrolysis of ions.

16.12 What concentration of F^- is necessary to start the precipitation of SrF_2 from a saturated solution of $SrSO_4$?

16.13 What concentration of $SO_4^{2-}(aq)$ is necessary to start the precipitation of $BaSO_4$ from a saturated solution of BaF_2?

16.14 What minimum concentration of NH_4^+ is necessary to prevent the formation of a $Fe(OH)_2$ precipitate from a solution that is $0.020\ M$ in Fe^{2+} and $0.020\ M$ in NH_3?

16.15 What minimum concentration of NH_4^+ is necessary to prevent formation of $Mn(OH)_2(s)$ from a solution that is $0.030\ M$ in $Mn^{2+}(aq)$ and $0.030\ M$ in NH_3?

16.16 A solution is $0.030\ M$ in Mn^{2+} and $0.025\ M$ in NH_4^+. What should the concentration of NH_3 be in order to cause $Mn(OH)_2$ to start to precipitate?

16.17 A solution is $0.090\ M$ in $Mg^{2+}(aq)$ and $0.33\ M$ in $NH_4^+(aq)$. What minimum concentration of NH_3 will cause $Mg(OH)_2$ to start to precipitate?

16.18 A solution is prepared by mixing 10 ml of $0.50\ M$ $CaCl_2$ with 10 ml of a solution that is $0.50\ M$ in NH_3 and $0.050\ M$ in $NH_4^+(aq)$. Will $Ca(OH)_2$ precipitate?

Precipitation of Sulfides

16.19 A solution that is $0.30\ M$ in $H^+(aq)$ and $0.15\ M$ in Ni^{2+} is saturated with H_2S. Should NiS precipitate?

16.20 A solution that is $0.25\ M$ in $H^+(aq)$ and $0.10\ M$ in $Co^{2+}(aq)$ is saturated with H_2S. Will CoS precipitate?

16.21 A solution that is $0.50\ M$ in $H^+(aq)$ and $0.030\ M$ in $Cd^{2+}(aq)$ is saturated with H_2S. Will CdS precipitate?

16.22 What should the $H^+(aq)$ concentration be in a solution that is $0.25\ M$ in Co^{2+} to prevent the precipitation of CoS when the solution is saturated with H_2S?

16.23 What is the lowest concentration of $H^+(aq)$ that must be present in a $0.50\ M$ solution of $Zn^{2+}(aq)$ to prevent the precipitation of ZnS when the solution is saturated with H_2S?

16.24 A solution that is $0.20\ M$ in $H^+(aq)$ and $0.20\ M$ in $Pb^{2+}(aq)$ is saturated with H_2S. What is the concentration of $Pb^{2+}(aq)$ after PbS has precipitated? Note that it is necessary to take into account the increase in acidity caused by the precipitation.

16.25 A solution that is $0.10\ M$ in $H^+(aq)$, $0.30\ M$ in $Cu^{2+}(aq)$, and $0.30\ M$ in $Fe^{2+}(aq)$ is saturated with H_2S. Calculate the concentrations of $H^+(aq)$, $S^{2-}(aq)$, $Cu^{2+}(aq)$, and $Fe^{2+}(aq)$. Note that it is necessary to take into account any increases in acidity caused by precipitation.

16.26 What concentration of $H^+(aq)$ should be present in a solution that is $0.20\ M$ in $Ni^{2+}(aq)$ and $0.20\ M$ in $Cd^{2+}(aq)$ so that when the solution is saturated with H_2S the maximum amount of CdS will precipitate but no NiS will precipitate at any stage?

16.27 (a) What concentration of $H^+(aq)$ should be present in a solution that is $0.20\ M$ in Pb^{2+} and $0.20\ M$ in Zn^{2+} so that upon saturation with H_2S, the maximum amount of PbS precipitates but no ZnS precipitates at any stage? **(b)** What concentration of Pb^{2+} remains in the solution after the PbS has precipitated? Ignore the hydrolysis of ions, but do not neglect the increase in acidity caused by the precipitation.

Complex Ions

16.28 Compare the molar solubilities of **(a)** $AgCl$, **(b)** $AgBr$, and **(c)** AgI in $0.50\ M$ NH_3 solution.

16.29 A $0.010\ M$ solution of $AgNO_3$ is made $0.50\ M$ in NH_3, thus forming the $Ag(NH_3)_2^+$ complex ion. Will $AgCl$ precipitate if sufficient $NaCl$ is added to make the solution $0.010\ M$ in Cl^-?

---

* The appendix contains answers to color-keyed problems.

ELEMENTS OF CHEMICAL THERMODYNAMICS

Thermodynamics is the study of the energy changes that accompany physical and chemical changes. An important aspect of the laws of chemical thermodynamics is that they enable us to predict whether a particular chemical reaction is theoretically possible under a given set of conditions. A reaction that has a natural tendency to occur of its own accord is said to be **spontaneous.**

Thermodynamic principles can also be used to determine the extent of a spontaneous reaction—the position of equilibrium. The maximum theoretical yield of the products of a reaction is limited by the equilibrium state of the reaction.

Thermodynamics, however, has nothing to say about the rate or mechanism of a spontaneous reaction. These questions are the concern of chemical kinetics (see Chapter 12). Some spontaneous changes occur extremely slowly. The stable form of carbon under ordinary conditions is graphite, not diamond. The change from diamond to graphite, therefore, is thermodynamically spontaneous. This change, however, is so extremely slow that it is not observed at ordinary temperatures and pressures.

17.1 First Law of Thermodynamics

Many scientists of the late-eighteenth and early-nineteenth centuries studied the relationship between work and heat. Thermodynamics had its origins in these studies. By the 1840s it became clear that

1. Work and heat are both forms of a larger classification called energy.
2. One form of energy can be converted into another form.
3. Energy cannot be created or destroyed.

The **first law of thermodynamics** is the law of conservation of energy: energy can be converted from one form into another but it cannot be created or destroyed. In other words, the total energy of the universe is a constant.

In applying thermodynamic concepts, we frequently confine our attention to the changes that occur within definite boundaries. That portion of nature that is included within these boundaries is called a **system.** The remainder is called the **surroundings.** A mixture of chemical compounds, for example, can constitute a system. The container and everything else around the system make up what is called the surroundings.

A system is assumed to have an **internal energy,** E, which includes all possible forms of energy attributable to the system. Important contributions to the internal energy of a system include the attractions and repulsions between the atoms, molecules, ions, and subatomic particles that make up the system and the kinetic energies of all of its parts.

According to the first law of thermodynamics, the internal energy of an *isolated* system is constant. The actual value of E for any system is not known and cannot be calculated. Thermodynamics, however, is concerned only with *changes* in internal energy, and these changes can be measured.

The state of a system can be defined by specifying the values of properties such as temperature, pressure, and composition. The internal energy of a system depends upon the state of the system and not upon how the system arrived at that state. Internal energy is therefore called a state function. Consider a sample of an ideal gas that occupies a volume of 1 liter at 100 K and 1 atm pressure (state A). At 200 K and 0.5 atm (state B), the sample occupies a volume of 4 liters. According to the first law, the internal energy of the system in state A, E_A, is a constant, as is the internal energy of the system in state B, E_B.

It follows that the difference in the internal energies of the two states, ΔE, is also a constant and is independent of the path taken between state A and state B. It makes no difference whether the gas is heated before the pressure change, whether the heating is done after the pressure change, or, indeed, whether the total change is brought about in several steps:

$$\Delta E = E_B - E_A$$

Suppose that we have a system in an *initial* state and that the internal energy of the system is E_i. If the system absorbs heat from the surroundings, q, the internal energy of the system will now be

$$E_i + q$$

If the system now uses some of its internal energy to do work on the surroundings, w, the internal energy of the system in the *final* state, E_f, will be

$$E_f = E_i + q - w$$
$$E_f - E_i = q - w$$
$$\Delta E = q - w \tag{17.1}$$

It is important to keep in mind the conventions regarding the signs of these quantities:

q, positive = heat *absorbed* by the system

q, negative = heat *evolved* by the system

w, positive = work done *by* the system

w, negative = work done *on* the system

The values of q and w involved in changing a system from an initial state to a final state depend upon the way in which the change is carried out. The value of $(q - w)$, however, is a constant, equal to ΔE, for the change no matter how it is brought about. If a system undergoes a change in which the internal energy of the system remains constant, the work done by the system equals the heat absorbed by the system.

17.2 Enthalpy

For ordinary chemical reactions, the work term generally arises as a consequence of pressure-volume changes. The work done against the pressure of the atmosphere if the system expands in the course of the reaction is an example of pressure-volume work. The term PV has the dimensions of work. Pressure, which is force per unit area, may be expressed in newtons per square meter (N/m^2). If volume is expressed in cubic meters (m^3), the product PV is

$$PV = (N/m^2)(m^3) = N \cdot m$$

The newton·meter (which is a joule) is a unit of work, since work is defined as force (the newton) times distance (the meter). In like manner, liter·atmospheres are units of work. If the pressure is held constant, the work done in expansion from V_A to V_B is

$$w = P(V_B - V_A) = P \, \Delta V \tag{17.2}$$

No pressure-volume work can be done by a process carried out at constant volume, and $w = 0$. Thus, at *constant volume* the equation

$$\Delta E = q - w$$

becomes

$$\Delta E = q_V \tag{17.3}$$

where q_V is the heat absorbed by the system at constant volume.

Processes carried out at constant pressure are far more common in chemistry than those conducted at constant volume. If we restrict our attention to pressure-volume work, the work done in constant pressure processes is $P \, \Delta V$. Thus, at *constant pressure* the equation

$$\Delta E = q - w$$

becomes

$$\Delta E = q_P - P \, \Delta V$$

Or, by rearranging,

$$q_P = \Delta E + P \, \Delta V \tag{17.4}$$

where q_P is the heat absorbed by the system at constant pressure.

The thermodynamic function **enthalpy**, H, is defined by the equation

$$H = E + PV \tag{17.5}$$

Therefore,

$$q_P = \Delta H \tag{17.6}$$

The heat absorbed by a reaction conducted at constant pressure is equal to the change in enthalpy. Enthalpy, like internal energy, is a function of the state of the system and is independent of the manner in which the state was achieved. The validity of the law of Hess rests on this fact (see Section 3.5).

When a bomb calorimeter is used to make a calorimetric determination (see Section 3.3), the heat effect is measured at constant volume. Ordinarily, reactions are run at constant pressure. The relationship between change in enthalpy and change in internal energy is used to convert heats of reaction at constant volume ($q_V = \Delta E$) to heats of reaction at constant pressure ($q_P = \Delta H$). The conversion is made by considering the change in volume of the products. The changes in the volumes of liquids and solids are so small that they are neglected.

For reactions involving gases, however, volume changes may be significant. Let us say that V_A is the total volume of gaseous reactants, V_B is the total volume of gaseous products, n_A is the number of moles of gaseous reactants, n_B is the number of moles of gaseous products, and the pressure and temperature are constant:

$$PV_A = n_A RT \text{ and } PV_B = n_B RT$$

Thus,

$$
\begin{aligned}
P \, \Delta V &= PV_B - PV_A \\
&= n_B RT - n_A RT \\
&= (n_B - n_A)RT \\
&= (\Delta n)RT
\end{aligned}
\tag{17.7}
$$

Since

$$\Delta H = \Delta E + P \, \Delta V \tag{17.8}$$

then,

$$\Delta H = \Delta E + (\Delta n)RT \tag{17.9}$$

where Δn is the number of moles of gaseous products minus the number of moles of gaseous reactants.

In order to solve problems using this equation, we must express the value of R in appropriate units. We have noted that liter atmospheres are units of energy. The value of R, 0.082056 liter·atm/(K·mol), may be converted to J/(K·mol) by use of factors derived from the relations

$$1 \text{ atm} = 1.01325 \times 10^5 \text{ N/m}^2$$

$$1 \text{ liter} = 1 \times 10^{-3} \text{ m}^3$$

$$1 \text{ J} = 1 \text{ N·m}$$

Thus,

$$8.2056 \times 10^{-2} \frac{\text{liter·atm}}{\text{K·mol}} \left(\frac{1.01325 \times 10^5 \text{ N/m}^2}{1 \text{ atm}} \right) \left(\frac{1 \text{ m}^3}{1 \text{ liter}} \right) \left(\frac{1 \text{ J}}{1 \text{ N·m}} \right)$$

$$= 8.3143 \text{ J/(K·mol)}$$

Various types of calculations involving enthalpy changes are a topic of Chapter 3. A list of standard enthalpies of formation is found in Table 3.1

Example 17.1

The heat of combustion at constant volume of $CH_4(g)$ is measured in a bomb calorimeter at 25°C and is found to be $-885,389$ J/mol. What is the ΔH?

Solution

For the reaction

$$CH_4(g) + 2O_2(g) \longrightarrow CO_2(g) + 2H_2O(l) \qquad \Delta E = -885,389 \text{ J}$$
$$\Delta n = 1 - (2 + 1) = -2$$

Therefore,

$$
\begin{aligned}
\Delta H &= \Delta E + (\Delta n)RT \\
&= -885,389 \text{ J} + (-2 \text{ mol})[8.3143 \text{ J/(K·mol)}](298.2 \text{ K}) \\
&= -885,389 \text{ J} - 4959 \text{ J} \\
&= -890,348 \text{ J} = -890.348 \text{ kJ}
\end{aligned}
$$

Example 17.2

Calculate $\Delta H°$ and $\Delta E°$ for the reaction

$$OF_2(g) + H_2O(g) \longrightarrow O_2(g) + 2HF(g)$$

The standard enthalpies of formation are $OF_2(g)$, $+23.0$ kJ/mol; $H_2O(g)$, -241.8 kJ/mol; and $HF(g)$, -268.6 kJ/mol.

Solution

The standard enthalpies of formation are used to calculate $\Delta H°$ for the reaction (see Section 3.6):

$$
\begin{aligned}
\Delta H° &= 2\Delta H_f°(HF) - [\Delta H_f°(OF_2) + \Delta H_f°(H_2O)] \\
&= 2(-268.6 \text{ kJ}) - [(+23.0 \text{ kJ}) + (-241.8 \text{ kJ})] \\
&= -537.2 \text{ kJ} + 218.8 \text{ kJ} = -318.4 \text{ kJ}
\end{aligned}
$$

This value of $\Delta H°$ is used to find $\Delta E°$. For the reaction $\Delta n = +1$ mol,

$$
\begin{aligned}
\Delta E° &= \Delta H° - (\Delta n)RT \\
&= -318.4 \text{ kJ} - (1 \text{ mol}) [(8.31 \text{ J/(K·mol)}] (298 \text{ K}) \\
&= -318.4 \text{ kJ} - 2476 \text{ J} \\
&= -318.4 \text{ kJ} - 2.5 \text{ kJ} = -320.9 \text{ kJ}
\end{aligned}
$$

Example 17.3

For the reaction

$$B_2H_6(g) + 3\,O_2(g) \longrightarrow B_2O_3(s) + 3\,H_2O(l)$$

ΔE° is -2143.2 kJ. (a) Calculate ΔH° for the reaction. (b) Determine the value of the standard enthalpy of formation of $B_2H_6(g)$. For $B_2O_3(s)$, $\Delta H_f^\circ = -1264.0$ kJ/mol and for $H_2O(l)$, $\Delta H_f^\circ = -285.9$ kJ/mol.

Solution

(a) $\Delta n = -4$. Therefore,

$$\begin{aligned}
\Delta H^\circ &= \Delta E^\circ + (\Delta n)RT \\
&= -2143.2 \text{ kJ} + (-4 \text{ mol})[8.314 \times 10^{-3} \text{ kJ/(K·mol)}](298 \text{ K}) \\
&= -2143.2 \text{ kJ} - 9.9 \text{ kJ} \\
&= -2153.1 \text{ kJ}
\end{aligned}$$

(b)
$$\Delta H^\circ = \Delta H_f^\circ(B_2O_3) + 3\,\Delta H_f^\circ(H_2O) - \Delta H_f^\circ(B_2H_6)$$

$$-2153.1 \text{ kJ} = (-1264.0 \text{ kJ}) + 3(-285.9 \text{ kJ}) - \Delta H_f^\circ(B_2H_6)$$

$$\Delta H_f^\circ(B_2H_6) = +31.4 \text{ kJ}$$

17.3 Second Law of Thermodynamics

The first law of thermodynamics puts only one restriction on chemical or physical changes—energy must be conserved. The first law, however, provides no basis for determining whether a proposed change will be spontaneous. The second law of thermodynamics establishes criteria for making this important prediction.

The thermodynamic function **entropy**, S, is central to the second law. Entropy may be interpreted as a measure of the randomness, or disorder, of a system. A highly disordered system is said to have a high entropy. Since a disordered condition is more probable than an ordered one, entropy may be regarded as a probability function. One statement of the **second law of thermodynamics** is: every spontaneous change is accompanied by an increase in entropy.

As an example of a spontaneous change, consider the mixing of two ideal gases. The two gases, which are under the same pressure, are placed in bulbs that are joined by a stopcock (see Figure 17.1). When the stopcock is opened, the gases spontaneously mix until each is evenly distributed throughout the entire apparatus. Why did this spontaneous change occur? The first law cannot help us answer this question. Throughout the mixing, the volume, total pressure, and temperature remain constant. Since the gases are ideal, no intermolecular forces exist, and neither the internal energy nor the enthalpy of the system is affected.

This change represents an increase in entropy. The final state is more random and hence more probable than the initial state. The random motion of the gas molecules has produced a more disordered condition. The fact that the gases mix spontaneously is not surprising; one would have predicted it from experience. Indeed, it would be surprising if the reverse were to be observed—a gaseous mixture spontaneously separating into two pure gases, each occupying one of the bulbs.

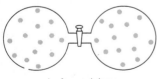
before mixing

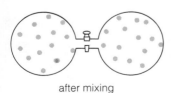
after mixing

Figure 17.1 Spontaneous mixing of two gases

For a given substance the solid, crystalline state is the state of lowest entropy (most ordered); the gaseous state is the state of highest entropy (most random); and the liquid state is intermediate between the other two. Hence, when a substance either melts or vaporizes, its entropy increases. The reverse changes, crystallization and condensation, are changes in which the entropy of the substance decreases. Why, then, should a substance spontaneously freeze at temperatures below its melting point, since this change represents a decrease in the entropy of the substance?

All the entropy effects that result from the proposed change must be considered. When two ideal gases mix by the process previously described, there is no exchange of matter or energy between the isolated system in which the change occurs and its surroundings. The only entropy effect is an increase in the entropy of the isolated system itself. Usually, however, a chemical reaction or a physical change is conducted in such a way that the system is not isolated from its surroundings. The total change in entropy is equal to the sum of the change in the entropy of the system (ΔS_{system}) and the change in entropy of the surroundings ($\Delta S_{surroundings}$):

$$\Delta S_{total} = \Delta S_{system} + \Delta S_{surroundings} \tag{17.10}$$

When a liquid freezes, the enthalpy of fusion is evolved by the liquid and absorbed by the surroundings. This energy increases the random motion of the surrounding molecules and therefore increases the entropy of the surroundings. The spontaneous freezing of a liquid at a temperature below the melting point occurs, therefore, because the decrease in entropy of the liquid (ΔS_{system}) is more than offset by the increase in entropy of the surroundings ($\Delta S_{surroundings}$) so that there is a net increase in entropy.

The total change in entropy should always be considered to determine spontaneity. When a substance melts, the entropy increases, but this effect alone does not determine whether or not the transformation is spontaneous. The entropy of the surroundings must also be considered, and spontaneity is indicated only if the total entropy of system and surroundings taken together increases.

The data of Table 17.1 pertain to the freezing of water. The meaning of the units of ΔS values will be discussed in later sections. For the moment, let us be concerned only with the numerical values listed in the table. At $-1°C$ the change is spontaneous; ΔS_{total} is positive. At $+1°C$, however, ΔS_{total} is negative, and freezing is not a spontaneous change. On the other hand, the reverse change, melting, is spontaneous at $+1°C$ (the signs of all ΔS values would be reversed).

At $0°C$, the melting point, ΔS_{total} is zero, which means that neither freezing nor melting is spontaneous. At this temperature a water-ice system would be in equilibrium, and no *net* change would be observed. Note, however, that freezing

Table 17.1 Entropy changes for the transformation $H_2O(l) \rightarrow H_2O(s)$ at 1 atm			
Temperature (°C)	ΔS_{system} [J/(K·mol)]	$\Delta S_{surroundings}$ (J/(K·mol))	ΔS_{total} [J/(K·mol)]
+1	−22.13	+22.05	−0.08
0	−21.99	+21.99	0
−1	−21.85	+21.93	+0.08

or melting can be made to occur at 0°C by removing or adding heat, but neither change will occur spontaneously.

Thus, the ΔS_{total} of a postulated change may be used as a criterion for whether the change will occur spontaneously. The entropy of the universe is steadily increasing as spontaneous changes occur. Rudolf Clausius summarized the first and second laws of thermodynamics as: "*The energy of the universe is constant; the entropy of the universe tends toward a maximum.*"

Entropy, like internal energy and enthalpy, is a state function. The entropy, or randomness, of a system in a given state is a definite value, and hence, ΔS for a change from one state to another is a definite value depending only on the initial and final states and not on the path between them.

It must be emphasized that, whereas thermodynamic concepts can be used to determine what changes are possible, thermodynamics has nothing to say about the rapidity of change. Some thermodynamically favored changes occur very slowly. Although reactions between carbon and oxygen, as well as between hydrogen and oxygen, at 25°C and 1 atm pressure are definitely predicted by theory, mixtures of carbon and oxygen and mixtures of hydrogen and oxygen can be kept for prolonged periods without significant reaction; such reactions are generally initiated by suitable means. Thermodynamics can authoritatively indicate postulated changes that will *not* occur and need not be attempted, and it can tell us how to alter the conditions of a presumably unfavored reaction in such a manner that the reaction will be thermodynamically possible.

17.4 Gibbs Free Energy

The type of change of primary interest to the chemist is, of course, the chemical reaction. The $\Delta S_{surroundings}$ for a reaction conducted at constant temperature and pressure may be calculated by means of the equation

$$\Delta S_{surroundings} = -\frac{\Delta H}{T} \tag{17.11}$$

where ΔH is the enthalpy change of the reaction and T is the absolute temperature. The change in the entropy of the surroundings is brought about by the heat transferred into or out of the surroundings because of the enthalpy change of the reaction. Since heat *evolved* by the reaction is *absorbed* by the surroundings (and *vice versa*), the sign of ΔH must be reversed. Hence, the larger the value of $-\Delta H$, the more disorder created in the surroundings and the larger the value of $\Delta S_{surroundings}$.

On the other hand, the change in the entropy of the surroundings is *inversely* proportional to the absolute temperature at which the change takes place. A given quantity of heat added to the surroundings at a low temperature (where the randomness is relatively low initially) will create a larger difference in the disorder of the surroundings than the same quantity of heat added at a high temperature (where the randomness is relatively high to begin with). Entropy is therefore measured in units of J/K.

In the last section we noted that

$$\Delta S_{total} = \Delta S_{system} + \Delta S_{surroundings}$$

If $-\Delta H/T$ is substituted for $\Delta S_{\text{surroundings}}$ and if the symbol ΔS (without a subscript) is used to indicate the entropy change of the system, the following equation is obtained:

$$\Delta S_{\text{total}} = \Delta S - \frac{\Delta H}{T} \tag{17.12}$$

Multiplication by T gives

$$T\Delta S_{\text{total}} = T\Delta S - \Delta H \tag{17.13}$$

By reversing the signs of the terms of this equation, we get

$$-T\Delta S_{\text{total}} = \Delta H - T\Delta S \tag{17.14}$$

We can use this equation to show how a function called the Gibbs free energy determines reaction spontaneity.

Gibbs free energy, G, is defined by the equation

$$G = H - TS \tag{17.15}$$

For a chemical reaction conducted at constant temperature and pressure,

$$\Delta G = \Delta H - T\Delta S \tag{17.16}$$

If we compare this equation to

$$-T\Delta S_{\text{total}} = \Delta H - T\Delta S$$

we see that

$$\Delta G = -T\Delta S_{\text{total}} \tag{17.17}$$

On the basis of this result, we note that

1. If ΔG is negative, the reaction is spontaneous. Since ΔS_{total} is greater than zero for a spontaneous change, $T\Delta S_{\text{total}}$ must also be greater than zero and $-T\Delta S_{\text{total}}$ must be less than zero. Therefore, for a spontaneous change at constant temperature and pressure,

$$\Delta G < 0$$

and when a spontaneous change occurs, the free energy of the system decreases.

The negative value of $T\Delta S_{\text{total}}$ is used so that the signs of ΔG values will be like the signs of other energy terms. A *negative* value indicates that energy is *liberated* by the system.

2. If ΔG is zero, the system is in equilibrium. For a system in equilibrium, ΔS_{total} is zero and, therefore,

$$\Delta G = 0$$

J. Willard Gibbs, 1839–1903.
*American Institute of Physics,
Niels Bohr Library.*

3. If ΔG is positive, the reaction is not spontaneous. The reverse of the reaction, however, will be spontaneous.

Notice that all the terms in the very important equation

$$\Delta G = \Delta H - T\Delta S$$

pertain to changes in the properties of the *system*. The use of free energy values, therefore, removes the need to consider changes in the surroundings. Notice also that each term in the equation is an energy term. Since ΔS is measured in J/K, the term $T\Delta S$ is expressed in J; ΔH and ΔG are also expressed in J.

Gibbs free energy is named for J. Willard Gibbs, who developed the application of thermodynamic concepts to chemistry. The change in Gibbs free energy, ΔG, combines two factors that influence reaction spontaneity:

$$\Delta G = \Delta H - T\Delta S$$

1. Reactions tend to seek a minimum in energy. Spontaneity is favored if the ΔH value is negative (the system liberates energy). A negative value of ΔH helps to make the ΔG value negative, which indicates a spontaneous reaction.

2. Reactions tend to seek a maximum in randomness. Spontaneity is favored if the value of ΔS is positive (randomness increases). Since ΔS appears in the term $-T\Delta S$, a positive value of ΔS helps to make the value of ΔG negative, which indicates a spontaneous reaction.

The most favorable circumstance for a negative value of ΔG, which indicates a spontaneous reaction, is a negative value of ΔH together with a positive value of ΔS [see reaction (a) in Table 17.2]. It is possible, however, for a large negative value of ΔH to outweigh an unfavorable entropy change, resulting in a negative value of ΔG [see reaction (b) in Table 17.2]. In addition, a large positive value of $T\Delta S$ can overshadow an unfavorable enthalpy change, giving rise to a negative value of ΔG [see reaction (c) in Table 17.2]. For most chemical reactions at 25°C and 1 atm, the absolute value of ΔH is much larger than the value of $T\Delta S$. Under these conditions exothermic reactions are usually spontaneous no matter how the entropy changes.

With increasing temperature, however, the value of $T\Delta S$ increases and the influence of the entropy effect on ΔG increases. Both ΔH and ΔS usually do not change greatly with increasing temperature. The term $T\Delta S$, however, includes the temperature itself so that at high temperatures $T\Delta S$ can be sufficiently large to be the dominant influence on ΔG. For example, consider reaction (d) in Table 17.2

Table 17.2 Thermodynamic values for some chemical reactions at 25°C and 1 atm (kJ)				
Reaction	ΔH	$-$	$(T\Delta S)$	$=$	ΔG
(a) $H_2(g) + Br_2(l) \longrightarrow 2HBr(g)$	-72.47	$-$	$(+34.02)$	$=$	-106.49
(b) $2H_2(g) + O_2(g) \longrightarrow 2H_2O(l)$	-571.70	$-$	(-97.28)	$=$	-474.42
(c) $Br_2(l) + Cl_2(g) \longrightarrow 2BrCl(g)$	$+29.37$	$-$	$(+31.17)$	$=$	-1.80
(d) $2Ag_2O(s) \longrightarrow 4Ag(s) + O_2(g)$	$+61.17$	$-$	$(+39.50)$	$=$	$+21.67$

(a typical decomposition reaction) for which both ΔH and ΔS are positive. At 25°C, ΔH is larger than $T\Delta S$ and therefore ΔG is positive. If we assume that ΔS does not change with increasing temperature (the actual change is small), at 300°C the value of $T\Delta S$ would be $+75.95$ kJ because of the increase in the value of T. Consequently, $T\Delta S$ would be larger than ΔH (which also does not change greatly with increasing temperature), and ΔG would be negative.

The values in Table 17.2 pertain to the differences between the free energy of the products and the free energy of the reactants at 25°C and 1 atm. However, in some cases [notably reaction (c)] the reaction goes to an intermediate, equilibrium state rather than to completion because the free energy of the equilibrium state is lower than the free energy of the state at which the reaction is complete. Until equilibrium is discussed (see Section 17.7), we should refrain from drawing conclusions as to *the degree of completion* of a reaction from such data.

Table 17.3 is a summary of how the signs of ΔH and ΔS for a given reaction determine reaction spontaneity. For a reaction to be spontaneous, ΔG must be negative, and the sign of ΔG depends upon the signs of ΔH and ΔS:

$$\Delta G = \Delta H - T\Delta S$$

At low temperatures, the sign of ΔG is determined by the sign of ΔH. At high temperatures, since T appears in the term $-T\Delta S$, this term becomes the dominant factor in the calculation.

Gibbs free energy, like the other thermodynamic functions we have discussed, is a state function. The value of ΔG for a process depends only upon the final and initial states of the system and not upon the path taken between those states. When a spontaneous reaction occurs, the free energy of the system declines.

17.5 Standard Free Energies

A standard free energy change, which is given the symbol $\Delta G°$, is the free energy change for a process at 25°C and 1 atm in which the reactants in their standard states are converted to the products in their standard states. The value of $\Delta G°$ for a reaction can be derived from standard free energies of formation in the same way that $\Delta H°$ values can be calculated from standard enthalpies of formation.

The **standard free energy of formation** of a compound, $\Delta G_f°$, is defined as the change in standard free energies when 1 mol of the compound is formed from its

Table 17.3	Effect of the signs of ΔH and ΔS on reaction spontaneity		
ΔH	ΔS	$\Delta G = \Delta H - T\Delta S$	Remarks
$-$	$+$	$-$	Reaction spontaneous at all temperatures
$+$	$-$	$+$	Reaction nonspontaneous at all temperatures
$-$	$-$	$-$ (at low T) $+$ (at high T)	Reaction spontaneous at low temperatures Reaction nonspontaneous at high temperatures
$+$	$+$	$+$ (at low T) $-$ (at high T)	Reaction nonspontaneous at low temperatures Reaction spontaneous at high temperatures

Table 17.4 Gibbs free energy of formation at 25°C and 1 atm

Compound	ΔG_f° (kJ/mol)	Compound	ΔG_f° (kJ/mol)
$H_2O(g)$	−228.61	$CO(g)$	−137.28
$H_2O(l)$	−237.19	$CO_2(g)$	−394.38
$HF(g)$	−270.7	$NO(g)$	+86.69
$HCl(g)$	−95.27	$NO_2(g)$	+51.84
$HBr(g)$	−53.22	$NaCl(s)$	−384.05
$HI(g)$	+1.30	$CaO(s)$	−604.2
$H_2S(g)$	−33.0	$Ca(OH)_2(s)$	−896.76
$NH_3(g)$	−16.7	$CaCO_3(s)$	−1128.76
$CH_4(g)$	−50.79	$BaO(s)$	−528.4
$C_2H_6(g)$	−32.89	$BaCO_3(s)$	−1138.9
$C_2H_4(g)$	+68.12	$Al_2O_3(s)$	−1576.41
$C_2H_2(g)$	+209.20	$Fe_2O_3(s)$	−741.0
$C_6H_6(l)$	+129.66	$AgCl(s)$	−109.70
$SO_2(g)$	−300.37	$ZnO(s)$	−318.19

constituent elements in their standard states. According to this definition, the standard free energy of formation of any element in its standard state is zero. The value of ΔG° for a reaction is equal to the sum of the standard free energies of formation of the products minus the sum of the standard free energies of formation of the reactants. Some standard free energies of formation are given in Table 17.4.

Example 17.4

Use ΔG_f° values from Table 17.4 to calculate ΔG° for the transformation

$$2\,NO(g) + O_2(g) \longrightarrow 2\,NO_2(g)$$

Solution

$$\begin{aligned}
\Delta G^\circ &= 2\Delta G_f^\circ(NO_2) - 2\Delta G_f^\circ(NO) \\
&= 2(+51.84 \text{ kJ}) - 2(+86.69 \text{ kJ}) \\
&= -69.70 \text{ kJ}
\end{aligned}$$

17.6 Absolute Entropies

The addition of heat to a substance results in an increase in molecular randomness. Hence, the entropy of a substance increases as the temperature increases. Conversely, cooling a substance makes it more ordered and decreases its entropy. At absolute zero the entropy of a perfect crystalline substance may be taken as zero. This statement is sometimes called the **third law of thermodynamics** and was first formulated by Walther Nernst in 1906. The entropy of an imperfect crystal, a glass, or a solid solution is not zero at 0 K.

Table 17.5 Absolute entropy at 25°C and 1 atm

Substance	$S°$ [J/(K·mol)]	Substance	$S°$ [J/(K·mol)]
$H_2(g)$	130.6	$HBr(g)$	198.5
$F_2(g)$	203.3	$HI(g)$	206.3
$Cl_2(g)$	223.0	$H_2S(g)$	205.6
$Br_2(l)$	152.3	$NH_3(g)$	192.5
$I_2(s)$	116.7	$CH_4(g)$	186.2
$O_2(g)$	205.03	$C_2H_6(g)$	229.5
S(rhombic)	31.9	$C_2H_4(g)$	219.5
$N_2(g)$	191.5	$C_2H_2(g)$	200.8
C(graphite)	5.69	$SO_2(g)$	248.5
Li(s)	28.0	$CO(g)$	197.9
Na(s)	51.0	$CO_2(g)$	213.6
Ca(s)	41.6	$NO(g)$	210.6
Al(s)	28.3	$NO_2(g)$	240.5
Ag(s)	42.72	NaCl(s)	72.38
Fe(s)	27.2	CaO(s)	39.8
Zn(s)	41.6	$Ca(OH)_2(s)$	76.1
Hg(l)	77.4	$CaCO_3(s)$	92.9
La(s)	57.3	$Al_2O_3(s)$	51.00
$H_2O(g)$	188.7	$Fe_2O_3(s)$	90.0
$H_2O(l)$	69.96	HgO(s)	72.0
HF(g)	173.5	AgCl(s)	96.11
HCl(g)	186.7	ZnO(s)	43.9

On the basis of the third law, absolute entropies can be calculated from heat capacity data. The **standard absolute entropy** of a substance, $S°$, is the entropy of the substance in its standard state at 25°C and 1 atm. Some $S°$ values are given in Table 17.5. The $\Delta S°$ value for a reaction is equal to the sum of the absolute entropies of the products minus the sum of the absolute entropies of the reactants. Note that the absolute entropy of an element is *not* equal to zero and that the absolute entropy of a compound is *not* the entropy change when the compound is formed from its constituent elements.

Example 17.5

(a) Use absolute entropies from Table 17.5 to calculate the standard entropy change, $\Delta S°$, for the formation of 1 mol of HgO(s) from its elements. (b) The standard enthalpy of formation, $\Delta H_f°$, of HgO(s) is -90.7 kJ/mol. Calculate the standard free energy of formation, $\Delta G_f°$, of HgO(s).

Solution

(a) $Hg(l) + \frac{1}{2}O_2(g) \longrightarrow HgO(s)$

$\Delta S° = S°(HgO) - [S°(Hg) + \frac{1}{2}S°(O_2)]$

$= (72.0 \text{ J/K}) - [(77.4 \text{ J/K}) + \frac{1}{2}(205.0 \text{ J/K})]$

$= -107.9 \text{ J/K}.$

(b) $\Delta G° = \Delta H° - T \Delta S°$
$= -90,700\ \text{J} - (298\ \text{K})(-107.9\ \text{J/K})$
$= -58,500\ \text{J} = -58.5\ \text{kJ}$

Many problems can be solved by use of $\Delta S°$ values (calculated from $S°$ values, see Table 17.5), $\Delta H_f°$ values (see Table 3.1), and $\Delta G_f°$ values (see Table 17.4). Approximate solutions for several types of problems can be found by making the assumption that the values of ΔH and ΔS do not change with a change in temperature. The assumption is usually a good one since observed changes in these values are comparatively small.

Example 17.6

For the reaction

$$NH_4Cl(s) \longrightarrow NH_3(g) + HCl(g)$$

$\Delta S° = +285\ \text{J/K}$, $\Delta H° = +177\ \text{kJ}$, and $\Delta G° = +91.9\ \text{kJ}$ at $25°C$. (a) Is the reaction spontaneous at $25°C$? (b) Assume that ΔH and ΔS do not change with an increase in temperature and calculate the value of ΔG at $500°C$ (which is $773\ \text{K}$). Is the reaction spontaneous at this temperature?

Solution

(a) Since $\Delta G°$ is a positive value at $25°C$ ($+91.9\ \text{kJ}$), the reaction is not spontaneous at this temperature. The reverse reaction

$$NH_3(g) + HCl(g) \longrightarrow NH_4Cl(s)$$

for which $\Delta G°$ is $-91.9\ \text{kJ}$ would be spontaneous at $25°C$.

(b) For the reaction as given in the problem,

$$NH_4Cl(s) \longrightarrow NH_3(g) + HCl(g)$$

$\Delta G = \Delta H° - T \Delta S°$
$= +177\ \text{kJ} - (773\ \text{K})(+0.285\ \text{kJ/K})$
$= +177\ \text{kJ} - 220\ \text{kJ} = -43\ \text{kJ}$

Notice that we express $\Delta S°$ in kJ/K so that the term $T \Delta S°$ will have the same units (kJ) as $\Delta H°$. Since ΔG is a negative value at $500°C$, the reaction is spontaneous at this temperature. This result is typical of decomposition reactions, which are endothermic and proceed to a greater and greater extent as the temperature is increased.

Example 17.7

For the vaporization of methyl alcohol,

$$CH_3OH(l) \rightleftharpoons CH_3OH(g)$$

$\Delta H° = +37.4$ kJ and $\Delta S°$ is $+111$ J/K at 25°C. Assume that these values do not change with an increase in temperature. When $\Delta G = 0$, the transformation is spontaneous in neither direction, the system is in equilibrium, and the temperature is the normal boiling point of methyl alcohol. Calculate the normal boiling point of CH_3OH.

Solution

$$\Delta G = \Delta H° - T\Delta S°$$

$$0 = +37.4 \text{ kJ} - T(+0.111 \text{ kJ/K})$$

$$T = \frac{+37.4 \text{ kJ}}{+0.111 \text{ kJ/K}} = 337 \text{ K}$$

The normal boiling point according to this calculation, therefore, is $337 - 273 = 64°C$. The handbook lists the normal boiling point of methyl alcohol as 64.96°C. Since $\Delta H°$ and $\Delta S°$ do change slightly with temperature, our calculated value is in error by a small amount.

17.7 Gibbs Free Energy and Equilibrium

Since ΔG is zero for a system in equilibrium, free energy changes can be used to evaluate the equilibrium state of a chemical reaction. The free energy of a substance in a state other than its standard state, G, is related to its standard free energy, $G°$, by the equation

$$G = G° + RT \ln a \tag{17.18}$$

where R is the ideal gas constant, 8.3143 J/(K·mol), T is the absolute temperature, and $\ln a$ is the natural logarithm of the activity of the substance.

The activity of a substance is its effective concentration. The activity of a substance in its standard state is equal to 1. Notice that since the natural logarithm of 1 is zero, the equation states that $G = G°$ when the substance is in its standard state. The standard state of an ideal gas is the gas at a pressure of 1 atm. If the partial pressure of the gas is 0.5 atm, the gas is said to have an activity of 0.5.

This equation can be used to derive an expression for the free energy change that accompanies a chemical reaction. The free energy change of the general equation

$$w\text{W} + x\text{X} \rightleftharpoons y\text{Y} + z\text{Z}$$

is given by the equation

$$\Delta G = \Delta G° + RT \ln \left(\frac{(a_Y)^y(a_Z)^z}{(a_w)^w(a_x)^x} \right) \tag{17.19}$$

The fraction contains the activities of the substances participating in the reaction with each activity raised to a power equal to the coficient in the balanced chemical equation. The logarithmic term of the equation corrects $\Delta G°$ to the more general condition in which the activities of the substances are not each equal to one.

At equilibrium, $\Delta G = 0$ and, therefore,

$$0 = \Delta G^\circ + RT \ln \frac{(a_Y)^y(a_Z)^z}{(a_w)^w(a_x)^x} \qquad (17.20)$$

Since the system is in equilibrium, the activities that appear in this equation are equilibrium activities, and the fraction is the thermodynamic equilibrium constant, K. Therefore,

$$\Delta G^\circ = -RT \ln K \qquad (17.21)$$

or

$$\Delta G^\circ = -2.303\, RT \log K \qquad (17.22)$$

The ideal gas constant, R, is 8.3143 J/(K·mol). For an equilibrium system at 25°C (which is 298.15 K), therefore,

$$
\begin{aligned}
\Delta G^\circ &= -2.303RT \log K \\
&= -2.303[8.3143 \text{ J/(K·mol)}](298.15 \text{ K}) \log K \\
&= -(5709 \text{ J/mol}) \log K \\
&= -(5.709 \text{ kJ/mol}) \log K
\end{aligned}
$$

Example 17.8

Calculate K_p for the following reaction at 25°C:

$$2\,SO_2(g) + O_2(g) \rightleftharpoons 2\,SO_3(g)$$

For $SO_2(g)$, ΔG_f° is -300.4 kJ/mol. For $SO_3(g)$, ΔG_f° is -370.4 kJ/mol.

Solution

The standard free energy change for the reaction is

$$
\begin{aligned}
\Delta G^\circ &= 2\Delta G_f^\circ(SO_3) - 2\Delta G_f^\circ(SO_2) \\
&= 2 \text{ mol}(-370.4 \text{ kJ/mol}) - 2 \text{ mol}(-300.4 \text{ kJ/mol}) \\
&= -140.0 \text{ kJ}
\end{aligned}
$$

We use this value to find K_p. Since

$$\Delta G^\circ = -(5.709 \text{ kJ/mol}) \log K$$

$$\log K = \frac{\Delta G^\circ}{-5.709 \text{ kJ/mol}}$$

$$= \frac{-140.0 \text{ kJ/mol}}{-5.709 \text{ kJ/mol}} = 24.52$$

Therefore,

$$K = \text{antilog } 24.52 = 3.3 \times 10^{24}$$

The type of K obtained by this method depends upon the definition of the standard states used in the determination of $\Delta G°$ values. For the reactions of gases, K_p's are obtained since the standard state of a gas is defined as the gas at a partial pressure of 1 atm.

For the reaction

$$N_2O_4(g) \rightleftharpoons 2\,NO_2(g)$$

$$\Delta G° = +5.40 \text{ kJ}$$

Since $\Delta G°$ is positive, we might predict that N_2O_4 in its standard state at 25°C and 1 atm would not dissociate into NO_2 at all and that the reverse reaction (the formation of N_2O_4 from NO_2) would go to completion. Both these predictions are incorrect.

If we use the value of $\Delta G°$ to calculate the equilibrium constant for this system at 25°C, we find

$$K_p = 0.113$$

At equilibrium, therefore, some of the N_2O_4 is dissociated.

The free energy of a system in which this reaction occurs at 25°C and 1 atm is plotted against the fraction of N_2O_4 dissociated in Figure 17.2. Point A represents the standard free energy of 1 mol of N_2O_4, point B represents the standard free energy of 2 mol of NO_2, and the intervening points on this curve represent the free energies of mixtures of N_2O_4 and NO_2. Absolute values of free energies

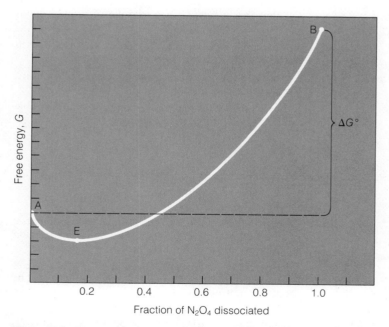

Figure 17.2 Free energy of a system that contains the equivalent of 1 mol of N_2O_4 as the reaction $N_2O_4(g) \rightleftharpoons NO_2(g)$ occurs (25°C and 1 atm)

Table 17.6 Values of K corresponding to $\Delta G°$ values according to the equation $\Delta G° = -2.303RT \log K$

$\Delta G°$ (kJ)	K
-200	1.1×10^{35}
-100	3.3×10^{17}
-50	5.7×10^{8}
-25	2.4×10^{4}
-5	7.5
0	1.0
+5	0.13
+25	4.2×10^{-5}
+50	1.7×10^{-9}
+100	3.0×10^{-18}
+200	9.3×10^{-36}

are not known, and no scale is indicated on the vertical axis of the diagram. Differences in free energies, however, can be calculated so that the shape of the curve is accurately represented.

The free energy curve exhibits a minimum at the equilibrium point, E, where 16.6% of the N_2O_4 is dissociated. The difference between the standard free energy of 2 mol of NO_2 (point B) and the standard free energy of 1 mol of N_2O_4 (point A) is $\Delta G°$ for the reaction ($+5.40$ kJ) and is indicated on the figure. However, ΔG for the preparation of the equilibrium mixture (point E) from 1 mol of N_2O_4 (point A) is -0.84 kJ, which indicates that N_2O_4 will spontaneously dissociate until equilibrium is reached.

The figure shows that equilibrium can be approached from either direction. Thus, $\Delta G = -6.23$ kJ for the preparation of the equilibrium mixture (point E) from 2 mol of pure NO_2 (point B). The negative values of ΔG for both changes (from A to E and from B to E) indicate that both changes are spontaneous.

We must be careful, therefore, when we interpret the meaning of $\Delta G°$ values in relation to reaction spontaneity. Values of K corresponding to various $\Delta G°$ values are listed in Table 17.6. A large *negative* value of $\Delta G°$ means that K for the reaction is a large *positive* value, and therefore the reaction from left to right will go virtually to completion. On the other hand, if $\Delta G°$ is a large positive value, K will be extremely small, thus indicating that the reverse reaction, from right to left, will go virtually to completion. Only if the value of $\Delta G°$ is neither very large nor very small (see Table 17.6) will the value of K indicate a situation in which the reaction will not go essentially to completion in one direction or the other.

Summary

The topics that have been discussed in this chapter are

1. The first law of thermodynamics: the law of conservation of energy. The internal energy of a system, E.

2. Enthalpy, H, and its relation to internal energy. Energy changes for processes conducted at constant volume and at constant pressure.

3. The second law of thermodynamics: the concept of entropy.

4. Gibbs free energy, G, as a measure of reaction spontaneity.

5. Standard free energies of formation and their use in calculating the change in free energy for a reaction.

6. Absolute entropies, calculated by means of the third law of thermodynamics.

7. The evaluation of the equilibrium state of a reversible reaction by use of free energy changes.

Key Terms

Some of the more important terms introduced in this chapter are listed below. Definitions for terms not included in this list may be located in the text by use of the index.

Enthalpy, H (Section 17.2) The heat content of a system; for a reaction run at constant pressure, the change in enthalpy, ΔH, is the heat transferred. Enthalpy is related to the internal energy of a system, E, by the equation: $H = E - PV$, where P is the pressure and V is the volume.

Entropy, S (Section 17.3) A measure of the randomness, or disorder, of a system; used as a criterion of reaction spontaneity; expressed in $J/(K \cdot mol)$.

First law of thermodynamics (Section 17.1) Energy can be converted from one form into another but it cannot be created or destroyed.

Gibbs free energy, G (Section 17.4) The thermodynamic function that takes into account both the enthalpy, H, and entropy, S, of a system; $G = H - TS$. For a reaction at constant temperature and pressure, $\Delta G = \Delta H - T\Delta S$.

Internal energy, E (Section 17.1)   The energy content of a system, which includes all possible forms of energy attributable to the system.

Second law of thermodynamics (Section 17.3)   Every spontaneous change is accompanied by an increase in entropy.

Spontaneous change (Section 17.4)   A change that has a natural tendency of its own to occur without work being done on the system. Some spontaneous changes are very slow, however.

Standard absolute entropy, $S°$ (Section 17.6)   The entropy of a substance in its standard state at 25°C and 1 atm; calculated on the basis of the third law of thermodynamics; expressed in J/(K·mol).

Standard free energy of formation, ΔG_f° (Section 17.5) The free energy change for the process, at 25°C and 1 atm,

in which 1 mol of a compound in its standard state is made from its constituent elements in their standard states.

State function (Section 17.1)   A function that depends upon the state of a system (defined by specifying properties such as temperature, pressure, and composition) and not upon how that system arrived at that state.

Surroundings (Section 17.1)   That part of nature not included within the boundaries of the system under consideration.

System (Section 17.1)   That part of nature that is under consideration.

Thermodynamics (Introduction)   The study of the energy changes that accompany chemical and physical changes.

Third law of thermodynamics (Section 17.6)   At 0 K, the entropy of a perfect crystalline substance is zero.

Problems*

The First Law, Internal Energy, Enthalpy

17.1 What is the first law of thermodynamics? What is the difference between internal energy, E, and enthalpy, H?

17.2 What is a state function? List the state functions described in this chapter.

17.3 The combustion of 1.000 g of ethyl alcohol, $C_2H_5OH(l)$, in a bomb calorimeter (at constant volume) evolves 29.62 kJ of heat at 25°C. The products of the combustion are $CO_2(g)$ and $H_2O(l)$. The molecular weight of ethyl alcohol is 46.06. (a) What is $\Delta E°$ for the combustion of 1 mol of ethyl alcohol? (b) Write the chemical equation for the reaction. What is $\Delta H°$ for the combustion? (c) Use values from Table 3.1 to calculate the enthalpy of formation of ethyl alcohol.

17.4 The combustion of 1.000 g of thiourea, $CS(NH_2)_2(s)$, in a bomb calorimeter (at constant volume) evolves 15.37 kJ of heat at 25°C. The products of the combustion are $CO_2(g)$, $SO_2(g)$, $N_2(g)$, and $H_2O(l)$. The molecular weight of thiourea is 76.12. (a) What is $\Delta E°$ for the combustion of 1 mol of thiourea? (b) Write the chemical equation for the reaction. What is $\Delta H°$ for the combustion? (c) Use values from Table 3.1 to calculate the enthalpy of formation of thiourea.

17.5 The combustion of 1.000 g of cyclohexane, $C_6H_{12}(l)$, in a bomb calorimeter (at constant volume) evolves 46.48 kJ of heat. The products of the combustion are $CO_2(g)$ and $H_2O(l)$. The molecular weight of cyclohexane is 84.16. (a) What is $\Delta E°$ for the combustion of 1 mol of cyclohexane? (b) Write the chemical equation for the

reaction. What is $\Delta H°$ for the combustion? (c) Use values from Table 3.1 to calculate the enthalpy of formation of cyclohexane.

17.6 Calculate $\Delta E°$ for the combustion of octane, $C_8H_{18}(l)$, to $CO_2(g)$ and $H_2O(l)$ at 25°C. The value of $\Delta H°$ for this reaction is -5470.71 kJ/mol of octane.

17.7 Calculate $\Delta E°$ for the combustion of ethylene, $C_2H_4(g)$, to $CO_2(g)$ and $H_2O(l)$ at 25°C. The value of $\Delta H°$ for this reaction is -1410.8 kJ/mol of ethylene.

17.8 For the reaction

$$3 NO_2(g) + H_2O(l) \longrightarrow 2 HNO_3(l) + NO(g)$$

$\Delta H°$ is -71.53 kJ. What is $\Delta E°$?

17.9 For the reaction

$$Ca_3P_2(s) + 6 H_2O(l) \longrightarrow 3 Ca(OH)_2(s) + 2 PH_3(g)$$

$\Delta H°$ is -721.70 kJ. What is $\Delta E°$?

17.10 For the reaction

$$CaNCN(s) + 3 H_2O(l) \longrightarrow CaCO_3(s) + 2 NH_3(g)$$

$\Delta E°$ is -261.75 kJ. Use values from Table 3.1 to calculate the enthalpy of formation of $CaNCN(s)$.

17.11 For the reaction

$$NH_4NO_3(s) \longrightarrow N_2O(g) + 2 H_2O(l)$$

$\Delta E°$ is -127.49 kJ. Use values from Table 3.1 to calculate the enthalpy of formation of $N_2O(g)$.

---

* The more difficult problems are marked with asterisks. The appendix contains answers to color-keyed problems.

The Second Law, Entropy, Gibbs Free Energy

17.12 How can entropy and Gibbs free energy be used as criteria for reaction spontaneity?

17.13 The standard enthalpy of formation of an element, ΔH_f°, is zero, the standard free energy of formation of an element, ΔG_f°, is zero, but the standard absolute entropy of an element, S°, is not zero. Explain.

17.14 For the reaction

$$2\,Li(s) + 2\,H^+(aq) \longrightarrow 2\,Li^+(aq) + H_2(g)$$

ΔH° is -556.9 kJ. The value of S° for $Li^+(aq)$ is 14.2 J/K mol, and that of $H^+(aq)$ is 0.0. Other values can be found in Table 17.5. Calculate **(a)** ΔS° and **(b)** ΔG° for the reaction.

17.15 For the reaction

$$2\,La(s) + 6\,H^+(aq) \longrightarrow 2\,La^{3+}(aq) + 3\,H_2(g)$$

ΔH° is -1474.4 kJ. The value of S° for $La^{3+}(aq)$ is -163.2 J/K mol, and that for $H^+(aq)$ is 0.0; other values can be found in Table 17.5. Calculate **(a)** ΔS° and **(b)** ΔG° for the reaction.

17.16 Use the data given in Table 17.4 to calculate ΔG° for the following proposed reactions:

$$SO_2(g) + H_2(g) \longrightarrow H_2S(g) + O_2(g)$$
$$SO_2(g) + 3\,H_2(g) \longrightarrow H_2S(g) + 2\,H_2O(l)$$

What would you predict the products of the reaction of $SO_2(g)$ and $H_2(g)$ at 25°C to be?

17.17 A chemist claims that the following reaction occurs at 25°C:

$$SF_6(g) + 8\,HI(g) \longrightarrow H_2S(g) + 6\,HF(g) + 4\,I_2(s)$$

Is it theoretically possible? Use values from Table 17.4 plus the fact that ΔG_f° for $SF_6(g)$ is -992 kJ/mol to calculate ΔG° for the reaction.

17.18 Will both $BF_3(g)$ and $BCl_3(l)$ hydrolyze at 25°C according to the following equation (where X is F or Cl)?

$$BX_3 + 3\,H_2O(l) \longrightarrow H_3BO_3(aq) + 3\,HX(aq)$$

The pertinent ΔG_f° values are: $H_3BO_3(aq)$, -963.32 kJ/mol; $HF(aq)$, -276.48 kJ/mol; $HCl(aq)$, -131.17 kJ/mol; $H_2O(l)$, -237.19 kJ/mol; $BF_3(g)$, -1093.28 kJ/mol; $BCl_3(l)$, -379.07 kJ/mol.

17.19 For oxygen difluoride, $OF_2(g)$, ΔG_f° is $+40.6$ kJ/mol. **(a)** Is the preparation of $OF_2(g)$ from its elements at 25°C a spontaneous reaction? **(b)** For ozone, $O_3(g)$, ΔG_f° is $+163.43$ kJ/mol. Is it theoretically possible to prepare $OF_2(g)$ at 25°C by the reaction

$$3\,F_2(g) + O_3(g) \longrightarrow 3\,OF_2(g)$$

17.20 The signs of the standard free energies of formation of all the oxides of nitrogen are positive. What implications does this fact have regarding **(a)** the prepara-

tion of these oxides from the elements at 25°C, and **(b)** the products of the combustion of a nitrogen-containing compound in oxygen?

17.21 For the reaction

$$HCOOH(l) \longrightarrow CO(g) + H_2O(l)$$

ΔH° is $+15.79$ kJ and ΔS° is $+215.27$ J/K. Calculate ΔG°. Is the decomposition of formic acid, $HCOOH(l)$ spontaneous at 25°C?

17.22 For the reaction

$$2\,CHCl_3(l) + O_2(g) \longrightarrow 2\,COCl_2(g) + 2\,HCl(g)$$

ΔH° is -366 kJ and ΔS° is $+340$ J/K. Calculate ΔG°. Is the formation of the poisonous gas phosgene, $COCl_2$, from chloroform, $CHCl_3$, and oxygen a spontaneous reaction at 25°C?

17.23 For the reaction

$$C_2H_4(g) + H_2(g) \longrightarrow C_2H_6(g)$$

ΔH° is -136.98 kJ. **(a)** Use values from Table 17.4 to calculate ΔG° for the reaction. **(b)** Use ΔG° and ΔH° to calculate ΔS° for the reaction. **(c)** Calculate the value of ΔS° for the reaction by using S° values from Table 17.5.

17.24 The standard enthalpy of formation, ΔH_f°, of $CS_2(l)$ is $+87.9$ kJ/mol. The absolute entropy, S°, of C(graphite) is 5.69 J/K mol, of S(rhombic) is 31.9 J/K mol, and of $CS_2(l)$ is 151.0 J/K mol. Calculate the standard free energy of formation, ΔG_f°, of $CS_2(l)$.

17.25 For the reaction

$$PCl_5(g) \longrightarrow PCl_3(g) + Cl_2(g)$$

at 25°C, $\Delta H^\circ = +92.5$ kJ and $\Delta S^\circ = +182$ J/K. **(a)** Calculate the value of ΔG° for 25°C. Is the reaction spontaneous at this temperature? **(b)** Assume that the value of ΔH° and ΔS° do not change with a change in temperature and calculate the value of ΔG for the reaction at 300°C. Is the reaction spontaneous at this temperature?

17.26 For the reaction

$$CaCO_3(s) \longrightarrow CaO(s) + CO_2(g)$$

at 25°C, $\Delta H^\circ = +178$ kJ and $\Delta S^\circ = +160$ J/K. **(a)** Calculate the value of ΔG° for 25°C. Is the reaction spontaneous at this temperature? **(b)** Assume that ΔH° and ΔS° are constant for changes in temperature and calculate the value of ΔG for the reaction at 1000°C. Is the reaction spontaneous at this temperature?

17.27 For the reaction

$$2\,Li(s) + H_2(g) \longrightarrow 2\,LiH(s)$$

at 25°C, $\Delta H^\circ = -180.8$ kJ and $\Delta S^\circ = -137.3$ J/K. **(a)** Calculate the value of ΔG° for 25°C. Is the reaction spontaneous at this temperature? **(b)** Assume that ΔH° and ΔS° are constant for changes in temperature and calculate the value of ΔG for the reaction at 1200°C. Is the reaction spontaneous at this temperature?

17.28 Calculate the normal boiling point of ammonia, $NH_3(l)$, given that $\Delta H = +23.3$ kJ/mol and $\Delta S = +97.2$ J/(K·mol) for the vaporization.

17.29 Calculate the normal boiling point of *n*-hexane, $C_6H_{14}(l)$, given that $\Delta H = +29.6$ kJ/mol and $\Delta S = +86.5$ J/(K·mol) for the vaporization.

17.30 What is the standard free energy of formation, ΔG_f°, of $PH_3(g)$? For $PH_3(g)$: ΔH_f° is $+9.25$ kJ/mol, and S° is 210.0 J/K mol. For the standard state of $P_4(s)$, S° is 177.6 J/K mol, and for $H_2(g)$, S° is 130.5 J/K mol.

17.31 For the reaction

$$CO(g) + 2H_2(g) \longrightarrow CH_3OH(l)$$

(a) Calculate ΔH° using values from Table 3.1. **(b)** The absolute entropy, S°, of $CH_3OH(l)$ is 126.78 J/K mol. Use this fact and values from Table 17.5 to calculate ΔS° for the reaction. **(c)** Use ΔH° and ΔS° to calculate ΔG° for the reaction. **(d)** Use the value of ΔG° for the reaction and the value of ΔG_f° for $CO(g)$ from Table 17.4 to calculate the standard free energy of formation of $CH_3OH(l)$.

17.32 Use the reaction

$$SO_2(g) + Cl_2(g) \longrightarrow SO_2Cl_2(l)$$

to find the absolute entropy, S°, of $SO_2Cl_2(l)$. For $SO_2Cl_2(l)$, ΔH_f° is -389.1 kJ/mol, and ΔG_f° is -313.8 kJ/mol. Thermodynamic values for $SO_2(g)$ and $Cl_2(g)$ are found in Tables 3.1, 17.4, and 17.5.

17.33 For $HgBr_2(s)$, ΔH_f° is -169.5 kJ/mol and ΔG_f° is -162.3 kJ/mol. The absolute entropies of $Hg(l)$ and $Br_2(l)$ are given in Table 17.5. What is S° for $HgBr_2(s)$?

17.34 Graphite is the standard state of carbon, and S° for graphite is 5.694 J/K mol. For diamond, ΔH_f° is $+1.895$ kJ/mol, and ΔG_f° is $+2.866$ kJ/mol. What is the absolute entropy, S°, of diamond? Which form of carbon is the more ordered?

17.35 The combustion of cyanogen, $C_2N_2(g)$, in oxygen forms $CO_2(g)$ and $N_2(g)$ and produces an extremely hot flame. For $C_2N_2(g)$, ΔG_f° is $+296.27$ kJ/mol and S° is 242.09 J/K mol. Use these data together with values from Tables 17.4 and 17.5 to calculate ΔH° for the reaction.

Gibbs Free Energy and Equilibrium

17.36 The value of ΔG_f° for $ZnCO_3(s)$ is -731.36 kJ/mol. Use this fact plus values from Table 17.4 to determine K_p for the reaction

$$ZnCO_3(s) \rightleftharpoons ZnO(s) + CO_2(g)$$

17.37 For urea, $CO(NH_2)_2(s)$, ΔG_f° is -197.15 kJ/mol. Use this value together with values from Table 17.4 to determine K_p for the reaction

$$CO_2(g) + 2NH_3(g) \rightleftharpoons H_2O(g) + CO(NH_2)_2(s)$$

17.38 For $SO_3(g)$, ΔG_f° is -370.37 kJ/mol. Use this value together with values from Table 17.4 to determine K_p for the reaction

$$SO_2(g) + NO_2(g) \rightleftharpoons SO_3(g) + NO(g)$$

17.39 For the reaction

$$Br_2(l) + Cl_2(g) \rightleftharpoons 2BrCl(g)$$

ΔG° is -1.80 kJ/mol. What is K_p for the reaction?

17.40 At 25°C, $K_p = 0.108$ for the reaction

$$NH_4HS(s) \rightleftharpoons NH_3(g) + H_2S(g)$$

Values of ΔG_f° for $NH_3(g)$ and $H_2S(g)$ are given in Table 17.4. What is the value of ΔG_f° for $NH_4HS(s)$?

17.41 For the equilibrium

$$PCl_5(g) \rightleftharpoons PCl_3(g) + Cl_2(g)$$

at 25°C, $K_p = 1.8 \times 10^{-7}$. What is ΔG° for the reaction?

17.42 For the reaction

$$H^+(aq) + OH^-(aq) \longrightarrow H_2O$$

$\Delta H^\circ = -55.9$ kJ and $\Delta S^\circ = +80.4$ J/K at 25°C. Calculate the value of ΔG° and the value of K_w at 25°C for the equilibrium

$$H_2O \rightleftharpoons H^+(aq) + OH^-(aq)$$

17.43 For the equilibrium

$$HC_2H_3O_2(aq) \rightleftharpoons H^+(aq) + C_2H_3O_2^-(aq)$$

K_a is 1.8×10^{-5} at 25°C. What is ΔG° for this reaction at 25°C?

*17.44 For the equilibrium system

$$H_2O(g) + CO(g) \rightleftharpoons H_2(g) + CO_2(g)$$

at 25°C, K_p is 9.78×10^4 and ΔH° is -41.20 kJ. **(a)** What is ΔG° at 25°C? **(b)** What is ΔS° at 25°C? **(c)** Assume that ΔH° and ΔS° do not change with changes in temperature, and calculate the value of ΔG for the reaction at 300°C. **(d)** What is the value of K_p for the system at 300°C?

CHAPTER ELECTROCHEMISTRY

18

All chemical reactions are fundamentally electrical in nature since electrons are involved (in various ways) in all types of chemical bonding. Electrochemistry, however, is primarily the study of oxidation-reduction phenomena.

The relations between chemical change and electrical energy have theoretical as well as practical importance. Chemical reactions can be used to produce electrical energy (in cells that are called either **voltaic** or **galvanic cells**). Electrical energy can be used to bring about chemical transformations (in **electrolytic cells**). In addition, the study of electrochemical processes leads to an understanding, as well as to the systemization, of oxidation-reduction phenomena that take place outside cells.

18.1 Metallic Conduction

An electric current is the flow of electric charge. In metals this charge is carried by electrons, and electrical conduction of this type is called **metallic conduction**. The current results from the application of an electric force supplied by a battery or some other source of electrical energy. A complete circuit is necessary to produce a current.

Metallic crystals may be described in terms of mobile electron clouds permeating relatively fixed lattices of positive metal ions (see Section 23.1). When electrons are forced into one end of a metal wire, the impressed electrons displace other electrons of the cloud at the point of entry. The displaced electrons, in turn, assume new positions by pushing neighboring electrons ahead, and this effect is transmitted down the length of the wire until electrons are forced out of the wire at the opposite end. The current source may be regarded as an electron pump, for it serves to force electrons into one end of the circuit and drain them off from the other end. At any position in the wire, electrical neutrality is preserved, since the rate of electrons in equals the rate of electrons out.

The analogy between the flow of electricity and the flow of a liquid is an old one. In earlier times electricity was described in terms of a current of "electric fluid." Conventions of long standing, which may be traced back to Benjamin Franklin (1747) and which were adopted before the electron was identified, ascribe a positive charge to this current. We shall interpret electrical circuits in terms of the movement of electrons. Remember, however, that conventional electric current is arbitrarily described as positive and as flowing in the opposite direction.

Electric current is measured in **amperes** (A). Quantity of electric charge is measured in **coulombs** (C); the coulomb is defined as the quantity of electricity carried in one second by a current of one ampere. Therefore,

$$1 \text{ A} = 1 \text{ C/s}$$

and

$$1 \text{ C} = 1 \text{ A} \cdot \text{s}$$

The current is forced through the circuit by an electrical potential difference, which is measured in **volts** (V). It takes one joule of work to move one coulomb from a lower to a higher potential when the potential difference is one volt. One volt, therefore, equals one joule/coulomb, and one volt·coulomb is a unit of energy and equals one joule:

$$1 \text{ V} = 1 \text{ J/C}$$
$$1 \text{ V} \cdot \text{C} = 1 \text{ J}$$

The higher the potential difference between two points in a given wire, the more current the wire will carry between those two points. George Ohm in 1826 expressed the quantitative relation between potential difference, $\mathscr{E}$, in volts, and current, I, in amperes, as

$$I = \mathscr{E}/R \quad \text{or} \quad \mathscr{E} = IR$$

where the proportionality constant, R, of **Ohm's law** is called the resistance. Resistance is measured in **ohms** (Ω). One volt is required to force a current of one ampere through a resistance of one ohm.

Resistance to the flow of electricity in metals is probably caused by the vibration of the metal ions about their lattice positions. These vibrations interfere with the motion of the electrons and retard the current. As the temperature is increased, the thermal motion of the metal ions is increased. Hence, the resistance of metals increases and the metals become poorer conductors.

18.2 Electrolytic Conduction

Electrolytic conduction, in which the charge is carried by ions, will not occur unless the ions of the electrolyte are free to move. Electrolytic conduction, therefore, is exhibited principally by molten salts and by aqueous solutions of electrolytes. Furthermore, a sustained current through an electrolytic conductor requires that chemical change accompany the movement of ions.

These principles of electrolytic conduction are best illustrated by reference to

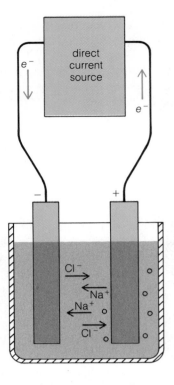

| cathode | anode | **Figure 18.1** Electrolysis of molten |
| $Na^+ + e^- \rightarrow Na$ | $2\,Cl^- \rightarrow Cl_2 + 2e^-$ | sodium chloride |

an electrolytic cell such as the one diagrammed in Figure 18.1 for the electrolysis of molten NaCl between inert electrodes.* The current source pumps electrons into the left-hand electrode, which therefore may be considered to be negatively charged. Electrons are drained from the right-hand, positive electrode. In the electric field thus produced, sodium ions (cations) are attracted toward the negative pole (cathode), and chloride ions (anions) are attracted toward the positive pole (anode). Electric charge in electrolytic conduction is carried by cations moving toward the **cathode** and anions moving in the opposite direction, toward the **anode.**

For a complete circuit, electrode reactions must accompany the movement of ions. At the cathode some chemical species (not necessarily the charge carrier) must accept electrons and be reduced. At the anode, electrons must be removed from some chemical species, which as a consequence is oxidized. The conventions relating to the terms anode and cathode are summarized in Table 18.1.

In the diagrammed cell, sodium ions are reduced at the cathode:

$$Na^+ + e^- \longrightarrow Na$$

and chloride ions are oxidized at the anode:

$$2\,Cl^- \longrightarrow Cl_2 + 2e^-$$

---

* Inert electrodes are not involved in electrode reactions.

Table 18.1 Electrode conventions		
	Cathode	Anode
ions attracted	cations	anions
direction of electron movement	into cell	out of cell
half-reaction	reduction	oxidation
sign		
electrolysis cell	negative	positive
galvanic cell	positive	negative

Proper addition of these two partial equations gives the reaction for the entire cell:

$$2\,NaCl(l) \xrightarrow{\text{electrolysis}} 2\,Na(l) + Cl_2(g)$$

In the actual operation of the commercial cell used to produce metallic sodium, calcium chloride is added to lower the melting point of sodium chloride, and the cell is operated at a temperature of approximately 600°C. At this temperature, sodium metal is a liquid.

We can trace the flow of negative charge through the circuit of Figure 18.1 as follows. Electrons leave the current source and are pumped into the cathode where they are picked up by and reduce sodium ions that have been attracted to this negative electrode. Chloride ions move away from the cathode toward the anode and thus carry negative charge in this direction. At the anode, electrons are removed from the chloride ions, thus oxidizing them to chlorine gas. These electrons are pumped out of the cell by the current source. In this manner the circuit is completed.

Electrolytic conduction, then, rests on the mobility of ions, and anything that inhibits the motion of these ions causes resistance to the current. Factors that influence the electrical conductivity of solutions of electrolytes include interionic attractions, solvation of ions, and viscosity of the solvent. These factors rest on solute-solute attractions, solute-solvent attractions, and solvent-solvent attractions, respectively. The average kinetic energy of the solute ions increases as the temperature is raised, and therefore, the resistance of electrolytic conductors generally decreases as the temperature is raised (that is, conduction increases). Furthermore, the effect of each of the three previously mentioned factors decreases as the temperature is increased.

At all times the solution is electrically neutral. The total positive charge of all of the cations equals the total negative charge of all of the anions.

18.3 Electrolysis

The electrolysis of molten sodium chloride serves as a commercial source of sodium metal and chlorine gas. Analogous procedures are used to prepare other very active metals (such as potassium and calcium). When certain aqueous solutions are electrolyzed, however, water is involved in the electrode reactions rather than the ions derived from the solute. Hence, the current-carrying ions are not necessarily discharged at the electrodes.

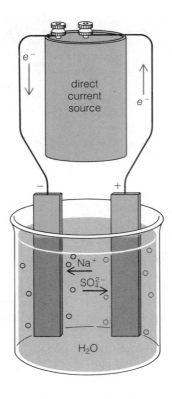

cathode   anode

$2e^- + 2H_2O \rightarrow H_2(g) + 2OH^-$ $2H_2O \rightarrow O_2(g) + 4H^+ + 4e^-$

Figure 18.2 Electrolysis of aqueous sodium sulfate

In the electrolysis of aqueous sodium sulfate, sodium ions move toward the cathode and sulfate ions move toward the anode (see Figure 18.2). Both these ions are difficult to discharge. When this electrolysis is conducted between inert electrodes, hydrogen gas is evolved at the cathode, and the solution surrounding the electrode becomes alkaline. Reduction occurs at the cathode; but rather than the reduction of the sodium ion:

$$e^- + Na^+ \longrightarrow Na$$

the *net change* that occurs is the reduction of water:

$$2e^- + 2H_2O \longrightarrow H_2(g) + 2OH^-$$

Water is an extremely weak electrolyte. Pure water is approximately $2 \times 10^{-7}\%$ ionized at 25°C:

$$2H_2O \rightleftharpoons H_3O^+ + OH^-$$

or, more briefly,

$$H_2O \rightleftharpoons H^+ + OH^-$$

The exact mechanism of the cathode reaction in the electrolysis of aqueous Na_2SO_4 is not known. It may be that the hydrogen ions from water are discharged and that the reaction proceeds as follows:

$$H_2O \rightleftharpoons H^+ + OH^-$$
$$2e^- + 2H^+ \longrightarrow H_2(g)$$

Multiplication of the first equation by 2 followed by addition of the two equations gives the net change

$$2e^- + 2H_2O \longrightarrow H_2(g) + 2OH^-$$

In general, water is reduced at the cathode (producing hydrogen gas and hydroxide ions) whenever the cation of the solute is difficult to reduce.

Oxidation occurs at the anode, and in the electrolysis of aqueous Na_2SO_4, the anions (SO_4^{2-}) that migrate toward the anode are difficult to oxidize:

$$2SO_4^{2-} \longrightarrow S_2O_8^{2-} + 2e^-$$

Therefore, the oxidation of water occurs preferentially. The mode of this reaction may be

$$H_2O \rightleftharpoons H^+ + OH^-$$
$$4OH^- \longrightarrow O_2(g) + 2H_2O + 4e^-$$

Multiplying the first equation by 4 and adding the equations, we get the net change

$$2H_2O \longrightarrow O_2(g) + 4H^+ + 4e^-$$

At the anode the evolution of oxygen gas is observed, and the solution surrounding the pole becomes acidic. In general, water is oxidized at the anode (producing oxygen gas and hydrogen ions) whenever the anion of the solute is difficult to oxidize.

The complete reaction for the electrolysis of aqueous Na_2SO_4 may be obtained by adding the cathode and anode reactions:

$$2[2e^- + 2H_2O \longrightarrow H_2(g) + 2OH^-]$$
$$2H_2O \longrightarrow O_2(g) + 4H^+ + 4e^-$$
$$\overline{6H_2O \longrightarrow 2H_2(g) + O_2(g) + 4H^+ + 4OH^-}$$

If the solution is mixed, the hydrogen and hydroxide ions that are produced neutralize one another, and the net change

$$2H_2O \xrightarrow{\text{electrolysis}} 2H_2(g) + O_2(g)$$

is merely the electrolysis of water. In the course of the electrolysis, the hydrogen ions migrate away from the anode, where they are produced, toward the cathode. In like manner, the hydroxide ions move toward the anode. These ions neutralize one another in the solution between the two electrodes.

Evaporators used to secure sodium hydroxide from the solution left after the electrolysis of aqueous sodium chloride. *Hooker Chemical Company.*

The electrolysis of an aqueous solution of NaCl between inert electrodes serves as an example of a process in which the anion of the electrolyte is discharged, but the cation is not:

$$
\begin{array}{ll}
\text{anode:} & 2\,Cl^- \longrightarrow Cl_2(g) + 2e^- \\
\text{cathode:} & \underline{2e^- + 2\,H_2O \longrightarrow H_2(g) + 2\,OH^-} \\
& 2\,H_2O + 2\,Cl^- \longrightarrow H_2(g) + Cl_2(g) + 2\,OH^-
\end{array}
$$

Since the sodium ion remains unchanged in the solution, the reaction may be indicated

$$2\,H_2O + 2\,Na^+ + 2\,Cl^- \xrightarrow{\ electrolysis\ } H_2(g) + Cl_2(g) + 2\,Na^+ + 2\,OH^-$$

This process is a commercial source of hydrogen gas, chlorine gas, and, by evaporation of the solution left after electrolysis, sodium hydroxide.

In the electrolysis of a solution of $CuSO_4$ between inert electrodes (see right portion of Figure 18.4, given subsequently), the current is carried by the Cu^{2+} and SO_4^{2-} ions. The current-carrying cations are discharged, but the anions are not:

$$
\begin{array}{ll}
\text{anode:} & 2\,H_2O \longrightarrow O_2(g) + 4\,H^+ + 4e^- \\
\text{cathode:} & \underline{2[2e^- + Cu^{2+} \longrightarrow Cu(s)]} \\
& 2\,Cu^{2+} + 2\,H_2O \longrightarrow O_2(g) + 2\,Cu(s) + 4\,H^+
\end{array}
$$

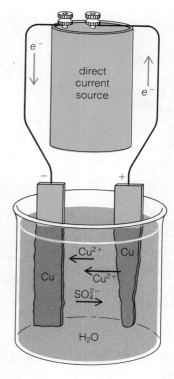

cathode	anode
$2e^- + Cu^{2+} \rightarrow Cu(s)$	$Cu(s) \rightarrow Cu^{2+} + 2e^-$

Figure 18.3 Electrolysis of aqueous cupric sulfate between copper electrodes

It is, of course, possible to have both ions of the solute discharged during the electrolysis of an aqueous solution. An example is the electrolysis of $CuCl_2$ between inert electrodes:

$$
\begin{aligned}
\text{anode:} \quad & 2\,Cl^- \longrightarrow Cl_2(g) + 2e^- \\
\text{cathode:} \quad & 2e^- + Cu^{2+} \longrightarrow Cu(s) \\
\hline
& Cu^{2+} + 2\,Cl^- \longrightarrow Cu(s) + Cl_2(g)
\end{aligned}
$$

It is also possible to have the electrode itself enter into an electrode reaction. If aqueous $CuSO_4$ is electrolyzed between copper electrodes (see Figure 18.3), Cu^{2+} ions are reduced at the cathode:

$$2e^- + Cu^{2+} \longrightarrow Cu$$

but of the *three* possible anode oxidations:

$$2\,SO_4^{2-} \longrightarrow 2\,S_2O_8^{2-} + 2e^-$$

$$2\,H_2O \longrightarrow O_2(g) + 4\,H^+ + 4e^-$$

$$Cu(s) \longrightarrow Cu^{2+} + 2e^-$$

(*Top*) About 90% of U.S. primary copper is produced from open-pit mines (such as shown here) in Arizona, Utah, New Mexico, Montana, and Nevada. The balance of U.S. primary copper comes from underground mines, chiefly in Arizona and Michigan. *Copper Development Association, Inc.*

(*Bottom*) Cathodes of 99.98% pure copper are lifted from an electrolytic refining tank. *The Anaconda Minerals Company.*

the oxidation of the copper metal of the electrode is observed to occur. Hence, at the anode, copper from the electrode goes into solution as Cu^{2+} ions, and at the cathode, Cu^{2+} ions plate out as $Cu(s)$ on the electrode. This process is used to refine copper. Impure copper is used as the anode of an electrolytic cell, and a solution of $CuSO_4$ is electrolyzed. Pure copper plates out on the cathode. Active electrodes are also used in electroplating processes. In silver plating, silver anodes are employed.

18.4 Stoichiometry of Electrolysis

The quantitative relationships between electricity and chemical change were first described by Michael Faraday in 1832 and 1833. Faraday's work is best understood by reference to the half-reactions that occur during electrolysis. The change at the cathode in the electrolysis of molten sodium chloride:

$$Na^+ + e^- \longrightarrow Na$$

shows that one electron is required to produce one sodium atom. One mole of electrons (Avogadro's number of electrons) is required to produce one mole of sodium metal (22.9898 g Na). The quantity of charge equivalent to one mole of electrons is called the **faraday** (F) and has been found to equal 96,485 coulombs (C), which for ordinary problem work is customarily rounded off to 96,500 C:

$$1 F = 96,500 C$$

If 2 F of electricity were used, 2 mol of Na would be produced.

In the same time that electrons equivalent to 1 F of electricity are added to the cathode, that same number of electrons is removed from the anode:

$$2 Cl^- \longrightarrow Cl_2(g) + 2e^-$$

The removal of 1 mol of electrons (1 F) from the anode would result in the discharge of 1 mol of Cl^- ions and the production of 0.5 mol of chlorine gas. If 2 F of electricity flow through the cell, 2 mol of Cl^- ions are discharged and 1 mol of Cl_2 gas liberated.

Electrode reactions, therefore, may be interpreted in terms of moles and faradays. The anode oxidation of the hydroxide ion, for example,

$$4 OH^- \longrightarrow O_2(g) + 2 H_2O + 4e^-$$

may be read as stating that 4 mol of OH^- ion produce 1 mol of O_2 gas and 2 mol of H_2O when 4 F of electricity are passed through the cell.

The relationships between moles of substances and faradays of electricity are the basis of stoichiometric calculations that involve electrolysis. Remember that one ampere (1 A) is equal to a current rate of one coulomb (1 C) per second:

$$1 A = 1 C/s$$

Example 18.1

The charge on a single electron is 1.6022×10^{-19} C. Calculate Avogadro's number from the fact that 1 F = 96,485 C.

Solution

$$? \text{ electrons} = 9.6485 \times 10^4 \text{ C} \left(\frac{1 \text{ electron}}{1.6022 \times 10^{-19} \text{ C}} \right)$$

$$= 6.0220 \times 10^{23} \text{ electrons}$$

Example 18.2

In the electrolysis of $CuSO_4$, how much copper is plated out on the cathode by a current of 0.750 A in 10.0 min?

Solution

The number of faradays used may be calculated as follows:

$$? \text{ F} = 10.0 \text{ min} \left(\frac{60 \text{ s}}{1 \text{ min}} \right) \left(\frac{0.75 \text{ C}}{1 \text{ s}} \right) \left(\frac{1 \text{ F}}{96,500 \text{ C}} \right)$$

$$= 0.00466 \text{ F}$$

The cathode reaction is $Cu^{2+} + 2e^- \rightarrow Cu(s)$, and therefore 2 F plate out 63.5 g of Cu(s):

$$? \text{ g Cu} = 0.00466 \text{ F} \left(\frac{63.5 \text{ g Cu(s)}}{2 \text{ F}} \right) = 0.148 \text{ g Cu(s)}$$

Example 18.3

(a) What volume of $O_2(g)$ at STP is liberated at the anode in the electrolysis of $CuSO_4$ described in Example 18.2? (b) If 100 ml of 1.00 M $CuSO_4$ is employed in the cell, what is the H^+ (aq) concentration at the end of the electrolysis? Assume that there is no volume change for the solution during the experiment and that the anode reaction is

$$2 H_2O \longrightarrow 4 H^+(aq) + O_2(g) + 4e^-.$$

Solution

(a) Four faradays produce 22.4 liter of $O_2(g)$ at STP:

$$? \text{ liter } O_2(g) = 0.00466 \text{ F} \left(\frac{22.4 \text{ liter } O_2(g)}{4 \text{ F}} \right)$$

$$= 0.0261 \text{ liter } O_2(g)$$

(b) Four faradays also produce 4 mol of $H^+(aq)$:

$$? \text{ mol } H^+(aq) = 0.00466 \text{ F} \left(\frac{1 \text{ mol } H^+(aq)}{1 \text{ F}} \right)$$

$$= 0.00466 \text{ mol } H^+(aq)$$

The small contribution of $H^+(aq)$ from the ionization of water may be ignored, and we may assume that there are 0.00466 mol $H^+(aq)$ in 100 ml of solution:

$$? \text{ mol } H^+(aq) = 1000 \text{ ml solution} \left(\frac{0.00466 \text{ mol } H^+(aq)}{100 \text{ ml solution}} \right)$$

$$= 0.0466 \text{ mol } H^+(aq)$$

The solution is therefore 0.0466 M in hydrogen ion.

In Figure 18.4, two electrolytic cells are set up in series. Electricity passes through one cell first and then through the other before returning to the current source. If silver nitrate is electrolyzed in one of the cells, the cathode reaction is

$$Ag^+ + e^- \longrightarrow Ag(s)$$

and metallic silver is plated out on the electrode used. By weighing this electrode before and after the electrolysis, one can determine the quantity of silver plated out and hence the number of coulombs that have passed through the cell. One faraday would plate out 107.868 g of silver. One coulomb, therefore, is equivalent to

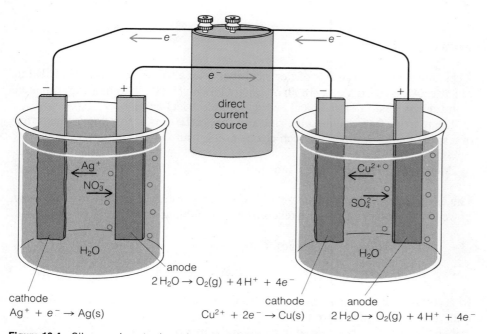

Figure 18.4   Silver coulometer in series with a cell for electrolysis

$$(107.868 \text{ g Ag})/(96,485 \text{ C}) = 1.1180 \times 10^{-3} \text{ g Ag/C}$$

The same number of coulombs pass through both cells in a given time when these cells are arranged in series. The number of coulombs used in an electrolysis, therefore, can be determined by the addition, in series, of this **silver coulometer** to the circuit of the experimental cell.

Example 18.4

(a) What mass of copper is plated out in the electrolysis of $CuSO_4$ in the same time that it takes to deposit 1.00 g of Ag in a silver coulometer that is arranged in series with the $CuSO_4$ cell? (b) If a current of 1.00 A is used, how many minutes are required to plate out this quantity of copper?

Solution

(a) From the electrode reactions we see that 2 F deposit 63.5 g Cu and 1 F deposits 107.9 g Ag:

$$? \text{ g Cu} = 1.00 \text{ g Ag} \left(\frac{1 \text{ F}}{107.9 \text{ g Ag}} \right) \left(\frac{63.5 \text{ g Cu}}{2 \text{ F}} \right) = 0.294 \text{ g Cu}$$

$$\text{(b) } ? \text{ min} = 1.00 \text{ g Ag} \left(\frac{96,500 \text{ C}}{107.9 \text{ g Ag}} \right) \left(\frac{1 \text{ s}}{1 \text{ C}} \right) \left(\frac{1 \text{ min}}{60 \text{ s}} \right)$$

$$= 14.9 \text{ min}$$

18.5 Voltaic Cells

A cell that is used as a source of electrical energy is called a voltaic cell or a galvanic cell after Alessandro Volta (1800) or Luigi Galvani (1780), who first experimented with the conversion of chemical energy into electrical energy.

The reaction between metallic zinc and copper(II) ions in solution is illustrative of a spontaneous change in which electrons are transferred:

$$Zn(s) + Cu^{2+}(aq) \longrightarrow Zn^{2+}(aq) + Cu(s)$$

The exact mechanism by which electron transfer occurs is not known. We may, however, represent the above reaction as a combination of two half-reactions:

$$Zn(s) \longrightarrow Zn^{2+}(aq) + 2e^-$$
$$2e^- + Cu^{2+}(aq) \longrightarrow Cu(s)$$

In a voltaic cell these half-reactions are made to occur at different electrodes so that the transfer of electrons takes place through the external electrical circuit rather than directly between zinc metal and copper(II) ions.

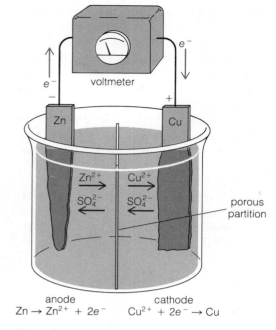

anode

cathode

$Zn \rightarrow Zn^{2+} + 2e^-$ $Cu^{2+} + 2e^- \rightarrow Cu$

Figure 18.5 The Daniell cell

The cell diagrammed in Figure 18.5 is designed to make use of this reaction to produce an electric current. The half-cell on the left contains a zinc metal electrode and $ZnSO_4$ solution. The half-cell on the right consists of a copper metal electrode in a solution of $CuSO_4$. The half-cells are separated by a porous partition that prevents the mechanical mixing of the solutions but permits the passage of ions under the influence of the flow of electricity. A cell of this type is called a **Daniell cell.**

When the zinc and copper electrodes are joined by a wire, electrons flow from the zinc electrode to the copper electrode. At the zinc electrode the zinc metal is *oxidized* to zinc ions. This electrode is the anode, and the electrons that are the product of the oxidation *leave* the cell from this pole (see Table 18.1). The electrons travel the external circuit to the copper electrode where they are used in the reduction of copper(II) ions to metallic copper. The copper thus produced plates out on the electrode. The copper electrode is the cathode. Here, the electrons *enter* the cell and *reduction* occurs.

Since electrons are produced at the zinc electrode, this anode is designated as the negative pole. Electrons travel from the negative pole to the positive pole in the external circuit of any voltaic cell when the cell is operating. The cathode, where electrons are used in the electrode reaction, is therefore the positive pole. Within the cell the movement of ions completes the electric circuit. At first glance, it is surprising that anions, which are negatively charged, should travel toward an anode that is the negative electrode. Conversely, cations, which carry a positive charge, travel toward the cathode, which is the positive pole.

Careful consideration of the electrode reactions provides the answer to this problem. At the anode, zinc ions are being produced and electrons left behind in the metal. At all times the electrical neutrality of the solution is maintained. In the solution surrounding the electrode there must be as much negative charge from anions as there is positive charge from cations. Hence, SO_4^{2-} ions move toward the anode to neutralize the effect of the Zn^{2+} ions that are being produced. At the same time, zinc ions move away from the anode toward the cathode. At the cathode, electrons are being used to reduce Cu^{2+} ions to copper metal. While the Cu^{2+} ions are being discharged, more Cu^{2+} ions move into the region surrounding the cathode to take the place of the ions being removed. If this did not occur, a surplus of SO_4^{2-} ions would build up around the cathode.

The porous partition is added to prevent mechanical mixing of the solutions of the half-cells. If Cu^{2+} ions came into contact with the zinc metal electrode, electrons would be transferred directly rather than through the circuit. In the normal operation of the cell, this "short circuit" does not occur because the Cu^{2+} ions move in a direction away from the zinc electrode.

Actually this cell would work if a solution of an electrolyte other than $ZnSO_4$ were used in the anode compartment, and if a metal other than copper were used for the cathode. The substitutes, however, must be chosen so that the electrolyte in the anode compartment does not react with the zinc electrode and the cathode does not react with Cu^{2+} ions.

18.6 Electromotive Force

If $1\,M$ $ZnSO_4$ and $1\,M$ $CuSO_4$ solutions are employed in the Daniell cell, the cell may be represented by the notation

$$Zn(s)|Zn^{2+}(1\,M)|Cu^{2+}(1\,M)|Cu(s)$$

in which the vertical lines represent phase boundaries. By convention, the substance forming the anode is listed first. The other materials of the cell are then listed in the order that one would encounter them leading from the anode to the cathode. The composition of the cathode is given last.

Electric current is produced by a voltaic cell as a result of the **electromotive force** (emf) of the cell, which is measured in volts. The greater the tendency for the cell reaction to occur, the higher the emf of the cell. The emf of a given cell, however, also depends upon the concentrations of the substances used to make the cell.

A **standard emf**, $\mathscr{E}°$, pertains to the electromotive force of a cell, at 25°C, in which all reactants and products are present in their standard states. The standard state of a solid or a liquid is, of course, the pure solid or pure liquid itself. The standard state of a gas or a substance in solution is a defined state of *ideal unit activity*—that is, corrections are applied for deviations from ideality caused by intermolecular and interionic attractions. For our discussion we shall make the assumption that the activity of ions may be represented by their molar concentrations and the activity of gases by their pressures in atmospheres. Hence, according to this approximation, a standard cell would contain ions at $1\,M$ concentrations and gases at 1 atm pressures. In the cell notations that follow, concentrations will be indicated only if they deviate from standard.

If the emf of a cell is to be used as a reliable measure of the tendency for the cell reaction to occur, the voltage must be the maximum value obtainable for the particular cell under consideration. If there is an appreciable flow of electricity during measurement, the voltage measured, $\mathscr{E}$, will be reduced because of the internal resistance of the cell. In addition, when the cell delivers current, the electrode reactions produce concentration changes that reduce the voltage.

The emf of a cell, therefore, must be measured with no appreciable flow of electricity through the cell. This measurement is accomplished by the use of a potentiometer. The circuit of a potentiometer includes a current source of variable voltage and a means of measuring this voltage. The cell being studied is connected to the potentiometer circuit in such a way that the emf of the cell is opposed by the emf of the potentiometer current source.

If the emf of the cell is larger than that of the potentiometer, electrons will flow in the normal direction for a spontaneously discharging cell of that type. On the other hand, if the emf of the potentiometer current source is larger than that of the cell, electrons will flow in the opposite direction, thus causing the cell reaction to be reversed. When the two emf's are exactly balanced, no electrons flow. This voltage is the **reversible emf** of the cell. The emf of a standard Daniell cell is 1.10 V.

Faraday's laws apply to the cell reactions of voltaic, as well as electrolytic, cells. One precaution must be observed, however. Electricity is generated by the simultaneous oxidation and reduction half-reactions that occur at the anode and cathode, respectively. Both must occur if the cell is to deliver current. Two faradays of electricity will be produced, therefore, by the oxidation of 1 mol of zinc at the anode *together with* the reduction of 1 mol of Cu^{2+} ions at the cathode. The partial equations

$$\text{anode:} \qquad Zn \longrightarrow Zn^{2+} + 2e^-$$

$$\text{cathode:} \quad 2e^- + Cu^{2+} \longrightarrow Cu$$

when read in terms of moles, represent the flow of two times Avogadro's number of electrons or the production of 2 F of electricity.

The quantity of electrical energy, in joules, produced by a cell is the product of the quantity of electricity delivered, in coulombs, and the emf of the cell, in volts (see Section 18.1). The electrical energy *produced* by the reaction between 1 mol of zinc metal and 1 mol of copper(II) ions may be calculated as follows:

$$2(96{,}500 \text{ C})(1.10 \text{ V}) = 212{,}000 \text{ J} = 212 \text{ kJ}$$

One volt·coulomb is a joule.

The emf used in the preceding calculation is the reversible emf ($\mathscr{E}°$) of the standard Daniell cell and hence the maximum voltage for this cell. Therefore, the value secured (212 kJ) is the maximum work that can be obtained from the operation of this type of cell. The maximum *net* work* that can be obtained from a chemical reaction conducted at a constant temperature and pressure is a measure of the *decrease* in the Gibbs free energy (see Section 17.4) of the system. Hence,

$$\Delta G = -nF\mathscr{E} \tag{18.1}$$

---

* Some reactions proceed with an increase in volume, and the system must do work to expand against the atmosphere in order to maintain a constant pressure. The energy for this pressure-volume work is not available for any other purpose; it must be expended in this way if the reaction is to occur at constant pressure. Pressure-volume work is not included in the potentiometric measurement of the electrical work of any cell. Net work (or available work) is work other than pressure-volume work.

where n is the number of moles of electrons transferred in the reaction (or the number of faradays produced), F is the value of the faraday in appropriate units, and $\mathscr{E}$ is the emf in volts. If F is expressed as 96,487 C, ΔG is obtained in joules. A change in free energy derived from a standard emf, $\mathscr{E}^\circ$, is given the symbol ΔG°.

The free energy change of a reaction is a measure of the tendency of the reaction to occur. If work must be done on a system to bring about a change, the change is not spontaneous. At constant temperature and pressure a spontaneous change is one from which net work can be obtained. Hence, for any spontaneous reaction the free energy of the system decreases; ΔG is negative. Since $\Delta G = -nF\mathscr{E}$, only if $\mathscr{E}$ is positive will the cell reaction be spontaneous and serve as a source of electrical energy.

18.7 Electrode Potentials

In the same way that a cell reaction may be regarded as the sum of two half-reactions, the emf of a cell may be thought of as the sum of two half-cell potentials. However, it is impossible to determine the absolute value of the potential of a single half-cell. A relative scale has been established by assigning a value of zero to the voltage of a standard reference half-cell and expressing all half-cell potentials relative to this reference electrode.

The reference half-cell used is the **standard hydrogen electrode,** which consists of hydrogen gas, at 1 atm pressure, bubbling over a platinum electrode (coated with finely divided platinum to increase its surface) that is immersed in an acid solution containing $H^+(aq)$ at unit activity. In Figure 18.6 a standard hydrogen electrode is shown connected by means of a salt bridge to a standard Cu^{2+}/Cu electrode. A **salt bridge** is a tube filled with a concentrated solution of a salt (usually KCl), which conducts the current between the half-cells but prevents the mixing of the solutions of the half-cells. The cell of Figure 18.6 may be diagrammed as

$$Pt|H_2|H^+||Cu^{2+}|Cu$$

A double bar indicates a salt bridge. The hydrogen electrode is the anode, the copper electrode is the cathode, the emf of the cell is 0.34 V.

The cell emf is considered to be the sum of the half-cell potential for the oxidation half reaction (which we shall give the symbol $\mathscr{E}^\circ_{ox}$) and the half-cell potential for the reduction half reaction (which we shall indicate as $\mathscr{E}^\circ_{red}$). For the cell of Figure 18.6,

anode:	$H_2 \longrightarrow 2H^+ + 2e^-$	$\mathscr{E}^\circ_{ox} = 0.00$ V
cathode:	$2e^- + Cu^{2+} \longrightarrow Cu$	$\mathscr{E}^\circ_{red} = +0.34$ V

Since the hydrogen electrode is arbitrarily assigned a potential of zero, the entire cell emf is ascribed to the standard Cu^{2+}/Cu electrode. The value $+0.34$ V is called the **standard electrode potential** of the Cu^{2+}/Cu electrode. Notice that *electrode potentials are given for reduction half-reactions*. If the symbol $\mathscr{E}^\circ$ (without a subscript) is used for an electrode potential, $\mathscr{E}^\circ_{red}$ is understood.

If a cell is constructed from a standard hydrogen electrode and a standard Zn^{2+}/Zn electrode, the zinc electrode is the anode, and the emf of the cell is 0.76 V. Thus,

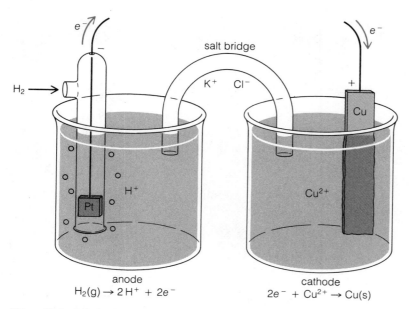

Figure 18.6 Standard hydrogen electrode and a Cu^{2+}/Cu electrode

anode: $\qquad\qquad$ $Zn \longrightarrow Zn^{2+} + 2e^-$ $\qquad$ $\mathscr{E}^\circ_{ox} = +0.76$ V

cathode: $\quad 2e^- + 2H^+ \longrightarrow H_2$ $\qquad$ $\mathscr{E}^\circ_{red} = \quad 0.00$ V

The value $+0.76$ V is sometimes called an **oxidation potential,** since it corresponds to an oxidation half-reaction. An electrode potential, however, is a **reduction potential.** To obtain the electrode potential of the Zn^{2+}/Zn couple, we must change the sign of the oxidation potential so that the potential corresponds to the reverse half-reaction, a reduction:

$$2e^- + Zn^{2+} \longrightarrow Zn \qquad \mathscr{E}^\circ_{red} = -0.76 \text{ V}$$

It is not necessary to use a cell containing a standard hydrogen electrode to obtain a standard electrode potential. For example, the standard potential of the Ni^{2+}/Ni electrode may be determined from the cell

$$Ni|Ni^{2+}||Cu^{2+}|Cu$$

The emf of this cell is 0.59 V, and the nickel electrode functions as the anode:

$$Ni + Cu^{2+} \longrightarrow Ni^{2+} + Cu \qquad \mathscr{E}^\circ_{cell} = +0.59 \text{ V}$$

The standard electrode potential of the Cu^{2+}/Cu electrode has been determined:

$$2e^- + Cu^{2+} \longrightarrow Cu \qquad \mathscr{E}^\circ_{red} = +0.34 \text{ V}$$

If we subtract the Cu^{2+}/Cu half-reaction from the cell reaction and subtract the half-cell potential from the cell emf, we obtain

$$Ni \longrightarrow Ni^{2+} + 2e^- \qquad \mathscr{E}^\circ_{ox} = +0.25 \text{ V}$$

Table 18.2 Standard electrode potentials at 25°C[a]

Half-Reaction	$\mathscr{E}°$ (Volts)
$Li^+ + e^- \rightleftharpoons Li$	−3.045
$K^+ + e^- \rightleftharpoons K$	−2.925
$Ba^{2+} + 2e^- \rightleftharpoons Ba$	−2.906
$Ca^{2+} + 2e^- \rightleftharpoons Ca$	−2.866
$Na^+ + e^- \rightleftharpoons Na$	−2.714
$Mg^{2+} + 2e^- \rightleftharpoons Mg$	−2.363
$Al^{3+} + 3e^- \rightleftharpoons Al$	−1.662
$2H_2O + 2e^- \rightleftharpoons H_2 + 2OH^-$	−0.82806
$Zn^{2+} + 2e^- \rightleftharpoons Zn$	−0.7628
$Cr^{3+} + 3e^- \rightleftharpoons Cr$	−0.744
$Fe^{2+} + 2e^- \rightleftharpoons Fe$	−0.4402
$Cd^{2+} + 2e^- \rightleftharpoons Cd$	−0.4029
$Ni^{2+} + 2e^- \rightleftharpoons Ni$	−0.250
$Sn^{2+} + 2e^- \rightleftharpoons Sn$	−0.136
$Pb^{2+} + 2e^- \rightleftharpoons Pb$	−0.126
$2H^+ + 2e^- \rightleftharpoons H_2$	0
$Cu^{2+} + 2e^- \rightleftharpoons Cu$	+0.337
$Cu^+ + e^- \rightleftharpoons Cu$	+0.521
$I_2 + 2e^- \rightleftharpoons 2I^-$	+0.5355
$Fe^{3+} + e^- \rightleftharpoons Fe^{2+}$	+0.771
$Ag^+ + e^- \rightleftharpoons Ag$	+0.7991
$Br_2 + 2e^- \rightleftharpoons 2Br^-$	+1.0652
$O_2 + 4H^+ + 4e^- \rightleftharpoons 2H_2O$	+1.229
$Cr_2O_7^{2-} + 14H^+ + 6e^- \rightleftharpoons 2Cr^{3+} + 7H_2O$	+1.33
$Cl_2 + 2e^- \rightleftharpoons 2Cl^-$	+1.3595
$MnO_4^- + 8H^+ + 5e^- \rightleftharpoons Mn^{2+} + 4H_2O$	+1.51
$F_2 + 2e^- \rightleftharpoons 2F^-$	+2.87

[a] Data from A. J. de Bethune and N. A. Swendeman Loud, "Table of Electrode Potentials and Temperature Coeffi-cients," pp. 414–424 in *Encyclopedia of Electrochemistry* (C. A. Hampel, editor), Van Nostrand Reinhold, New York, 1964, and from A. J. de Bethune and N. A. Swendeman Loud, *Standard Aqueous Electrode Potentials and Temperature Coefficients*, 19 pp., C. A. Hampel, publisher, Skokie, Illinois, 1964.

The electrode potential is therefore

$$2e^- + Ni^{2+} \longrightarrow Ni \qquad \mathscr{E}°_{red} = -0.25 \text{ V}$$

Standard electrode potentials are listed in Table 18.2, and a more complete list is found in the appendix. The table is constructed with the most positive electrode potential (greatest tendency for reduction) at the bottom. Hence, if a pair of electrodes is combined to make a voltaic cell, the reduction half-reaction (cathode) of the cell will be that listed for the electrode that stands lower in the table, and the oxidation half-reaction (anode) will be the *reverse* of that shown for the electrode that stands higher in the table.

For example, consider a cell constructed from standard Ni^{2+}/Ni and Ag^+/Ag electrodes. The table entries for these electrodes are

$$Ni^{2+} + 2e^- \rightleftharpoons Ni \qquad \mathscr{E}°_{red} = -0.250 \text{ V}$$

$$Ag^+ + e^- \rightleftharpoons Ag \qquad \mathscr{E}°_{red} = +0.799 \text{ V}$$

Of the two ions, the Ag^+ ion shows the greater tendency for reduction. The Ag^+/Ag electrode is therefore the cathode, and the Ni^{2+}/Ni electrode is the anode. The half-reaction that takes place at an anode is an oxidation, and the half-cell potential is an oxidation potential. The sign of the table entry for the Ni^{2+}/Ni half-cell, therefore, must be reversed to give an $\mathscr{E}°_{ox}$:

anode: $\qquad Ni \longrightarrow Ni^{2+} + 2e^- \qquad \mathscr{E}°_{ox} = +0.250 \text{ V}$

cathode: $\quad 2e^- + 2Ag^+ \longrightarrow 2Ag \qquad \mathscr{E}°_{red} = +0.799 \text{ V}$

The cell reaction and cell emf may be obtained by addition:

$$Ni + 2Ag^+ \longrightarrow Ni^{2+} + 2Ag \qquad \mathscr{E}°_{cell} = +1.049 \text{ V}$$

Notice that the half-reaction for the reduction of Ag^+ must be multiplied by 2 before the addition so that the electrons lost and gained in the half-reactions will cancel. The $\mathscr{E}°$ for the Ag^+/Ag electrode, however, is *not* multiplied by 2. The magnitude of an electrode potential depends upon the temperature and the concentrations of materials used in the construction of the half-cell. These variables are fixed for *standard* electrode potentials. Indication of the stoichiometry of the cell reaction does not imply that a concentration change has been made.

Actually the half-reactions implied by the half-cell potentials are

$$\begin{array}{l} 2H^+ + Ni \longrightarrow H_2 + Ni^{2+} \\ \underline{H_2 + 2Ag^+ \longrightarrow 2H^+ + 2Ag} \\ Ni + 2Ag^+ \longrightarrow Ni^{2+} + 2Ag \end{array}$$

Notice, however, that in the addition of these half-reactions, the H_2 molecules and H^+ ions cancel.

Electrode potentials are also useful for the evaluation of oxidation-reduction reactions that take place outside of electrochemical cells. An oxidizing agent is a substance that brings about an oxidation and in the process is itself reduced. A strong *oxidizing* agent, therefore, has a high positive *reduction* potential, $\mathscr{E}°_{red}$. The strongest oxidizing agent given in Table 18.2 is $F_2(g)$ since the highest $\mathscr{E}°_{red}$ given in the table is

$$F_2(g) + 2e^- \longrightarrow 2F^-(aq) \qquad \mathscr{E}°_{red} = +2.87 \text{ V}$$

The best oxidizing agents listed in the table are F_2, MnO_4^- in acid, Cl_2, and $Cr_2O_7^{2-}$ in acid.

A reducing agent is itself oxidized in bringing about a reduction. A strong *reducing* agent, therefore, has a high, positive *oxidation* potential. Remember that $\mathscr{E}°_{ox}$ values are obtained by changing the signs of the table values; the corresponding oxidation half-reactions are derived by reversing the partial equations shown. The strongest reducing agent given in Table 18.2 is Li metal since the highest $\mathscr{E}°_{ox}$ derived from the table values is

$$Li(s) \longrightarrow Li^+(aq) + e^- \qquad \mathscr{E}°_{ox} = +3.045 \text{ V}$$

The best reducing agents given in the table are the active metals Li, K, Ba, Ca, and Na.

Whether or not a proposed reaction will be spontaneous *with all substances present at unit activity* can be determined by use of electrode potentials. A spontaneous reaction is indicated only if the emf of the reaction is positive.

Example 18.5

Use electrode potentials to determine whether the following proposed reactions are spontaneous with all substances present at unit activity:
(a) $Cl_2(g) + 2I^-(aq) \longrightarrow 2Cl^-(aq) + I_2(s)$
(b) $2Ag(s) + 2H^+(aq) \longrightarrow 2Ag^+(aq) + H_2(g)$

Solution

(a) In the proposed reaction, the Cl_2 is reduced to Cl^- (and we need an $\mathscr{E}^\circ_{red}$ for this half-reaction) and the I^- is oxidized to I_2 (and we need an $\mathscr{E}^\circ_{ox}$ for this half-reaction):

$$
\begin{array}{ll}
2e^- + Cl_2(g) \longrightarrow 2Cl^-(aq) & \mathscr{E}^\circ_{red} = +1.360\ V \\
2I^-(aq) \longrightarrow I_2(s) + 2e^- & \mathscr{E}^\circ_{ox} = -0.536\ V \\
\hline
Cl_2(g) + 2I^-(aq) \longrightarrow 2Cl^-(aq) + I_2(s) & emf = +0.824\ V
\end{array}
$$

Since the overall emf is positive, the reaction is spontaneous.
(b) In this reaction, Ag is oxidized ($\mathscr{E}^\circ_{ox}$ needed) and H^+ is reduced ($\mathscr{E}^\circ_{red}$ needed):

$$
\begin{array}{ll}
2Ag(s) \longrightarrow 2Ag^+(aq) + 2e^- & \mathscr{E}^\circ_{ox} = -0.799\ V \\
2H^+(aq) + 2e^- \longrightarrow H_2(g) & \mathscr{E}^\circ_{red} = 0.000\ V \\
\hline
2Ag(s) + 2H^+(aq) \longrightarrow 2Ag^+(aq) + H_2(g) & emf = -0.799\ V
\end{array}
$$

The reaction is *not* spontaneous as written. The reverse reaction (between Ag^+ and H_2) would be spontaneous (emf = +0.799 V).

There are several factors that must be kept in mind when using a table of electrode potentials to predict the course of a chemical reaction. Because $\mathscr{E}$ changes with changes in concentration, many presumably unfavored reactions can be made to occur by altering the concentrations of the reacting species. In addition, some theoretically favored reactions proceed at such a slow rate that they are of no practical consequence.

Correct use of the table also demands that all pertinent half-reactions of a given element be considered before making a prediction. On the basis of the half-reactions

$$
\begin{array}{ll}
3e^- + Fe^{3+} \rightleftharpoons Fe & \mathscr{E}^\circ_{red} = -0.036\ V \\
2e^- + 2H^+ \rightleftharpoons H_2 & \mathscr{E}^\circ_{red} = 0.000\ V
\end{array}
$$

one might predict that the products of the reaction of iron with H^+ would be hydrogen gas and Fe^{3+} ions (emf for the complete reaction, $+0.036$ V). The oxidation state iron(II), however, lies between metallic iron and the oxidation state iron(III). Once an iron atom has lost two electrons and becomes an Fe^{2+} ion, further oxidation is opposed, as may be seen from the reverse of the following:

$$e^- + Fe^{3+} \rightleftharpoons Fe^{2+} \qquad \mathscr{E}^\circ_{red} = +0.771 \text{ V}$$

Thus, the reaction yields Fe^{2+} ions only. This fact could have been predicted by an examination of the half-reaction

$$2e^- + Fe^{2+} \rightleftharpoons Fe \qquad \mathscr{E}^\circ_{red} = -0.440 \text{ V}$$

The $\mathscr{E}_{ox}$ for the production of Fe^{2+} ions from the reaction of iron metal and H^+ ions (+0.440 V) is greater than that for the production of Fe^{3+} ions (+0.036 V), and hence the former is favored.

We may summarize the electrode potentials for iron and its ions as follows:

$$Fe^{3+} \xrightarrow{+0.771 \text{ V}} Fe^{2+} \xrightarrow{-0.440 \text{ V}} Fe$$
$$\underline{\qquad\qquad -0.036 \text{ V} \qquad\qquad}$$

The preceding predictions are immediately evident from this diagram, if we remember that oxidation is the reverse of the relation corresponding to an electrode potential.

Occasionally an oxidation state of an element is unstable toward disproportionation (auto oxidation-reduction, see Section 11.3). The electrode potentials for copper and its ions may be summarized as follows:

$$Cu^{2+} \xrightarrow{+0.153 \text{ V}} Cu^+ \xrightarrow{+0.521 \text{ V}} Cu$$
$$\underline{\qquad\qquad +0.337 \text{ V} \qquad\qquad}$$

From this we see that the Cu^+ ion is not a very stable one. In water, Cu^+ ions are disproportionate to copper metal and Cu^{2+} ions:

$$2\,Cu^+(aq) \longrightarrow Cu(s) + Cu^{2+}(aq)$$

The emf for this reaction is $+0.521 - 0.153 = +0.368$ V. Species instable toward disproportionation may be readily recognized by the fact that the electrode potential for the reduction to the next lower oxidation state is more positive than the electrode potential for the couple with the next higher oxidation state. An inspection of the diagram for iron and its ions shows that the Fe^{2+} ion is stable toward such disproportionation.

18.8 Gibbs Free Energy Change and Electromotive Force

The reversible emf of a cell, $\mathscr{E}^\circ_{cell}$, is a measure of the decrease in Gibbs free energy for the cell reaction:

$$\Delta G^\circ = -nF\mathscr{E}^\circ \tag{18.1}$$

We can, therefore, use standard electrode potentials to calculate ΔG° values.

Example 18.6

Use electrochemical data to calculate the value of $\Delta G°$ for the reaction

$$2\,Ag(s) + Cl_2(g) \longrightarrow 2\,AgCl(s)$$

Solution

We use the standard electrode potentials listed in the appendix to calculate the standard potential for the reaction:

$$
\begin{array}{ll}
2\,Ag(s) + 2\,Cl^-(aq) \longrightarrow 2\,AgCl(s) + 2e^- & \mathscr{E}°_{ox} = -0.222\ V \\
\underline{2e^- + Cl_2(g) \longrightarrow 2\,Cl^-(aq)} & \mathscr{E}°_{red} = +1.359\ V \\
2\,Ag(s) + Cl_2(g) \longrightarrow 2\,AgCl(s) & \mathscr{E}°_{cell} = +1.137\ V
\end{array}
$$

Since $n = 2$ (2 mol of electrons transferred),

$$
\begin{aligned}
\Delta G° &= -nF\mathscr{E}° \\
&= -2(96{,}500\ C)(+1.137\ V) \\
&= -219{,}400\ J = -219.4\ kJ
\end{aligned}
$$

Note that $1\ J = 1\ V \cdot C$.

The $\Delta G°$ values obtained in this way can be used, together with standard enthalpy changes ($\Delta H°$ values), to calculate standard changes in entropy ($\Delta S°$ values).

Example 18.7

Calculate the value of $\Delta S°$ for the reaction

$$2\,Ag(s) + Cl_2(g) \longrightarrow 2\,AgCl(s) \qquad \Delta H° = -254.0\ kJ$$

Solution

We are given $\Delta H°$ for the reaction and we note from Example 18.6 that $\Delta G° = -219.4\ kJ$. Therefore,

$$\Delta G° = \Delta H° - T\Delta S°$$

$$-219.4\ kJ = -254.0\ kJ - T\,\Delta S°$$

$$T\,\Delta S° = -34.6\ kJ$$

Since $T = 298\ K$,

$$
\begin{aligned}
\Delta S° &= \frac{T\Delta S°}{T} \\
&= \frac{-(34{,}600\ J)}{298\ K} \\
&= -116\ J/K
\end{aligned}
$$

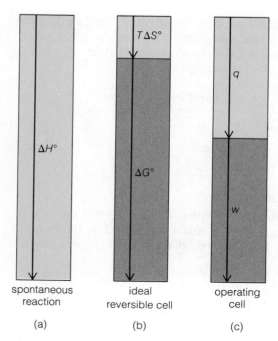

Figure 18.7 Relations between thermodynamic functions for the reaction $2\,Ag(s) + Cl_2(g) \longrightarrow 2\,AgCl(s)$ at 25°C and 1 atm

The fact that $\Delta S°$ is negative means that the system becomes more ordered (less random) as the reaction proceeds. Notice that a mole of gas is consumed during the course of the reaction; the decrease in the entropy of the system is therefore not surprising.

The results of Examples 18.6 and 18.7 are summarized in Figure 18.7. If the reaction is conducted outside the cell (a), heat equivalent to $\Delta H°$ is evolved. In the case of the ideal, reversible cell (b), the maximum amount of useful work is obtained ($\Delta G°$) and heat equivalent to $T\Delta S°$ is evolved. Since $\Delta G° = \Delta H° - T\Delta S°$,

$$\Delta H° = \Delta G° + T\Delta S°$$

as the figure illustrates. The ideal, reversible cell is an abstraction, not an operating device. When the reversible emf of a cell is measured, the emf of the cell is balanced against an external emf in such a way that no current flows. In this way the maximum voltage that the cell is capable of producing is measured. In an operating cell (c), less than the maximum amount of work is done (w) and an amount of heat greater than $T\Delta S°$ is evolved (q).

The relationship between $\Delta G°$ and emf is useful in several other ways. An $\mathscr{E}°_{cell}$, for example, can be calculated from a $\Delta G°$ value for a cell reaction. In addition, free energy changes provide an answer to the problem of combining two $\mathscr{E}^\circ_{red}$ values to obtain a third $\mathscr{E}^\circ_{red}$ value. Even though an $\mathscr{E}^\circ_{ox}$ and an $\mathscr{E}^\circ_{red}$ may be combined to give an emf of a *cell*, two electrode potentials ($\mathscr{E}^\circ_{red}$ values) may *not* be combined directly to give a third *electrode* potential (an $\mathscr{E}^\circ_{red}$ value). For example, the sum of the two partial equations

$$2e^- + Fe^{2+} \longrightarrow Fe \qquad \mathscr{E}^\circ_{red} = -0.440 \text{ V}$$

$$e^- + Fe^{3+} \longrightarrow Fe^{2+} \qquad \mathscr{E}^\circ_{red} = +0.771 \text{ V}$$

is the partial equation

$$3e^- + Fe^{3+} \longrightarrow Fe$$

The electrode potential for the final half-reaction, however, is *not* the sum of the other two $\mathscr{E}^\circ_{red}$ values. A solution to the problem is provided by the fact that free energy changes are additive in the same way that enthalpy changes are (the law of Hess).

Example 18.8

Use $\mathscr{E}^\circ_{red}$ values to find the $\mathscr{E}^\circ_{red}$ for the half-reaction

$$3e^- + Fe^{3+} \longrightarrow Fe$$

Solution

Whereas two $\mathscr{E}^\circ_{red}$ values may not be added to give a third, the ΔG° values for two half-reactions may be added to give a ΔG° value for a third half-reaction:

	$\mathscr{E}^\circ_{red}$	$\Delta G^\circ = -nF\mathscr{E}^\circ$
$2e^- + Fe^{2+} \longrightarrow Fe$	-0.440 V	$= -2(-0.440)F = +0.880F$
$e^- + Fe^{3+} \longrightarrow Fe^{2+}$	$+0.771$ V	$= -1(+0.771)F = \underline{-0.771F}$
$3e^- + Fe^{3+} \longrightarrow Fe$		$+0.109F$

Notice that we make no attempt to multiply $+0.109$ V by the value of the faraday. The $\mathscr{E}^\circ_{red}$ for the third half-reaction is found from the ΔG° value ($+0.109F$). Since three electrons are gained, $n = 3$:

$$\Delta G^\circ = -nF\mathscr{E}^\circ_{red}$$

$$+0.109F = -3F\mathscr{E}^\circ_{red}$$

$$\mathscr{E}^\circ_{red} = \frac{+0.109F}{-3F} = -0.036 \text{ V}$$

Electrode potentials may also be used to determine equilibrium constants. In Section 17.7, we noted that

$$\Delta G^\circ = -2.303RT \log K \qquad (18.2)$$

Since

$$\Delta G^{\circ} = -nF\mathscr{E}^{\circ}$$

$$nF\mathscr{E}^{\circ} = 2.303RT \log K$$

$$\mathscr{E}^{\circ} = \frac{2.303RT}{nF} \log K \tag{18.3}$$

When $T = 298.15$ K ($25°C$), substitution for R, F, and T gives

$$\mathscr{E}^{\circ} = \frac{0.05916 \text{ V}}{n} \log K \tag{18.4}$$

Example 18.9

Use electrochemical data to calculate the equilibrium constant K for the reaction at $25°C$:

$$Fe^{2+}(aq) + Ag^{+}(aq) \rightleftharpoons Fe^{3+}(aq) + Ag(s)$$

Solution

The half-reactions are

$$Fe^{2+} \longrightarrow Fe^{3+} + e^{-} \qquad \mathscr{E}^{\circ}_{ox} = -0.771 \text{ V}$$

$$e^{-} + Ag^{+} \longrightarrow Ag(s) \qquad \mathscr{E}^{\circ}_{red} = +0.799 \text{ V}$$

The $\mathscr{E}^{\circ}$ value for the reaction, therefore, is $+0.028$ V and $n = 1$:

$$\mathscr{E}^{\circ} = \frac{0.0592 \text{ V}}{n} \log K$$

$$+0.028 \text{ V} = \frac{0.0592 \text{ V}}{1} \log K$$

$$\log K = 0.47$$

$$K = 3.0$$

18.9 Effect of Concentration on Cell Potentials

In Section 17.7, we considered the relationship

$$\Delta G = \Delta G^{\circ} + 2.303RT \log Q \tag{18.5}$$

where ΔG is the free energy change for a chemical reaction, ΔG° is the *standard* free energy change for that reaction, R is 8.3143 J/(K·mol), T is the absolute temperature, and Q is the reaction quotient, a fraction derived from the activities of the substances involved in the reaction. For the hypothetical reaction

$$w\text{W} + x\text{X} \longrightarrow y\text{Y} + z\text{Z}$$

where the lower-case letters represent the coefficients of the balanced chemical equation

$$Q = \frac{(a_\text{Y})^y (a_\text{Z})^z}{(a_\text{W})^w (a_\text{X})^x} \qquad (18.6)$$

the activity of each substance is raised to a power equal to the coefficient of that substance in the balanced chemical equation. The *numerator* of Q is the product of the activity terms for the substances on the *right* of the chemical equation. The *denominator* of Q is the product of the activity terms for the substances on the *left* of the chemical equation. Since the activity of a pure solid is assumed to be unity at all times, the activity term for a solid is always equal to 1. For our work, we will assume that the activity of a substance in solution is given by the molar concentration of the substance and the activity of a gas is equal to the partial pressure of the gas in atmospheres.

Since $\Delta G = -nF\mathscr{E}$ and $\Delta G^\circ = -nF\mathscr{E}^\circ$,

$$\Delta G = \Delta G^\circ + 2.303 RT \log Q$$

$$-nF\mathscr{E} = -nF\mathscr{E}^\circ + 2.303 RT \log Q$$

$$\mathscr{E} = \mathscr{E}^\circ - \frac{2.303 RT}{nF} \log Q \qquad (18.7)$$

If we substitute 298.15 K for T (which is 25°C), 8.3143 J/(K·mol) for R, 96,485 C/mol for F, we get

$$\mathscr{E} = \mathscr{E}^\circ - \frac{0.05916\text{ V}}{nF} \log Q \qquad (18.8)$$

which is called the Nernst equation, named for Walther Nernst, who developed it in 1889. When the activities of all substances are unity (standard states), $\log Q = 0$ and $\mathscr{E} = \mathscr{E}^\circ$. The Nernst equation may be used to determine the emf of a cell constructed from nonstandard electrodes or to calculate the electrode potential of a half-cell in which all species are not present at unit activity.

Walther Nernst, 1864–1941.
American Institute of Physics,
Niels Bohr Library, Sawyer
Collection.

Example 18.10

What is the electrode potential of a Zn^{2+}/Zn electrode in which the concentration of Zn^{2+} ions is $0.1 M$?

Solution

The partial equation

$$2e^- + \text{Zn}^{2+} \longrightarrow \text{Zn}$$

shows 2 electrons gained. If the symbol $[Zn^{2+}]$ is used to designate the molar concentration of Zn^{2+} ions:

$$\mathscr{E} = \mathscr{E}^\circ - \frac{0.0592}{2} \log\left(\frac{1}{[Zn^{2+}]}\right)$$

$\mathscr{E}^\circ_{red}$ for the Zn^{2+}/Zn electrode is -0.76 V:

$$\mathscr{E} = -0.76 - \frac{0.0592}{2} \log\left(\frac{1}{0.1}\right)$$

$$\mathscr{E} = -0.76 - 0.0296(1) = -0.79 \text{ V}$$

Example 18.11

What is the potential for the cell

$$Ni\,|\,Ni^{2+}(0.01\ M)\,\|\,Cl^-(0.2\ M)\,|\,Cl_2(1\ atm)\,|\,Pt$$

Solution

Oxidation occurs at the Ni^{2+}/Ni electrode since it is the anode of the cell. The two half-reactions of the cell are

$$Ni \longrightarrow Ni^{2+} + 2e^- \qquad \mathscr{E}^\circ_{ox} = +0.25 \text{ V}$$

$$2e^- + Cl_2 \longrightarrow 2\,Cl^- \qquad \mathscr{E}^\circ_{red} = +1.36 \text{ V}$$

The cell reaction and $\mathscr{E}^\circ$ for the cell, therefore, are

$$Ni + Cl_2 \longrightarrow Ni^{2+} + 2\,Cl^- \qquad \mathscr{E}^\circ = +1.61 \text{ V}$$

Since $n = 2$,

$$\mathscr{E} = \mathscr{E}^\circ - \frac{0.0592}{2} \log\left(\frac{[Cl^-]^2[Ni^{2+}]}{p_{Cl_2}}\right)$$

$$\mathscr{E} = +1.61 - \frac{0.0592}{2} \log\left(\frac{(0.2)^2(0.01)}{(1)}\right)$$

$$\mathscr{E} = +1.61 - 0.0296 \log(0.0004)$$

$$\mathscr{E} = +1.61 + 0.10 = +1.71 \text{ V}$$

Example 18.12

What is the $\mathscr{E}$ of the cell

$$Sn\,|\,Sn^{2+}(1.0\ M)\,\|\,Pb^{2+}(0.0010\ M)\,|\,Pb$$

Solution

The following may be obtained from a table of standard electrode potentials:

$$\text{Sn} \longrightarrow \text{Sn}^{2+} + 2e^- \qquad \mathscr{E}^\circ_{ox} = +0.136 \text{ V}$$

$$2e^- + \text{Pb}^{2+} \longrightarrow \text{Pb} \qquad \mathscr{E}^\circ_{red} = -0.126 \text{ V}$$

Thus the reaction in a standard cell is

$$\text{Sn} + \text{Pb}^{2+} \longrightarrow \text{Sn}^{2+} + \text{Pb} \qquad \mathscr{E}^\circ = +0.010 \text{ V}$$

For the cell as diagrammed in the problem,

$$\mathscr{E} = \mathscr{E}^\circ - \frac{0.0592}{2} \log \left(\frac{[\text{Sn}^{2+}]}{[\text{Pb}^{2+}]} \right)$$

$$\mathscr{E} = +0.010 - \frac{0.0592}{2} \log \left(\frac{1.0}{0.0010} \right)$$

$$\mathscr{E} = +0.010 - 0.0296(3)$$

$$= +0.010 - 0.089 = -0.079 \text{ V}$$

This result means that the cell will not function in the manner implied by the diagram. Instead, it would operate in the reverse order and in a direction opposite to that of the standard cell. The cell is properly diagrammed

$$\text{Pb} | \text{Pb}^{2+}(0.0010 \ M) \| \text{Sn}^{2+}(1.0 \ M) | \text{Sn}$$

and the reaction of the cell is

$$\text{Pb} + \text{Sn}^{2+} \longrightarrow \text{Pb}^{2+} + \text{Sn} \qquad \mathscr{E} = +0.079 \text{ V}$$

From this example we see that concentration effects can sometimes reverse the direction that a reaction is expected to take.

Notice that the results of the preceding examples are qualitatively in agreement with predictions based on Le Chatelier's principle. Increasing the concentrations of reactants and decreasing the concentrations of products would be expected to increase the driving force of the reaction and lead to a more positive $\mathscr{E}$. On the other hand, decreasing the concentrations of reactants and increasing the concentrations of products would be expected to retard a reaction and result in a more negative $\mathscr{E}$.

Example 18.13

Consider a cell based on the reaction

$$\text{Mg(s)} + 2\,\text{H}^+ \longrightarrow \text{Mg}^{2+}(\text{aq}) + \text{H}_2(\text{g}) \qquad \mathscr{E}^\circ_{cell} = +2.363 \text{ V}$$

What is the concentration of $\text{H}^+(\text{aq})$ in a cell in which $[\text{Mg}^{2+}] = 1.00 \ M$ and $p_{\text{H}_2} = 1.00$ atm, if the emf of the cell is $+2.099$ V?

Solution

$$\mathscr{E} = \mathscr{E}° - \frac{0.0592}{n} \log\frac{[Mg^{2+}](p_{H_2})}{[H^+]^2}$$

$$+2.099 = +2.363 - \frac{0.0592}{2} \log\frac{1}{[H^+]^2}$$

$$-0.264 = -\frac{0.0592}{2}(-2\log[H^+])$$

$$\log[H^+] = -4.46$$

$$[H^+] = 3.5 \times 10^{-5}\ M$$

Notice that the pH (which is $-\log[H^+]$) of the solution in the cathode of the cell is 4.46. This example illustrates how a voltaic cell can be used to measure the concentration of an ion. A pH meter makes use of this concept with the instrument calibrated to read in pH units rather than volts.

18.10 Concentration Cells

Since an electrode potential depends upon the concentration of the ions used in the electrode, a cell may be constructed from two half-cells composed of the same materials but differing in concentration of ions. For example:

$$Cu|Cu^{2+}(0.010\ M)||Cu^{2+}(0.10\ M)|Cu$$

From the partial equation

$$2e^- + Cu^{2+} \rightleftharpoons Cu \qquad \mathscr{E}° = +0.34\ V$$

and Le Chatelier's principle, we can predict that *increasing* the concentration of Cu^{2+} ions would drive the reaction to the right and raise the reduction potential, whereas *decreasing* the concentration of Cu^{2+} ions would drive the reaction to the left and raise the oxidation potential (or lower the reduction potential). Hence, of the two electrodes in the cell diagrammed, the left electrode (with the lower concentration of Cu^{2+} ions) would have the stronger tendency for oxidation and the right (with the higher concentration of Cu^{2+} ions) would have the stronger tendency for reduction. The "reaction" of the cell is

$$Cu + Cu^{2+}(0.10\ M) \longrightarrow Cu^{2+}(0.010\ M) + Cu$$

$\mathscr{E}°$ for this cell is zero since the same electrode is involved in each half-cell. Then

$$\mathscr{E} = 0.00 - \frac{0.0592}{2}\log\left(\frac{0.010}{0.10}\right)$$

$$\mathscr{E} = -0.0296(-1) = +0.0296\ V$$

18.11 Electrode Potentials and Electrolysis

An emf of a cell calculated from electrode potentials is the maximum voltage that the cell can develop. For the reverse process, an electrolysis, the emf is the minimum voltage required to bring about the electrolysis. In theory, we should be able to use $\mathscr{E}°$ values to determine what electrode reaction will occur in an electrolysis when a choice of several is possible.

Consider the electrolysis of an aqueous solution of $CuCl_2$. There are two possible cathode reactions:

$$2e^- + Cu^{2+}(aq) \longrightarrow Cu(s) \qquad \mathscr{E}°_{red} = +0.34 \text{ V}$$

$$2e^- + 2H_2O \longrightarrow H_2(g) + 2OH^-(aq) \qquad \mathscr{E}_{red} = -0.414 \text{ V}$$

The electrode potential for the reduction of water has been adjusted by use of the Nernst equation for the fact that in neutral aqueous solutions $[OH^-] = 10^{-7} M$, not $1.0 M$. Clearly, the reduction of Cu^{2+} is easier to bring about than the reduction of water.

Two possible anode reactions should be considered:

$$2Cl^-(aq) \longrightarrow Cl_2(g) + 2e^- \qquad \mathscr{E}°_{ox} = -1.36 \text{ V}$$

$$2H_2O \longrightarrow O_2(g) + 4H^+(aq) + 4e^- \qquad \mathscr{E}_{ox} = -0.82 \text{ V}$$

The electrode potential for the oxidation of water has been adjusted by use of the Nernst equation for the fact that in neutral aqueous solutions $[H^+] = 10^{-7} M$, not $1.0 M$. From these values it would seem that the oxidation of water would occur. In fact, the oxidation of Cl^- is the half-reaction that is observed.

Frequently, the voltage required for an electrolysis is higher than the value calculated by use of electrode potentials by an amount called the **overvoltage.** Overvoltage is thought to be caused by a slow rate of reaction at the electrodes. Excess applied voltage is required to make the electrolysis proceed at an appreciable rate. Overvoltages for the deposition of metals are low, but those required for the liberation of hydrogen gas or oxygen gas are usually appreciable. In the electrolysis of an aqueous solution of $CuCl_2$ the overvoltage of chlorine is less than the overvoltage of oxygen so that Cl_2 is liberated at the anode, not O_2.

The minimum voltage required for the electrolysis, therefore, is

$$\mathscr{E} = \mathscr{E}°_{red} + \mathscr{E}°_{ox}$$
$$= +0.34 \text{ V} - 1.36 \text{ V} = -1.02 \text{ V}$$

The value has a negative sign since it represents voltage *required*. A higher voltage than this would have to be used to take care of the overvoltage as well as to overcome the internal resistance of the cell.

The products of an electrolysis vary with the concentrations of ions in solution since the half-cell emf's are dependent upon concentrations. For example, the electrolysis of *dilute* aqueous solutions of chlorides yields oxygen gas at the anode rather than chlorine. Furthermore, after the primary electrode reactions occur, in which electrons are transferred, secondary reactions may occur. If chlorine is liberated in an alkaline solution, for example, ClO^- or ClO_3^- may be formed by the reaction of Cl_2 with OH^- ions.

18.12 The Corrosion of Iron

The corrosion of iron has serious economic consequences. The annual world-wide cost of replacing rusted iron objects runs to billions of dollars. The process itself is electrochemical in nature.

The corrosion of iron, rusting, occurs only in the presence of oxygen and water. At one place on the surface of the iron object, *oxidation* of the iron takes place:

$$\text{anode:} \quad Fe(s) \longrightarrow Fe^{2+}(aq) + 2e^-$$

At another spot on the surface, a reduction occurs, which involves $O_2(g)$ and H_2O:

$$\text{cathode:} \quad 4e^- + O_2(g) + 2\,H_2O \longrightarrow 4\,OH^-(aq)$$

In effect, therefore, a miniature voltaic cell is set up. The electrons produced at the anodic region move through the iron toward the cathodic region.

The *cations*, Fe^{2+} ions, produced at the anode, move through the water on the surface of the object toward the *cathode*. The *anions*, OH^- ions, produced at the cathode, move toward the *anode*. Somewhere between these two regions, the ions meet and form $Fe(OH)_2$. Iron(II) hydroxide, however, is not stable in the presence of moisture and oxygen. The hydroxide is oxidized to iron(III) hydroxide, which in reality is hydrated iron(III) oxide, $Fe_2O_3 \cdot x\,H_2O$, or rust.

The places where a rusted iron object becomes pitted are the anodic regions where iron goes into solution as Fe^{2+} ions. The cathodic regions are those that are most open to moisture and air, since $O_2(g)$ and H_2O are involved in the cathode reaction. The rust always forms at spots somewhat removed from those places where pitting occurs (between the anodic and cathodic regions).

When a painted iron object rusts, for example, the cathodic regions are spots where the paint has been broken away and bare iron metal is exposed to moisture and oxygen. The anodic regions, where the iron becomes pitted, are spots beneath the painted surface. The pitting causes more paint to flake off, which accelerates the rusting. The rust itself forms at spots between these two regions, usually closer to the cathodic region than to the anodic region—the transformation of $Fe(OH)_2$ to rust requires $O_2(g)$ and H_2O.

Salt water accelerates rusting because the ions present in the water help to carry the current in the miniature voltaic cells that are set up on the surface of the iron. Some ions, Cl^- for example, appear to catalyze the electrode reactions.

Impurities in the iron also enhance rusting. Very pure iron does not rust rapidly. Some types of impurities, strains, and crystal defects present in the iron attract electrons away from regions in the iron that become, therefore, anodic sites.

Iron or steel objects can be prevented from rusting by applying protective coatings (such as grease, paint, or other metals) which keep air and moisture away from the iron. Metal coatings are applied by electrolysis (Cr, Ni, and Cd are examples) or by dipping the object into molten metal (Zn and Sn are examples).

Galvanized iron is iron that has been coated with zinc. The iron is protected from rusting even if the zinc coating is broken. In such a case, the zinc, rather than the iron, serves as the anode and is oxidized since zinc is a more reactive metal than iron. For tin coatings, such as are used in "tin cans," the reverse is

Twin ribbons of zinc wire (about $\frac{1}{2}$ inch in diameter) are buried as sacrificial anodes alongside a section of the trans-Alaska pipeline to prevent electrochemical corrosion of the pipe. The ribbons are connected to the pipe at 500 or 1000-foot intervals. *The Alyeska Pipeline Service Company.*

true. If a tin coating is broken, the corrosion of the iron beneath is enhanced since iron is a more reactive metal than tin.

Metals that are more reactive than iron can be used as sacrificial anodes. To protect an underground iron tank, pipeline, or cable from rusting, pieces of reactive metals (such as Mg or Zn) are buried beside the iron object and connected to it by wires. By this means, the iron is not oxidized; it becomes the cathode and the more reactive metal becomes the anode. The reactive metal anodes are sacrificed to protect the iron. They oxidize away rapidly and must be replaced from time to time.

18.13 Some Commercial Voltaic Cells

Several voltaic cells are of commercial importance. The **dry cell** (see Figure 18.8) consists of a zinc metal container (which serves as the anode) that is filled

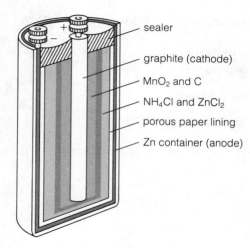

sealer

graphite (cathode)

MnO_2 and C

NH_4Cl and $ZnCl_2$

porous paper lining

Zn container (anode)

Figure 18.8 The dry cell

with a moist paste of ammonium chloride and zinc chloride and contains a graphite electrode (the cathode) surrounded by manganese dioxide. The electrode reactions are complex, but they may be approximately represented by

anode:
$$Zn \longrightarrow Zn^{2+} + 2e^-$$

cathode: $2e^- + 2\,MnO_2 + 2\,NH_4^+ \longrightarrow Mn_2O_3 \cdot H_2O + 2\,NH_3$

The dry cell generates a voltage of approximately 1.25 to 1.50 V.

A newer type of dry cell which has found use in small electrical devices (such as hearing aids) consists of a zinc container as the anode, a carbon rod as the cathode, and moist mercury(II) oxide mixed with potassium hydroxide as the electrolyte. A lining of porous paper keeps the electrolyte separated from the zinc anode. The cell has a potential of approximately 1.35 V:

anode:
$$Zn + 2\,OH^- \longrightarrow Zn(OH)_2 + 2e^-$$

cathode: $2e^- + HgO + H_2O \longrightarrow Hg + 2\,OH^-$

The **lead storage cell** consists of a lead anode and a grid of lead packed with lead dioxide as the cathode. The electrolyte is sulfuric acid, and the half-cell reactions are

anode:
$$Pb(s) + SO_4^{2-} \longrightarrow PbSO_4(s) + 2e^-$$

cathode: $2e^- + PbO_2(s) + SO_4^{2-} + 4\,H^+ \longrightarrow PbSO_4(s) + 2\,H_2O$

In practice, the current obtainable from a lead storage cell is increased by constructing the cell from a number of cathode plates joined together and arranged alternately with a number of anode plates, which are also joined together. The potential difference of one cell is approximately 2 V. A storage battery consists of three or six such cells joined in series to produce a 6- or a 12-volt battery.

The electrode reactions of the storage battery can be reversed by the application of an external current source and in this manner the battery can be recharged. Since sulfuric acid is consumed as the storage battery delivers current, the state of charge of the battery can be determined by measuring the density of the battery electrolyte.

The **nickel-cadmium storage cell** has a longer life than the lead storage cell but is more expensive to manufacture:

anode: $\qquad$ $Cd(s) + 2\,OH^- \longrightarrow Cd(OH)_2 + 2e^-$

cathode: $\quad 2e^- + NiO_2(s) + 2\,H_2O \longrightarrow Ni(OH)_2 + 2\,OH^-$

The potential of each cell of a nickel-cadmium battery is approximately 1.4 V, and the battery is rechargeable.

18.14 Fuel Cells

In the generation of electrical energy, heat from the combustion of a fuel (coal, oil, or natural gas) is used to convert water into steam. The steam is used to run a turbine which in turn drives a generator and produces electric current. At every step in the process, energy is lost in the form of heat. As a result, only about 30% to 40% of the energy obtained from the combustion of the fuel ends up as electrical energy.

Electrical cells that are designed to convert the energy from the combustion of fuels such as hydrogen, carbon monoxide, or methane directly into electrical energy are called fuel cells. Since in theory 100% of the free energy released by a combustion (ΔG) should be obtainable from an efficient fuel cell, extensive research into their development is currently being undertaken. Although approximately only 60% to 70% efficiency has been realized as yet, present fuel cells are about twice as efficient as processes in which the heat of combustion is used to generate electricity by mechanical means.

In a typical fuel cell, hydrogen and oxygen are bubbled through porous carbon electrodes into concentrated aqueous sodium hydroxide or potassium hydroxide. Catalysts are incorporated in the electrodes:

$$C|H_2(g)|OH^-|O_2(g)|C$$

The gaseous materials are consumed and are continuously supplied. The electrode reactions are

anode: $\qquad$ $2\,H_2(g) + 4\,OH^- \longrightarrow 4\,H_2O + 4e^-$

cathode: $\quad 4e^- + O_2(g) + 2\,H_2O \longrightarrow 4\,OH^-$

The complete cell reaction is

$$2\,H_2(g) + O_2(g) \longrightarrow 2\,H_2O(l)$$

The cell is maintained at an elevated temperature, and the water produced by the cell reaction evaporates as it is formed.

Although hydrogen-oxygen fuel cells have been used to supply electricity in spacecraft, existing fuel cells are expensive and not commercially practical at the present time. Problems in their design include the development of electrode

Demonstration fuel cell designed to generate electrical power by the air oxidation of hydrogen-enriched hydrocarbon fuel. *U.S. Department of Energy.*

catalysts that will cause the electrode reactions to occur more rapidly, the design of cells that will function at lower temperatures than those that must be used currently, and the improvement of methods used to handle corrosive liquids (such as the KOH electrolyte) and gases under pressure.

Summary

The topics that have been discussed in this chapter are

1. Electrical conduction by metals and by electrolytes.

2. Electrolysis; the use of electricity to bring about chemical change.

3. The quantitative relationships between electricity and chemical change.

4. Voltaic cells, in which chemical reactions are used to generate electricity.

5. Electromotive force, the force that produces electric current in a voltaic cell and is a measure of the tendency for the cell reaction to occur.

6. Standard electrode potentials, which are used to systematize and interpret oxidation-reduction phenomena.

7. The relationship between Gibbs free energy changes and electromotive force; the use of electrochemical data to determine ΔG values, ΔS values, and equilibrium constants.

8. The effect of concentration changes on the emf's of voltaic cells; the Nernst equation.

9. Concentration cells, in which the emf is produced by the combination of two half-cells of the same type that differ only in the concentrations of ions employed.

10. The use of electrode potentials to predict the products of an electrolysis.

11. The corrosion of iron, an electrochemical process.

12. Some commercial voltaic cells.

13. Fuel cells, devices that convert the energy from a combustion directly into electrical energy.

Key Terms

Some of the more important terms introduced in this chapter are listed below. Definitions for terms not included in this list may be located in the text by use of the index.

Ampere, A (Section 18.1)   The SI base unit for electric current; a current of one coulomb per second.

Anode (Section 18.2)   An electrode at which oxidation occurs.

Cathode (Section 18.2)   An electrode at which reduction occurs.

Concentration cell (Section 18.10)   A voltaic cell constructed from two half-cells that are composed of the same substances but that differ in the concentrations of ions.

Coulomb, C (Section 18.1)   A unit of electrical charge; the quantity of electricity carried in one second by a current of one ampere.

Daniell cell (Section 18.5)   A voltaic cell in which Zn metal is oxidized to Zn^{2+} ions at the anode and Cu^{2+} ions are reduced to Cu metal at the cathode.

Electrode (Section 18.2)   An anode or a cathode.

Electrolysis (Section 18.3)   The use of electric current to bring about chemical changes.

Electrolytic conduction (Section 18.2)   The conduction of electricity by the movement of ions through a solution or a molten salt. A sustained current requires that chemical changes at the electrodes also occur.

Electromotive force, emf (Section 18.6)   The potential difference between two electrodes of a voltaic cell, measured in volts; a measure of the tendency for an oxidation-reduction reaction to occur.

Faraday, F (Section 18.4)   The total charge of one mole of electrons; 9.64846×10^4 C.

Faraday's laws (Section 18.4)   The laws developed by Michael Faraday that describe the quantitative relationships between amount of electricity used and chemical change in an electrolysis.

Fuel cell (Section 18.14)   Voltaic cells that are designed to convert the energy obtained from a combustion of a fuel directly into electrical energy.

Gibbs free energy change, ΔG (Sections 18.6 and 18.8)   For a voltaic cell, a measure of the maximum work that can be obtained from the cell; $\Delta G = -nF\mathscr{E}$.

Half-cell (Sections 18.5 and 18.7)   Half of a voltaic cell in which an oxidation *or* a reduction occurs.

Metallic conduction (Section 18.1)   The conduction of electricity through a metal by electron displacement.

Nernst equation (Section 18.9)   The equation used to determine the emf of a cell in which the constituents are present in concentrations other than standard.

Overvoltage (Section 18.11)   An excess voltage (over that theoretically calculated as necessary) that must be applied in certain electrolyses so that they proceed at appreciable rates.

Oxidation potential, $\mathscr{E}_{ox}$ (Section 18.7)   A potential that corresponds to an oxidation half-reaction; an $\mathscr{E}_{ox}$ is the reverse of an electrode potential, $\mathscr{E}_{red}$ (which is a reduction potential) and the signs of the two values are different.

Salt bridge (Section 18.7)   A tube filled with a concentrated solution of an electrolyte and connecting two half-cells of a voltaic cell; it conducts electric current between the half-cells while preventing their contents from mixing.

Silver coulometer (Section 18.4)   An electrolytic cell in which silver is plated out on a cathode placed in an electrical circuit to determine the number of coulombs that have passed through the circuit (by weighing the deposited silver).

Standard electrode potential, $\mathscr{E}^\circ_{red}$ (Section 18.7)   A half-cell potential (measured in volts) for a reduction relative to a standard hydrogen electrode, which is assigned a potential of zero; measured at 25°C with all substances present in their standard states.

Standard emf (Section 18.6)   The emf of a voltaic cell in which all reactants and products are in their standard states; the standard state for an ion is approximately a 1 M concentration and for a gas is approximately 1 atm pressure.

Standard hydrogen electrode (Section 18.7)   A reference electrode in which hydrogen gas, at 1 atm pressure, is bubbled over a Pt electrode that is immersed in an acid solution containing H^-(aq) ions at unit activity.

Voltaic cell (Section 18.5)   A cell that uses a chemical reaction to produce electrical energy; also called a **galvanic cell.**

Problems*

Conduction

18.1 Describe the mechanism of metallic conduction and that of electrolytic conduction.

18.2 What is the effect of an increase in temperature on the conductivity of metals and on the conductivity of solutions of electrolytes? Discuss the causes of resistance to the flow of electricity in these two types of conductors.

18.3 In an electrolytic cell: **(a)** What type of ions move toward the anode? **(b)** What type of half-reaction occurs at the anode? **(c)** What is the sign of the anode? **(d)** Do electrons enter the cell or leave the cell at the anode?

Electrolytic Cells, Quantitative Relationships

18.4 Sketch a cell for the electrolysis of molten $MgCl_2$ between inert electrodes. On the sketch indicate **(a)** the signs of the electrodes, **(b)** the cathode and the anode, **(c)** the directions that the ions move, **(d)** the direction in which electrons move, **(e)** the electrode reactions.

18.5 Sketch a cell for the electrolysis of a $CuSO_4$ solution between Cu electrodes. On the sketch indicate **(a)** the signs of the electrodes, **(b)** the cathode and the anode, **(c)** the directions that the ions move, **(d)** the direction in which electrons move, **(e)** the electrode reactions.

18.6 Write the partial equations for the electrode reactions that occur in the electrolysis of **(a)** $Na_2SO_4(aq)$, **(b)** $NaCl(aq)$, **(c)** $CuCl_2(aq)$, **(d)** $CuSO_4(aq)$.

18.7 A Ni^{2+} solution is electrolyzed using a current of 1.25 A. How many grams of Ni plate out in 25.0 min?

18.8 An acidic solution containing the BiO^+ ion is electrolyzed using a current of 0.750 A. How many grams of Bi plate out in 30.0 min? Write the equation for the reaction at the cathode.

18.9 An acidic solution containing Pb^{2+} ions is electrolyzed and $PbO_2(s)$ plated out on the *anode*. **(a)** Write the chemical equation for the anode reaction. **(b)** If a current of 0.500 A is used for 15.0 min, how many grams of PbO_2 plate out? **(c)** If a solution contains 1.50 g of Pb^{2+}, how many minutes will it take to plate out all of the lead, as PbO_2, using a current of 0.500 A?

18.10 A Ag^+ solution is electrolyzed using a current of 2.50 A. How many grams of Ag plate out in 60.0 min?

18.11 How many minutes will it take to plate out 3.00 g of Cd from a Cd^{2+} solution using a current of 3.00 A?

18.12 How many minutes will it take to plate out 5.00 g of Fe from a Fe^{3+} solution using a current of 1.50 A?

18.13 **(a)** If 0.872 g of Ag is deposited on the cathode of a silver coulometer, how many coulombs have passed through the circuit? **(b)** If the process takes 15.0 min, what was the rate of the current?

18.14 **(a)** How many grams of Ni are deposited in the electrolysis of a solution of $NiSO_4$ in the same time that it takes to deposit 0.575 g of Ag in a silver coulometer arranged in series with the $NiSO_4$ cell? **(b)** If a current of 2.00 A is used, how many minutes will the procedure require?

18.15 In the Hall process, aluminium is produced by the electrolysis of molten Al_2O_3. The electrode reactions are

$$\text{anode:} \quad C + 2O^{2-} \longrightarrow CO_2 + 4e^-$$

$$\text{cathode:} \quad 3e^- + Al^{3+} \longrightarrow Al$$

In the process, the carbon of which the anode is composed is gradually consumed by the anode reaction. How many grams of C are lost from the anode in the time that it takes to plate out 1.00 kg of Al?

18.16 How long would it take to produce enough aluminum by the Hall process (see problem 18.15) to make a case of aluminum soft-drink cans (24 cans) if each can uses 5.00 g of Al, a current of 50,000 amp is employed, and the efficiency of the cell is 90.0%?

18.17 In the electrolysis of molten $MgCl_2$, how many liters of $Cl_2(g)$, measured at STP, are produced in the same time that it takes to plate out 10.0 g of Mg?

18.18 In the electrolysis of a $CuSO_4$ solution, how many grams of Cu are plated out on the cathode in the time that it takes to liberate 5.00 liter of $O_2(g)$, measured at STP, at the anode?

18.19 What is the molarity of H^+ in the solution after the electrolysis described in problem 18.18? The final volume of the solution is 750 ml.

18.20 **(a)** In the electrolysis of aqueous NaCl, what volume of $Cl_2(g)$ is produced in the time that it takes to liberate 5.00 liter of $H_2(g)$? Assume that both gases are measured at STP. **(b)** The volume of the solution used in the electrolysis is 500 ml. What is the molarity of NaOH after the process?

18.21 If 125 ml of 0.750 M $CuCl_2$ solution is electrolyzed using a current of 3.50 A for 45.0 min, what are the final concentrations of Cu^{2+} and Cl^- ions? Assume that the volume of the solution does not change during the course of the electrolysis.

18.22 How many hours will it take to plate out all of the Ni in 200 ml of a 0.350 M Ni^{2+} solution using a current of 0.650 A?

Voltaic Cells, Electrode Potentials

18.23 Sketch a voltaic cell in which the cell reaction is

$$Mg(s) + 2Ag^+(aq) \longrightarrow Mg^{2+}(aq) + 2Ag(s)$$

---

* The more difficult problems are marked with asterisks. The appendix contains answers to color-keyed problems.

On the sketch indicate (a) the signs of the electrodes, (b) the cathode and anode, (c) the directions in which the ions move, (d) the direction in which electrons move, (e) the electrode reactions, (f) the cell voltage.

18.24 Sketch a voltaic cell in which the reaction is

$$Cd + Cl_2 \longrightarrow Cd^{2+} + 2\ Cl^-$$

On the sketch indicate: (a) the signs of the electrodes, (b) the cathode and anode, (c) the directions in which the ions move, (d) the direction in which the electrons move, (e) the electrode reactions, (f) the cell voltage.

18.25 (a) What is $\mathcal{E}°$ for the cell

$$Mg|Mg^{2+}||Sn^{2+}|Sn$$

(b) Write the equation for the cell reaction. **(c)** Which electrode is positive?

18.26 (a) What is $\mathcal{E}°$ for the cell

$$Ni|Ni^{2+}||Cu^{2+}|Cu$$

(b) Write the equation for the cell reaction. **(c)** Which electrode is positive?

18.27 (a) Give the notation for a cell that utilizes the reaction

$$Cl_2(g) + 2\ I^-(aq) \longrightarrow 2\ Cl^-(aq) + I_2(s)$$

(b) What is $\mathcal{E}°$ for the cell? **(c)** Which electrode is the cathode?

18.28 (a) Give the notation for a cell that utilizes the reaction

$$H_2(g) + Br_2(l) \longrightarrow 2\ H^+(aq) + 2\ Br^-(aq)$$

(b) What is $\mathcal{E}°$ for the cell? **(c)** Which electrode is the anode?

18.29 For the cell

$$U|U^{3+}||Ag^+|Ag$$

$\mathcal{E}°$ is $+2.588$ V. Use the emf of the cell and $\mathcal{E}°$ for the Ag^+/Ag couple to calculate $\mathcal{E}°$ for the U^{3+}/U half-reaction.

18.30 For the cell

$$Cu|Cu^{2+}||Pd^{2+}|Pd$$

$\mathcal{E}°$ is $+0.650$ V. Use the emf of the cell and $\mathcal{E}°$ for the Cu^{2+}/Cu couple to calculate $\mathcal{E}°$ for the Pd^{2+}/Pd half-reaction.

18.31 For the cell

$$Sn|Sn^{2+}||H^+, BiO^+|Bi$$

$\mathcal{E}°$ is $+0.456$ V. Use the emf of the cell and $\mathcal{E}°$ for the Sn^{2+}/Sn couple to calculate $\mathcal{E}°$ for the BiO^+/Bi half-reaction.

18.32 From the table of electrode potentials (acid solution) that appears in the appendix, select a suitable substance for each of the following transformations

(assume that all soluble substances are present at $1\ M$ concentrations): **(a)** an oxidizing agent capable of oxidizing Fe to Fe^{2+} but not Tl to Tl^+, **(b)** an oxidizing agent capable of oxidizing Mn^{2+} to MnO_4^- but not MnO_2 to MnO_4^-, **(c)** a reducing agent capable of reducing Fe^{2+} to Fe but not Mn^{2+} to Mn, **(d)** a reducing agent capable of reducing PbO_2 to Pb^{2+} but not MnO_2 to Mn^{2+}.

18.33 From the table of electrode potentials (acid solution) that appears in the appendix, select a suitable substance for each of the following transformations (assume that all soluble substances are present in $1\ M$ concentrations): **(a)** an oxidizing agent capable of oxidizing Hg to Hg_2^{2+} but not Hg to Hg^{2+}, **(b)** an oxidizing agent capable of oxidizing Mn^{2+} to MnO_2 but not Cr^{3+} to $Cr_2O_7^{2-}$, **(c)** a reducing agent capable of reducing Sn^{4+} to Sn^{2+} but not Sn^{2+} to Sn, **(d)** a reducing agent capable of reducing I_2 to I^- but not Cu^{2+} to Cu.

18.34 Use $\mathcal{E}°$ values to predict whether each of the following skeleton equations represents a reaction that will occur in acid solution with all soluble substances present in $1\ M$ concentration. Complete and balance the equation for each reaction that is predicted to occur:

(a) $H_2O_2 + Cu^{2+} \longrightarrow Cu + O_2$

(b) $H_2O_2 + Ag^+ \longrightarrow Ag + O_2$

(c) $Ag^+ + Fe^{2+} \longrightarrow Ag + Fe^{3+}$

(d) $Au + Cl_2 \longrightarrow Au^{3+} + Cl^-$

(e) $H_2SO_3 + H_2S \longrightarrow S$

18.35 Use $\mathcal{E}°$ values to predict whether each of the following skeleton equations represents a reaction that will occur in acid solution with all soluble substances present in $1\ M$ concentration. Complete and balance the equation for each reaction that is predicted to occur:

(a) $PbO_2 + Cl^- \longrightarrow Pb^{2+} + Cl_2$

(b) $Co + Cd^{2+} \longrightarrow Co^{2+} + Cd$

(c) $I^- + NO_3^- \longrightarrow I_2 + NO$

(d) $Mn^{2+} + Cr_2O_7^{2-} \longrightarrow MnO_4^- + Cr^{3+}$

(e) $MnO_4^- + Mn^{2+} \longrightarrow MnO_2$

18.36 Use $\mathcal{E}°$ values to predict whether each of the following skeleton equations represents a reaction that will occur in acid solution with all soluble substances present in $1\ M$ concentration. Complete and balance the equation for each reaction that is predicted to occur:

(a) $Hg + Hg^{2+} \longrightarrow Hg_2^{2+}$

(b) $Mn^{2+} \longrightarrow MnO_2 + Mn$

(c) $SO_4^{2-} + S \longrightarrow H_2SO_3$

(d) $Mn^{3+} + Mn \longrightarrow Mn^{2+}$

(e) $Sn^{4+} + Sn \longrightarrow Sn^{2+}$

18.37 Given the following standard electrode potential diagram (acid solution):

$$In^{3+} \xrightarrow{-0.434\ V} In^+ \xrightarrow{-0.147\ V} In$$
$$\underset{-0.338\ V}{\underline{\hspace{4cm}}}$$

(a) Is the In^+ ion stable toward disproportionation in water solution? **(b)** Which ion is produced when In metal reacts with $H^+(aq)$? (c) The $\mathscr{E}°$ for Cl_2/Cl^- is $+1.36$ V. Will In react with Cl_2? What is the product? (d) Write balanced chemical equations for all reactions.

18.38 Given the following standard electrode potential diagram (acid solution):

$$Tl^{3+} \xrightarrow{+1.25 \text{ V}} Tl^+ \xrightarrow{-0.34 \text{ V}} Tl$$
$$\xrightarrow{+0.72 \text{ V}}$$

(a) Is the Tl^+ ion stable toward disproportionation in water solution? (b) Which ion is produced when Tl metal reacts with $H^+(aq)$? (c) The $\mathscr{E}°$ for Cl_2/Cl^- is $+1.36$ V. Will Tl react with Cl_2? What is the product? (d) Write balanced chemical equations for all reactions.

18.39 The following reactions occur at 25°C with all soluble substances present in 1 M concentrations:

$$Zn + Pb^{2+} \longrightarrow Zn^{2+} + Pb$$
$$Ti + Zn^{2+} \longrightarrow Ti^{2+} + Zn$$
$$2\,Lu + 3\,Ti^{2+} \longrightarrow 2\,Lu^{3+} + 3\,Ti$$

From this information alone, predict whether the following reactions will occur under similar conditions:
(a) $Pb + Ti^{2+} \longrightarrow Pb^{2+} + Ti$
(b) $2\,Lu + 3\,Pb^{2+} \longrightarrow 2\,Lu^{3+} + 3\,Pb$
(c) $2\,Lu^{3+} + 3\,Zn \longrightarrow 3\,Zn^{2+} + 2\,Lu$

18.40 (a) Given

$$Ti^{3+} + e^- \longrightarrow Ti^{2+} \qquad \mathscr{E}° = -0.369 \text{ V}$$
$$Ti^{2+} + 2e^- \longrightarrow Ti \qquad \mathscr{E}° = -1.628 \text{ V}$$

calculate $\mathscr{E}°$ for the half-reaction

$$Ti^{3+} + 3e^- \longrightarrow Ti$$

(b) Will the Ti^{2+} ion disproportionate in aqueous solution? (c) Will Ti metal react with $H^+(aq)$? If so, which ion is produced?

18.41 (a) Given

$$Co^{2+} + 2e^- \longrightarrow Co \qquad \mathscr{E}° = -0.277 \text{ V}$$
$$Co^{3+} + 3e^- \longrightarrow Co \qquad \mathscr{E}° = +0.418 \text{ V}$$

calculate $\mathscr{E}°$ for the half-reaction

$$Co^{3+} + e^- \longrightarrow Co^{2+}$$

(b) Will the Co^{2+} ion disporportionate in aqueous solution? (c) Will Co metal react with $H^+(aq)$? If so, what ion is produced?

18.42 (a) Given

$$UO_2^+ + 4H^+ + e^- \longrightarrow U^{4+} + 2H_2O$$
$$\mathscr{E}° = +0.620 \text{ V}$$
$$UO_2^{2+} + 4H^+ + 2e^- \longrightarrow U^{4+} + 2H_2O$$
$$\mathscr{E}° = +0.330 \text{ V}$$

calculate $\mathscr{E}°$ for the half-reaction

$$UO_2^{2+} + e^- \longrightarrow UO_2^+$$

(b) Will the UO_2^+ ion disproportionate in aqueous solution? If so, write the chemical equation for the reaction.

18.43 (a) Given

$$Au^+ + e^- \longrightarrow Au \qquad \mathscr{E}° = +1.691 \text{ V}$$
$$Au^{3+} + 3e^- \longrightarrow Au \qquad \mathscr{E}° = +1.495 \text{ V}$$

calculate $\mathscr{E}°$ for the half-reaction

$$2e^- + Au^{3+} \longrightarrow Au^+$$

(b) Will the Au^+ ion disproportionate in aqueous solution? (c) Will Au metal react with $H^+(aq)$? If so, what ion is produced?

18.44 (a) Given

$$Eu^{3+} + 3e^- \longrightarrow Eu \qquad \mathscr{E}° = -2.407 \text{ V}$$
$$Eu^{3+} + e^- \longrightarrow Eu^{2+} \qquad \mathscr{E}° = -0.429 \text{ V}$$

calculate $\mathscr{E}°$ for the half-reaction

$$Eu^{2+} + 2e^- \longrightarrow Eu$$

(b) Will the Eu^{2+} ion disproportionate in aqueous solution? (c) Will Eu^{3+} and Eu react to give Eu^{2+}? (d) Will Eu metal react with $H^+(aq)$? If so, what ion is produced?

Gibbs Free Energy and Emf

18.45 Given the following:

$$PbSO_4 + 2e^- \rightleftharpoons Pb + SO_4^{2-} \qquad \mathscr{E}° = -0.359 \text{ V}$$
$$Pb^{2+} + 2e^- \rightleftharpoons Pb \qquad \mathscr{E}° = -0.126 \text{ V}$$

(a) Write the notation for a cell utilizing these half-reactions. (b) Write the equation for the cell reaction. (c) Calculate $\mathscr{E}°$ for the cell. (d) Determine $\Delta G°$.

18.46 Given the following:

$$AgI + e^- \rightleftharpoons Ag + I^- \qquad \mathscr{E}° = -0.152 \text{ V}$$
$$Ag^+ + e^- \rightleftharpoons Ag \qquad \mathscr{E}° = +0.799 \text{ V}$$

(a) Write the notation for a cell utilizing these half-reactions. (b) Write the equation for the cell reaction. (c) Calculate $\mathscr{E}°$ for the cell. (d) Determine $\Delta G°$.

*18.47 (a) Diagram a cell for which the cell reaction is

$$H^+ + OH^- \longrightarrow H_2O$$

(b) Calculate $\mathscr{E}°$ for the cell and $\Delta G°$ for the reaction.

18.48 (a) Diagram a cell for which the cell reaction in acid solution is

$$2H_2 + O_2 \longrightarrow 2H_2O$$

(b) Calculate $\mathscr{E}°$ for the cell and $\Delta G°$ for the reaction.

18.49 (a) Use electrode potentials (see appendix) to calculate the emf of a standard cell utilizing the reaction

$$Cl_2(g) + 2 Br^-(aq) \longrightarrow 2 Cl^-(aq) + Br_2(l)$$

(b) What is $\Delta G°$ for the reaction? (c) If $\Delta H°$ for the reaction is -93.09 kJ, what is $\Delta S°$?

18.50 (a) Use electrode potentials (see appendix) to calculate the emf of a standard cell utilizing the reaction

$$2 H_2O_2(aq) \longrightarrow O_2(g) + 2 H_2O$$

(b) What is $\Delta G°$ for the reaction? (c) If $\Delta H°$ for the reaction is -189.44 kJ, what is $\Delta S°$?

18.51 (a) Use electrode potentials (see appendix) to calculate the emf of a standard cell utilizing the reaction

$$I_2(s) + H_2S(aq) \longrightarrow S(rhombic) + 2 H^+(aq) + 2 I^-(aq)$$

(b) What is $\Delta G°$ for the reaction? (c) If $\Delta S°$ for the reaction is $+11.7$ J/K, what is $\Delta H°$?

18.52 For the half-reaction

$$XeF_2(aq) + 2 H^+(aq) + 2e^- \longrightarrow Xe(g) + 2 HF(aq)$$

$\mathscr{E}°$ has been reported to equal $+2.64$ V. (a) Calculate $\Delta G°$ for the reaction

$$XeF_2(aq) + H_2(g) \longrightarrow Xe(g) + 2 HF(aq)$$

(b) The free energy of formation, $\Delta G_f°$ of HF(aq) is -276.48 kJ/mol. What is $\Delta G_f°$ of $XeF_2(aq)$?

18.53 Given the following electrode potentials

$$H_3BO_3(s) + 3H^+(aq) + 3e^- \longrightarrow B(s) + 3 H_2O$$
$$\mathscr{E}° = -0.869 \text{ V}$$

$$4 H^+(aq) + O_2(g) + 4e^- \longrightarrow 2 H_2O$$
$$\mathscr{E}° = +1.229 \text{ V}$$

calculate the value of $\Delta G_f°$ for $H_3BO_3(s)$. For $H_2O(l)$, $\Delta G_f°$ is -237.19 kJ/mol.

18.54 Perbromates have only recently been prepared. For the half-reaction

$$BrO_4^-(aq) + 8 H^+(aq) + 7e^- \longrightarrow \tfrac{1}{2} Br_2(l) + 4 H_2O$$

$\mathscr{E}°$ is reported to be $+1.59$ V. (a) Calculate $\Delta G_f°$ for $BrO_4^-(aq)$. For $H_2O(l)$, $\Delta G_f°$ is -237.19 kJ/mol, and for $H^+(aq)$, $\Delta G_f°$ is 0.0. (b) Calculate $\mathscr{E}°$ for the half-reaction

$$BrO_4^-(aq) + 2 H^+(aq) + 2e^- \longrightarrow BrO_3^-(aq) + H_2O$$

The value of $\Delta G_f°$ for $BrO_3^-(aq)$ is $+21.71$ kJ/mol. (c) What oxidizing agent, listed in the table in the appendix, *should* be capable of oxidizing $BrO_3^-(aq)$ to $BrO_4^-(aq)$?

18.55 Use electrode potentials (see appendix) to calculate the equilibrium constant for the reaction

$$Ni(s) + Sn^{2+}(aq) \rightleftharpoons Ni^{2+}(aq) + Sn(s)$$

18.56 Use standard electrode potentials (given in the appendix) to determine the equilibrium constant at 25°C for the reaction

$$Cl_2(g) + H_2O \rightleftharpoons H^+(aq) + Cl^-(aq) + HOCl(aq)$$

18.57 Given the following

$$2 H_2O + 2e^- \longrightarrow H_2(g) + 2 OH^-(aq)$$
$$\mathscr{E}° = -0.828 \text{ V}$$

calculate the value of the water constant, K_w:

$$H_2O \rightleftharpoons H^+(aq) + OH^-(aq)$$

18.58 Use standard electrode potentials (given in the appendix) to calculate the equilibrium constant at 25°C for the reaction

$$4 H^+(aq) + 4 Br^-(aq) + O_2(g) \rightleftharpoons 2 Br_2(l) + 2 H_2O$$

18.59 Calculate $\mathscr{E}°$ for the half-reaction

$$PbBr_2(s) + 2e^- \longrightarrow Pb(s) + 2 Br^-(aq)$$

from the following data:

$$PbBr_2(s) \rightleftharpoons Pb^{2+}(aq) + 2 Br^-(aq)$$

$$K = 4.60 \times 10^{-6}$$

$$Pb^{2+}(aq) + 2e^- \longrightarrow Pb(s)$$

$$\mathscr{E}° = -0.126 \text{ V}$$

The Nernst Equation

18.60 Suggest ways to increase the emf of a cell that is based on the reaction

$$Fe(s) + 2 H^+(aq) \longrightarrow Fe^{2+}(aq) + H_2(g)$$

18.61 (a) Calculate the emf of a cell formed from a Mg^{2+}/Mg half-cell in which $[Mg^{2+}]$ is 0.0500 M and a Ni^{2+}/Ni half-cell in which $[Ni^{2+}]$ is 1.50 M. (b) Write the equation for the cell reaction. (c) Which electrode is positive?

18.62 (a) Calculate the emf of a cell formed from a Zn^{2+}/Zn half-cell in which $[Zn^{2+}]$ is 0.0500 M and a Cl_2/Cl^- half-cell in which $[Cl^-]$ is 0.0500 M and the pressure of $Cl_2(g)$ is 1.25 atm. (b) Write the equation for the cell reaction. (c) Which electrode is negative?

18.63 (a) Calculate the emf of a cell formed from a Pb^{2+}/Pb half-cell in which $[Pb^{2+}]$ is 6.00 M and a H^+/H_2 half-cell in which $[H^+]$ is 0.0200 M and the pressure of $H_2(g)$ is 2.00 atm. (b) Write the equation for the cell reaction and the notation for the cell. (c) Which electrode is positive?

18.64 What is the concentration of Cd^{2+} in the cell

$$Zn|Zn^{2+}(0.0900 \text{ } M)||Cd^{2+}(? \text{ } M)|Cd$$

if the emf of the cell is 0.4000 V?

18.65 What is the concentration of Ag^+ in the cell

$$Cu|Cu^{2+}(3.50\ M)||Ag^+(?\ M)|Ag$$

if the emf of the cell is 0.350 V?

18.66 **(a)** According to the $\mathscr{E}^\circ$ value, should the following reaction proceed spontaneously at 25°C with all soluble substances present at 1.00 M concentration?

$$MnO_2(s) + 4\ H^+(aq) + 2\ Cl^-(aq) \longrightarrow$$
$$Mn^{2+}(aq) + 2\ H_2O + Cl_2(g)$$

(b) Would the reaction be spontaneous if the concentrations of H^+ and Cl^- were each increased to 10.0 M?

18.67 For the half-reaction

$$Cr_2O_7^{2-} + 14\ H^+ + 6e^- \rightleftharpoons 2\ Cr^{3+} + 7\ H_2O$$

$\mathscr{E}^\circ$ is + 1.33 V. What would the potential be if the concentration of $H^+(aq)$ were reduced to 0.100 M?

18.68 For a half-reaction of the form

$$M^{2+} + 2e^- \rightleftharpoons M$$

what would be the effect on the electrode potential if **(a)** the concentration of M^{2+} were doubled, **(b)** the concentration of M^{2+} were cut in half?

18.69 Consider the following standard cell:

$$Sn|Sn^{2+}||Pb^{2+}|Pb$$

(a) Calculate the value of $\mathscr{E}^\circ$ for the cell. **(b)** When the cell operates, the concentration of Sn^{2+} increases and the concentration of Pb^{2+} decreases. What are the concentrations of the ions when the emf of the cell is zero?

18.70 **(a)** Determine the emf of a cell prepared from two H^+/H_2 half-cells, one in which $[H^+]$ is 0.0250 M and the other in which $[H^+]$ is 5.00 M. The pressure of $H_2(g)$ in both half-cells is 1.00 atm. **(b)** Write an equation for the "cell reaction." **(c)** Determine the emf of the cell if the pressure of $H_2(g)$ is changed to 2.00 atm in the anode half-cell and to 0.100 atm in the cathode half-cell.

18.71 **(a)** Determine the emf of a cell prepared from two Ga^{3+}/Ga half-cells, one in which $[Ga^{3+}]$ is 2.00 M and the other in which $[Ga^{3+}]$ is 0.300 M. **(b)** Write an equation for the "cell reaction." Which electrode is negative?

18.72 Why does measurement of the density of the electrolyte in a lead storage battery give an indication of the state of charge of the battery?

CHAPTER 19

THE NONMETALS, PART I: HYDROGEN AND THE HALOGENS

Within a period of the periodic table, metallic character *decreases* from left to right; within a group, metallic character *increases* from top to bottom. The stepped diagonal line shown in the periodic table is the approximate division between the metals (on the left) and the nonmetals (on the right). The first element, hydrogen, is also classified as a nonmetal. In this chapter, the chemistry of hydrogen and of the halogens (group VII A) is discussed.

HYDROGEN

Hydrogen does not fit well into any group of the periodic table. The hydrogen atom has only one valence electron and in this regard is like the atoms of group I A. The group I A elements, however, are metals and hydrogen is a nonmental. The hydrogen atom is one electron short of a noble-gas configuration like the atoms of the group VII A elements. Hydrogen, however, is less electronegative than the group VII A elements and its chemical properties deviate in important ways from the properties of those elements. The reason for the unique character of hydrogen is found in its very small atomic radius.

19.1 Occurrence and Properties of Hydrogen

Hydrogen atoms constitute about 15% of all the atoms present in the crust, bodies of water, and atmosphere of the earth. On the basis of mass, however, the percentage drops to less than 1% since hydrogen has a very low atomic mass. The most important compound of hydrogen found in nature is water, which is the principal source of the element. Hydrogen is also found in combined form in the hydrocarbons (compounds of carbon and hydrogen found in coal, natural gas, and petroleum), a few minerals (such as clay and certain hydrates), and the organic compounds that constitute the principal part of all plant and animal matter (see Chapters 26 and 27). Free hydrogen occurs in nature only in negligible amounts (for example, in volcanic gases).

Hydrogen is a colorless, odorless, tasteless gas. It has the lowest density of any chemical substance; at STP one liter of hydrogen weighs 0.0899 g. The two atoms of hydrogen of the H_2 molecule are joined by a single covalent bond, giving each atom a stable helium electronic configuration. The molecule is nonpolar; the weak nature of the intermolecular forces of attraction is indicated by the low normal boiling point ($-252.7°C$), the normal melting point ($-259.1°C$) and the critical temperature ($-240°C$ at a critical pressure of 12.8 atm). Hydrogen is virtually insoluble in water; approximately 2 ml of hydrogen will dissolve in 1 liter of water at room temperature and atmospheric pressure.

There are three isotopes of hydrogen. The most abundant isotope $_1^1H$, constitutes 99.985% of naturally occurring hydrogen; deuterium $_1^2H$ (also indicated as $_1^2D$) constitutes 0.015%; and the radioactive tritium, $_1^3H$ (also indicated as $_1^3T$) occurs only in trace amounts.

19.2 Industrial Production of Hydrogen

Large quantities of hydrogen are used by the chemical industry. The principal industrial sources are

1. Steam reformer process. This process is the one most widely used for the production of hydrogen in large quantities. A hydrocarbon, such as methane (CH_4), and steam are passed over a nickel catalyst at 900°C. The reactions that occur are

$$CH_4(g) + 2\,H_2O(g) \longrightarrow CO_2(g) + 4\,H_2(g)$$
$$CH_4(g) + H_2O(g) \longrightarrow CO(g) + 3\,H_2(g)$$

The gas emerging from the reformer furnace consists of H_2, CO, CO_2, and excess steam. This gas mixture is passed into a *shift converter* at 450°C, in which the CO is converted into CO_2:

$$CO(g) + H_2O(g) \longrightarrow CO_2(g) + H_2(g)$$

The CO_2 is removed from the resulting mixture of CO_2 and H_2 by passing the mixture through cold water under pressure or through a solution of an amine (a basic compound with which CO_2 reacts). In each case, the CO_2 dissolves but the H_2 does not.

2. Water gas. Coke is impure carbon that has been obtained from coal by heating it in the absence of air to drive off volatile components. Coke and steam react at high temperatures (1000°C) to produce a gaseous mixture known as water gas:

$$C(s) + H_2O(g) \longrightarrow CO(g) + H_2(g)$$

Since both CO and H_2 will burn, the mixture is used as a fuel as well as a source for H_2.

Hydrogen is obtained from water gas by liquefying the CO (high pressure, low temperature) and separating the gaseous H_2 from the liquid CO. As an alternative, the water gas can be passed into a shift converter, the CO converted into CO_2, and the CO_2 removed as described for the steam reformer process.

3. Iron and steam. Iron and steam react at temperatures of 650°C or higher:

$$3\,Fe(s) + 4\,H_2O(g) \longrightarrow Fe_3O_4(s) + 4\,H_2(g)$$

A catalytic cracking unit. *Mobil Oil Corporation.*

4. Cracking. Hydrogen is obtained by the catalytic decomposition of hydrocarbons at high temperatures. In petroleum refining, high molecular weight hydrocarbons are "cracked" into lower molecular weight compounds that are more valuable. Hydrogen is a by-product.

5. Electrolysis of sodium chloride brine. Hydrogen is a by-product (as is chlorine) in the industrial preparation of sodium hydroxide by the electrolysis of concentrated solutions of sodium chloride:

$$2\,Na^+(aq) + 2\,Cl^-(aq) + 2\,H_2O \xrightarrow{\text{electrolysis}}$$
$$2\,Na^+(aq) + 2\,OH^-(aq) + H_2(g) + Cl_2(g)$$

6. Electrolysis of water. Very pure but relatively expensive hydrogen is obtained by the electrolysis of water that contains a small amount of sulfuric acid or sodium hydroxide:

$$2\,H_2O \xrightarrow{\text{electrolysis}} 2\,H_2(g) + O_2(g)$$

19.3 Hydrogen from Displacement Reactions

Reactions in which one element (or a group of elements) displaces another element (or a group of elements) from a compound are called **displacement reactions.** Hydrogen is a product of several types of displacement reactions:

1. The very reactive elements (Ca, Ba, Sr, and the group I A metals) react vigorously with water at room temperature to produce hydrogen and solutions of hydroxides. For example,

$$2\,Na(s) + 2\,H_2O \longrightarrow 2\,Na^+(aq) + 2\,OH^-(aq) + H_2(g)$$

$$Ca(s) + 2\,H_2O \longrightarrow Ca^{2+}(aq) + 2\,OH^-(aq) + H_2(g)$$

These reactions may be dangerous.

2. A long list of metals react with aqueous solutions of acids to produce hydrogen. For example,

$$Zn(s) + 2\,H^+(aq) \longrightarrow Zn^{2+}(aq) + H_2(g)$$

$$Fe(s) + 2\,H^+(aq) \longrightarrow Fe^{2+}(aq) + H_2(g)$$

$$2\,Al(s) + 6\,H^+(aq) \longrightarrow 2\,Al^{3+}(aq) + 3\,H_2(g)$$

By convention, the standard electrode potential of the H^+/H_2 electrode is zero:

$$2e^- + 2\,H^+(aq) \rightleftharpoons H_2(g) \qquad \mathscr{E}^\circ = 0.000 \text{ V}$$

Any metal that has a $\mathscr{E}^\circ_{ox}$ that is a positive value (the corresponding $\mathscr{E}^\circ_{red}$ would be a negative value) should, therefore, be capable of displacing hydrogen from an aqueous solution of an acid. The very reactive metals (Ca, Ba, Sr, and the group I A metals) react violently and such reactions are not run. Other metals (Pb, for example) react very slowly. A common laboratory preparation of hydrogen involves the reaction of zinc (a metal with a satisfactory order of reactivity) with hydrochloric acid.

3. Certain metals and nonmetals displace hydrogen from alkaline solutions. Examples are

$$Zn(s) + 2\,OH^-(aq) + 2\,H_2O \longrightarrow H_2(g) + Zn(OH)_4^{2-}(aq)$$
zincate ion

$$2\,Al(s) + 2\,OH^-(aq) + 6\,H_2O \longrightarrow 3\,H_2(g) + 2\,Al(OH)_4^-(aq)$$
aluminate ion

$$Si(s) + 2\,OH^-(aq) + H_2O \longrightarrow 2\,H_2(g) + SiO_3^{2-}(aq)$$
silicate ion

Hydrogen is also conveniently prepared in the laboratory by means of the reactions of the metal hydrides with water. These reactions are discussed in the next section.

19.4 Reactions of Hydrogen

The bond energy of the H—H bond is 431 kJ/mol. In the course of most reactions this bond must be broken so that the H atoms can form new bonds with other atoms. Since this bond energy is relatively high, most reactions of hydrogen take place at elevated temperatures.

Hydrogen reacts with the metals of group I A and the heavier metals of group II A (Ba, Sr, and Ca) to form **saltlike hydrides.** The hydride ion (H^-) of these compounds is isoelectronic with helium and achieves this stable configuration by the addition of an electron from the metal. Since the electron affinity of hydrogen is low (-73 kJ/mol), hydrogen reacts in this manner only with the most reactive metals:

$$2\,Na(s) + H_2(g) \longrightarrow 2\,NaH(s)$$

$$Ca(s) + H_2(g) \longrightarrow CaH_2(s)$$

The saltlike hydrides react with water to form hydrogen:

$$H^- + H_2O \longrightarrow H_2(g) + OH^-(aq)$$

With certain other metals, such as platinum, palladium, and nickel, hydrogen forms **interstitial hydrides.** Many of these hydrides are nonstoichiometric; their apparent formulas depend upon the conditions under which they are prepared. Palladium can absorb up to 900 times its own volume of hydrogen. The interstitial hydrides resemble the metals from which they are derived. The crystal structures of the metals do not change (or change only slightly) as hydrogen is added. Whether these substances should be regarded as compounds is debatable. The name, interstitial hydrides, arises from the view that the hydrogen is only absorbed into the interstices of the metallic crystal.

Magnetic studies, however, indicate that electron pairing of some sort is involved in the formation of these hydrides. Many of the metals have unpaired electrons and, hence, are paramagnetic. A gradual loss of paramagnetism is observed as the hydrides form from the metal. Since the electrons of molecular hydrogen are paired, these studies imply that the hydrogen is incorporated into the crystal in atomic form, which would account for the catalytic activity of Pt, Pd, Ni, and other metals for many reactions of hydrogen. If the H—H bond is broken or even only weakened, the absorbed hydrogen should be much more reactive than ordinary H_2 gas.

Complex hydrides of boron and aluminum are important and useful compounds for chemical syntheses. Important examples are sodium borohydride and lithium aluminum hydride:

$$Na^+ \begin{bmatrix} H \\ H\!:\!\overset{\cdot\cdot}{\underset{\cdot\cdot}{B}}\!:\!H \\ H \end{bmatrix}^- \qquad Li^+ \begin{bmatrix} H \\ H\!:\!\overset{\cdot\cdot}{\underset{\cdot\cdot}{Al}}\!:\!H \\ H \end{bmatrix}^-$$

Since these compounds, as well as the saltlike hydrides, react with water to liberate hydrogen, they are prepared by reactions in ether:

$$4\,LiH + AlCl_3 \longrightarrow Li[AlH_4] + 3\,LiCl$$

Hydrogen forms covalent compounds with most of the nonmetals. It reacts with the halogens to form colorless, polar covalent gases:

$$H_2(g) + Cl_2(g) \longrightarrow 2\,HCl(g)$$

Reactions with fluorine or chlorine will take place at room temperature or below, but the reactions with the less reactive bromine or iodine require higher temperatures ($400°C$ to $600°C$).

The reaction of hydrogen and oxygen in which H_2O is formed is highly exothermic and is the basis of the high temperature (ca. $2800°C$) produced by the oxyhydrogen torch. The reaction of hydrogen with sulfur is more difficult and requires high temperatures:

$$H_2(g) + S(g) \longrightarrow H_2S(g)$$

Hydrogen reacts with nitrogen at high pressures (300 to 1000 atm), at high temperatures (400°C to 600°C), and in the presence of a catalyst to produce ammonia (Haber process):

$$3\,H_2(g) + N_2(g) \rightleftharpoons 2\,NH_3(g)$$

Hydrogen does not react readily with carbon, but at high temperatures, hydrogen can be made catalytically to react with finely divided carbon (Bergius process); hydrocarbons are produced in these reactions:

$$2\,H_2(g) + C(g) \longrightarrow CH_4(g)$$

The ionization potential of hydrogen is relatively high (1312 kJ/mol). In none of the compounds formed between hydrogen and the nonmetals does hydrogen exist as positive ions; the bonding in all these compounds is polar covalent. Because of its large charge-to-radius ratio, the unassociated proton (H^+) does not exist in ordinary chemical systems. In pure acids the hydrogen atom is covalently bonded to the rest of the molecule, and in aqueous solutions the proton is hydrated.

Hydrogen reacts with many metal oxides to produce water and the free metal:

$$CuO(s) + H_2(g) \longrightarrow Cu(s) + H_2O(g)$$

$$WO_3(s) + 3\,H_2(g) \longrightarrow W(s) + 3\,H_2O(g)$$

$$FeO(s) + H_2(g) \longrightarrow Fe(s) + H_2O(g)$$

Some metals (W, for example) are commercially produced by the hydrogen reduction of oxide ores. The method is expensive, however, and is not used if an alternative, less-expensive method is available.

Carbon monoxide and hydrogen react at high temperatures and high pressures in the presence of a catalyst to produce methyl alcohol, CH_3OH:

$$CO(g) + 2\,H_2(g) \longrightarrow CH_3OH(g)$$

19.5 Industrial Uses of Hydrogen

The principal industrial uses of hydrogen are

1. Production of ammonia from N_2 and H_2 by the Haber process.
2. Production of hydrogen chloride from Cl_2 and H_2.
3. Synthesis of methyl alcohol from CO and H_2.
4. Refining of petroleum.
5. Hydrogenation of edible oils (corn, cotton seed, soy bean, peanut, and others) to produce shortenings and other foods (see Section 27.3).

6. Reduction of oxide ores to produce certain metals.

7. As a rocket fuel.

8. As a fuel in oxyhydrogen welding, atomic hydrogen welding, annealing furnaces, and electronic component fabrication.

THE HALOGENS

The elements of group VII A—fluorine, chlorine, bromine, iodine, and astatine— are called the **halogens.** The name "halogen" is derived from Greek and means "salt former." These elements, with the exception of astatine, occur extensively in nature in the form of halide salts. Astatine probably occurs in nature, in extremely small amounts, as a short-lived intermediate of natural radioactive decay processes. Most of our meager information about the chemistry of astatine, however, comes from the study of the small amounts of a radioactive isotope of this element prepared by nuclear reactions.

19.6 Properties of the Halogens

The electronic configurations of the halogens are listed in Table 19.1. Each halogen atom has one electron less than the noble gas that follows it in the periodic classification. There is, therefore, a marked tendency on the part of a halogen atom to attain a noble-gas configuration by the formation of a uninegative ion or a single covalent bond. Positive oxidation states exist for all the elements except fluorine.

Some properties of the halogens are summarized in Table 19.2; the symbol X stands for any halogen. Notice the following:

1. Physical state. Under ordinary conditions, the halogens exist as diatomic molecules with a single covalent bond joining the atoms of a molecule. The molecules are held together in the solid and liquid states by London forces. Of all the halogen molecules, I_2 is the largest, has the most electrons, and is the most polarizable. It is not surprising, therefore, that the intermolecular attractions between I_2 molecules are the strongest and that I_2 has the highest melting point and boiling point. At ordinary temperatures and pressures, I_2 is a solid, Br_2 is a liquid, and Cl_2 and F_2 are gases.

Table 19.1 Electronic configurations of the halogens

Element	Z	1s	2s	2p	3s	3p	3d	4s	4p	4d	4f	5s	5p	5d	6s	6p
F	9	2	2	5												
Cl	17	2	2	6	2	5										
Br	35	2	2	6	2	6	10	2	5							
I	53	2	2	6	2	6	10	2	6	10		2	5			
At	85	2	2	6	2	6	10	2	6	10	14	2	6	10	2	5

2. First ionization energy. Within the *group*, ionization energy decreases with increasing atomic radius in the expected manner. The first ionization energy of fluorine is the highest of the group and of iodine is the lowest of the group. The halogen of each *period* has a relatively high ionization energy, second only to that of the noble gas of the period. There is, therefore, little tendency for a halogen atom to form a positive ion (although such ions as I_2^+, Br_2^+, Cl_2^+, and I_3^+ have been identified).

3. Electronegativity. Each halogen is the most reactive nonmetal of its period, and fluorine is the most reactive of all the nonmetals. Fluorine has the highest electronegativity of any element and is one of the strongest oxidizing agents known. The electronegativity of the halogens decreases in the order F > Cl > Br > I, and the oxidizing power of the halogens decreases in the same order.

4. Bond energy. Bond energy decreases from Cl_2 to Br_2 to I_2 since the increasing size of the halogen atom makes it easier and easier to break the bond between the atoms of the X_2 molecule. The bond dissociation energy of the F_2 molecule is, however, unusually low and out of line in comparison to the other values. The reason for this relatively low value is not completely understood but is ascribed to the effect of the nonbonding electrons in the F_2 molecule. A repulsion between the highly dense electron clouds of the small fluorine atoms is believed to weaken the bond and lower the energy required to break it.

The bond formed between fluorine and an element other than itself is always stronger than the bonds formed by any of the other halogens with the same element. The bond energies of the hydrogen halides, for example, are HF, 565 kJ/mol; HCl, 431 kJ/mol; HBr, 364 kJ/mol; and HI, 297 kJ/mol. The high order of chemical reactivity of F_2 in reactions with other nonmetals is the result, therefore, of the low bond energy of the F_2 molecule (energy *required*), coupled with the high bond energies of the new bonds (energy *released*).

5. Electrode potentials. The values for the X_2/X^- electrode potentials fall in the same order as the electronegativity values. Fluorine is the most reactive halogen—iodine the least. Since electrode potentials refer to processes that occur in aqueous solution and since fluorine reacts with water, the F_2/F^- electrode potential is obtained by calculation rather than by direct measurement.

Table 19.2   Some properties of the halogens

Property	F_2	Cl_2	Br_2	I_2
color	pale yellow	yellow-green	red-brown	violet-black
melting point (°C)	−218	−101	−7	+113
boiling point (°C)	−188	−35	+59	+183
atomic radius (pm)	72	99	114	133
ionic radius, X^- (pm)	136	181	195	216
first ionization energy (kJ/mol)	1.68×10^3	1.25×10^3	1.14×10^3	1.00×10^3
electronegativity	4.0	3.2	3.0	2.7
bond energy (kJ/mol)	155	243	193	151
standard electrode potential (V) $2e^- + X_2 \rightleftharpoons 2X^-$	+2.87	+1.36	+1.07	+0.54

The relative oxidizing ability of the halogens may be observed in displacement reactions. Thus, fluorine can displace chlorine, bromine, and iodine from their salts; chlorine can displace bromine and iodine from their salts; and bromine can displace iodine from iodides:

$$F_2(g) + 2\,NaCl(s) \longrightarrow 2\,NaF(s) + Cl_2(g)$$

$$Cl_2(g) + 2\,Br^-(aq) \longrightarrow 2\,Cl^-(aq) + Br_2(l)$$

$$Br_2(l) + 2\,I^-(aq) \longrightarrow 2\,Br^-(aq) + I_2(s)$$

Since fluorine actively oxidizes water (producing O_2), displacement reactions involving F_2 cannot be run in water solution.

19.7 Occurrence and Industrial Preparation of the Halogens

The principal natural sources of the halogens are listed in Table 19.3. The halogens are too reactive to occur in elemental form. The most common state in which they occur is as halide ions.

The halogens are produced commercially in the following ways:

1. Fluorine. Fluorine must be prepared by an electrochemical process because no suitable chemical agent is sufficiently powerful to oxidize fluoride ion to fluorine. Furthermore, since the oxidation of water is easier to accomplish than the oxidation of the fluoride ion, the electrolysis must be carried out under anhydrous conditions. In actual practice, a solution of potassium fluoride in anhydrous hydrogen fluoride is electrolyzed. Pure HF does not conduct electric current; the KF reacts with HF to produce ions (K^+ and HF_2^-) that can act as charge carriers. The HF_2^- ion is formed from a F^- ion hydrogen-bonded to a HF molecule ($F-H\cdots F^-$); this hydrogen bond is so strong that the H atom is exactly midway between the two F atoms. The hydrogen fluoride used in the electrolysis is commercially derived from fluorospar, CaF_2 (see Section 19.10):

$$2\,KHF_2(l) \xrightarrow[\text{heat}]{electrolysis} H_2(g) + F_2(g) + 2\,KF(l)$$

Table 19.3	Occurrence of the halogens	
Element	Percent of Earth's Crust	Occurrence
fluorine	6.5×10^{-2}	CaF_2 (fluorospar), Na_3AlF_6 (cryolite), $Ca_5(PO_4)_3F$ (fluorapatite)
chlorine	5.5×10^{-2}	Cl^- (sea water and underground brines) NaCl (rock salt)
bromine	1.6×10^{-4}	Br^- (sea water, underground brines, solid salt beds)
iodine	3.0×10^{-5}	I^- (oil-well brines, sea water) $NaIO_3$, $Na O_4$ (impurities in Chilean saltpeter, $NaNO_3$)

Cells used for the electrolysis of sodium chloride brine. *Hooker Chemical Company.*

2. Chlorine. The principal industrial source of chlorine is the electrolysis of aqueous sodium chloride, from which sodium hydroxide and hydrogen are also products:

$$2\,Na^+(aq) + 2\,Cl^-(aq) + 2\,H_2O \xrightarrow{\text{electrolysis}}$$
$$H_2(g) + Cl_2(g) + 2\,Na^+(aq) + 2\,OH^-(aq)$$

Chlorine is also obtained, as a by-product, from the industrial processes in which the reactive metals Na, Ca, and Mg are prepared. In each of these processes an anhydrous molten chloride is electrolyzed and Cl_2 gas is produced at the anode. For example,

$$2\,NaCl(l) \xrightarrow{\text{electrolysis}} 2\,Na(l) + Cl_2(g)$$

3. Bromine. Bromine is commercially prepared by the oxidation of the bromide ion of salt brines or sea water with chlorine as the oxidizing agent:

$$Cl_2(g) + 2\,Br^-(aq) \longrightarrow 2\,Cl^-(aq) + Br_2(l)$$

The liberated Br_2 is removed from the solution by a stream of air and, in subsequent steps, collected from the air and purified.

4. Iodine. In the United States the principal source of iodine is the iodide ion found in oil-well brines. Free iodine is obtained by chlorine displacement:

$$Cl_2(g) + 2I^-(aq) \longrightarrow 2Cl^-(aq) + I_2(s)$$

In addition, iodine is commercially obtained from the iodate impurity found in Chilean nitrates. Sodium bisulfite is used to reduce the iodate ion:

$$2IO_3^-(aq) + 5HSO_3^-(aq) \longrightarrow I_2(s) + 5SO_4^{2-}(aq) + 3H^+(aq) + H_2O$$

19.8 Laboratory Preparation of the Halogens

With the exception of fluorine (which must be prepared electrochemically), the free halogens are usually prepared in the laboratory by the action of oxidizing agents on aqueous solutions of the hydrogen halides or on solutions containing the sodium halides and sulfuric acid.

From a table of standard electrode potentials we can get an approximate idea of what oxidizing agents will satisfactorily oxidize a given halide ion. Thus, any couple with a standard electrode potential more positive than $+1.36$ V should oxidize the chloride ion ($\mathscr{E}^\circ_{red} = +1.36$ V) as well as the bromide ion ($\mathscr{E}^\circ_{red} = +1.07$ V) and the iodide ion ($\mathscr{E}^\circ_{red} = +0.54$ V). Recall, however, that standard electrode potentials are listed for half-reactions at $25°C$ with all materials in their standard states. Thus, even though the standard electrode potential of $MnO_2 \rightarrow Mn^{2+}$ is only $+1.23$ V, MnO_2 is capable of oxidizing the chloride ion if concentrated HCl is used (rather than HCl at unit activity) and if the reaction is heated. In actual practice, $KMnO_4$, $K_2Cr_2O_7$, PbO_2, and MnO_2 are frequently used to prepare the free halogens from halide ions:

$$MnO_2(s) + 4H^+(aq) + 2Cl^-(aq) \longrightarrow Mn^{2+}(aq) + Cl_2(g) + 2H_2O$$

$$2MnO_4^-(aq) + 16H^+(aq) + 10Br^-(aq) \longrightarrow 2Mn^{2+}(aq) + 5Br_2(l) + 8H_2O$$

$$Cr_2O_7^{2-}(aq) + 14H^+(aq) + 6I^-(aq) \longrightarrow 2Cr^{3+}(aq) + 3I_2(s) + 7H_2O.$$

19.9 The Interhalogen Compounds

Some of the reactions of the halogens are summarized in Table 19.4. The halogens react with each other to produce a number of interhalogen compounds. All the compounds with the formula XX' (such as BrCl) are known except IF. Four XX'_3 compounds have been prepared (ClF_3, BrF_3, ICl_3, and IF_3), and three XX'_5 compounds are known (ClF_5, BrF_5, and IF_5). The only XX'_7 compound that has been made is IF_7.

With the exception of ICl_3, all the molecules for which n of the formula XX'_n is greater than 1 are halogen fluorides in which fluorine atoms (the smallest and most electronegative of all the halogen atoms) surround a Cl, Br, or I atom. The stability of these compounds increases as the size of the central atom increases.

Table 19.4 Some reactions of the halogens ($X_2 = F_2$, Cl_2, Br_2, or I_2)

General Reaction	Remarks
$nX_2 + 2M \longrightarrow 2MX_n$	F_2, Cl_2 with practically all metals; Br_2, I_2 with all except noble metals
$X_2 + H_2 \longrightarrow 2HX$	
$3X_2 + 2P \longrightarrow 2PX_3$	with excess P; similar reactions with As, Sb, and Bi
$5X_2 + 2P \longrightarrow 2PX_5$	with excess X_2, but not with I_2; SbF_5, $SbCl_5$, AsF_5, $AsCl_5$, and BiF_5 may be similarly prepared
$X_2 + 2S \longrightarrow S_2X_2$	with Cl_2, Br_2
$X_2 + H_2O \longrightarrow H^+ + X^- + HOX$	not with F_2
$2X_2 + 2H_2O \longrightarrow 4H^+ + 4X^- + O_2$	F_2 rapidly; Cl_2, Br_2 slowly in sunlight
$X_2 + H_2S \longrightarrow 2HX + S$	
$X_2 + CO \longrightarrow COX_2$	Cl_2, Br_2
$X_2 + SO_2 \longrightarrow SO_2X_2$	F_2, Cl_2
$X_2 + 2X'^- \longrightarrow X_2' + 2X^-$	$F_2 > Cl_2 > Br_2 > I_2$
$X_2 + X_2' \longrightarrow 2XX'$	formation of the interhalogen compounds (all except IF)

Hence, neither BrF_7 nor ClF_7 has been prepared, although IF_7 is known; ClF_5 readily decomposes into ClF_3 and F_2, whereas BrF_5 and IF_5 are stable at temperatures above 400°C.

The structures of the higher interhalogen compounds have received considerable attention because the bonding of the central atom in each of these molecules violates the octet principle (see Figure 19.1). The central atom of each XX_3' molecule has three bonding pairs and two nonbonding pairs of electrons in its valence shell, and the molecules, therefore, are T-shaped. The XX_5' molecules are square pyramidal since each of the central atoms has five bonding pairs and one nonbonding pair of electrons in its valence shell. The nonbonding electron pairs of the central atoms of these two types of molecules introduce some distortion. The I atom of the IF_7 molecule has seven bonding pairs of electrons in its valence shell. The IF_7 molecule is pentagonal bipyramidal.

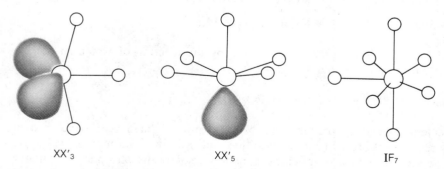

XX_3' XX_5' IF_7

Figure 19.1 Structures of XX_3', XX_5', and IF_7. (The unshared pairs in XX_3' and XX_5' molecules are responsible for some distortion from the regular T-shaped and square pyramidal structures.)

19.10 The Hydrogen Halides

Each of the hydrogen halides may be prepared by the direct reaction of hydrogen with the corresponding free halogen:

$$H_2 + X_2 \longrightarrow 2\,HX$$

The vigor of the reaction decreases markedly from fluorine to iodine. The reactions serve as important industrial sources of HCl, HBr, and HI.

Both HF and HCl can be prepared by the action of warm concentrated sulfuric acid on the corresponding natural halide, CaF_2 and NaCl. Both reactions serve as important industrial sources of the gases:

$$CaF_2(s) + H_2SO_4(l) \longrightarrow CaSO_4(s) + 2\,HF(g)$$

$$NaCl(s) + H_2SO_4(l) \longrightarrow NaHSO_4(s) + HCl(g)$$

All hydrogen halides are colorless gases at room temperature; sulfuric acid, on the other hand, is a high-boiling liquid. Thus, the foregoing reactions are examples of a general method for the preparation of a volatile acid from its salts by means of a nonvolatile acid. At higher temperatures (about 500°C), further reaction occurs between $NaHSO_4$ and NaCl:

$$NaCl(s) + NaHSO_4(l) \longrightarrow HCl(g) + Na_2SO_4(s)$$

Hydrogen bromide and hydrogen iodide cannot be made by the action of concentrated sulfuric acid on bromides and iodides because hot, concentrated sulfuric acid oxidizes these anions to the free halogens. The bromide and iodide ions are easier to oxidize than the fluoride and chloride ions:

$$2\,NaBr(s) + 2\,H_2SO_4(l) \longrightarrow Br_2(g) + SO_2(g) + Na_2SO_4(s) + 2\,H_2O(g)$$

Since the iodide ion is a stronger reducing agent (more easily oxidized) than the bromide ion, S and H_2S, as well as SO_2, are obtained as reduction products from the reaction of NaI with hot concentrated sulfuric acid.

Pure HBr or HI can be obtained by the action of phosphoric acid on NaBr or NaI; phosphoric acid is an essentially nonvolatile acid and is a poor oxidizing agent:

$$NaBr(s) + H_3PO_4(l) \longrightarrow HBr(g) + NaH_2PO_4(s)$$

$$NaI(s) + H_3PO_4(l) \longrightarrow HI(g) + NaH_2PO_4(s)$$

The hydrogen halides may be prepared by the reaction of water on the appropriate phosphorus trihalide:

$$PX_3 + 3\,H_2O \longrightarrow 3\,HX(g) + H_3PO_3(aq)$$

Convenient laboratory preparations of HBr and HI have been developed in which red phosphorus, bromine or iodine, and a limited amount of water are employed and in which no attempt is made to isolate the phosphorus trihalide intermediate.

Hydrogen fluoride molecules associate with each other through hydrogen bonding. The vapor consists of aggregates up to $(HF)_6$ at temperatures near the boiling point (19.4°C) but is less highly associated at higher temperatures. Gaseous HCl, HBr, and HI consist of single molecules. Liquid HF and solid HF are more highly hydrogen bonded than gaseous HF, and the boiling point and melting point of HF are abnormally high in comparison with those of the other hydrogen halides.

All hydrogen halides are very soluble in water; water solutions are called hydrohalic acids. Aqueous HI, for example, is called hydroiodic acid. The H—F bond is stronger than any other H—X bond; HF is a weak acid in water solution, whereas HCl, HBr, and HI are completely dissociated:

$$HF(aq) \rightleftharpoons H^+(aq) + F^-(aq)$$

The F^- ions from this dissociation are largely associated with HF molecules:

$$F^-(aq) + HF(aq) \rightleftharpoons HF_2^-(aq)$$

Concentrated HF solutions are more strongly ionic than dilute solutions and contain high concentrations of ions of the type HF_2^-, $H_2F_3^-$, and higher.

Hydrofluoric acid reacts with silica, SiO_2, and glass, which is made from silica:

$$SiO_2(s) + 6\,HF(aq) \longrightarrow 2\,H^+(aq) + SiF_6^{2-}(aq) + 2\,H_2O$$

When warmed, the reaction is

$$SiO_2(g) + 4\,HF(aq) \longrightarrow SiF_4(g) + 2\,H_2O(g)$$

For this reason, hydrofluoric acid must be stored in wax or plastic containers instead of glass bottles.

The fluorosilicate ion, SiF_6^{2-}, is an example of a large group of complex ions formed by the halide ions. Halo complexes are formed by most metals (with the notable exceptions of the group I A, group II A, and lanthanide metals) and with some nonmetals (for example, BF_4^-). The formulas of these complex ions are most commonly of the types $(MX_4)^{(4-)+n}$ and $(MX_6)^{(6-)+n}$, where n is the oxidation number of the central atom of the complex.

19.11 The Metal Halides

Metal halides can be prepared by direct interaction of the elements, by the reactions of the hydrogen halides with hydroxides or oxides, and by the reactions of the hydrogen halides with carbonates:

$$K_2CO_3(s) + 2\,HF(l) \longrightarrow 2\,KF(s) + CO_2(g) + H_2O$$

The character of the bonding in metal halides varies widely as do the physical properties of these compounds. A metal that has a low ionization energy generally forms halides that are highly ionic and consequently have high melting and boiling points. On the other hand, metals that have comparatively high ionization energies react, particularly with bromine and iodine, to form halides in which the bonding has a high degree of covalent character. These compounds have comparatively low melting points and boiling points.

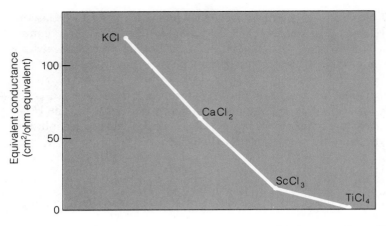

Figure 19.2 Equivalent conductances of some molten chlorides near the melting point

In general, the halides of the I A metals, the II A metals (with the exception of Be), and most of the inner-transition metals are largely ionic; halides of the remaining metals are covalent to a varying degree. Within a series of halides of metals of the same period, ionic character decreases from compound to compound as the oxidation number of the metal increases and the size of the cation undergoes a concomitant decrease. The ionic character of a compound is reflected in its ability to conduct electric current as measured by its conductance. The equivalent conductances of the molten chlorides of some fourth-period metals are plotted in Figure 19.2. The "cations" of the compounds (K^+, Ca^{2+}, Sc^{3+}, and Ti^{4+}) are isoelectronic. Potassium chloride is a completely ionic solid, and molten KCl has the highest conductance of the four compounds. Titanium(IV) chloride, $TiCl_4$, is a covalent liquid and is nonconducting.

If a metal exhibits more than one oxidation state, the halides of the highest oxidation state are the most covalent. Thus, the second member of each of the following pairs is a highly covalent, volatile liquid (melting points are given in parentheses): $SnCl_2$ (246°C), $SnCl_4$ (−33°C); $PbCl_2$ (501°C), $PbCl_4$ (−15°C); $SbCl_3$ (73.4°C), $SbCl_5$ (2.8°C).

Since fluorine is the most electronegative halogen, fluorides are the most ionic of the halides; ionic character decreases in the order: fluoride > chloride > bromide > iodide. The halides of aluminum are an excellent example of the relationship between ionic or covalent character and the size of the halide ion. Aluminum fluoride is an ionic substance. Aluminum chloride is semicovalent and crystallizes in a layer lattice in which electrically neutral layers are held together by London forces. Aluminum bromide and aluminum iodide are essentially covalent; the crystals consist of Al_2Br_6 and Al_2I_6 molecules, respectively.

The water solubility of fluorides is considerably different from that of the chlorides, bromides, and iodides. The fluorides of lithium, the group II A metals, and the lanthanides are only slightly soluble, whereas the other halides of these metals are relatively soluble.

Most chlorides, bromides, and iodides are soluble in water. The cations that form slightly soluble compounds with these halide ions include silver(I), mercury(I), lead(II), copper(I), and thallium(I).

The insolubility of the silver salts of Cl^-, Br^-, and I^- is the basis of a common test for these halide ions; AgCl is white, AgBr is cream, and AgI is yellow. The

Table 19.5 Oxyhalogen acids

Oxidation State of the Halogen	Formula of Acid				Name of Acid	Name of Anion Derived from Acid
1+	HOF	HOCl	HOBr	HOI	hypohalous acid	hypohalite ion
3+	–	$HClO_2$	–	–	halous acid	halite ion
5+	–	$HClO_3^a$	$HBrO_3^a$	HIO_3^a	halic acid	halate ion
7+	–	$HClO_4^a$	$HBrO_4^a$	$\begin{cases} HIO_4 \\ H_4I_2O_9 \\ H_5IO_6 \end{cases}$	perhalic acid	perhalate ion

^{a} Strong acids.

silver halide precipitates may be formed by the addition of a solution of silver nitrate to a solution containing the appropriate halide ion. Silver iodide is insoluble in excess ammonia; however, AgCl readily dissolves to form the $Ag(NH_3)_2^+$ complex ion and AgBr dissolves with difficulty. Silver fluoride is soluble. Precipitates of MgF_2 or CaF_2 are usually used to confirm the presence of the fluoride ion in a solution.

If the iodide ion in an aqueous solution is oxidized to iodine (the usual procedure employs chlorine), the I_2 can be extracted by cyclohexane, which forms a two-liquid-layer system with water. The solution of I_2 in cyclohexane is violet colored. In the corresponding test for the bromide ion, the Br_2-cyclohexane solution is brown.

19.12 Oxyacids of the Halogens

The oxyacids of the halogens are listed in Table 19.5. The only oxyacid of fluorine that has been prepared is hypofluorous acid, HOF, a thermally unstable compound. The acids of chlorine and their salts are the most important of these compounds.

Lewis formulas for the oxychlorine acids are shown in Figure 19.3; removal of H^+ from each of these structures gives the electronic formula of the corresponding anion. However, Lewis structures, in which each Cl and O atom has a valence-electron octet, do not show that the Cl—O bonds in these compounds can have a considerable amount of double-bond character due to $p\pi$-$d\pi$ bonding (see Section 7.6).

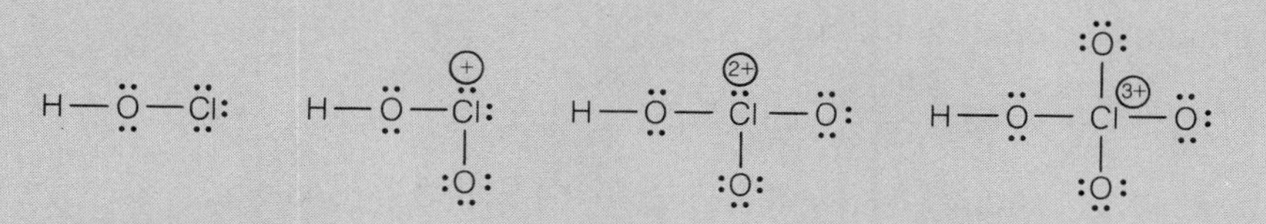

Figure 19.3 Lewis formulas for the oxychlorine acids

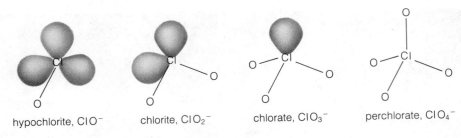

hypochlorite, ClO^- chlorite, ClO_2^- chlorate, ClO_3^- perchlorate, ClO_4^-

Figure 19.4 Structures of the oxychlorine anions

The oxybromine and oxyiodine anions have configurations similar to the analogous oxychlorine anions shown in Figure 19.4. The H_5IO_6 molecule is octahedral with five OH groups and one O atom in the six positions surrounding the central I atom.

The standard electrode potentials for Cl_2, Br_2, I_2, and the compounds of these elements in acidic and alkaline solution are summarized in Figure 19.5. The number shown over each arrow is the $\mathscr{E}^\circ_{red}$, in volts, for the reduction of the species on the left to that on the right. An oxidation potential for a transformation from right to left may, of course, be obtained by changing the sign of $\mathscr{E}^\circ_{red}$.

Perchloric acid ($HClO_4$), perbromic acid ($HBrO_4$), and the halic acids ($HClO_3$, $HBrO_3$, and HIO_3) are strong acids, but the remaining oxyacids are incompletely dissociated in water solution and exist in solution largely in molecular form. Hence, molecular formulas are shown in Figure 19.5 for the weak acids in acidic solution. In general, acid strength increases with increasing oxygen content.

Much of the chemistry of these compounds is effectively correlated by means of these electrode potential diagrams. Remember, however, that standard electrode potentials refer to reductions that take place at 25°C with all substances present in their standard states. Concentration changes and temperature changes alter $\mathscr{E}^\circ$ values. Furthermore, a cell emf tells nothing about the speed of the transformation to which it applies; some reactions for which the cell emfs are positive occur so slowly that they are of no practical importance. According to $\mathscr{E}^\circ_{red}$ values, the oxidation of water (in acid) and of the OH^- ion (in base) should be readily accomplished by many oxyhalogen compounds. These reactions, however, occur so slowly that it is possible to observe all the oxyhalogen compounds in solution.

One of the outstanding characteristics of the oxyhalogen compounds is their ability to function as oxidizing agents; all the $\mathscr{E}^\circ_{red}$ values are positive. The $\mathscr{E}^\circ_{red}$ values also show that, in general, these compounds are stronger oxidizing agents in acidic solution than in alkaline solution.

Many of these oxyhalogen compounds are unstable toward disproportionation. Such materials are readily identified from the diagrams of Figure 19.5. For example, in alkaline solution:

$$ClO^- \xrightarrow{+0.40} Cl_2 \xrightarrow{+1.36} Cl^-$$

Since $+1.36$ is more positive than $+0.40$, Cl_2 will disproportionate:

$$
\begin{array}{ll}
4\,OH^- + Cl_2 \longrightarrow 2\,ClO^- + 2\,H_2O + 2e^- & \mathscr{E}^\circ_{ox} = -0.40\ V \\
\underline{2e^- + Cl_2 \longrightarrow 2\,Cl^-} & \underline{\mathscr{E}^\circ_{red} = +1.36\ V} \\
2\,OH^- + Cl_2 \longrightarrow ClO^- + Cl^- + H_2O & \mathscr{E}_{emf} = +0.96\ V
\end{array}
$$

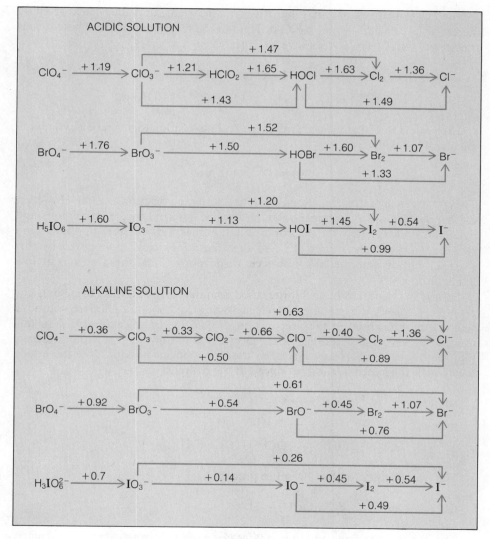

Figure 19.5 Standard electrode potential diagrams for chlorine, bromine, iodine, and their compounds

The $\mathscr{E}^\circ_{red}$ values indicate that HOCl, HOBr, HOI, HClO$_2$, and ClO$_3^-$ should disproportionate in acidic solution. In alkaline solution all species should disproportionate except the halide ions, ClO$_4^-$, BrO$_4^-$, BrO$_3^-$, H$_3$IO$_6^{2-}$, and IO$_3^-$.

1. Hypohalous acids and hypohalites. The hypohalous acids (HOX) are the weakest halogen oxyacids. They exist in solution but cannot be prepared in pure form. Each of the three halogens is slightly soluble in water and reacts to produce a low concentration of the corresponding hypohalous acid:

$$X_2 + H_2O \rightleftharpoons H^+(aq) + X^-(aq) + HOX(aq)$$

All standard potentials for these reactions are negative, and the reactions proceed to a very limited extent. Thus, at 25°C saturated solutions of the halogens contain the following HOX concentrations: HOCl, 3×10^{-2} M; HOBr, 1×10^{-3} M; HOI, 6×10^{-6} M.

The X_2-H_2O reactions may be driven to the right, and the yields of HOX increased, by the addition of Ag_2O or HgO to the reaction mixture. These oxides precipitate the X^- ion and remove $H^+(aq)$:

$$2X_2 + Ag_2O(s) + H_2O \longrightarrow 2AgX(s) + 2HOX(aq)$$

$$2X_2 + 4HgO(s) + H_2O \longrightarrow HgX_2 \cdot 3HgO(s) + 2HOX(aq)$$

Each of the three halogens dissolves in alkaline solution to produce a hypohalite ion and a halide ion:

$$X_2 + 2OH^-(aq) \longrightarrow X^-(aq) + XO^-(aq) + H_2O$$

The emf for each of these reactions is positive, and each reaction is rapid. However, the hypohalite ions disproportionate in alkaline solution to the halate ions and halide ions. Fortunately the disproportionation reactions of ClO^- and BrO^- are slow at temperatures around 0°C, so that these ions can be prepared by this method. The hypoiodite ion, however, disproportionates rapidly even at low temperatures.

Solutions of sodium hypochlorite, used commercially in cotton bleaching (for example, Clorox), are prepared by electrolyzing cold sodium chloride solutions such as the ones used in the preparation of chlorine (see Section 19.7). In this process, however, the products of the electrolysis are not kept separate. Rather, the electrolyte is vigorously mixed so that the chlorine produced at the anode reacts with the hydroxide ion produced at the cathode:

anode: $\qquad\qquad 2Cl^- \longrightarrow Cl_2 + 2e^-$

cathode: $\quad 2e^- + 2H_2O \longrightarrow 2OH^- + H_2$

$\qquad\qquad\quad Cl_2 + 2OH^- \longrightarrow OCl^- + Cl^- + H_2O$

The overall equation for the entire process is

$$Cl^- + H_2O \xrightarrow[cold]{electrolysis} OCl^- + H_2$$

The hypohalous acids and hypohalites are good oxidizing agents, particularly in acidic solution. The compounds decompose not only by disproportionation but also by the liberation of O_2 from the solutions. The reactions in which O_2 is liberated are slow, however, and are catalyzed by metal salts.

2. Halous acids and halites. The only known halous acid is chlorous acid ($HClO_2$). This compound cannot be isolated in pure form, and even in aqueous solution it decomposes rapidly. Chlorous acid is weak acid, but it is stronger than hypochlorous acid. The disproportionation of HOCl in acidic solution does not yield $HClO_2$. The electrode potentials indicate, however, that it should be possible to make the chlorite ion by the disproportionation of ClO^- in alkaline solution. The disproportionation of ClO^- into ClO_3^- and Cl^-, however, is more favorable, and ClO_2^- cannot be made from ClO^-.

Chlorites are comparatively stable in alkaline solution and may be prepared from ClO_2 gas. Chlorine dioxide, a very reactive odd-electron molecule, is prepared by the reduction of chlorates in aqueous solution using sulfur dioxide gas as the reducing agent:

$$2ClO_3^-(aq) + SO_2(g) \longrightarrow ClO_2(g) + SO_4^{2-}(aq)$$

The chlorite ion is prepared by the reaction of ClO_2 gas with an alkaline solution of sodium peroxide (which forms the hydroperoxide ion HO_2^- in alkaline solution):

$$2\,ClO_2(g) + HO_2^-(aq) + OH^-(aq) \longrightarrow 2\,ClO_2^-(aq) + O_2(g) + H_2O$$

Solid chlorites are dangerous chemicals. They detonate when heated and are explosive in contact with combustible material.

3. Halic acids and halates. We have mentioned the disproportionation reactions of the hypohalite ions:

$$3\,XO^- \longrightarrow XO_3^- + 2\,X^-$$

Thus if a free halogen is added to a *hot*, concentrated solution of alkali, the corresponding halide and halate ions are produced rather than the halide and hypohalite ions:

$$3\,X_2 + 6\,OH^-(aq) \longrightarrow 5\,X^-(aq) + XO_3^-(aq) + 3\,H_2O$$

The chlorates are commercially prepared by the electrolysis of hot, concentrated solutions of chlorides (instead of the cold solutions used for the electrochemical preparation of the hypochlorites). The electrolyte is stirred vigorously so that the chlorine produced at the anode reacts with the hydroxide ion that is a product of the reduction at the cathode:

anode: $\qquad\qquad 2\,Cl^- \longrightarrow Cl_2 + 2e^-$

cathode: $\quad 2e^- + 2\,H_2O \longrightarrow 2\,OH^- + H_2$

$$3\,Cl_2 + 6\,OH^- \longrightarrow 5\,Cl^- + ClO_3^- + 3\,H_2O$$

If the three foregoing equations are added after the first two equations have each been multiplied through by 3, the equation for the overall process is obtained:

$$Cl^- + 3\,H_2O \xrightarrow[\text{hot}]{\textit{electrolysis}} ClO_3^- + 3\,H_2$$

The chlorate crystallizes from the concentrated solution employed as the electrolyte of the cell. Although chlorates are generally water soluble, they are much less soluble than the corresponding chlorides.

Solutions of a halic acid can be prepared by adding sulfuric acid to a solution of the barium salt of the acid:

$$Ba^{2+}(aq) + 2\,XO_3^-(aq) + 2\,H^+(aq) + SO_4^{2-}(aq) \longrightarrow$$
$$BaSO_4(s) + 2\,H^+(aq) + 2\,XO_3^-(aq)$$

Pure $HBrO_3$ or $HClO_3$ cannot be isolated from aqueous solutions because of decomposition. Iodic acid, HIO_3, however, can be obtained as a white solid; this acid is generally prepared by oxidizing iodine with concentrated nitric acid. All halic acids are strong. The halates, as well as the halic acids, are strong oxidizing agents. The reactions of chlorates with easily oxidized materials may be explosive.

Chlorates decompose upon heating in a variety of ways. At high temperatures, and particularly in the presence of a catalyst, chlorates decompose into chlorides and oxygen:

$$2\,KClO_3(s) \xrightarrow[MnO_2]{heat} 2\,KCl(s) + 3\,O_2(g)$$

At more moderate temperatures, in the absence of a catalyst, the decomposition yields perchlorates and chlorides:

$$4\,KClO_3(s) \xrightarrow{heat} 3\,KClO_4(s) + KCl(s)$$

4. Perhalic acids and perhalates. Perchlorate salts are made by the controlled thermal decomposition of a chlorate or by the electrolysis of a cold solution of a chlorate. The free acid, a clear hygroscopic liquid, may be prepared by distilling a mixture of a perchlorate salt with concentrated sulfuric acid. A number of crystalline hydrates of perchloric acid are known. The compound $HClO_4 \cdot H_2O$ is of interest; the lattice positions of the crystal are occupied by H_3O^+ and ClO_4^- ions, and the solid is isomorphous (of the same crystalline structure) with NH_4ClO_4. Perchloric acid is a strong acid and a strong oxidizing agent. Concentrated $HClO_4$ may react violently when heated with organic substances; the reactions are frequently explosive.

Perbromates are prepared by the oxidation of bromates in alkaline solution using F_2 as the oxidizing agent. Several periodic acids have been prepared; the most common one is H_5IO_6. Periodates are made by the oxidation of iodates, usually by means of Cl_2 in alkaline solution. Salts of the anions IO_4^-, $I_2O_9^{4-}$, IO_6^{5-}, and IO_5^{3-} have been obtained.

19.13 Industrial Uses of the Halogens

The most important industrial uses of the halogens and halogen compounds are

1. Fluorine. The most important fluorides produced commercially are synthetic cryolite and the fluorocarbons. Cryolite, Na_3AlF_6, is used as a supporting electrolyte in the electrolysis of molten Al_2O_3 for the production of aluminum (Hall process). The fluorocarbons are noted for their chemical inertness. The Freons (for example, CCl_2F_2) are used as refrigerants and aerosol propellants. Polytetrafluoroethylene (polymerized $F_2C{=}CF_2$, Teflon) is a solid fluorocarbon polymer that is highly resistant to chemical attack. Liquid fluorocarbons are used as chemically resistant lubricants.

The principal use of elemental fluorine is in the separation of ^{235}U (the isotope of uranium that undergoes atomic fission) from natural uranium (which is mostly ^{238}U). Uranium metal (which contains both isotopes) is converted into UF_6 (which sublimes at $56°C$). Since $^{235}UF_6$ has a lower molecular weight than $^{238}UF_6$, the $^{235}UF_6$ vapor effuses through a porous barrier more rapidly than the $^{238}UF_6$ does and a separation of the two isotopes is effected by this means.

Minor but well publicized uses of fluorides (principally NaF), are their addition to water supplies and toothpaste to prevent tooth decay.

2. Chlorine. A large number of chlorine-containing compounds are produced commercially. Most of these compounds are organic compounds that are made by use of chlorine or hydrogen chloride. They are used, for example, as plastics, solvents, pesticides, herbicides, pharmaceuticals, refrigerants, and dyes. Large quantities of HCl are produced to be used not only in the synthesis of organic products, but also in petroleum technology, metallurgy, metal cleaning (to remove metal oxides from metals), food processing, and in the manufacture of inorganic chlorides. Chlorine is used in the manufacture of paper, rayon, hydrogen chloride, bromine, iodine, sodium hypochlorite, and metal chlorides, in the disinfection of water, and in the bleaching of textiles.

3. Bromine. The principal use of bromine, at present, is in the manufacture of ethylene dibromide (CH_2BrCH_2Br) which is used with the anti-knock agent tetraethyl lead in leaded gasolines. Ethylene dibromide supplies bromine to convert the lead into $PbBr_2$ which is volatile at cylinder combustion temperatures and is swept out of the engine in the exhaust. The importance of this use of bromine is declining since antipollution laws forbid the use of leaded gasoline in new cars.

Bromine-containing organic compounds are used as intermediates in industrial syntheses, dyes, pharmaceuticals, fumigants, and fire-proofing agents. Inorganic bromides are used in medicine, in bleaching, and in photography (AgBr).

4. Iodine. Iodine and its compounds are not used as extensively as the other halogens and halides. Significant uses include the production of pharmaceuticals, dyes, and silver iodide (for photography).

Summary

The topics that have been discussed in this chapter are

1. The occurrence and physical properties of hydrogen.
2. The industrial production of hydrogen.
3. Displacement reactions in which hydrogen is liberated.
4. The reactions of hydrogen; the hydrides.
5. Commercial uses of hydrogen.
6. The halogens: physical properties, occurrence, industrial and laboratory preparations.
7. Compounds of the halogens: hydrogen halides, metal halides, oxyacids and their salts, electrode-potential diagrams.
8. Industrial uses of the halogens and their compounds.

Key Terms

Some of the more important terms introduced in this chapter are listed below. Definitions for terms not included in this list may be located in the text by use of the index.

Cracking (Section 19.2) A process in which high molecular weight hydrocarbons are broken down into lower molecular weight compounds; used in petroleum refining.

Displacement reaction (Section 19.3) A reaction in which one element (or group of elements) displaces another element (or group of elements) from a compound

Halogen (Section 19.6) A group VII A element; F, Cl, Br, I, or At.

Interhalogen compound (Section 19.9) A compound that is composed of two different halogens.

Steam reformer process (Section 19.2) An industrial process used to prepare hydrogen by the reaction of steam with a hydrocarbon.

Water gas (Section 19.2) A mixture of carbon monoxide and hydrogen produced industrially by the reaction of steam and coke.

Problems*

Hydrogen

19.1 What are the principal ways in which hydrogen occurs in nature? Discuss three methods by which hydrogen is prepared from water industrially.

19.2 Write chemical equations for the preparation of hydrogen from **(a)** Na(s) and H_2O, **(b)** Fe(s) and steam, **(c)** Zn(s) and H^+(aq), **(d)** Zn(s) and OH^-(aq), **(e)** C(s) and steam, **(f)** CH_4(g) and steam, **(g)** CaH_2(s) and H_2O.

19.3 Write chemical equations for the reactions of hydrogen with **(a)** Na(s), **(b)** Ca(s), **(c)** Cl_2(g), **(d)** N_2(g), **(e)** Cu_2O(s), **(f)** CO(g), **(g)** WO_3(s).

19.4 Describe the properties and structures of **(a)** saltlike hydrides, **(b)** interstitial hydrides, **(c)** complex hydrides, **(d)** covalent hydrides.

19.5 How do the physical properties of hydrogen reflect the nature of the London forces between H_2 molecules?

19.6 What accounts for the catalytic activity of Pd, Pt, and Ni in many reactions that involve hydrogen?

19.7 What mass of hydrogen can be obtained by the reaction of excess water with **(a)** 10.0 g of Ca(s), **(b)** 10.0 g of CaH_2(s)?

19.8 (a) What is the mass of 22.4 liters of H_2(g) at STP? **(b)** Assume that air is 22.0% O_2(g) and 78.0% N_2(g) and calculate the mass of 22.4 liters of air at STP. **(c)** If a 22.4-liter balloon were filled with hydrogen at STP, the difference between the mass of 22.4 liters of air and the mass of 22.4 liters of hydrogen would be the approximate value of the lifting power of the balloon. Calculate this value. Approximately how many times its own mass can a sample of hydrogen lift?

19.9 Calculate the mass of **(a)** Al, **(b)** Zn, and **(c)** Sn required to liberate 1.000 g of hydrogen from excess acid.

19.10 Use the bond energies found in Table 3.2 to calculate the standard enthalpy of formation of **(a)** HF(g), **(b)** H_2O(g), **(c)** NH_3(g).. **(d)** Compare your values with those listed in Table 3.1.

19.11 Use the following data to calculate the lattice energy of NaH(s): The enthalpy of formation of NaH(s) is -57.3 kJ/mol. The enthalpy of sublimation of Na(s) is $+108$ kJ/mol, and the first ionization energy of Na(g) is $+496$ kJ/mol. The bond energy of H_2(g) is $+435$ kJ/mol, and the first electron affinity of H(g) is -73 kJ/mol. Compare the value you get with the lattice energy of NaCl(s), which is -789 kJ/mol.

19.12 (a) What mass of hydrogen would theoretically be required to reduce 1.00 kg of WO_3(s) to yield W(s)? **(b)** What volume would this mass of H_2(g) occupy at STP?

The Halogens

19.13 Write chemical equations to show how Cl_2(g) may be prepared from Cl^-(aq) by use of **(a)** MnO_2(s), **(b)** PbO_2(s), **(c)** MnO_4^-(aq), **(d)** $Cr_2O_7^{2-}$(aq). **(e)** Can analogous reactions be used to prepare fluorine, bromine, or iodine? Justify your answer.

19.14 Write chemical equations to show how to prepare the following: **(a)** F_2 from CaF_2, **(b)** Cl_2 from NaCl, **(c)** Br_2 from sea water, **(d)** I_2 from $NaIO_3$, **(e)** HBr from PBr_3.

19.15 Write chemical equations for the reactions of Cl_2 with **(a)** H_2, **(b)** Zn, **(c)** P, **(d)** S, **(e)** H_2S, **(f)** CO, **(g)** SO_2, **(h)** I^-(aq), **(i)** cold H_2O.

19.16 Write chemical equations for the reactions of HF with **(a)** SiO_2, **(b)** Na_2CO_3, **(c)** KF, **(d)** CaO.

19.17 Write equations to show why concentrated HF(aq) is more strongly ionic than dilute HF(aq).

19.18 Since HCl(g) can be prepared from NaCl and H_2SO_4, why is it that the reaction of NaBr and H_2SO_4 cannot be used to prepare HBr(g)?

19.19 From each of the following pairs of compounds, select the compound that would have the higher electrical conductivity in the molten state. Explain the basis for your prediction in each case. **(a)** $FeCl_2$ or $FeCl_3$, **(b)** RbCl or $SrCl_2$, **(c)** BeF_2 or $BeCl_2$.

19.20 Discuss the molecular geometry of the interhalogen compounds XX_3', XX_5', and XX_7'.

19.21 Write chemical equations for the electrolysis of **(a)** dry, molten NaCl, **(b)** cold NaCl solution, **(c)** cold NaCl solution with the electrolyte stirred, **(d)** hot, concentrated NaCl solution with the electrolyte stirred, **(e)** cold $NaClO_3$ solution.

19.22 Which is larger: **(a)** the acid strength of HF or of HCl, **(b)** the melting point of F_2 or of Cl_2, **(c)** the boiling point of HF or of HCl, **(d)** the first ionization energy of F_2 or of I_2, **(e)** the bond energy of F_2 or of Cl_2, **(f)** the bond energy of Cl_2 or of I_2?

19.23 Write chemical equations for the reactions used to identify each of the halide ions in aqueous solution.

19.24 How many faradays of electricity are needed to produce 1.000 kg of Cl_2(g) by the electrolysis of dry, molten NaCl?

19.25 How many grams of Cl_2(g) are produced in the same time that it takes to prepare 1.000 kg of NaOH by the electrolysis of an aqueous solution of NaCl?

19.26 How many grams of Cl_2(g) are produced in one hour by the electrolysis of an aqueous solution of NaCl if 1000 A of electricity are supplied?

---

* The appendix contains answers to color-keyed problems.

THE NONMETALS, PART II: THE GROUP VI A ELEMENTS

CHAPTER 20

Group VI A includes oxygen, sulfur, selenium, tellurium, and polonium. Oxygen is the most important and abundant of the group. Since the chemistry of oxygen is different from that of the other members, oxygen is considered separately before the others. Polonium is the product of the radioactive disintegration of radium. The most abundant isotope of polonium, ^{210}Po, has a half-life of only 138.7 days.

20.1 Properties of the Group VI A Elements

The electronic configurations of the group VI A elements are listed in Table 20.1. Each element is two electrons short of a noble-gas structure. Hence, these elements attain a noble-gas electronic configuration in the formation of ionic compounds by accepting two electrons per atom:

$$2\,Na^+ \qquad :\!\ddot{\underset{..}{S}}\!:^{2-}$$

The elements also acquire noble-gas configurations through covalent-bond formation:

$$H\!:\!\underset{\underset{H}{..}}{\overset{..}{Se}}\!:$$

Certain properties of the group VI A elements are summarized in Table 20.2. Each member of the group is a less active nonmetal than the halogen of its period. The electronegatives of the elements decrease, in the expected manner, with increasing atomic number. Oxygen is the second most electronegative element

Table 20.1	Electronic configurations of the group VI A elements																
Elements	Z	1s	2s	2p	3s	3p	3d	4s	4p	4d	4f	5s	5p	5d	6s	6p
O	8	2	2	4												
S	16	2	2	6	2	4										
Se	34	2	2	6	2	6	10	2	4							
Te	52	2	2	6	2	6	10	2	6	10		2	4			
Po	84	2	2	6	2	6	10	2	6	10	14	2	6	10	2	4

Table 20.2 Some properties of the group VI A elements				
Property	Oxygen	Sulfur	Selenium	Tellurium
color	colorless	yellow	red to black	silver-white
molecular formula	O_2	S_8 rings	Se_8 rings Se_n chains	Te_n chains
melting point (°C)	−218.4	119	217	452
boiling point (°C)	−182.9	444.6	688	1390
atomic radius (pm)	74	104	117	137
ionic radius (2− ion) (pm)	140	184	198	221
first ionization energy (kJ/mol)	1312	1004	946	870
electronegativity	3.4	2.6	2.6	2.1
bond energy (single bonds) (kJ/mol)	138	213	184	138
$\mathscr{E}°$ for reduction of element to H_2X in acid solution (V)	+1.23	+0.14	−0.40	−0.72

(fluorine is first); sulfur is about as electronegative as iodine. Thus, the oxides of most metals are ionic, whereas the sulfides, selenides, and tellurides of only the most active metals (such as the I A and II A metals) are truly ionic compounds.

The group VI A elements are predominantly nonmetallic in chemical behavior; however, metallic characteristics appear in the heavier members of the group. The trend in increasing metallic character parallels, as expected, increasing atomic number, increasing atomic radius, and decreasing ionization potential. Polonium is the most metallic member of the group; it appears to be capable of forming cations that exist in aqueous solution, and the 2− state of polonium (in H_2Po, for example) is unstable. Whereas tellurium is essentially nonmetallic in character, unstable salts of tellurium with anions of strong acids have been reported. The ordinary form of tellurium is metallic. Selenium exists in both metallic and non-metallic crystalline modifications.

Sulfur, selenium, and tellurium exist in positive oxidation states in compounds in which they are combined with more electronegative elements (such as oxygen and the halogens). Oxygen is considered to have a positive oxidation number only in the few compounds that it forms with fluorine. For sulfur, selenium, and tellurium, the oxidation states of 4+ and 6+ are particularly important.

The electrode potentials listed in Table 20.2 give an idea of the strength of the group VI A elements as oxidizing agents. Oxygen is a strong oxidizing agent, but there is a striking decrease in this property from oxygen to tellurium. In fact, H_2Te and H_2Se are better *reducing* agents than hydrogen. Compare the $\mathscr{E}°$ values listed in Table 20.2 with those given for the halogens in Table 19.2.

20.2 Occurrence and Industrial Production of Oxygen

Oxygen is the most abundant element (see Table 1.2). Free oxygen makes up about 21.0% by volume or 23.2% by mass of the atmosphere. Most minerals

Table 20.3 Composition of dry air

Substance	Percent by Volume	Substance	Percent by Volume
N_2	78.00	CH_4	2×10^{-4}
O_2	20.95	Kr	1×10^{-4}
Ar	0.93	N_2O	5×10^{-5}
CO_2	0.03	H_2	5×10^{-5}
Ne	0.0018	Xe	8×10^{-6}
He	0.0005	O_3	1×10^{-6}

contain combined oxygen. Silica, SiO_2, is a common ingredient of many minerals and the chief constituent of sand. Silicon is second to oxygen in the order of natural abundance because of the widespread occurrence of silica. Other oxygen-containing minerals are oxides, sulfates, and carbonates. Oxygen is a constituent of the compounds that make up plant and animal matter. The human body is more than 60% oxygen.

Three isotopes of oxygen occur in nature: ^{16}O (99.759%), ^{18}O (0.204%), and ^{17}O (0.037%). The isotopes ^{14}O, ^{15}O, ^{19}O, and ^{20}O are artificial and unstable.

The principal commercial source of oxygen is the atmosphere. Air is a mixture. The composition of air varies with altitude and to a lesser extent with location. The analysis of air is made after water and solid particles (such as dust and spores) have been removed. Some of the components of air are listed in Table 20.3. The percentages given (by volume) are for clean, dry air at sea level.

Over 99% of the oxygen produced industrially is obtained from the liquefaction and fractional distillation of air. In the process, filtered, dry air from which the CO_2 has been removed is liquefied by compression and cooling. When the air is allowed to warm, nitrogen (boiling point, $-196°C$) boils away from the oxygen (boiling point, $-183°C$). The noble gases are obtained from the nitrogen and oxygen fractions by repeated distillations and other separation techniques.

A small amount of very pure but relatively expensive oxygen is produced commercially by the electrolysis of water:

$$2 H_2O \xrightarrow{\text{electrolysis}} 2 H_2(g) + O_2(g)$$

20.3 Laboratory Preparation of Oxygen

Oxygen is usually prepared in the laboratory by the thermal decomposition of certain oxygen-containing compounds. The following are used:

1. Oxides of metals of low reactivity. The oxides of silver (Ag_2O), mercury (HgO), and gold (Au_2O_3) decompose on heating to give oxygen gas and the free metal:

$$2 HgO(s) \longrightarrow 2 Hg(l) + O_2(g)$$

2. Peroxides. Oxygen and the oxide ion are produced when the peroxide ion, O_2^{2-}, is heated:

$$2\left[:\ddot{O}:\ddot{O}:\right]^{2-} \longrightarrow 2\left[:\ddot{O}:\right]^{2-} + O_2$$

Thus,

$$2\,Na_2O_2(s) \longrightarrow 2\,Na_2O(s) + O_2(g)$$
$$2\,BaO_2(s) \longrightarrow 2\,BaO(s) + O_2(g)$$

Oxygen can be obtained from the reaction of sodium peroxide and water at room temperature. The other product is an aqueous solution of sodium hydroxide:

$$2\,O_2^{2-}(aq) + 2\,H_2O \longrightarrow 4\,OH^-(aq) + O_2(g)$$

3. Nitrates and chlorates. Certain other compounds release all or part of their oxygen upon heating. Nitrates of the I A metals form nitrites:

$$2\,NaNO_3(l) \longrightarrow 2\,NaNO_2(l) + O_2(g)$$

Potassium chlorate loses all its oxygen; a catalyst (MnO_2) is generally used to lower the temperature required for this decomposition:

$$2\,KClO_3(s) \longrightarrow 2\,KCl(s) + 3\,O_2(g)$$

20.4 Reactions of Oxygen

The reactions of oxygen are often more sluggish than would be predicted from the fact that oxygen has a high electronegativity (3.4), second in this property only to fluorine (4.0). The reason for this slowness is that the bond energy of oxygen is high (494 kJ/mol); therefore, reactions that require the oxygen-to-oxygen bond to be broken occur only at high temperatures. Many of these reactions are relatively highly exothermic and produce sufficient heat to sustain themselves after having once been initiated by external heating. Whether self-sustaining or not, most oxygen reactions occur at temperatures considerably higher than room temperature.

Oxygen forms four different anions: the superoxide, peroxide, oxide, and ozonide ions. Molecular orbital diagrams for oxygen, the superoxide ion, and the peroxide ion are given in Figure 20.1. The **superoxide ion**, O_2^-, can be considered to arise from the addition of one electron to the $\pi*2p$ orbital of the O_2 molecule, which reduces the number of unpaired electrons to 1 and the bond order to $1\frac{1}{2}$. The **peroxide ion**, O_2^{2-}, contains two more electrons (in the $\pi*2p$ orbitals) than the O_2 molecule; hence, the bond order is reduced to 1, and the ion is diamagnetic. The **oxide ion**, O^{2-}, is isoelectronic with neon and is diamagnetic. The **ozonide ion**, O_3^-, is paramagnetic with one unpaired electron and is produced by reactions of ozone, O_3, with hydroxides of K, Rb, and Cs (see Section 20.6).

All metals except the less-reactive metals (for example, Ag and Au) react with oxygen. Oxides of all metals are known but some must be made indirectly. The most reactive metals of group I A (and those with largest atomic radii)—Cs, Rb, and K—react with oxygen at atmospheric pressure to produce superoxides. For example,

$$Cs(s) + O_2(g) \longrightarrow CsO_2(s)$$

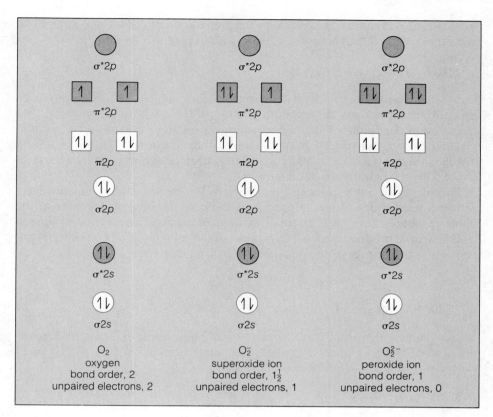

Figure 20.1 Molecular orbital energy-level diagrams for oxygen, the superoxide ion, and the peroxide ion

Sodium peroxide is produced by the reaction of sodium with oxygen:

$$2\,Na(s) + O_2(g) \longrightarrow Na_2O_2(s)$$

Lithium metal forms an ordinary oxide with O_2 rather than a peroxide or a superoxide because the small Li^+ ion cannot form a stable lattice with the larger O_2^{2-} or O_2^- ions:

$$4\,Li(s) + O_2(g) \longrightarrow 2\,Li_2O(s)$$

Generally, oxides form at much higher temperatures than either peroxides or superoxides. The ordinary oxides of Na, K, Rb, and Cs can be obtained by heating oxygen with an excess of the metal.

With the exception of barium (which reacts with oxygen to yield barium peroxide), the remaining metals generally produce normal oxides in their reactions with oxygen:

$$2\,Mg(s) + O_2(g) \longrightarrow 2\,MgO(s)$$

$$4\,Al(s) + 3\,O_2(g) \longrightarrow 2\,Al_2O_3(s)$$

Analogous reactions can be written for the preparation of CaO, CuO, ZnO, PbO, and other oxides. The reaction of mercury and oxygen is reversible:

$$2\,Hg(l) + O_2(g) \rightleftharpoons 2\,HgO(s)$$

For metals that have more than one electrovalence number, the oxide produced generally depends upon the quantity of oxygen, the quantity of the metal, and the reaction conditions. Thus, the reaction of iron and oxygen can be made to yield FeO (low pressure of oxygen, temperature above 600°C), Fe_3O_4 (finely divided iron, heated in air at 500°C), or Fe_2O_3 (iron heated in air at temperatures above 500°C). Hydrated Fe_2O_3 is iron rust.

Except for the noble gases and the group VII A elements, all nonmetals react with oxygen. Oxides of the halogens and those of the heavier members of the noble-gas family have been prepared by indirect means. The reaction of oxygen with hydrogen produces water. The product of the reaction of carbon with oxygen depends upon the proportion of carbon to oxygen employed:

$$2\,C(s) + O_2(g) \longrightarrow 2\,CO(g)$$
$$C(s) + O_2(g) \longrightarrow CO_2(g)$$

In like manner, the product of the reaction of phosphorus and oxygen depends upon whether phosphorus is reacted in a limited oxygen supply (P_4O_6) or in excess oxygen (P_4O_{10}). Sulfur reacts to produce SO_2:

$$S(s) + O_2(g) \longrightarrow SO_2(g)$$

The reaction of nitrogen with oxygen requires extremely high temperatures. The following reaction occurs in a high-energy electric arc:

$$N_2(g) + O_2(g) \longrightarrow 2\,NO(g)$$

Additional oxides of sulfur (for example, SO_3) and nitrogen (for example, NO_2 and N_2O_5) are prepared by means other than the direct combination of the elements. Lower oxides can be reacted with oxygen to produce higher oxides. For example,

$$2\,Cu_2O(s) + O_2(g) \longrightarrow 4\,CuO(s)$$
$$2\,CO(g) + O_2(g) \longrightarrow 2\,CO_2(g)$$

Most reactions of compounds with oxygen yield the same products that would be obtained if the individual elements that make up the compounds were reacted directly. Thus,

$$2\,H_2S(g) + 3\,O_2(g) \longrightarrow 2\,H_2O(g) + 2\,SO_2(g)$$
$$CS_2(l) + 3\,O_2(g) \longrightarrow CO_2(g) + 2\,SO_2(g)$$
$$2\,C_2H_2(g) + 5\,O_2(g) \longrightarrow 4\,CO_2(g) + 2\,H_2O(g)$$
$$C_2H_6O(l) + 3\,O_2(g) \longrightarrow 2\,CO_2(g) + 3\,H_2O(g)$$

The reaction of zinc sulfide with oxygen illustrates a metallurgical process known as roasting. Many sulfide ores are subjected to this procedure (see Section 23.5):

$$2\,ZnS(s) + 3\,O_2(g) \longrightarrow 2\,ZnO(s) + 2\,SO_2(g)$$

The products of the reaction of a hydrocarbon with oxygen depend upon the amount of oxygen supplied. Thus, when natural gas (methane, CH_4) is burned in air, $H_2O(g)$, $C(s)$, $CO(g)$, and $CO_2(g)$ are produced by the oxidation.

In addition to water, hydrogen and oxygen form a compound called hydrogen peroxide, H_2O_2, which is colorless liquid that boils at $150.2°C$ and freezes at $-0.41°C$. Hydrogen peroxide can be made by treating peroxides with acids:

$$BaO_2(s) + 2\,H^+(aq) + SO_4^{2-}(aq) \longrightarrow H_2O_2(aq) + BaSO_4(s)$$

In this preparation the barium sulfate, which is insoluble, can be removed by filtration.

The peroxide linkage (—O—O—) exists in covalent compounds and ions in addition to H_2O_2 and the O_2^{2-} ion. For example, the peroxydisulfate ion, written $S_2O_8^{2-}$ or $[O_3SOOSO_3]^{2-}$, contains this linkage (see Section 20.12). The $S_2O_8^{2-}$ ion is produced by the electrolysis of sulfuric acid under suitable conditions; the reaction of this ion with water serves as a commercial preparation of hydrogen peroxide:

$$S_2O_8^{2-}(aq) + 2\,H_2O \longrightarrow H_2O_2(aq) + 2\,HSO_4^-(aq)$$

Hydrogen peroxide is a weak, diprotic acid in water solution. One or two hydrogens can be neutralized by sodium hydroxide to produce either sodium hydroperoxide ($NaHO_2$) or sodium peroxide (Na_2O_2). In the laboratory, H_2O_2 is used as an oxidizing or reducing agent.

20.5 Industrial Uses of Oxygen

Most of the commercial uses of oxygen stem from its ability to support combustion and sustain life. In many applications, the use of oxygen or oxygen-enriched air in place of atmospheric air increases the intensity and speed of reaction and thereby lowers costs and improves yields. The principal uses of oxygen are:

1. Production of steel

2. Processing and fabrication of metals

3. Production of oxygen-containing compounds such as sodium peroxide and organic compounds

4. Oxidizer for rocket fuels

5. The oxyacetylene torch

6. Biological treatment of waste water

7. Life support systems in medicine, in air and space travel, and in submarines

The fuel used to propel this Delta space vehicle was oxidized by liquid oxygen. *NASA.*

20.6 Ozone

The existence of an element in more than one form in the same physical state is called **allotropy**, and the forms are called **allotropes**. A number of elements exhibit allotropy, for example, carbon, sulfur, and phosphorus. Oxygen exists in a triatomic form, ozone, in addition to the common diatomic modification.

The ozone molecule is diamagnetic and has an angular structure. Both oxygen-to-oxygen bonds have the same length (128 pm), which is intermediate between the double-bond distance (110 pm) and the single-bond distance (148 pm). The molecule may be represented as a resonance hybrid:

Ozone is a pale blue gas with a characteristic odor; predictably, its density is $1\frac{1}{2}$ times that of O_2. The normal boiling point of ozone is $-112°C$, and the normal melting point is $-193°C$. It is slightly more soluble in water than is O_2.

Ozone is produced by passing a silent electric discharge through oxygen gas. The reaction proceeds through the dissociation of an O_2 molecule into oxygen atoms and the combination of an O atom with a second O_2 molecule:

$$\tfrac{1}{2}O_2 \longrightarrow O \qquad \Delta H = +247 \text{ kJ}$$

$$O + O_2 \longrightarrow O_3 \qquad \Delta H = -105 \text{ kJ}$$

The energy released in the second step, in which a new bond is formed, is not sufficient to compensate for the energy required by the first step, in which a bond is broken. Hence, the overall reaction for the preparation of ozone is endothermic:

$$\tfrac{3}{2}O_2 \longrightarrow O_3 \qquad \Delta H_f° = +142 \text{ kJ}$$

Ozone is highly reactive; it is explosive at temperatures above 300°C or in the presence of substances that catalyze its decomposition. Ozone will react with many substances at temperatures that are not high enough to produce reaction with O_2. The higher reactivity of O_3 in comparison to O_2 is consistent with the higher energy content of O_3.

20.7 Air Pollution

Several oxides, found in air in variable amounts, are air pollutants. Modern technological civilization is introducing foreign substances into the atmosphere at an ever-increasing rate. The principal air pollutants in terms of quantities present are the following:

1. *Carbon monoxide* is produced by the incomplete combustion of fuels. The automobile's internal-combustion engine is the principal source of this pollutant. The mass of CO produced by this source is about equal to half the mass of gasoline consumed.

Carbon monoxide is toxic because it combines with the hemoglobin of the blood and prevents the hemoglobin from carrying oxygen to the body tissues (see Section 24.1). In other ways, however, CO is not very reactive. Reaction with O_2 of the air to form CO_2 does occur, but very slowly.

2. *Oxides of sulfur* (SO_2 and SO_3) result from the combustion of coal, metallurgical processes, and petroleum combustion and refining. The major source is the

combustion of coal (which contains from 0.5% to 3.0% S) in the generation of electricity. The roasting of sulfide ores (see Section 23.5) is also a significant source of SO_2 pollution:

$$2\ PbS(s) + 3\ O_2(g) \longrightarrow 2\ PbO(s) + 2\ SO_2(g)$$

Sulfur dioxide from these sources is slowly oxidized to SO_3 by O_2 in the air. The SO_3 forms sulfuric acid with atmospheric water, and SO_2 forms sulfurous acid. These substances are, of course, extremely corrosive.

The oxides of sulfur are in some ways the most serious air pollutants. They are toxic, cause respiratory ailments, and are a serious health threat. They damage plant life, corrode metals, and erode marble and limestone. Ancient monuments (such as the Parthenon in Athens) which have stood for centuries are crumbling because of the pollution of modern civilization.

3. *Oxides of nitrogen* (NO and NO_2) are produced from the N_2 and O_2 of the air at the high temperatures characteristic of some combustions. Significant amounts of NO are formed in the combustions carried out in automobile engines and in electric generating plants. Nitrogen dioxide is formed in the air by the oxidation of NO.

Small amounts of NO and NO_2 ordinarily occur in air and form a minor part of the nitrogen cycle. Nitrogen dioxide is considerably more toxic than NO. The gases, however, usually occur at relatively low concentrations so that the *direct* effect of these pollutants is not serious. The significance of these oxides lies in the role that they play in the formation of other, more serious pollutants (see *Hydrocarbons*, below).

4. *Hydrocarbons* are compounds that contain carbon and hydrogen. They are found in petroleum, natural gas, and coal. The compounds are released into the atmosphere by evaporation, petroleum refining, and incomplete combustion of fuels. The unburned hydrocarbons in automobile exhaust constitute a major source of this type of contamination.

A few hydrocarbons are carcinogenic (cancer-producing). The principal danger associated with hydrocarbon pollution, however, lies in the pollutants that are produced from hydrocarbons in the air. Nitrogen dioxide decomposes in sunlight to give O atoms:

$$NO_2(g) \longrightarrow NO(g) + O(g)$$

These O atoms react with O_2 to produce ozone, O_3:

$$O(g) + O_2(g) \longrightarrow O_3(g)$$

Ozone is highly reactive and reacts with some hydrocarbons to produce oxygen-containing organic compounds. Since the process is initiated by sunlight, the products are sometimes called **photochemical pollutants.**

These substances are toxic and very irritating to the eyes, skin, and respiratory tract. They cause extensive crop damage and the deterioration of materials. They constitute what is called photochemical smog.

5. *Small particles* suspended in air are another important contributor to air pollution. The particles may consist of liquids or solids and vary in diameter from about 0.01 μm to 100 μm (the micrometer, μm, previously called the micron, is 10^{-4} cm). A principal source is the smoke from the combustion of coal. In-

Pediment of the Parthenon in Athens, Greece, showing deterioration from air pollution. *Wide World Photos*.

dustrial processes (for example, the manufacture of cement) also introduce particulate matter into the air. Automobile exhaust contains suspended particles.

Particles reduce visibility and pose a threat to health. They cause lung damage and may be toxic. Carcinogenic substances may be contained in soot particles. Particles in automobile exhaust contain lead, a toxic substance which comes from the lead compounds that are added to gasoline to improve engine performance.

20.8 Allotropic Modifications of S, Se, and Te

All members of this group exist in more than one allotropic modification. For a given element the difference may be in molecular complexity (O_2 and O_3 for oxygen, see Section 20.6), in crystalline form, or in both.

The most important solid modifications of sulfur belong to the rhombic and monoclinic crystal systems (see Figure 20.2). The crystals of both allotropes are built from S_8 molecules. These molecules are in the form of puckered, eight-membered rings of S atoms in which the S atoms are bonded to each other by single covalent bonds and the S—S—S bond angle is about 105° (see Figure 20.3). In chemical equations, therefore, elementary sulfur should be indicated by the formula S_8. The usual practice, however, is to designate sulfur by the symbol S. This practice leads to less complicated equations that are nevertheless stoichiometrically valid.

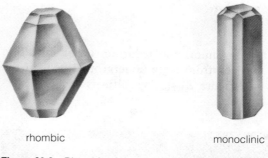

rhombic   monoclinic

Figure 20.2 Rhombic and monoclinic crystals of sulfur

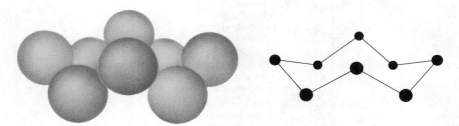

Figure 20.3 Structure of the S_8 molecule

Liquid sulfur undergoes a series of changes as its temperature is increased. At the melting point liquid sulfur is a light yellow, mobile liquid consisting principally of S_8 molecules. Upon continued heating the sulfur changes into a red-brown, highly viscous material. The viscosity reaches a maximum between 160°C and 200°C; upon further heating (up to the boiling point at 444.6°C), the viscosity of the liquid decreases. It is believed that the viscosity effect is caused by the dissociation of the S_8 rings and the formation of long chains of S atoms; at temperatures approaching the boiling point, it appears that these chains break into fragments. If sulfur is heated to approximately 200°C and then poured into cold water, a red-brown rubbery mass called plastic sulfur is obtained. It is assumed that plastic sulfur consists mainly of long chains of sulfur atoms; X-ray analysis of plastic sulfur has shown that it has the molecular structure characteristic of fibers. At room temperature, plastic sulfur, which is a supercooled liquid, slowly crystallizes and the S_8 rings re-form. Sulfur vapor has been shown to consist of S_8, S_6, S_4, and S_2 molecules; S_2 is paramagnetic, like O_2.

The stable modification of selenium at room temperature is a gray, metallic, hexagonal form, the crystals of which are constructed of zigzag chains of selenium atoms. It is this form that is used in photoelectric cells; the normally low electrical conductivity of hexagonal selenium is increased about 1000 times by exposure of the element to light. There are two monoclinic forms of selenium, both red; one of these has been shown to be made up of Se_8 rings that are similar to S_8 rings. In addition, amorphous forms of selenium have been described.

The common form of tellurium consists of silver-white, metallic hexagonal crystals built from zigzag chains of tellurium atoms. A black, amorphous form of tellurium exists.

The two modifications of polonium that have been reported belong to the cubic and rhombohedral systems.

20.9 Occurrence and Industrial Preparation of S, Se, and Te

The principal forms in which sulfur, selenium, and tellurium occur in nature are listed in Table 20.4. Sulfur is obtained from large underground beds of the free element by the **Frasch process** (see Figure 20.4). The sulfur is melted under-

Table 20.4	Occurrence of sulfur, selenium, and tellurium	
Element	Percent of Earth's Crust	Occurrence
sulfur	0.05	native FeS_2 (pyrite), PbS (galena), HgS (cinnabar), ZnS (sphalerite), Cu_2S (chalcocite), $CuFeS_2$ (chalcopyrite) $CaSO_4 \cdot 2H_2O$ (gypsum), $BaSO_4$ (barite), $MgSO_4 \cdot 7H_2O$ (epsomite)
selenium	9×10^{-6}	small amounts of Se in some S deposits rare minerals: Cu_2Se, $PbSe$, Ag_2Se low concentrations in sulfide ores of Cu, Fe, Pb, Ni
tellurium	2×10^{-7}	small amounts of Te in some S deposits rare minerals: $AuTe_2$, $PbTe$, Ag_2Te, Au_2Te, Cu_2Te low concentrations in sulfide ores of Cu, Fe

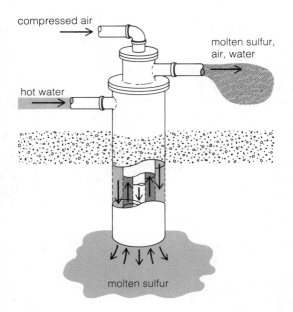

Figure 20.4 The Frasch process

Sulfur, in liquid form from the Frasch process, is sprayed into pools to solidify. *Freeport-McMoRan Inc.*

ground by water that is heated to approximately 170°C under pressure and forced down to the deposits. A froth of sulfur, air, and water is forced to the surface by hot, compressed air. The sulfur thus obtained is about 99.5% pure.

The principal commercial source of both selenium and tellurium is the anode sludge obtained from the electrolytic refining of copper (see Section 23.7).

20.10 Hydrogen Compounds of S, Se, and Te

Some reactions of sulfur, selenium, and tellurium are summarized in Table 20.5. The hydrogen compounds of sulfur, selenium, and tellurium can be prepared by the direct combination of the elements at elevated temperatures. Direct combination, however, is not a satisfactory source of the compounds. In addition to the inconvenience of the method, H_2S, H_2Se, and H_2Te are unstable at high temperatures, and the products are contaminated by starting materials.

The hydrogen compounds are readily obtained by the action of dilute acid on sulfides, selenides, and tellurides. For example,

$$FeS(s) + 2 H^+(aq) \longrightarrow Fe^{2+}(aq) + H_2S(g)$$

Table 20.5   Some reactions of sulfur, selenium, and tellurium

Reaction of Sulfur	Remarks
$nS + mM \longrightarrow M_mS_n$	Se, Te react similarly with many metals (not noble metals)
$nS + S^{2-} \longrightarrow S_{n+1}^{2-}$	for S and Te, $n = 1$ to 5; for Se, $n = 1$ to 4
$S + H_2 \longrightarrow H_2S$	S > Se > Te; elevated temperatures; compounds are better prepared by actions of dilute HCl on sulfides, selenides, or tellurides
$S + O_2 \longrightarrow SO_2$	S > Se > Te; dioxides of Se and Te are easier to prepare with a mixture of $O_2 + NO_2$
$S + 3F_2 \longrightarrow SF_6$	S, Se, Te with excess F_2
$S + 2F_2 \longrightarrow SF_4$	S, Se; TeF_4 is made indirectly ($TeF_6 + Te$)
$2S + X_2 \longrightarrow S_2X_2$	S, Se; $X_2 = Cl_2$ or Br_2
$S + 2X_2 \longrightarrow SX_4$	S, Se, Te with excess Cl_2; Se, Te with excess Br_2; Te with excess I_2
$S_2Cl_2 + Cl_2 \longrightarrow 2SCl_2$	SCl_2 only, SBr_2 unknown; $SeCl_2$, $SeBr_2$ (only in vapor state); $TeCl_2$, $TeBr_2$ are made by thermal decomposition of higher halides
$S + 4HNO_3 \longrightarrow SO_2 + 4NO_2 + 2H_2O$	hot, concentrated nitric acid; S yields mixtures of SO_2 and SO_4^{2-}; Se yields H_2SeO_3 ($SeO_2 \cdot H_2O$); Te yields $2TeO_2 \cdot HNO_3$

Rhomic crystals of sulfur. *Fundamental Photographs, New York.*

The reaction of thioacetamide with water is a convenient laboratory source of H_2S:

$$\underset{\text{thioacetamide}}{\overset{\displaystyle H \quad\quad S}{H-\underset{\underset{\displaystyle H}{|}}{\overset{\overset{\displaystyle H}{|}}{C}}-\underset{\underset{\displaystyle H}{|}}{\overset{\displaystyle \parallel}{C}}}}\ \text{(aq)} + H_2O \longrightarrow \underset{\text{acetamide}}{\overset{\displaystyle H \quad\quad O}{H-\underset{\underset{\displaystyle H}{|}}{\overset{\overset{\displaystyle H}{|}}{C}}-\underset{\underset{\displaystyle H}{|}}{\overset{\displaystyle \parallel}{C}}}}\ \text{(aq)} + H_2S\text{(g)}$$

Hydrogen sulfide, hydrogen selenide, and hydrogen telluride are colorless, unpleasant-smelling, highly poisonous gases. They are composed of angular molecules, similar to the water molecule.

Hydrogen sulfide reacts with oxygen to yield water and either SO_2 or free S depending upon the amount of oxygen employed. The combustion of H_2Se or H_2Te in oxygen produces water and either Se or Te.

The compounds H_2S, H_2Se, and H_2Te are all moderately soluble in water and are weak acids in aqueous solution. The trend in acid strength parallels that of the hydrogen halides—the elements of highest atomic number form the strongest acids. Thus, H_2Te is the strongest acid, and H_2S is the weakest acid of the group. The acids are diprotic and ionize in two steps:

$$H_2S\text{(aq)} \rightleftharpoons H^+\text{(aq)} + HS^-\text{(aq)}$$
$$HS^-\text{(aq)} \rightleftharpoons H^+\text{(aq)} + S^{2-}\text{(aq)}$$

The aqueous hydrogen sulfide system is discussed in Section 15.7.

The sulfides of the group I A and group II A metals are water soluble. The sulfides of the remaining metals are either insoluble or decompose in the presence of water to form insoluble hydroxides:

$$Al_2S_3\text{(s)} + 6\,H_2O \longrightarrow 2\,Al(OH)_3\text{(s)} + 3\,H_2S\text{(g)}$$

Precipitation of the insoluble sulfides, under varying conditions, is extensively used in analytical procedures for the separation and identification of cations in solution (see Section 16.3). An analytical test for the sulfide ion consists of generating H_2S gas by adding acid to the sulfide and then identifying the H_2S by the insoluble, black PbS that forms on a filter paper wet with a solution of a soluble Pb^{2+} salt that is held in the gas:

$$Pb^{2+}\text{(aq)} + H_2S\text{(g)} \longrightarrow PbS\text{(s)} + 2\,H^+\text{(aq)}$$

Sulfur dissolves in solutions of soluble sulfides and forms a mixture of polysulfide anions:

$$S^{2-}\text{(aq)} + n\text{S(s)} \longrightarrow S_{n+1}^{2-}\text{(aq)}$$

Polyselenide and polytelluride ions can be prepared by analogous reactions. For sulfur, ions varying in complexity from S_2^{2-} to S_6^{2-} have been prepared; Se_5^{2-} and Te_6^{2-} are the highest polyselenides and polytellurides known. The atoms of a polysulfide ion are joined into chains by single covalent bonds:

The structure of the disulfide ion, S_2^{2-}, is similar to that of the peroxide ion (see Section 20.4). The mineral pyrite, FeS_2, is iron(II) disulfide.

At room temperature, polysulfides decompose in acid solution to yield mainly H_2S and free S; however, careful treatment of a polysulfide solution with concentrated HCl at $-15°C$ yields H_2S_2, H_2S_3, and small quantities of higher homologs. The hydrogen polysulfides are unstable, yellow oils.

20.11 The 4+ Oxidation State of S, Se, and Te

The important compounds in which sulfur appears in a 4+ oxidation state are sulfur dioxide (SO_2), sulfurous acid (H_2SO_3), and the salts of sulfurous acid—the sulfites. Both selenium and tellurium form compounds analogous to those of sulfur.

Sulfur dioxide is commercially obtained by burning sulfur:

$$S(s) + O_2(g) \longrightarrow SO_2(g)$$

or by roasting sulfide ores (such as ZnS, PbS, Cu_2S, and FeS_2) in air (see Section 23.5):

$$2 ZnS(s) + 3 O_2(g) \longrightarrow 2 ZnO(s) + 2 SO_2(g)$$

Sulfur dioxide is a colorless gas. It has a sharp, irritating odor and is poisonous. The molecules of SO_2 are angular, and the structure of the compound may be represented as a resonance hybrid:

The S—O bonds have additional double-bond character from $p\pi$-$d\pi$ bonding.

The polar nature of the SO_2 molecule is reflected in the ease with which sulfur dioxide may be liquefied. Sulfur dioxide liquefies at $-10°C$ (the normal boiling point) under a pressure of 1 atm, and at 20°C a pressure of about 3 atm will liquefy the gas. It is this property of SO_2 that makes the compound useful as a refrigerant.

Sulfur dioxide is moderately soluble in water, producing solutions of sulfurous acid, H_2SO_3. The acid is not very stable, and pure H_2SO_3 cannot be isolated. The extent of the reaction between dissolved SO_2 molecules and water is not known. Probably both H_2SO_3 and SO_2 molecules exist in the solution in equilibrium:

$$SO_2(aq) + H_2O \rightleftharpoons H_2SO_3(aq)$$

Sulfurous acid is a weak, diprotic acid. The first dissociation is

$$H_2SO_3(aq) \rightleftharpoons H^+(aq) + HSO_3^-(aq)$$

which is sometimes written as

$$SO_2(aq) + H_2O \rightleftharpoons H^+(aq) + HSO_3^-(aq)$$

The second dissociation is

$$HSO_3^-(aq) \rightleftharpoons H^+(aq) + SO_3^{2-}(aq)$$

Sulfurous acid, therefore, forms two series of salts: normal salts (for example, Na_2SO_3, sodium sulfite) and acid salts (for example, $NaHSO_3$, sodium bisulfite or sodium hydrogen sulfite).

Sulfites are often prepared by bubbling SO_2 gas through a solution of a hydroxide:

$$2\,OH^- + SO_2(g) \longrightarrow SO_3^{2-}(aq) + H_2O$$

If the addition is continued, acid sulfites are produced:

$$H_2O + SO_3^{2-}(aq) + SO_2(g) \longrightarrow 2\,HSO_3^-(aq)$$

The salts of sulfurous acid can be isolated from solution.

Since sulfurous acid is unstable, the addition of an acid to a sulfite or to an acid sulfite liberates SO_2 gas, which is a convenient way to prepare the gas in the laboratory:

$$SO_3^{2-}(aq) + 2\,H^+(aq) \longrightarrow SO_2(g) + H_2O$$

Sulfur dioxide, sulfurous acid, and sulfites can function as mild oxidizing agents, but reactions in which these compounds react as reducing agents (and are oxidized to the sulfate ion, SO_4^{2-}) are more numerous and more important. Many substances (such as potassium permanganate, potassium dichromate, chlorine, and bromine) oxidize sulfite to sulfate. In fact, sulfites are usually contaminated by traces of sulfates because of oxidation by the oxygen of the air:

$$2\,SO_3^{2-} + O_2(g) \longrightarrow 2\,SO_4^{2-}$$

Sulfites may be identified by the production of SO_2 gas upon acidification and by oxidation to the sulfate ion,

$$6\,H^+(aq) + 5\,SO_3^{2-}(aq) + 2\,MnO_4^-(aq) \longrightarrow$$
$$5\,SO_4^{2-}(aq) + 2\,Mn^{2+}(aq) + 3\,H_2O$$

followed by the precipitation of the sulfate as the insoluble barium salt:

$$Ba^{2+}(aq) + SO_4^{2-}(aq) \longrightarrow BaSO_4(s)$$

The dioxides of selenium and tellurium may be prepared by direct combination of the elements with oxygen; however, SeO_2 and TeO_2 are usually made by heating the product obtained from the oxidation of selenium or tellurium by concentrated nitric acid (see Table 20.5). Both SeO_2 and TeO_2 are white solids.

Selenous acid, H_2SeO_3, is formed when the very soluble SeO_2 is dissolved in water; it is a weak, diprotic acid and may be obtained in pure form by the evaporation of a solution of SeO_2. Tellurium dioxide is only slightly soluble in water, and pure H_2TeO_3 has never been prepared. Consequently, only very dilute solutions of tellurous acid have been studied.

Tellurium dioxide, as well as selenium dioxide, will dissolve in aqueous solutions of hydroxides to produce tellurites and selenites. If an excess of the dioxide is employed, the corresponding acid salt is obtained.

The selenium and tellurium compounds of the 4+ oxidation state are better oxidizing agents and poorer reducing agents than the corresponding sulfur compounds. Selenium dioxide, SeO_2, is a good oxidizing agent and is employed in certain organic syntheses.

20.12 The 6+ Oxidation State of S, Se, and Te

Sulfur trioxide is produced when sulfur dioxide reacts with atmospheric oxygen. Since the reaction is very slow at ordinary temperatures, the commercial preparation is conducted at elevated temperatures (400 to 700°C) and in the presence of a catalyst (such as divanadium pentoxide or spongy platinum):

$$2 SO_2(g) + O_2(g) \longrightarrow 2 SO_3(g)$$

Sulfur trioxide is a volatile material (boiling point, 44.8°C). In the gas phase it consists of single molecules that are triangular planar in form with O—S—O bond angles of 120°. The electronic structure of the molecule may be represented as a resonance hybrid:

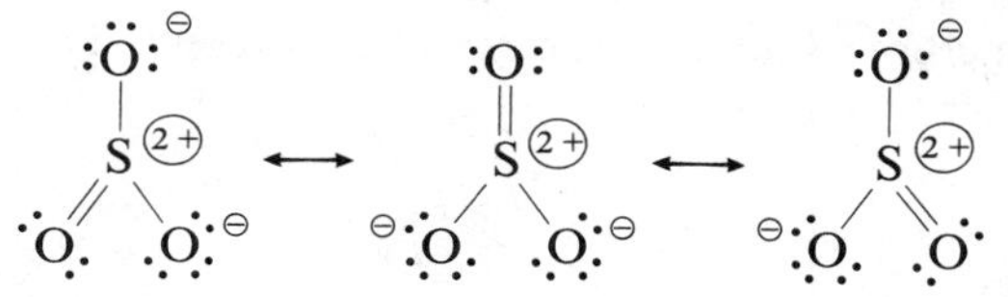

Additional double bonding, in excess of that represented by the resonance forms, arises from $p\pi$-$d\pi$ bonding. The compound exists in at least three solid modifications that are formed by the condensation of SO_3 units into larger molecules called polymers.

Sulfur trioxide is an extremely reactive substance and a strong oxidizing agent. It is the anhydride of sulfuric acid, H_2SO_4, and reacts vigorously with water to produce the acid and with metallic oxides to produce sulfates (see Section 11.5).

Sulfuric acid, an important industrial chemical, is made by the **contact process** in which SO_2 is catalytically oxidized to SO_3 in the manner previously described. The SO_3 vapor is bubbled through H_2SO_4 and pyrosulfuric acid ($H_2S_2O_7$) is formed:

$$SO_3(g) + H_2SO_4(l) \longrightarrow H_2S_2O_7(l)$$

Water is then added to the pyrosulfuric acid to make sulfuric acid of the desired concentration:

$$H_2S_2O_7(l) + H_2O \longrightarrow 2 H_2SO_4(l)$$

This procedure, in which pyrosulfuric acid is formed, is easier to control than the direct reaction of SO_3 with water.

A sulfuric acid plant. *Texasgulf, Inc.*

Sulfuric acid is a colorless, oily liquid that freezes at 10.4°C and begins to boil at approximately 290°C with decomposition into water and sulfur trioxide. The electronic structure of sulfuric acid may be represented as

$$H-\overset{\cdot\cdot}{\underset{\cdot\cdot}{O}}-\overset{\overset{\displaystyle :\overset{\cdot\cdot}{O}:^{\ominus}}{|}}{\underset{\underset{\displaystyle :\overset{\cdot\cdot}{O}:^{\ominus}}{|}}{S}^{2+}}-\overset{\cdot\cdot}{\underset{\cdot\cdot}{O}}-H$$

However, the S—O bonds have a considerable degree of double-bond character from $p\pi\text{-}d\pi$ bonding, and the electronic formula

$$H-\overset{\cdot\cdot}{\underset{\cdot\cdot}{O}}-\overset{\overset{\displaystyle :O:}{\parallel}}{\underset{\underset{\displaystyle :O:}{\parallel}}{S}}-\overset{\cdot\cdot}{\underset{\cdot\cdot}{O}}-H$$

is sometimes used. The H_2SO_4 molecule, as well as the SO_4^{2-} ion, is tetrahedral.

When concentrated sulfuric acid is added to water, a large amount of heat is evolved. Sulfuric acid has a strong affinity for water and forms a series of hydrates (such as $H_2SO_4 \cdot H_2O$, $H_2SO_4 \cdot 2\,H_2O$, and $H_2SO_4 \cdot 4\,H_2O$). Sulfuric acid, therefore, is used as a drying agent. Gases that do not react with H_2SO_4 may be dried by being bubbled through the acid. The dehydrating power of sulfuric acid is also seen in the charring action of the acid on carbohydrates:

$$C_{12}H_{22}O_{11}(s) \xrightarrow{\;H_2SO_4\;} 12\,C(s) + 11\,H_2O(g)$$
sucrose

Table 20.6 Standard electrode potentials of sulfuric, selenic, and telluric acids	
Reaction	Standard Electrode Potential
$2e^- + 4H^+ + SO_4^{2-} \rightleftharpoons H_2SO_3 + H_2O$	$\mathscr{E}° = +0.17 \text{ V}$
$2e^- + 4H^+ + SeO_4^{2-} \rightleftharpoons H_2SeO_3 + H_2O$	$\mathscr{E}° = +1.15 \text{ V}$
$2e^- + 2H^+ + H_6TeO_6 \rightleftharpoons TeO_2 + 4H_2O$	$\mathscr{E}° = +1.02 \text{ V}$

In aqueous solution, sulfuric acid ionizes in two steps:

$$H_2SO_4 \longrightarrow H^+(aq) + HSO_4^-(aq)$$

$$HSO_4^-(aq) \rightleftharpoons H^+(aq) + SO_4^{2-}(aq)$$

Sulfuric acid is a strong electrolyte as far as the first dissociation is concerned. The second dissociation, however, is not complete. The acid forms two series of salts: normal salts (such as sodium sulfate, Na_2SO_4) and acid salts (such as sodium bisulfate, $NaHSO_4$). Most sulfates are soluble in water. However, barium sulfate ($BaSO_4$), strontium sulfate ($SrSO_4$), lead sulfate ($PbSO_4$), and mercury(I) sulfate (Hg_2SO_4) are practically insoluble; calcium sulfate ($CaSO_4$) and silver sulfate (Ag_2SO_4) are but slightly soluble. The formation of white, insoluble barium sulfate, the least soluble of the substances listed, is commonly used as a laboratory test for the sulfate ion.

Since sulfuric acid has a relatively high boiling point (or decomposition temperature), it is used to secure more volatile acids from their salts. This use is illustrated by the preparations of HF and HCl (see Section 19.10), as well as by the preparation of HNO_3 (see Section 21.7).

Sulfuric acid at unit concentration and 25°C is not a particularly good oxidizing agent (note the relatively low standard electrode potential in Table 20.6). Hot, concentrated sulfuric acid, however, is a moderately effective oxidizing agent. The oxidizing ability of hot, concentrated H_2SO_4 on bromides and iodides has already been noted (see Section 19.10). This reagent will also oxidize many nonmetals:

$$C(s) + 2H_2SO_4(l) \longrightarrow CO_2(g) + 2SO_2(g) + 2H_2O(g)$$

Most metals are oxidized by hot, concentrated sulfuric acid, including those metals of relatively low reactivity that are not oxidized by hydronium ion. For example, copper will not displace hydrogen from aqueous acids. Copper metal is, however, oxidized by hot, concentrated sulfuric acid, although hydrogen gas is not a product of the reaction:

$$Cu(s) + 2H_2SO_4(l) \longrightarrow CuSO_4(s) + SO_2(g) + 2H_2O(g)$$

Selenic acid, H_2SeO_4, is prepared by the oxidization of selenous acid, H_2SeO_3. Selenates may be prepared by the oxidation of the corresponding selenites. Selenium trioxide, the acid anhydride of selenic acid, is not very stable and decomposes to SeO_2 and O_2 on warming. Low yields of SeO_3 mixed with SeO_2 are produced when an electric discharge is passed through selenium vapor and oxygen. The trioxide may also be prepared by the reaction of potassium selenate, K_2SeO_4, and SO_3.

Selenic acid is very similar to sulfuric acid. In aqueous solution the first dissociation of H_2SeO_4 is strong and the second dissociation is weak. The acid forms normal and acid salts. Like the sulfate ion, the selenate ion is tetrahedral.

Telluric acid is prepared by the action of vigorous oxidizing agents on elemental tellurium. Unlike H_2SO_4 and H_2SeO_4, the formula of telluric acid is H_6TeO_6, which may be regarded as a hydrated form of the nonexistent H_2TeO_4. No compound of formula H_2TeO_4 has ever been prepared, although salts of an acid corresponding to this formula exist. Tellurium, like iodine, is a large atom and can accommodate six oxygen atoms, and telluric acid, like IO_6^{5-}, is octahedral. Telluric acid is not very similar to sulfuric acid, and H_6TeO_6 functions in aqueous solution as a very weak diprotic acid. If H_6TeO_6 is heated to approximately $350°C$, water is driven off and solid tellurium trioxide results. This compound is not very water soluble but reacts with alkalies to produce tellurates.

Both selenic and telluric acids are stronger oxidizing agents than sulfuric acid (see Table 20.6), although the reactions of H_2SeO_4 and H_6TeO_6 often proceed slowly.

There are other acids of sulfur in which S is in a 6+ oxidation state. Pyrosulfuric acid ($H_2S_2O_7$, also called disulfuric acid):

$$\begin{array}{ccccc}
& O & & O & \\
& \| & & \| & \\
H-O- & S & -O- & S & -O-H \\
& \| & & \| & \\
& O & & O &
\end{array}$$

has been mentioned as the product of the reaction of SO_3 with H_2SO_4 in a $1:1$ molar ratio. The pyrosulfate ion has been shown to have a structure in which two SO_4 tetrahedra are joined by an O atom common to both tetrahedra: $[O_3SOSO_3]^{2-}$. Pyrosulfuric acid is a stronger oxidizing agent and a stronger dehydrating agent than sulfuric acid.

A peroxy acid is an acid that contains a peroxide group (—O—O—) somewhere in the molecule. Two peroxy acids of sulfur exist: peroxymonosulfuric acid (H_2SO_5) and peroxydisulfuric acid ($H_2S_2O_8$):

$$\begin{array}{ccccccccc}
& & & O & & & & O & \\
& & & \| & & & & \| & \\
H-O-O- & S & -O-H & & H-O- & S & -O-O- & S & -O-H \\
& & \| & & & \| & & \| & \\
& & O & & & O & & O &
\end{array}$$

The structure of the peroxydisulfate ion consists of two complete SO_4 tetrahedra joined by an O—O bond.

Peroxydisulfuric acid is prepared by the electrolysis of moderately concentrated solutions of sulfuric acid (50 to 70%) at temperatures below room temperature (5 to 10°C). Potassium and ammonium salts of this acid are prepared by the electrolysis of the corresponding acid sulfates. The anode reaction in these electrolyses may be represented by the partial equation:

$$2\,HSO_4^- \longrightarrow S_2O_8^{2-} + 2\,H^+ + 2e^-$$

The reaction of peroxydisulfuric acid with water yields peroxymonosulfuric acid:

$$H_2S_2O_8 + H_2O \longrightarrow H_2SO_5 + H_2SO_4$$

Upon further hydrolysis, H_2SO_5 is decomposed into hydrogen peroxide and H_2SO_4:

$$H_2SO_5 + H_2O \longrightarrow H_2SO_4 + H_2O_2$$

Both peroxymonosulfuric acid and peroxydisulfuric acid are low melting solids. The first ionization of peroxydisulfuric acid is strong. The peroxydisulfate ion is one of the strongest oxidizing agents known:

$$2e^- + S_2O_8^{2-} \rightleftharpoons 2SO_4^{2-} \qquad \mathscr{E}^\circ_{red} = +2.01 \text{ V}$$

The oxidations, however, are slow and are usually catalyzed by Ag^+ ions. In both peroxy acids, sulfur is assumed to be in its highest oxidation state (6+). The oxygen atoms of the peroxide grouping, however, are each assigned an oxidation number of $1-$. In the course of an oxidation, it is the peroxide oxygens that change oxidation state (from $1-$ to $2-$).

There is another significant group of sulfur-containing anions. The most important representative of the group is the thiosulfate ion, $S_2O_3^{2-}$:

$$\left[\begin{array}{c} O \\ | \\ O-S-O \\ | \\ S \end{array} \right]^{2-}$$

This tetrahedral ion may be regarded as a sulfate ion in which one oxygen atom has been replaced by a sulfur atom. In fact, the prefix "thio-" is used to name any species that may be considered to be derived from another compound by replacing an oxygen atom by a sulfur atom; the prefix is placed before the base name of the oxygen-containing compound. Thus, OCN^- is the cyanate ion and SCN^- is the thiocyanate ion; $CO(NH_2)_2$ is urea and $CS(NH_2)_2$ is thiourea.

Thiosulfates may be prepared by the reaction of sulfur with sulfites in aqueous solution:

$$SO_3^{2-}(aq) + S(s) \longrightarrow S_2O_3^{2-}(aq)$$

The corresponding acid does not exist. Upon acidification, thiosulfates decompose to elementary sulfur and SO_2 gas:

$$S_2O_3^{2-}(aq) + 2H^+(aq) \longrightarrow S(s) + SO_2(g) + H_2O$$

The two sulfur atoms of the thiosulfate ion are not equivalent. This has been shown by the reactions of a compound derived from a sulfite and radioactive sulfur, $^{35}_{16}S$. When a compound thus prepared is decomposed by acidification, all the activity ends in the elementary sulfur:

$$SO_3^{2-}(aq) + {}^{35}S(s) \longrightarrow {}^{35}SSO_3^{2-}(aq)$$

$${}^{35}SSO_3^{2-}(aq) + 2H^+(aq) \longrightarrow {}^{35}S(s) + SO_2(g) + H_2O$$

Hence, the central sulfur atom is generally assigned an oxidation number of $6+$ (as in the sulfate ion), and the coordinated sulfur is usually assigned an oxidation number of $2-$ (corresponding to the oxidation number of the oxygen it replaces). The average oxidation state of sulfur in the ion is $2+$.

The thiosulfate ion is readily oxidized to the tetrathionate ion, $S_4O_6^{2-}$. The tetrathionate ion may be regarded as an analog of the peroxydisulfate ion, $S_2O_8^{2-}$, in which the peroxide group (—O—O—) is replaced by a disulfide group (—S—S—). Other thionate ions exist—for example, the dithionate ion, $S_2O_6^{2-}$, and the trithionate ion, $S_3O_6^{2-}$; the latter is structurally similar to the pyrosulfate ion, $S_2O_7^{2-}$, with a sulfur atom replacing the central oxygen atom.

20.13 Electrode Potential Diagrams for S

Electrode potential diagrams for sulfur and its compounds appear in Figure 20.5. On the basis of the $\mathscr{E}_{red}^{\circ}$ values, several substances shown in the diagrams should be able to disproportionate. In alkaline solution (but not in acid), sulfur itself forms S^{2-} and $S_2O_3^{2-}$. The thiosulfate ion is stable in alkaline solution but disproportionates to S and SO_2 in acid. The electrode potentials also show that both SO_2, in acid, and SO_3^{2-}, in base, are unstable toward disproportionation. The latter two reactions, however, are slow under ordinary conditions.

At standard concentrations and in acidic solution, the oxysulfur compounds shown in the diagram are only moderately strong oxidizing agents. In alkaline solution the oxysulfur compounds are poor oxidants. In fact, all the ions with the exception of SO_4^{2-} are easily oxidized in base and can function as reducing agents. The thiosulfate ion is easily oxidized to the tetrathionate ion $S_4O_6^{2-}$; $\mathscr{E}_{ox}^{\circ}$ for this transformation is only -0.08 V.

20.14 Industrial Uses of S, Se, and Te

The commercial uses of sulfur, selenium, and tellurium include the following:

1. Sulfur. Over 80% of the sulfur mined is used in the manufacture of sulfuric acid. The industrial importance of H_2SO_4 is reflected in the significance that is attached to figures for the annual consumption of the acid. These figures are used to rate the state of industrialization, standard of living, and economic well-being of a country.

Sulfuric acid is used in many industrial processes: in the production of other chemicals, fertilizers, pigments, iron, steel, and in petroleum refining. It is the electrolyte used in lead storage batteries. Elemental sulfur is used in the vulcanization of rubber, in the production of pigments, paints, paper, fungicides, insecticides, and in pharmaceuticals.

2. Selenium. Selenium has been used in photocells, devices which transmit electric current in proportion to the intensity of incident light. The electrical response to light of gray selenium also accounts for its use in xerography, the dry photocopying process. Selenium is also used in the production of colored glass, ceramics, pigments, alloys, steel, oxidation inhibitors for lubricating oils, and in the vulcanization of rubber.

3. Tellurium. Tellurium finds fewer uses than the other elements of this group.

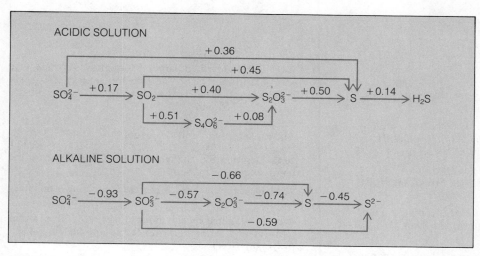

Figure 20.5 Electrode potential diagrams for sulfur and its compounds ($\mathscr{E}^{\circ}_{red}$ values given in volts)

Like selenium, tellurium is used in the vulcanization of rubber and in the manufacture of glass, ceramics, alloys, and enamel pigments.

Summary

The topics that have been discussed in this chapter are

1. The properties of the group VI elements.

2. Oxygen: occurrence, industrial and laboratory preparations.

3. Reactions of oxygen: oxides, superoxides, peroxides, and ozonides.

4. Industrial uses of oxygen.

5. Ozone, an allotrope of oxygen.

6. Air pollution.

7. Sulfur, selenium, and tellurium: allotropes, occurrence, and industrial preparation.

8. Compounds of sulfur, selenium, and tellurium: hydrogen compounds, compounds with the group VI A element in a 4+ oxidation state, compounds with the group VI A element in a 6+ oxidation state, and electrode potential diagrams of S.

9. Industrial uses of sulfur, selenium, and tellurium and their compounds.

Key Terms

Some of the more important terms introduced in this chapter are listed below. Definitions for terms not included in this list may be located in the text by use of the index.

Allotropes (Section 20.6)   Two or more forms of the same element in the same physical state.

Contact process (Section 20.12)   A process for the manufacture of sulfuric acid in which SO_2 is catalytically oxidized to SO_3, the SO_3 vapor dissolved in H_2SO_4, and the resulting $H_2S_2O_7$ diluted with water to give H_2SO_4.

Frasch process (Section 20.9)   A process in which molten sulfur is obtained from underground deposits.

Peroxy acid (Section 20.12)   An acid that contains a peroxide group (—O—O—) somewhere in the molecule.

Photochemical pollutants (Section 20.7)   Air pollutants produced by a sequence of reactions that is initiated by sunlight.

Problems*

Oxygen

20.1 List the forms in which oxygen occurs in nature. How is oxygen produced industrially?

20.2 Write chemical equations for the preparation of oxygen from **(a)** $HgO(s)$, **(b)** $Na_2O_2(s)$ and H_2O, **(c)** $NaNO_3(s)$, **(d)** $KClO_3(s)$, **(e)** H_2O.

20.3 Write chemical equations for the reactions of oxygen with **(a)** $K(s)$, **(b)** $Na(s)$, **(c)** $Li(s)$, **(d)** $Mg(s)$, **(e)** $Hg(l)$, **(f)** $Ba(s)$, **(g)** $C(s)$, **(h)** $S(s)$, **(i)** $Cu_2O(s)$, **(j)** $P_4(s)$.

20.4 Write chemical equations for the complete combustion in oxygen of **(a)** C_4H_{10}, **(b)** $C_5H_{12}S$, **(c)** C_3H_8O, **(d)** ZnS, **(e)** PbS.

20.5 The products of the combustion of a hydrocarbon in oxygen depend upon the amount of oxygen supplied. Write chemical equations for the reactions of methane, $CH_4(g)$, with oxygen that yield **(a)** $C(s)$, **(b)** $CO(g)$, **(c)** $CO_2(g)$.

20.6 Compare the oxides of the I A elements with those of the VI A elements in regard to **(a)** their physical state, **(b)** their melting point, **(c)** the nature of the bonding, **(d)** the products of their reactions with water.

20.7 Draw molecular-orbital energy-level diagrams for **(a)** O_2, **(b)** O_2^-, **(c)** O_2^{2-}. State the number of unpaired electrons and the bond order of each.

20.8 Compounds that contain the dioxygenyl ion, O_2^+, are known. Draw molecular-orbital energy-level diagrams for the dioxygenyl ion, O_2^+, and the superoxide ion, O_2^-. Compare the two ions as to **(a)** bond order, **(b)** number of unpaired electrons.

20.9 Because the oxygen of H_2O_2 can be either oxidized (to O_2) or reduced (to H_2O), hydrogen peroxide can function as a reducing agent or as an oxidizing agent. Using the ion-electron method, write balanced chemical equations for the following reactions of H_2O_2 **(a)** the oxidation of PbS to $PbSO_4$ in acid solution, **(b)** the oxidation of $Cr(OH)_3$ to CrO_4^{2-} in alkaline solution, **(c)** the reduction of MnO_4^- to Mn^{2+} in acid solution, **(d)** the reduction of Ag_2O to Ag in alkaline solution.

20.10 Draw the resonance forms of the ozone molecule, O_3. What is the bond order? What is the shape of the molecule?

20.11 The standard enthalpy of formation of $H_2O(l)$ is -285.9 kJ/mol and of $O_3(g)$ is $+142.3$ kJ/mol. What is the enthalpy change when one mole of $H_2O(l)$ is prepared from **(a)** $H_2(g)$ and $O_2(g)$, **(b)** $H_2(g)$ and $O_3(g)$? In general, how do enthalpy changes for the reactions of ozone compare with those of oxygen?

*20.12 The standard enthalpy of formation of ozone, O_3, is $+142$ kJ/mol. The bond dissociation energy of $O_2(g)$ is $+494$ kJ/mol. What is the *average* bond energy of the *two* bonds in ozone?

Sulfur, Selenium, and Tellurium

20.13 Describe the changes in sulfur that occur as the temperature is increased.

20.14 Describe the Frasch process for mining elementary sulfur.

20.15 Write chemical equations for the reactions of sulfur with **(a)** O_2, **(b)** $S^{2-}(aq)$, **(c)** $SO_3^{2-}(aq)$, **(d)** Fe, **(e)** F_2, **(f)** Cl_2, **(g)** HNO_3.

20.16 Write equations for the reactions of H_2SO_4 with **(a)** $C_{12}H_{22}O_{11}$, **(b)** $NaNO_3$, **(c)** Cu, **(d)** Zn, **(e)** ZnS, **(f)** Fe_2O_3.

20.17 Write a sequence of equations representing reactions that could be used to make each of the following and that start with elemental S **(a)** H_2S (not by direct union of the elements), **(b)** H_2SO_3, **(c)** $Na_2S_2O_3$, **(d)** $NaHSO_4$, **(e)** $H_2S_2O_7$.

20.18 Write equations for the reactions of $SO_2(g)$ with **(a)** O_2 (Pt catalyst) **(b)** $Cl_2(g)$, **(c)** H_2O, **(d)** $ClO_3^-(aq)$, **(e)** $OH^-(aq)$, **(f)** $SO_3^{2-}(aq)$ and H_2O.

20.19 Draw Lewis structures for and describe the geometric structure of **(a)** S_3^{2-}, **(b)** SO_3^{2-}, **(c)** SO_4^{2-}, **(d)** $S_2O_3^{2-}$, **(e)** $S_4O_6^{2-}$, **(f)** $H_2S_2O_7$, **(g)** $H_2S_2O_8$.

20.20 Write an equation for the reaction of water with **(a)** $CH_3C(NH_2)S$, **(b)** SO_2, **(c)** SO_3, **(d)** Al_2S_3, **(e)** SeO_3, **(f)** TeO_3, **(g)** H_2SO_5, **(h)** $H_2S_2O_7$.

20.21 Write chemical equations for the reactions of $O_2(g)$ with **(a)** H_2S, **(b)** H_2Te, **(c)** PbS, **(d)** Na_2SO_3.

20.22 Write equations for the reactions of HCl(aq) with **(a)** Na_2SO_3, **(b)** Na_2S, **(c)** $Na_2S_2O_3$.

20.23 Write balanced chemical equations for the disproportionation reaction of **(a)** S (to S^{2-} and $S_2O_3^{2-}$) in alkaline solution, **(b)** $S_2O_3^{2-}$ (to S and SO_2) in acid solution.

*20.24 According to standard electrode potentials, both SO_2 (in acid solution) and SO_3^{2-} (in alkaline solution) should disproportionate. These reactions, however, are slow. What products should be obtained in each of these disproportionation reactions? Note that it is necessary to take all possibilities given in Figure 20.5 into account.

20.25 Explain why OF_4 cannot be prepared but SF_4 can be.

---

* The more difficult problems are marked with asterisks. The appendix contains answers to color-keyed problems.

20.26 What is the difference in meaning between the prefixes *per-* and *peroxy-* as applied in the naming of acids?

20.27 Describe a laboratory test for **(a)** $S^{2-}(aq)$, **(b)** $SO_3^{2-}(aq)$, **(c)** $SO_4^{2-}(aq)$, **(d)** $S_2O_3^{2-}(aq)$.

20.28 Chlorosulfonic acid, $HOSO_2Cl$, is the product of the reaction of SO_3 and HCl. The peroxysulfuric acids (H_2SO_5 and $H_2S_2O_8$) can be prepared by the reaction of one mole of hydrogen peroxide, H_2O_2, with either one or two moles of chlorosulfonic acid. Write chemical equations for the reactions.

20.29 Describe the geometric shapes of the following: **(a)** H_2S, **(b)** SO_2, **(c)** SO_3, **(d)** SO_3^{2-}, **(e)** $S_2O_3^{2-}$, **(f)** SO_4^{2-}, **(g)** SF_4, **(h)** SF_6, **(i)** H_6TeO_6.

20.30 Explain why **(a)** H_2Te is a stronger acid than H_2S, **(b)** H_2SO_4 is a stronger acid than H_6TeO_6.

C H A P T E R

21

THE NONMETALS, PART III: THE GROUP V A ELEMENTS

Group V A includes nitrogen, phosphorus, arsenic, antimony, and bismuth. Collectively, these elements show a wider range of properties than is exhibited by either the group VI A elements or the group VII A elements.

21.1 Properties of the Group V A Elements

Within any A family of the periodic classification, metallic character increases (and nonmetallic character decreases) with increasing atomic number, atomic weight, and atomic size. This trend is particularly striking in group V A. The first ionization energies of the elements in the group, which are listed in Table 21.1, decrease from values typical of a nonmetal (N) to those characteristic of a metal (Bi). Nitrogen and phosphorus are generally regarded as nonmetals, arsenic and antimony as semimetals or metalloids, and bismuth as a metal.

The electronic configurations of the elements are listed in Table 21.2. Each element has three electrons less than the noble gas of its period, and the formation of trinegative ions might be expected. Nitrogen forms the nitride ion, N^{3-}, in combination with certain reactive metals, and phosphorus forms the phosphide ion, P^{3-}, less readily. The remaining elements of the group (As, Sb, and Bi), however, are more metallic than N and P and have no tendency to form comparable anions.

The loss of electrons and consequent formation of cations, which is characteristic of metals, is observed for the heavier members of the group. High ionization energies prohibit the loss of all five valence electrons by any element. Consequently, 5+ ions do not exist, and the 5+ oxidation state is attained only through covalent bonding. In addition, most of the compounds in which the group V A elements appear in the 3+ oxidation state are covalent. Antimony and bismuth, however, can form $d^{10}s^2$ ions, Sb^{3+} and Bi^{3+}, through loss of the p electrons of their valence levels. The compounds $Sb_2(SO_4)_3$, BiF_3, and $Bi(ClO_4)_3 \cdot 5H_2O$ are ionic. The 3+ ions of antimony and bismuth react with water to form antimonyl and bismuthyl ions (SbO^+ and BiO^+), as well as hydrated forms of these ions (for example, $Bi(OH)_2^+$):

$$Bi^{3+}(aq) + H_2O \rightleftharpoons BiO^+(aq) + 2H^+(aq)$$

Nitrogen, phosphorus, and arsenic do not form simple cations.

The oxides of the group V A elements become less acidic and more basic as the metallic character of the element increases. Thus N_2O_3, P_4O_6, and As_4O_6

Table 21.1 Some properties of the group V A elements

Property	Nitrogen	Phosphorus	Arsenic	Antimony	Bismuth
color	colorless	white, red, black	gray metallic, yellow	gray metallic, yellow	gray metallic
molecular formula	N_2	P_4 (white) P_n (black)	As_n (metallic) As_4 (yellow)	Sb_n (metallic) Sb_4 (yellow)	Bi_n
melting point (°C)	−210	44.1 (white)	814 (36 atm) (metallic)	630.5 (metallic)	271
boiling point (°C)	−195.8	280	633 (sublimes)	1325	1560
atomic radius (pm)	74	110	121	141	152
ionic radius (pm)	140 (N^{3-})	185 (P^{3-})		92 (Sb^{3+})	108 (Bi^{3+})
first ionization energy (kJ/mol)	1399	1061	965	830	772
electronegativity	3.0	2.2	2.2	2.1	2.0

Table 21.2 Electronic configurations of the group V A elements

ELEMENT	Z	1s	2s	2p	3s	3p	3d	4s	4p	4d	4f	5s	5p	5d	6s	6p
N	7	2	2	3												
P	15	2	2	6	2	3										
As	33	2	2	6	2	6	10	2	3							
Sb	51	2	2	6	2	6	10	2	6	10		2	3			
Bi	83	2	2	6	2	6	10	2	6	10	14	2	6	10	2	3

are acidic oxides; they dissolve in water to form acids, and they dissolve in solutions of alkalies to form salts of these acids. The compound Sb_4O_6 is amphoteric; it will dissolve in hydrochloric acid as well as in sodium hydroxide. The comparable oxide of bismuth is strictly basic; Bi_2O_3 is not soluble in alkalies, but the compound will dissolve in acids to produce bismuth salts.

All the oxides in which the elements exhibit a 5+ oxidation state are acidic, but the acidity declines markedly from N_2O_5 to Bi_2O_5. In addition, the stability of the 5+ oxidation state decreases with increasing atomic number; Bi_2O_5 is extremely unstable and has never been prepared in a pure state.

Many of the properties of nitrogen are anomalous in comparison to those of the other V A elements. This departure is characteristic of the first members of the groups of the periodic classification. Free nitrogen is surprisingly unreactive, partly because of the great strength of the bonding in the N_2 molecule:

$$:N{\equiv}N:$$

According to the molecular orbital theory, two π bonds and one σ bond join the atoms of a N_2 molecule, and the bond order is 3. The energy required to dissociate molecular N_2 into atoms is very high (941 kJ/mol).

Since nitrogen has no d orbitals in its valence level ($n = 2$), the maximum number of covalent bonds formed by nitrogen is four (for example, in NH_4^+). In

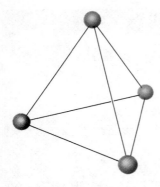

Figure 21.1 Structure of the P₄ molecule

the valence levels of the other V A elements, there are empty d orbitals which may be utilized in covalent bond formation. Hence, P, As, Sb, and Bi form as many as six covalent bonds in such species as PCl_5, PCl_6^-, AsF_5, $SbCl_6^-$, and $BiCl_5^{2-}$.

For the group as a whole, the $3-$, $3+$, and $5+$ oxidation states are most common. The importance and stability of the $5+$ and $3-$ states decline from the lighter to the heavier elements. Nitrogen, however, appears in every oxidation state from $3-$ to $5+$. Nitrogen also has a tendency toward the formation of multiple bonds (for example, in the cyanide ion, $C{\equiv}N^-$). The other V A elements do not form π bonds with p orbitals, but some multiple bond character can arise in the compounds of these elements (particularly those of P) from $p\pi$-$d\pi$ bonding.

Phosphorus, arsenic, and antimony occur in allotropic modifications. There are three important forms of phosphorus: white, red, and black. White phosphorus, a waxy solid, is obtained by condensing phosphorus vapor. Crystals of white phosphorus are formed from P₄ molecules (see Figure 21.1) in which each phosphorus atom has an unshared pair of electrons and completes its octet by forming single covalent bonds with the other three phosphorus atoms of the molecule.

White phosphorus is soluble in a number of nonpolar solvents (for example, benzene and carbon disulfide). In such solutions, in liquid white phosphorus, and in phosphorus vapor, the element exists as P₄ molecules. At temperatures above 800°C a slight dissociation of the P₄ molecules of the vapor into P₂ molecules is observed; these latter molecules are assumed to have a structure similar to that of the N₂ molecule. White phosphorus is the most reactive form of the element and is stored under water to protect it from atmospheric oxygen with which it spontaneously reacts.

Red phosphorus may be prepared by heating white phosphorus to about 250°C in the absence of air. It is a polymeric material in which many phosphorus atoms are joined in a network, but the details of the structure of red phosphorus are not known. Red phosphorus is not soluble in common solvents and is considerably less reactive than the white variety. It does not react with oxygen at room temperature.

Black phosphorus, a less common allotrope, is made by subjecting the element to very high pressures or by a slow crystallization of liquid white phosphorus in the presence of mercury as a catalyst and a seed of black phosphorus. Crystalline black phosphorus consists of layers of phosphorus atoms covalently joined into a network (see Figure 21.2). The distance between P atoms of adjacent layers is much greater than the distance between P atoms of the same layer since the

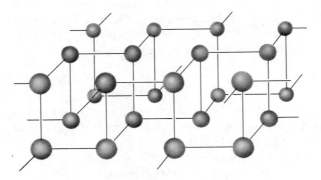

Figure 21.2 Structure of a layer of the black phosphorus crystal

P atoms of a given layer are covalently bonded to one another and the layers are held together by comparatively weak London forces. Hence, black phosphorus is a flaky material much like graphite (which also has a layer-type crystal, see Section 22.2), and like graphite, black phosphorus is an electrical conductor. Black phosphorus is the least soluble and least reactive form of the element.

Arsenic and antimony exist in soft, yellow, nonmetallic modifications which are thought to be formed from tetrahedral As_4 and Sb_4 molecules analogous to the P_4 molecules of white phosphorus. These yellow forms may be obtained by the rapid condensation of vapors and are soluble in carbon disulfide. They are unstable and are readily converted into stable, gray, metallic modifications.

Bismuth commonly occurs as a light gray metal with a reddish cast; the element does not exist in other modifications. The metallic modifications of arsenic, antimony, and bismuth are comparatively soft and brittle and have a metallic luster. Their crystalline structures are similar to the structure of black phosphorus, and they are electrical conductors.

21.2 The Nitrogen Cycle

In nature, nitrogen is constantly being removed from the atmosphere and returned to the atmosphere by several natural and artificial processes. These processes, taken together, constitute what is called the **nitrogen cycle.**

Nitrogen is a constituent element of all plant and animal protein. Since nitrogen is a comparatively unreactive element, the cells of living systems cannot directly assimilate the nitrogen of the air to use in the synthesis of proteins. The nitrogen of the air, however, is converted by several **nitrogen-fixation processes** into compounds that can be used by plants. These nitrogen-fixation processes constitute the first part of the nitrogen cycle.

During storms, lightning flashes cause some nitrogen and oxygen of the air to form nitrogen oxide:

$$N_2(g) + O_2(g) \longrightarrow 2\,NO(g)$$

Nitrogen dioxide is produced by the reaction of NO with additional O_2 from the air:

$$2\,NO(g) + O_2(g) \longrightarrow 2\,NO_2(g)$$

The NO_2 reacts with water to form nitric acid:

$$3\,NO_2(g) + H_2O(l) \longrightarrow 2\,HNO_3(l) + NO(g)$$

The nitric acid is washed to the earth where it forms nitrates in the soil, which can be used by plants as nutrients.

Certain soil bacteria, as well as nitrogen-fixing bacteria in the root nodules of leguminous plants (such as peas, beans, and alfalfa) fix atmospheric nitrogen into compounds that plants can assimilate. Fertilizers are used to augment the fixed nitrogen in the soil. Nitrogen-containing fertilizers are made from ammonia, which is itself produced commercially by a nitrogen-fixation process, the **Haber process:**

$$N_2(g) + 3\,H_2(g) \rightleftharpoons 2\,NH_3(g)$$

In the second stage of the nitrogen cycle, plants use the fixed nitrogen in the soil to make plant protein. The plant protein is in turn eaten by animals and used to make animal protein. Indeed, humans obtain their fixed nitrogen from the ingestion of both plant and animal protein.

In the third part of the nitrogen cycle, the cycle is completed. The decay of the waste products of animal metabolism and the death and decay of plants and animals liberates nitrogen as an end product. The N_2, therefore, is returned to the air.

21.3 Occurrence and Preparation of the Group V A Elements

The principal natural sources of the group V A elements are listed in Table 21.3. Nitrogen, like oxygen (see Section 20.2) and the noble gases (see Section 22.9), is produced commercially by the fractionation of liquid air. Nitrogen from this source, in cylinders, is usually employed when the gas is needed in the laboratory. On occasion, however, small amounts for laboratory use may be obtained by heating an aqueous solution saturated with ammonium chloride and sodium nitrite,

$$NH_4^+(aq) + NO_2^-(aq) \longrightarrow N_2(g) + 2H_2O$$

or by heating either sodium azide or barium azide,

$$2NaN_3(s) \longrightarrow 2Na(1) + 3N_2(g)$$

Phosphorus is the only member of group V A that does not occur in nature as an uncombined element. It is prepared industrially by heating a mixture of phosphate rock, sand, and coke in an electric furnace:

Table 21.3	Occurrence of the group V A elements	
Element	Percent of Earth's crust	Occurrence
nitrogen	0.0046 (0.03 including atmosphere)	N_2 (atmosphere) $NaNO_3$ (Chilean saltpeter)
phosphorus	0.12	$Ca_3(PO_4)_2$ (phosphate rock), $Ca_5(PO_4)_3F$ and $Ca_5(PO_4)_3Cl$ (apatite)
arsenic	5×10^{-4}	$FeAsS$(arsenopyrite), As_4S_4 (realgar), As_2S_3 (orpiment), As_4O_6 (arsenolite); native As; in ores of Cu, Pb, Co, Ni, Zn, Sn, Ag, and Au
antimony	5×10^{-5}	Sb_2S_3 (stibnite), Sb_4O_6 (senarmontite); native Sb; in ores of Cu, Pb, Ag, and Hg
bismuth	1×10^{-5}	Bi_2S_3 (bismuthinite), Bi_2O_3 (bismite); native Bi; in ores of Cu, Pb, Sn, Co, Ni, Ag, and Au

Mining phosphate rock. *W. R. Grace & Co.*

$$2 \, Ca_3(PO_4)_2(s) + 6 \, SiO_2(s) \longrightarrow 6 \, CaSiO_3(l) + P_4O_{10}(g)$$

$$P_4O_{10}(g) + 10 \, C(s) \longrightarrow P_4(g) + 10 \, CO(g)$$

The calcium silicate is withdrawn as a molten slag from the bottom of the furnace, and the product gases are passed through water, which condenses the phosphorus vapor into a white solid.

Arsenic, antimony, and bismuth are obtained by carbon reduction of their oxides at elevated temperatures:

$$As_4O_6(s) + 6 \, C(s) \longrightarrow As_4(g) + 6 \, CO(g)$$

An important industrial source of the oxides is the flue dust obtained from the processes used in the production of certain metals, notably copper and lead. In addition, the oxides are obtained by roasting the sulfide ores of the elements in air; for example,

$$2 \, Sb_2S_3(s) + 9 \, O_2(g) \longrightarrow Sb_4O_6(g) + 6 \, SO_2(g)$$

Although arsenic, antimony, and bismuth all occur as native ores, only the deposits of native bismuth are sufficiently large to be of commercial importance.

21.4 Nitrides and Phosphides

Elementary nitrogen reacts with a number of metals at high temperatures to form **ionic nitrides,** high-melting, white, crystalline solids that contain the N^{3-} ion. The group II A metals, cadmium, and zinc form ionic nitrides with the formula

M_3N_2 (where M is Be, Mg, Ca, Sr, Ba, Cd, or Zn) and lithium forms Li_3N. Ionic nitrides react with water to yield ammonia and hydroxides:

$$Ca_3N_2(s) + 6\,H_2O \longrightarrow 3\,Ca^{2+}(aq) + 6\,OH^-(aq) + 2\,NH_3(g)$$

Interstitial nitrides are made at elevated temperatures from many transition metals, in powdered form, and nitrogen or ammonia. A crystal of an interstitial nitride (VN, Fe_4N, W_2N, and TiN are examples) consists of metal atoms arranged in a lattice with nitrogen atoms occupying lattice holes (the interstices). These substances, therefore, frequently deviate from exact stoichiometry. They resemble metals and are hard, extremely high melting, good electrical conductors, and chemically unreactive.

Covalent nitrides include such compounds as S_4N_4, P_3N_5, Si_3N_4, Sn_3N_4, BN, and AlN. Some of these compounds are molecular in form. Others, such as BN and AlN, are substances in which a large number of atoms of the two elements are covalently bonded together into a network crystal. Both BN and AlN are made by reacting the elements at high temperatures. Two C atoms taken together have the same number of valence electrons (8) as one B atom (3 valence electrons) and one N atom (5 valence electrons) combined. The compound BN, therefore, may be considered to be isoelectronic with carbon. Indeed, BN is known in two crystalline modifications, one resembling graphite and another extremely hard form resembling diamond.

Many metals react with white phosphorus to form phosphides. The group II A elements form phosphides with the formula M_3P_2 (where M is Be, Mg, Ca, Sr, or Ba), lithium forms Li_3P and sodium forms Na_3P. These compounds readily react with water to form phosphine. For example,

$$Ca_3P_2(s) + 6\,H_2O \longrightarrow 3\,Ca^{2+}(aq) + 6\,OH^-(aq) + PH_3(g)$$

The phosphides of the group III A elements (such as BP, AlP, and GaP) form covalent network crystals similar to silicon, and like silicon these substances are semiconductors. Many phosphides of the transition metals are known (FeP, Fe_2P, Co_2P, RuP, and OsP_2 are examples). These substances are gray-black, semimetallic crystals that are electrical conductors.

The reactions of metals with arsenic, antimony, and, to a lesser extent, bismuth yield arsenides, stibnides, and bismuthides. These compounds become progressively more difficult to prepare as the atomic number of the group V A element increases.

21.5 Hydrogen Compounds

The group V A elements all form hydrogen compounds, the most important of which is ammonia, NH_3. Large quantities of ammonia are commercially prepared by the direct union of the elements (**Haber process**):

$$N_2(g) + 3\,H_2(g) \rightleftharpoons 2\,NH_3(g)$$

Ammonia is the only hydrogen compound of the V A elements that can be prepared directly. The reaction is conducted under high pressures (from 100 to 1000 atm), at 400° to 550°C, and in the presence of a catalyst. One catalyst, so employed, consists of finely divided iron and Fe_3O_4 containing small amounts of K_2O and Al_2O_3.

Smaller quantities of ammonia are produced as a by-product in the manufacture of coke by the destructive distillation of coal. Ammonia was formerly produced commercially by the reaction of calcium cyanamide, CaNCN, with steam under pressure:

$$CaNCN(s) + 3\,H_2O(g) \longrightarrow CaCO_3(s) + 2\,NH_3(g)$$

However, the Haber process has largely displaced this method as a commercial source of ammonia, and calcium cyanamide is produced chiefly as a fertilizer and as a raw material in the manufacture of certain nitrogen-containing organic compounds. Calcium cyanamide is produced in a two-step process. Calcium carbide, CaC_2, is made by the reaction of CaO and coke in an electric furnace:

$$CaO(s) + 3\,C(s) \longrightarrow CaC_2(s) + CO(g)$$

and the calcium carbide is reacted with relatively pure nitrogen at approximately 1000°C to produce calcium cyanamide:

$$CaC_2(s) + N_2(g) \longrightarrow CaNCN(s) + C(s)$$

In the laboratory, ammonia is conveniently prepared by the hydrolysis of nitrides (see Section 21.4) or by heating an ammonium salt with a strong alkali, such as NaOH or $Ca(OH)_2$, either dry or in solution:

$$NH_4^+(aq) + OH^-(aq) \longrightarrow NH_3(g) + H_2O$$

The ammonia molecule,

$$H-\overset{\displaystyle ..}{N}-H$$
$$|$$
$$H$$

is trigonal pyramidal with the nitrogen atom at the apex; this compound is associated through hydrogen bonding in the liquid and solid states.

Aqueous solutions of ammonia are alkaline:

$$NH_3(aq) + H_2O \rightleftharpoons NH_4^+(aq) + OH^-(aq)$$

In solution or as a dry gas, ammonia reacts with acids to produce ammonium salts:

$$NH_3(g) + HCl(g) \longrightarrow NH_4Cl(s)$$

The ammonium ion is tetrahedral.

Nitrogen is formed when ammonia is burned in pure oxygen:

$$4\,NH_3(g) + 3\,O_2(g) \longrightarrow 2\,N_2(g) + 6\,H_2O(g)$$

However, when a mixture of ammonia and air is passed over platinum gauze at 1000°C, nitric oxide, NO, is produced:

$$4\,NH_3(g) + 5\,O_2(g) \longrightarrow 4\,NO(g) + 6\,H_2O(g)$$

This catalyzed oxidation of NH_3 is a part of the Ostwald process for the manufacture of nitric acid (see Section 21.7).

Hydrazine, N_2H_4, may be considered to be derived from NH_3 by the replacement of a H atom by a $-NH_2$ group:

$$H-\overset{..}{N}-\overset{..}{N}-H$$
$$\quad\ \ |\quad\ |$$
$$\quad\ \ H\quad H$$

The compound, which is a liquid, may be prepared by oxidizing NH_3 with NaOCl. Hydrazine is less basic than NH_3 but does form cations in which one proton or two protons are bonded to the free electron pairs of the molecule. For example,

$$\left[\begin{array}{c} H \\ | \\ H-\overset{..}{N}-N-H \\ |\quad\ | \\ H\quad H \end{array}\right]^+ Cl^- \qquad \left[\begin{array}{c} H\ \ \ H \\ |\ \ \ | \\ H-N-N-H \\ |\ \ \ | \\ H\ \ \ H \end{array}\right]^{2+} 2\,Cl^-$$

Hydrazine is a strong reducing agent and has found some use in rocket fuels.

Hydroxylamine, NH_2OH, is another compound which, like hydrazine, may be considered to be derived from the NH_3 molecule:

$$H-\overset{..}{N}-O-H$$
$$\quad\ \ |$$
$$\quad\ \ H$$

Like hydrazine, hydroxylamine is a weaker base than NH_3, but salts containing the $[NH_3OH]^+$ ion may be prepared.

Hydrazoic acid, HN_3, is another hydrogen compound of nitrogen. The structure of hydrazoic acid may be represented as a resonance hybrid:

$$H-\overset{..}{N}=\overset{\oplus}{N}=\overset{..}{N}\overset{\ominus}{:} \longleftrightarrow H-\overset{..}{\underset{..}{N}}\overset{\ominus}{-}\overset{\oplus}{N}\equiv N:$$

The two N—N bond distances of the molecule are not the same; the distance from the central N to the N bearing the H atom (124 pm) is longer than the other (113 pm). Hydrazoic acid may be made by reacting hydrazine (which forms the $[N_2H_5]^+$ ion in acidic solution) with nitrous acid (HNO_2):

$$N_2H_5^+(aq) + HNO_2(aq) \longrightarrow HN_3(aq) + H^+(aq) + 2\,H_2O$$

The free acid, a low-boiling liquid, may be obtained by distillation of the water solution. Hydrazoic acid is a weak acid. The heavy metal salts of the acid, such as lead azide, $Pb(N_3)_2$, explode upon being struck and are used in detonation caps.

Phosphine (PH_3) is a very poisonous, colorless gas that is prepared by the hydrolysis of phosphides or by the reaction of white phosphorus with concentrated solutions of alkalies:

$$P_4(s) + 3\,OH^-(aq) + 3\,H_2O \longrightarrow PH_3(g) + \quad 3\,H_2PO_2^-(aq)$$
$$\qquad\qquad\qquad\qquad\qquad\qquad\qquad\qquad \textit{hypophosphite ion}$$

The PH_3 molecule is pyramidal, similar to the NH_3 molecule. Unlike NH_3, however, the compound is not associated by hydrogen bonding in the liquid state.

Phosphine is much less basic than ammonia. Phosphonium compounds, such as $PH_4I(s)$ which can be made from dry $PH_3(g)$ and $HI(g)$, are unstable. They decompose at relatively low temperatures, or in aqueous solution, to yield the component gases.

Arsine (AsH_3), stibine (SbH_3), and bismuthine (BiH_3) are extremely poisonous gases that may be produced by the hydrolysis of arsenides, stibnides, and bismuthides (for example, Na_3As, Zn_3Sb_2, and Mg_3Bi_2). The yields of the hydrogen compounds become poorer with increasing molecular weight. Very poor yields of bismuthine are obtained by this method. The stability of the hydrogen compounds declines in the series from NH_3 to BiH_3. Bismuthine is very unstable and decomposes to the elements at room temperature. Arsine and stibine may be similarly decomposed by warming.

Arsine, stibine, and bismuthine have no basic properties and do not form salts with acids.

21.6 Halogen Compounds

The most important halides of the V A elements are the trihalides (for example, NF_3) and the pentahalides (for example, PF_5). All four binary trihalides of each of the V A elements have been made, but the tribromide and triiodide of nitrogen can be isolated only in the form of ammoniates ($NBr_3 \cdot 6\,NH_3$ and $NI_3 \cdot x\,NH_3$). The nitrogen trihalides are prepared by the halogenation of ammonia gas (NF_3, NBr_3), of an ammonium salt in acidic solution (NCl_3, NBr_3), or of concentrated aqueous ammonia (NI_3). Each of the trihalides of P, As, Sb, and Bi is prepared by direct halogenation of the V A element using a stoichiometric excess of the V A element to prevent pentahalide formation.

Bismuth trifluoride is an ionic compound, but the other trihalides are covalent. In the gaseous state the covalent trihalides exist as trigonal pyramidal molecules (see Figure 21.3). This molecular form persists in the liquid state and in all of the solids except AsI_3, SbI_3, and BiI_3, which crystallize in covalent layer lattices.

Nitrogen trifluoride is a very stable colorless gas, whereas the other trihalides of nitrogen are explosively unstable. The trihalides undergo hydrolysis:

$$NCl_3(l) + 3\,H_2O \longrightarrow NH_3(g) + 3\,HOCl(aq)$$

$$PCl_3(l) + 3\,H_2O \longrightarrow H_3PO_3(aq) + 3\,H^+(aq) + 3\,Cl^-(aq)$$

$$AsCl_3(l) + 3\,H_2O \longrightarrow H_3AsO_3(aq) + 3\,H^+(aq) + 3\,Cl^-(aq)$$

$$SbCl_3(s) + H_2O \longrightarrow SbOCl(s) + 2\,H^+(aq) + 2\,Cl^-(aq)$$

$$BiCl_3(s) + H_2O \longrightarrow BiOCl(s) + 2\,H^+(aq) + 2\,Cl^-(aq)$$

Nitrogen is more electronegative than Cl, and in the hydrolysis of NCl_3, N and Cl appear in $3-$ and $1+$ oxidation states, respectively. In each of the other hydrolysis reactions, the V A element appears in a $3+$ oxidation state and Cl in a $1-$ state. The metallic character of the V A elements increases with increasing atomic number, and Sb and Bi occur as oxo cations (SbO^+ and BiO^+) in the hydrolysis products of $SbCl_3$ and $BiCl_3$.

The pentahalide series is not so complete as the trihalide series. Since N has no d orbitals in its valence level, N can form no more than four covalent bonds.

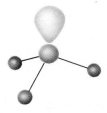

Figure 21.3 Molecular structure of the covalent trihalides of group V A elements

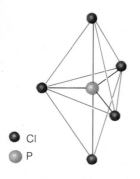

Figure 21.4 Structure of the gaseous PCl_5 molecule

Pentahalides of N, therefore, do not exist. All pentahalides of phosphorus are known with the exception of the pentaiodide; presumably there is not sufficient room around a phosphorus atom to accommodate five large iodine atoms. In addition, AsF_5, SbF_5, BiF_5, and $SbCl_5$ have been prepared.

The pentahalides may be prepared by direct reaction of the elements using an excess of the halogen and by the reaction of the halogen with the trihalide:

$$PCl_3(l) + Cl_2(g) \rightleftharpoons PCl_5(s)$$

The preceding reaction is reversible. In the gas phase the pentahalides dissociate to varying degrees.

The pentahalides are trigonal bipyramidal molecules in the gaseous and liquid state (see Figure 21.4). The crystal lattice of $SbCl_5$ consists of such molecules. Solid PCl_5 and PBr_5, however, form ionic lattices composed of PCl_4^+ and PCl_6^- and PBr_4^+ and Br^-, respectively. Apparently it is impossible to pack six bromine atoms around a phosphorus atom since PBr_6^- does not form. The cations are tetrahedral and the PCl_6^- ion is octahedral (see Figure 21.5).

The phosphorus pentahalides undergo hydrolysis in two steps. For example,

$$PCl_5(s) + H_2O \longrightarrow POCl_3(l) + 2H^+(aq) + 2Cl^-(aq)$$

$$POCl_3(l) + 3H_2O \longrightarrow H_3PO_4(aq) + 3H^+(aq) + 3Cl^-(aq)$$

The phosphoryl halides, POX_3 (X = F, Cl, or Br) can be prepared by the hydrolysis of the appropriate pentahalide in a limited amount of water or by the reaction of the trihalide with oxygen:

$$2PCl_3(l) + O_2(g) \longrightarrow 2POCl_3(l)$$

Molecules of the phosphoryl halides have a PX_3 grouping arranged as a trigonal pyramid (see Figure 21.3) with an oxygen atom bonded to the phosphorus atom, thus forming a distorted tetrahedron.

A number of mixed trihalides (for example, NF_2Cl, $PFBr_2$, and $SbBrI_2$) and mixed pentahalides (for example, PCl_2F_3, $PClF_4$, and $SbCl_3F_2$) have been prepared. In addition, halides are known that conform to the general formula E_2X_4: N_2F_4, P_2Cl_4, P_2I_4, and As_2I_4. These compounds have molecular structures similar to the structure of hydrazine, N_2H_4.

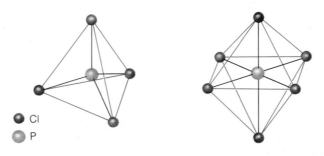

Figure 21.5 Structures of PCl_4^+ and PCl_6^-

Isomers are substances that have the same molecular formula but differ in the way the constituent atoms are arranged into molecules. Dinitrogen difluoride exists in two isomeric forms:

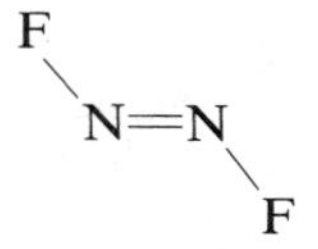

$$
\underset{trans}{
\begin{array}{c}
\text{F} \\
\diagdown \\
\text{N}\!=\!\text{N} \\
\diagdown \\
\text{F}
\end{array}
}
\qquad
\underset{cis}{
\begin{array}{c}
\text{N}\!=\!\text{N} \\
\diagup \quad \diagdown \\
\text{F} \qquad \text{F}
\end{array}
}
$$

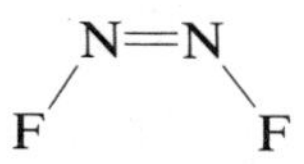

The double bond, composed of a σ bond and a π bond, between the two nitrogen atoms prevents free rotation about the nitrogen-nitrogen axis. In the *cis* isomer both fluorine atoms are on the same side of the double-bonded nitrogen atoms, whereas in the *trans* isomer the fluorine atoms are on opposite sides. Both molecules are planar.

21.7 Oxides and Oxyacids of Nitrogen

Oxides are known for every oxidation state of nitrogen from $1+$ to $5+$:

1. The 1+ oxidation state. Dinitrogen oxide (also called nitrous oxide), N_2O, is prepared by gently heating molten ammonium nitrate:

$$NH_4NO_3(l) \longrightarrow N_2O(g) + 2\,H_2O(g)$$

It is a colorless gas and is relatively unreactive. At temperatures around $500°C$, however, dinitrogen oxide decomposes to nitrogen and oxygen; hence, N_2O supports combustion. Molecules of N_2O are linear, and the electronic structure of the compound may be represented as a resonance hybrid:

$$:\!\overset{\ominus}{\underset{..}{\overset{..}{N}}}\!=\!\overset{\oplus}{N}\!=\!\overset{..}{\underset{..}{O}}\!: \longleftrightarrow :\!N\!\equiv\!\overset{\oplus}{N}\!-\!\overset{..}{\underset{..}{O}}\!:^{\ominus}$$

Dinitrogen oxide is commonly called "laughing gas" because of the effect it produces when breathed in small amounts. The gas is used as a general anesthetic, and because of its solubility in cream, it is the gas used to charge whipped cream aerosol cans.

2. The 2+ oxidation state. Nitrogen oxide (also called nitric oxide), NO, may be prepared by the direct reaction of the elements at high temperatures:

$$N_2(g) + O_2(g) \longrightarrow 2\,NO(g)$$

The reaction is endothermic ($\Delta H = +90.4$ kJ/mol), but even at $3000°C$ the yield of NO is only approximately 4%. In a successful preparation the hot gases from the reaction must be rapidly cooled to prevent the decomposition of NO back into nitrogen and oxygen. By this reaction, atmospheric nitrogen is fixed during lightning storms. This reaction also serves as the basis of the **arc process** of nitrogen fixation in which an electric arc is used to provide the high temperatures necessary for the direct combination of nitrogen and oxygen. As a commercial source of NO, the arc process has been supplanted by the catalytic oxidation of ammonia from the Haber process (see Section 21.5).

The NO molecule contains an odd number of electrons, which means that one electron must be unpaired; for this reason, NO is paramagnetic. The electronic structure of the molecule may be represented by the resonance forms:

$$:\!N\!=\!\ddot{O}\!: \longleftrightarrow :\!\overset{\ominus}{N}\!=\!\overset{\oplus}{\ddot{O}}\!:$$

The structure, however, is best described by molecular orbital theory, which assigns a bond order of $2\frac{1}{2}$ to the molecule and indicates that the odd electron is in a π^* orbital. The loss of an electron from NO produces the nitrosonium ion, NO^+. Since the electron is lost from an antibonding orbital, the NO^+ ion has a bond order of 3, and the bond distance in NO^+ (106 pm) is shorter than the bond distance in the NO molecule (114 pm). Ionic compounds of NO^+ are known (for example, $NO^+\ HSO_4^-$, $NO^+\ ClO_4^-$, and $NO^+\ BF_4^-$).

Whereas odd-electron molecules are generally very reactive and highly colored, nitric oxide is only moderately reactive and is a colorless gas (condensing to a blue liquid and blue solid at low temperatures). In addition, NO shows little tendency to associate into N_2O_2 molecules by electron pairing. Nitrogen oxide reacts instantly with oxygen at room temperature to form nitrogen dioxide:

$$2\,NO(g) + O_2(g) \longrightarrow 2\,NO_2(g)$$

3. The 3+ oxidation state. Dinitrogen trioxide, N_2O_3, forms as a blue liquid when an equimolar mixture of nitric oxide and nitrogen dioxide is cooled to $-20°C$:

$$NO(g) + NO_2(g) \longrightarrow N_2O_3(l)$$

The compound is unstable under ordinary conditions and decomposes into NO and NO_2. Both NO and NO_2 are odd-electron molecules; N_2O_3 is formed by electron pairing, and the N_2O_3 molecule is thought to contain a N—N bond. Dinitrogen trioxide is the anhydride of nitrous acid, HNO_2, and dissolves in aqueous alkali to produce the nitrite ion, NO_2^-.

4. The 4+ oxidation state. Nitrogen dioxide, NO_2, and dinitrogen tetroxide, N_2O_4, exist in equilibrium:

$$2\,NO_2 \rightleftharpoons N_2O_4$$

Nitrogen dioxide consists of odd-electron molecules, is paramagnetic, and is brown in color. The dimer, in which the electrons are paired, is diamagnetic and colorless. In the solid state the oxide is colorless and consists of pure N_2O_4. The liquid is yellow in color and consists of a dilute solution of NO_2 in N_2O_4. As the temperature is raised, the gas contains more and more NO_2 and becomes deeper and deeper brown in color. At $135°C$ approximately 99% of the mixture is NO_2.

Nitrogen dioxide molecules are angular:

The structure of N_2O_4 is thought to be planar with two NO_2 units joined by a N—N bond.

Nitrogen dioxide is produced by the reaction of nitric oxide with oxygen. In the laboratory the compound is conveniently prepared by heating lead nitrate:

$$2\,Pb(NO_3)_2(s) \longrightarrow 2\,PbO(s) + 4\,NO_2(g) + O_2(g)$$

5. The 5+ oxidation state. Dinitrogen pentoxide, N_2O_5, is the acid anhydride of nitric acid; N_2O_5 may be prepared from this acid by dehydration using phosphorus (V) oxide:

$$4\,HNO_3(g) + P_4O_{10}(s) \longrightarrow 4\,HPO_3(s) + 2\,N_2O_5(g)$$

The compound is a colorless, crystalline material that sublimes at $32.5°C$. The vapor consists of N_2O_5 molecules which are thought to be planar with the two nitrogen atoms joined through an oxygen atom (O_2NONO_2).

The electronic structure of the molecule may be represented as a resonance hybrid of

and other equivalent structures that have different arrangements of the double bonds. The compound is unstable in the vapor state and decomposes according to the equation

$$2\,N_2O_5(g) \longrightarrow 4\,NO_2(g) + O_2(g)$$

Crystals of N_2O_5 are composed of nitronium, NO_2^+, and nitrate, NO_3^-, ions. The compound is dissociated into these two ions in solutions in anhydrous sulfuric acid, nitric acid, and phosphoric acid. The nitrate ion is triangular planar. The nitronium ion is linear; it is isoelectronic with CO_2 and may be considered as a nitrogen dioxide molecule minus the odd electron. The ion is probably a reaction intermediate in certain reactions of nitric acid in the presence of sulfuric acid (nitrations). Ionic nitronium compounds have been prepared (for example, $NO_2^+ClO_4^-$, $NO_2^+BF_4^-$, $NO_2^+PF_6^-$).

The most important oxyacid of nitrogen is nitric acid, HNO_3, in which nitrogen exhibits an oxidation number of 5+. Commercially, nitric acid is produced by the **Ostwald process.** Nitric oxide from the catalytic oxidation of ammonia is reacted with oxygen to form nitrogen dioxide. This gas, together with excess oxygen, is passed into a tower where it reacts with warm water:

$$3\,NO_2(g) + H_2O \longrightarrow 2\,H^+(aq) + 2\,NO_3^-(aq) + NO(g)$$

The excess oxygen converts the NO into NO_2; this NO_2 then reacts with water as before. In this cyclic manner the nitric oxide is eventually completely converted into nitric acid. The product of the Ostwald process is about 70% HNO_3 and is known as concentrated nitric acid; more concentrated solutions may be prepared from it by distillation.

Pure nitric acid is a colorless liquid that boils at $83°C$. It may be prepared in the laboratory by heating sodium nitrate with concentrated sulfuric acid:

$$NaNO_3(s) + H_2SO_4(l) \longrightarrow NaHSO_4(s) + HNO_3(g)$$

This preparation (from Chilean saltpeter) is a minor commercial source of the acid.

The HNO_3 molecule is planar and may be represented as a resonance hybrid:

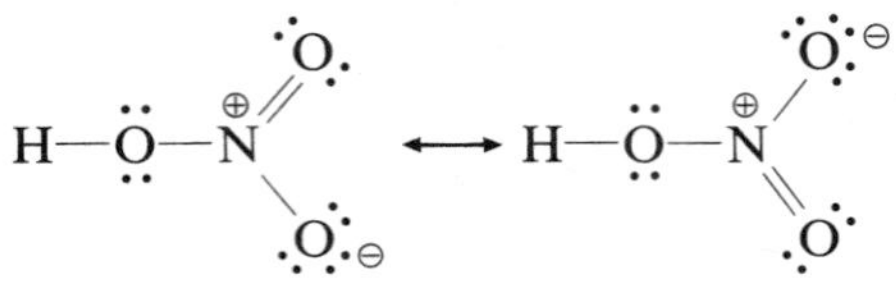

Nitric acid is a strong acid and is almost completely dissociated in aqueous solution. Most salts of nitric acid, which are called nitrates, are very soluble in water. The nitrate ion is triangular planar:

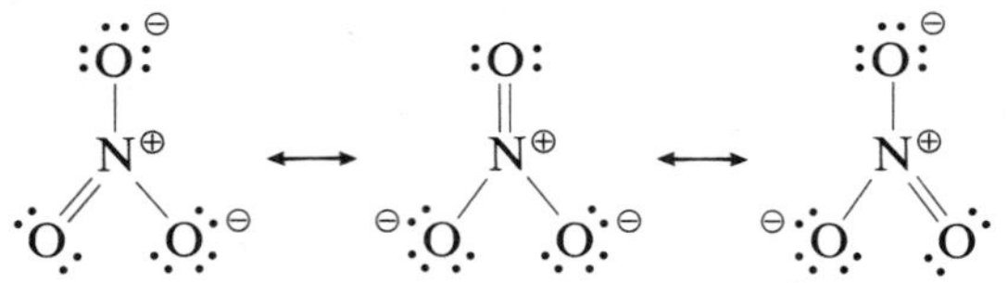

Nitric acid is a powerful oxidizing agent; it oxidizes most nonmetals (generally to oxides or oxyacids of their highest oxidation state) and all metals with the exception of a few of the noble metals. Many unreactive metals, such as silver and copper, that do not react to yield hydrogen with nonoxidizing acids, such as HCl, dissolve in nitric acid.

In nitric acid oxidations, hydrogen is almost never obtained; instead, a variety of nitrogen-containing compounds, in which nitrogen is in a lower oxidation state, is produced (see Table 21.4). The product to which HNO_3 is reduced depends upon the concentration of the acid, the temperature, and the nature of the material being oxidized. A mixture of products is usually obtained, but the principal product in many cases is NO when dilute HNO_3 is employed and the nitrogen(IV) oxides when concentrated HNO_3 is used.

Dilute:

$$3\,Cu(s) + 8\,H^+(aq) + 2\,NO_3^-(aq) \longrightarrow 3\,Cu^{2+}(aq) + 2\,NO(g) + 4\,H_2O$$

Concentrated:

$$Cu(s) + 4\,H^+(aq) + 2\,NO_3^-(aq) \longrightarrow Cu^{2+}(aq) + 2\,NO_2(g) + 2\,H_2O$$

Table 21.4	**Standard electrode potentials for reductions of the nitrate ion**
Half Reaction	Standard Electrode Potential ($\mathscr{E}^\circ_{red}$)
$e^- + 2\,H^+ + NO_3^- \rightleftharpoons NO_2 + H_2O$	+0.80 V
$8e^- + 10\,H^+ + NO_3^- \rightleftharpoons NH_4^+ + 3\,H_2O$	+0.88 V
$2e^- + 3\,H^+ + NO_3^- \rightleftharpoons HNO_2 + H_2O$	+0.94 V
$3e^- + 4\,H^+ + NO_3^- \rightleftharpoons NO + 2\,H_2O$	+0.96 V
$8e^- + 10\,H^+ + 2\,NO_3^- \rightleftharpoons N_2O + 5\,H_2O$	+1.12 V
$10e^- + 6\,H^+ + 2\,NO_3^- \rightleftharpoons N_2 + 6\,H_2O$	+1.25 V

In some instances, however, strong reducing agents are known to produce almost pure compounds of nitrogen in lower oxidation states. (For example, the reaction of zinc and dilute nitric acid yields NH_3 as the reduction product of HNO_3.)

The half-reactions for the reduction of the nitrate ion in acid solution (see Table 21.4) show the $\mathscr{E}^\circ_{red}$ values to be strongly dependent upon the $H^+(aq)$ concentration. This concentration dependence is experimentally observed— below a concentration of 2 M, nitric acid has little more oxidizing power than solutions of HCl of corresponding concentration.

The oxyacid of nitrogen in which nitrogen has an oxidation number of 3+ is nitrous acid, HNO_2. The compound is unstable toward disproportionation (particularly when warmed):

$$3\,HNO_2(aq) \longrightarrow H^+(aq) + NO_3^-(aq) + 2\,NO(g) + H_2O$$

As a result, pure HNO_2 has never been prepared. Instead, aqueous solutions of the acid are usually prepared by adding a strong acid (such as HCl) to a *cold* aqueous solution of a nitrite (such as $NaNO_2$):

$$H^+(aq) + NO_2^-(aq) \longrightarrow HNO_2(aq)$$

Such solutions are used directly, without attempting to isolate HNO_2, in laboratory procedures that require HNO_2. Aqueous solutions of HNO_2 may also be prepared by adding an equimolar mixture of $NO(g)$ and $NO_2(g)$ to water:

$$NO(g) + NO_2(g) + H_2O \rightleftharpoons 2\,HNO_2(aq)$$

The reaction is exothermic and reversible. Note that the acid anhydride of nitrous acid, N_2O_3, is prepared by the reaction of $NO(g)$ and $NO_2(g)$ at $-20°C$.

The acid is weak and may function as an oxidizing agent or a reducing agent. The electronic structure may be represented as:

Nitrites are prepared by adding $NO(g)$ and $NO_2(g)$ to solutions of alkalies:

$$NO(g) + NO_2(g) + 2\,OH^-(aq) \longrightarrow 2\,NO_2^-(aq) + H_2O$$

The nitrites of the I A metals are formed when the nitrates are heated. They may also be prepared by heating the nitrate with a reducing agent, such as lead, iron or coke:

$$NaNO_3(l) + C(s) \longrightarrow NaNO_2(l) + CO(g)$$

The nitrite ion is angular:

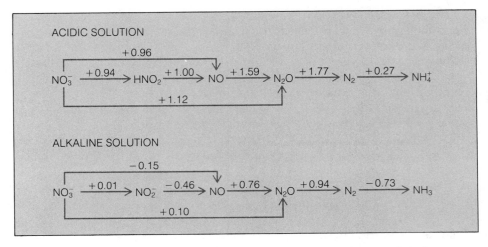

Figure 21.6 Electrode potential diagrams for nitrogen and some of its compounds (values given in volts)

Electrode potential diagrams for N_2 and some of its compounds appear in Figure 21.6. The compounds of nitrogen shown in the diagrams are much stronger oxidizing agents in acidic solution than in alkaline solution. In fact, the NO_3^- ion is a very weak oxidant in alkaline solution. Conversely, a given compound of nitrogen is easier to oxidize in alkaline solution than in acidic solution with the exception of NO_3^- which cannot be oxidized. The potentials also indicate that HNO_2 is unstable toward disproportionation to NO and NO_3^- in acid, whereas in alkaline solution the NO_2^- ion is stable toward such disproportionation.

21.8 Oxides and Oxyacids of Phosphorus

The two important oxides of phosphorus contain phosphorus in oxidation states of 3+ (P_4O_6) and 5+ (P_4O_{10}). Phosphorus(III) oxide, P_4O_6, is frequently called phosphorus trioxide, a name which dates from a time when only the empirical formula of the compound, P_2O_3, was known. The compound is a colorless substance that melts at 23.9°C and is the principal product when white phosphorus is burned in a limited supply of air. The structure of the P_4O_6 molecule (shown in Figure 21.7) is based on a P_4 tetrahedron (see Figure 21.1) and has an O atom inserted in each of the six edges of the tetrahedron.

Phosphorus(V) oxide, P_4O_{10}, is often called phosphorus pentoxide (based on the empirical formula, P_2O_5). The oxide is the product of the combustion of white phosphorus in an excess of oxygen. It is a white powder that sublimes at 360°C. The molecular structure of P_4O_{10} (see Figure 21.7) can be derived from the structure of P_4O_6 by adding an extra O atom to each P atom.

Phosphorus(V) oxide has a great affinity for water and is a very effective drying agent. Many different phosphoric acids may be prepared by the addition of water to P_4O_{10}. The most important acid of phosphorus in the 5+ state, orthophosphoric acid (usually called phosphoric acid) results from the complete hydration of the oxide:

$$P_4O_{10} + 6H_2O \longrightarrow 4H_3PO_4$$

Phosphoric acid is obtained commercially by this means or by treating phosphate rock with sulfuric acid:

$$Ca_3(PO_4)_2(s) + 3H_2SO_4(l) \longrightarrow 2H_3PO_4(l) + 3CaSO_4(s)$$

The compound is a colorless solid but is generally sold as an 85% solution. The electronic structure of H_3PO_4 may be represented as

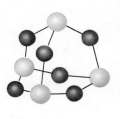

P_4O_6

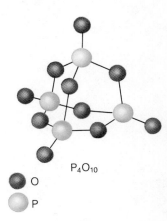

P_4O_{10}

○ O
○ P

Figure 21.7 Structures of P_4O_6 and P_4O_{10}

since the P—O bond has double-bond character from $p\pi$-$d\pi$ bonding. The H_3PO_4 molecule and the ions derived from it are tetrahedral (Figure 21.8).

Phosphoric acid is a weak, triprotic acid without effective oxidizing power:

$$H_3PO_4(aq) \rightleftharpoons H^+(aq) + H_2PO_4^-(aq)$$

$$H_2PO_4^-(aq) \rightleftharpoons H^+(aq) + HPO_4^{2-}(aq)$$

$$HPO_4^{2-}(aq) \rightleftharpoons H^+(aq) + PO_4^{3-}(aq)$$

Three series of salts may be derived from H_3PO_4 (the sodium salts, for example, are: NaH_2PO_4, Na_2HPO_4, and Na_3PO_4). The product of a given neutralization depends upon the stoichiometric ratio of H_3PO_4 to alkali.

Phosphates are important ingredients of commercial fertilizers. Phosphate rock is too insoluble in water to be used directly for this purpose. The more soluble calcium dihydrogen phosphate is a satisfactory fertilizer ingredient, however, and may be obtained by treatment of phosphate rock with an acid:

$$Ca_3(PO_4)_2(s) + 2H_2SO_4(l) \longrightarrow Ca(H_2PO_4)_2(s) + 2CaSO_4(s)$$

The mixture of $Ca(H_2PO_4)_2$ and $CaSO_4$ is called **superphosphate fertilizer.** A higher yield of the dihydrogen phosphate is obtained if phosphoric acid is employed in the reaction instead of sulfuric acid:

$$Ca_3(PO_4)_2(s) + 4H_3PO_4(l) \longrightarrow 3Ca(H_2PO_4)_2(s)$$

Since nitrates are also important constituents of fertilizers, the mixture obtained by treatment of phosphate rock with nitric acid is a highly effective fertilizer:

$$Ca_3(PO_4)_2(s) + 4HNO_3(l) \longrightarrow Ca(H_2PO_4)_2(s) + 2Ca(NO_3)_2(s)$$

Condensed phosphoric acids have more than one P atom per molecule. The members of one group of condensed phosphoric acids, the **polyphosphoric acids,** conform to the general formula $H_{n+2}P_nO_{3n+1}$ where n is 2 to 10. Examples are

PO_4^{3-}

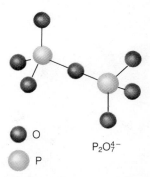

○ O
○ P

$P_2O_7^{4-}$

Figure 21.8 Structures of PO_4^{3-} and $P_2O_7^{4-}$

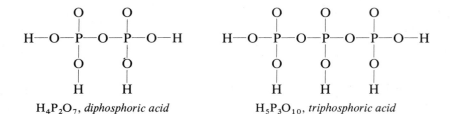

$H_4P_2O_7$, *diphosphoric acid* $H_5P_3O_{10}$, *triphosphoric acid*

The polyphosphoric acids and polyphosphates have chain structures based on PO_4 tetrahedra which are joined through O atoms that are common to adjacent tetrahedra (Figure 21.8).

The **metaphosphoric acids** constitute another group of condensed phosphoric acids. These compounds have the general formula $H_nP_nO_{3n}$ where n is 3 to 7. Some of the metaphosphoric acids are cyclic. For example:

$H_3P_3O_9$, *trimetaphosphoric acid* $H_4P_4O_{12}$, *tetrametaphosphoric acid*

In addition, there are high-molecular-weight, long-chain metaphosphoric acids which are always obtained as complex mixtures that are assigned the formula $(HPO_3)_n$. The molecules of these mixtures are based on long chains of

$$\begin{array}{c} O \\ \| \\ -P-O- \\ | \\ OH \end{array}$$

units joined in such a way that each P atom is tetrahedrally bonded to four O atoms, but the complete structures are very complicated and involve phosphorus-oxygen units that link two chains together.

The condensed phosphoric acids may be obtained by the controlled addition of water to P_4O_{10}. For example,

$$P_4O_{10}(s) + 4\,H_2O \longrightarrow 2\,H_4P_2O_7(s)$$
diphosphoric acid

$$P_4O_{10}(s) + 2\,H_2O \xrightarrow{\ 0°C\ } H_4P_4O_{12}(s)$$
tetrametaphosphoric acid

The dehydration of H_3PO_4 by heating also yields condensed acids. For example,

$$2\,H_3PO_4(l) \xrightarrow{\ 215°C\ } H_4P_2O_7(l) + H_2O(g)$$
diphosphoric acid

$$n\,H_3PO_4(l) \xrightarrow{\ 325°C\ } (HPO_3)_n(l) + n\,H_2O(g)$$
metaphosphoric acid mixture

On standing in water, all condensed phosphoric acids revert to H_3PO_4.

The oxyacid in which phosphorus has an oxidation number of $3+$ is phosphorous acid, H_3PO_3. Note that the *-ous* ending of the name of the acid differs from the *-us* ending of the name of the element. Phosphorous acid can be made by adding P_4O_6 to cold water,

$$P_4O_6(s) + 6 H_2O \longrightarrow 4 H_3PO_3(aq)$$

Condensed phosphorous acids also exist. Even though the H_3PO_3 molecule contains three hydrogen atoms, phosphorous acid is a weak, *diprotic* acid and is probably better formulated as $H_2(HPO_3)$:

$$H_2(HPO_3)(aq) \rightleftharpoons H^+(aq) + H(HPO_3)^-(aq)$$
$$H(HPO_3)^-(aq) \rightleftharpoons H^+(aq) + HPO_3^{2-}(aq)$$

The sodium salts NaH_2PO_3 and Na_2HPO_3 are known, but it is impossible to prepare Na_3PO_3.

Phosphorus has an oxidation number of $1+$ in hypophosphorous acid, H_3PO_2. Solutions of salts of the acid may be prepared by boiling white phosphorus with solutions of alkalies:

$$P_4(s) + 3 OH^-(aq) + 3 H_2O \longrightarrow PH_3(g) + 3 H_2PO_2^-(aq)$$

The acid, which is a colorless crystalline material, may be obtained by treating a solution of barium hypophosphite with sulfuric acid:

$$Ba^{2+}(aq) + 2 H_2PO_2^-(aq) + 2 H^+(aq) + SO_4^{2-}(aq) \longrightarrow$$
$$BaSO_4(s) + 2 H_3PO_2(aq)$$

removing the precipitated $BaSO_4$ by filtration, and evaporating the solution. Hypophosphorous acid is a weak monoprotic acid. The formula of the compound, therefore, is sometimes written $H(H_2PO_2)$:

$$H(H_2PO_2)(aq) \rightleftharpoons H^+(aq) + H_2PO_2^-(aq)$$

The number of protons released by phosphoric (three), phosphorous (two), and hypophosphorous (one) acids may be explained on the basis of the molecular structures of these compounds:

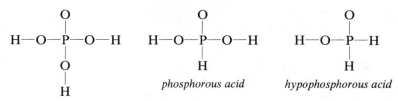

phosphoric acid

phosphorous acid

hypophosphorous acid

Only the hydrogen atoms bonded to oxygen atoms are acidic; those bonded to phosphorus atoms do not dissociate as $H^+(aq)$ in water. In each of these molecules the P—O bonds have *pπ-dπ* double-bond character.

Electrode potential diagrams for phosphorus and some of its compounds appear in Figure 21.9. The most striking feature of the diagrams is that all the

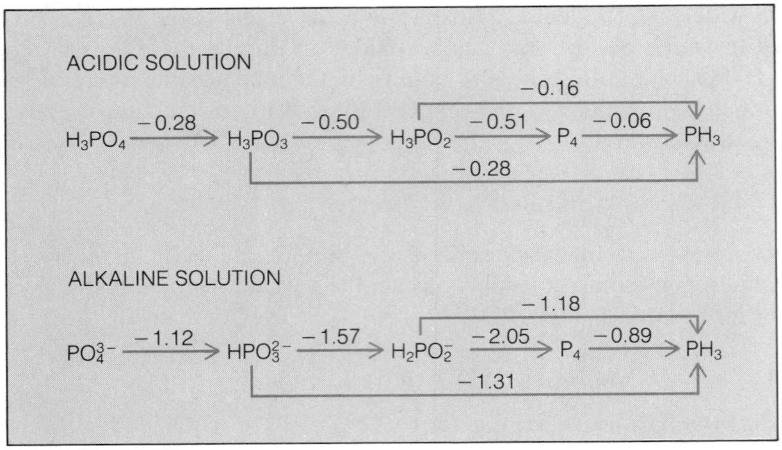

Figure 21.9 Electrode potential diagrams for phosphorus and some of its compounds

potentials are negative—in contrast to the potentials of the nitrogen diagram. Thus none of the substances is a good oxidizing agent, particularly in alkaline solution. Rather, H_3PO_3, H_3PO_2, and the salts of these acids have strong reducing properties; the acids are readily oxidized to H_3PO_4 and the anions are readily oxidized to PO_4^{3-}. With the exception of HPO_3^{2-} in alkaline solution, all species of intermediate oxidation state disproportionate; PH_3 is one of the products of each disproportionation.

21.9 Oxides and Oxyacids of As, Sb, and Bi

When arsenic, antimony, or bismuth is heated in air, a 3+ oxide is formed: As_4O_6, Sb_4O_6, or Bi_2O_3. These oxides serve as an excellent example of the change in metallic character that is observed within a group of elements of the periodic table. The lightest element of a group is the most nonmetallic, and its oxide, therefore, is the most acidic. The heaviest member of a group is the most metallic, and its oxide, therefore, is the most basic.

1. The oxide of the lightest element of the three, As_4O_6, is the most acidic of the three. Aqueous solutions of As_4O_6 are acidic and thought to contain arsenious acid, H_3AsO_3. When As_4O_6 dissolves in aqueous alkalies, arsenites (salts of AsO_3^{3-} and other forms of this anion) form. The oxide is exclusively acidic and will not dissolve in aqueous acids.

2. The oxide of the intermediate element of the three, Sb_4O_6, is amphoteric (it has both acidic and basic properties). The oxide will not dissolve in water, but it will dissolve in alkaline solutions (to give antimonites, salts of SbO_2^-) and also in acidic solutions (to give salts of SbO^+ or Sb^{3+}).

3. The oxide of the heaviest element of the three, Bi_2O_3, is exclusively basic. It will not dissolve in water or in aqueous alkalies, but it will dissolve in acids (to give salts of BiO^+ or Bi^{3+}). Bismuth(III) hydroxide, $Bi(OH)_3$, is the only true hydroxide of the group V A elements. When $OH^-(aq)$ is added to a solution of a Bi^{3+} salt, $Bi(OH)_3$ precipitates.

The molecular structures of the 5+ oxides of arsenic and antimony are not accurately known, and empirical formulas (As_2O_5 and Sb_2O_5) are employed. These oxides are prepared by the action of concentrated HNO_3 on the elements or the 3+ oxides, followed by the dehydration of the products of these reactions ($2 H_3AsO_4 \cdot H_2O$ and $Sb_2O_5 \cdot x H_2O$). The 5+ oxides are exclusively acidic. Orthoarsenic acid (H_3AsO_4, a triprotic acid) forms when As_2O_5 is dissolved in water. Arsenates (salts of AsO_4^{3-}) or antimonates (salts of $Sb(OH)_6^-$) can be prepared by dissolving As_2O_5 or Sb_2O_5 in aqueous alkalies. Sodium antimonate, $NaSb(OH)_6$, is one of the least soluble sodium salts, and its formation is often used as a test for Na^+.

Bismuth(V) oxide has never been prepared in the pure state; it is unstable and readily loses oxygen. The red-brown product obtained by the action of strong oxidizing agents (such as Cl_2, OCl^-, and $S_2O_8^{2-}$) on a suspension of Bi_2O_3 in an alkaline solution is thought to be impure Bi_2O_5.

21.10 Industrial Uses of the Group V A Elements

The most important industrial uses of the group V A elements and their compounds are:

1. Nitrogen. The manufacture of ammonia constitutes the principal use of elemental nitrogen. Smaller amounts of N_2 are used in the production of calcium cyanamid ($CaCN_2$). Because N_2 is relatively unreactive, gaseous nitrogen is used as an inert atmosphere in place of air for chemical and metallurgical processes that must be run in the absence of oxygen. The gas is used in food processing and packaging (coffee, for example) to prevent the spoilage and deterioration that is brought about by exposure to atmospheric oxygen. Liquid nitrogen has replaced liquid air in cryogenic work (low-temperature work) to avoid the danger associated with contact between the oxygen of the air and combustible materials. Liquid nitrogen is used for the preparation and transportation of frozen food as well as the transportation of perishable food.

The metal nitrides are high-melting, very hard, chemically unreactive, and electrical conductors, and their uses reflect these properties. They are used in the fabrication of refractory materials (heat-resistant materials), abrasives, grinding and cutting tools, and semiconductors.

The major industrial use of ammonia is the manufacture of nitric acid. Large amounts of ammonia are converted into various ammonium salts, used principally as fertilizers, and ammonia itself is used directly as a fertilizer. The compound is used to make hydrazine (H_2NNH_2, a component of rocket fuels) and urea (H_2NCONH_2, a fertilizer and an ingredient in the manufacture of plastic resins). Ammonia is employed in the Solvay process (for the manufacture of sodium carbonate), in petroleum, paper, rubber, and textile technology, and in the manufacture of dyes, drugs, and explosives.

The major use of nitric acid is in the production of nitrates, which are principally used in fertilizers and explosives. The reactions of nitric acid with certain organic compounds produce a number of commercial explosives (nitroglycerine, nitrocellulose, and trinitrotoluene are examples). Nitric acid has a large number of minor applications; all the nitrogen compounds of commerce are produced from nitric acid and/or ammonia.

2. Phosphorus. Most elemental phosphorus is used to make phosphorus(V) oxide, phosphoric acids, and the salts of these acids. Some phosphorus is used to make matches, warfare agents, and rodent poisons. Metal phosphides, and phosphorus itself, are used as alloying ingredients in the metallurgy of steel and copper. Some phosphides (GaP, BP, AlP, and InP) are semiconductors.

Phorphorus(V) oxide is used to make phosphoric acids, phosphates, and flame-retardant materials. The oxide functions as a catalyst for some reactions, is a desiccant (drying agent), and is used as a dehydrating agent in some organic reactions.

Phosphoric acid is used in the manufacture of fertilizers as well as directly as a fertilizer. The acid is used in phosphatizing, a process that produces a corrosion-resistant coating on iron objects and is applied prior to painting the objects. Phosphoric acid is used in polishing aluminum objects, in electropolishing steel articles, and in several ways in food technology.

Phosphates are used as fertilizers and in food processing, drugs, detergents, scouring powders, and toothpastes. Ammonium phosphate functions as a flame retardant and is used in textile technology.

3. Arsenic, Antimony, and Bismuth. These elements are not used in the amounts nor to the extent that nitrogen and phosphorus are. Arsenic, antimony, and bismuth are used in the production of a wide range of alloys. Either antimony or bismuth is a usual component of the alloys used as type metal. Molten type-metal alloys expand upon freezing. When they are used to make type for printing, the casts obtained have sharp edges. Low-melting alloys of antimony and bismuth are used to make safety plugs for boilers, fire sprinklers, solders, and electrical fuses. Arsenic is used in lead and copper alloys as a hardener and arsenic compounds find use as insecticides, rodent poisons, and weed killers. An important use of all three elements is the fabrication of semiconductors.

Summary

The topics that have been discussed in this chapter are

1. The group V A elements: their physical properties, allotropic forms, occurrence, and industrial preparation.

2. The nitrogen cycle.

3. Compounds of the group V A elements: nitrides, phosphides, hydrogen compounds, halogen compounds, oxides, oxyacids and their salts, electrode-potential diagrams.

4. Industrial uses of the group V A elements and their compounds.

Key Terms

Some of the more important terms introduced in this chapter are listed below. Definitions for terms not found in this list may be located in the text by use of the index.

Arc process (Section 21.7) A nitrogen-fixation process in which nitrogen oxide, NO, is produced by the reaction of nitrogen and oxygen in an electric arc.

Cyanamid process (Section 21.5) A nitrogen-fixation in which calcium cyanamid, CaNCN, is prepared from calcium carbide, CaC_2, and nitrogen.

Haber process (Sections 21.2 and 21.5) A nitrogen-fixation process for the preparation of ammonia from nitrogen and hydrogen.

Isomers (Section 21.6) Substances that have the same molecular formula but differ in the way the constituents are arranged into a molecule.

Nitrogen cycle (Section 21.2) A group of natural and artificial processes by which nitrogen is constantly being removed from the atmosphere and returned to it.

Nitrogen fixation (Section 21.2) A process in which elemental nitrogen is converted into a nitrogen-containing compound.

Ostwald process (Section 21.7) A commercial process for the manufacture of nitric acid; ammonia is catalytically oxidized to NO, the NO is reacted with O_2 to form NO_2, and the NO_2 is reacted with water to form HNO_3.

Superphosphate fertilizer (Section 21.8) A mixture of $Ca(H_2PO_4)_2$ and $CaSO_4$ used as a fertilizer and produced by the reaction of sulfuric acid and phosphate rock.

Problems*

21.1 Discuss how the properties of the elements of group V A and their compounds change with increasing atomic number.

21.2 Write all the equations that you can for the reactions of elementary nitrogen. Explain the low chemical reactivity of N_2.

21.3 Why is it advantageous for a farmer to rotate the crops sown in a given field?

21.4 In 1892 Lord Rayleigh observed that the density of nitrogen prepared from air (by the removal of H_2O, CO_2, and O_2) was different from the density of nitrogen prepared from NH_4Cl and $NaNO_2$. On the basis of this observation, Rayleigh and William Ramsey isolated argon (1894). Assume that air is 1.00% Ar, 78.00% N_2, and 21.00% O_2 (by volume). How should the densities of the two "nitrogen" samples compare?

21.5 The normal boiling point of the compound ethylene diamine, $H_2NCH_2CH_2NH_2$, is 117°C and that of propyl amine, $CH_3CH_2CH_2NH_2$, is 49°C. The molecules, however, are similar in size and molecular weight. What reason can you give for the difference in boiling point?

21.6 Discuss the comparative reactivities, solubilities, and electrical conductivities of the three allotropic forms of phosphorus in terms of molecular structure.

21.7 Write chemical equations for the separate reactions of each of the following with H_2O, with aqueous NaOH, and with aqueous HCl: **(a)** P_4O_6, **(b)** Sb_4O_6, **(c)** Bi_2O_3.

21.8 What are the characteristics of the three types of nitrides?

21.9 Boron nitride, BN, exists in two crystalline modifications—one resembling diamond and the other graphite. Describe, using drawings, the structures of these substances.

21.10 Discuss the preparation, structure, and properties of ammonia. Why is the boiling point of NH_3 high in comparison to the boiling points of PH_3, AsH_3, and SbH_3?

21.11 Write equations for the following reactions of ammonia:
(a) $NH_3(aq) + Ag^+(aq) \longrightarrow$
(b) $NH_3(aq) + H^+(aq) \longrightarrow$
(c) $NH_3(aq) + H_2O + CO_2(aq) \longrightarrow$
(d) $NH_3(g) + O_2(g) \xrightarrow{heat}$
(e) $NH_3(g) + O_2(g) \xrightarrow{Pt, heat}$
(f) $NH_3(g) + HCl(g) \longrightarrow$
(g) $NH_3(g) + V(s) \xrightarrow{heat}$
(h) $NH_3(l) + Na(s) \longrightarrow$

21.12 Write a series of equations for the preparation of nitric acid starting with N_2 as the source of nitrogen.

21.13 Write an equation for the reaction of HNO_3 with **(a)** Cu, **(b)** Zn, **(c)** P_4O_{10}, **(d)** NH_3, **(e)** $Ca(OH)_2$.

21.14 (a) List the oxides of nitrogen. **(b)** Write a chemical equation for the preparation of each of them. **(c)** Diagram the electronic structure of each oxide.

21.15 State the products of the thermal decomposition of **(a)** NH_4NO_3, **(b)** NH_4NO_2, **(c)** $Pb(NO_3)_2$, **(d)** $NaNO_3$, **(e)** NaN_3.

21.16 Write equations for the following oxidation-reduction reactions: **(a)** the reaction of NH_3 with OCl^- that produces N_2H_4 and Cl^-; **(b)** the reaction of NO_2 with H_2 in HCl solution that produces $[NH_3OH]Cl$; **(c)** the electrolysis of NH_4F in anhydrous HF that produces NF_3 and H_2.

21.17 Draw electronic structures for orthophosphoric, phosphorous, and hypophosphorous acids. Tell how these structures explain the number of $H^+(aq)$ dissociated per mole when each of these compounds is dissolved in water.

21.18 Write equations for the preparations of all acids that may be obtained by the hydrolysis of P_4O_{10}.

21.19 Write an equation for the reaction of each of the following with water: **(a)** Li_3N, **(b)** AlN, **(c)** Ca_3P_2, **(d)** CaNCN, **(e)** NCl_3, **(f)** PCl_3, **(g)** PCl_5, **(h)** $H_4P_2O_7$, **(i)** NO_2.

---

* The appendix contains answers to color-keyed problems.

21.20 What is the acid anhydride of each of the following: **(a)** $H_2N_2O_2$, **(b)** HNO_2, **(c)** HNO_3, **(d)** $H_4P_2O_7$, **(e)** $H_3P_3O_9$, **(f)** H_3PO_3, **(g)** H_3AsO_4?

21.21 Diagram the resonance forms of **(a)** NO_2^-, **(b)** NO_2, **(c)** NO, **(d)** NO_3^-, **(e)** HNO_3, **(f)** HN_3.

21.22 Write equations for the reactions of HCl with **(a)** H_2NNH_2, **(b)** H_2NOH, **(c)** $NaNO_2$, **(d)** NH_3, **(e)** $(NH_4)_2CO_3$.

21.23 Write equations for the reactions of O_2 with **(a)** As, **(b)** P_4, **(c)** PCl_3, **(d)** NO, **(e)** Sb_2S_3.

21.24 Write equations for the reactions of water with **(a)** Na_3As, **(b)** As_2O_5, **(c)** Mg_3Bi_2, **(d)** As_4O_6, **(e)** $SbCl_3$, **(f)** PH_4I.

21.25 Draw Lewis structures for *cis*- and *trans*-N_2F_2.

21.26 Describe the shape of each of the following: **(a)** NH_3, **(b)** NH_2^-, **(c)** PCl_4^+, **(d)** PCl_6^-, **(e)** $SbCl_5$, **(f)** $Sb(OH)_6^-$.

21.27 Describe the commercial processes used in the manufacture of **(a)** P_4, **(b)** NH_3, **(c)** H_3PO_4, **(d)** HNO_3, **(e)** superphosphate fertilizer, **(f)** As_4.

21.28 Write equations for the reactions of aqueous NaOH with **(a)** NH_4Cl, **(b)** P_4, **(c)** As_4O_6, **(d)** Sb_2O_5.

21.29 Describe the shape of the following molecules: **(a)** P_4, **(b)** $POCl_3$, **(c)** P_4O_6, **(d)** P_4O_{10}.

THE NONMETALS, PART IV: CARBON, SILICON, BORON, AND THE NOBLE GASES

C H A P T E R

22

In this chapter, the nonmetals of group IV A (carbon and silicon), group III A (boron alone), and group 0 (the noble gases) are considered. These discussions complete the survey of the nonmetals that was begun in Chapter 19.

CARBON AND SILICON

Carbon, silicon, germanium, tin, and lead make up group IV A. The compounds of carbon are more numerous than the compounds of any other element with the possible exception of hydrogen. In fact, approximately ten compounds that contain carbon are known for every compound that does not. The chemistry of the compounds of carbon (most of which also contain hydrogen) is the subject of **organic chemistry** (see Chapter 26).

22.1 Properties of the Group IV A Elements

The transition from nonmetallic character to metallic character with increasing atomic number that is exhibited by the elements of group V A is also evident in the chemistry of the IV A elements. Carbon is strictly a nonmetal (although graphite is an electrical conductor). Silicon is essentially a nonmetal in its chemical behavior, but its electrical and physical properties are those of a semimetal. Germanium is a semimetal; its properties are more metallic than nonmetallic. Tin and lead are truly metallic, although some vestiges of nonmetallic character remain (the oxides and hydroxides of tin and lead, for example, are amphoteric).

Carbon exists in network crystals with the atoms of the crystal held together by covalent bonds (see Section 22.2). A large amount of energy is required to rupture some or all of these bonds in fusion or vaporization. Carbon, therefore, has the highest melting point and boiling point of the family (see Table 22.1). The heaviest member of the family, lead, exists in a typical metallic lattice. The crystalline forms of the intervening members show a transition between the two extremes displayed by carbon and lead, and this accounts for the trend in melting points and boiling points of the elements (see Table 22.1).

The crystalline forms of silicon and germanium are similar to the diamond. The bonds are not so strong as those in the diamond, however, and silicon and

Table 22.1 Some properties of the group IV A elements

Property	Carbon[a]	Silicon	Germanium	Tin	Lead
melting point (°C)	3570	1420	959	232	327
boiling point (°C)	4827	2355	2700	2360	1755
atomic radius (pm)	77	117	122	141	154
ionization energy (kJ/mol)					
first	1090	782	782	704	714
second	2350	1570	1530	1400	1450
third	4620	3230	3290	2940	3090
fourth	6220	4350	4390	3800	4060
electronegativity	2.6	1.9	2.0	2.0	2.3

[a] Diamond

germanium are semiconductors (the diamond is not). Silicon and germanium may be used to prepare impurity semiconductors (see Section 23.2) which are employed in transistors. Although one modification of tin has a diamond-type structure, the principal form of tin is metallic.

The electronic configurations of the elements are listed in Table 22.2. With the possible exception of carbon, the assumption of a noble-gas configuration through the formation of a $4-$ ion by electron gain is not observed. The electronegativities of the group IV A elements are generally low (see Table 22.1).

The ionization energies of the elements (see Table 22.1) show that the energy required for the removal of all four valence electrons from any given element is extremely high. Consequently, simple $4+$ ions of group IV A elements are unknown. Germanium, tin, and lead appear to be able to form $d^{10}s^2$ ions by the loss of two electrons. Of these $2+$ ions, however, only some of the compounds of Pb^{2+} (such as PbF_2 and $PbCl_2$) are ionic, the compounds of Ge^{2+} and Sn^{2+} are predominantly covalent, and those of Ge^{2+} are relatively unstable toward disproportionation to the $4+$ and 0 states.

In the majority of their compounds, the IV A elements are covalently bonded. Through the formation of four covalent bonds per atom, a IV A element attains the electronic configuration of the noble gas of its period. Compounds of the type AB_4 are tetrahedral. All the IV A elements can form such compounds, but only a few compounds of this type are known for lead (PbF_4, $PbCl_4$, and PbH_4) and they, with the exception of PbF_4, are thermally unstable.

In the case of carbon, the formation of four covalent bonds saturates the valence level. However, the other members of the group have empty d orbitals available in their valence levels and can form species in which the atom of the IV A element exhibits a covalence greater than four. The ions SiF_6^{2-}, $GeCl_6^{2-}$, $SnBr_6^{2-}$, $Sn(OH)_6^{2-}$, $Pb(OH)_6^{2-}$, and $PbCl_6^{4-}$ are octahedral.

The most important way in which carbon differs from the remaining elements of group IV A (as well as from all other elements) is the pronounced ability of carbon to form compounds in which many carbon atoms are bonded to each other in chains or rings. This property, called catenation, is exhibited by other elements near carbon in the periodic classification (such as boron, nitrogen, phosphorus, sulfur, oxygen, silicon, germanium, and tin) but to a much lesser extent than carbon; this property of carbon accounts for the large number of organic compounds.

Table 22.2 Electronic configurations of the group IV A elements

Element	Z	1s	2s	2p	3s	3p	3d	4s	4p	4d	4f	5s	5p	5d	6s	6p
C	6	2	2	2												
Si	14	2	2	6	2	2										
Ge	32	2	2	6	2	6	10	2	2							
Sn	50	2	2	6	2	6	10	2	6	10		2	2			
Pb	82	2	2	6	2	6	10	2	6	10	14	2	6	10	2	2

In group IV A the tendency for self-linkage diminishes markedly with increasing atomic number. The hydrides, which have the general formula E_nH_{2n+2} (where E is a group IV A element), illustrate this trend. There appears to be no limit to the number of carbon atoms that can bond together to form chains, and a very large number of hydrocarbons are known. For the other IV A elements, the most complex hydrides that have been prepared are Si_6H_{14}, Ge_9H_{20}, Sn_2H_6, and PbH_4. The carbon-carbon single bond energy (347 kJ/mol) is much greater than that of the silicon-silicon bond (226 kJ/mol), the germanium-germanium bond (188 kJ/mol), or the tin-tin bond (151 kJ/mol).

In addition, the C—C bond is about as strong as any bond that carbon forms with any other element. Typical bond energies for carbon bonds are: C—C, 347 kJ/mol; C—O, 335 kJ/mol; C—H, 414 kJ/mol; and C—Cl, 326 kJ/mol. In comparison, the Si—Si bond is much weaker than the bonds that silicon forms with other elements. Some bond energies for silicon bonds are: Si—Si, 226 kJ/mol; Si—O, 368 kJ/mol; Si—H, 328 kJ/mol; and Si—Cl, 391 kJ/mol. Silicon, therefore, has more of a tendency to bond to other elements than to bond to itself.

Another characteristic of carbon is its pronounced ability to form multiple bonds with itself and with other nonmetals. Groupings such as

$$>C=C<, \quad -C\equiv C-, \quad -C\equiv N, \quad >C=O, \quad \text{and} \quad >C=S$$

are frequently encountered. No other IV A element uses p orbitals to form π bonds.

Only the truly nonmetallic members of the family, carbon and silicon, are treated in the sections that follow.

22.2 Occurrence and Preparation of Carbon and Silicon

Carbon constitutes approximately 0.03% of the earth's crust. In addition, the atmosphere contains 0.03% CO_2 by volume and carbon is also an important constituent of all plant and animal matter. The allotropes of carbon, diamond and graphite, as well as impure forms of the element such as coal occur in nature. In combined form the element occurs in compounds with hydrogen (which are called **hydrocarbons** and are found in natural gas and petroleum), in the atmosphere as CO_2, and in carbonate minerals such as limestone ($CaCO_3$), dolomite ($CaCO_3 \cdot MgCO_3$), siderite ($FeCO_3$) witherite ($BaCO_3$), and malachite ($CuCO_3 \cdot Cu(OH)_2$).

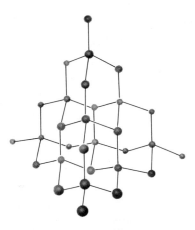

Figure 22.1 Arrangement of atoms in a diamond crystal

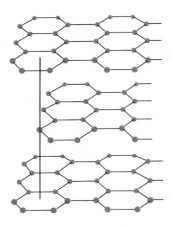

Figure 22.2 Arrangement of atoms in a graphite crystal

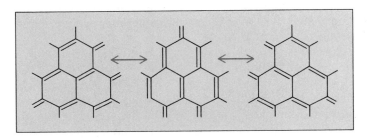

Figure 22.3 Resonance forms for a fragment of a graphite layer

In diamond each carbon atom is bonded through sp^3 hybrid orbitals to four other carbon atoms arranged tetrahedrally (see Figure 22.1). Strong bonds hold this network crystal together. Furthermore, all valence electrons of each carbon atom are paired in bonding orbitals—the valence level of each carbon atom can hold no more than eight electrons. Thus the diamond is extremely hard, high melting, stable, and a nonconductor of electricity.

Whereas the diamond is a colorless, transparent material with a high refractivity, graphite is a soft, black solid with a slight metallic luster. The graphite crystal is composed of layers formed from hexagonal rings of carbon atoms (see Figure 22.2). The layers are held together by relatively weak London forces; the distance from carbon atom to carbon atom in adjacent planes is 335 pm as compared with a distance of 141.5 pm between bonded carbon atoms of a plane. Since it is easy for the layers to slide over one another, graphite is soft and has a slippery feel. It is less dense than diamond.

The nature of the bonding in the layers of the graphite crystal accounts for some of the properties of this substance. Each carbon atom is bonded to three other carbon atoms, and all the bonds are perfectly equivalent. The C—C bond distance in graphite (141.5 pm) compared with that in the diamond (154 pm) suggests that a degree of multiple bonding exists in the former. Graphite may be represented as a resonance hybrid (see Figure 22.3) in which each bond is a $1\frac{1}{3}$ bond.

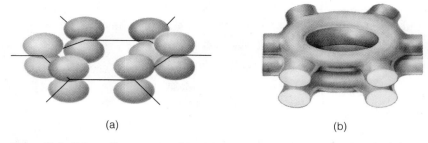

(a)　　　　　　　　　　(b)

Figure 22.4 Schematic representation of the formation of the multicenter bonding system of a graphite fragment (σ bonds are shown as solid lines)

Each C atom in graphite forms σ bonds with three other C atoms through the use of sp^2 hybrid orbitals. The structure, therefore, is planar with the three σ bonds of each atom directed to the corners of equilateral triangles. This bonding accounts for three of the four valence electrons of each C atom; the fourth, a p electron, is not involved in σ bond formation (see Figure 22.4a).

If only two adjacent atoms in a molecule had this electron arrangement, the additional p electrons would pair to form a localized π bond, and thus the atoms would be joined by a double bond. However, in graphite *each* C atom has an additional p electron, and the resonance forms depict the possibility of forming conventional double bonds in three ways. Since each p orbital overlaps with more than one other p orbital, an extended π bonding system that encompasses the entire structure forms (see Figure 22.4b). The electrons in this π bonding system are not localized between two atoms but are free to move throughout the entire layer. Hence, graphite has a metallic luster and is an electrical conductor. The conductivity is fairly large in a direction parallel to the layers but is small in a direction perpendicular to the planes of the crystal.

Silicon, which constitutes approximately 28% of the earth's crust, is the second most abundant element (oxygen is first). The element does not occur free in nature; rather, it is found as silicon dioxide (sometimes called silica) and in an enormous variety of silicate minerals.

Silicon is prepared by the reduction of silicon dioxide by coke at high temperatures in an electric furnace:

$$SiO_2(l) + 2\ C(s) \longrightarrow Si(l) + 2\ CO(g)$$

If a larger quantity of carbon is employed, silicon carbide (SiC, which is called "carborundum") is produced rather than silicon.

The only known modification of silicon has a structure similar to diamond. Crystalline silicon is a gray, lustrous solid that is a semiconductor. The bonds in silicon are not so strong as those in diamond, and the bonding electrons are not so firmly localized. Evidently, in silicon, electrons may be thermally excited to a conductance band that is energetically close to the valence band (see Section 23.2).

Very pure silicon, which is used in transistors, is prepared by a series of steps. First, impure silicon is reacted with chlorine to produce $SiCl_4$. The tetrachloride, a volatile liquid, is purified by fractional distillation and then reduced by hydrogen to elementary silicon. This product is further purified by zone refining (see Figure 23.10). In this process a short section of one end of a silicon rod is melted, and this melted zone is caused to move slowly along the rod to the other end by the movement of the heater. Pure silicon crystallizes from the melt, and the impurities are swept along in the melted zone to one end of the rod which is subsequently sawed off and discarded.

22.3 Carbides and Silicides

A large number of carbides are known. These compounds may be made by heating the appropriate metal or its oxide with carbon, carbon monoxide, or a hydrocarbon.

The **saltlike carbides** are made up of metal cations together with anions that contain carbon alone. The I A and II A metals, as well as Cu^+, Ag^+, Au^+, Zn^{2+}, and Cd^{2+}, form carbides that are sometimes called *acetylides* because they contain the *acetylide* ion, C_2^{2-}, which has the structure

$$[:C\equiv C:]^{2-}$$

Upon hydrolysis, acetylides yield acetylene, C_2H_2:

$$H-C\equiv C-H$$

For example,

$$CaC_2(s) + 2 H_2O \longrightarrow Ca(OH)_2(s) + C_2H_2(g)$$

Calcium carbide is used in the commercial production of calcium cyanamid, $CaNCN$ (see Section 21.5).

Beryllium carbide, Be_2C, and aluminum carbide, Al_4C_3, contain the *methanide* ion, C^{4-}, which upon hydrolysis yields methane, CH_4:

$$\begin{array}{c} H \\ | \\ H-C-H \\ | \\ H \end{array}$$

For example,

$$Al_4C_3(s) + 12 H_2O \longrightarrow 4 Al(OH)_3(s) + 3 CH_4(g)$$

Interstitial carbides are formed by the transition metals and consist of metallic crystals with carbon atoms in the holes between the metal atoms of the crystal structure (the interstices). Examples of this type of carbide include TiC, TaC, W_2C, VC, and Mo_2C. Because interstitial carbides are very hard, high-melting, and chemically unreactive, they are used in the fabrication of cutting tools. They resemble metals in appearance and electrical conductivity.

The bonding in the **covalent carbides** SiC and Be_4C is completely covalent. These compounds are very hard, chemically unreactive, and do not melt even at high temperatures. Because of these properties, they are used as abrasives (in the place of industrial diamonds). Silicon carbide (carborundum) is produced by the reaction of SiO_2 and C in an electric furnace. The SiC crystal consists of a diamond-like tetrahedral network formed from alternating Si and C atoms. Boron carbide, B_4C, which is harder than SiC, is made by the reduction of B_2O_3 by carbon in an electric furnace.

Silicon dissolves in almost all molten metals, and in many of these instances, definite compounds called silicides are produced (Mg_2Si, $CaSi_2$, Li_3Si, and $FeSi$ are examples). Although probably none of the silicides is truly ionic, some of them react with water to produce hydrogen-silicon compounds that are called *silicon hydrides* or *silanes*.

The **silanes** are compounds that conform to the general formula Si_nH_{2n+2}; compounds in which n equals 1 to 6 are known. They resemble structurally the hydrocarbons that conform to the general formula C_nH_{2n+2} and are called the alkanes (see Figure 22.5). Although there is presumably no limit to the number of C atoms that can join together to form alkanes, the number of Si atoms that can bond together to form silanes appears to be limited because of the comparative weakness of the Si—Si bond (see Section 22.1). In addition, the alkanes are much less reactive than the silanes, which are spontaneously flammable in air:

$$2\,Si_2H_6(g) + 7\,O_2(g) \longrightarrow 4\,SiO_2(s) + 6\,H_2O(g)$$

The C—C bonds in the alkanes are all single bonds, but there are hydrocarbons (acetylene for example) which contain multiple bonds between C atoms. Silicon hydrides that contain multiple Si to Si bonds are unknown. The hydrocarbons are discussed in Chapter 26.

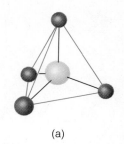

(a)

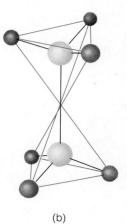

(b)

Figure 22.5 Arrangement of atoms in (a) CH_4 and SiH_4 and (b) C_2H_6 and Si_2H_6. (Larger spheres, C or Si; smaller spheres, H.)

22.4 Oxides and Oxyacids of C and Si

Carbon monoxide is formed by the combustion of carbon in a limited supply of oxygen at high temperatures (approximately 1000°C):

$$2\,C(s) + O_2(g) \longrightarrow 2\,CO(g)$$

It is also produced by the reaction of CO_2 and C at high temperatures and is a constituent of water gas (see Section 19.2). Carbon monoxide is isoelectronic with nitrogen:

$$^{\ominus}\!:\!C\!\equiv\!O\!:^{\oplus}$$

and has two π bonds and one σ bond joining the atoms.

Carbon monoxide burns in air:

$$2\,CO(g) + O_2(g) \longrightarrow 2\,CO_2(g) \qquad \Delta H = -283\text{ kJ}$$

Since this reaction is highly exothermic, carbon monoxide can be used as a fuel. The compound reacts with the halogens in sunlight to produce the carbonyl halides (COX_2) and with sulfur vapor at high temperatures to form carbonyl sulfide (COS).

Carbon monoxide is used as a reducing agent in metallurgical processes. At high temperatures, it reacts with many metal oxides to yield the free metal and CO_2:

$$FeO(s) + CO(g) \longrightarrow Fe(l) + CO_2(g)$$

The catalyzed reactions of carbon monoxide and hydrogen are commercially important for the production of hydrocarbons and methanol (see Section 26.6).

Certain transition metals and transition metal salts react with CO to produce **metal carbonyls** in which the CO molecule uses the unshared electron pair of the C atom for bond formation. Examples of metal carbonyls include $Ni(CO)_4$, a tetrahedral molecule; $Fe(CO)_5$, a trigonal bipyramidal molecule; and $Cr(CO)_6$, an octahedral molecule. Carbon monoxide is poisonous because it combines with the iron atom of the hemoglobin of the blood, thus preventing the hemoglobin from combining with oxygen (see Section 24.1 for a more complete discussion).

Carbon dioxide is formed by the complete combustion of carbon or compounds of carbon (notably the hydrocarbons). It is also produced by the reaction of carbonates or hydrogen carbonates with acids,

$$CaCO_3(s) + 2H^+(aq) \longrightarrow Ca^{2+}(aq) + CO_2(g) + H_2O$$

and by heating carbonates,

$$CaCO_3(s) \longrightarrow CaO(s) + CO_2(g)$$

The molecule is linear and nonpolar:

$$:\ddot{O}{=}C{=}\ddot{O}:$$

In the **oxygen–carbon-dioxide cycle** in nature, oxygen is removed from the air and CO_2 is added to the air by the respiration of animals, the combustion of fuels, and the decay of organic matter. The reverse occurs in **photosynthesis.** In the formation of carbohydrates by plants, CO_2 is removed from the air and O_2 is released to the air. The energy for photosynthesis is supplied by sunlight and the process is catalyzed by the green coloring matter of plants, chlorophyll. The geochemical equilibrium between CO_2, H_2O, and limestone, $CaCO_3$, also serves to control the level of CO_2 in the air:

$$CaCO_3(s) + CO_2(g) + H_2O \rightleftharpoons Ca^{2+}(aq) + 2HCO_3^-(aq)$$

At the present time, the rate at which CO_2 is being added to the atmosphere exceeds the rate at which it is being removed by other processes. The concentration of CO_2 in the atmosphere, therefore, is steadily increasing. The concentration today is about 10% higher than it was a century ago.

Carbon dioxide is moderately soluble in water. It is the acid anhydride of carbonic acid, H_2CO_3. The acid, however, has never been obtained in the pure state. Solutions of CO_2 in water consist mainly of dissolved CO_2 molecules—less than 1% of the dissolved CO_2 is in the form of H_2CO_3 molecules. Carbonic acid is a weak, diprotic acid, and equations for its ionizations are probably best written

$$CO_2(aq) + H_2O \rightleftharpoons H^+(aq) + HCO_3^-(aq)$$
$$HCO_3^-(aq) \rightleftharpoons H^+(aq) + CO_3^{2-}(aq)$$

Two series of salts are formed: the normal carbonates, such as Na_2CO_3 and $CaCO_3$, and the hydrogen carbonates, such as $NaHCO_3$ and $Ca(HCO_3)_2$ The structure of the carbonate ion has been discussed in Sections 7.2 and 7.6 (see Figure 7.19).

Mining trona ore in Green River, Wyoming. The ore is the principal source of sodium carbonate, which is used in the manufacture of glass, soap, detergents, paper, and other chemicals. *Allied Corporation*.

The only well-characterized oxide of silicon is SiO_2. In contrast to the oxides of carbon, which are volatile molecular species held together by London forces in the solid state, SiO_2 forms very stable, nonvolatile, three-dimensional network crystals (melting point, $\sim 1700°C$). One of the three crystal modifications of SiO_2 has a lattice that may be considered to be derived from the diamond lattice with silicon atoms replacing carbon atoms and an oxygen atom midway between each pair of silicon atoms.

The chemistry of silicon is dominated by compounds that contain the Si—O linkage. Silicon dioxide is the product of the reaction of the elements; it is also produced by reaction of the spontaneously flammable silicon hydrides with air. Hydrous SiO_2 is the product of the hydrolysis of many silicon compounds. Silicon dioxide occurs in several forms in nature; among them are sand, flint, agate, jasper, onyx, and quartz.

Silicon dioxide is an acidic oxide; however, no acids of silicon have ever been isolated. The oxide does not react directly with water; acidification of a water solution of a soluble silicate yields only hydrous SiO_2. Silicates may be made by heating metal oxides or metal carbonates with SiO_2. Certain silicates of the I A metals (with a molar ratio of silicon dioxide to metal oxide of not more than 2 to 1) are water soluble.

A large number of silicates of various types occurs in nature. The basic unit of all silicates is the tetrahedral SiO_4 group (see Figure 22.6). The simple ion SiO_4^{4-} occurs in certain minerals (zircon, $ZrSiO_4$, is an example):

Quartz crystals. *U.S. Geological Survey*.

$$\left[\begin{array}{c} :\overset{\cdot\cdot}{O}: \\ | \\ :\overset{\cdot\cdot}{O}\!-\!\underset{}{Si}\!-\!\overset{\cdot\cdot}{O}: \\ | \\ :\overset{\cdot\cdot}{O}: \end{array} \right]^{4-}$$

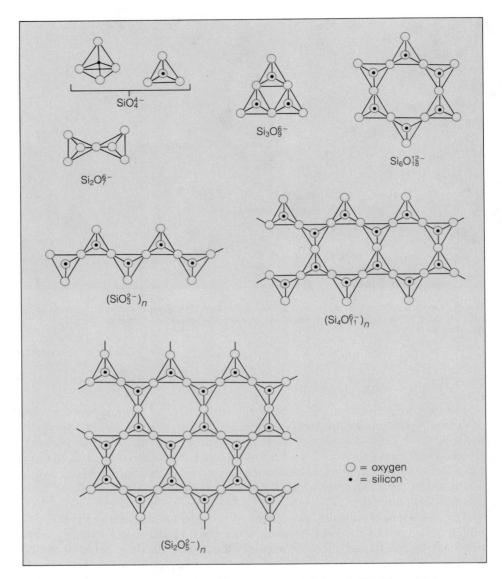

Figure 22.6 Schematic representation of the arrangement of atoms in the silicate ions

The four bond pairs of the Si atom are arranged in a tetrahedral manner.

The other silicate anions contain more than one SiO_4 tetrahedron joined together by bridge oxygen atoms (that is, oxygen atoms that are shared by two tetrahedra). The mineral thortveitite ($Sc_2Si_2O_7$) contains the $Si_2O_7^{6-}$ ion, which is formed by two SiO_4 tetrahedra joined by a single bridge oxygen atom (see Figure 22.6):

$$
\left[\begin{array}{c} \ddot{O} \quad\quad \ddot{O} \\ \ddot{O}-Si-\ddot{O}-Si-\ddot{O} \\ \ddot{O} \quad\quad \ddot{O} \end{array} \right]^{6-}
$$

Cyclic silicate anions are known in which tetrahedra are joined in a circle with each tetrahedron sharing two bridge oxygen atoms. The $Si_3O_9^{6-}$ anion in bentonite ($BaTiSi_3O_9$) consists of three SiO_4 tetrahedra joined in a circle, and in the $Si_6O_{18}^{12-}$ ion found in beryl ($Be_2Al_2Si_6O_{18}$) there are six (see Figure 22.6).

If three oxygen atoms of each SiO_4 tetrahedron are used as bridge atoms, a sheetlike anion results. The anion $(Si_2O_5^{2-})_n$ occurs in talc, $Mg_3(Si_2O_5)(OH)_2$. Because of the layer structure, this material feels slippery. Occasionally, aluminum atoms take the places of some of the silicon atoms in certain anions. The hypothetical AlO_4 tetrahedron would have a 5− charge; consequently, such substitutions increase the negative charge of the anion. Muscovite, $KAl_3Si_3O_{10}(OH)_2$ contains a sheetlike alumino-silicate anion with one-fourth of the silicon atoms of the $(Si_2O_5^{2-})_n$ structure replaced by aluminum atoms.

If all four oxygen atoms of each SiO_4 tetrahedron are used as bridge atoms, the three-dimensional, diamond-like network crystal of SiO_2 results. If some of the Si atoms of this SiO_2 structure are replaced with Al atoms, an aluminosilicate anion is produced. Since Si atoms (each with four valence electrons) are replaced by aluminum atoms (each with three valence electrons), additional electrons (which provide the charge on the anion) are required to satisfy the bonding requirements of the structure. Framework aluminosilicates, such as the feldspars and zeolites, are of this type.

Glass is a mixture of silicates made by fusing SiO_2 with metal oxides and carbonates. Common soda-lime glass is made from Na_2CO_3, $CaCO_3$, and SiO_2. Special glasses may be made by the addition of other acidic and basic oxides (such as Al_2O_3, B_2O_3, PbO, and K_2O). Cement is a complex aluminosilicate mixture made from limestone ($CaCO_3$) and clay ($H_4Al_2Si_2O_9$).

22.5 Sulfur, Halogen, and Nitrogen Compounds of Carbon

Carbon disulfide, CS_2, is a volatile liquid prepared commercially by heating carbon and sulfur together in an electric furnace. The structure of the molecule is similar to that of CO_2:

$$:\overset{..}{S}=C=\overset{..}{S}:$$

The compound is a good solvent for nonpolar substances such as waxes and grease. Its use, however, is limited by its toxicity and flammability. It is used commercially in the manufacture of rayon and in the production of carbon tetrachloride.

Carbon tetrachloride is made commercially by heating carbon disulfide with chlorine:

$$CS_2(g) + 3\,Cl_2(g) \longrightarrow CCl_4(g) + S_2Cl_2(g)$$

Carbon tetrachloride, which is a liquid under ordinary conditions, is not flammable and is a good solvent for many nonpolar materials. It was formerly used in fire extinguishers and in dry cleaning, but these uses are now illegal because of the toxicity of the compound.

Carbon tetrafluroride, CF_4, which can be obtained by the fluorination of almost any carbon-containing compound, is a very stable gas. Mixed chlorine-fluorine compounds of carbon (in particular CCl_2F_2) are called the "freons." They are very stable, odorless, nontoxic gases that are used principally as refrigerants.

Carbon tetrabromide, CBr_4, and carbon tetraiodide, CI_4, are solids. They are thermally unstable, presumably because of the difficulty of the carbon atom in accommodating four large atoms around itself. Other halogen-containing compounds of carbon are discussed in Section 26.5.

There are many compounds in which carbon is bonded to nitrogen. Hydrogen cyanide, HCN, is commercially prepared by the catalyzed reaction of methane (CH_4), ammonia, and oxygen at high temperatures:

$$2\,CH_4(g) + 2\,NH_3(g) + 3\,O_2(g) \longrightarrow 2\,HCN(g) + 6\,H_2O(g)$$

The compound is a highly poisonous, low-boiling liquid (boiling point, 26.5°C) and is used in the manufacture of plastics.

Hydrogen cyanide readily dissolves in water to produce solutions of hydrocyanic acid, a weak acid. Cyanides, such as sodium cyanide (NaCN), are produced by neutralization of this acid. The cyanide ion is isoelectronic with N_2 and CO:

$$[:C\equiv N:]^-$$

The ion has a strong tendency to form covalent complexes with metal cations—such as $Ag(CN)_2^-$, $Cd(CN)_4^{2-}$, $Ni(CN)_4^{2-}$, $Hg(CN)_4^{2-}$, $Fe(CN)_6^{4-}$, $Fe(CN)_6^{3-}$, and $Cr(CN)_6^{3-}$.

Mild oxidation of the cyanide ion in aqueous solution produces the cyanate ion, OCN^-. The corresponding acid, cyanic acid (HOCN), is not stable in aqueous solution and decomposes to CO_2 and NH_3. An analogous ion, the thiocyanate ion, SCN^-, is prepared by melting a mixture of a I A cyanide with sulfur.

BORON

Of the group III A elements (boron, aluminum, gallium, indium, and thallium), boron alone is a nonmetal.

22.6 Properties of the Group III A Elements

The electronic configurations of the elements are listed in Table 22.3 and properties of the elements are given in Table 22.4. The boron atom is much smaller than those of the other elements of the group. This difference accounts for the sharp distinction in properties between the nonmetallic boron and the other group members, which are metallic. In general, metals have atomic radii that are greater than 120 pm and nonmetals have atomic radii that are less than 120 pm. The atomic radii of Ga, In, and Tl are influenced by the electronic inner-building of elements that immediately precede them in the periodic table (particularly in the case of Tl, which follows the lanthanides). Atomic radius, therefore, does not rapidly and regularly increase with increasing atomic number in the series from Al to Tl.

None of the III A elements shows the slightest tendency to form simple anions. The most important oxidation state of this group is 3+, as would seem reasonable from the $ns^2\,np^1$ electron configuration of the valence level. The first three ionization energies of boron are relatively high because of the small size of the boron atom. As a result, the B^{3+} ion is never formed. The energy required to remove

Table 22.3 Electronic configurations of the group III A elements

Element	Z	1s	2s	2p	3s	3p	3d	4s	4p	4d	4f	5s	5p	5d	6s	6p
B	5	2	2	1												
Al	13	2	2	6	2	1										
Ga	31	2	2	6	2	6	10	2	1							
In	49	2	2	6	2	6	10	2	6	10		2	1			
Tl	81	2	2	6	2	6	10	2	6	10	14	2	6	10	2	1

Table 22.4 Some properties of the group III A elements

Property	Boron	Aluminum	Gallium	Indium	Thallium
melting point (°C)	2300	659	30	155	304
boiling point (°C)	2550	2500	2070	2100	1457
atomic radius (pm)	80	125	125	150	155
ionic radius, M^{3+} (pm)	—	52	62	81	95
ionization energy (kJ/mol)					
first	801	579	579	560	589
second	2422	1814	1968	1814	1959
third	3657	2740	2953	2692	2866
electrode potential, $\mathscr{E}^{\circ}_{red}$, ($M^{3+}/M$)(V)	—	−1.67	−0.52	−0.34	+0.72

three electrons from the boron atom cannot be supplied by either lattice energies or hydration enthalpies. All the compounds of boron are covalent. The compounds of the others in which the III A element is in a $3+$ oxidation state are either ionic or covalent.

Gallium, indium, and thallium can exist in a $1+$ oxidation state (a $d^{10}s^2$ configuration). Within a group, the importance and stability of a lower oxidation state increases as the group is descended. For Ga and In the $1+$ state is less important and less stable than the $3+$ state. For Tl, the largest member of the group, the reverse is true; Tl^+ compounds resemble compounds of the I A metals.

The oxides and hydroxides exhibit the usual trend in decreasing acidic character with increasing atomic number. Thus B_2O_3 and $B(OH)_3$ (which is usually written H_3BO_3) are acidic, the compounds of aluminum and gallium are amphoteric, and In_2O_3 and Tl_2O_3 are exclusively basic.

Boron has a slight tendency for catenation (for example, the atoms of B_2Cl_4 are arranged Cl_2BBCl_2), but no other member of this group displays this characteristic.

22.7 Boron

Boron (which makes up $3 \times 10^{-4}\%$ of the earth's crust) does not occur free in nature. The principal ores of boron are borates such as kernite ($Na_2B_4O_7 \cdot 4H_2O$), borax ($Na_2B_4O_7 \cdot 10H_2O$), colemanite ($Ca_2B_6O_{11} \cdot 5H_2O$), and

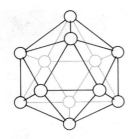

Figure 22.7 Arrangement of atoms in a B_{12} icosahedron

ulexite ($NaCaB_5O_9 \cdot 8H_2O$). Pure, crystalline boron may be obtained by reducing BBr_3 by hydrogen on a hot tungsten filament (at approximately 1500°C):

$$2\,BBr_3(g) + 3\,H_2(g) \longrightarrow 2\,B(s) + 6\,HBr(g)$$

The crystals, which condense on the filament, are black with a metallic luster.

The structures of three types of boron crystals have been determined, and at least one other allotrope is known to exist. Groups of 12 B atoms arranged in regular icosahedra (solid figures, each with 20 equilateral-triangular faces) occur in each of these crystalline modifications (see Figure 22.7). In the α-rhombohedral form, each B atom participates not only in the bonding of its icosahedron but also in bonding that icosahedron to others. Ordinary electron-pair bonds cannot account for the large number of boron-boron interactions of the structure since a boron atom has only three electrons and four orbitals in its valence level.

In metallic crystals there are too few valence electrons to form covalent bonds between neighboring metal atoms of the crystal lattice; the electrons belong to the crystal as a whole and serve to bind many nuclei together. Boron, however, is a nonmetal, and the properties of crystalline boron (low electrical conductivity, extreme hardness, brittleness, and high melting point) are more typical of a covalent, network type of crystal than of a metallic crystal.

The bonding in elementary boron has been interpreted as involving **three-center bonds** as well as the more common two-center bonds. In a three-center bond, three atoms are held together by an electron pair in a single molecular orbital. The molecular orbital is assumed to arise from the overlap of three atomic orbitals, one from each of the three bonded atoms. Several types of three-center bonds have been postulated, including the BBB bond found in crystalline boron (see Figure 22.8) and a BHB bond found in the boron hydrides (see Section 22.8).

22.8 Compounds of Boron

At very high temperatures(about 2000°C) boron reacts with many metals to form borides. These substances are very hard, are chemically stable, and have metallic conductivity. In the crystals of some metallic borides, the boron atoms are interstitial; in others, chains, octahedra, or layers of boron atoms are present. Magnesium boride, MgB_2, unlike the other borides, is readily hydrolyzed to produce a mixture of boron hydrides.

Figure 22.8 Formation of a three-center B—B—B bond

Boron reacts with ammonia or nitrogen at elevated temperatures to produce boron nitride, BN. This material is isoelectronic with carbon and has a crystal structure similar to graphite but with alternating boron and nitrogen atoms. At very high temperatures and pressures this modification of BN is converted to another form that has a diamond-type lattice and is almost as hard as diamond.

At high temperatures, boron reacts with the halogens to yield the trihalides. BF_3 and BCl_3 are gases, BBr_3 is a liquid, and BI_3 is a solid. The molecules of the trihalides are triangular planar:

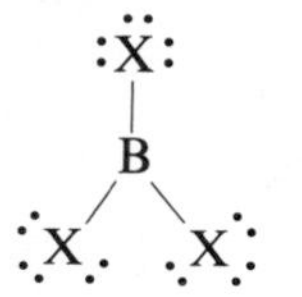

Since the boron atom does not have an octet of electrons in each of these molecules, the trihalides react as electron acceptors (Lewis acids):

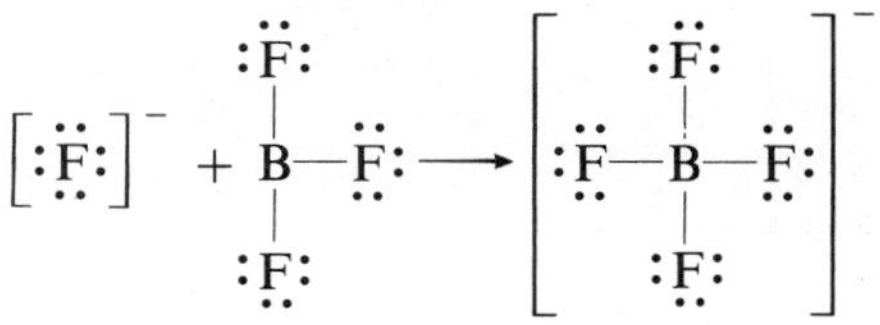

$$NH_3 + BF_3 \longrightarrow H_3N{:}BF_3$$

$$H_2O + BF_3 \longrightarrow H_2O{:}BF_3$$

The fluoborate ion, BF_4^-, is tetrahedral.

Many borates occur in nature. Some may be prepared by fusion of metallic oxides with B_2O_3 or boric acid. Hydrated borates may be obtained by crystallization of the solution resulting from the neutralization of boric acid with aqueous alkali. The borate anions may be considered to be built up from triangular BO_3 units, with or without BO_4 tetrahedra, by the sharing of oxygen atoms between units in much the same way as the silicates are constructed. Borax, $Na_2B_4O_5(OH)_4 \cdot 8H_2O$, which can be prepared by neutralizing boric acid with NaOH in aqueous solution, contains the $B_4O_5(OH)_4^{2-}$ ion.

The acidification of an aqueous solution of any borate precipitates orthoboric acid (also called boric acid), H_3BO_3 or $B(OH)_3$, as soft, white crystals. This acid, which may also be prepared by the complete hydration of B_2O_3, has a layer-type crystal lattice in which there is extensive hydrogen bonding between $B(OH)_3$ molecules (see Figure 22.9). The crystal structure of the material accounts for its cleavage into sheets and its slippery feel. Boric acid and the borates are toxic.

Boric acid is a very weak, *monobasic* acid in solution. Its ionization in fairly dilute solution may be represented as

$$B(OH)_3(aq) + H_2O \rightleftharpoons H^+(aq) + B(OH)_4^-(aq)$$

or

$$(H_2O)B(OH)_3(aq) \rightleftharpoons H^+(aq) + B(OH)_4^-(aq)$$

The $B(OH)_4^-$ ion is tetrahedral. In more concentrated solutions, polymeric anions, such as $B_3O_3(OH)_4^-$, exist.

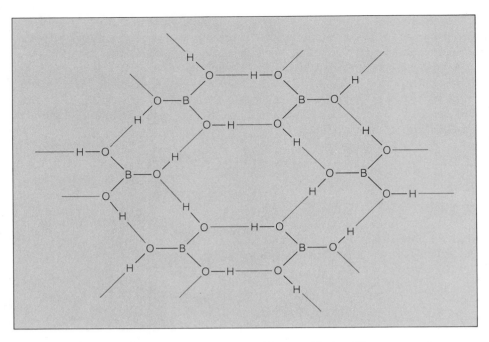

Figure 22.9 Arrangement of atoms in a layer of H_3BO_3 crystal (colored lines represent hydrogen bonds)

When boric acid is heated to about 170°C, metaboric acid, HBO_2, results. The meta- acid reverts to the more highly hydrated ortho- form in water. Although salts of many hypothetical boric acids are known, these are the only two acids that exist in the pure condition.

Metaboric acid may be dehydrated by heating into boric oxide, B_2O_3. The oxide may also be prepared by heating boron in air. Crystals of B_2O_3 consist of interconnected spiral chains of BO_4 tetrahedra. The material, however, is usually obtained as a glass.

Boron forms two series of hydrides: B_nH_{n+4}, where n = 2, 5, 6, 8, 10, or 18, and B_nH_{n+6}, where n = 4, 5, 6, 9, or 10. A few compounds that do not conform to either of these formulas have been reported. The simplest hydride of boron is diborane, B_2H_6. The molecule BH_3 cannot be isolated but is thought to be a transitory intermediate in some of the reactions of the boron hydrides.

Diborane, which is a gas under ordinary conditions, can be prepared by the reaction of BF_3 and lithium hydride in ether:

$$6\,LiH + 8\,BF_3 \longrightarrow 6\,LiBF_4 + B_2H_6$$

A mixture of higher boranes is obtained by heating diborane.

The compounds that conform to the general formula B_nH_{n+4} are thermally more stable (with regard to decomposition to the elements) than the second group, B_nH_{n+6}. The compounds B_2H_6, B_5H_9, and B_5H_{11} are spontaneously flammable in air. All boranes hydrolyze, at various rates, to yield boric acid and hydrogen.

The molecular structure of diborane is shown in Figure 22.10. In the molecule, each B atom has two H atoms (called terminal hydrogen atoms) bonded to it by

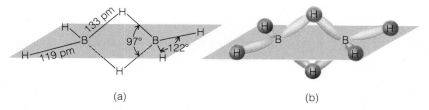

(a)   (b)

Figure 22.10   (a) Structure of diborane, (b) Formation of bonding orbitals (shapes of orbitals simplified)

regular two-center, electron-pair bonds, and the resulting BH_2 fragments are joined together by two hydrogen bridges (three-center BHB bonds). The two B atoms and the four terminal H atoms lie in a plane, and the bridge H atoms lie above and below this plane.

Each B atom may be considered to use sp^3 hybrid orbitals (tetrahedral) for the formation of the molecule. Two of the sp^3 orbitals of each B atom are used to form ordinary two-center bonds with terminal H atoms, and the other two hybrid orbitals are used to form three-center BHB bonds. Each three-center bond consists of a pair of electrons in a bonding molecular orbital formed by the overlap of the $1s$ orbital of the H atom with orbitals of the two B atoms (see Figure 22.10). Of the twelve valence electrons in the molecule (six from the two B atoms and six from the six H atoms), four pairs are utilized in the two-center bonds to the four terminal H atoms, and two pairs are used in the two BHB three-center bonds. The molecule is diamagnetic. The bond distance between a B atom and a terminal H atom is shorter than the distance between a B atom and a bridge H atom, thus indicating a stronger bond in the former case.

The structures of the higher boranes involve BBB as well as BHB three-center bonds. Several of the higher boranes have skeletons of boron atoms that are fragments of the B_{12} icosahedron found in crystalline boron.

Boron forms a series of borohydride ions. The most important one is BH_4^-, a tetrahedral ion that may be prepared by the reaction of lithium hydride and B_2H_6 in ether:

$$2\,LiH\ +\ B_2H_6\ \longrightarrow\ 2\,LiBH_4$$

Lithium borohydride, and the alkali borohydrides in general, are ionic materials that are important reducing agents. Other borohydride ions are known, such as $B_3H_8^-$, $B_{10}H_{13}^-$, and $B_{12}H_{12}^{2-}$. The boron skeleton of the $B_{12}H_{12}^{2-}$ ion is a B_{12} icosahedron.

THE NOBLE GASES

The outstanding characteristic of the noble gases (group 0) is their low order of chemical reactivity. Until 1962 no true compounds of these elements were known, and they were called the "inert gases." Since 1962, more than 30 compounds of the heavier members of this group have been prepared.

22.9 Properties of the Noble Gases

The properties of the noble gases reflect their very stable electronic configurations. The atoms have no tendency to combine with each other to form molecules; each element occurs as a colorless, monatomic gas. Each noble gas has the highest first ionization potential of any element in its period (see Table 22.5). The low melting points and boiling points of these elements are evidence of the weak nature of the London forces of attraction that operate between the atoms.

With increasing atomic number, the atomic size increases, and the outer electrons become slightly less tightly held. Therefore, the ionization potential decreases regularly from helium to radon. This factor (that is, increasing size of the electron cloud) also accounts for the increasing strength of London forces and consequently the increasing boiling point and melting point from helium to radon.

All the noble gases occur in the atmosphere (see Table 22.5), and Ne, Ar, Kr, and Xe are by-products of the fractionation of liquid air. Certain natural gas deposits (located principally in Kansas, Oklahoma, and Texas) contain a higher percentage of helium than is found in air; these deposits constitute the major commercial source of helium. Isotopes of radon are produced in nature by the decay of certain radioactive elements. All the isotopes of radon are themselves radioactive; $^{222}_{86}Rn$ is produced by the radioactive decay of $^{226}_{88}Ra$ and has a half-life of 3.82 days, the longest of any radon isotope.*

Argon is used to fill electric light bulbs. The gas does not react with the hot filament but conducts heat away from it, thus prolonging its life. Argon is also used as an inert atmosphere in welding and high-temperature metallurgical processes. The gas protects the hot metals from air oxidation. Helium is used in lighter-than-air craft (its density is about 14% of the density of air) and in low-temperature work (it has the lowest boiling point of any known substance). Neon signs are made from discharge tubes containing neon gas at a low pressure. Radon has been used as a source of α particles in cancer therapy.

The first chemical reaction of a noble gas to be observed, the reaction of xenon with platinum hexafluoride (PtF_6), was reported by Neil Bartlett in 1962.

Table 22.5	Some properties of the noble gases			
Gas	Melting Point (°C)	Boiling Point (°C)	Ionization Energy (kJ/mol)	Abundance in Atmosphere (Volume %)[a]
He	—[b]	−268.9	2.37×10^3	5×10^{-4}
Ne	−248.6	−245.9	2.08×10^3	2×10^{-3}
Ar	−189.3	−185.8	1.52×10^3	0.93
Kr	−157	−152.9	1.35×10^3	1×10^{-4}
Xe	−112	−107.1	1.17×10^3	8×10^{-6}
Rn	−71	−61.8	1.04×10^3	trace

[a] Dry air at sea level.

[b] −272.2°C at 26 atm pressure.

---

* The half-life of a radioactive isotope is the time that it takes for half of a sample of the isotope to disappear.

A jet of argon being used to stir molten steel. *American Iron and Steel Institute.*

Platinum hexafluoride is a powerful oxidizing agent; it reacts with oxygen to give $[O_2^+][PtF_6^-]$. Since the first ionization energy of *molecular* oxygen,

$$O_2 \longrightarrow O_2^+ + e^- \qquad \Delta H = +1180 \text{ kJ}$$

is close to the first ionization energy of xenon,

$$Xe \longrightarrow Xe^+ + e^- \qquad \Delta H = +1170 \text{ kJ}$$

Bartlett reasoned that xenon should react with PtF_6. Experiment verified this prediction, and a yellow-orange crystalline solid, originally reported as $[Xe]^+ [PtF_6]^-$ and now thought to be $[XeF]^+ [Pt_2 F_{11}]^-$, is produced by the reaction.

The best characterized noble-gas compounds are the xenon fluorides: XeF_2, XeF_4, and XeF_6. Each of these compounds may be prepared by direct reaction of Xe and F_2. The choice of reaction conditions, particularly the proportion of Xe to F_2 used, determines which fluoride is obtained. The xenon fluorides are colorless, crystalline solids (see Table 22.6).

Table 22.6 Properties of some compounds of xenon			
Oxidation State	Compound	Form	Melting Point (°C)
2+	XeF_2	colorless crystals	129
4+	XeF_4	colorless crystals	117
6+	XeF_6	colorless crystals	50
	$XeOF_4$	colorless liquid	−46
	XeO_3	colorless crystals	−
8+	XeO_4	colorless gas	−
	$Na_4XeO_6 \cdot 8H_2O$	colorless crystals	−

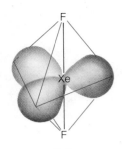

XeF$_2$

Figure 22.11 Structure of XeF$_2$

Oxygen-containing compounds of xenon are produced by reactions of the xenon fluorides with water. Partial hydrolysis of XeF$_6$ yields XeOF$_4$, a colorless liquid:

$$XeF_6(s) + H_2O \longrightarrow XeOF_4(l) + 2 HF(g)$$

Complete hydrolysis of XeF$_6$ or hydrolysis of XeOF$_4$ produces a solution that yields solid XeO$_3$ upon evaporation:

$$XeF_6(s) + 3 H_2O \longrightarrow XeO_3(aq) + 6 HF(aq)$$
$$XeOF_4(l) + 2 H_2O \longrightarrow XeO_3(aq) + 4 HF(aq)$$

The compounds of xenon in its highest oxidation state, 8+, include XeO$_4$ and the perxenates (compounds of the XeO$_6^{4-}$ ion). Pure perxenic acid, H$_4$XeO$_6$, has never been prepared. Salts of this acid may be made by passing ozone gas through alkaline solutions of XeO$_3$:

$$12 OH^-(aq) + 3 XeO_3(aq) + O_3(g) \longrightarrow 3 XeO_6^{4-}(aq) + 6 H_2O$$

The acid anhydride of perxenic acid, XeO$_4$, is made by reacting barium perxenate, Ba$_2$XeO$_6$, with concentrated sulfuric acid at $-5°C$:

$$Ba_2XeO_6(s) + H_2SO_4(l) \longrightarrow 2 BaSO_4(s) + XeO_4(g) + 2 H_2O$$

The order of the group 0 elements according to decreasing reactivity (increasing ionization energy; see Table 22.5) should be: Rn > Xe > Kr > Ar > Ne > He. Thus, radon should be the most reactive noble gas. There is evidence that Rn reacts with fluorine. The radioactive disintegration of Rn isotopes, however, makes the chemistry of Rn difficult to assess. Krypton is not so reactive as Xe, but a few compounds of Kr have been prepared. The most important of these compounds is KrF$_2$, which is made by subjecting a mixture of Kr and F$_2$ to an electric discharge. No compounds of He, Ne, or Ar have been prepared as yet.

The structures of most xenon compounds have been determined and may be interpreted by a valence bond approach (see Figures 22.11 and 22.12).

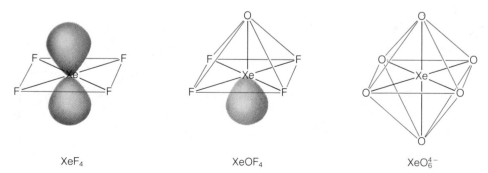

XeF$_4$ XeOF$_4$ XeO$_6^{4-}$

Figure 22.12 Structures of XeF$_4$, XeOF$_4$, and XeO$_6^{4-}$

Summary

The topics that have been discussed in this chapter are

1. The properties of the group IV A elements.

2. The occurrence, preparation, and allotropic forms of carbon and silicon.

3. Certain compounds of carbon and silicon: carbides and silicides; oxides; oxyacids and their salts; sulfur, halogen, and nitrogen compounds of carbon.

4. The properties of the group III A elements.

5. Boron: its occurrence, allotropic forms, compounds, and the three-center bonds found in its hydrides.

6. The noble gases: their occurrence, physical properties, compounds, uses.

Key Terms

Some of the more important terms introduced in this chapter are listed below. Definitions for terms not included in this list may be located in the text by use of the index.

Carbonyl (Section 22.4) A compound that contains a carbonyl group (CO) such as carbonyl sulfide (COS), the carbonyl halides (COX_2), and the metal carbonyls.

Catenation (Section 22.1) The formation of chains or rings by atoms of the same element bonded to each other.

Interstitial carbides (Section 22.3) Substances in which carbon atoms occupy holes between the atoms of a metallic crystal.

Noble gases (Section 22.9) The elements of group 0: He, Ne, Ar, Kr, Xe, and Rn.

Organic chemistry (Section 22.1) The chemistry of the hydrocarbons (compounds of carbon and hydrogen) and their derivatives.

Oxygen–carbon-dioxide cycle (Section 22.4) A group of natural and artificial processes by which O_2 and CO_2 are constantly being removed from the atmosphere and returned to it.

Photosynthesis (Section 22.4) The process by which plants make carbohydrates; CO_2 is removed from the air and O_2 is returned to it; the energy for the process is supplied by sunlight.

Silanes (Section 22.3) Compounds that contain only silicon and hydrogen.

Three-center bond (Section 22.7) A bond in which three atoms are bonded together by an electron pair in a single molecular orbital; found in boron and the hydrides of boron.

Problems*

Carbon and Silicon

22.1 Discuss the ways in which the chemistry of carbon differs from the chemistry of the other elements of group IV A.

22.2 What explanations may be offered for the fact that carbon has a greater tendency toward catenation than any other element?

22.3 Describe the crystal structure of **(a)** diamond, **(b)** graphite, **(c)** SiC.

22.4 Discuss the three types of carbides.

22.5 Name the following: **(a)** HCN(g), **(b)** HCN(aq), **(c)** KCN, **(d)** KOCN, **(e)** KSCN, **(f)** $Fe(CO)_5$, **(g)** $NaHCO_3$, **(h)** Na_2CO_3.

22.6 Write chemical equations for the reactions of carbon monoxide with **(a)** Cl_2, **(b)** S, **(c)** O_2, **(d)** FeO, **(e)** Ni.

22.7 Write chemical equations for the reaction of **(a)** CaC_2 and H_2O, **(b)** Al_4C_3 with H_2O.

22.8 Draw Lewis structures for **(a)** N_2, **(b)** CO, **(c)** CN^-, **(d)** OCN^-, **(e)** CS_2, **(f)** C_2H_2, **(g)** CCl_4, **(h)** CO_3^{2-}.

---

* The appendix contains answers to color-keyed problems.

22.9 Write chemical equations for the following reactions: **(a)** $CaCO_3(s)$ and $SiO_2(s)$ (heated), **(b)** $CaCO_3(s)$ (heated), **(c)** $CaCO_3(s)$ and $H^+(aq)$.

22.10 In alkaline solution, $\mathscr{E}°$ for the reduction of OCN^- to CN^- is -0.970 V and $\mathscr{E}°$ for the reduction of PbO to Pb is -0.580 V. Write the partial equations for these half-reactions. Will PbO oxidize CN^- to OCN^-? If so, write the chemical equation for the reaction.

Boron

22.11 Describe the structure of B_2H_6.

22.12 Write equations to show how orthoboric acid ionizes in aqueous solution.

22.13 Write chemical equations for the following reactions: **(a)** B_2O_3 and Mg (heated), **(b)** BBr_3 and H_2 (heated), **(c)** B and N_2 (heated), **(d)** B and Mg (heated), **(e)** BF_3 and F^-, **(f)** B_2O_3 and H_2O, **(g)** $B(OH)_3$ and $OH^-(aq)$, **(h)** $B(OH)_3$ (mild heating), **(i)** $B(OH)_3$ (strong heating), **(j)** LiH and B_2H_6.

22.14 What is a three-center bond?

22.15 A boron hydride is 81.2% boron. A 25.0 ml sample of the gas (measured at STP) weighs 0.0594 g. What is the molecular formula of the compound?

The Noble Gases

22.16 Diagram the structures of the following using dots to indicate valence electrons. Use VSEPR theory to predict their geometric configurations: **(a)** XeF_2, **(b)** XeF_4, **(c)** XeO_3, **(d)** $XeOF_4$, **(e)** XeO_4, **(f)** XeO_6^{4-}.

22.17 Write equations to show how to prepare **(a)** XeF_2, **(b)** XeF_4, **(c)** XeF_6, **(d)** $XeOF_4$, **(e)** XeO_3.

22.18 What reasons can you give for the fact that the only binary noble-gas compounds known are the fluorides and oxides of Kr, Xe, and Rn?

22.19 **(a)** Use the ion-electron method to write an equation for the oxidation, in acid solution, of Mn^{2+} to MnO_4^- by XeO_3, which is reduced to Xe(g). **(b)** In a reaction between XeO_3 and Mn^{2+} that occurs in 200 ml of solution, the Xe(g) produced occupied a volume of 448 ml at STP. What is the molarity of the MnO_4^- solution resulting from the reaction? What weight of XeO_3 was consumed by the reaction?

22.20 What reason can you give for the fact that the boiling point of helium is lower than the boiling point of hydrogen?

METALS AND METALLURGY

C H A P T E R

23

Several physical and chemical properties that are typical of metals are used to define this classification of elements—although the extent of each of these properties varies widely from metal to metal. Metals have superior electrical and thermal conductivities, characteristic luster, and the ability to be deformed under stress without cleaving. The tendency of these elements toward the formation of cations through electron loss and their formation of basic oxides are among the chemical characteristics of metals.

More than three-quarters of all known elements are metals. The stepped line that appears in the periodic table marks an approximate division between metals and nonmetals—the nonmetals appear in the upper right corner of the table. The division, however, is slightly arbitrary; elements that appear close to the line have intermediate properties.

23.1 The Metallic Bond

Metals have comparatively low ionization energies and electronegativities. Consequently, the outer electrons of metal atoms are relatively loosely held. In a metallic crystal, positive ions (which consist of metal atoms minus the outer electrons) occupy positions in the crystal structure. The outer electrons move freely throughout this structure and bond the crystal together.

The **band theory** describes this bonding in terms of molecular orbitals that extend over the entire crystal. A general idea of the band model for the lithium crystal can be developed in the following way. In the Li_2 molecule, the $2s$ orbital of one Li atom overlaps the $2s$ orbital of another, and $\sigma 2s$ and $\sigma^* 2s$ molecular orbitals are formed (see Section 7.4 and Table 7.3). In the ground state of the Li_2 molecule, only the $\sigma 2s$ molecular orbital is occupied by electrons. Nevertheless, the combination of *two* atomic orbitals produces *two* molecular orbitals even if one of them is unoccupied.

The $2s$ orbitals of two separate Li atoms have the same orbital energy and are said to be **degenerate.** When two Li atoms are brought together to form the Li_2 molecule, the degeneracy is split. The $\sigma 2s$ molecular orbital has a lower orbital energy than the $\sigma^* 2s$ has.

Now consider that a number of Li atoms are brought together into the arrangement of the Li crystal. The $2s$ orbitals of these Li atoms overlap and produce delocalized molecular orbitals that extend throughout the entire structure. *The number of molecular orbitals produced equals the number of atomic orbitals used to produce them.*

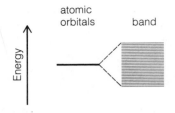

Figure 23.1 Formation of a band by the interaction of the 2s orbitals of lithium

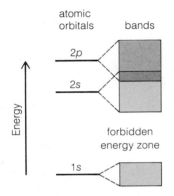

Figure 23.2 Overlap of the 2s and 2p-bands in metallic lithium

In even a very small sample of Li metal there is an extremely large number of atoms. Since each Li atom contributes one 2s orbital toward the formation of the delocalized molecular orbitals, an extremely large number of molecular orbitals are formed. The 2s atomic orbitals of the isolated Li atoms are degenerate, but the molecular orbitals of the Li crystal are not (see Figure 23.1). The molecular orbitals, taken together, make up what is called a **band**; each orbital constitutes an **energy level** within the band. These levels are very closely spaced in terms of energy and for all practical purposes form a continuous energy band. Even at low temperatures, electrons have enough energy to move from level to level within a band.

The 2s electron in a Li atom (electron configuration, $1s^2 2s^1$) is a valence electron, and the 2s-band of Li (see Figure 23.1) is sometimes called the **valence band.** Each Li atom has *one* electron in *one* 2s orbital. If N Li atoms formed a 2s-band, the band would contain N electrons and consist of N molecular orbitals. A molecular orbital, like any orbital, can hold *two* electrons with opposed spins. The 2s-band, therefore, can hold a maximum of $2N$ electrons. Since it contains only N electrons, the valence band of the Li crystal is one-half occupied. The band derived from the 2p orbitals of Li must also be considered, however.

In an isolated Li atom, there are three empty 2p orbitals that are close in energy to the 2s orbital. In metallic Li, a 2p-band, derived from these 2p orbitals, overlaps the 2s-band (see Figure 23.2). A 2p-band formed by N Li atoms would consist of $3N$ energy levels (since each Li atom has three 2p orbitals) and would be empty. This 2p-band is sometimes called the **conduction band** since electrons can move freely throughout it and thereby conduct electricity. In the case of the Li crystal, the 2s- and 2p-bands overlap so that the two bands can be considered to form one.

The shading in Figure 23.2 is designed to show band overlap, not electron occupancy. In fact, the combined bands formed by N Li atoms would consist of $4N$ energy levels (with a maximum capacity of $8N$ electrons) and would contain N electrons. The combined band, therefore, would be one-eighth filled.

If band overlap did not occur, beryllium would not conduct electricity. The electron configuration of Be is $1s^2 2s^2$, and consequently the 2s-band (the valence band) of Be is filled. The 2s-band of Be, however, is overlapped by an empty 2p-band (a conduction band). Beryllium is a metallic conductor because the combined band, resulting from the overlap of the 2s- and 2p-bands, is only one-quarter filled.

Figure 23.2 also shows a narrow band produced by the interaction of the 1s orbitals of Li. Since the 1s orbital is filled in the Li atom, the 1s-band is filled. These electrons do not appreciably affect the bonding. They are closely held by the Li nuclei and form a part of the positive ions that make up the crystal structure. The 1s-band is not overlapped by another band in the Li crystal. Instead, it is separated from the 2s-2p-band by a **forbidden energy zone.** No electron in Li has an energy that would place it in this forbidden zone. Electrons in the filled 1s-band do not have energies high enough to traverse the forbidden energy zone and enter the 2s-2p-band where they could contribute to the bonding and conductivity of Li.

In a metallic crystal, therefore, the valence electrons are held by all the atoms of the crystal and are highly mobile. The valence band, which may be only partially filled, is overlapped by an unfilled conduction band. Transference of an electron to a higher level within a band requires the addition of very little energy since the levels are close together. Thus the valence electrons of a metal can move to higher levels by absorbing light of a wide range of wavelengths. When these electrons fall to lower energy levels, light is radiated. The lustrous appearance of metals is caused by these electron transitions.

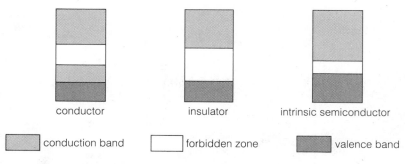

conductor   insulator   intrinsic semiconductor

conduction band   forbidden zone   valence band

Figure 23.3   Energy-level band diagrams for three types of solids

The freely moving electrons of metallic crystals account for the high thermal and electrical conductivities of metals. Valence electrons of a metal absorb heat as kinetic energy and transfer it rapidly to all parts of the metal since their motion is relatively unrestricted.

Energy-level diagrams for conductors, insulators, and intrinsic semiconductors are shown in Figure 23.3. In the diagram for a **conductor** (such as Li), the valence band (from the 2s orbitals) is overlapped by an empty conduction band (from the 2p orbitals). These two bands are separated from an empty, upper conduction band (derived from orbitals of the third shell) by a forbidden energy zone. Electrical conduction takes place by means of the motion of electrons within the lower conduction band. There is no need to supply the energy required to bridge the forbidden zone and use the upper conduction band.

The diagram for the **insulator,** however, shows a completely filled valence band that is widely separated from a conduction band by a forbidden energy zone. Electron motion, and hence electrical conductivity, is possible only if energy is provided to promote electrons across the comparatively large forbidden zone to the conduction band. Normally such promotion does not occur, and hence the conductivities of insulators are extremely low.

An **intrinsic semiconductor** is a material that has a low electrical conductivity that is intermediate between that of a conductor and an insulator. This conductivity increases markedly with increasing temperature. For a semiconductor, the forbidden zone is sufficiently narrow that electrons can be promoted from the valence band to the conduction band by heat (see Figure 23.3). The vacancies left by the removal of electrons from the valence band permit the electrons remaining in the valence band to move under the influence of an electric field. Conduction takes place through the motion of electrons in the valence, as well as in the conduction, band.

The electrical conductivity of a metal is not dependent upon thermal excitation of electrons. Whereas the conductivity of an intrinsic semiconductor increases with increasing temperature, the electrical conductivity of a metal decreases with increasing temperature. Presumably the increased temperature causes increased vibration of the metal ions of the crystal lattice which, in turn, impedes the flow of conduction electrons.

23.2 Semiconductors

Pure silicon and germanium function as intrinsic semiconductors. Both these elements crystallize in the diamond-type network structure in which each atom

A semiconductor chip of silicon. *Intel Corporation.*

in bonded to four other atoms (see Figure 9.12). At room temperature the conductivity of either silicon or germanium is extremely low, since the electrons are fixed by the bonding of the crystal. At higher temperatures, however, the crystal bonding begins to break down; electrons are freed and are able to move through the structure, thereby causing the conductivity to increase.

The addition of small traces of certain impurities to either silicon or germanium enhances the conductivity of these materials and produces what are called **extrinsic** (or **impurity**) **semiconductors.** For example, if boron is added to pure silicon at the rate of one B atom per million Si atoms, the conductivity is increased by a factor of approximately one hundred thousand (from $4 \times 10^{-6}/(\Omega \cdot cm)$ to $0.8/(\Omega \cdot cm)$ at room temperature). Each Si atom has four valence electrons that are used in the bonding of the network lattice. A boron atom, however, has only three valence electrons. Hence, a B atom that assumes a position of a Si atom in the crystal can form only three of the four bonds required for a perfect lattice, and an electron vacancy (or hole) is introduced. An electron from a nearby bond can move into this vacancy, thus completing the four bonds on the B atom but, at the same time, leaving a vacancy at the original site of the electron. In this way electrons move through the structure, and the holes move in a direction opposite to that of the conduction electrons.

This type of extrinsic semiconductor, in which holes enable electron motion, is called a *p*-type semiconductor; the *p* stands for positive. The term is somewhat misleading, however, since the crystal, which is formed entirely from neutral atoms, is electrically neutral. The structure is electron deficient only with regard to the covalent bonding requirements of the lattice; it never has an excess of positive charge.

In addition to boron, other III A elements (Al, Ga, or In) can be added in small amounts to pure silicon or germanium to produce *p*-type semiconductors.

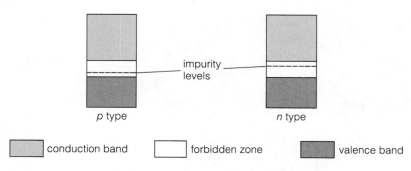

impurity levels

p type   n type

☐ conduction band   ☐ forbidden zone   ☐ valence band

Figure 23.4   Energy-level band diagrams for extrinsic semiconductors

Impurities that owe their effect to their ability to accept electrons by means of the holes they produce in the crystal lattice are known as **acceptor impurities.**

The energy-level band diagram for *p*-type semiconductors is shown in Figure 23.4. Electron vacancies are responsible for the addition of *empty* impurity levels, just above the filled valence band, to a diagram that otherwise would be characteristic of an intrinsic semiconductor. Promotion of electrons from the valence band to the unoccupied impurity levels requires the addition of comparatively little energy, and conduction takes place in this manner.

Traces of group V A elements (P, As, Sb, or Bi) when added to silicon or germanium produce a second type of extrinsic semiconductor known as an *n*-**type semiconductor**; the *n* stands for negative. In this instance, each impurity atom has five valence electrons—one more than required by the bonding of the host crystal. The extra electron can move through the structure and function as a conduction electron. Impurities of this type are known as **donor impurities,** since they provide conduction electrons. The *n*-type semiconductor is negative only in the peculiar sense that more electrons are present than are required by the bonding scheme of the crystal; the substance is electrically neutral.

The energy-level diagram for the *n*-type semiconductor is shown in Figure 23.4. *Filled* impurity levels are introduced just below the empty conduction band by the addition of donor impurities. The electrons from these levels are easily excited into the conduction band by heating, and conduction occurs by this means. Semiconductors find use in photocells and transistors.

23.3 Physical Properties of Metals

The densities of metals show a wide variation. Of all the *solid* elements, lithium has the lowest density and osmium has the highest. The majority of metals, however, have relatively high densities in comparison with the densities of the solid nonmetals. The metals of groups I A and II A, which are called the *light metals*, are notable exceptions to this generalization. The close-packed arrangement of the atoms of most metallic crystals helps to explain the relatively high densities of most metals.

In Figure 23.5, the densities of the metals of the fourth, fifth, and sixth periods are plotted against group number. Each curve reaches a maximum approximately at group VIII (Fe, Ru, and Os). This trend reflects the trend in atomic radius.

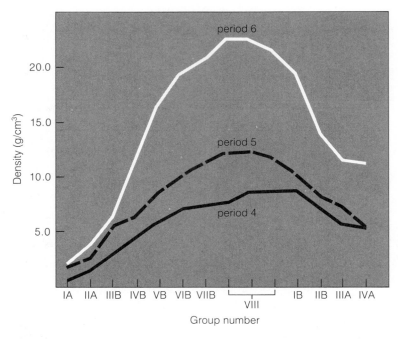

Figure 23.5 Densities of the metals of the fourth, fifth, and sixth periods

The atomic radii of the metals of each period reach a *minimum* at this point and the densities reach a **maximum.** The I A metals have the largest atomic radii and the smallest atomic masses of their periods; these factors coupled with their crystal structure give them comparatively low densities.

The melting points and the boiling points of the nonmetals show extreme variation—from the very low values of the gaseous elements (such as hydrogen and helium) to the extremely high values of the elements that crystallize in covalent network structures (such as the diamond and boron). There is considerable variation in the melting and boiling points of the metals, but, of course, the low values typical of the gaseous nonmetals are not observed. In general, most metals have comparatively high melting and boiling points.

In Figure 23.6, the melting points of the metals of the fourth, fifth, and sixth periods are plotted against group number. The striking feature of the curves is the maximum that appears at approximately the center of each one. Curves for the boiling point, enthalpy of fusion, enthalpy of vaporization, and hardness have approximately the same appearance. For a given period, the strength of the metallic bonding must therefore reach a maximum around the center of the transition series.

The strength of the metallic bonding is, of course, related to the number of delocalized electrons per atom used in the bonding. If we count only the *ns*, *np*, and *unpaired* $(n - 1)d$ electrons of a metal as bonding electrons, we get an order that approximates that of the properties we are considering. This analysis is a decided oversimplification. Other factors, such as atomic radius, nuclear charge, and crystal form, are involved.

The deformability, luster, thermal conductivity, and electrical conductivity are the properties most characteristic of metallic bonding and the metallic state. High electrical conductivity, which is measured in units of $1/(\Omega \cdot cm)$, is character-

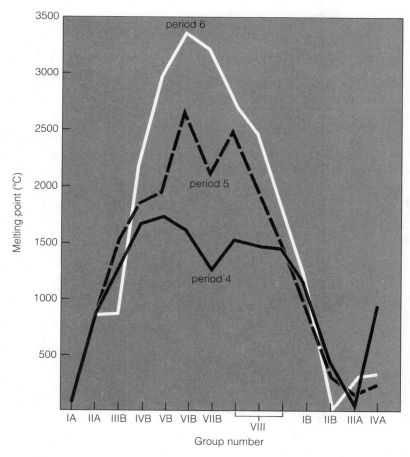

Figure 23.6 Melting points of the metals of the fourth, fifth, and sixth periods

istic of metals.* The following ranges of electrical conductivities, measured at 25°C, are observed:

1. *Metals:* 10^4 to $10^6/(\Omega \cdot cm)$
2. *Semiconductors:* 10^{-5} to $10/(\Omega \cdot cm)$
3. *Nonmetals, or insulators:* 10^{-12} to $10^{-10}/(\Omega \cdot cm)$

---

* According to Ohm's law, $\mathscr{E} = IR$, where $\mathscr{E}$ is the potential difference in volts (V), I is the current in amperes (A), and R is the resistance in ohms (Ω). Hence, $I = \mathscr{E}/R$, and when the potential difference is 1 V, the electrical current is equal to $1/R$, a value called the conductance.

$$conductance = 1/R$$

The conductance of a metal wire varies *directly* with the cross-sectional area of the wire, a (at a constant voltage, more electricity will flow through a thick wire than a thin one) and *inversely* with the length of the wire, l:

$$conductance = ka/l$$

where k is a proportionality constant called the *conductivity* and is equal to the conductance of a 1 cm³ cube of the metal (1 cm² in cross-section and 1 cm long). Thus:

$$conductance = 1/R = ka/l$$
$$k = l/Ra$$

If l is measured in cm and a in cm², k has the units of $1/(\Omega \cdot cm)$.

Table 23.1 Electrical conductivities of the metals at 0°C in units of $10^4/(\text{ohm}\cdot\text{cm})$

Li 11.8	Be 18													
Na 23	Mg 25											Al 40		
K 15.9	Ca 23	Sc	Ti 1.2	V 0.6	Cr 6.5	Mn 20	Fe 11.2	Co 16	Ni 16	Cu 65	Zn 18	Ga 2.2		
Rb 8.6	Sr 3.3	Y	Zr 2.4	Nb 4.4	Mo 23	Tc	Ru 8.5	Rh 22	Pd 10	Ag 66	Cd 15	In 12	Sn 10	Sb 2.8
Cs 5.6	Ba 1.7	La 1.7	Hf 3.4	Ta 7.2	W 20	Re 5.3	Os 11	Ir 20	Pt 10	Au 49	Hg 4.4	Tl 7.1	Pb 5.2	Bi 1

The conductivities of some metals are listed in Table 23.1. The metals of group I B (copper, silver, and gold) are outstanding electrical conductors. Except for similarities in the relative standing of elements of the same group within a period, periodic trends are difficult to discern. In general, thermal conductivities parallel electrical conductivities.

Most metals are highly deformable. They are **malleable,** capable of being pounded into new shapes, and **ductile,** capable of being drawn into wire. The nondirectional character of metallic bonding accounts for the ease with which planes of metal atoms slide across one another under stress and thus explains why a metal crystal may be deformed without shattering. When the planes of an ionic crystal are displaced (see Figure 9.11), the new alignment brings ions of the same charge into proximity and results in the cleavage of the crystal. Covalent network crystals are deformed only by breaking the covalent bonds of the crystal, a process which also results in the fragmentation of the crystal.

23.4 Natural Occurrence of Metals

An ore is a naturally occurring material from which one or more metals can be profitably extracted. The principal types of ores and examples of each are listed in Table 23.2. A few of the less reactive metals occur in nature in elemental form; for many of these metals, these **native ores** constitute the most important source. The greatest tonnage of metals is derived from oxides—either oxide ores or metal oxides that are produced by the roasting of carbonate or sulfide ores.

Silicate minerals are abundant in nature. However, the extraction of metals from silicates is difficult, and the cost of such processes may be prohibitive. Consequently, only the less common metals are commercially derived from silicate ores. Phosphate minerals are, in general, rare and occur in low concentrations.

A number of metals occur as impurities in the ores of other metals so that both metals are derived from the same commercial operation. For example, cadmium metal is obtained as a by-product in the production of zinc.

Ores, as mined, generally contain variable amounts of unwanted materials (such as silica, clay, and granite), which are called **gangue.** The concentration of

Table 23.2 Occurrence of metals

Type of Ore	Examples
native metals	Cu, Ag, Au, As, Sb, Bi, Pd, Pt
oxides	Al_2O_3, Fe_2O_3, Fe_3O_4, SnO_2, MnO_2, TiO_2, $FeO\cdot Cr_2O_3$, $FeO\cdot WO_3$, Cu_2O, ZnO
carbonates	$CaCO_3$, $CaCO_3\cdot MgCO_3$, $MgCO_3$, $FeCO_3$, $PbCO_3$, $BaCO_3$, $SrCO_3$, $ZnCO_3$, $MnCO_3$, $CuCO_3\cdot Cu(OH)_2$, $2\,CuCO_3\cdot Cu(OH)_2$
sulfides	Ag_2S, Cu_2S, CuS, PbS, ZnS, HgS, $FeS\cdot CuS$, FeS_2, Sb_2S_3, Bi_2S_3, MoS_2, NiS, CdS
halides	NaCl, KCl, AgCl, $KCl\cdot MgCl_2\cdot 6\,H_2O$, NaCl and $MgCl_2$ in sea water
sulfates	$BaSO_4$, $SrSO_4$, $PbSO_4$, $CaSO_4\cdot 2\,H_2O$, $CuSO_4\cdot 2\,Cu(OH)_2$
silicates	$Be_3AlSi_6O_{18}$, $ZrSiO_4$, $Sc_2Si_2O_7$, $(NiSiO_3, MgSiO_3)$[a]
phosphates	$[CePO_4, LaPO_4, NdPO_4, PrPO_4, Th_3(PO_4)_4]$[a], $LiF\cdot AlPO_4$

[a] Occur in a single mineral but not in fixed proportions.

the desired metal must be sufficiently high to make its extraction chemically feasible and economically competitive. Ores of low concentration are worked only if they can be processed comparatively easily and inexpensively or if the metal product is scarce and valuable. The required concentration varies greatly from metal to metal. For aluminum or iron it should be 30% or more; for copper it may be 1% or less.

23.5 Metallurgy: Preliminary Treatment of Ores

Metallurgy is the science of extracting metals from their ores and preparing them for use. Metallurgical processes may be conveniently divided into three principal operations: (1) **preliminary treatment,** in which the desired component of the ore is concentrated, specific impurities removed, and/or the mineral is put into a suitable form for subsequent treatment; (2) **reduction,** in which the metal compound is reduced to the free metal; and (3) **refining,** in which the metal is purified and, in some cases, substances added to give desired properties to the final product. Since the problems encountered in each step vary from metal to metal, many different metallurgical procedures exist.

The processing of many ores requires, as a first step, that most of the gangue be removed. Such **concentration** procedures, which are usually carried out on ores that have been crushed and ground, may be based on physical or chemical properties. **Physical separations** are based on differences between the physical properties of the mineral and the gangue. Thus, through washing with water the particles of rocky impurities may often be separated from the heavier mineral particles. This separation may be accomplished by shaking the crushed ore in a stream of water on an inclined table; the heavier mineral particles settle to the bottom and are collected. Adaptations of this process exist.

Flotation is a method of concentration applied to many ores, especially those of copper, lead, and zinc. The finely crushed ore is mixed with a suitable oil

Froth flotation used to
concentrate a copper ore.
Freeport-McMoRan Inc.

and water in large tanks. The mineral particles are wetted by the oil, whereas the gangue particles are wetted by the water. Agitation of the mixture with air produces a froth which contains the oil and mineral particles. This froth floats on the top of the water and is skimmed off.

Magnetic separation is employed to separate Fe_3O_4 (the mineral magnetite) from gangue. The ore is crushed, and electromagnets are used to attract the particles of Fe_3O_4, which is ferromagnetic.

The removal of free metals from native ores may be considered to be a type of concentration. Certain ores, such as native bismuth and native copper, are heated to a temperature just above the melting point of the metal, and the liquid metal is poured away from the gangue. The process is called **liquation.**

Mercury will dissolve silver and gold to form what are called **amalgams.** Hence, the native ores of silver and gold are treated with mercury, the resulting liquid amalgam collected, and the free silver or gold recovered from the amalgam by distilling the mercury away.

The chemical properties of minerals are the basis of **chemical separations.** An important example is the **Bayer method** of obtaining pure aluminum oxide from the ore bauxite. Since aluminum hydroxide is amphoteric, the aluminum oxide is dissolved away from the ore impurities (principally ferric oxide and silicates) by treatment of the crushed ore with hot sodium hydroxide solution:

$$Al_2O_3(s) + 2\,OH^-(aq) + 3\,H_2O \longrightarrow 2\,Al(OH)_4^-(aq)$$

The solution of sodium aluminate is filtered and cooled; its pH is adjusted downward by dilution and/or neutralization with carbon dioxide so that aluminum hydroxide precipitates. Pure aluminum oxide, ready for reduction, is obtained by heating the aluminum hydroxide:

$$2\,Al(OH)_3(s) \longrightarrow Al_2O_3(s) + 3\,H_2O(g)$$

The first step in the extraction of magnesium from sea water is a chemical concentration. The Mg^{2+} in sea water (approximately 0.13%) is precipitated as magnesium hydroxide by treatment of the sea water with a slurry of calcium hydroxide. The calcium hydroxide is obtained from CaO that is produced by roasting oyster shells ($CaCO_3$):

$$CaCO_3(s) \xrightarrow{\text{heat}} CaO(s) + CO_2(g)$$

$$CaO(s) + H_2O \longrightarrow Ca(OH)_2(s)$$

$$Mg^{2+}(aq) + Ca(OH)_2(s) \longrightarrow Mg(OH)_2(s) + Ca^{2+}(aq)$$

The magnesium hydroxide is converted to magnesium chloride by reaction with hydrochloric acid, and the magnesium chloride solution is evaporated to produce the dry $MgCl_2$ necessary for electrolytic reduction.

The metal component of some ores may be obtained by **leaching.** Thus, low-grade carbonate and oxide ores of copper may be leached with dilute sulfuric acid:

$$CuO(s) + 2\,H^+(aq) \longrightarrow Cu^{2+}(aq) + H_2O$$

$$CuCO_3(s) + 2\,H^+(aq) \longrightarrow Cu^{2+}(aq) + CO_2(g) + H_2O$$

and the resulting copper sulfate solutions may be directly subjected to electrolysis.

Silver and gold ores are leached with solutions of sodium cyanide in the presence of air since these metals form very stable complex ions with the cyanide ion. For native silver, argenitite (Ag_2S), and cerargyrite (AgCl), the reactions are

$$4\,Ag(s) + 8\,CN^-(aq) + O_2(g) + 2\,H_2O \longrightarrow 4\,Ag(CN)_2^-(aq) + 4\,OH^-(aq)$$

$$Ag_2S(s) + 4\,CN^-(aq) \longrightarrow 2\,Ag(CN)_2^-(aq) + S^{2-}(aq)$$

$$AgCl(s) + 2\,CN^-(aq) \longrightarrow Ag(CN)_2^-(aq) + Cl^-(aq)$$

After concentration many ores are **roasted** in air. The sulfide ores of the less reactive metals are directly reduced to the free metal by heating (see Section 23.6). However, the majority of the sulfide ores, as well as the carbonate ores, are converted into oxides by roasting:

$$2\,ZnS(s) + 3\,O_2(g) \longrightarrow 2\,ZnO(s) + 2\,SO_2(g)$$

$$PbCO_3(s) \longrightarrow PbO(s) + CO_2(g)$$

The free metals are more readily obtained from oxides than from sulfides or carbonates.

23.6 Metallurgy: Reduction

By far the largest quantity of metals, as well as the largest number of metals, are produced by **smelting** operations—high-temperature reduction processes in which the metal is usually secured in a molten state. In most of these processes a **flux** (such as limestone, $CaCO_3$) is used to remove the gangue that remains after ore concentration. The flux forms a **slag** with silicon dioxide and silicate impurities; for limestone and silicon dioxide the simplified equations are

$$CaCO_3(s) \longrightarrow CaO(s) + CO_2(g)$$

$$CaO(s) + SiO_2(s) \longrightarrow CaSiO_3(l)$$

The slag, which is a liquid at the smelting temperatures, generally floats on top of the molten metal and is readily separated from the metal.

The reducing agent employed for a given smelting operation is the least expensive material that is capable of yielding a product of the required purity. For the ores of the metals of low reactivity—for example, the sulfide ores of mercury, copper, and lead—no chemical agent is required. Mercury is produced by roasting cinnabar, HgS, in air:

$$HgS(s) + O_2(g) \longrightarrow Hg(g) + SO_2(g)$$

The mercury vapor is condensed in a receiver and requires no further purification.

Copper sulfide ores, after concentration by flotation, are smelted to a material called **matte,** which is essentially cuprous sulfide, Cu_2S. The matte is then reduced by blowing air through the molten material:

$$Cu_2S(l) + O_2(g) \longrightarrow 2\,Cu(l) + SO_2(g)$$

Any cuprous oxide that forms in this process is reduced by stirring the molten metal with poles of green wood. The copper produced by this method (blister copper) is about 99% pure and is refined by electrolysis.

Lead sulfide ores (galena) are subjected to a roasting operation in which a portion of the PbS is converted into lead oxide and lead sulfate:

$$2\,PbS(s) + 3\,O_2(g) \longrightarrow 2\,PbO(s) + 2\,SO_2(g)$$

$$PbS(s) + 2\,O_2(g) \longrightarrow PbSO_4(s)$$

The resulting PbS, PbO, and $PbSO_4$ mixture, together with fluxes, is smelted in the absence of air. The reactions, in which sulfur is oxidized and lead is reduced, are

$$PbS(s) + 2\,PbO(s) \longrightarrow 3\,Pb(l) + SO_2(g)$$

$$PbS(s) + PbSO_4(s) \longrightarrow 2\,Pb(l) + 2\,SO_2(g)$$

Some lead is also produced by smelting PbO with coke. The lead oxide is obtained by roasting the sulfide ore in an excess of air so that virtually complete conversion to the oxide results.

Iron, zinc, tin, cadmium, antimony, nickel, cobalt, molybdenum, lead, and other metals are produced by carbon reduction of oxides, which are obtained as ores or as the products of roasting operations.

The reactions that occur in a high-temperature carbon reduction are complex. In most cases the reduction is effected principally by carbon monoxide, not carbon. Since both the mineral and coke are solids that are not readily fusible, contact between them is poor and direct reaction slow:

$$MO(s) + C(s) \longrightarrow M(l) + CO(g)$$

(M stands for a metal.) However, a gas and a solid make better contact and react more readily:

$$2\,C(s) + O_2(g) \longrightarrow 2\,CO(g)$$

$$MO(s) + CO(g) \longrightarrow M(l) + CO_2(g)$$

The carbon dioxide that is produced is converted into carbon monoxide by reaction with coke:

$$C(s) + CO_2(g) \longrightarrow 2\,CO(g)$$

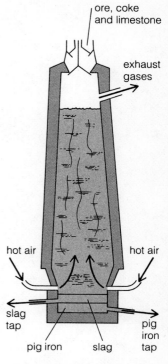

Figure 23.7 Diagram of a blast furnace (schematic)

The most important commercial metal, iron, is produced by carbon reduction in a blast furnace designed to operate continuously (see Figure 23.7). The ore, coke, and a limestone flux are charged into the top of the furnace, and heated air, which is sometimes enriched with oxygen, is blown in at the bottom. The incoming air reacts with carbon to form carbon monoxide and to liberate considerable amounts of heat; at this point the temperature of the furnace is the highest (approximately 1500°C).

The iron ore (usually Fe_2O_3) is reduced in stages depending upon the temperature. Near the top of the furnace, where the temperature is lowest, Fe_3O_4 is the reduction product:

$$3\,Fe_2O_3(s) + CO(g) \longrightarrow 2\,Fe_3O_4(s) + CO_2(g)$$

The descending Fe_3O_4 is reduced to FeO in a lower, hotter zone:

$$Fe_3O_4(s) + CO(g) \longrightarrow 3\,FeO(s) + CO_2(g)$$

In the hottest zone, reduction to metallic iron occurs:

$$FeO(s) + CO(g) \longrightarrow Fe(l) + CO_2(g)$$

The molten iron collects in the bottom of the furnace. Molten slag, principally calcium silicate produced by the ultimate action of the flux on the gangue, also collects on the bottom. The slag floats on top of the molten iron, thus protecting the metal from being oxidized by the incoming air. The slag and iron are periodically drawn off. The hot exhaust gases are used to heat incoming air.

The impure iron produced by the blast furnace (**pig iron**) contains up to 4% carbon, up to 2% silicon, some phosphorus, and a trace of sulfur. In the manufacture of steel (see Section 23.7), these impurities are removed or their concentrations are adjusted; in addition, certain metallic ingredients are added. The presence of some carbon in steel is desirable, and the amount of carbon in the final product is readily controlled in the refining process.

The carbon reduction of some oxides (for example, those of chromium, manganese, and tungsten) yields products that contain an appreciable quantity of carbon. If these metals are to be used in the manufacture of steel, the carbon content does not matter. However, since carbon impurities are difficult to remove from many metals, carbon reduction is not a satisfactory method for preparing these metals in a pure form. In such instances, as well as those in which carbon is incapable of effecting a reduction, reactive metals (such as Na, Mg, and Al) may be used as reducing agents. Processes that use metals as reducing agents are more expensive than those that employ carbon because the cost of the production of the metal used as a reducing agent must be taken into account.

The reduction of an oxide by aluminum is called a **Goldschmidt process** or a **thermite process**:

$$Cr_2O_3(s) + 2\,Al(s) \longrightarrow 2\,Cr(l) + Al_2O_3(l)$$

$$3\,MnO_2(s) + 4\,Al(s) \longrightarrow 3\,Mn(l) + 2\,Al_2O_3(l)$$

$$3\,BaO(s) + 2\,Al(s) \longrightarrow 3\,Ba(l) + Al_2O_3(l)$$

The reactions are highly exothermic, and molten metals are produced. The reaction of Fe_2O_3 and aluminum is used at times for the production of molten iron in welding operations and as the basis of incendiary (thermite) bombs. Other oxides commercially reduced by metals include: UO_3 (by Al or Ca), V_2O_5 (by Al), Ta_2O_5 (by Na), MoO_3 (by Al), ThO_2 (by Ca), and WO_3 (by Al).

The **Kroll process** involves the reduction of metal halides (such as $TiCl_4$, $ZrCl_4$, UF_4, and $LaCl_3$) by magnesium, sodium, or calcium. In the production of titanium, titanium tetrachloride is prepared by the reaction of titanium dioxide, carbon, and chlorine. The chloride is a liquid and is readily purified by distillation. Purified $TiCl_4$ is passed into molten magnesium or sodium at approximately 700°C under an atmosphere of argon or helium (which prevents oxidation of the product):

$$TiCl_4(g) + 2\,Mg(l) \longrightarrow Ti(s) + 2\,MgCl_2(l)$$

A **metal displacement** reaction employing zinc is used to obtain silver and gold from the solutions obtained from cyanide leaching operations:

$$2\,Ag(CN)_2^-(aq) + Zn(s) \longrightarrow 2\,Ag(s) + Zn(CN)_4^{2-}(aq)$$

Hydrogen reduction is used for the preparation of some metals (at high temperatures) when carbon reduction yields an unsatisfactory product. Examples are

$$GeO_2(s) + 2\,H_2(g) \longrightarrow Ge(s) + 2\,H_2O(g)$$

$$WO_3(s) + 3\,H_2(g) \longrightarrow W(s) + 3\,H_2O(g)$$

$$MoO_3(s) + 3\,H_2(g) \longrightarrow Mo(s) + 3\,H_2O(g)$$

Cells used for the production of magnesium metal by the electrolysis of molten $MgCl_2$. The cells operate at 700°C and greater than 100,000 A of direct current. The heavy copper busbars are required to deliver electrical power to the cells and the large black graphite rods are anodes. *Dow Chemical U.S.A.*

The metals are produced as powders by this method. Germanium is melted and cast into ingots, but tungsten (wolfram) and molybdenum have high melting points. They are heated and pounded into compact form. Hydrogen cannot be used to reduce the oxides of some metals because of the formation of undesirable hydrides.

The most reactive metals are prepared by **electrolytic reduction** of molten compounds. Sodium and magnesium (as well as the other I A and II A metals) are prepared by electrolysis of the fused chlorides. Sodium chloride is electrolyzed in a **Downs cell** (see Figure 23.8), which is designed to keep the products separate from one another so that they do not react:

anode: $\qquad 2\,Cl^- \longrightarrow Cl_2(g) + 2e^-$

cathode: $\quad e^- + Na^+ \longrightarrow Na(l)$

Purified Al_2O_3 is electrolyzed in the **Hall process** for the production of aluminum (see Figure 23.9). Aluminum oxide dissolved in molten cryolite, Na_3AlF_6, conducts electric current. The carbon lining of the cell serves as the cathode, and carbon anodes are used. The primary reaction at the anode may be considered to be the discharge of oxygen from oxide ions. However, the oxygen reacts with the carbon anodes so that the cell reactions are

anode: $\quad C(s) + 2\,O^{2-} \longrightarrow CO_2(g) + 4e^-$

cathode: $\quad 3e^- + Al^{3+} \longrightarrow Al(l)$

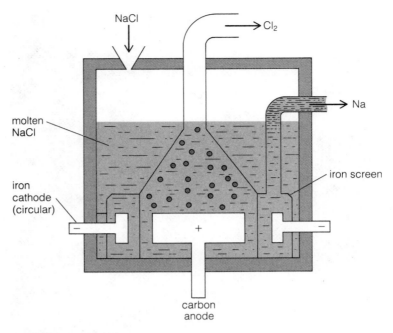

Figure 23.8 Schematic diagram of the Downs cell for the production of sodium and chlorine.

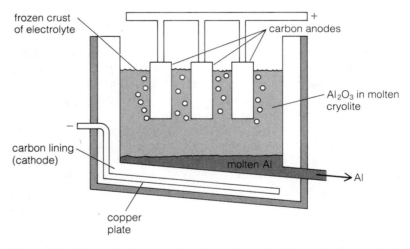

Figure 23.9 Schematic diagram of an electrolytic cell for the production of aluminum

The molten aluminum collects at the bottom of the cell and is periodically withdrawn.

Some metals are prepared by the electrolysis of aqueous solutions of their salts. Very pure zinc is produced by the electrolysis of zinc sulfate solutions. The zinc sulfate is obtained by treating zinc oxide (obtained from zinc sulfide ores by roasting) with sulfuric acid. The sulfuric acid for the process is made from the sulfur dioxide derived from the roasting of the sulfide ore:

Aluminum production by electrolytic reduction: A craneman is manipulating a giant crucible into position to tap one of the electrolytic cells containing molten aluminum. *Aluminum Company of America.*

$$\text{anode:} \qquad 2\,H_2O \longrightarrow O_2(g) + 4\,H^+(aq) + 4e^-$$

$$\text{cathode:} \quad 2e^- + Zn^{2+}(aq) \longrightarrow Zn(s)$$

Aqueous solutions of salts of copper, cadmium, chromium, cobalt, gallium, indium, manganese, and thorium are electrolyzed to produce the corresponding metals. Electrolytically prepared metals generally require no further refining.

23.7 Metallurgy: Refining

Most of the metals obtained from reduction operations require refining to rid them of objectionable impurities. Refining processes vary widely from metal to metal, and for a given metal the method employed may vary with the proposed use of the final product. Along with the removal of materials that impart undesirable properties to the metal, the refining step may include the addition of substances to give the product desired characteristics. Some refining processes are designed to recover valuable metal impurities, such as gold, silver, and platinum.

Crude tin, lead, and bismuth are purified by **liquation.** In this process, ingots of the impure metals are placed at the top of a sloping hearth that is maintained at a temperature slightly above the melting point of the metal. The metal melts and flows down the inclined hearth into a well, leaving behind the solid impurities. Some low-boiling metals, such as zinc and mercury, are purified by **distillation.**

The **Parkes process** for refining lead, which also is a concentration method for silver, relies upon the selective dissolution of silver in molten zinc. A small amount of zinc (1 to 2%) is added to molten lead that contains silver as an impurity. Silver is much more soluble in zinc than in lead; lead and zinc are insoluble in each other. Hence, most of the silver concentrates in the zinc, which comes to the top of the molten lead. The zinc layer solidifies first upon cooling and is removed. The silver is obtained by remelting the zinc layer and distilling away the zinc, which is collected and used over again.

The **Van Arkel process** is based upon the thermal decomposition of a metal compound. The method, which is used for the purification of titanium, hafnium, and zirconium, involves the decomposition of a metal iodide on a hot metal filament. For example, impure zirconium is heated with a limited amount of iodine in an evacuated glass apparatus:

$$Zr(s) + 2 I_2(g) \longrightarrow ZrI_4(g)$$

The gaseous zirconium tetraiodide is decomposed upon contact with a hot filament, and pure zirconium metal is deposited upon the filament:

$$ZrI_4(g) \longrightarrow Zr(s) + 2 I_2(g)$$

The regenerated iodine reacts with more zirconium. The process is very expensive and is employed for the preparation of limited amounts of very pure metals for special uses.

Another process capable of producing metals of very high purity is **zone refining.** A circular heater is fitted around a rod of an impure metal, such as germanium (see Figure 23.10). The heater, which is slowly moved down the rod, melts a band of the metal. As the heater moves along, pure metal recrystallizes out of the melt, and impurities are swept along in the molten zone to one end of the rod, which is subsequently discarded. More than one pass of the heater may be made on the same rod.

Electrolytic refining is an important and widely used method of purification. Many metals, including copper, tin, lead, gold, zinc, chromium, and nickel, are refined electrolytically. Plates of the impure metal are used as anodes, and the electrolyte is a solution of a salt of the metal. The pure metal plates out on the cathode. For copper, copper sulfate is employed as the electrolyte, and the electrode reactions are

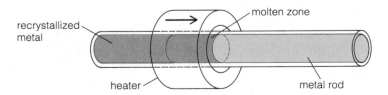

Figure 23.10 Schematic diagram of zone refining

anode: $$Cu(s) \longrightarrow Cu^{2+}(aq) + 2e^-$$

cathode: $$2e^- + Cu^{2+}(aq) \longrightarrow Cu(s)$$

The more reactive metals in the crude copper anode, such as iron, are oxidized and pass into solution where they remain; they are not reduced at the cathode. The less reactive metals, such as silver, gold, and platinum, are not oxidized. As the copper anode dissolves away, they fall to the bottom of the cell from where they are recovered as a valuable "anode sludge."

There are two important ways in which pig iron is refined into steel. The principal impurities in pig iron are carbon (from the coke added in the reduction of the iron ore) as well as silicon, phosphorus, and sulfur (from the ore itself). In a refining process, these impurities are oxidized. The $CO(g)$ and $CO_2(g)$ produced from the C escape and the oxides of Si, P, and S (which are acidic oxides) are removed by slag formation with calcium oxide or other basic oxides:

$$CaO(s) + SiO_2(s) \longrightarrow CaSiO_3(l)$$

Molten iron from a blast furnace being poured into a basic oxygen furnace. *American Iron and Steel Institute.*

Calcium oxide is formed by the decomposition of limestone, $CaCO_3$, which is added in the refining process.

In the **open hearth process,** pig iron, scrap iron, Fe_2O_3, and limestone are heated in a shallow hearth lined with CaO or MgO. A blast of hot air and burning fuel is directed on the molten charge. The impurities are oxidized by the iron oxide and the hot air. Since it takes about 8 to 10 hours to produce a batch of steel, the quality of the product is easily controlled. Alloying metals (such as Mn, Cr, Ni, W, Mo, and V) may be added before the charge is poured.

Most of the steel produced today is made by the **basic oxygen process.** Molten pig iron, scrap iron, and powdered $CaCO_3$ are placed in a converter that is lined with basic oxides. The impurities are oxidized by pure oxygen under a pressure of 10 to 12 atm, which is blown onto the surface of the molten charge through a lance. The stream of oxygen and the escape of gaseous oxidation products keeps the charge mixed. The reactions are rapid and highly exothermic so that no external heat is required to keep the mass molten. The process is complete in about 20 to 50 minutes and gives a high-quality product.

23.8 The Group I A Metals

The I A metals, also called the **alkali metals,** constitute the most reactive group of metals. None of these elements is found free in nature, and all may be prepared by the electrolysis of dry, molten salts. The element francium, $Z = 87$, is formed in certain natural radioactive processes. All the isotopes of francium are radioactive with short half-lives, and the element is extremely rare.

The group I A elements are silvery metals; cesium has a slight golden yellow cast. The elements are comparatively soft (they can be cut by a knife) and have low melting points and boiling points (see Table 23.3). Melting point, boiling point, and hardness decrease with increasing atomic number. The metals are good conductors of heat and electricity. They have very low densities. Compare the densities given in Table 23.3 with the densities of the metals of the first transition series, which range from 2.5 g/cm^3 for $_{21}$Sc to 8.9 g/cm^3 for $_{30}$Cu.

The alkali metals emit electrons when irradiated (the photoelectric effect). Cesium, which ejects electrons most readily, is used in the manufacture of photo-

Table 23.3 Some properties of the I A metals

	Lithium	Sodium	Potassium	Rubidium	Cesium
outer electronic configuration	$2s^1$	$3s^1$	$4s^1$	$5s^1$	$6s^1$
melting point (°C)	179	97.5	63.7	39.0	28.5
boiling point (°C)	1336	880	760	700	670
density (g/cm^3)	0.53	0.97	0.86	1.53	1.90
atomic radius (pm)	123	157	203	216	235
ionic radius, M$^+$ (pm)	60	95	133	148	169
ionization energy (kJ/mol)					
first	520	496	419	403	376
second	7296	4563	3069	2640	2258
electrode potential, $\mathscr{E}^{\circ}_{red}$, (M$^+$/M) (V)	−3.05	−2.71	−2.93	−2.93	−2.92

cells (employed in light meters and electric eyes), which convert light signals into electric signals.

The electronic configuration of each of the alkali metals is that of the preceding noble gas (a noble gas core) plus a single s valence electron in the outer shell. These valence electrons are easily lost to give $1+$ ions that are isoelectronic with noble gases. Thus, an element of group I A has the lowest first ionization energy of any element of its period. The ease with which these elements lose electrons makes them extremely strong reducing agents.

The second ionization energies of the I A metals are so much higher than the first ionization energies that the $1+$ oxidation state is the only one observed for these metals. With few exceptions, alkali metal compounds are ionic.

In general, the reactivity of the elements increases with increasing atomic number—paralleling a decrease in first ionization potential. Thus, in most cases cesium is the most reactive element of the group and lithium is the least reactive. This order of reactivity is expected since a large atom holds its valence electron less tightly than a small atom; in a large atom the electron is farther from the nucleus and is screened from the positive nuclear charge by a large number of underlying electron shells.

Each I A element has the largest atomic radius and largest ionic radius of any element of its period. This comparatively large size combined with the low charge of the I A ions leads to species that have little polarizing ability (and hence form strongly ionic compounds) and that do not readily form complex ions.

All the hydroxides of the alkali metals are strong electrolytes. Alkali metal compounds are generally very soluble in water. However, the hydroxide, carbonate, phosphate, and fluoride of lithium are much less soluble than the corresponding salts of the other I A metals. Sodium ion may be precipitated from solution as sodium zinc uranyl acetate, $NaZn(UO_2)_3(C_2H_3O_2)_9 \cdot 6\,H_2O$. The perchlorate ion, ClO_4^-, the hexachloroplatinate ion, $[PtCl_6]^{2-}$, and the cobaltinitrite ion, $[Co(NO_2)_6]^{3-}$, may be used to precipitate K^+, Rb^+, and Cs^+.

The unipositive ammonium ion has a radius that lies between those of K^+ and Rb^+, and ammonium salts have solubilities that resemble those of the alkali metal salts. The ionic radius of the thallous ion, Tl^+, is similar to that of the rubidium ion. Not only do thallous compounds resemble rubidium compounds in solubility, but $TlOH$ is also a strong electrolyte.

Table 23.4 Reaction of the I A metals[a]

Reaction	Remarks
$2\,M + X_2 \longrightarrow 2\,MX$	X_2 = all halogens
$4\,Li + O_2 \longrightarrow 2\,Li_2O$	excess oxygen
$2\,Na + O_2 \longrightarrow Na_2O_2$	
$M + O_2 \longrightarrow MO_2$	M = K, Rb, Cs
$2\,M + S \longrightarrow M_2S$	also with Se and Te
$6\,Li + N_2 \longrightarrow 2\,Li_3N$	Li only
$12\,M + P_4 \longrightarrow 4\,M_3P$	also with As, Sb
$2\,M + 2\,C \longrightarrow M_2C_2$	M = Li and Na; other I A metals give non-stoichiometric interstitial compounds
$2\,M + H_2 \longrightarrow 2\,MH$	
$2\,M + 2\,H_2O \longrightarrow 2\,MOH + H_2$	room temperature
$2\,M + 2\,H^+ \longrightarrow 2\,M^+ + H_2$	violent reaction
$2\,M + 2\,NH_3 \longrightarrow 2\,MNH_2 + H_2$	liquid NH_3 presence of catalysts such as Fe; gaseous NH_3, heated

[a] M = any I A metal, except where noted.

Some of the reactions of the I A metals are summarized in Table 23.4. Whereas the properties of lithium are similar in most respects to those of the other members of group I A, there are some dissimilarities which may be traced to the comparatively small size of Li and Li^+. The first members of the groups of the periodic table often deviate slightly in character from the remaining members of their groups. In addition, similarities may be observed between these first members and elements that adjoin them diagonally on the periodic table (diagonal relationships):

$$\begin{array}{cccc} Li & Be & B & C \\ Na & Mg & Al & Si \end{array}$$

Thus, lithium resembles magnesium in some of its properties. The magnesium ion has a slightly larger ionic radius (65 pm) than the lithium ion (60 pm), but the charge on the magnesium ion is $2+$.

Lithium resembles magnesium and differs from its congeners in the following ways. The carbonate, phosphate, and fluoride of lithium are only slightly soluble in water. Lithium forms an ordinary oxide rather than a peroxide or a superoxide upon burning in oxygen. Lithium ions are more strongly hydrated than those of any other I A element. Lithium reacts directly with nitrogen to give a nitride. Lithium hydroxide may be decomposed to Li_2O and water upon heating, and lithium carbonate decomposes to Li_2O and CO_2 upon ignition. Lithium nitrate decomposes upon heating:

$$4\,LiNO_3(s) \longrightarrow 2\,Li_2O(s) + 4\,NO_2(g) + O_2(g)$$

The nitrates of the other alkali metals form nitrites upon heating:

$$2\,MNO_3(s) \longrightarrow 2\,MNO_2(s) + O_2(g)$$

and the hydroxides and carbonates of Na, K, Rb, and Cs are thermally stable.

23.9 The Group II A Metals

The group II A metals, called the **alkaline earth metals,** are highly electropositive and constitute the second most reactive group of metals. They are not found free in nature and are commonly produced by the electrolysis of molten chlorides. Radium is a comparatively scarce element and all its isotopes are radioactive.

Because of its larger nuclear charge, each II A metal has a smaller atomic radius than that of the I A metal of its period. Since atoms of the II A metals are smaller and have two valence electrons instead of one, the II A metals have higher melting and boiling points and greater densities than the I A metals (see Table 23.5). In addition, the alkaline earth metals are harder than the alkali metals. Beryllium is hard enough to scratch glass and has a tendency toward brittleness; the degree of hardness declines with increasing atomic number. The alkaline earth metals are white metals with a silvery luster. They are good conductors of electricity.

Each of the II A metals has an electronic configuration consisting of a noble gas core plus two s electrons in the outer, valence level. The loss of the two valence electrons produces an ion that is isoelectronic not only with a noble gas but also with the I A metal ion of the same period. However, the alkaline earth metal ions have a larger nuclear charge than the alkali metal ions and are therefore considerably smaller. The beryllium ion is an exceptionally small cation.

Compared with the I A metal ions, the II A metal ions have considerably higher ratios of ionic charge to ionic radius. There are several important consequences of this fact. The hydration energy of a given alkaline earth ion is about five times that of the alkali metal ion of the same period. Furthermore, the compounds of the smaller members of the group, magnesium and beryllium, have appreciable covalent character because the cations of these metals have a strong polarizing effect on anions. This tendency for covalent bond formation is particularly pronounced for beryllium. All compounds of this metal, even those with the most electronegative elements such as oxygen and fluorine, have significant covalent character. Beryllium has the greatest tendency of the group toward the formation of complex ions; examples are $[Be(H_2O)_4]^{2+}$, $[Be(NH_3)_4]^{2+}$, BeF_4^{2-}, and $[Be(OH)_4]^{2-}$. Beryllium hydroxide is the only amphoteric hydroxide formed by a group II A metal.

Table 23.5 Some properties of the II A metals

	Beryllium	Magnesium	Calcium	Strontium	Barium
outer electronic configuration	$2s^2$	$3s^2$	$4s^2$	$5s^2$	$6s^2$
melting point (°C)	1280	651	851	800	850
boiling point (°C)	1500	1107	1440	1366	1537
density (g/cm³)	1.86	1.75	1.55	2.6	3.59
atomic radius (pm)	89	136	174	191	198
ionic radius, M^{2+} (pm)	31	65	99	113	135
ionization energy (kJ/mol)					
first	899	738	590	540	503
second	1757	1450	1145	1059	960
electrode potential, $\mathscr{E}^\circ_{red}$, (M^{2+}/M) (V)	−1.85	−2.36	−2.87	−2.89	−2.91

The covalent nature of beryllium halides is indicated by the poor electrolytic conduction of the anhydrous molten compounds; NaCl is generally added to anhydrous molten $BeCl_2$ in the electrolytic preparation of beryllium. Gaseous $BeCl_2$ molecules are linear and presumably of the sp hybrid type. In most solid compounds beryllium is tetrahedrally bonded through sp^3 hybrid orbitals. This configuration is observed in the complex ions of beryllium. Beryllium chloride is polymeric; each beryllium atom is joined to four chlorine atoms:

The BeO crystal may be considered to be composed of BeO_4 tetrahedra.

The ionization potentials of the II A metals are higher than those of the I A metals because of differences in atomic size and nuclear charge. The hydration energies of the II A cations, however, are also larger than those of the I A cations because of the smaller size and higher charge of the II A cations. Consequently, the electrode potentials of the II A metals are, in general, similar in magnitude to those of the I A metals.

The hydration energy is largest for Be^{2+} and smallest for Ba^{2+}, but the electrode potentials fall in the same order as the ionization potentials; Ba is the strongest reducing agent of the group and Be is the weakest. This order of reactivity is illustrated by the reactions of the alkaline earth metals with water. Beryllium fails to react at red heat. Magnesium will react with boiling water or steam. Calcium, strontium, and barium react vigorously with cold water. Some chemical reactions of the group II A metals are summarized in Table 23.6.

Table 23.6   Reactions of the II A metals[a]

Reaction	Remarks
$M + X_2 \longrightarrow MX_2$	X_2 = all halogens
$2M + O_2 \longrightarrow 2MO$	Ba also gives BaO_2
$M + S \longrightarrow MS$	also with Se and Te
$3M + N_2 \longrightarrow M_3N_2$	at high temperatures
$6M + P_4 \longrightarrow 2M_3P_2$	at high temperatures
$M + 2C \longrightarrow MC_2$	all except Be which forms Be_2C; high temperatures
$M + H_2 \longrightarrow MH_2$	M = Ca, Sr, Ba; high temperatures; Mg with H_2 under pressure
$M + 2H_2O \longrightarrow M(OH)_2 + H_2$	M = Ca, Sr, Ba; room temperature steam; Be does not react at red heat
$Mg + H_2O \longrightarrow MgO + H_2$	
$M + 2H^+ \longrightarrow M^{2+} + H_2$	
$Be + 2OH^- + 2H_2O \longrightarrow Be(OH)_4^{2-} + H_2$	Be only
$M + 2NH_3 \longrightarrow M(NH_2)_2 + H_2$	M = Ca, Sr, Ba; liquid ammonia in presence of catalysts
$3M + 2NH_3 \longrightarrow M_3N_2 + 3H_2$	gaseous NH_3; high temperatures

[a] M = any II A metal, except where noted.

Table 23.7 Solubility product constants of some II A metal salts

	OH^-	SO_4^{2-}	CO_3^{2-}	$C_2O_4^{2-}$	F^-	CrO_4^{2-}
Be^{2+}	1.6×10^{-26}	–	–	–	–	–
Mg^{2+}	8.9×10^{-12}	–	10^{-5}	8.6×10^{-5}	8×10^{-8}	–
Ca^{2+}	1.3×10^{-6}	2.4×10^{-5}	4.7×10^{-9}	1.3×10^{-9}	1.7×10^{-10}	7.1×10^{-4}
Sr^{2+}	3.2×10^{-4}	7.6×10^{-7}	7×10^{-10}	5.6×10^{-8}	7.9×10^{-10}	3.6×10^{-5}
Ba^{2+}	5.0×10^{-3}	1.5×10^{-9}	1.6×10^{-9}	1.5×10^{-8}	2.4×10^{-5}	8.5×10^{-11}

Almost all the salts of the I A metals are very soluble in water. In contrast, a number of II A metal compounds are not appreciably water soluble; the solubility products of some of these insoluble compounds are listed in Table 23.7.

The solubility of a salt depends upon the lattice energy of the salt (energy absorbed in the solution process):

$$MA(s) \longrightarrow M^{2+}(g) + A^{2-}(g)$$

and the hydration energies of the ions (energy evolved):

$$M^{2+}(g) \longrightarrow M^{2+}(aq)$$
$$A^{2-}(g) \longrightarrow A^{2-}(aq)$$

When the solubilities of salts containing the same anion are compared, the hydration energy of the anion may be neglected and solubility differences may be attributed to the combination of the other two factors.

The solubility of the alkaline earth sulfates decreases with increasing cation size; $BeSO_4$ is very soluble, and $BaSO_4$ is very insoluble. The lattice energies of the sulfates do not change greatly in the sequence from $BeSO_4$ to $BaSO_4$, presumably because the anion is so much larger than any II A cation. The trend in the solubility of the sulfates therefore parallels the trend in hydration energies of the ions. The hydration of the small Be^{2+} ion is by far the most exothermic of any ion of the group, and $BeSO_4$ is by far the most soluble sulfate formed by any ion of the group.

The trend in the solubility of hydroxides is the reverse of that of the sulfates; $Be(OH)_2$ is the least soluble alkaline earth hydroxide, and the solubility increases down the group. For hydroxides the lattice energy is dependent upon cation size. The strength of the crystal forces decreases with increasing cation size; $Be(OH)_2$ has the largest lattice energy of any alkaline earth hydroxide. Apparently the trend in lattice energy (energy required) overshadows the trend in hydration energy (energy released).

Solubility data cannot always be so simply interpreted. In the preceding solubility considerations we have ignored entropy effects. In addition, even though the lattice energies and hydration energies of a group of salts may vary regularly, the sum of the two energy effects and hence the solubilities of the salts may vary in an irregular manner.

The lattice energies of the oxides of the II A metals decrease regularly from BeO to BaO, and the effect of this trend is seen in the series of reactions of the alkaline earth oxides with water. Beryllium oxide is insoluble in and unreactive toward water. Magnesium oxide reacts very slowly with water; MgO that has been ignited at high temperatures, however, is practically inert. The oxides of calcium, strontium, and barium readily react with water to form hydroxides:

$$MO + H_2O \longrightarrow M(OH)_2$$

The carbonates of the alkaline earth metals decompose upon heating:

$$MCO_3(s) \longrightarrow MO(s) + CO_2(g)$$

The thermal stability of the carbonates varies directly with the size of the cation. Beryllium carbonate is very unstable and can be prepared only in an atmosphere of carbon dioxide presumably because of the enhanced stability of BeO over $BeCO_3$.

The tendency of the II A metal ions toward hydration causes a number of the compounds of these ions to hydrate with ease; $Mg(ClO_4)_2$, $CaCl_2$, $CaSO_4$, and $Ba(ClO_4)_2$ are used as desiccants.

The beryllium ion hydrolyzes in solution:

$$[Be(H_2O)_4]^{2+} + H_2O \rightleftharpoons [Be(H_2O)_3(OH)]^+ + H_3O^+$$

but the other cations do not. The hydroxides of beryllium and magnesium may be regarded as insoluble (see Table 23.7). Even though the hydroxides of calcium, strontium, and barium are limitedly soluble in water, these compounds are completely dissociated in aqueous solution.

The sulfides of the group are water soluble, and solutions of these compounds are alkaline due to the hydrolysis of the sulfide ion. In quantitative determinations barium ion is usually precipitated as barium sulfate; strontium ion, as strontium sulfate or strontium oxalate; and calcium ion, as calcium oxalate. Magnesium ion is precipitated as $Mg(NH_4)PO_4 \cdot 6H_2O$ by the HPO_4^{2-} ion in the presence of ammonia. Beryllium may be determined as the hydroxide.

Because of the extremely small size of Be and Be^{2+}, beryllium is even more exceptional with regard to the other members of group II A than lithium is with regard to the other members of group I A. The diagonal relationship between beryllium and aluminum is particularly striking. Although Al^{3+} is larger than Be^{2+}, the two ions have similar electric fields because of the higher charge of Al^{3+}.

Both elements have a strong tendency toward the formation of covalent compounds. The halides are largely covalent in nature, are soluble in organic solvents, and are strong Lewis acids. Both aluminum hydroxide and beryllium hydroxide are amphoteric, and both metals dissolve in solutions of hydroxides to give hydrogen. The standard electrode potential of beryllium is similar in magnitude to that of aluminum but much smaller than those of the other II A metals.

The carbides of beryllium and aluminum yield methane, CH_4, on hydrolysis:

$$Be_2C(s) + 4H_2O \longrightarrow 2Be(OH)_2(s) + CH_4(g)$$

$$Al_4C_3(s) + 12H_2O \longrightarrow 4Al(OH)_3(s) + 3CH_4(g)$$

The hydrolysis of the other II A metal carbides yields acetylene, C_2H_2.

Aluminum and beryllium (as well as magnesium) have thin protective oxide coatings, which make them resistant to attack by dilute nitric acid. Beryllium oxide and aluminum oxide are hard, extremely high melting, insoluble solids. Beryllium and aluminum form the stable fluoro complex anions BeF_4^{2-} and AlF_6^{3-}. The other II A metals do not form fluoro complexes that are stable in solution. Normal beryllium carbonate is unstable, and aluminum carbonate cannot be made at all.

23.10 The Transition Metals

The transition metals are found in groups III B to II B in the periodic table. Some chemists do not include group II B (the zinc group) in this classification, and some exclude group I B (the copper group) as well. However, there are advantages in classifying the copper- and zinc-group elements as transition elements, and we shall follow this practice.

In general, the transition metals have high melting and boiling points, high heats of fusion, and high heats of vaporization. The group II B elements (zinc, cadmium, and mercury) are exceptions to this generalization. Mercury is a liquid under ordinary conditions, and all of the II B elements are comparatively low melting and relatively easy to volatilize (distillation is a method of refining the metals). Most transition metals are good conductors of heat and electricity; the I B elements are outstanding in these respects (see Table 23.1).

The significant feature of the electronic configurations of the transition elements (see Table 23.8) is the gradual build-up of the d subshell of the shell adjacent to the outer shell. Included in the sixth and seventh periods are the lanthanides and actinides, respectively. These are the inner-transition elements in which inner f orbitals are being filled (see Section 23.11).

The transition elements exhibit a wide variation in chemical properties. Many compounds of the transition elements are colored and paramagnetic because of unpaired electrons. Since the inner d orbitals are energetically close to the s orbital of the outer level, both ns and $(n-1)$ d electrons are involved in compound formation. With the exceptions of zinc, cadmium, and the elements of group III B, all transition elements exhibit more than one oxidation state in compound formation. The largest number of oxidation states, as well as the maximum oxidation number, is observed for the fifth transition element of period 4 (Mn) and the sixth elements of periods 5 and 6 (Ru and Os, respectively). For these elements and the ones preceding them in each transition series, the maximum oxidation number corresponds to the total number of ns and $(n-1)$ d electrons (see Table 23.9). The higher oxidation numbers of a given element are most frequently seen in compounds containing the more electronegative elements: fluorine, oxygen, and chlorine.

Table 23.8 Postulated electronic configurations of the valence subshells of the transition metals

Sc	$3d^1$	$4s^2$	Y	$4d^1$	$5s^2$	La	$5d^1$	$6s^2$
Ti	$3d^2$	$4s^2$	Zr	$4d^2$	$5s^2$	Hf	$5d^2$	$6s^2$
V	$3d^3$	$4s^2$	Nb	$4d^4$	$5s^1$	Ta	$5d^3$	$6s^2$
Cr	$3d^5$	$4s^1$	Mo	$4d^5$	$5s^1$	W	$5d^4$	$6s^2$
Mn	$3d^5$	$4s^2$	Tc	$4d^6$	$5s^1$	Re	$5d^5$	$6s^2$
Fe	$3d^6$	$4s^2$	Ru	$4d^7$	$5s^1$	Os	$5d^6$	$6s^2$
Co	$3d^7$	$4s^2$	Rh	$4d^8$	$5s^1$	Ir	$5d^7$	$6s^2$
Ni	$3d^8$	$4s^2$	Pd	$4d^{10}$		Pt	$5d^9$	$6s^1$
Cu	$3d^{10}$	$4s^1$	Ag	$4d^{10}$	$5s^1$	Au	$5d^{10}$	$6s^1$
Zn	$3d^{10}$	$4s^2$	Cd	$4d^{10}$	$5s^2$	Hg	$5d^{10}$	$6s^2$

Table 23.9 Oxidation states of the transition elements (less common or unstable states in parentheses)

Sc	Ti	V	Cr	Mn	Fe	Co	Ni	Cu	Zn
		(1+)	(1+)	(1+)		(1+)	(1+)	1+	(1+)
	(2+)	(2+)	2+	2+	2+	2+	2+	2+	2+
3+	3+	3+	3+	(3+)	3+	3+	(3+)	(3+)	
	4+	4+	(4+)	4+	(4+)	(4+)	(4+)		
		5+	(5+)	(5+)	(5+)				
			6+	(6+)	(6+)				
				7+					

Y	Zr	Nb	Mo	Tc	Ru	Rh	Pd	Ag	Cd
	(1+)	(1+)	(1+)		(1+)	(1+)		1+	(1+)
	(2+)	(2+)	(2+)	(2+)	2+	(2+)	2+	(2+)	2+
3+	(3+)	(3+)	3+	(3+)	3+	3+	(3+)	(3+)	
	4+	(4+)	4+	4+	4+	4+	4+		
		5+	5+	5+	(5+)	(5+)			
			6+	(6+)	(6+)	(6+)			
				7+	(7+)				
					(8+)				

La	Hf	Ta	W	Re	Os	Ir	Pt	Au	Hg
				(1+)	(1+)	(1+)		1+	1+
		(2+)	(2+)	(2+)	(2+)	(2+)	2+		2+
3+	(3+)	(3+)	(3+)	3+	(3+)	3+	3+	3+	
	4+	(4+)	4+	4+	4+	4+	4+		
		5+	5+	5+	(5+)	(5+)	(5+)		
			6+	(6+)	6+	(6+)	(6+)		
				7+					
					8+				

After the maximum oxidation state of a period is reached in each transition series, there is a decrease in the highest oxidation number observed for each subsequent element; these oxidation states are, in general, difficult to obtain and unstable. Less common and unstable oxidation states are enclosed in parentheses in Table 23.9.

Compounds of Cu^+, Ag^+, Au^+, and Hg_2^{2+} are the only compounds in which the $1+$ oxidation state is important. The mercury(I) ion is unique; evidence for its dimeric structure includes the following. Mercurous salts are diamagnetic in agreement with the formula Hg_2^{2+}; the ion Hg^+ would have one unpaired electron. X-ray studies of mercurous salts indicate discrete Hg_2^{2+} ions. The interpretations of certain equilibria involving the mercurous ion are in agreement with experimentally derived values for the equilibrium constants only if the formula Hg_2^{2+} is used.

Within each group the higher oxidation states become more stable and more important with increasing atomic number; a corresponding decline in the stability and importance of the lower oxidation states is also observed. The gain in stability of the higher states with increasing atomic number may be attributed to the increasing atomic size, which makes the d electrons more available for compound formation.

Thus, the $2+$ ions of many of the fourth period transition elements are stable and characteristic of the elements, whereas the $2+$ states of the heavier elements are not particularly common or stable. The most important oxidation states of iron are $2+$ and $3+$; for osmium, the heaviest member of the same group, the most important states are $4+$, $6+$, and $8+$.

The maximum oxidation number of manganese (which is the first member of its group) is attained in the permanganate ion, MnO_4^-. This ion is a strong oxidizing agent in acidic solution:

$$5e^- + 8H^+ + MnO_4^- \longrightarrow Mn^{2+} + 4H_2O \qquad \mathscr{E}^{\circ}_{red} = +1.51 \text{ V}$$

In contrast, the analogous perrhenate ion, ReO_4^-, formed by the heaviest member of the group, does not react in this way under comparable conditions.

The transition elements of group VI B form series of oxy anions in which the central metal ion is in an oxidation state of $6+$; the simplest members of this series are the chromate ion, CrO_4^{2-}, the molybdate ion, MoO_4^{2-}, and the tungstate ion, WO_4^{2-}. Only a few polynuclear complexes of chromium have been reported; the most important of these is the dichromate ion, $Cr_2O_7^{2-}$, which results when chromate solutions are acidified:

$$2H^+ + CrO_4^{2-} \longrightarrow Cr_2O_7^{2-} + H_2O$$

The dichromate ion is easily reduced in acid and consequently is a strong oxidizing agent:

$$6e^- + 14H^+ + Cr_2O_7^{2-} \longrightarrow 2Cr^{3+} + 7H_2O \qquad \mathscr{E}^{\circ}_{red} = +1.33 \text{ V}$$

In contrast, molybdenum and tungsten form very extensive series of polynuclear molybdates and tungstates, and these compounds, as well as the corresponding acids, are very stable and are without significant oxidizing power. The formation of such polynuclear oxy anions is not a common phenomenon in transition metal chemistry.

With increasing oxidation number the oxides of a given element become less basic and more acidic. Thus, CrO is an exclusively basic oxide and dissolves in acids to form Cr^{2+} salts; Cr_2O_3 is amphoteric and forms Cr^{3+} salts with acids and the chromite ion, $Cr(OH)_4^-$, in alkaline solution; CrO_3 is entirely acidic in character and gives rise to chromates, CrO_4^{2-}, and dichromates, $Cr_2O_7^{2-}$.

The atomic and ionic radii of the transition elements are, in general, smaller than those of the representative elements of the same period since, for the transition elements, electrons are being added to an inner d sublevel where the effect of the nuclear charge is greater than it is in the outer shell. This relatively small size causes the transition element ions to have comparatively high densities. Small size plus the availability of d orbitals for bonding account for the pronounced tendency of most transition elements to form numerous stable complex ions (see Chapter 24).

There are horizontal (or period) similarities as well as vertical (or group) similarities between the transition metals. Since the elements Cr through Cu, which are a part of the first transition series, have very similar atomic radii and ionic radii for ions of the same charge (see Table 23.10), horizontal similarities occur (such as the formation of compounds of formula MO, MS, MCl_2, and so on). Among the heavier elements, horizontal similarities are also evident (for example, the formation of MO_2 compounds). Similarities of this type are particularly striking among the transition triads of group VIII (Fe, Co, and Ni; Ru, Rh, and Pd; and Os, Ir, and Pt). The elements of the two heavier triads are similar both physically and chemically.

Table 23.10	Atomic and ionic radii of the transition elements of the fourth period									
	Sc	Ti	V	Cr	Mn	Fe	Co	Ni	Cu	Zn
atomic radii (pm)	144	132	122	117	117	117	116	115	117	125
ionic radii (pm)										
M^{2+}	–	90	88	84	80	76	74	72	72	74
M^{3+}	81	76	74	69	66	64	63	–	–	–

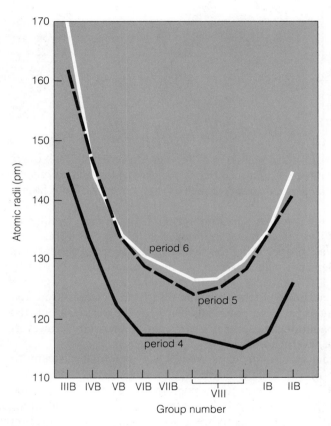

Figure 23.11 Atomic radii of the transition elements

The lanthanides occur at the beginning of the transition series of the sixth period. For the lanthanides, electrons are being added to the 4*f* sublevel without significant change in the 5*s*, 5*p*, 5*d*, and 6*s* sublevels. The nuclear charge increases as this 4*f* sublevel (which is deep within the atom) is being filled, and there is a consequent decrease in the atomic and ionic radii of the lanthanides, which is known as the **lanthanide contraction.**

The lanthanide contraction has an important effect on the properties of the transition elements that follow the lanthanides in the sixth period. Because of the lanthanide contraction, the atomic radii of these elements are not much different from those of the corresponding elements of the fifth period (see Figure 23.11).

Table 23.11	Standard electrode potentials of the transition elements (volts)									
	Sc	Ti	V	Cr	Mn	Fe	Co	Ni	Cu	Zn
M^+/M	–	–	–	–	–	–	–	–	+0.52	–
M^{2+}/M	–	–1.63	–1.19	–0.91	–1.18	–0.44	–0.28	–0.25	+0.34	–0.76
M^{3+}/M	–2.08	–1.21	–0.88	–0.74	–0.28	–0.04	+0.42	–	–	–
	Y	Zr	Nb	Mo	Tc	Ru	Rh	Pd	Ag	Cd
M^+/M	–	–	–	–	–	–	+0.6	–	+0.80	–
M^{2+}/M	–	–	–	–	+0.4	+0.45	+0.6	+0.99	–	–0.40
M^{3+}/M	–2.37	–	–1.1	–0.2	–	–	+0.8	–	–	–
	La	Hf	Ta	W	Re	Os	Ir	Pt	Au	Hg
M^+/M	–	–	–	–	–	–	–	–	+1.69	+0.79
M^{2+}/M	–	–	–	–	–	+0.85	–	+1.2	–	+0.85
M^{3+}/M	–2.52	–	–	–0.11	+0.3	–	+1.15	–	+1.50	–

Consequently, the second and third members of each group resemble each other much more than they resemble the first member of the group; there is no regular increase in size with increasing atomic number such as is observed for the A family elements.

Even though 32 elements intervene, hafnium ($Z = 72$) is notably similar to zirconium ($Z = 40$). Both elements have approximately the same atomic and ionic radii, and the chemical properties of the two elements are very similar; the separation of the two elements, which occur together in nature, is extremely difficult.

Some of the standard electrode potentials of the transition elements are listed in Table 23.11. For the most part, the potentials listed for the heavier elements, since they are for lower oxidation states, are not particularly important; however, they do serve to illustrate several trends.

Many of the metals react with dilute acids, as well as with water or steam, to liberate hydrogen. Some of the metals, however, are poor reducing agents—notably, mercury, the group I B elements (copper, silver, and gold), and the transition triads of the fifth and sixth periods (ruthenium, rhodium, palladium; osmium, iridium, and platinum). Frequently these six elements of the transition triads, silver, and gold are referred to as noble metals.

Thus, the elements of low reactivity appear to be concentrated toward the end of the transition series, particularly those of the fifth and sixth periods. In general, there is a decrease in electropositive character (or strength of the metals as reducing agents) across a period and down a group. The decreasing size of the atoms across a period causes the electrons to be held more tightly. For *transition* elements the increase in nuclear charge on the atoms down a group is not accompanied by a large increase in atomic size, and the electrons are held more tightly. In the case of the transition elements of the sixth period that follow the lanthanides, the unusually small sizes of the atoms cause the electrons to be very tightly held. With the exceptions of La and Hf, the least electropositive member of each group is the sixth period element.

Although a given electrode potential may indicate a thermodynamic tendency for a reaction to occur, at times the rates at which some transition elements react

Table 23.12 Products of some reactions of the transition elements of the fourth period					
	Sc	Ti	V	Cr	Mn
O_2	Sc_2O_3	TiO_2	V_2O_5, VO_2	Cr_2O_3	Mn_3O_4
X_2	ScX_3	TiX_4	VF_5, VCl_4, VBr_3, VI_3	CrX_3 (except CrI_2)	MnX_2
S	Sc_2S_3	TiS_2	V_2S_5, VS_2	CrS	MnS
N_2	ScN	TiN	VN	CrN	Mn_3N_2
HCl	$Sc^{3+} + H_2$	$Ti^{3+} + H_2$	—	$Cr^{2+} + H_2$	$Mn^{2+} + H_2$
H_2O	$Sc(OH)_3 + H_2$	$TiO_2 + H_2^a$	—	$Cr_2O_3 + H_2^a$	$Mn(OH)_2 + H_2$
NaOH	—	—	—	$Cr(OH)_6^{3-} + H_2$	—
	Fe	Co	Ni	Cu	Zn
O_2	Fe_3O_4, Fe_2O_3	Co_3O_4	NiO	Cu_2O, CuO	ZnO
X_2	FeX_3 (except FeI_2)	CoX_2	NiX_2	CuX_2 (except CuI)	ZnX_2
S	FeS	CoS	NiS	Cu_2S	ZnS
N_2	—	—	—	—	—
HCl	$Fe^{2+} + H_2$	$Co^{2+} + H_2$	$Ni^{2+} + H_2$	—	$Zn^{2+} + H_2$
H_2O	$Fe_3O_4 + H_2^a$	$CoO + H_2^a$	$NiO + H_2^a$	—	$ZnO + H_2^a$
NaOH	—	—	—	—	$[Zn(OH)_4]^{2-} + H_2$

^{a} Steam.

are extremely slow. Thus, chromium is a moderately strong reducing agent and reacts with nonoxidizing acids, such as HCl, to liberate hydrogen. Chromium does not, however, react with nitric acid, a strong oxidizing agent. It is said that the metal is made passive by the nitric acid. The phenomenon of passivity, which is displayed by many transition elements, is not well understood. In some instances it may be that the metal is protected by a thin, transparent, impervious oxide coating.

The products of some of the reactions of the transition metals of the fourth period, which are the most common of the transition elements, are given in Table 23.12.

23.11 The Lanthanides

Postulated electronic configurations for the lanthanides are given in Table 23.13. The differentiating electrons of the lanthanides, or inner-transition elements, are lodged in the third shell from the outside, in $4f$ orbitals. This factor accounts for some of the unusual aspects of the chemistry of these elements. The atomic and ionic radii of the elements show a regular decrease with increasing atomic number (the lanthanide contraction).

The outstanding feature of the chemistry of the lanthanides is their similarity. The differentiating $4f$ electrons lie deep within the lanthanide atoms; thus, these f electrons are shielded from the surroundings of the atom and do not cause strong variations in properties. In contrast, the properties of the transition elements vary widely. The d electrons project from transition element atoms and ions and interact with the surroundings (see Section 23.10). Hence, the number and arrange-

Table 23.13 Some properties of the lanthanides

	Z	Postulated Electronic Configurations of Valence Subshells	Oxidation States	Atomic Radius, pm	Ionic Radius, M^{3+}, pm	$\mathscr{E}^{\circ}_{red}$ $3e^- + M^{3+} \rightarrow M$ (volts)
La	57	$5d^1\,6s^2$	$3+$	169	106	-2.52
Ce	58	$4f^2\,6s^2$	$3+, 4+$	165	103	-2.48
Pr	59	$4f^3\,6s^2$	$3+, 4+$	165	101	-2.46
Nd	60	$4f^4\,6s^2$	$2+, 3+, 4+$	164	100	-2.43
Pm	61	$4f^5\,6s^2$	$3+$	–	98	-2.42
Sm	62	$4f^6\,6s^2$	$2+, 3+$	166	96	-2.41
Eu	63	$4f^7\,6s^2$	$2+, 3+$	185	95	-2.41
Gd	64	$4f^7\,5d^1 6s^2$	$3+$	161	94	-2.40
Tb	65	$4f^9\,6s^2$	$3+, 4+$	159	92	-2.39
Dy	66	$4f^{10}\,6s^2$	$3+, 4+$	159	91	-2.35
Ho	67	$4f^{11}\,6s^2$	$3+$	158	89	-2.32
Er	68	$4f^{12}\,6s^2$	$3+$	157	88	-2.30
Tm	69	$4f^{13}\,6s^2$	$2+, 3+$	156	87	-2.28
Yb	70	$4f^{14}\,6s^2$	$2+, 3+$	170	86	-2.27
Lu	71	$4f^{14}\,5d^1 6s^2$	$3+$	156	85	-2.26

Table 23.14 Reactions of the lanthanides[a]

Reaction	Remarks
$2M + 3X_2 \longrightarrow 2MX_3$	X_2 = all halogens; Ce gives CeF_4 with F_2
$4M + 3O_2 \longrightarrow 2M_2O_3$	Ce gives CeO_2
$2M + 3S \longrightarrow M_2S_3$	not Eu; reactions with Se similar
$2M + N_2 \longrightarrow 2MN$	high temperatures; reactions with P_4, As, Sb, and Bi similar
$2M + 6H^+ \longrightarrow 2M^{3+} + 3H_2$	
$2M + 6H_2O \longrightarrow 2M(OH)_3 + 3H_2$	slow with cold water

[a] M = any lanthanide, except where noted.

ment of the d electrons of the transition elements are significant, whereas the number and arrangement of the f electrons of the inner-transition elements cause only minor variations in properties.

All lanthanides form ions in the characteristic group III B oxidation state, $3+$ (see Table 23.14). For some of the elements other oxidation states are known, but these are less stable. The $3+$ ions are formed through the loss of the two $6s$ electrons and one $4f$ electron (or one $5d$ electron, if available).

There is some evidence that f^0, f^7 (half-filled), and f^{14} (filled) ions have stable configurations. Thus, La^{3+} (f^0), Gd^{3+} (f^7), and Lu^{3+} (f^{14}) are the only ions that these elements form. The most stable $2+$ and $4+$ ions are Eu^{2+} (f^7), Yb^{2+} (f^{14}), Ce^{4+} (f^0), and Tb^{4+} (f^7). The ceric ion, Ce^{4+}, is a good oxidizing agent.

The elements occur together in nature because of their great chemical similarity. Promethium, which is radioactive and probably occurs only in trace amounts, is

Table 23.15 Reactions of aluminum, gallium, indium, and thallium

Reaction	Remarks
$2M + 3X_2 \longrightarrow 2MX_3$	X_2 = all halogens; Tl also gives TlX; no iodide of Tl^{3+}
$4M + 3O_2 \longrightarrow 2M_2O_3$	high temperatures; Tl also gives Tl_2O
$2M + 3S \longrightarrow M_2S_3$	high temperatures; Tl also gives Tl_2S; also with Se and Te
$2Al + N_2 \longrightarrow 2AlN$	Al only; GaN and InN may be prepared indirectly
$2M + 6H^+ \longrightarrow 2M^{3+} + 3H_2$	M = Al, Ga, and In; Tl gives Tl^+
$2M + 2OH^- + 6H_2O \longrightarrow 2M(OH)_4^- + 3H_2$	M = Al and Ga

M = Al, Ga, In, and Tl, except where noted.

an exception. The elements are extremely difficult to separate; repeated fractional crystallization and ion-exchange techniques have been employed to effect separations.

All the lanthanides are silvery white, very reactive metals. The electrode potentials for the reduction of the 3+ ions are strikingly similar (see Table 23.13). Some reactions of the lanthanides are given in Table 23.14; a number of the metals are also known to react with carbon to produce saltlike carbides and with hydrogen to give saltlike hydrides. The M_2O_3 oxides react with water to form insoluble hydroxides, $M(OH)_3$, that are not amphoteric. Insoluble carbonates, $M_2(CO_3)_3$, may be produced by the reactions of carbon dioxide with the oxides or hydroxides. The halides (with the exception of fluorides), nitrates, chlorates, acetates, and sulfates of the 3+ ions are water soluble; the phosphates, fluorides, oxalates, hydroxides, and carbonates are insoluble. The precipitation of the elements as oxalates serves as the basis of an analytical determination. In general, the compounds are paramagnetic and highly colored.

23.12 The Metals of Group III A

The characteristics of the group III A elements have been discussed in Section 22.6, and some of the properties of the metals are listed in Table 22.4.

Aluminum, the most abundant metal of the earth's crust (approximately 8%), is obtained by the electrolysis of molten Al_2O_3 (the Hall process, see Section 23.6). Gallium, indium, and thallium are widely distributed in nature but occur only in trace amounts; they may be prepared by the electrolysis of aqueous solutions of salts of the metals. They are soft, white metals with relatively low melting points (see Figure 23.6). Gallium has an unusually low melting point (30°C). Since its boiling point is not abnormally low (2070°C), gallium has an exceptional liquid range and has found use as a thermometer fluid.

The metals are fairly reactive (see Table 23.15). Alluminum, gallium, and indium (but not thallium) have protective oxide coatings and are passive toward nitric acid. However, all the metals react with nonoxidizing acids to liberate hydrogen. As expected from their $ns^2 np^1$ electronic configurations, the most important oxidation state is $3+$. Most of the compounds of the metals in the $3+$ oxidation state are covalent. In water, however, the M^{3+} ions are stabilized through hydration, and the enthalpies of hydration are high.

The sulfates, nitrates, and halides are water soluble, but the M^{3+} ions hydrolyze readily:

$$M(H_2O)_6^{3+} + H_2O \longrightarrow M(H_2O)_5(OH)^{2+} + H_3O^+$$

For this reason, salts of weak acids (such as acetates, carbonates, sulfides, and cyanides) do not exist in water solution. Such salts are completely hydrolyzed:

$$M(H_2O)_6^{3+} + 3\,C_2H_3O_2^- \longrightarrow M(OH)_3(s) + 3\,H_2O + 3\,HC_2H_3O_2$$

In addition, complexes with ammonia do not exist in water:

$$M(H_2O)_6^3 + NH_3 \longrightarrow M(H_2O)_5(OH)^{2+} + NH_4^+$$

The hydroxides, $M(OH)_3$, are insoluble in water. Aluminum hydroxide and gallium hydroxide are amphoteric:

$$Al(OH)_3 + OH^- \longrightarrow Al(OH)_4^-$$
$$Ga(OH)_3 + OH^- \longrightarrow Ga(OH)_4^-$$

Aluminum sulfate forms an important series of double salts called alums, $MAl(SO_4)_2 \cdot 12\,H_2O$ (or $M_2SO_4 \cdot Al_2(SO_4)_3 \cdot 24\,H_2O$), where M may be almost any univalent cation (Na^+, K^+, Rb^+, Cs^+, NH_4^+, Ag^+, and Tl^+) except Li^+ (which is too small). In addition to Al^{3+}, other M^{3+} species form series of such double sulfates: Fe^{3+}, Cr^{3+}, Mn^{3+}, Ti^{3+}, Co^{3+}, Ga^{3+}, In^{3+}, Rh^{3+}, and Ir^{3+}.

Hydrides analogous to those of boron are not formed by aluminum, gallium, indium, and thallium. However, MH_4^- ions, which are analogous to the borohydride ion, are formed.

Among the heavier members of the group, compounds with the metal in a $1+$ oxidation state are known. The np^1 electron is more readily removed than the ns^2 electrons, which are sometimes called an inert pair. This effect is also seen for the M^{2+} ions of group IV A. The low reactivity of mercury has been ascribed to the $6s^2$ inert pair of its valence level.

Unlike the transition elements, however, the $1+$ state of the III A elements (their lowest state) is most stable among the heavier members of the group. No $1+$ compounds of aluminum are known. Gallium and indium form a few compounds (such as Ga_2S, Ga_2O at high temperatures, $InCl$, $InBr$, and In_2O). The Ga^+ and In^+ ions are not stable in water solution.

For thallium, however, the $1+$ state is important and stable. In fact, Tl^+ is more stable in water than Tl^{3+} (which is a strong oxidizing agent):

$$2e^- + Tl^{3+} \rightleftharpoons Tl^+ \qquad \mathscr{E}_{red}^\circ = +1.25 \text{ V}$$

The oxide Tl_2O dissolves in water to form the soluble hydroxide $TlOH$. Whereas the Tl^{3+} ion is extensively hydrolyzed in aqueous solution, the Tl^+ ion is not. A wide variety of Tl^+ compounds are known. The sulfate, nitrate, acetate, and fluoride are water soluble; the chloride, bromide, iodide, sulfide, and chromate are insoluble.

23.13 The Metals of Group IV A

Of the elements of group IV A, germanium, tin, and lead are classified as metals. The characteristics of the group as a whole are discussed in Section 22.1, and some of the properties of the metals are listed in Table 22.1. The metals are not abundant in nature; germanium is a rare element.

Germanium is a semiconductor and is used in the manufacture of transistors. It is a hard, brittle, white metal and has the highest melting point of the metals of the group. Tin and lead have relatively low melting points and low tensile strengths. Lead is especially soft and malleable.

The metals are fairly reactive (see Table 23.16). Lead frequently appears less reactive than the electrode potential,

$$2e^- + Pb^{2+} \rightleftharpoons Pb \qquad \mathscr{E}^{\circ}_{red} = -0.13 \text{ V}$$

would indicate because of the formation of surface coatings. Thus, the reaction of lead with sulfuric acid or hydrochloric acid is impeded by the formation of insoluble lead sulfate or lead chloride on the surface of the metal.

Since the elements have $ns^2\,np^2$ valence shell configurations, two oxidation states are observed: 4+ and 2+ (inert pair species). The 4+ state declines in importance and the 2+ state becomes increasingly important down the series: Ge, Sn, Pb. Thus, only a few 2+ germanium compounds are important (GeO, GeS, $GeCl_2$, $GeBr_2$, and GeI_2), and only a few 4+ lead compounds are important [PbO_2, $Pb(C_2H_3O_2)_4$, PbF_4, and $PbCl_4$]. The Sn^{2+} [stannous, or tin(II) ion] and Pb^{2+} [plumbous, or lead(II) ion] are the only cationic species of the group that exist in water; 4+ ions probably do not exist. Most of the pure 2+ and 4+ compounds are covalent, although PbF_2 is known to be ionic.

All the metals form dioxides; GeO_2 and SnO_2 are the products of the reactions of the metals with oxygen. Lead dioxide may be prepared by the oxidation of PbO or Pb^{2+} salts in alkaline solution:

$$Pb(OH)_3^-(aq) + OCl^-(aq) \longrightarrow PbO_2(s) + Cl^-(aq) + OH^-(aq) + H_2O$$

Table 23.16 Reactions of germanium, tin, and lead[a]	
Reaction	Remarks
$M + 2X_2 \longrightarrow MX_4$	X_2 = any halogen; M = Ge and Sn; Pb yields PbX_2
$M + O_2 \longrightarrow MO_2$	M = Ge and Sn; high temperatures; Pb yields PbO or Pb_3O_4
$M + 2S \longrightarrow MS_2$	M = Ge and Sn; high temperatures; Pb yields PbS
$M + 2H^+ \longrightarrow M^{2+} + H_2$	M = Sn and Pb
$3M + 4H^+ + 4NO_3^- \longrightarrow 3MO_2 + 4NO + 2H_2O$	M = Ge and Sn
$3Pb + 8H^+ + 2NO_3^- \longrightarrow 3Pb^{2+} + 2NO + 4H_2O$	
$M + OH^- + 2H_2O \longrightarrow M(OH)_3^- + H_2$	M = Sn and Pb; slow
$Ge + 2OH^- + 4H_2O \longrightarrow Ge(OH)_6^{2-} + 2H_2$	

[a] M = Ge, Sn, and Pb, except where noted.

Lead dioxide is a strong oxidizing agent:

$$2e^- + 4H^+ + PbO_2 \longrightarrow Pb^{2+} + 2H_2O \qquad \mathscr{E}^{\circ}_{red} = +1.46 \text{ V}$$

In acid solution PbO_2 oxidizes Mn^{2+} to MnO_4^- and Cl^- to Cl_2. In contrast to GeO_2 and SnO_2, PbO_2 is thermally unstable. Upon gentle heating PbO_2 decomposes to Pb_3O_4 (red lead); stronger heating gives PbO (litharge). The compound Pb_3O_4 contains lead in two oxidation states and may be represented as $Pb_2^{II}Pb^{IV}O_4$ like the compounds Fe_3O_4 and Co_3O_4, both of which conform to the formula $M^{II}(M^{III}O_2)_2$.

Germanates, stannates, and plumbates may be derived from the dioxides by reactions with aqueous alkali or in the case of PbO_2 by fusion with alkali metal oxides or alkaline earth oxides. The stannates and plumbates form trihydrates in which the anion is an octahedral hydroxy complex: $[M(OH)_6]^{2-}$; a few germanates of similar structure are known (for example, $Fe[Ge(OH)_6]$). There are no hydroxides of formula $M(OH)_4$, but hydrous MO_2 oxides may be prepared.

Treatment of the Sn^{2+} and Pb^{2+} ions in water solution with OH^- ion gives hydrous-oxide precipitates that are commonly assigned the formulas $Sn(OH)_2$ and $Pb(OH)_2$. By heating these products, SnO and PbO can be prepared; PbO is also the product of the reaction of lead and oxygen as well as of the decompositions of Pb_3O_4 and PbO_2. The compound GeO may be prepared by the hydrolysis of $GeCl_2$.

All the monoxides, as well as the hydrous oxides (or hydroxides), are amphoteric and dissolve in either acid or alkaline solution. In excess alkali, germanites, stannites, and plumbites are produced. For example:

$$PbO(s) + OH^-(aq) + H_2O \longrightarrow Pb(OH)_3^-(aq)$$

The stannite ion is a strong reducing agent:

$$2e^- + Sn(OH)_6^{2-} \longrightarrow Sn(OH)_3^- + 3OH^- \qquad \mathscr{E}^{\circ}_{red} = -0.93 \text{ V}$$

All the metals form monosulfides, MS, but only germanium and tin form disulfides, MS_2. The disulfides of germanium and tin may be prepared by the reactions of the metals and sulfur and may be precipitated from solutions of Ge^{IV} or Sn^{IV} compounds by H_2S. The disulfides, like the dioxides, are amphoteric. In solutions of alkali metal sulfides or ammonium sulfide, GeS_2 and SnS_2 dissolve to form thioanions; compounds of the SnS_3^{2-} and SnS_4^{4-} ions have been isolated from such solutions, but thiogermanates have not been obtained in pure form:

$$SnS_2(s) + S^{2-}(aq) \longrightarrow SnS_3^{2-}(aq)$$

Insoluble PbS and SnS may be precipitated from solutions by means of H_2S. Lead(II) sulfide is also the product of the direct union of the elements; SnS may be prepared by the thermal decomposition of SnS_2. Germanium(II) sulfide may be derived from the disulfide of germanium by the reaction

$$GeS_2 + Ge \longrightarrow 2GeS$$

None of the monosulfides are soluble in sulfide solutions.

Complete series of the tetrahalides of germanium, tin, and lead are known except for $PbBr_4$ and PbI_4. Lead in a 4+ state has strong oxidizing power and cannot exist in a compound with bromine and iodine in $1-$ states (which have reducing abilities). Lead(IV) chloride readily decomposes at $100°C$:

$$PbCl_4(l) \longrightarrow PbCl_2(s) + Cl_2(g)$$

The tetrahalides, in general, are volatile covalent substances. They react readily with water to produce hydrous dioxides.

All the dihalides of germanium, tin, and lead are known. Germanium(II) halides, as well as tin(II) halides, may be prepared by the reaction of the appropriate tetrahalide and metal:

$$GeCl_4 + Ge \longrightarrow 2\,GeCl_2$$

The reactions of tin and the hydrohalic acids yield tin(II) halides. Lead(II) halides may be produced by the direct reaction of the elements. Since all PbX_2 compounds are insoluble in water, they may be precipitated from solutions containing Pb^{2+} by the addition of halide ions.

The dihalides are much less volatile than the tetrahalides, a fact which indicates an increased degree of ionic character; PbF_2, $PbCl_2$, and $PbBr_2$ are ionic in the solid state. The dihalides of germanium are not particularly stable. Complex anions, such as $GeCl_3^-$, $SnCl_4^{2-}$, $PbCl_3^-$, $PbCl_4^{2-}$, and $PbCl_6^{4-}$, are known. Both Sn^{2+} and Pb^{2+} form such ions in aqueous solution in the presence of halide ions; in general, the ions MX^+ and MX_3^- are most important. Insoluble lead(II) halides dissolve in solutions containing excess halide ions because of the formation of such complexes.

All three metals form covalent, volatile hydrides: GeH_4, germane; SnH_4, stannane; and PbH_4, plumbane. The hydride of lead is thermally unstable and decomposes to the elements at $0°C$; SnH_4 decomposes at approximately $150°C$. Continuing the trend established by carbon and silicon, germanium forms catenated hydrides of formula Ge_nH_{2n+2}, where n is any number from 2 to 8. Tin or lead form only the simple hydrides.

The Sn^{2+} and Pb^{2+} ions are extensively hydrolyzed in water:

$$[Sn(H_2O)_6]^{2+} + H_2O \rightleftharpoons [Sn(H_2O)_5(OH)]^+ + H_3O^+ \qquad K_a \cong 10^{-2}$$

$$[Pb(H_2O)_6]^{2+} + H_2O \rightleftharpoons [Pb(H_2O)_5(OH)]^+ + H_3O^+ \qquad K_a \cong 10^{-8}$$

The reaction with water is particularly pronounced in the case of tin(II) compounds; excess acid is generally added to aqueous solutions of such compounds to inhibit this reaction and prevent the precipitation of basic salts, such as $Sn(OH)Cl$.

Lead(II) nitrate and lead(II) acetate are soluble in water; the acetate is only slightly dissociated. The sulfate, chromate, carbonate, sulfide, and all of the halides of Pb^{2+} are only slightly soluble in water.

Summary

The topics that have been discussed in this chapter are

1. The band theory of the bonding in metals; conductors, insulators, and semiconductors.

2. Extrinsic semiconductors.

3. The physical properties and natural occurrence of metals.

4. Metallurgical operations, which may be grouped into processes for preliminary treatment of ores, for reduction, and for refining.

5. Surveys of the chemistry of the metals found in groups I A, II A, III A, IV A, the transition elements, and the lanthanides.

Key Terms

Some of the more important terms introduced in this chapter are listed below. Definitions for terms not included in this list may be located in the text by use of the index.

Alkali metal (Section 23.8)   An element of group I A: Li, Na, K, Rb, or Cs.

Alkaline earth metal (Section 23.9)   An element of group II A: Be, Mg, Ca, Sr, or Ba.

Band theory (Section 23.1)   A theory that explains the bonding in metals in terms of molecular orbitals that extend over the entire crystal and together make up what are called bands.

Basic oxygen process (Section 23.7)   A process in which pig iron is refined into steel. The impurities in the pig iron are oxidized by a blast of pure oxygen.

Bayer process (Section 23.5)   A process for obtaining pure aluminum oxide (for use in the production of aluminum metal) from the ore bauxite. Advantage is taken of the amphoteric nature of aluminum oxide to dissolve it away from ore impurities by use of a hot solution of NaOH.

Blast furnace (Section 23.6)   A furnace for the carbon reduction of iron ore; pig iron, an impure form of iron, is produced.

Concentration of ores (Section 23.5)   Processes in which the desired components of ores are separated from gangue.

Conduction band (Section 23.1)   An empty band in a metallic crystal through which electrons are free to move and thereby conduct electricity.

Conductor (Section 23.1)   A metal in which an empty conduction band overlaps the valence band and the valence electrons are free to conduct electricity.

Downs cell (Section 23.6)   An electrolytic cell in which a reactive metal is produced by the electrolysis of the molten chloride of the metal.

Electrical conductivity (Section 23.3)   A measure of the ability of a substance to conduct electricity; measured in units of $1.0/(ohm \cdot cm)$.

Extrinsic (or impurity) semiconductor (Section 23.2)   A crystalline semiconductor to which impurities have been added to enhance its conductivity.

Flotation (Section 23.5)   A method of ore concentration; the finely crushed ore is mixed with a suitable oil and water in a large tank. Agitation of the mixture with air produces a froth which contains the mineral particles and which is skimmed off.

Flux (Section 23.6)   A substance (usually limestone, $CaCO_3$) used to form a slag in a smelting operation and thus remove the gangue.

Forbidden energy zone (Section 23.1)   A zone in an energy diagram for the bands of a crystal; no electron can have an energy that would place it in this zone.

Gangue (Section 23.4)   Unwanted material included with ores as mined.

Hall process (Section 23.6)   The process by which aluminum is produced by the electrolysis of a solution of aluminum oxide in molten cryolite, Na_3AlF_6.

Insulator (Section 23.1)   A substance in which a completely filled valence band is widely separated from a conduction band by a forbidden energy zone so that the valence electrons are normally not able to conduct electricity.

Intrinsic semiconductor (Sections 23.1 and 23.2)   A pure, crystalline substance that functions as a semiconductor.

Kroll process (Section 23.6)   The reduction of a metal halide by Mg, Na, or Ca.

Lanthanides (Section 23.11)   The 14 inner-transition elements (with atomic numbers from 58 to 71) that follow lanthanum in the periodic table.

Leaching (Section 23.5)   The removal of the metal component of an ore by use of a solution that contains a substance with which the desired component reacts to form a soluble species.

Matte (Section 23.6)   Impure cuprous sulfide, Cu_2S, obtained by smelting copper sulfide ores.

Metallurgy (Section 23.5)   The science of extracting metals from their ores and preparing these metals for use.

Native ore (Section 23.4)   An ore in which a metal or a nonmetal (for example, Ag, Au, Bi, and S) occurs in elemental form.

n-type semiconductor (Section 23.2)   An extrinsic semiconductor produced by the addition of a donor impurity (which has a larger number of valence electrons than the atoms of the host crystal) to an intrinsic semiconductor.

Open hearth process (Section 23.7)   A process in which pig iron is refined into steel. The oxidation of the impurities in the pig iron is accomplished by reaction with iron oxide and air.

Ore (Section 23.4)   A naturally occurring material from which one or more metals (or other chemical substances) can be profitably extracted.

Parkes process (Section 23.7)   A process for removing silver from impure lead; the silver is selectively dissolved in molten zinc.

Pig iron (Section 23.6)   An impure form of iron produced by the blast furnace.

p-type semiconductor (Section 23.2)   An extrinsic semiconductor produced by adding an acceptor impurity (which has fewer valence electrons than the atoms of the host crystal) to an intrinsic semiconductor.

Reduction of an ore (Sections 23.5 and 23.6)   A process in which a free metal is obtained from a metal compound found in an ore.

Refining (Sections 23.5 and 23.7)   A process in which an impure metal is purified; in some cases, substances are added to give desired properties to the final product.

Roasting of ores (Section 23.5)   The process in which an ore is heated in air; used to convert sulfides and carbonates to oxides, which are more readily reduced to the free metal.

Semiconductor (Sections 23.1 and 23.2)   A material with a low electrical conductivity that increases markedly with increasing temperature. In crystals of semiconductors, the forbidden energy zone is sufficiently narrow that electrons can be promoted from the valence band to the conduction band by heat.

Slag (Section 23.6)   A material that is formed from a flux and gangue in a smelting operation and therefore serves to remove the gangue.

Smelting (Section 23.6)   A high-temperature reduction process in which a free metal is secured from its concentrated and purified ore.

Thermite process (Section 23.6)   The reduction of an oxide of a metal by aluminum; also called the Goldschmidt process.

Valence band (Section 23.1)   A band in a metallic crystal that contains the valence electrons of the metal.

Van Arkel process (Section 23.7)   A method for the purification of certain metals by the formation and subsequent thermal decomposition of an iodide of the metal.

Zone refining (Section 23.7)   A process used to purify certain metals in which a heated zone is caused to move along a rod of the impure metal. The impurities are swept along in the molten zone to one end of the rod.

Problems*

The Metallic Bond

23.1 How does the band theory of metallic bonding explain the luster, thermal conductivity, and electrical conductivity of metals?

23.2 Explain why the electrical conductivity of a metal *decreases* when the temperature is increased and the electrical conductivity of a semiconductor *increases* when the temperature is increased.

23.3 Use energy-level diagrams to explain the difference between conductors, insulators, and semiconductors.

23.4 Explain clearly the origin of the bands in a metallic crystal. What is a forbidden energy zone?

Physical Properties, Occurrence of Metals

23.5 What properties distinguish metals from nonmetals?

23.6 What evidence supports the belief that the strength of the metallic bonding of the metals of the fourth and subsequent periods reaches a maximum around the center of the transition series? How can the trend be explained?

*23.7 What diameter of iron wire must be used so that a given length will have an electrical conductivity equal to that of the same length of copper wire that is 1.0 cm in diameter? Electrical conductivities are listed in Table 23.1.

23.8 What are the principal types of ores? Give an example of each type.

23.9 Explain why it is not surprising that the following ores are found in nature: **(a)** native ores of platinum, gold, and silver, **(b)** sulfate ores of barium, strontium, and lead, **(c)** Na^+ and Mg^{2+} in sea water, **(d)** sulfide ores of lead, bismuth, and nickel.

Metallurgy

23.10 What types of ores are roasted? Write a chemical equation for the roasting of an ore of each type. Why is this procedure used?

23.11 Describe the following metallurgical processes. Use equations when possible. **(a)** Froth flotation, **(b)** Parkes process, **(c)** Van Arkel process, **(d)** Kroll process, **(e)** liquation, **(f)** zone refining, **(g)** thermite process.

23.12 Identify the following: **(a)** ore, **(b)** gangue, **(c)** alum, **(d)** amalgam, **(e)** flux, **(f)** slag, **(g)** smelting.

23.13 Give equations for reactions that occur in the blast furnace for the production of pig iron.

---

* The more difficult problems are marked with asterisks. The appendix contains answers to color-keyed problems.

23.14 What processes are used to refine pig iron into steel? How do they differ?

23.15 Why is carbon reduction not used to obtain some metals from their ores?

23.16 Write equations for the thermite reduction of **(a)** UO_3 (by Al), **(b)** V_2O_5 (by Al), **(c)** Ta_2O_5 (by Na), **(d)** ThO_2 (by Ca), **(e)** WO_3 (by Al).

23.17 Describe, with equations, the steps in the production of aluminum metal from bauxite.

23.18 Use equations to show how magnesium is obtained from sea water.

23.19 What metals are produced by the electrolysis of molten salts? by the electrolysis of aqueous salt solutions?

23.20 Compare the metallurgical processes used to obtain copper from **(a)** low-grade $CuCO_3$ ores, **(b)** CuS ores.

23.21 Explain, using equations, how impure copper can be refined electrolytically. What is the "anode sludge?"

23.22 Explain, using chemical equations, how silver is obtained from low-grade ores of native silver.

23.23 Use equations to compare the processes used to obtain lead from **(a)** $PbCO_3$ by roasting and carbon reduction, **(b)** PbS by incomplete roasting and smelting in the absence of air.

The Representative Metals

23.24 Write an equation for the reaction that occurs between Na and **(a)** H_2, **(b)** N_2, **(c)** O_2, **(d)** Cl_2, **(e)** S, **(f)** P_4, **(g)** C, **(h)** H_2O, **(i)** NH_3.

23.25 According to first ionization potentials, the most reactive I A metal is cesium. According to electrode potentials, the most reactive I A metal is lithium. Reconcile these observations.

23.26 Discuss the ways in which the chemistry of lithium and its compounds differs from the chemistry of sodium and its compounds.

23.27 Write an equation for the reaction that occurs between Ca and **(a)** H_2, **(b)** N_2, **(c)** O_2, **(d)** Cl_2, **(e)** S, **(f)** P_4, **(g)** C, **(h)** H_2O, **(i)** NH_3.

23.28 The second ionization potentials of the II A metals are much larger than the first ionization potentials. Why do these elements not form $1+$ ions instead of $2+$ ions in their reactions?

23.29 Explain why the compounds of beryllium are far more covalent than those of any other II A metal.

23.30 Write an equation for the reaction that occurs between Al and **(a)** Cl_2, **(b)** O_2, **(c)** S, **(d)** N_2, **(e)** OH^-(aq), **(f)** H^+(aq).

23.31 Why do Al and Pb appear at times to be less reactive than electrode potentials would indicate?

23.32 A "diagonal relationship" exists between Al and Be. Explain the atomic-ionic properties that cause such a relationship and cite evidence for its existence.

23.33 Why is it impossible to precipitate Al_2S_3 from aqueous solutions containing Al^{3+}(aq) ions?

23.34 Write an equation for the reaction that occurs between Sn and **(a)** Cl_2, **(b)** O_2, **(c)** S, **(d)** H^+(aq), **(e)** OH^-(aq), **(f)** HNO_3.

23.35 Complete the following equations:

(a) $PbO_2(s) + H_2O + OH^-(aq) \longrightarrow$

(b) $PbO_2(s) \xrightarrow[heating]{mild}$

(c) $PbO_2(s) \xrightarrow[heating]{strong}$

(d) $PbO(s) + H_2O + OH^-(aq) \longrightarrow$

(e) $SnS(s) + S^{2-}(aq) \longrightarrow$

(f) $SnS_2(s) + S^{2-}(aq) \longrightarrow$

(g) $PbCl_2(s) + Cl^-(aq) \longrightarrow$

(h) $SnF_4 + F^-(aq) \longrightarrow$

The Transition Metals and Inner-transition Metals

23.36 Write equations to compare the reactions of Fe, Cr, and Zn with **(a)** Cl_2, **(b)** O_2, **(c)** S, **(d)** N_2, **(e)** H_2O, **(f)** H^+(aq), **(g)** OH^-(aq).

23.37 In what ways do the properties of the transition metals differ from those of the A family metals?

23.38 Group IV B consists of Ti, Zr, and Hf. Why do Zr and Hf resemble each other chemically much more than either of them resembles Ti?

23.39 Write equations for the reactions of CrO, Cr_2O_3, and CrO_3 with H^+(aq) and with OH^-(aq).

23.40 What reasons can you give to explain the low order of reactivity of the noble metals?

*23.41 The standard electrode potential, $\mathscr{E}_{red}^\circ$, for the $Hg_2^{2+}/$ Hg half-reaction is $+0.788$ V. For the cell

$$Pt|Hg|Hg_2^{2+}(10^{-3}\,M)||Hg_2^{2+}(10^{-2}\,M)|Hg|Pt$$

$\mathscr{E}$ equals $+0.0296$ V. **(a)** What would be the emf of this concentration cell if the mercurous ion had the formula Hg^+? Notice that the ion concentrations given apply to the formula Hg_2^{2+}. **(b)** What would be the standard electrode potential of the mercurous ion/mercury couple if the formula of the mercurous ion were Hg^+? Notice that a liter of a $1\,M$ Hg^+ solution would contain one-half the weight of "mercurous" ion that a liter of $1\,M$ Hg_2^{2+} solution would contain. **(c)** Explain why measurement of the emf of the concentration cell proves that the formula of the mercurous ion is Hg_2^{2+}, whereas measurement of the $\mathscr{E}_{red}^\circ$ for the mercurous ion/mercury couple does *not* establish the formula of the ion.

23.42 Write an equation for each reaction that occurs between La and **(a)** Cl_2, **(b)** O_2, **(c)** S, **(d)** N_2, **(e)** H_2O, **(f)** H^+(aq).

23.43 The chemical properties of the transition metals vary widely from element to element. In contrast, the inner-transition metals markedly resemble each other chemically. Explain why this difference exists.

COMPLEX COMPOUNDS

C H A P T E R

24

Chemists of the late-nineteenth century had difficulty in understanding how "molecular compounds" or "compounds of higher order" are bonded. The formation of a compound such as $CoCl_3 \cdot 6\,NH_3$ was baffling—particularly in this case since simple $CoCl_3$ does not exist. In 1893 Alfred Werner proposed a theory to account for compounds of this type. Werner wrote the formula of the cobalt compound as $[Co(NH_3)_6]Cl_3$. He assumed that the six ammonia molecules are symmetrically "coordinated" to the central cobalt atom by "subsidiary valencies" of cobalt while the "principal valencies" of cobalt are satisfied by the chloride ions. Werner spent over twenty years preparing and studying coordination compounds and perfecting and proving his theory. Although modern work has amplified his theory, it has required little modification.

Many practical applications have been derived from the study of complex compounds. Advances have resulted in such fields as metallurgy, analytical chemistry, biochemistry, water purification, textile dyeing, electrochemistry, and bacteriology. In addition, the study of these compounds has enlarged our understanding of chemical bonding, certain physical properties (for example, spectral and magnetic properties), minerals (many minerals are complex compounds), and metabolic processes (both heme of blood and chlorophyll of plants are complex compounds).

Alfred Werner, 1866–1919.
Edgar Fahs Smith Collection.

24.1 Structure

A complex ion or complex compound consists of a central metal cation to which several anions and/or molecules (called ligands) are bonded. With few exceptions, *free* ligands have at least one electron pair that is not engaged in bonding:

$$:\!\overset{\displaystyle ..}{\underset{\displaystyle ..}{Cl}}\!:^{-} \qquad :C\!\equiv\!N\!:^{-} \qquad H\!-\!\overset{\displaystyle ..}{\underset{\displaystyle |}{O}}\!: \qquad H\!-\!\overset{\displaystyle ..}{\underset{\displaystyle |}{N}}\!-\!H$$
$$\qquad\qquad\qquad\qquad\qquad\qquad H \qquad\quad H$$

These electron pairs may be considered to be donated to the electron-deficient metal ions in the formation of complexes. Ligands, therefore, are substances that are capable of acting as Lewis bases. The bonding of complexes, however, shows a wide variation in character—from strongly covalent to predominantly ionic (see Section 24.5).

The ligands are said to be coordinated around the central cation in a first coordination sphere. In the formulas of complex compounds, such as

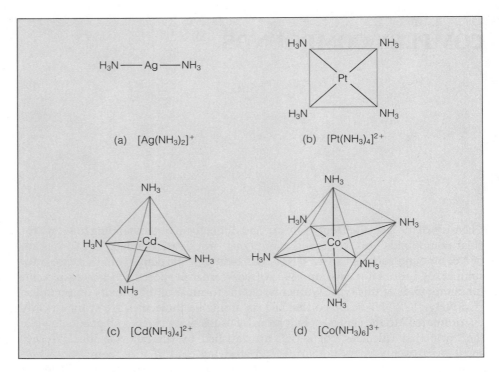

Figure 24.1 Common configurations of complex ions: (a) linear, (b) square-planar, (c) tetrahedral, (d) octahedral

$K_3[Fe(CN)_6]$ and $[Cu(NH_3)_4]Cl_2$, the first coordination sphere is indicated by square brackets. The ligands are arranged about the central ion in a regular geometric manner (see Figure 24.1). The number of atoms *directly* bonded to the central metal ion, or the number of coordination positions, is called the co-ordination number of the central ion.

The charge of a complex is the sum of the charges of the constituent parts. Complexes may be cations, anions, or neutral molecules. In each of the complexes of platinum(IV)—$[Pt(NH_3)_5Cl]^{3+}$, $[Pt(NH_3)_2Cl_4]^0$, and $[PtCl_6]^{2-}$—the platinum contributes $4+$, each chlorine contributes $1-$, and the coordinated ammonia molecules do not contribute to the charge of the complex.

An interesting series of platinum (IV) complexes appears in Table 24.1. The list is headed by the chloride of the $[Pt(NH_3)_6]^{4+}$ ion, and in each subsequent entry an ammonia molecule of the coordination sphere is replaced by a chloride ion. The *coordinated* ammonia molecules and chloride ions are tightly held and do not dissociate in water solution. Those chloride ions of the compound that are *not coordinated* to the platinum are ionizable. Aqueous solutions of the last three compounds of the table do not precipitate silver chloride upon the addition of silver nitrate, whereas solutions of the first four precipitate AgCl in amounts proportional to 4/4, 3/4, 2/4, and 1/4, respectively, of their total chlorine content. In each case, the total number of ions per formula unit derived from conductance data agrees with the formula listed in Table 24.1.

In general, the most stable complexes are formed by metal ions that have a high positive charge and a small ionic radius. The transition elements and the metals immediately following (notably the III A and IV A metals) have a marked tendency to form complexes. Few complexes are known for the lanthanides and

Table 24.1   Some platinum (IV) complex compounds

	Molar Conductance 0.001 M Solution[a] (25°C)	Number of Ions per Formula Unit	Number of Cl^- Ions per Formula Unit
$[Pt(NH_3)_6]Cl_4$	523	5	4
$[Pt(NH_3)_5Cl]Cl_3$	404	4	3
$[Pt(NH_3)_4Cl_2]Cl_2$	228	3	2
$[Pt(NH_3)_3Cl_3]Cl$	97	2	1
$[Pt(NH_3)_2Cl_4]$	0	0	0
$K[Pt(NH_3)Cl_5]$	108	2	0
$K_2[PtCl_6]$	256	3	0

[a] $cm^2/(\Omega \cdot mol)$

the I A and II A metals (with the exception of beryllium). The bonding of transition-metal complexes involves the d orbitals of the central metal atom.

Complexes containing metal ions with coordination numbers ranging from two to twelve are known. The majority of complexes, however, are two-, four-, and six-fold coordinate (see Figure 24.1), and the coordination number six is by far the most common.

Six-coordinate complexes are octahedral. The regular octahedral arrangement of atoms is frequently represented:

This representation is a convenient way to create a three-dimensional illusion. No difference between the bonds of the vertical axis and the other bonds should be inferred from this drawing. *All* the bonds are equivalent. Tetragonal geometry is a distorted form of the octahedral in which the bond distances along one of the axes are longer or shorter than the remaining bonds:

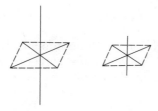

Four-coordinate complexes are known in tetrahedral and square-planar configurations. The square-planar configuration is the usual one for Pt^{II}, Pd^{II}, and Au^{III} and is assumed by many complexes of Ni^{II} and Cu^{II}. Examples include $[Pt(NH_3)_4]^{2+}$, $[PdCl_4]^{2-}$, $[AuCl_4]^-$, $[Ni(CN)_4]^{2-}$, and $[Cu(NH_3)_4]^{2+}$. In some instances there is evidence that square complexes may be tetragonal forms with two groups, along a vertical axis, located at greater distances from the

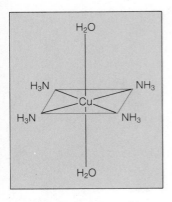

Figure 24.2 Configuration of $[Cu(NH_3)_4(H_2O)_2]^{2+}$

central ion than the ligands of the plane. Thus, $[Cu(NH_3)_4]^{2+}$ in water solution may have two water molecules coordinated in the manner shown in Figure 24.2. Hence, square-planar and octahedral geometries may be considered to merge.

For four-coordinate complexes the tetrahedral configuration is encountered more frequently than the square-planar and is particularly common for complexes of the nontransition elements. Certain complexes of Cu^I, Ag^I, Au^I, Be^{II}, Zn^{II}, Cd^{II}, Hg^{II}, Al^{III}, Ga^{III}, In^{III}, Fe^{III}, Co^{II}, and Ni^0 are tetrahedral. Examples include $[Cu(CN)_4]^{3-}$, $[BeF_4]^{2-}$, $[AlF_4]^-$, $[FeCl_4]^-$, $[Cd(CN)_4]^{2-}$, $[ZnCl_4]^{2-}$, and $[Ni(CO)_4]^0$. The oxyanions of certain transition metals (such as VO_4^{3-}, CrO_4^{2-}, FeO_4^{2-}, and MnO_4^-) are tetrahedral, resembling the tetrahedral oxyanions of nonmetals (such as SiO_4^{4-}, PO_4^{3-}, AsO_4^{3-}, SO_4^{2-}, and ClO_4^-). Although a number of tetrahedral complexes of transition elements are known, the majority of complexes of these elements are octahedral.

Linear, two-coordinate complexes are not so common as the other forms previously mentioned. However, well-characterized complexes of this type are known for Cu^I, Ag^I, and Hg^{II}; examples are $[CuCl_2]^-$, $[Ag(NH_3)_2]^+$, $[Au(CN)_2]^-$, $[Hg(NH_3)_2]^{2+}$, and $[Hg(CN)_2]^0$.

In general, each metal exhibits more than one coordination number and geometry in its complexes. Although all known complexes of Co^{III} are octahedral, most cations form more than one type of complex. For example, Al^{III} forms tetrahedral and octahedral complexes, Cu^I forms linear and tetrahedral complexes, and Ni^{II} forms square-planar, tetrahedral, and octahedral complexes.

Complexes of the transition elements are frequently highly colored. Examples of complexes that exhibit a wide range of colors are $[Co(NH_3)_6]^{3+}$ (yellow), $[Co(NH_3)_5(H_2O)]^{3+}$ (pink), $[Co(NH_3)_5Cl]^{2+}$ (violet), $[Co(H_2O)_6]^{3+}$ (purple), and $[Co(NH_3)_4Cl_2]^+$ (a violet form and a green form—see Section 24.4).

The ligands we have discussed thus far have been capable of forming only one bond with the central ion. They are referred to as **unidentate** (from Latin, meaning *one-toothed*) ligands. Certain ligands are capable of occupying more than one coordination position of a metal ion. Ligands that coordinate through two bonds from different parts of the molecule or anion are called **bidentate.** Examples are

carbonate ion   *oxalate ion*   *ethylenediamine*

The carbonate and oxalate ions each coordinate through two oxygen atoms. The ethylenediamine molecule (abbreviated *en*) coordinates through both nitrogen atoms. These positions are marked with arrows. Bidentate ligands form rings with the central metal ions:

$$CH_2—CH_2$$
$$NH_2 \qquad NH_2$$
$$M$$

The resulting metal complexes are called **chelates** (from Greek, meaning *claw*). The formation of five- or six-membered rings is generally favored.

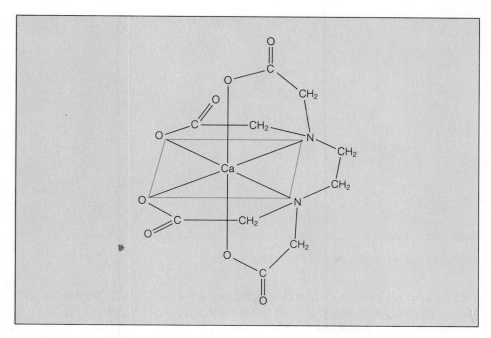

Figure 24.3 Ethylenediaminetetraacetate complex of Ca^{2+} ($[Ca(EDTA)]^{2-}$)

Multidentate ligands have been prepared that can coordinate at 2, 3, 4, 5, or 6 positions. In general, chelates are more stable than complexes containing only unidentate ligands. Thus, the sexadentate complexing agent ethylenediaminetetraacetate ion (EDTA),

$$^-O_2C—CH_2\qquad\qquad\qquad CH_2—CO_2^-$$
$$\diagdown\qquad\qquad\qquad\qquad\diagup$$
$$N—CH_2—CH_2—N$$
$$\diagup\qquad\qquad\qquad\qquad\diagdown$$
$$^-O_2C—CH_2\qquad\qquad\qquad CH_2—CO_2^-$$

is capable of forming a very stable complex with the calcium ion—an ion with one of the least tendencies toward the formation of complexes (see Figure 24.3).

Heme of hemoglobin is a chelate of Fe^{2+}, and chlorophyll is a chelate of Mg^{2+}. In both these substances the metal atom is coordinated to a quadridentate ligand, which may be considered to be derived from a porphin structure (see Figure 24.4) by the substitution of various groups for the H atoms of the porphin. The substituted porphins are called porphyrins.

Coordination around the iron atom of heme is octahedral. Four of the coordination positions are utilized in the formation of the heme (which is essentially planar), the fifth is used to bond the heme to a protein molecule (globin), and the sixth is used to coordinate either H_2O (hemoglobin) or O_2 (oxyhemoglobin). Coordination about this sixth position is reversible:

$$\text{hemoglobin} + O_2 \rightleftharpoons \text{oxyhemoglobin} + H_2O$$

and dependent upon the pressure of O_2. Hemoglobin picks up O_2 in the lungs and releases it in the body tissues, where it is used for the oxidation of food.

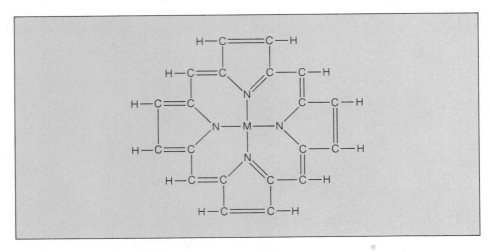

Figure 24.4 Metal porphin (M indicates a metal cation)

Hemoglobin reacts with carbon monoxide to form a complex that is more stable than oxyhemoglobin, and therefore CO is toxic.

Chlorophyll, a porphyrin of Mg^{2+}, the green pigment of plants, serves as a catalyst for the process of photosynthesis in which CO_2 and H_2O are converted into a carbohydrate (glucose) and O_2. The energy for photosynthesis comes from sunlight, and the chlorophyll molecule initiates this process by absorbing a quantum of light.

24.2 Labile and Inert Complexes

Certain complexes (which are called labile) rapidly undergo reactions in which ligands are replaced; other complexes (nonlabile, or inert, complexes) do not undergo these substitution reactions or do so slowly. This distinction applies to the *rate* of attainment of equilibrium and has no bearing on the position of equilibrium. With the exception of the complexes of Cr^{III} and Co^{III}, most octahedral complexes of the fourth period transition elements are very labile; exchange reactions come to equilibrium almost as fast as the reagents are mixed. The reason that the complexes of Co^{III} have been studied more than any other group of complexes is that these substances undergo ligand exchange reactions at a slower, more convenient rate.

The inertness of a complex should not be confused with its thermodynamic stability. Although $[Co(NH_3)_6]^{3+}$ is stable in aqueous solution:

$$[Co(NH_3)_6]^{3+} + 6\,H_2O \rightleftharpoons [Co(H_2O)_6]^{3+} + 6\,NH_3 \qquad K \cong 10^{-34}$$

the complex is unstable in aqueous acid:

$$[Co(NH_3)_6]^{3+} + 6\,H_3O^+ \rightleftharpoons [Co(H_2O)_6]^{3+} + 6\,NH_4^+ \qquad K \cong 10^{22}$$

Nevertheless, the $[Co(NH_3)_6]^{3+}$ ion can exist in dilute acid for weeks; the latter reaction must have a high activation energy. Thus, the complex is thermodynamically unstable and at the same time inert.

Examples of the reverse situation are also known. The stable complex

$[FeCl_6]^{3-}$ is very labile and undergoes rapid exchange with radioactive chloride in aqueous solution.

In many cases complex formation results in the stabilization of metal ions toward oxidation or reduction. Electrode potentials for Zn^{2+} and some complexes of this ion are

$$2e^- + Zn^{2+}(aq) \rightleftharpoons Zn(s) \qquad \mathscr{E}^\circ_{red} = -0.763 \text{ V}$$

$$2e^- + [Zn(NH_3)_4]^{2+} \rightleftharpoons Zn(s) + 4\,NH_3 \qquad \mathscr{E}^\circ_{red} = -1.04 \text{ V}$$

$$2e^- + [Zn(CN)_4]^{2-} \rightleftharpoons Zn(s) + 4\,CN^- \qquad \mathscr{E}^\circ_{red} = -1.26 \text{ V}$$

The increasing stability toward reduction observed in this series parallels increasing stability of the complexes toward dissociation in aqueous solution. The instability constant of $[Zn(NH_3)_4]^{2+}$ is approximately 10^{-10} and that of the $[Zn(CN)_4]^{2-}$ ion is about 10^{-18}.

In many instances complex formation results in the stabilization of a metal in a rare or otherwise unknown oxidation state. A classic example is afforded by the complexes of Co^{III}. Simple compounds containing cobalt in an oxidation state of $3+$ are rare. The electrode potential

$$e^- + [Co(H_2O)_6]^{3+} \rightleftharpoons [Co(H_2O)_6]^{2+} \qquad \mathscr{E}^\circ_{red} = +1.81 \text{ V}$$

indicates that the hydrated Co^{3+} ion is a strong oxidizing agent which is capable of oxidizing water to oxygen and hence incapable of prolonged existence in water.

In the presence of many complexing agents, such as NH_3, the $3+$ oxidation state of cobalt is much more stable toward reduction:

$$e^- + [Co(NH_3)_6]^{3+} \rightleftharpoons [Co(NH_3)_6]^{2+} \qquad \mathscr{E}^\circ_{red} = +0.11 \text{ V}$$

The $[Co(NH_3)_6]^{3+}$ ion can exist in aqueous solution. It does not oxidize water to oxygen, and, in fact, the reverse reaction—the oxidation of Co^{II} complexes by a stream of air—is used to prepare Co^{III} complexes. The ammonia complex of Co^{III}, which has an instability constant of about 10^{-34}, is much more stable toward dissociation than the ammonia complex of Co^{II}, which has an instability constant of approximately 10^{-5}.

The formation of a complex can prevent a metal cation from disproportionating. The copper(I) ion, for example, is unstable in water solution:

$$2\,Cu^+(aq) \longrightarrow Cu^{2+}(aq) + Cu(s)$$

The ammonia complex of this ion, $[Cu(NH_3)_2]^+$ is stable toward disproportionation. A similar situation is observed for the ions of gold: Au^+ is theoretically unstable toward disproportionation into Au^{3+} and Au metal. In addition, both gold ions are strong oxidizing agents and are capable of oxidizing water. However, stable complexes of both Au^I and Au^{III} are known.

24.3 Nomenclature

Since thousands of complexes are known and the number is constantly expanding, a system of nomenclature has been adopted for these compounds. The following list summarizes the important rules of the system. They are adequate for naming the simple and frequently encountered complexes.

1. If the complex compound is a salt, the cation is named first whether or not it is the complex ion.

2. The constituents of the complex are named in the following order: anions, neutral molecules, central metal ion.

3. Anionic ligands are given -*o* endings. Examples are: OH^-, hydroxo; O^{2-}, oxo; S^{2-}, thio; Cl^-, chloro; F^- fluoro; CO_3^{2-}, carbonato; CN^-, cyano; CNO^-, cyanato; $C_2O_4^{2-}$, oxalato; NO_3^-, nitrato; NO_2^-, nitro; SO_4^{2-}, sulfato; and $S_2O_3^{2-}$, thiosulfato.

4. The names of neutral ligands are not changed. Exceptions to this rule are: H_2O, aquo; NH_3, ammine; CO, carbonyl; and NO, nitrosyl.

5. The number of ligands of a particular type is indicated by a prefix: di-, tri-, tetra-, penta-, and hexa- (for two to six). For complicated ligands (such as ethylenediamine), the prefixes bis-, tris-, and tetrakis- (two to four) are employed.

6. The oxidation number of the central ion is indicated by a Roman numeral, which is set off by parentheses and placed after the name of the complex.

7. If the complex is an anion, the ending -*ate* is employed. If the complex is a cation or a neutral molecule, the name is not changed.

Examples of the rules of nonmenclature are

$[Ag(NH_3)_2]Cl$	Diamminesilver(I) chloride
$[Co(NH_3)_3Cl_3]$	Trichlorotriamminecobalt(III)
$K_4[Fe(CN)_6]$	Potassium hexacyanoferrate(II)
$[Ni(CO)_4]$	Tetracarbonylnickel(0)
$[Cu(en)_2]SO_4$	Bis(ethylenediamine)copper(II) sulfate
$[Pt(NH_3)_4][PtCl_6]$	Tetraammineplatinum(II) hexachloroplatinate(IV)
$[Co(NH_3)_4(H_2O)Cl]Cl_2$	Chloroaquotetraamminecobalt(III) chloride

Common names are frequently employed when they are clearly more convenient than the systematic name (for example, ferrocyanide rather than hexacyanoferrate(II) for $[Fe(CN)_6{}^{4-}]$) or when the structure of the complex is not certain (for example, the aluminate ion).

24.4 Isomerism

Two compounds with the same molecular formula but different arrangements of atoms are called isomers. Such compounds differ in their chemical and physical properties. Structural isomerism is displayed by compounds that have different ligands within their coordination spheres. Several types of structural isomers may be identified.

The following pair of compounds of Co^{III} serve as an example of ionization isomers:

(a) $[Co(NH_3)_5(SO_4)]Br$

  red

(b) $[Co(NH_3)_5Br]SO_4$

  violet

Conductance data show that both compounds dissociate into two ions in aqueous solution. In the first compound, the SO_4^{2-} ion is a part of the coordination sphere, and the Br^- ion is ionizable. An aqueous solution of compound (a) gives an immediate precipitate of AgBr upon the addition of $AgNO_3$, but since the SO_4^{2-} ion is not free, no precipitate forms upon the addition of $BaCl_2$. For compound (b), the reverse is true. An aqueous solution of this compound gives a precipitate of $BaSO_4$ but not AgBr since the SO_4^{2-} is ionizable and the Br^- is coordinated. Note that SO_4^{2-} functions as a unidentate ligand and that the charge on the complex ion of compound (a) is $1+$ and that of compound (b) is $2+$. There are numerous additional examples of ionization isomers, such as

  $[Pt(NH_3)_4Cl_2]Br_2$ $[Pt(NH_3)_4Br_2]Cl_2$

Hydrate isomerism is analogous to ionization isomerism and is probably best illustrated by the following series of compounds that have the formula $CrCl_3 \cdot 6 H_2O$:

(a) $[Cr(H_2O)_6]Cl_3$

  violet

(b) $[Cr(H_2O)_5Cl]Cl_2 \cdot H_2O$

  green

(c) $[Cr(H_2O)_4Cl_2]Cl \cdot 2 H_2O$

  green

In a mole of each of these compounds there are six moles of water. However, in compound (a), six water molecules are coordinated; in compound (b), five; and in compound (c), four. The uncoordinated water molecules occupy separate positions in the crystals and are readily lost when compounds (b) and (c) are exposed to desiccants. The coordinated water, however, is not so easily removed. Further evidence for the structures of the compounds is afforded by conductance data (the compounds are composed of four, three, and two ions, respectively) and the quantity of AgCl precipitated (the compounds have three, two, and one ionizable chloride ions, respectively).

Another example of hydrate isomerism is given by the following pair of compounds:

  $[Co(NH_3)_4(H_2O)Cl]Cl_2$ $[Co(NH_3)_4Cl_2]Cl \cdot H_2O$

Coordination isomers may exist in compounds that have two or more centers of coordination. Isomers arise through the exchange of ligands between these coordination centers. In simple examples, which involve only two complex ions per compound, the coordinated metal ions may be the same:

  $[Cr(NH_3)_6][Cr(NCS)_6]$ $[Cr(NH_3)_4(NCS)_2][Cr(NH_3)_2(NCS)_4]$

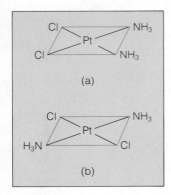

Figure 24.5 (a) *Cis* and (b) *trans* isomers of dichlorodiammineplatinum(II)

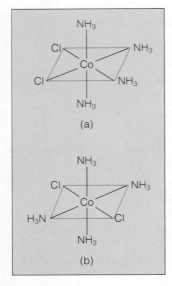

Figure 24.6 (a) *Cis* and (b) *trans* isomers of the dichlorotetraamminecobalt(III) ion

or different:

$$[Cu(NH_3)_4][PtCl_4] \qquad [Pt(NH_3)_4][CuCl_4]$$

and the oxidation state of the metals may vary:

$$[Pt(NH_3)_4][PtCl_6] \qquad\qquad [Pt(NH_3)_4Cl_2][PtCl_4]$$
tetraammineplatinum(II)   *dichlorotetraammineplatinum(IV)*
hexachloroplatinate(IV)   *tetrachloroplatinate(II)*

Linkage isomerism arises when ligands are capable of coordinating in two ways. The nitrite ion, NO_2^-, for example, can coordinate through an oxygen atom (—ONO, nitrito compounds) or through the nitrogen atom (—NO_2, nitro compounds):

$$[Co(NH_3)_5(NO_2)]Cl_2 \qquad\qquad [Co(NH_3)_5(ONO)]Cl_2$$
(yellow)   *(red)*
nitropentaamminecobalt(III)   *nitritopentaamminecobalt(III)*
chloride   *chloride*

Some other ligands that are capable of forming linkage isomers are: CN^- which can coordinate either through the C atom or the N atom; SCN^-, through N or S; and CO, through C or O.

Stereoisomerism is a second general classification of isomers. Compounds are stereoisomers when they both contain the same ligands in their coordination spheres but differ in the way that these ligands are arranged in space. One type of stereoisomerism is **geometric,** or *cis-trans*, **isomerism.**

An example of geometric isomerism is afforded by the *cis* and *trans* isomers of the square-planar dichlorodiammineplatinum (II); see Figure 24.5. In the *cis* isomer the chlorine atoms are situated on adjacent corners of the square (along an edge), whereas in the *trans* isomer they occupy opposite corners (along the diagonal).

Since all the ligands of a tetrahedral complex have the same relationship to one another, *cis-trans* isomerism does not exist for this geometry. Many geometric isomers are known for octahedral complexes, however. There are two isomers of the dichlorotetraamminecobalt(III) ion: a violet *cis* form and a green *trans* form (see Figure 24.6).

A second type of stereoisomerism is **optical isomerism.** Some molecules and ions can exist in two forms that are not superimposable and that bear the same relationship that a right hand bears to a left hand. That the hands are not superimposable is readily demonstrated by attempting to put a left-handed glove on a right hand. Such molecules and ions are spoken of as being **chiral** or **dissymmetric,** and the forms are called **enantiomorphs** (from Greek, meaning *opposite forms*) or **mirror images** (since the one may be considered to be a mirror reflection of the other).

Enantiomorphs have identical physical properties except for their effects on plane-polarized light. Light that has been passed through a polarizer consists of waves that vibrate in a single plane. One enantiomorph (the **dextro** form), whether pure or in solution, will rotate the plane of the plane-polarized light to the right; the other (the **levo** form) will rotate the plane an equal extent to the left (see Figure 24.7).

For this reason enantiomorphs are called optical isomers or **optical antipodes.** An equimolar mixture of enantiomorphs, called a **racemic modification,** has no

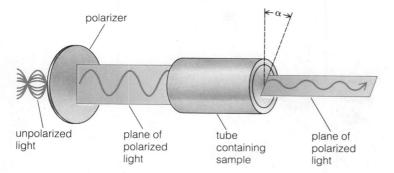

Figure 24.7 Rotation of the plane-polarized light. (The plane is rotated by an angle after the light passes through a solution of an optically active isomer. In this case, rotation is to the right and the isomer is said to be dextrorotatory.)

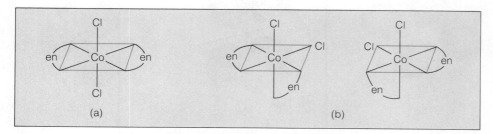

Figure 24.9 Isomers of the dichlorobis(ethylenediamine)cobalt(III) ion: (a) *trans* isomer, (b) optical isomers of the *cis* form

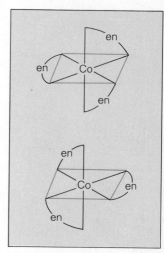

Figure 24.8 Dextro and levo forms of the tris(ethylenediamine)cobalt(III) ion

effect on plane-polarized light since it contains an equal number of *dextro*rotatory and *levo*rotatory forms.

The tris(ethylenediamine)cobalt(III) ion exists in enantiomorphic forms. Examination of the diagrams of Figure 24.8 confirms that the enantiomorphs are not superimposable and that the ions are chiral. Note that bidentate chelating agents can span only *cis* positions.

Both types of stereoisomerism—geometric and optical—are illustrated by the isomers of the dichlorobis(ethylenediamine)cobalt(III) ion (see Figure 24.9). The *trans* configuration of this ion is optically inactive, and a mirror image would be identical to the original. The *cis* arrangement, however, is chiral and exists in *dextro* and *levo* forms. The *trans* modification is said to be a *diastereoisomer* of either the *dextro* or *levo cis* modification.

24.5 The Bonding in Complexes

The first theory used to explain the bonding in complexes assumed quite simply that the ligands donate electron pairs to form covalent bonds with the central metal ions. Bonds formed in this way were called coordinate covalent bonds. During the years that followed, the **valence bond theory** grew out of this simple concept. The central metal ion is assumed to supply empty hybrid orbitals toward the formation of the complex (for example, $d^2 sp^3$ hybrid orbitals for the formation of an octahedral complex), and the ligands supply electron pairs to fill these orbitals and form covalent bonds. The valence bond theory fits many experimental observations well, but it fails to explain others in a satisfactory manner (notably,

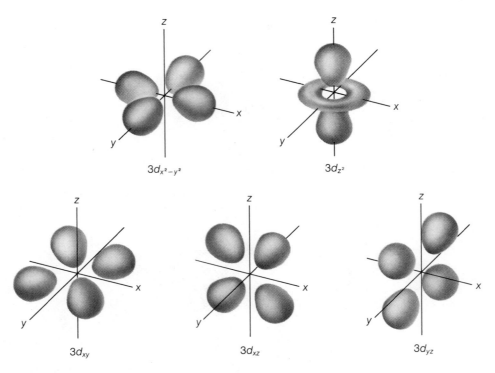

Figure 24.10 Boundary surface diagrams for the *d* orbitals

the numbers of unpaired electrons in certain complexes), and it has nothing at all to say about still other observations (notably, the colors of complexes).

Modern theories of the bonding in complexes are derived from the **crystal field theory.** In its simplest form, this theory explains the bonding in terms of electrostatic attractions between the positive charge of the central metal cation and the negative charges of the electron pairs of the ligands. An important aspect of this theory is the affect that the electrostatic field of the ligands has on the *d* orbitals of the central metal cation.

The outer electrons of transition-metal ions are *d* electrons (see Table 5.4). In a free transition-metal ion, all five of the *d* orbitals have equal orbital energies; they are degenerate. Not all of the *d* orbitals are energy equivalent, however, when they are surrounded by the negative field created by the ligands.

Consider the relation of the *d* orbitals (see Figure 24.10) to the arrangement of the ligands of an octahedral complex (see Figure 24.11a). The d_{z^2} and $d_{x^2-y^2}$ orbitals have lobes that point toward ligands, whereas the lobes of the d_{xy}, d_{xz}, and d_{yz} orbitals lie between ligands. Thus, in the complex two sets of *d* orbitals exist. The d_{xy}, d_{xz}, and d_{yz} orbitals (or t_{2g} orbitals) are equivalent to each other, and the d_{z^2} and $d_{x^2-y^2}$ orbitals (or e_g orbitals) are equivalent to each other and different from the first three. The symbols t_{2g} and e_g are applied to threefold degenerate and twofold degenerate sets of orbitals, respectively.

It is not immediately obvious that the d_{z^2} orbital is perfectly equivalent to the $d_{x^2-y^2}$ orbital. The d_{z^2} orbital may be regarded as a combination, in equal parts, of two hypothetical orbitals, $d_{z^2-y^2}$ and $d_{z^2-x^2}$, which have shapes exactly like that of the $d_{x^2-y^2}$ orbital (see Figure 24.12). Since the number of *d* orbitals is limited to five, the $d_{z^2-y^2}$ and $d_{z^2-x^2}$ orbitals have no independent existence.

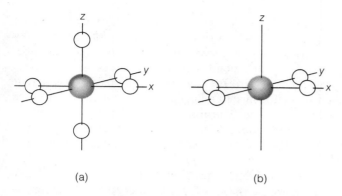

Figure 24.11 Arrangement of ligands of (a) octahedral and (b) square-planar complexes in relation to sets of Cartesian coordinates

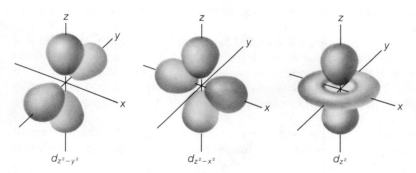

$d_{z^2-y^2}$ $d_{z^2-x^2}$ d_{z^2}

Figure 24.12 Diagram showing that the d_{z^2} orbitals may be considered to be a combination of the $d_{z^2-y^2}$ and $d_{z^2-x^2}$ orbitals

In the crystal field theory the assumption is made that an electrostatic field surrounding the central metal ion is produced by the negative ends of the dipolar molecules or by the anions that function as ligands. A metal-ion electron in a d orbital that has lobes directed toward ligands has a higher energy (owing to electrostatic repulsion) than an electron in an orbital with lobes that point between ligands. In octahedral complexes the orbitals of the e_g group, therefore, have higher orbital energies than those of the t_{2g} group. The difference between the orbital energies of the t_{2g} and e_g orbitals in an octahedral complex is given the symbol Δ_o.

In a square-planar complex (see Figure 24.11b), the d orbitals exhibit four different relationships. The lobes of the $d_{x^2-y^2}$ orbital point toward ligands, and this orbital has the highest orbital energy. The lobes of the d_{xy} orbital lie between orbitals but are coplanar with them, and hence this orbital is next highest in orbital energy. The lobes of the d_{z^2} orbital point out of the plane of the complex, but the belt around the center of the orbital (which contains about a third of the electron density) lies in the plane. The d_{z^2} orbital, therefore, is next highest in orbital energy. The d_{xz} and d_{yz} orbitals, which are degenerate, are least affected by the electrostatic field of the ligands since their lobes point out of the plane of the complex. These orbitals are lowest in orbital energy.

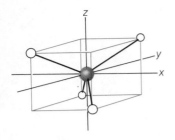

Figure 24.13 The relation of tetrahedrally arranged ligands to a set of Cartesian coordinates

The order of splitting of the d orbitals in a tetrahedral complex may be derived from an examination of Figure 24.13, which shows the relation of tetrahedrally arranged ligands to a system of Cartesian coordinates and a hypothetical cube. The lobes of the d_{xy}, d_{xz}, and d_{yz} orbitals point toward cube edges, and the lobes of the d_{z^2} and $d_{x^2-y^2}$ orbitals point toward the centers of cube faces. Notice that the distance from the center of a cube face to a ligand is farther than the distance from the center of a cube edge to a ligand. The order of orbital energies, therefore, is the reverse of that for octahedral coordination, and the threefold degenerate set, t_{2g}, is of higher energy than the twofold degenerate set, e_g. The difference in energy is given the symbol Δ_t.

The splitting of d-orbital energies in tetrahedral, octahedral, and square-planar complexes is summarized in Figure 24.14. A defect of the crystal field theory is that is centers on the ionic aspects of the bonding and fails to take into account the covalent character of the bonding. The **molecular orbital theory** is at the opposite extreme; it describes the bonding in terms of the formation of bonding and antibonding molecular orbitals. From either point of view, however, the same conclusions are reached regarding the distribution of electrons in the d orbitals of the central metal ion. The orders of d-orbital splitting are those shown in Figure 24.14. The importance of these splitting diagrams is discussed later in this section.

In an octahedral complex, the $3d_{z^2}$ and $3d_{x^2-y^2}$ orbitals along with the $4s$ and three $4p$ orbitals are assumed to overlap six orbitals from the ligands with the attendant formation of six bonding molecular orbitals and six antibonding molecular orbitals. The d_{xy}, d_{xz}, and d_{yz} orbitals (the t_{2g} set), which do not overlap the σ orbitals of the ligands, are essentially nonbonding. (The t_{2g} set can be used in π bonding, however.)

A bonding molecular orbital concentrates electron density between the atoms and is of relatively low energy in comparison with an antibonding molecular

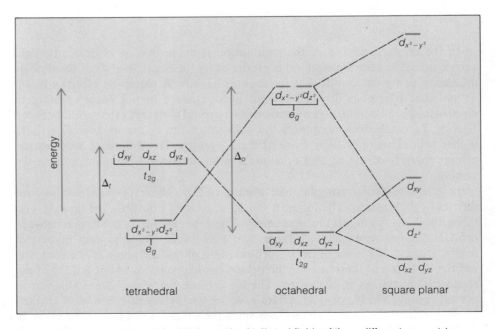

Figure 24.14 The splitting of d-orbital energies by ligand fields of three different geometries

orbital, which has a low electron density between the atoms and acts as a disruptive force. A molecular orbital energy-level diagram for an octahedral complex with no π bonding is given in Figure 24.15.

Whenever two atomic orbitals of different energies combine, the character of the resulting bonding molecular orbital is predominantly that of the atomic orbital of lower energy, and the antibonding molecular orbital has mainly the character of the higher energy atomic orbital. In an octahedral complex the bonding orbitals have predominantly the character of ligand orbitals. The antibonding orbitals resemble metal orbitals more than ligand orbitals. The t_{2g} set, which are nonbonding, may be considered as purely metal orbitals.

In an octahedral complex, the six electron pairs from the ligands completely occupy the bonding molecular orbitals. The d electrons of the central metal ion are accommodated in the t_{2g} nonbonding orbitals and the $(e_g)_a$ antibonding orbitals. The difference between the energies of these sets is Δ_o. The four remaining antibonding orbitals are never occupied in the ground states of any known complex.

The conclusion reached by this treatment is much the same as that postulated by the crystal field theory. In an octahedral complex the degeneracy of the metal d orbitals may be considered to be split into a threefold degenerate set, t_{2g}, and a higher energy, metal-like, twofold degenerate set, which may be labeled $(e_g)_a$ or simply e_g.

The molecular orbital treatment may be applied to tetrahedral and square-planar complexes, but the applications are more complicated. The conclusions reached are in essential agreement with the splittings diagrammed in Figure 24.14.

For the octahedral complexes of a given metal ion, the magnitude of Δ_o is different for each set of ligands, and the electronic configurations of many complexes depend upon the size of Δ_o.

For complexes of transition-element ions with one, two, or three d electrons (which are referred to as d^1, d^2, or d^3 ions), the orbital occupancy is certain and

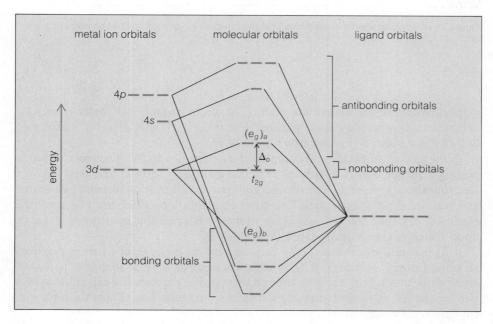

Figure 24.15 Diagram indicating the formation of molecular orbitals for an octahedral complex with no π bonding

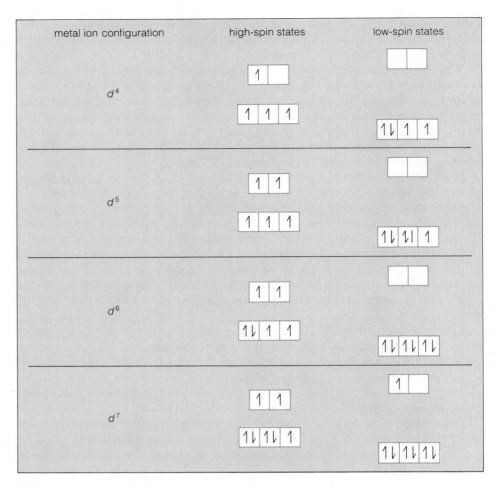

Figure 24.16 Arrangements of electrons in high-spin and low-spin octahedral complexes of d^4, d^5, d^6, and d^7 ions

is independent of the magnitude of Δ_o. The electrons enter the lower energy t_{2g} orbitals singly with their spins parallel. For d^4, d^5, d^6, and d^7 ions, a choice of two configurations is possible (see Figure 24.16).

In the case of an octahedral complex of a d^4 ion, the fourth electron can singly occupy a higher energy e_g orbital or it can enter a t_{2g} orbital and pair with an electron already present. The former configuration has four unpaired electrons and is called the **high-spin state**. The latter configuration, which has two unpaired electrons, is called the **low-spin state**. Which configuration is assumed depends upon which is energetically more favorable.

If Δ_o is small, the electron may be promoted to the e_g level where it occupies an orbital singly. However, if Δ_o is large, the electron may be forced to pair with an electron in a t_{2g} orbital even though this pairing requires the expenditure of a pairing energy, P, to overcome the inter-electronic repulsion. Thus, the high-spin configuration results when

$$\Delta_o < P$$

and the low-spin configuration results when

$$\Delta_o > P$$

The value of P depends upon the metal ion; Δ_o is different for each complex.

This conclusion is also valid for complexes of d^5, d^6, and d^7 ions. For a complex of a d^8 ion there is only one possible configuration: six electrons paired in the t_{2g} orbitals and two unpaired electrons in the e_g orbitals. Likewise, complexes of d^9 and d^{10} ions exist in only one configuration.

Values of Δ_o can be obtained from studies of the absorption spectra of complexes. In the complex $[Ti(H_2O)_6]^{3+}$, there is only one electron to be accommodated in either the t_{2g} or e_g orbitals. In the ground state this electron occupies a t_{2g} orbital. Excitation of the electron to an e_g orbital is possible when the energy required for this transition, Δ_o, is supplied. The absorption of light by the complex can bring about such excitations. The wavelength of light absorbed most strongly by the $[Ti(H_2O)_6]^{3+}$ ion is approximately 490 nm, which corresponds to a Δ_o of about 243 kJ/mol. The single absorption band of this complex spreads out over a considerable portion of the visible spectrum. Most of the red and violet light, however, is not absorbed, and this causes the red-violet color of the complex.

The interpretation of the absorption spectra of complexes with more than one d electron is considerably more complicated, since more than two arrangements of d electrons are then possible. In general, for a given metal ion the replacement of one set of ligands by another causes a change in the energy difference between the t_{2g} and e_g orbitals, Δ_o, which gives rise to different light-absorption properties. In many instances a striking color change is observed when the ligands of a complex are replaced by other ligands.

Ligands may be arranged in a **spectrochemical series** according to the magnitude of Δ_o they bring about. From the experimental study of the spectra of many complexes, it has been found that the order is generally the same for the complexes of all of the transition elements in their common oxidation states with only occasional inversions of order between ligands that stand near to one another on the list. The order of some common ligands is

$$I^- < Br^- < Cl^- < F^- < OH^- < C_2O_4^{2-} < H_2O < NH_3$$
$$< en < NO_2^- < CN^-$$

The values of Δ_o induced by the halide ions are generally low, and complexes of these ligands usually have high-spin configurations. The cyanide ion, which stands at the opposite end of the series from the halide ions, induces the largest d-orbital splittings of any ligand listed. Cyano complexes generally have low-spin configurations.

A given ligand, however, does not always produce complexes of the same spin type. Thus, the hexaammine complex of Fe^{II} has a high-spin configuration, whereas the hexaammine complex of Co^{III} (which is isoelectronic with Fe^{II}) has a low-spin configuration.

For each metal ion there is a point in the series that corresponds to the change from ligands that produce high-spin complexes to ligands that form low-spin complexes. For example, Co^{II} forms high-spin complexes with NH_3 and ethylenediamine, but the NO_2^- and CN^- complexes of Co^{II} have low-spin configurations. The actual position in the series at which this change from high- to low-spin complex formation occurs depends upon the electron-pairing energy, P, for the metal ion, as well as the values of Δ_o for the complexes under consideration.

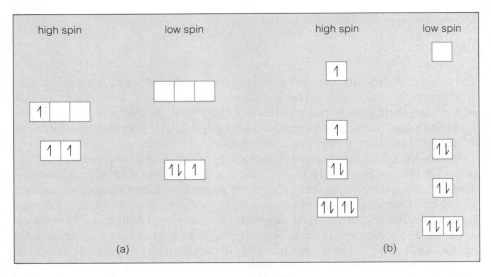

Figure 24.17 Arrangements of d electrons in high-spin and low-spin states in (a) a tetrahedral complex of a d^3 ion, and (b) a square-planar complex of a d^8 ion

In the case of tetrahedral complexes, the two orbitals of the e_g set have lower orbital energies than the three orbitals of the t_{2g} set (see Figure 24.14). Only one spin configuration is possible for a d^1 or a d^2 ion, each with electrons entered singly into the e_g orbitals. In theory, however, a high-spin state and a low-spin state should be possible for a d^3 ion (see Figure 24.17a), as well as for a d^4, d^5, or d^6 ion. No low-spin tetrahedral complexes are known, however.

In a typical tetrahedral complex, the energy difference between the t_{2g} and e_g sets, Δ_t, is low, about one-half of a typical Δ_o value. For all tetrahedral complexes:

$$\Delta_t < P$$

Consequently, after two electrons have been introduced singly into the two orbitals of the e_g set, it is easier to place the third electron singly into an orbital of the t_{2g} set than it is to supply the pairing energy, P, and pair this third electron with one already present in an orbital of the e_g set. This conclusion is also valid for the tetrahedral complexes of d^4, d^5, and d^6 ions. Consequently, all known tetrahedral complexes are high-spin complexes. Only one spin configuration is theoretically possible for the tetrahedral complexes of d^7, d^8, d^9, and d^{10} ions.

Square-planar complexes are sometimes formed by d^8 ions. In theory, a high-spin state and a low-spin state should be possible for the square-planar complexes of a d^8 ion (see Figure 24.17b). Actually, all known d^8 square-planar complexes have low-spin configurations (and are diamagnetic since all electrons are paired). The energy separation between the highest and next highest two orbitals is so large that a high-spin state cannot be attained. For the square-planar complexes of d^7 ions, the configuration is similar to that of the low-spin state shown in Figure 24.17b with the exception that only one electron (instead of two) is present in the highest orbital occupied.

Summary

The topics that have been discussed in this chapter are

1. The structure of complex compounds; chelates.

2. The thermodynamic stability of complexes and their stability toward reactions in which ligands are replaced.

3. The nomenclature of complex compounds.

4. Isomers of complexes, compounds that have the same molecular formula but different arrangements of atoms. The types of isomerism exhibited by complex compounds.

5. The theories used to explain the bonding in complexes.

Key Terms

Some of the more important terms introduced in this chapter are listed below. Definitions for terms not included in this list may be located in the text by use of the index.

Chelate (24.1) A metal complex that contains coordinated multidentate ligands.

Chelating agent (Section 24.1) A ligand capable of occupying more than one coordination position on a central metal ion in a complex; a **multidentate ligand.**

Chiral molecule or ion (Section 24.4) A molecule or an ion that can exist in two forms that are mirror images that are not superimposable.

Coordination number (Section 24.1) The number of atoms directly bonded to the central metal cation of a complex.

Coordination isomerism (Section 24.4) A type of isomerism in which two complex compounds, each with more than one coordination center, vary in the way that the same set of ligands is distributed between the coordination centers.

Coordination sphere (Section 24.1) The group of anions or molecules that are directly coordinated to the central metal ion of a complex.

Crystal field theory (Section 24.5) A theory that explains the bonding of the central metal ion to the ligands of a complex in terms of electrostatic attractions between the positive charge of the metal ion and the negative charge of the electron pairs of the ligands.

e_g **orbitals** (Section 24.5) A degenerate set of two d orbitals of the central metal ion of a complex.

Enantiomorphs (Section 24.4) Two different compounds that are mirror images of one another; one may be considered to be the mirror reflection of the other. The **dextro** form rotates the plane of polarized light to the right; the **levo** form, to the left.

Geometric isomerism (Section 24.4) A type of isomerism in which the isomers differ in the geometric arrangement of the ligands around the central metal ion.

High-spin state (Section 24.5) A state in which the electron configuration of the d orbitals of the central metal ion of a complex has a maximum number of unpaired electrons.

Hydrate isomerism (Section 24.4) A type of isomerism in which two complex compounds differ as to whether water molecules are coordinated or occupy separate positions in the crystal lattice of the compound.

Inert complex (Section 24.2) A complex that does not undergo reactions in which its ligands are replaced, or does so only slowly.

Ionization isomerism (Section 24.4) A type of isomerism in which two complexes differ as to whether a given anion is inside or outside the coordination sphere of the complex.

Labile complex (Section 24.2) A complex that rapidly undergoes reactions in which its ligands are replaced.

Linkage isomerism (Section 24.4) A type of isomerism in which two complexes differ as to the way a given ligand (that is present in both) coordinates to the central metal ion.

Low-spin state (Section 24.5) A state in which the electron configuration of the d orbitals of the central metal ion of a complex has a minimum number of unpaired electrons.

Optical isomerism (Section 24.4) A type of isomerism in which the two isomers are mirror images that are not superimposable.

Porphyrins (Section 24.1) Chelates derived from a quadridentate ligand that is a derivative of the porphin structure (see Figure 24.4); examples include hemoglobin and chlorophyll.

Spectrochemical series (Section 24.5) A series of ligands arranged in increasing order of the size of the splitting of d orbital energies, Δ_o, that they bring about.

Stereoisomers (Section 24.4) Compounds that have the same molecular formula and are bonded in the same way but that differ in the way the constituent atoms are arranged in space.

Structural isomers (Section 24.4) Compounds that have the same molecular formula but differ in the way the atoms are bonded together.

t_{2g} **orbitals** (Section 24.5) A degenerate set of three d orbitals of the central metal ion of a complex.

Problems*

Structure of Complexes

24.1 Define the following terms: **(a)** coordination number, **(b)** ligand, **(c)** chelate, **(d)** enantiomorph, **(e)** inert complex, **(f)** low-spin state.

24.2 State the oxidation number of the central atom in each of the following complexes:

(a) $[Co(NO_2)_6]^{3-}$

(b) $[Au(CN)_4]^-$

(c) $[V(CO)_6]$

(d) $[Co(NH_3)_4Br_2]^+$

(e) $[Co(en)Cl_4]^{2-}$

24.3 What types of metal ions form complexes most readily? What configurations of complexes are most common?

24.4 Interpret the formation of $Ag(NH_3)_2^+$ from Ag^+ and NH_3 in terms of the Lewis acid-base theory.

*24.5 Given the following data, calculate the instability constant, K_{inst}, of the $Al(OH)_4^-$ ion:

$$3e^- + Al(OH)_4^- \rightleftharpoons Al + 4OH^- \quad \mathscr{E}_{red}^\circ = -2.330 \text{ V}$$

$$3e^- + Al^{3+} \rightleftharpoons Al \quad \mathscr{E}_{red}^\circ = -1.662 \text{ V}$$

*24.6 Given:

$$3e^- + AuBr_4^- \rightleftharpoons Au + 4Br^- \quad \mathscr{E}_{red}^\circ = +0.870 \text{ V}$$

$$3e^- + Au^{3+} \rightleftharpoons Au \quad \mathscr{E}_{red}^\circ = +1.498 \text{ V}$$

Calculate the instability constant of the $AuBr_4^-$ ion.

24.7 Write formulas for

(a) potassium pentachloroaquorhodate(III)

(b) sulfatotetraamminecobalt(III) nitrate

(c) sodium dioxotetracyanorhenate(V)

(d) diamminebis-(ethylenediamine)-cobalt(II) chloride

(e) potassium tetracyanonickelate(0)

(f) potassium hexacyanonickelate(II)

(g) tetraamminecopper(II) hexachlorochromate-(III)

24.8 Write formulas for

(a) hexacarbonylvanadium(0)

(b) zinc hexachloroplatinate(IV)

(c) chloronitrotetraaminecobalt(III) sulfate

(d) sodium dithiosulfatoargentate(I)

(e) tetraamineplatinum(II) trichloroammineplatinate(II)

(f) tetachlorodiammineplatinum(IV)

(g) potassium hexabromoaurate(III)

(h) hexaamminenickel(II) hexanitrocobaltate(III)

24.9 Name the following compounds:

(a) $K_4[Ni(CN)_4]$

(b) $K_2[Ni(CN)_4]$

(c) $(NH_4)_2[Fe(H_2O)Cl_5]$

(d) $[Cu(NH_3)_4][PtCl_4]$

(e) $[Ir(NH_3)_5(ONO)]Cl_2$

(f) $[Co(NH_3)_6]_2[Ni(CN)_4]_3$

24.10 Name the following compounds:

(a) $[Co(NO)(CO)_3]$

(b) $[Pt(NH_3)_2Cl_4]$

(c) $K_4[Pt(CN)_4]$

(d) $[Co(NH_3)_6]_4[Co(NO_2)_6]_3$

(e) $Na[Au(CN)_2]$

(f) $[Co(en)_2(SCN)Cl]Cl$

Isomerism of Complexes

24.11 Addition of a solution of potassium hexacyanoferrate(II) to $Fe^{3+}(aq)$ yields a precipitate called Prussian blue. Addition of a solution of potassium hexacyanoferrate(III) to $Fe^{2+}(aq)$ also yields a blue precipitate (Turnbull's blue). The blue precipitates are now known to be identical and are called potassium iron(III) hexacyanoferrate(II). Write the formula for **(a)** Prussian blue, **(b)** iron(III) hexacyanoferrate(III), a brown precipitate, **(c)** copper(II) hexacyanoferrate(II), a purple precipitate, **(d)** potassium iron(II) hexacyanoferrate(II), a white precipitate.

24.12 Two compounds have the same empirical formula: $Co(NH_3)_3(H_2O)_2ClBr_2$. One mole of compound A readily loses one mole of water in a desiccator, whereas compound B does not lose any water under the same conditions. An aqueous solution of A has a conductivity equivalent to that of a compound with two ions per formula unit. The conductivity of an aqueous solution of B corresponds to that of a compound with three ions per formula unit. When $AgNO_3$ is added to a solution of compound A, one mole of AgBr is precipitated per mole of A. A solution of compound B yields two moles of AgBr per mole of B. **(a)** Write the formulas of A and B. **(b)** What type of isomers are A and B?

24.13 Write a formula for an example of **(a)** an ionization isomer of $[Co(NH_3)_5(NO_3)]SO_4$, **(b)** a linkage isomer of $[Mn(CO)_5(SCN)]$ that involves the SCN^- ion, **(c)** a coordination isomer of $[Pt(NH_3)_4][PtCl_4]$, **(d)** a hydrate isomer of $[Co(en)_2(H_2O)_2]Br_3$.

24.14 Write a formula for an example of **(a)** an ionization isomer of $[Pt(NH_3)_4(OH)_2]SO_4$, **(b)** a linkage isomer of $[Pd(dipy)(SCN)_2]$ (dipy is a bidentate ligand—2,2'-dipyridine), **(c)** a hydrate isomer of $[Co(NH_3)_4(H_2O)Cl]Cl_2$, and **(d)** a coordination isomer of $[Cr(NH_3)_4(C_2O_4)][Cr(NH_3)_2(C_2O_4)]$.

---

* The more difficult problems are marked with asterisks. The appendix contains answers to color-keyed problems.

24.15 Identify all the possible isomers (stereoisomers as well as coordination isomers) of the compound $[Pt(NH_3)_4][PtCl_6]$.

24.16 Diagram the structures of all of the possible stereoisomers of each of the following octahedral complexes. Classify the structures as geometric or optical isomers.

(a) $[Cr(NH_3)_2(NCS)_4]^-$

(b) $[Co(NH_3)_3(NO_2)_3]$

(c) $[Co(en)(NH_3)_2Cl_2]^+$

(d) $[Co(en)Cl_4]^-$

(e) $[Co(en)_2ClBr]^+$

(f) $[Cr(NH_3)_2(C_2O_4)_2]^-$

(g) $[Cr(C_2O_4)_3]^{3-}$

24.17 Diagram all the stereoisomers of each of the following molecules. In the formulas, py stands for pyridine, C_5H_5N, a unidentate ligand.

(a) $[Pt(NH_3)(py)ClBr]$ (square-planar)

(b) $[Pt(NH_3)(py)Cl_2]$ (square-planar)

(c) $[Pt(py)_2Cl_2]$ (square-planar)

(d) $[Co(py)_2Cl_2]$ (tetrahedral)

24.18 How can dipole-moment measurement distinguish between the *cis-* and *trans-*isomers of the square-planar $[Pt(NH_3)_2Cl_2]$?

24.19 Diagram the four stereoisomers of $[Pt(en)(NO_2)_2Cl_2]$.

Bonding in Complexes

24.20 The complex $[Ni(CN)_4]^{2-}$ is square-planar, and the complex $[NiCl_4]^{2-}$ is tetrahedral. Refer to Figure 24.17 and the text and predict the number of unpaired electrons in each complex.

24.21 The octahedral complexes $[Fe(CN_6)]^{3-}$ and $[FeF_6]^{3-}$ have one and five unpaired electrons, respectively. Offer an explanation for these observations.

24.22 For Fe^{2+} the electron-pairing energy, P, is approximately 210 kJ/mol. Approximate values of Δ_o for the complexes $[Fe(H_2O)_6]^{2+}$ and $[Fe(CN)_6]^{4-}$ are 120 kJ/mol and 390 kJ/mol, respectively. **(a)** Do these complexes have high-spin or low-spin configurations? **(b)** Draw a *d*-orbital splitting diagram for each.

24.23 For $[Mn(H_2O)_6]^{3+}$, which is a high-spin complex, Δ_o is approximately 250 kJ/mol. For $[Mn(CN)_6]^{3-}$, which is a low-spin complex, Δ_o is approximately 460 kJ/mol. **(a)** What conclusion can you reach in regard to the magnitude of the electron-pairing energy, P, for Mn^{3+}? **(b)** Would you expect the complex $[Mn(C_2O_4)_3]^{3-}$ to be a high-spin or a low-spin complex?

24.24 Draw *d*-orbital splitting diagrams for $[Co(NH_3)_6]^{2+}$ and $[Co(NH_3)_6]^{3+}$. The Δ_o values for these two complexes are approximately 120 kJ/mol and 270 kJ/mol, respectively. The pairing energy of Co^{2+} is approximately 270 kJ/mol and of Co^{3+} is approximately 210 kJ/mol.

24.25 Draw *d*-orbital splitting diagrams for the high-spin and low-spin octahedral complexes of **(a)** Zn^{2+}, **(b)** Cr^{2+}, **(c)** Ni^{2+}, **(d)** Ni^{3+}, **(e)** Mn^{2+}, **(f)** Fe^{3+}, **(g)** V^{2+}, **(h)** V^{3+}, **(i)** V^{4+}, **(j)** V^{5+}.

24.26 Draw *d*-orbital splitting diagrams for **(a)** the octahedral complexes of Cr^{3+}, **(b)** the octahedral complexes of Ni^{2+}, **(c)** the square-planar complexes of Ni^{2+} (all of which are diamagnetic), **(d)** the square-planar complexes of Co^{2+} (all of which have one unpaired electron), **(e)** the tetrahedral complexes of Co^{2+}, **(f)** the octahedral complexes of Ru^{2+} (all of which are diamagnetic), **(g)** the octahedral complexes of Ir^{4+} (all of which have one unpaired electron).

*24.27 The $[Ti(H_2O)_6]^{3+}$ complex is red-violet. What change in color would be expected if the ligands of this complex were replaced by ligands that induce a larger Δ_o? Note that the color of the complex corresponds to light transmitted, not absorbed.

CHAPTER 25

NUCLEAR CHEMISTRY

Ordinary chemical reactions involve only the electrons. In such reactions the nucleus is important only insofar as it influences the electrons. However, matter does undergo important transformations that involve the nucleus directly. The study of nuclear transformations has enlarged our understanding of the nature of matter and the courses of many chemical and biological processes; many technological applications have resulted from these investigations.

When Wilhelm Röntgen discovered X rays in 1895, he noticed that these invisible rays expose photographic plates and cause fluorescent salts to glow. Henri Becquerel in 1896 investigated the hypothesis that salts of this type emit invisible radiation independent of any external stimulus. Becquerel found that a double sulfate of potassium and uranium (which he happened to have on hand) emits radiation capable of exposing a photographic plate that is well protected from light. He subsequently identified uranium as the source of the "radioactive" rays. Following Becquerel's discovery, other radioactive elements were identified and isolated (notably by Marie and Pierre Curie), and the nature of the rays was elucidated (principally by Ernest Rutherford). Radioactive rays originate from transformations that take place within the nucleus.

25.1 The Nucleus

The nuclei of atoms are thought to contain protons and neutrons, particles that are collectively called **nucleons.** The number of protons in a specific atomic nucleus corresponds to the atomic number (or nuclear charge), Z, and the total number of nucleons is given by the mass number, A. Thus, the number of neutrons equals $(A - Z)$. This information is indicated on the chemical symbol of a given nuclide by appending the atomic number as a subscript and the mass number as a super-script. The symbols for the two naturally occurring isotopes of lithium are

$$^6_3\text{Li} \qquad \text{and} \qquad ^7_3\text{Li}$$

Much is known about the structure of the nucleus, but much more remains to be learned. Determination of the radii of a large number of atomic nuclei shows that the radius of a given nucleus, r, is directly related to the cube root of its mass number:

$$r = (1.3 \times 10^{-13} \text{ cm})A^{1/3} \tag{25.1}$$

The volume of a sphere is $\frac{4}{3}\pi r^3$, and if we assume a spherical nucleus, it follows that nuclear volume varies directly with mass number. In other words, the mass of a nucleus determines its volume; nuclear density is therefore approximately constant for all atomic nuclei. This density is about 2.44×10^{14} g/cm^3, an amazingly high value; 1 cm^3 of this nuclear matter would weigh more than 250 million tons.

The density of a liquid is constant and independent of the size of any drop considered. Since all nuclei have approximately the same density, a fluid-droplet model of the nucleus has been proposed. The nature of the cohesive forces holding the nucleus into a nuclear fluid is far from being completely understood.

It is clear, however, that powerful cohesive forces between nucleons exist and that they effectively overcome the electrostatic forces of repulsion between the protons of the nucleus. The attractive forces between nucleons are charge independent, that is, proton-proton, proton-neutron, and neutron-neutron attractions are identical. The range of the force is believed to be on the order of 2×10^{-13} cm, the diameter of a nucleon.

In 1935 Hidekei Yukawa postulated that neutrons and protons are bound together by the very rapid exchange of a nuclear particle, which was identified as a π meson, or pion. Three types of pions are known to exist: positive, π^+, negative, π^-, and neutral, π^0. According to the exchange theory, a neutron, n_A^0, is converted into a proton, p_A^+, by the emission of a negative pion. The emitted π^- is accepted by a proton, p_B^+, which is converted into a neutron, n_B^0:

$$n_A^0 \longrightarrow p_A^+ + \pi^-$$
$$\pi^- + p_B^+ \longrightarrow n_B^0$$

If we consider the deuterium nucleus, which consists of one proton and one neutron, this exchange then reverses, the neutron n_B^0 becoming a proton, p_B^+, and the proton p_A^+ becoming a neutron, n_A^0. The exchange is extremely rapid; approximately 10^{24} transfers occur in one second.

The same type of interaction can be postulated for the positive pion:

$$p_A^+ \longrightarrow n_A^0 + \pi^+$$
$$\pi^+ + n_B^0 \longrightarrow p_B^+$$

or the neutral pion:

$$p_A^+ \longrightarrow p_A^+ + \pi^0$$
$$\pi^0 + n_B^0 \longrightarrow n_B^0$$

The exchange of a neutral pion can also be used to account for the existence of neutron-neutron and proton-proton forces.

In Figure 25.1 the number of neutrons is plotted against the number of protons for the naturally occurring, nonradioactive nuclei. The points, which represent stable combinations of protons and neutrons, lie in what may be called a zone of stability. Nuclei that have compositions represented by points that lie outside of this zone spontaneously undergo radioactive transformations that tend to bring their compositions into or closer to this zone (see Section 25.2).

The stable nuclei of the lighter elements contain approximately equal numbers of neutrons and protons, a neutron/proton ratio of 1. The heavier nuclei contain more neutrons than protons. With increasing atomic number more and more protons are packed into a tiny nucleus, and the electrostatic forces of repulsion

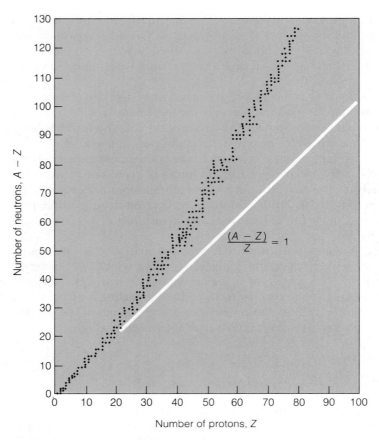

Figure 25.1 Neutron/proton ratio

increase sharply. A larger and larger excess of neutrons is required to diminish the effect of these repulsion forces, and the neutron/proton ratio increases with increasing atomic number until the ratio is approximately 1.5 at the end of the curve of Figure 25.1. There appears to be an upper limit to the number of protons that can be packed into a nucleus, no matter how many neutrons are present. The largest stable nucleus is $^{209}_{83}$Bi; nuclei that are larger than this exist, but all of them are radioactive.

Most naturally occurring stable nuclides have an even number of protons and an even number of neutrons; only five ($^{2}_{1}$H, $^{6}_{3}$Li, $^{10}_{5}$B, $^{14}_{7}$N, and $^{180}_{73}$Ta) have an odd number of protons and an odd number of neutrons (see Table 25.1). For each odd atomic number there are never more than two stable nuclides, whereas for an even atomic number as many as ten stable nuclides may occur. The two elements of atomic number less than 83 that have never been proven to be naturally existing ($_{43}$Tc and $_{61}$Pm) have odd atomic numbers. Empirical observations such as these suggest that there is a periodicity in nuclear structure similar to the periodicity of atomic structure. A nuclear shell model, as yet incompletely developed, has been suggested.

Periodic variations are observed in many nuclear properties. Certain nuclides have relatively high binding energies (which are indicative of comparatively great stability; see Section 25.6) compared with nuclides of close atomic number or mass number. Exceptional nuclear stability is also shown by certain nuclides that have a poor ability to capture neutrons (see Section 25.5). Comparison of

Table 25.1	Distribution of naturally occurring stable nuclides	
Protons	Neutrons	Number of Nuclides
even	even	157 ⎫ 209
even	odd	52 ⎭
odd	even	50 ⎫ 55
odd	odd	5 ⎭

data from these and other studies of nuclear properties indicates that unusual nuclear stability is associated with nuclides having either a number of protons or a number of neutrons equal to a magic number: 2, 8, 20, 28, 50, 82, and 126. It is thought that the magic numbers indicate closed nuclear shells in the same way that the atomic numbers of the noble gases, 2, 10, 18, 36, 54, and 86, indicate stable electronic configurations. In general, elements that have an atomic number equal to a magic number have a larger number of stable isotopes than neighboring elements. The nuclides with a magic number of protons as well as a magic number of neutrons, ^4_2He, $^{16}_8O$, $^{40}_{20}Ca$, and $^{208}_{82}Pb$, have notably high stabilities in terms of neutron capture (see Section 25.5), binding energy (see Section 25.6), and relative abundance.

25.2 Radioactivity

Unstable nuclei spontaneously undergo certain changes that result in the attainment of more stable nuclear compositions. Some unstable nuclides are naturally occurring, others are produced synthetically. Certain synthetic radioactive nuclides undergo some types of radioactive decay that have not been observed for any naturally occurring unstable nuclide.

Alpha emission consists of the ejection of α particles, which have an atomic number of 2 and a mass number of 4 and may be considered to be ^4_2He nuclei. Both synthetic and natural nuclides undergo α decay, which is common only for nuclides of mass number greater than 209 and atomic number greater than 82. Nuclides such as these have too many protons for stability. Points indicating the composition of these nuclides fall outside the plot of Figure 25.1 to the upper right, beyond the zone of stability. The emission of α particles reduces the number of protons by two and the number of neutrons by two and adjusts the composition of the nucleus downward, closer to the stability zone.

An example of α decay is

$$^{210}_{84}Po \longrightarrow {}^{206}_{82}Pb + {}^4_2He$$

Notice that the equation indicates the conservation of mass number (superscripts) and atomic number (subscripts). Equations such as these are written to indicate nuclear changes only; the electrons are customarily ignored. An α particle is emitted as a ^4_2He nucleus without electrons and with a 2+ charge. Subsequent to its emission, however, the particle attracts electrons from other atoms (which become cations), and the α particle becomes a neutral atom of ^4_2He. The $^{206}_{82}Pb$ is left therefore with a surplus of two electrons and a 2− charge; these excess electrons are rapidly lost to surrounding cations.

The energy released by this process is equivalent to the difference in mass between the reactant nucleus ($^{210}_{84}$Po) and the products of the transformation (the product nucleus, $^{206}_{82}$Pb, and an α particle). The masses of neutral *atoms*, rather than the masses of *nuclei*, are generally recorded, but this causes no trouble in the calculation. If the mass of the $^{210}_{84}$Po *atom* is used, the mass of 84 electrons is included; for the products the mass of the $^{206}_{82}$Pb *atom* includes the mass of 82 electrons, and the mass of the $^{4}_{2}$He *atom* includes the mass of two electrons. Thus, the masses of the electrons cancel when atomic masses are employed for the calculation:

(mass $^{210}_{84}$Po) − (mass $^{206}_{82}$Pb + mass $^{4}_{2}$He)
  209.9829 u − (205.9745 u + 4.0026 u) = 0.0058 u

The energy equivalent of a mass difference can be calculated by means of Einstein's equation, $E = mc^2$. Since 1 u is 1.6605×10^{-27} kg and the speed of light (c) is 2.9979×10^8 m/s, the energy equivalent of 1 u is

$$E = mc^2$$
$$= (1.6605 \times 10^{-27} \text{ kg})(2.9979 \times 10^8 \text{ m/s})^2$$
$$= 1.4924 \times 10^{-10} \text{ kg m}^2/\text{s}^2$$
$$= 1.4924 \times 10^{-10} \text{ J}$$

The energy unit customarily employed is the MeV (a megaelectron volt, which is 10^6 electron volts). An electron volt is the energy acquired by an electron when it is accelerated through a potential difference of 1 V:

$$1 \text{ eV} = (\text{charge of electron})(1 \text{ V})$$
$$= (1.6022 \times 10^{-19} \text{ coulomb})(1 \text{ V})$$
$$= 1.6022 \times 10^{-19} \text{ J}$$

$$1 \text{ MeV} = 1.6022 \times 10^{-13} \text{ J}$$

The energy equivalent of 1 u, therefore, is

$$\frac{1.4924 \times 10^{-10} \text{ J/u}}{1.6022 \times 10^{-13} \text{ J/MeV}} = 931.47 \text{ MeV/u}$$

The energy released by the α decay of $^{210}_{84}$Po, therefore, is

$$0.0058 \text{ u} \times 931 \text{ MeV/u} = 5.4 \text{ MeV}$$

If the product nucleus in an α-decay process is left in the ground state, the kinetic energy of the emitted α particle accounts for most of the energy released, and the kinetic energy of the recoiling product nucleus accounts for the remainder. If all the product nuclei are left in the ground state, all the α particles emitted have the same energy. In some α-decay processes, however, several energy groups of α particles are emitted. In these cases the α particles of highest energy correspond to product nuclei in the ground state, and the α particles of lower energies originate from nuclei left in excited states. A nucleus in an excited state subsequently emits energy in the form of γ radiation to reach the ground state. The sum of the kinetic energy of the α particle, the recoil energy of the product nucleus, and the energy of the γ radiation equals the decay energy.

Gamma radiation is electromagnetic radiation of very short wavelength; its emission is caused by energy changes within the nucleus. Its emission alone does not cause changes in the mass number or in the atomic number of the nucleus. At times, nuclides are produced in excited states by nuclear reactions (see Section

25.5), and such nuclides revert to their ground states by the emission of the excess energy in the form of γ radiation:

$$[^{125}_{52}\text{Te}]^* \longrightarrow \quad ^{125}_{52}\text{Te} \quad + \gamma$$

excited state   *ground state*

The γ rays emitted by a specific nucleus have a definite energy value or set of energy values because they correspond to transitions between discrete energy levels of the nucleus. Thus, an emission spectrum of γ radiation is analogous to the line spectrum that results from transitions of electrons between energy levels in an excited atom.

Gamma radiation frequently accompanies all other types of radioactive decay. The following α-decay process is an example:

$$^{240}_{94}\text{Pu} \longrightarrow [^{236}_{92}\text{U}]^* + {}^4_2\text{He}$$

$$[^{236}_{92}\text{U}]^* \longrightarrow {}^{236}_{92}\text{U} + \gamma$$

In cases such as this, deductions can be made concerning the energy levels of the product nucleus. The emission of a 5.16 MeV α particle results directly in the production of $^{236}_{92}\text{U}$ in the ground state; no γ radiation accompanies such α particles. However, for the process outlined in the set of equations, a 5.12 MeV α particle is emitted along with a 0.04 MeV γ ray. It is assumed that the energy of the γ ray corresponds to the energy difference between the ground state and the first excited state of $^{236}_{92}\text{U}$. Alpha particles of other energies are also emitted, and the complete analysis of these energies results in a more detailed picture of the energy levels of the $^{236}_{92}\text{U}$ nucleus.

Beta emission is observed for nuclides that have too high a neutron/proton ratio for stability; points representing such nuclides lie to the left of the zone of stability of Figure 25.1. The β particle is an electron, indicated by the symbol $_{-1}^{0}e$, which may be considered to result from the transformation of a nuclear neutron into a nuclear proton. Electrons, as such, do not exist in the nucleus. The net effect of β emission is that the number of neutrons is decreased by 1 and the number of protons is increased by 1. Thus, the neutron/proton ratio is decreased; the mass number does not change.

Beta decay is a very common mode of radioactive disintegration and is observed for both natural and synthetic nuclides. Examples include

$$^{186}_{73}\text{Ta} \longrightarrow {}^{186}_{74}\text{W} + {}_{-1}^{0}e$$

$$^{82}_{35}\text{Br} \longrightarrow {}^{82}_{36}\text{Kr} + {}_{-1}^{0}e$$

$$^{27}_{12}\text{Mg} \longrightarrow {}^{27}_{13}\text{Al} + {}_{-1}^{0}e$$

$$^{14}_{6}\text{C} \longrightarrow {}^{14}_{7}\text{N} + {}_{-1}^{0}e$$

Notice that the sums of the subscripts and superscripts on the right side of the equation equal the subscript and superscript of the parent nucleus on the left.

The last equation represents the mode of decay of the radioactive carbon isotope that is present in small amount in the atmosphere. The energy released by this process may be calculated by using atomic masses instead of nuclear masses. If we add six orbital electrons to both sides of the equation, we have the equivalent of the *atomic* mass of $^{14}_{6}\text{C}$ (which includes the mass of six electrons) on the left and, taking into account the electron ejected as a β particle, the equivalent of the *atomic* mass $^{14}_{7}\text{N}$ (which includes the mass of seven electrons) on the right. Therefore,

$$(\text{mass } {}^{14}_{6}\text{C}) - (\text{mass } {}^{14}_{7}\text{N})$$
$$14.00324 \text{ u} - 14.00307 \text{ u} = 0.00017 \text{ u} = 0.16 \text{ MeV}$$

In β decay the recoil energy of the product nucleus is negligible since the ejected electron has a small mass. One would expect that the decay energy would be taken up by the kinetic energy of the β particle and that all the β particles emitted would have energies corresponding to this value. Instead, however, a continuous spectrum of β-particle energies is observed with the highest energy almost equal to the decay energy. It is postulated that when a β particle of less than maximum energy is ejected, a **neutrino,** v, that carries off the excess energy is emitted at the same time:

$$ {}^{14}_{6}\text{C} \longrightarrow {}^{14}_{7}\text{N} + {}^{0}_{-1}e + v $$

The neutrino is assumed to be an uncharged particle of a vanishingly small mass.

Unstable nuclides that have neutron/proton ratios below those required for stability (points that fall to the right and below the zone of stability of Figure 25.1) do not occur in nature. Many such artificial nuclides are known, however. Two types of radioactive processes are observed that increase the neutron/proton ratio of this type of nuclide: positron emission and electron capture.

Positron emission, or β^{+} emission, consists of the ejection of a positive electron, which is called a positron and indicated by the symbol ${}^{0}_{1}e$, from the nucleus. A positron has the same mass as an electron but an opposite charge. It arises from the conversion of a nuclear proton into a neutron. Positron emission results in decrease of *one* in the number of protons and an increase of *one* in the number of neutrons; no change in mass number occurs. Hence, positron emission raises the numerical value of the neutron/proton ratio.

Examples of this mode of radioactive decay include

$$ {}^{122}_{53}\text{I} \longrightarrow {}^{122}_{52}\text{Te} + {}^{0}_{1}e $$

$$ {}^{38}_{19}\text{K} \longrightarrow {}^{38}_{18}\text{Ar} + {}^{0}_{1}e $$

$$ {}^{23}_{12}\text{Mg} \longrightarrow {}^{23}_{11}\text{Na} + {}^{0}_{1}e $$

$$ {}^{15}_{8}\text{O} \longrightarrow {}^{15}_{7}\text{N} + {}^{0}_{1}e $$

We can use atomic masses to calculate the energy released by the process described by the last equation. If we add eight orbital electrons to both sides of the equation, the equivalent of the atomic mass of ${}^{15}_{8}\text{O}$ (eight electrons) would be indicated on the left. The equivalent of the atomic mass of ${}^{15}_{7}\text{N}$ (seven electrons) plus the mass of one orbital electron (making up the eight added) plus the mass of the ejected positron would be indicated on the right. Thus, the mass difference is

$$(\text{mass } {}^{15}_{8}\text{O}) - (\text{mass } {}^{15}_{7}\text{N} + \text{mass } {}^{0}_{-1}e + \text{mass } {}^{0}_{1}e)$$

Since the mass of a positron is identical to that of an electron:

$$(\text{mass } {}^{15}_{8}\text{O}) - [\text{mass } {}^{15}_{7}\text{N} + 2(\text{mass } {}^{0}_{-1}e)]$$
$$15.00308 \text{ u} - [15.00011 \text{ u} + 2(0.00055 \text{ u})]$$
$$15.00308 \text{ u} - 15.00121 \text{ u} = 0.00187 \text{ u} = 1.74 \text{ MeV}$$

For spontaneous positron emission the atomic mass of the reactant nuclide must exceed the atomic mass of the product nuclide by at least 0.00110 u, the mass of two electrons.

In positron emission a spectrum of β^+ energies is observed similar to that observed for β emission, and the simultaneous ejection of neutrinos is postulated:

$$^{15}_{8}\text{O} \longrightarrow {}^{15}_{7}\text{N} + {}^{0}_{1}e + \nu$$

The positron was the first antiparticle to be observed. It is similar to the electron in all respects except charge. When a positron and an electron collide, they annihilate each other, and γ radiation equivalent to the masses of the two particles is produced. It is believed that antiparticles exist for all particles except the neutral pion, π^0, and the photon. The negative proton (antiproton) has been detected. The annihilation of proton with antiproton usually produces several high-energy pions. Since the antineutron, like the neutron, has no charge, its detection rests on annihilation phenomena.

Electron capture (ec), sometimes called K capture, is another process through which the neutron/proton ratio of an unstable, proton-rich nuclide may be increased. Unlike positron emission, electron capture can occur when the mass difference between reactant and product nuclides does not exceed 0.00110 u. In this process the nucleus captures an orbital electron from the K or L shell, and the captured electron converts a nuclear proton into a neutron. The transformation results in a product nuclide with *one* less proton and *one* more neutron. Consequently, the atomic number of the product nuclide is *one* less than that of the reactant nuclide, and the mass number does not change. Examples are

$$^{0}_{-1}e + {}^{197}_{80}\text{Hg} \xrightarrow{ec} {}^{197}_{79}\text{Au}$$

$$^{0}_{-1}e + {}^{106}_{47}\text{Ag} \xrightarrow{ec} {}^{106}_{46}\text{Pd}$$

$$^{0}_{-1}e + {}^{37}_{18}\text{Ar} \xrightarrow{ec} {}^{37}_{17}\text{Cl}$$

$$^{0}_{-1}e + {}^{7}_{4}\text{Be} \xrightarrow{ec} {}^{7}_{3}\text{Li}$$

$$^{0}_{-1}e + {}^{55}_{26}\text{Fe} \xrightarrow{ec} {}^{55}_{25}\text{Mn}$$

The energy released can be calculated directly from the atomic masses of the reactant and product nuclides. For the last example,

(mass ${}^{55}_{26}\text{Fe}$) − (mass ${}^{55}_{25}\text{Mn}$)
54.93830 u − 54.93805 u = 0.00025 u = 0.23 MeV

The recoil energy of the product nucleus is negligible, and if the product nucleus is left in the ground state, all the available energy is carried away by a neutrino that is ejected in the process. In addition, electron capture is accompanied by the production of X rays. The capture of an orbital electron leaves a vacancy in the K or L shell, and when an outer electron falls into this vacancy, X-ray emission follows.

25.3 Rate of Radioactive Decay

Many techniques are employed to study the emissions of radioactive substances. Radiations from these materials affect photographic film in the same way that

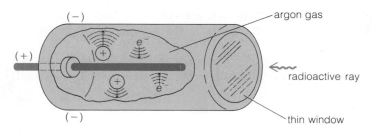

Figure 25.2   Essential features of a Geiger-Müller counter tube

ordinary light does. Photographic techniques for the qualitative and quantitative detection of radiation are employed, but they are not very accurate nor are they suitable for rapid analysis.

The energy of emissions from radioactive sources is absorbed by some materials (for example, zinc sulfide) and transformed into radiant energy of visible wavelength. The zinc sulfide is said to **fluoresce,** and a little flash of light may be observed from the impact of each particle from the radioactive source. This property has been put to use in an instrument known as a **scintillation counter.** The window of a sensitive photoelectric tube is coated with ZnS, and the flashes of light emitted by the ZnS when it is struck by radioactive emissions cause pulses of electric current to pass through the photoelectric tube. These signals are amplified and made to operate various kinds of counting devices.

The **Wilson cloud chamber** enables the path of ionizing radiation to be seen. The chamber contains air saturated with water vapor. By the movement of a piston the air in the chamber is suddenly expanded and cooled; this causes droplets of water to condense on the ions that are formed by the particles as they move through the vapor, and makes the paths of the particles visible. Photographs of these cloud tracks may be made and studied. Such photographs provide information on the length of the paths, collisions undergone by the particles, the speed of the particles, and the effects of external forces on the behavior of the particles.

The essential features of a **Geiger-Müller counter** are diagrammed in Figure 25.2. The radiation enters the tube through a thin window. As a particle or a γ ray traverses the tube, which contains argon gas, it knocks electrons off the argon atoms in its path and forms Ar^+ ions. A potential of about 1000 to 1200 V is applied between the electrodes of the tube, and the electrons and Ar^+ ions cause a pulse of electric current to flow through the circuit. This pulse is amplified to cause a clicker to sound or an automatic counting device to operate.

The rates of decay of all radioactive substances have been found to be first order (see Section 12.2) and to be independent of temperature. This lack of temperature dependence implies that the activation energy of any radioactive-decay process is zero. The rate of decay, therefore, depends only upon the amount of radioactive material present. If we let N equal the number of atoms of radioactive material, ΔN is the number of atoms that disintegrate in a time interval, Δt:

$$-\frac{\Delta N}{\Delta t} = kN \tag{25.2}$$

where k is the rate constant. The rate expression is negative because it represents the disappearance of the radioactive substance.

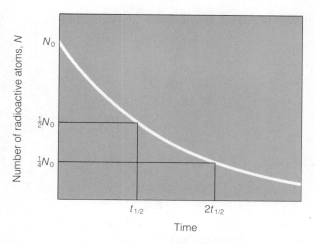

Figure 25.3   Curve showing the rate of a radioactive decay

Rearrangement of the rate expression gives

$$-\frac{\Delta N}{N} = k\Delta t \tag{25.3}$$

which states that the fraction lost $(-\Delta N/N)$ in a given time interval (Δt) is directly proportional to the length of the interval. The time required for half of the sample to decay depends upon a time interval known as the **half-life**, $t_{1/2}$ which is a constant.

The curve of Figure 25.3, which shows number of radioactive atoms versus time, is typical of first-order processes. Let N_0 equal the number of radioactive atoms present at the start. After a single half-life period has elapsed, one-half of the original number of atoms remain ($\frac{1}{2}N_0$). This number is reduced by half (to $\frac{1}{4}N_0$) by the time another half-life has passed. Each radioactive isotope has a characteristic half-life, and these vary widely. For example, $_3^5\text{Li}$ has a half-life estimated to be 10^{-21} s, and $_{92}^{238}\text{U}$ has a half-life of 4.51×10^9 years.

The rate equation may be written in its differential form:

$$-\frac{dN}{dt} = kN \tag{25.4}$$

and by means of the calculus this equation may be integrated:

$$-2.303 \log\left(\frac{N}{N_0}\right) = kt \tag{25.5}$$

where N_0 is the amount present at time zero, and N is the amount present at time t. The last equation may be rearranged to give

$$\log\left(\frac{N_0}{N}\right) = \frac{kt}{2.303} \tag{25.6}$$

After a half-life period the number of radioactive atoms left is equal to $\frac{1}{2}N_0$:

$$N = \tfrac{1}{2}N_0$$

Therefore,

$$\frac{N_0}{N} = 2$$

Substitution into Equation 25.6 gives

$$\log\left(\frac{N_0}{N}\right) = \frac{kt}{2.303}$$

$$\log 2 = \frac{kt_{1/2}}{2.303}$$

$$t_{1/2} = \frac{2.303 \log 2}{k}$$

$$t_{1/2} = \frac{0.693}{k} \qquad (25.7)$$

According to the rate law, a radioactive substance never completely disappears. The rate law describes the behavior of a sample containing a large number of atoms. When the number of atoms in the sample declines to a very small value, the rate law is no longer followed exactly. How can a fraction of an atom disintegrate?

Example 25.1

The isotope $^{60}_{27}\text{Co}$ has a half-life of 5.27 years. What amount of a 0.0100 g sample of $^{60}_{27}\text{Co}$ remains after 1.00 year?

Solution

The rate constant for this disintegration is

$$k = \frac{0.693}{t_{1/2}}$$

$$= \frac{0.693}{5.27 \text{ years}} = 0.132/\text{year}$$

The fraction remaining undecomposed at the end of 1.00 year may be found in the following way:

$$\log\left(\frac{N_0}{N}\right) = \frac{kt}{2.30}$$

$$= \frac{(0.132/\text{year})(1.00 \text{ year})}{2.30} = 0.0574$$

$$\left(\frac{N_0}{N}\right) = \text{antilog}\, 0.0574 = 1.14$$

$$\frac{N}{N_0} = \frac{1}{1.14} = 0.877$$

Therefore, the amount remaining after 1.00 year is

$$0.877 \times 0.0100\ \text{g} = 0.00877\ \text{g}$$

The radioactive isotope $^{14}_{6}C$ is produced in the atmosphere by the action of cosmic-ray neutrons on $^{14}_{7}N$:

$$^{14}_{7}N + ^{1}_{0}n \longrightarrow ^{14}_{6}C + ^{1}_{1}H$$

The $^{14}_{6}C$ is oxidized to CO_2, and this radioactive CO_2 mixes with nonradioactive CO_2. The radiocarbon disappears through radioactive decay, but it is also constantly being made. The steady state is reached when the proportion is one atom of $^{14}_{6}C$ to 10^{12} ordinary carbon atoms, which represents 15.3 ± 0.1 disintegrations per minute per gram of carbon.

The CO_2 of the atmosphere is absorbed by plants through the process of photosynthesis, and the ratio of $^{14}_{6}C$ to ordinary carbon in plant materials that are alive and growing is the same ratio as that in the atmosphere. When the plant dies, however, the amount of $^{14}_{6}C$ diminishes through radioactive decay and is not replenished by the assimilation of atmospheric CO_2 by the plant.

The half-life of $^{14}_{6}C$, a β emitter, is 5770 years. The age of a wooden object can be determined by comparing the radiocarbon activity of the object with that of growing trees. This method of **radiocarbon dating** has been applied to many archeological finds and objects of historical interest. By this means the Dead Sea scrolls were determined to be 1917 ± 200 years old.

Example 25.2

A sample of carbon from a wooden artifact is found to give 7.00 $^{14}_{6}C$ counts per minute per gram of carbon. What is the approximate age of the artifact? The $^{14}_{6}C$ from wood recently cut down decays at the rate of 15.3 disintegrations per minute per gram of carbon.

Solution

The half-life of $^{14}_{6}C$ is 5770 years. Therefore,

$$k = \frac{0.693}{t_{1/2}} = \frac{0.693}{5770\ \text{years}} = 1.20 \times 10^{-4}/\text{year}$$

Since the $^{14}_{6}C$ from wood recently cut down decays at the rate of 15.3 disintegrations per minute per gram of carbon,

$$\log\left(\frac{N_0}{N}\right) = \frac{kt}{2.30}$$

$$\log\left(\frac{15.3 \text{ disintegrations/min}}{7.00 \text{ disintegrations/min}}\right) = \frac{(1.20 \times 10^{-4}/\text{year})t}{2.30}$$

$$t = \frac{2.30 \log 2.19}{1.20 \times 10^{-4}/\text{year}}$$

$$= 6520 \text{ years}$$

The amount of radiation emanating from a source per unit time is termed the **activity** of the source:

$$\text{activity} = -\frac{dN}{dt} = kN \tag{25.8}$$

Activities are generally expressed in curies; 1 curie (Ci) is defined as 3.70×10^{10} disintegrations per second, and 1 microcurie (μCi) is 3.70×10^4 disintegrations per second.

Example 25.3

The half-life of $^{100}_{43}\text{Tc}$, a β emitter, is 16 s. How many atoms of $^{100}_{43}\text{Tc}$ are present in a sample with an activity of 0.200 μCi? What is the mass of the sample?

Solution

For this radioactive decay the rate constant, k, is

$$k = \frac{0.693}{t_{1/2}}$$

$$= \frac{0.693}{16 \text{ s}} = 0.0433/\text{s}$$

The activity of the sample in terms of disintegrations per second is

$$0.200(3.70 \times 10^4 \text{ disintegrations/s}) = 7.40 \times 10^3 \text{ disintegrations/s}$$

This value represents the decay of 7.40×10^3 atoms/s and

$$\text{activity} = kN$$

$$(7.40 \times 10^3 \text{ atoms/s}) = (4.33 \times 10^{-2}/\text{s})N$$

$$N = 1.71 \times 10^5 \text{ atoms}$$

The mass of the sample can be derived from the fact that the atomic weight of ^{100}Tc to three significant figures is 100; thus,

$$? \text{ g Tc} = 1.71 \times 10^5 \text{ atoms Tc}\left(\frac{100 \text{ g Tc}}{6.02 \times 10^{23} \text{ atoms Tc}}\right)$$

$$= 2.84 \times 10^{-19} \text{ g Tc}$$

25.4 Radioactive Disintegration Series

The examples of α, β^-, and β^+ emission and of electron capture given in Section 25.2 are one-step processes that lead to stable nuclides. Frequently, however, the nucleus produced by a radioactive process is itself radioactive. The repetition of this situation creates a chain, or series, of disintegration processes involving many radioactive nuclides and leading ultimately to the production of a stable nuclide.

Three such disintegration series, which involve only α and β emission, occur in nature: the $^{232}_{90}\text{Th}$, $^{238}_{92}\text{U}$, and $^{235}_{92}\text{U}$ series. The $^{238}_{92}\text{U}$ series, which leads finally to the stable nuclide $^{206}_{82}\text{Pb}$, is diagrammed in Figure 25.4. Branching occurs in several places in the series, and the series proceeds by two different routes. The branches, however, always rejoin at a later point, and commonly one branch is preferred over the other (note the percentage figures in Figure 25.4). By any given route the $^{238}_{92}\text{U}$ series consists of 14 steps—8 involving α decay and 6 involving β decay.

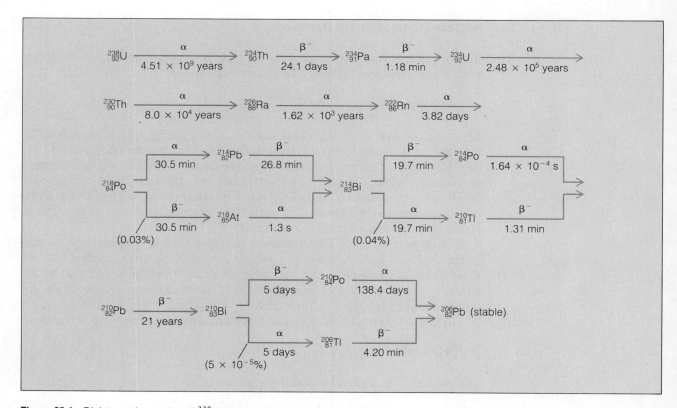

Figure 25.4 Disintegration series of $^{238}_{92}\text{U}$ (half-lives of isotopes are indicated)

The radioactive nuclides that occur in nature are those that have very long half-lives or those that are constantly being produced by the disintegration of other nuclides. This last type of naturally occurring radioactive nuclide eventually exists in a steady state at which time the amount of the nuclide remains essentially constant because the material is being produced at the same rate that it is decomposing.

The natural disintegration series serve as the basis of a method of geological dating. A sample of rock may be analyzed for its $^{206}_{82}Pb$ and $^{238}_{92}U$ content, and the length of time to produce this ratio of lead to uranium may be calculated from the decay constants of the series. Results from studies such as these place the age of some rocks at from 3 to 3.5 billion years; the age of the earth is estimated to be 4.5 billion years.

Decay series, some more elaborate than others, are known for artificial nuclides. Position emission and K capture are observed in some of these series as well as α and β emission. Following are three examples of simple two-step disintegration series of artificial nuclides:

$$^{20}_{8}O \xrightarrow[14\,s]{\beta^-} {}^{20}_{9}F \xrightarrow[11\,s]{\beta^-} {}^{20}_{10}Ne$$

$$^{30}_{16}S \xrightarrow[1.4\,s]{\beta^+} {}^{30}_{15}P \xrightarrow[2.6\,min]{\beta^+} {}^{30}_{14}Si$$

$$^{76}_{36}Kr \xrightarrow[10\,hours]{ec} {}^{76}_{35}Br \xrightarrow[16.5\,hours]{\beta^+\ or\ ec} {}^{76}_{34}Se$$

25.5 Nuclear Bombardment Reactions

In 1915 Ernest Rutherford reported that the following transformation occurs when α particles from $^{214}_{84}Po$ are passed through nitrogen:

$$^{14}_{7}N + {}^{4}_{2}He \longrightarrow {}^{17}_{8}O + {}^{1}_{1}H$$

This was the first artificial transmutation of one element into another to be reported; in the years following, thousands of such nuclear transformations have been studied. It is assumed that the projectile (in this case, an α particle) forms a compound nucleus with the target ($^{14}_{7}N$) and that the compound nucleus very rapidly ejects a subsidiary particle ($^{1}_{1}H$) to form the product nucleus ($^{17}_{8}O$). Other projectiles, such as neutrons, deuterons ($^{2}_{1}H$), protons, and ions of low atomic number, are used in addition to α particles.

Particle-particle reactions are usually classified according to the type of projectile employed and the subsidiary particle ejected. Thus, the preceding reaction is called an (α, p) reaction, and the complete transformation is indicated by the notation $^{14}_{7}N\ (\alpha, p)\ ^{17}_{8}O$. Examples of several of the more common types of nuclear transformations are listed in Table 25.2.

The first artificial, radioactive nuclide produced was made by the (α, n) reaction:

$$^{27}_{13}Al + {}^{4}_{2}He \longrightarrow {}^{30}_{15}P + {}^{1}_{0}n$$

The product, $^{30}_{15}P$, decays by positron emission:

$$^{30}_{15}P \longrightarrow {}^{30}_{14}Si + {}^{0}_{1}e$$

Table 25.2 Examples of nuclear reactions

Type	Reaction	Radioactivity of Product Nuclide
(α, n)	$^{75}_{33}\text{As} + {}^{4}_{2}\text{He} \longrightarrow {}^{78}_{35}\text{Br} + {}^{1}_{0}\text{n}$	β^{+}
(α, p)	$^{106}_{46}\text{Pd} + {}^{4}_{2}\text{He} \longrightarrow {}^{109}_{47}\text{Ag} + {}^{1}_{1}\text{H}$	stable
(p, n)	$^{7}_{3}\text{Li} + {}^{1}_{1}\text{H} \longrightarrow {}^{7}_{4}\text{Be} + {}^{1}_{0}\text{n}$	ec
(p, γ)	$^{14}_{7}\text{N} + {}^{1}_{1}\text{H} \longrightarrow {}^{15}_{8}\text{O} + \gamma$	β^{+}
(p, α)	$^{9}_{4}\text{Be} + {}^{1}_{1}\text{H} \longrightarrow {}^{6}_{3}\text{Li} + {}^{4}_{2}\text{He}$	stable
(d, p)	$^{31}_{15}\text{P} + {}^{2}_{1}\text{H} \longrightarrow {}^{32}_{15}\text{P} + {}^{1}_{1}\text{H}$	β^{-}
(d, n)	$^{209}_{83}\text{Bi} + {}^{2}_{1}\text{H} \longrightarrow {}^{210}_{84}\text{Po} + {}^{1}_{0}\text{n}$	α
(n, γ)	$^{59}_{27}\text{Co} + {}^{1}_{0}\text{n} \longrightarrow {}^{60}_{27}\text{Co} + \gamma$	β^{-}
(n, p)	$^{45}_{21}\text{Sc} + {}^{1}_{0}\text{n} \longrightarrow {}^{45}_{20}\text{Ca} + {}^{1}_{1}\text{H}$	β^{-}
(n, α)	$^{27}_{13}\text{Al} + {}^{1}_{0}\text{n} \longrightarrow {}^{24}_{11}\text{Na} + {}^{4}_{2}\text{He}$	β^{-}

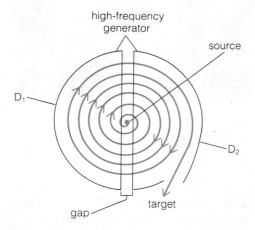

Figure 25.5 Path of a particle in a cyclotron

Except for the nature of the product there is no difference between nuclear reactions that produce stable nuclides and those that yield radioactive nuclides.

Projectile particles that bear a positive charge are repelled by target nuclei; this is particularly true of the heavier nuclei, which have high charges. Consequently, only a small number of nuclear transformations can be brought about by the positive particles emitted by radioactive sources. Various particle accelerators are used to give protons, deuterons, α particles, and other cationic projectiles sufficiently high kinetic energies to overcome the electrostatic repulsions of the target nuclei. The cyclotron (see Figure 25.5) is one such instrument.

The ion source is located between two hollow D-shaped plates (D_1 and D_2), called dees, that are separated by a gap. The dees are enclosed in an evacuated chamber located between the poles of a powerful electromagnet (not shown in the figure). A high-frequency generator keeps the dees oppositely charged. Under the influence of the magnetic and electrical fields the ions move from the source in a circular path. Each time they reach the gap between the dees, the polarity of the dees is reversed. Thus, the positively charged particles are pushed out of a

An overview of the Super-HILAC (Heavy Ion Linear Accelerator) at the Lawrence Berkeley Laboratory. This instrument will accelerate ions of all the elements in the periodic chart. *University of California, Lawrence Berkeley Laboratory.*

positive dee and attracted into a negative dee. Each time they traverse the gap, therefore, they are accelerated. Because of this, the particles travel an ever-increasing spiral path; eventually they penetrate a window in the instrument and, moving at extremely high speed, strike a target.

The **linear accelerator** (see Figure 25.6) operates in much the same way except that no magnetic field is employed. The particles are accelerated through a series of tubes enclosed in an evacuated chamber. A positive ion from the source is attracted into tube 1, which is negatively charged. At this time, the odd-numbered tubes have negative charges and the even-numbered tubes have positive charges. As the particle emerges from tube 1, the charges of the tubes are reversed so that the even-numbered tubes are now negatively charged. The particle is repelled out of tube 1 (now positive) and attracted into tube 2 (now negative); as a result, it is accelerated.

Each time the particle leaves one tube to enter another, the charges of the tubes are reversed. Since the polarity of the tubes is reversed at a constant time interval, and since the speed of the particle increases constantly, each tube must be longer than the preceding one. The accelerated particles leave the last tube at high speed and strike the target.

Neutrons are particularly important projectiles because they bear no charge and therefore are not repelled by the positive charge of the target nuclei. A mixture of beryllium and an α emitter (such as $^{222}_{86}\text{Rn}$) is a convenient neutron source:

$$^{9}_{4}\text{Be} + {}^{4}_{2}\text{He} \longrightarrow {}^{12}_{6}\text{C} + {}^{1}_{0}\text{n}$$

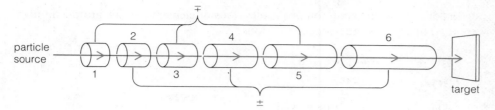

Figure 25.6 Schematic representation of a linear accelerator

This reaction was used by James Chadwick in his experiments that characterized the neutron (1932). The bombardment of beryllium by accelerated deuterons from a cyclotron is a more intense source of neutrons:

$$\ce{^{9}_{4}Be + ^{2}_{1}H -> ^{10}_{5}B + ^{1}_{0}n}$$

A very important source of neutrons is the nuclear reactor (see Section 25.6).

Neutrons from nuclear reactions are known as **fast neutrons**; they cause reactions in which a subsidiary particle is ejected [such as (n, α) and (n, p) reactions]. **Slow neutrons,** or **thermal neutrons,** are produced when the neutrons derived from a nuclear reaction are passed through a moderator (such as carbon, paraffin, hydrogen, deuterium, or oxygen). Through collisions with the nuclei of the moderator, the kinetic energies of the neutrons are decreased to values approximating those of ordinary gas molecules. Bombardments using slow neutrons bring about (n, γ) reactions, which are also called **neutron-capture reactions** since no subsidiary particle is ejected:

$$\ce{^{34}_{16}S + ^{1}_{0}n -> ^{35}_{16}S + \gamma}$$

Isotopes of practically every element have been prepared by this type of reaction.

Nuclear reactions have been used to prepare isotopes belonging to elements that do not exist in nature or that exist in extremely minute concentrations. Thus, isotopes of technetium and astatine have been prepared by the reactions

$$\ce{^{96}_{42}Mo + ^{2}_{1}H -> ^{97}_{43}Tc + ^{1}_{0}n}$$

$$\ce{^{209}_{83}Bi + ^{4}_{2}He -> ^{211}_{85}At + 2 ^{1}_{0}n}$$

The elements following uranium in the periodic classification are called the **transuranium elements**; none of these elements are naturally occurring, but many have been made by nuclear reactions. Some of these reactions use targets of artificial nuclides, and in these cases the final products are therefore the result of syntheses consisting of several steps. Examples of these preparations are

$$\ce{^{238}_{92}U + ^{1}_{0}n -> ^{239}_{92}U + \gamma}$$

$$\ce{^{239}_{92}U -> ^{239}_{93}Np + ^{0}_{-1}e}$$

$$\ce{^{239}_{93}Np -> ^{239}_{94}Pu + ^{0}_{-1}e}$$

$$\ce{^{239}_{94}Pu + ^{2}_{1}H -> ^{240}_{95}Am + ^{1}_{0}n}$$

$$\ce{^{239}_{94}Pu + ^{4}_{2}He -> ^{242}_{96}Cm + ^{1}_{0}n}$$

A 3×10^{-7} g sample of californium oxychloride, the first pure californium compound to be isolated in the laboratory (magnified 170 times). The compound gives off its own light as a result of radioactive decay. *University of California, Lawrence Berkeley Laboratory.*

In addition to the common projectiles, ions of elements of low atomic number are used in some bombardment reactions:

$$^{238}_{92}U + ^{12}_{6}C \longrightarrow ^{244}_{98}Cf + 6\,^{1}_{0}n$$

$$^{238}_{92}U + ^{14}_{7}N \longrightarrow ^{246}_{99}Es + 6\,^{1}_{0}n$$

$$^{238}_{92}U + ^{16}_{8}O \longrightarrow ^{255}_{100}Fm + 4\,^{1}_{0}n$$

$$^{252}_{98}Cf + ^{10}_{5}B \longrightarrow ^{257}_{103}Lr + 5\,^{1}_{0}n$$

25.6 Nuclear Fission and Fusion

Nuclear fission reactions are more famous and infamous than particle-particle or particle-capture transformations. In a fission process a heavy nucleus is split into nuclei of lighter elements and several neutrons. The fission of a given nuclide results in more than one set of products. Two possible reactions for the slow neutron-induced fission of $^{235}_{92}U$ are

$$^{235}_{92}U + ^{1}_{0}n \longrightarrow ^{93}_{36}Kr + ^{140}_{56}Ba + 3\,^{1}_{0}n$$

$$^{235}_{92}U + ^{1}_{0}n \longrightarrow ^{90}_{38}Sr + ^{144}_{54}Xe + 2\,^{1}_{0}n$$

Heavy nuclei have much larger neutron/proton ratios than nuclei of moderate mass (see Figure 25.1). The neutrons released in the fission process lower the neutron/proton ratios of the product nuclei; nevertheless, these nuclei are generally radioactive, and further adjustment of the neutron/proton ratios occurs, usually through β emission. Certain fissions can be induced by protons, deuterons, or α particles, but the most important are those that are brought about by neutrons.

The **binding energy** of a nucleus may be considered to be the energy *required* to pull the nucleons of the nucleus apart or the energy *released* by the hypothetical formation of the nucleus by the condensation of the individual nucleons. The binding energy of a nucleus may be calculated from the difference between the sum of the masses of the constituent nucleons and the mass of the corresponding nucleus. The sum of the masses of the nucleons of the $^{35}_{17}Cl$ nucleus (17 protons plus 18 neutrons) is 0.320 u greater than the actual mass of this nucleus. The energy equivalent of this mass is

$$0.320 \text{ u} \times 931 \text{ MeV/u} = 298 \text{ MeV}$$

The magnitude of the binding energy of a given nucleus indicates the stability of that nucleus toward radioactive decay. For the purpose of comparison, the values are usually given in terms of binding energy per nucleon, and the largest values are characteristic of the most stable nuclei. The binding energy per nucleon in the case of $^{35}_{17}Cl$ is

$$\frac{298 \text{ MeV}}{35 \text{ nucleons}} = 8.51 \text{ MeV/nucleon}$$

In Figure 25.7 mass number is plotted against binding energy per nucleon for the nuclides. Inspection of the curve shows that nuclides of intermediate mass have larger values of binding energy per nucleon than the heavier nuclides. Thus,

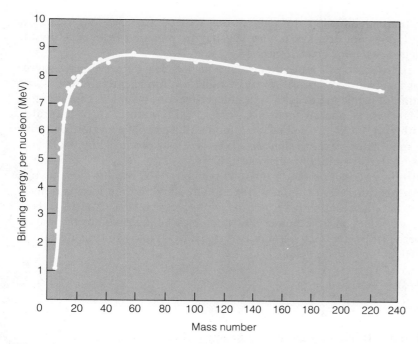

Figure 25.7 Binding energy per nucleon versus mass number

the fission of $^{235}_{92}U$ produces lighter nuclei with higher binding energy/nucleon values, and energy is liberated. The sum of the masses of the products of a fission reaction is less than the sum of the masses of the reactants. A typical fission of a single $^{235}_{92}U$ nucleus releases approximately 200 MeV.

The fission of each $^{235}_{92}U$ nucleus is induced by a single neutron and produces several neutrons; in the overall process an average of approximately 2.5 neutrons are produced per fission. If each neutron causes the fission of another $^{235}_{92}U$ nucleus, an explosively rapid chain reaction ensues. If a reproduction factor of 2 is used for simplicity, the first fission would cause 2 first-generation fissions. Each of these would cause 2 fissions—a total of 4 second-generation fissions; these would be followed by 8, 16, 32, . . . , fissions in succeeding generations. In the nth generation, 2^n fissions would occur. Since each fission is extremely rapid, an explosion results. If 200 MeV is released by each fission, a tremendous amount of energy is liberated.

In a small amount of $^{235}_{92}U$ undergoing fission, many of the neutrons produced are lost from the surface of the mass before they can bring about nuclear reactions. If the size of the fissionable material exceeds a certain **critical mass,** however, the neutrons are captured before they can leave the mass and an explosive chain reaction results. An atomic bomb is detonated by bringing two pieces of fissionable material, each of subcritical size, together into one piece of supercritical size. The reaction is started by a stray neutron.

A number of nuclei, such as $^{238}_{92}U$, $^{231}_{91}Pa$, $^{237}_{93}Np$, and $^{232}_{90}Th$, undergo fission with fast, but not with slow, neutrons. Slow neutrons, however, can induce fission in $^{233}_{92}U$, $^{235}_{92}U$ and $^{239}_{94}Pu$. The latter two nuclides are those commonly employed as nuclear fuels. The nuclide $^{235}_{92}U$, which constitutes only about 0.7% of natural uranium, was employed in the construction of the first atomic bomb. The separation of $^{235}_{92}U$ from the principal uranium isotope, $^{238}_{92}U$, was carried out in many ways, the most successful being the separation of gaseous $^{235}UF_6$ from

$^{238}UF_6$ by thermal diffusion through porous barriers (see Section 8.12). The nuclide $^{239}_{94}Pu$ does not occur in nature but may be prepared from $^{238}_{92}U$ by the series of nuclear reactions given in Section 25.5.

The controlled fission of nuclear fuels in a **nuclear reactor** serves as a source of energy or as a source of neutrons and γ radiation for scientific research and the production of artificial nuclides. In a nuclear reactor, cylinders containing at least a critical mass of nuclear fuel ($^{235}_{92}U$, $^{233}_{92}U$, $^{239}_{94}Pu$, natural U, or natural U enriched with $^{235}_{92}U$) are surrounded by a moderator (graphite, water, or heavy water). The moderator serves to slow down the neutrons produced by the fission, thus increasing the possibility of their capture.

The neutron reproduction factor must be maintained at a value close to 1. If it falls below this value, the reaction will eventually stop, and if it increases much above this value, the reaction may become explosively violent. Control rods of cadmium, boron steel, or other materials, which may be inserted into the reactor to any desired depth, serve to capture excess neutrons and control the rate of the reaction:

$$^{113}_{48}Cd + ^1_0n \longrightarrow ^{114}_{48}Cd + \gamma$$

$$^{10}_5B + ^1_0n \longrightarrow ^{11}_5B + \gamma$$

The $^{114}_{48}Cd$ and $^{11}_5B$ nuclides are not radioactive. Provision is made for the insertion of research samples into the reactor.

Early reactors that used natural uranium as a fuel and graphite blocks as a moderator were large—cubes on the order of 20 to 25 feet on an edge. The use of enriched fuels and moderators other than graphite has reduced the size required considerably. **Breeder reactors** produce a fissionable nuclide (for example, $^{239}_{94}Pu$ from $^{238}_{92}U$) within the reactor and thus provide for the maintenance of a supply of nuclear fuel. Nuclear reactors that are designed as sources of energy employ a circulating coolant to remove the heat from the reactor to the outside where it is used in power production. All nuclear reactors must be heavily shielded.

Nuclear **fusion** is a process in which very light nuclei are fused into a heavier nucleus. The curve of Figure 25.7 shows that such processes should liberate more energy than fission processes do. The hydrogen bomb is based on nuclear fusion. Extremely high activation energies are required for fusion reactions, and in actual practice, a fission bomb is used to supply the high temperatures required. The energy of the sun is believed to be derived from the conversion of hydrogen nuclei into helium by nuclear fusion. Reactions such as the following are postulated:

$$^1_1H + ^1_1H \longrightarrow ^2_1H + ^0_1e$$

$$^2_1H + ^1_1H \longrightarrow ^3_2He + \gamma$$

$$^3_2He + ^3_2He \longrightarrow ^4_2He + 2^1_1H$$

The liquid drop model of the nucleus interprets fusion and fission in terms of the forces of repulsion and attraction between the nucleons in the nucleus. The size of the nucleus is critical in determining the total effect of these two opposing forces. The nucleons on the surface of the nucleus are subjected to fewer forces of attraction than the nucleons that are surrounded on all sides by other nucleons in the center of the nucleus. Therefore, the total effect of the attractive forces is increased when the surface area is decreased. One large drop has a smaller surface area than several small drops comprising the same amount of nuclear fluid.

The world's largest superconductivity magnets to be used in atomic fusion research. The strong magnetic field produced by these magnets (150,000 times that of the earth) will be used to confine the superhot fusion fuel. The magnets are constructed of superconducting niobium-titanium cable cooled by liquid helium to about 4 K. They consume less than 0.003% of the electric power that conventional water-cooled copper magnets of this size would consume. *University of California, Lawrence Livermore National Laboratory.*

At the same time, the electrostatic forces of repulsion between protons tend to disrupt a large drop into smaller ones. Whether the process of fusion or fission occurs depends upon the balance between these two effects, which do not vary with the size of the drop in a parallel manner. Since the surface area of a sphere is $4\pi r^2$, the increase or decrease in the energy of attraction brought about by a change in surface area is proportional to the square of the radius, r^2 (or $A^{2/3}$). The coulombic forces of repulsion, however, are proportional to the square of the nuclear charge divided by the radius, Z^2/r (or $Z^2/A^{1/3}$). For light nuclei the increase in the effect of the attractive forces brought about by the reduction in nuclear surface area more than offsets the increase in the effect of the forces of repulsion, and fusion occurs. For heavy nuclei the reverse applies; fission occurs since it leads to a reduction in the effect of the forces of repulsion that is of more significance than the accompanying reduction in the effect of the forces of attraction.

25.7 Uses of Radioactive Nuclides

A large number of uses have been developed for the nuclides that are the products of the processes run in nuclear reactors. Thickness gauges have been developed in which a radioactive source is placed on one side of the material to be tested (cigarettes, metal plates, and so forth) and a counting device on the other. The amount of radiation reaching the counter is a measure of the thickness of the material.

The effectiveness of lubricating oils is measured in an engine constructed from metal into which radioactive nuclides have been incorporated. After the engine has been run for a fixed time period, the oil is withdrawn and tested for the presence of radioactive particles accumulated through engine wear.

When a single pipeline is used to transfer more than one petroleum derivative, a small amount of a radioactive nuclide is placed in the last portion of one substance to signal its end and the start of another. The radiation may be used to activate an automatic valve system so that the liquids are diverted into different tanks.

The use of radiocarbon in determining the age of materials of plant origin has been mentioned. Photosynthesis has been studied by tracing the absorption of CO_2 containing $^{14}_{6}C$ by plants. In this way much has been learned about the conversion of CO_2 into sugars, starches, and cellulose.

Radioactive nuclides have found many uses in medical research, therapy, and diagnosis. The radiations from $^{60}_{27}Co$, a β and γ emitter, are used in cancer therapy. The nuclide $^{131}_{53}I$ is used in the diagnosis and treatment of thyroid disorders since this gland concentrates ingested iodine.

Radioactive tracers have found wide use in chemical studies. The following structure of the thiosulfate ion,

$$\left[\begin{array}{c} S \\ | \\ O-S-O \\ | \\ O \end{array} \right]^{2-}$$

is indicated by studies using $^{35}_{16}S$. The thiosulfate ion is prepared by heating a sulfite with the radioactive sulfur:

$$^{35}S(s) + SO_3^{2-}(aq) \longrightarrow [^{35}SSO_3]^{2-}(aq)$$

Upon the acidification of this thiosulfate solution, the radioactive sulfur is quantitatively precipitated; none is found in the resulting SO_2 gas:

$$[^{35}SSO_3]^{2-}(aq) + 2\,H^+(aq) \longrightarrow {}^{35}S(s) + SO_2(g) + H_2O$$

In addition, upon the decomposition of silver thiosulfate derived from this ion, all the radioactive sulfur ends in the silver sulfide:

$$Ag_2[^{35}SSO_3](s) + H_2O \xrightarrow{heat} Ag_2{}^{35}S(s) + SO_4^{2-}(aq) + 2\,H^+(aq)$$

These results indicate that the two sulfur atoms of the thiosulfate ion are not equivalent and that the proposed structure is likely.

Radioactive nuclides have been used to study reaction rates and mechanisms as well as the action of catalysts. By using radioactive tracers it becomes possible to follow the progress of tagged molecules and atoms through a chemical reaction.

For example, the mechanism of the reaction between the sulfite ion and the chlorate ion:

$$SO_3^{2-} + ClO_3^- \longrightarrow SO_4^{2-} + ClO_2^-$$

has been shown to proceed by the exchange of an oxygen atom. The SO_3^{2-} ion used contained ordinary oxygen atoms, but the ClO_3^- ion was prepared with

$^{18}_8O$, which is not radioactive but which may be detected by a mass spectrograph. The SO_4^{2-} ion produced by the reaction contained $^{18}_8O$ in an amount that would indicate that one $^{18}_8O$ atom had been added to each SO_3^{2-} ion.

The mechanism proposed, therefore, is a Lewis nucleophilic displacement of ClO_2^- by SO_3^{2-} on one of the oxygen atoms of the ClO_3^- ion. The $^{18}_8O$ atoms are marked with asterisks:

$$\begin{bmatrix} & O & \\ & \| & \\ O- & S & \\ & \| & \\ & O & \end{bmatrix}^{2-} + \begin{bmatrix} & O^* & \\ & \| & \\ O^*- & Cl & -O^* \end{bmatrix}^{-} \longrightarrow \begin{bmatrix} & O & & O^* & \\ & \| & & \| & \\ O- & S & \cdots O^* \cdots & Cl & -O^* \\ & \| & & & \\ & O & & & \end{bmatrix}^{3-} \longrightarrow$$

$$\begin{bmatrix} & O & \\ & \| & \\ O- & S & -O^* \\ & \| & \\ & O & \end{bmatrix}^{2-} + \begin{bmatrix} & O^* & \\ & \| & \\ & Cl & -O^* \end{bmatrix}^{-}$$

Activation analysis is the determination of the quantity of an element in a sample by bombarding the sample with suitable nuclear projectiles and measuring the intensity of the radioactivity induced in the element being investigated. The induced activity is not influenced by the chemical bonding of the element. Neutrons are the most frequently employed projectiles. Techniques have been developed for the determination of more than 50 elements. This method is particularly valuable for the determination of elements present in extremely low concentrations.

Determinations of very low vapor pressures and very low solubilities may be made conveniently by using tagged materials. For example, the solubility of water in benzene may be determined by using water that has been enriched with the radioactive ^3_1H and measuring the activity of the water-saturated benzene.

Summary

The topics that have been discussed in this chapter are

1. Some properties of the nucleus, the exchange theory of nuclear bonding, nuclear stability.

2. Types of radioactive decay of both natural and synthetic radioactive nuclides and the energy changes that accompany these processes.

3. The detection and mesurement of radioactive emissions, the rate of radioactive decay, half-life of a radioactive nuclide, radiocarbon dating.

4. Natural and artificial radioactive decay series.

5. Nuclear reactions, the preparation of artificial nuclides.

6. The uses of radioactive nuclides.

Key Terms

Some of the more important terms introduced in this chapter are listed below. Definitions for terms not included in this list may be located in the text by use of the index.

Alpha emission (Section 25.2) A type of radioactivity in which alpha particles (each of which consists of two neutrons and two protons, given the symbol ^4_2He) are ejected.

Antiparticle (Section 25.2) One part of a particle-antiparticle pair, which destory each other when they come into contact and are converted into a form of energy. The antiparticle of a charged particle has the same mass as the particle but a charge with the opposite sign (for example, the proton, p^+, and antiproton, p^-). Some particle-antiparticle pairs are uncharged (for example, the neutron and antineutron).

Beta emission (Section 25.2) A type of radioactivity in which electrons (symbol, $_{-1}^{0}e$) are ejected.

Breeder reactor (Section 25.6) A type of nuclear reactor that manufactures more nuclear fuel than it uses.

Critical mass (Section 25.6) The amount of a fissionable material required to sustain a nuclear chain reaction.

Curie, Ci (Section 25.3) A unit used in the measurement of the activity of a radioactive source (the number of disintegrations per unit time). One curie is 3.70×10^{10} disintegrations per second.

Cyclotron (Section 25.5) An instrument in which a charged particle is accelerated along a spiral path and caused to strike a target nucleus.

Electron capture (Section 25.2) A radioactive decay process exhibited by some artificial nuclides in which a nucleus captures an electron from an inner shell and the captured electron converts a proton into a neutron; also called K capture.

Fast neutron (Section 25.5) A fast-moving neutron produced by a nuclear reaction.

Gamma radiation (Section 25.2) Electromagnetic radiation of very short wavelength.

Geiger-Müller counter (Section 25.3) A device that is used for the quantitative detection of radioactive emissions and that functions by counting the electrical impulses caused by the ionization of a gas in a chamber through which the emissions pass.

Half-life (Section 25.3) The time that it takes for one-half of a sample of a radioactive nuclide to decay.

Linear accelerator (Section 25.5) An instrument in which a charged particle is accelerated along a linear path and caused to strike a target nucleus.

Neutrino (Section 25.2) An uncharged particle of extremely small mass that is ejected in the course of some radioactive processes from which it carries off energy.

Neutron-capture reaction (Section 25.5) A nuclear reaction in which a target nucleus is bombarded with slow neutrons. No subsidiary particle is ejected and the net result is that the target nucleus captures an additional neutron.

Nuclear fission (Section 25.6) A process in which heavy nuclei are split into lighter nuclei.

Nuclear fusion (Section 25.6) A process in which very light nuclei are fused into heavier nuclei.

Nuclear reactor (Section 25.6) A reactor in which the controlled fisson of a nuclear fuel serves as a source of energy.

Nuclide (Section 25.1) A term used to refer to a species of atom characterized by its atomic number and mass number; the term isotope refers to one species of atom out of two or more that are of the same element and that have the same atomic number but different mass numbers.

Particle-particle reaction (Section 25.5) A nuclear reaction in which a target nucleus and a projectile particle form a compound nucleus that rapidly ejects a subsidiary particle and forms a product nucleus. Reactions are indicated: target nucleus (projectile, ejected particle) product nucleus; for example, $_7^{14}N\ (\alpha, p)\ _8^{17}O$.

Pion, π (Section 25.1) A particle that is thought to be rapidly exchanged by nucleons, thereby binding them together into a nucleus. Three types of pions are known: positive, π^+, negative, π^-, and neutral, π^0.

Positron emission (Section 25.2) A type of radioactivity exhibited by some artificial nuclides in which positrons (positive electrons, given the symbol $_1^0e$) are ejected.

Radioactive disintegration series (Section 25.4) A series of radioactive disintegrations that successively produce new radioactive nuclides until a nonradioactive nuclide is obtained.

Radiocarbon dating (Section 25.3) A method of dating a carbon-containing object by measuring the number of radioactive disintegrations of the $_6^{14}C$ present in a sample and calculating, by means of the half-life of this nuclide, the length of time for the object to achieve its present condition.

Scintillation counter (Section 25.3) A device that is used for the quantitative detection of radioactive emissions and that functions by counting the flashes of light given off by a fluorescent substance struck by these emissions.

Slow neutron (Section 25.5) A neutron, derived from a nuclear reaction, that has been slowed down by passage through a moderator; also called a thermal neutron.

Transuranium element (Section 25.5) The elements following uranium (atomic number, 92) in the periodic table.

Wilson cloud chamber (Section 25.3) A device that enables the path of ionizing radiation to be seen as a cloud track formed by condensation of water vapor on the ions.

Zone of stability (Section 25.1) A zone in a graph of number of neutrons versus number of protons, in which points that represent stable nuclides fall.

Problems*

Radioactivity

25.1 Briefly describe the exchange theory of nuclear bonding.

25.2 Define the following terms: **(a)** nuclide, **(b)** nucleon, **(c)** critical mass, **(d)** thermal neutron, **(e)** fission, **(f)** fusion, **(g)** transuranium element.

25.3 Write equations for the following examples of radioactive decay: **(a)** alpha emission by $^{218}_{85}At$, **(b)** beta emission by $^{198}_{79}Au$, **(c)** positron emission by $^{25}_{13}Al$, **(d)** electron capture by $^{108}_{47}Ag$.

25.4 Write equations for the following examples of radioactive decay: **(a)** alpha emission by $^{181}_{78}Pt$, **(b)** beta emission by $^{25}_{11}Na$, **(c)** positron emission by $^{62}_{29}Cu$, **(d)** electron capture by $^{37}_{18}Ar$.

25.5 Write equations for the following examples of radioactive decay: **(a)** alpha emission by $^{221}_{87}Fr$, **(b)** beta emission by $^{66}_{29}Cu$, **(c)** positron emission by $^{18}_{9}F$, **(d)** electron capture by $^{133}_{56}Ba$.

25.6 Write equations for the following examples of radioactive decay: **(a)** alpha emission by $^{227}_{91}Pa$, **(b)** beta emission by $^{114}_{47}Ag$, **(c)** positron emission by $^{39}_{20}Ca$, **(d)** electron capture by $^{55}_{26}Fe$.

25.7 The nuclide $^{192}_{78}Pt$ decays to $^{188}_{76}Os$ by alpha emission. The mass of $^{192}_{78}Pt$ is 191.9614 u, and the mass of $^{188}_{76}Os$ is 187.9560 u. Calculate the energy released in this process.

25.8 The nuclide $^{213}_{85}At$ decays to $^{209}_{83}Bi$ by alpha emission. The mass of $^{213}_{85}At$ is 212.9931 u, and the mass of $^{209}_{83}Bi$ is 208.9804 u. Calculate the energy released in this process.

25.9 When $^{221}_{87}Fr$ decays, 6.34 MeV and 6.12 MeV α particles are emitted. What are the energy of and the source of the γ radiation accompanying this alpha emission?

25.10 When $^{222}_{88}Ra$ decays, 6.56 MeV and 6.23 MeV α particles are emitted. What are the energy of and the source of the γ radiation accompanying this α emission?

25.11 The disintegration energy is 7.03 MeV for the β decay of $^{20}_{9}F$ to $^{20}_{10}Ne$ (mass, 19.99244 u). What is the mass of $^{20}_{9}F$?

25.12 The disintegration energy is 11.5 MeV for the β decay of $^{11}_{4}Be$ to $^{11}_{5}B$ (mass, 11.00931 u). What is the mass of $^{11}_{4}Be$?

25.13 The nuclide $^{21}_{11}Na$ (mass, 20.99883 u) decays to $^{21}_{10}Ne$ (mass, 20.99395 u) by positron emission. What is the energy released by this radioactive decay process?

25.14 The nuclide $^{18}_{10}Ne$ (mass, 18.00572 u) decays to $^{18}_{9}F$ (mass, 18.00095 u) by positron emission. What is the energy released by this radioactive decay process?

25.15 The energy released by the decay of $^{31}_{16}S$ by positron emission is 5.40 MeV. The nuclide produced, $^{31}_{15}P$, has a mass of 30.97376 u. What is the mass of $^{31}_{16}S$?

25.16 The energy released by the decay of $^{27}_{14}Si$ by positron emission is 3.80 MeV. The nuclide produced, $^{27}_{13}Al$, has a mass of 26.98154 u. What is the mass of $^{27}_{14}Si$?

25.17 The nuclide $^{37}_{18}Ar$ (mass, 36.96678 u) decays to $^{37}_{17}Cl$ (mass, 36.96590 u). Is it energetically possible for the process to occur by positron emission, or must it occur by electron capture?

25.18 Is it energetically possible for $^{41}_{20}Ca$ (mass, 40.96227 u) to decay to $^{41}_{19}K$ (mass, 40.96183 u) by positron emission, or must the process occur by electron capture?

25.19 The disintegration energy for the decay of $^{51}_{24}Cr$ by electron capture is 0.75 MeV, and the mass of the nuclide produced, $^{51}_{23}V$, is 50.9440 u. What is the mass of $^{51}_{24}Cr$?

25.20 Why is electron capture accompanied by the production of X rays?

Rate of Radioactive Decay

25.21 The nuclide $^{76}_{35}Br$ has a half-life of 16.5 hours. How much of a 0.0100 g sample remains at the end of 1.00 day?

25.22 The nuclide $^{198}_{79}Au$ has a half-life of 64.8 hours. How much of a 0.0100 g sample remains at the end of 1.00 day?

25.23 The half-life of $^{35}_{16}S$ is 86.7 days. How long will it take for 90.0% of a sample to disappear?

25.24 The half-life of $^{112}_{47}Ag$ is 3.20 hours. How long will it take for 75.0% of a sample to disappear?

25.25 The rate of decay of $^{18}_{9}F$ is such that 10.0% of the original quantity remains after 369 min. **(a)** What is the rate constant for this radioactive disintegration? **(b)** What is the half-life of $^{18}_{9}F$?

25.26 The nuclide $^{22}_{11}Na$ decays by positron emission. The rate is such that 76.6% of the original quantity remains after 1.00 year. **(a)** What is the rate constant? **(b)** What is the half-life of $^{22}_{11}Na$?

25.27 A sample of a radioactive material initially gives 2500 counts/min and 15 min later gives 2400 counts/min. What is the half-life of the nuclide?

25.28 The carbon from the heart-wood of a giant sequoia tree gives 10.8 $^{14}_{6}C$ counts per minute per gram of carbon, whereas the wood from the outer portion of the tree gives 15.3 $^{14}_{6}C$ counts per minute per gram of carbon. How old is the tree?

---

* The appendix contains answers to color-keyed problems.

25.29 The half-life of $^{237}_{93}Np$ is 2.1×10^6 years, and the age of earth is approximately 4.5×10^9 years. What fraction of the quantity of $^{237}_{93}Np$ that was formed when the earth was created is now present?

25.30 The half-life of $^{65}_{30}Zn$ is 243 days. What is the activity in curies of a 0.000100 g sample of $^{65}_{30}Zn$? The atomic mass of $^{65}_{30}Zn$ to three significant figures is 64.9 u.

25.31 The half-life of $^{59}_{26}Fe$ is 45 days. **(a)** How many atoms are in a sample that has an activity of 0.75 Ci? **(b)** What is the weight of the sample?

Disintegration Series

25.32 Starting with $^{154}_{68}Er$, the successive steps in an artificial decay chain are: α, β^+, ec, α, α. What are the members of the chain?

25.33 One of the naturally occurring decay series is that of the nuclide $^{232}_{90}Th$. The particles successively emitted in one route are: α, β, β, α, α, α, β, α, β, α. Determine in order the members of the chain.

25.34 It is believed that a $^{237}_{93}Np$ disintegration series existed in nature at one time. The members of the series (with the exception of the stable nuclide that ends the series), however, have virtually disappeared through radioactive decay in the time since the earth was created. The particles successively emitted in the $^{237}_{93}Np$ series are: α, β, α, α, β, α, α, α, β, α, β. Determine in order the members of the chain.

Nuclear Reactions

25.35 Write equations for the following induced nuclear reactions: **(a)** $^{82}_{35}Br\,(n, \gamma)$, **(b)** $^{10}_{5}B\,(n, \alpha)$, **(c)** $^{35}_{17}Cl\,(n, p)$, **(d)** $^{7}_{3}Li\,(p, n)$, **(e)** $^{130}_{52}Te\,(d, 2n)$, **(f)** $^{43}_{20}Ca\,(\alpha, p)$, **(g)** $^{237}_{93}Np\,(\alpha, n)$, **(h)** $^{238}_{92}U\,(^{22}_{10}Ne, 4n)$.

25.36 Write equations for the following induced nuclear reactions: **(a)** $^{10}_{5}B\,(\alpha, n)$, **(b)** $^{6}_{3}Li\,(n, \alpha)$, **(c)** $^{7}_{3}Li\,(d, n)$, **(d)** $^{12}_{6}C\,(p, \gamma)$, **(e)** $^{96}_{42}Mo\,(p, n)$, **(f)** $^{75}_{33}As\,(d, p)$, **(g)** $^{45}_{21}Sc\,(\alpha, p)$, **(h)** $^{51}_{23}V\,(d, 2n)$.

25.37 Some transuranium nuclides and the types of induced nuclear reactions that have been used to prepare them are listed. In each case what isotope was used as the starting material? **(a)** $(d, 6n)\ ^{231}_{93}Np$, **(b)** $(d, 3n)$ $^{243}_{97}Bk$, **(c)** $(\alpha, {}^{3}_{1}H)\ ^{250}_{99}Es$, **(d)** $(^{12}_{6}C, 4n)\ ^{245}_{99}Es$, **(e)** $(^{16}_{8}O, 5n)\ ^{249}_{100}Fm$, **(f)** $(\alpha, 2n)\ ^{255}_{101}Md$, **(g)** $(^{22}_{10}Ne, 4n)\ ^{256}_{102}No$, **(h)** $(^{10}_{5}B, 3n)\ ^{257}_{103}Lr$.

25.38 An isotope of element 105 was prepared by bombarding a $^{249}_{98}Cf$ target with $^{15}_{7}N$ nuclei. Four neutrons were emitted when a $^{249}_{98}Cf$ nucleus absorbed a $^{15}_{7}N$ nucleus. Write an equation for the transformation.

Nuclear Fission and Fusion

25.39 Calculate the binding energy per nucleon of $^{235}_{92}U$ (mass, 235.0439 u). The masses of the proton, neutron, and electron are 1.007277 u, 1.008665 u, and 0.0005486 u, respectively.

25.40 **(a)** Calculate the energy released by the fission

$$^{235}_{92}U + {}^{1}_{0}n \longrightarrow {}^{94}_{38}Sr + {}^{139}_{54}Xe + 3{}^{1}_{0}n$$

The atomic masses are: $^{235}_{92}U$, 235.0439 u; $^{94}_{38}Sr$, 93.9154 u; $^{139}_{54}Xe$, 138.9179 u. The mass of the neutron is 1.0087 u. **(b)** What percent of the total mass of the starting materials is converted into energy?

25.41 **(a)** Calculate the energy released by the fusion

$$^{1}_{1}H + {}^{2}_{1}H \longrightarrow {}^{3}_{2}He$$

The atomic masses are: $^{1}_{1}H$, 1.0078 u; $^{2}_{1}H$, 2.0141 u; $^{3}_{2}He$, 3.0160 u. **(b)** What percent of the total mass of the starting materials is converted into energy?

25.42 Each of the following nuclides represents one of the products of a neutron-induced fission of $^{235}_{92}U$. In each case another nuclide and three neutrons are also produced. Identify the other nuclides. **(a)** $^{148}_{58}Ce$, **(b)** $^{95}_{37}Rb$, **(c)** $^{134}_{52}Te$.

ORGANIC CHEMISTRY

CHAPTER

26

The name organic chemistry derives from the early concept that substances of plant or animal origin (organic substances) were different from those of mineral origin (inorganic substances). In the middle of the nineteenth century, however, the idea that organic substances could only be synthesized by living organisms was gradually discounted. Not only has a large number of natural products been synthesized in the laboratory, but also countless related materials have been made that do not occur in nature. All these compounds contain carbon. Over a million carbon compounds are known.

Because it represents a convenient division of chemistry, the term organic chemistry is retained, but it is now commonly defined as the chemistry of carbon and its compounds. However, since some carbon compounds (such as carbonates, carbides, and cyanides) are traditionally classed as inorganic compounds, organic chemistry is probably better defined as the chemistry of the hydrocarbons (compounds containing only carbon and hydrogen) and their derivatives.

Carbon forms an unusually large number of compounds because of its exceptional ability to catenate (see Section 22.1). In addition, the carbon atom can form four very stable, single-covalent bonds; it also has the ability to form multiple bonds with other carbon atoms or with atoms of other elements.

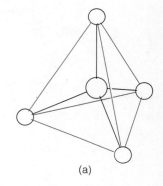

(a)

26.1 The Alkanes

In the first four sections of this chapter various types of hydrocarbons are characterized. The reactions of the hydrocarbons is the topic of Section 26.5. The simplest hydrocarbon is methane, CH_4. This molecule contains four equivalent carbon-hydrogen bonds arranged tetrahedrally (see Figure 26.1). The bonding may be considered to arise through the use of sp^3 hybrid orbitals of the carbon atom. Each of the bonds of methane is of the same length (109.5 pm), and each of the H—C—H bond angles is 109° 28′ (the so-called tetrahedral angle). The commonly employed structural formula of methane:

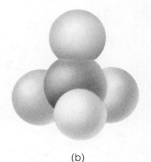

(b)

$$\begin{array}{c} H \\ | \\ H-C-H \\ | \\ H \end{array}$$

does not accurately represent the tetrahedral arrangement of the molecule.

Figure 26.1 Representations of the structure of methane, CH_4. (The bonds in (a) are of exaggerated length in comparison with the atomic sizes.)

The **alkanes** are hydrocarbons in which all the carbon-carbon bonds are single bonds; they may be considered to be derived from methane by the successive addition of $-CH_2-$ units. Such a series of compounds is said to be **homologous** and the compounds are said to be **homologs**. Thus, the formula of the second member of the family, ethane:

$$H-\underset{\underset{H}{|}}{\overset{\overset{H}{|}}{C}}-\underset{\underset{H}{|}}{\overset{\overset{H}{|}}{C}}-H$$

may be formally derived by the introduction of a $-CH_2-$ unit between the carbon and a hydrogen of methane. This molecule is also represented by the formula CH_3-CH_3.

The alkanes conform to the general formula C_nH_{2n+2}, where n is the number of carbon atoms in the compound. A few subsequent members of the family, and their names, are

$$H-\underset{\underset{H}{|}}{\overset{\overset{H}{|}}{C}}-\underset{\underset{H}{|}}{\overset{\overset{H}{|}}{C}}-\underset{\underset{H}{|}}{\overset{\overset{H}{|}}{C}}-H \qquad H-\underset{\underset{H}{|}}{\overset{\overset{H}{|}}{C}}-\underset{\underset{H}{|}}{\overset{\overset{H}{|}}{C}}-\underset{\underset{H}{|}}{\overset{\overset{H}{|}}{C}}-\underset{\underset{H}{|}}{\overset{\overset{H}{|}}{C}}-H$$

$$CH_3CH_2CH_3 \qquad\qquad CH_3CH_2CH_2CH_3$$
$$\textit{propane} \qquad\qquad\qquad \textit{butane}$$

$$H-\underset{\underset{H}{|}}{\overset{\overset{H}{|}}{C}}-\underset{\underset{H}{|}}{\overset{\overset{H}{|}}{C}}-\underset{\underset{H}{|}}{\overset{\overset{H}{|}}{C}}-\underset{\underset{H}{|}}{\overset{\overset{H}{|}}{C}}-\underset{\underset{H}{|}}{\overset{\overset{H}{|}}{C}}-H$$

$$CH_3CH_2CH_2CH_2CH_3$$
$$\textit{pentane}$$

These compounds are spoken of as **straight-chain** compounds even though the carbon chains are far from linear (see Figure 26.2); all the bond angles are approximately tetrahedral, and axial rotation of any carbon atom around a single bond of the chain is possible.

There is only one compound for each of the formulas CH_4, C_2H_6, and C_3H_8. However, there are two compounds with the formula C_4H_{10}: the compound with a straight chain (butane) and one with a **branched chain**:

$$CH_3\underset{\underset{CH_3}{|}}{CH}CH_3$$
$$\textit{methylpropane}$$

The two compounds of formula C_4H_{10} are **structural isomers** (compounds with the same molecular formula but different structural formulas); they have different properties.

There are three structural isomers of C_5H_{12}: pentane (the straight-chain compound),

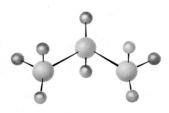

Figure 26.2 Representation of the structure of propane, $CH_3CH_2CH_3$

Table 26.1 Simple alkyl radicals

Formula	Name
CH_3-	methyl
CH_3CH_2-	ethyl
$CH_3CH_2CH_2-$	*normal*-propyl or *n*-propyl
$CH_3\overset{\displaystyle |}{\underset{\displaystyle CH_3}{CH}}-$	isopropyl
$CH_3CH_2CH_2CH_2-$	*normal*-butyl or *n*-butyl
$CH_3\overset{\displaystyle |}{\underset{\displaystyle CH_3}{CH}}CH_2-$	isobutyl
$CH_3CH_2\overset{\displaystyle |}{\underset{\displaystyle CH_3}{CH}}-$	*secondary*-butyl or *sec*-butyl
$CH_3\overset{\displaystyle CH_3}{\overset{\displaystyle |}{\underset{\displaystyle |}{\underset{\displaystyle CH_3}{C}}}}-$	*tertiary*-butyl or *tert*-butyl

$$CH_3CH_2\overset{\displaystyle |}{\underset{\displaystyle CH_3}{CH}}CH_3 \qquad and \qquad CH_3-\overset{\displaystyle CH_3}{\overset{\displaystyle |}{\underset{\displaystyle |}{\underset{\displaystyle CH_3}{C}}}}-CH_3$$

methylbutane   *dimethylpropane*

In the alkane series the number of possible structural isomers increases rapidly: there are five structural isomers for C_6H_{14}, nine for C_7H_{16}, and 75 for $C_{10}H_{22}$. It has been calculated that more than 4×10^9 isomers are possible for $C_{30}H_{62}$.

A branched-chain compound may be named in terms of the longest straight chain in the molecule. The side chains are named as alkyl radicals, and their positions on the straight chain are usually indicated by a numbering system. Alkyl radicals are fragments of alkane molecules from which a hydrogen atom has been removed; their names are derived from the name of the parent alkane with the ending changed to -*yl*. A list of common alkyl radicals appears in Table 26.1.

For example, the longest straight chain in the molecule:

$$CH_3\overset{\displaystyle CH_3}{\overset{\displaystyle |}{\underset{\displaystyle |}{\underset{\displaystyle CH_3}{C}}}}-CH\overset{}{\underset{\displaystyle |}{\underset{\displaystyle CH_2}{\underset{\displaystyle |}{\underset{\displaystyle CH_3}{}}}}}CH_2CH_3$$

consists of five carbon atoms, and the compound is named as a derivative of pentane. The pentane chain is numbered starting at the end that will give the side chains the lowest numbers:

Table 26.2 Physical properties of some straight-chain alkanes

Compound	Formula	Melting Point (°C)	Boiling Point (°C)
methane	CH_4	−183	−162
ethane	C_2H_6	−172	−88
propane	C_3H_8	−187	−42
butane	C_4H_{10}	−135	1
pentane	C_5H_{12}	−131	36
hexane	C_6H_{14}	−94	69
heptane	C_7H_{16}	−91	99
octane	C_8H_{18}	−57	126
nonane	C_9H_{20}	−54	151
decane	$C_{10}H_{22}$	−30	174
hexadecane	$C_{16}H_{34}$	20	288
heptadecane	$C_{17}H_{36}$	23	303

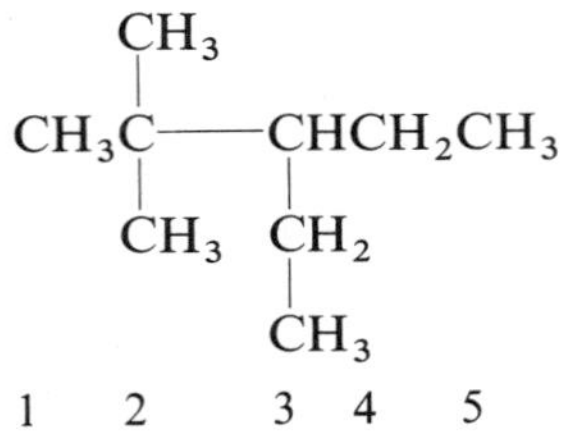

$$CH_3C\!\!-\!\!CHCH_2CH_3$$

with CH_3 above CH_3C; CH_3 and CH_2 below (and CH_3 below CH_2); numbered 1 2 3 4 5.

In the name a number is used to indicate the position of each substituent radical. Thus, the name of the compound is

2,2-dimethyl-3-ethylpentane

No numbers appear in the names of our earlier examples of structural isomers. The name methylpropane requires no number to indicate the position of the methyl radical with regard to the propane chain since there is only one possible position for the methyl group (number 2). If the methyl group were placed on either terminal carbon atom (number 1 or number 3), the compound would be the straight-chain isomer, butane. In like manner, the names methylbutane and dimethylpropane do not require the use of numbers; there is only one possible structure corresponding to each name.

Names for some of the higher straight-chain homologs may be obtained from Table 26.2, which lists the melting points and boiling points of some of the straight-chain alkanes. The melting point and boiling point increase as the length of the chain (and hence the molecular weight) of the hydrocarbon increases. The first four compounds are gases under ordinary conditions; higher homologs are liquids (C_5H_{12} to $C_{15}H_{32}$) and solids (from $C_{16}H_{34}$ on).

In addition to open-chain hydrocarbons, cyclic hydrocarbons are known. The cycloalkanes are ring structures that contain only single carbon-carbon bonds; they have the general formula C_nH_{2n}. Examples are

Table 26.3 Petroleum fractions

Fraction	Boiling Range (°C)	Carbon Content of Compounds	Use
gas	below 20	C_1–C_4	fuel
petroleum ether	20–90	C_5–C_7	solvent
gasoline	35–220	C_5–C_{12}	motor fuel
kerosene	200–315	C_{12}–C_{16}	jet fuel
fuel oil	250–375	C_{15}–C_{18}	diesel fuel
lubricating oils, greases	350 up	C_{16}–C_{20}	lubrication
paraffin wax	50–60 (m.p.)	C_{20}–C_{30}	candles
asphalt	viscous liquid		paving
residue	solid		fuel

cyclopropane   *cyclohexane*

The cyclohexane ring is not planar but is puckered in such a way that the C—C—C bond angles are 109° 28′, the tetrahedral angle. Three- and four-membered rings are strained because the C—C—C bond angles deviate significantly from the tetrahedral angle. The strain is greatest in cyclopropane, in which the C atoms form an equilateral triangle and the bond angles are 60°. Thus, the cyclopropane ring is readily broken by H_2 (Ni catalyst) to give propane. Cyclobutane reacts with H_2 in an analogous manner (to give butane), but a higher temperature is required.

Petroleum is the principal source of alkanes and cycloalkanes. In the refining of petroleum the crude oil is first separated into fractions by distillation (see Table 26.3). Some higher-boiling fractions are subjected to **cracking processes** in which large molecules are broken down into smaller molecules, thus increasing the yield of the most valuable fraction, gasoline. In addition, small molecules are converted into large ones in what are called **alkylation processes.** Gases that are by-products of these processes, together with other substances that are obtained from the petroleum fractions, are important starting materials for the manufacture of many chemical products. More than 2000 such products (such as rubber, detergents, plastics, and textiles) are derived from the substances obtained from petroleum and natural gas.

26.2 The Alkenes

The alkenes, which are also called **olefins,** are hydrocarbons that have a carbon-carbon double bond somewhere in their molecular structure. The first member of the series is ethene (also called ethylene):

Night view of a fractional distillation column at a modern oil refinery. *American Petroleum Institute.*

$$H{\diagdown}{\diagup}H$$

$$\underset{H}{\overset{}{\diagup}} C = C \underset{H}{\overset{}{\diagdown}}$$

Each carbon atom of ethene uses sp^2 hybrid orbitals to form three sigma bonds: one with the other C atom and two with H atoms. Thus, all six atoms lie in the same plane, and all bond angles are approximately 120°. In addition, each carbon atom has an electron in a p orbital that is not engaged in the formation of sp^2 σ bonds. These two p orbitals are coplanar and pependicular to the plane of the molecule; they overlap to form a π bonding orbital with regions of charge density above and below the plane of the molecule (see Section 7.5, Figure 7.17).

The p orbitals can overlap and form a π bonding orbital only when they lie in the same plane. Consequently, free rotation around the carbon-carbon double bond is not possible without breaking the π bond. The carbon-carbon bond distance in ethene (133 pm) is shorter than that in ethane (154 pm) because of the double bond. The π bond is not so strong as the σ bond; the carbon-carbon bond energy in ethane is approximately 340 kJ/mol, and the bond energy of the carbon-carbon double bond in ethene is about 610 kJ/mol.

Alkanes are called **saturated** hydrocarbons because all the valence electrons of the carbon atoms are engaged in single bond formation, and no more hydrogen atoms or atoms of other elements can be accommodated by the carbon atoms of the chain. **Unsaturated** hydrocarbons have one or more carbon-carbon multiple bonds and can undergo **addition reactions** (see Section 26.5) because of the availability of the π electrons of the multiple bond. In one such reaction hydrogen adds to ethene in the presence of a catalyst to form ethane:

$$CH_2 = CH_2 + H_2 \xrightarrow{\text{Ni}} CH_3 - CH_3$$

Alkenes conform to the general formula C_nH_{2n}. The name of an alkene is derived from the name of the corresponding alkane by changing the ending from *-ane* to *-ene*; a number is used, when necessary, to indicate the position of the double bond. The second member of the series is propene:

$$CH_2 = CH - CH_3$$

New types of isomerism arise in the alkene series. The type of structural isomerism displayed by the alkanes (for example, butane and methylpropane) is known as **chain isomerism**. In addition to chain isomerism, **position isomerism** occurs in the olefin series. There is only one straight-chain butane. There are, however, two straight-chain butenes, and they differ in the position of the double bond:

$$CH_2 = CH - CH_2 - CH_3 \qquad CH_3 - CH = CH - CH_3$$
$$\textit{1-butene} \qquad\qquad\qquad \textit{2-butene}$$

In the first compound the double bond is located between carbon atoms number 1 and number 2, and the lower number (1) is used to indicate its position. In like manner, the lower number is used to indicate the position of the double bond (which occurs between carbon atoms number 2 and number 3) in the name of the second compound. In each case the chain is numbered from the end that gives the lowest possible number for the name. The name of a branched-chain alkene is derived from that of the longest straight chain that contains the double bond.

The alkenes also exhibit **geometric (or cis-trans) isomerism,** which is a type of **stereoisomerism.** Stereoisomers have the same structural formula but differ in the arrangement of the atoms in space. An example is provided by the isomers of 2-butene. Because of the restricted rotation around the double bond, one isomer exists with both methyl groups on the same side of the double bond (the *cis* isomer), and another isomer exists with the methyl groups on opposite sides of the double bond (the *trans* isomer). All carbon atoms of the molecule lie in the same plane:

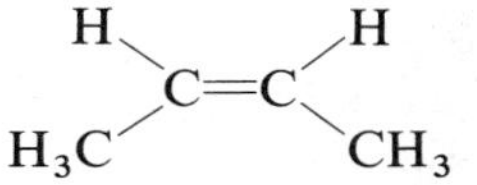

$$H_3C\diagdown C=C\diagup CH_3$$

cis-2-butene
boiling point, 1°C

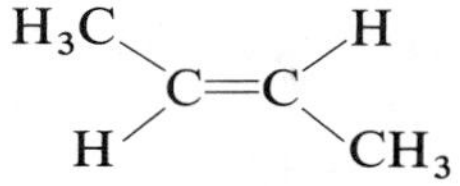

trans-2-butene
boiling point, 2.5°C

The properties of the 2-butenes do not differ greatly. Other *cis-trans* isomers exhibit wide variation in their physical properties:

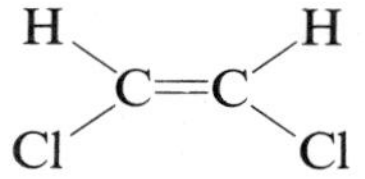

cis-1,2-dichloroethene
boiling point, 60.1°C

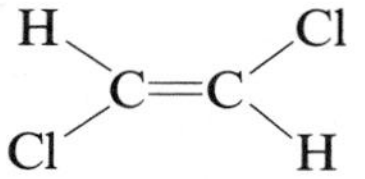

trans-1,2-dichloroethene
boiling point, 48.4°C

The physical properties of the alkenes are very similar to those of the alkanes. The C_2H_4, C_3H_6, and C_4H_8 compounds are gases under ordinary conditions; the C_5H_{10} to $C_{18}H_{36}$ compounds are liquids; and the higher alkenes are solids. All the compounds are only slightly soluble in water.

Molecules can contain more than one double bond—for example:

$$CH_2=CH-CH=CH_2$$
1,3-*butadiene*

an **alkadiene,** which is used to produce a type of synthetic rubber (see Section 26.10). **Cycloalkenes** are known—for example:

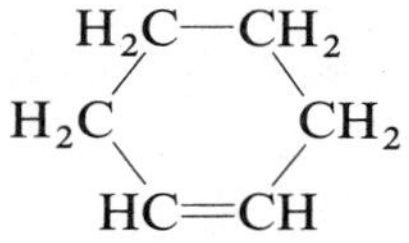

$$H_2C-CH_2$$
$$H_2C \qquad CH_2$$
$$HC=CH$$
cycloxhexene

26.3 The Alkynes

Molecules that contain carbon-carbon triple bonds are known as alkynes. These unsaturated hydrocarbons have the general formula C_nH_{2n-2} and constitute the **alkyne,** or **acetylene, series.** The first member of the series is ethyne, or acetylene:

$$H-C\equiv C-H$$

In acetylene each carbon atom uses sp hybrid orbitals to form a σ bond with a hydrogen atom and a σ bond with the other carbon atom. Consequently, the four atoms of the molecule lie in a straight line. In addition to the carbon-carbon σ bond, two π bonds are formed between the carbon atoms by the overlap of p orbitals (see Section 7.5, Figure 7.18). The resultant carbon-carbon bond distance (121 pm) is shorter than the carbon-carbon double-bond distance (133 pm). The π bonds are weaker than the C—C σ bond; the bond energy of the triple bond is approximately 830 kJ/mol (compared with the single bond energy of 340 kJ/mol). Alkynes readily undergo addition reactions across the triple bond because of the availability of the four electrons of the π bonds.

Alkynes are named in a manner analogous to that employed in naming alkenes; the ending *-yne* is used in place of *-ene*. Their physical properties are similar to those of the alkanes and alkenes. Both chain and position isomerism occur in the alkyne series, but *cis-trans* isomerism is not possible because of the linear geometry of the group X—C≡C—X, where X is a carbon atom or an atom of another element.

Notice that a branched-chain isomer of C_4H_6 is impossible; only two isomers of C_4H_6 exist:

$$HC{\equiv}CCH_2CH_3 \qquad CH_3C{\equiv}CCH_3$$
1-butyne   *2-butyne*

Both branched-chain and straight-chain isomers of C_5H_8 are known:

$$HC{\equiv}CCH_2CH_2CH_3 \qquad CH_3C{\equiv}CCH_2CH_3 \qquad HC{\equiv}CCHCH_3$$

$$\underset{\text{CH}_3}{|}$$

1-pentyne   *2-pentyne*   *methyl-1-butyne*

26.4 Aromatic Hydrocarbons

Aromatic hydrocarbons are compounds that have molecular structures based on that of benzene, C_6H_6. The six carbon atoms of benzene are arranged in a ring from which the hydrogen atoms are radially bonded (see Figure 26.3). The entire structure is planar, and all of the bond angles are 120°. The carbon-carbon bond distance is 139 pm, which is between the single-bond distance of 154 pm and the double-bond distance of 133 pm.

The electronic structure of benzene may be represented as a resonance hybrid:

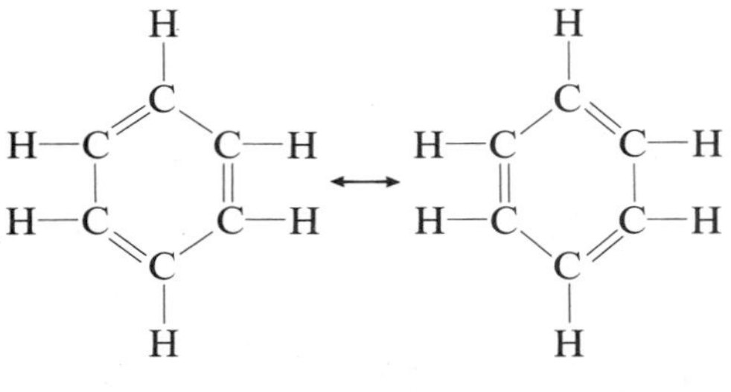

This representation correctly shows that all of the carbon-carbon bonds are equivalent and that each one is intermediate between a single bond and a double bond.

Each carbon atom of the ring uses sp^2 hybrid orbitals to form σ bonds with two adjacent carbon atoms and with a hydrogen atom. Therefore, the resulting framework of the molecule is planar with bond angles of 120° (see Figure 26.3).

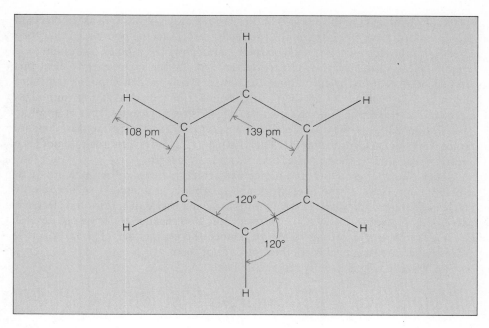

Figure 26.3 Geometry of the benzene molecule

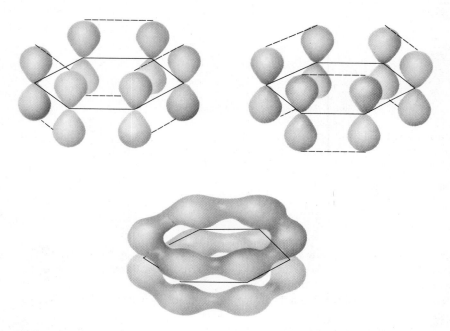

Figure 26.4 The π bonding system of benzene; σ bonds are indicated by the solid lines

A $2p$ electron of each carbon atom is not employed in the formation of the sp^2 hybrid σ bonds. The axes of these p orbitals are perpendicular to the plane of the molecule, and each p orbital can overlap two adjacent p orbitals. The result is the formation of a system of π orbitals with regions of charge density above and below the plane of the ring (see Figure 26.4).

The electron pairs of the σ bonds of the benzene molecule (as well as the electron pair of an ordinary π bond) are localized with respect to the two nuclei that they serve to bond. However, the six electrons that occupy the π bonding system (one contributed by each carbon atom) are delocalized. On the average, the π electrons tend to be distributed evenly about the benzene ring, which explains the bond length and equivalence of all the C to C bonds. Since delocalization minimizes electron-electron repulsion, the structure, which is reminiscent of that of graphite (see Section 22.2), is very stable and is responsible for the properties that are typical of aromatic compounds. The benzene ring does not readily undergo addition reactions in the way that alkenes and alkynes do.

Notice that the resonance and molecular-orbital pictures of benzene are qualitatively equivalent. If the overlap of the *p* orbitals is imagined to occur only between orbitals of adjacent carbon atoms in such a way that traditional electron-pair bonds between only two atoms result, the two resonance forms are derived (see Figure 26.4). According to the resonance theory, neither of these forms is correct; the true structure is a hybrid of both of them.

The benzene ring is usually represented as

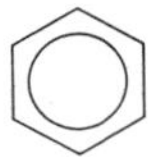

The carbon and hydrogen atoms of the ring are not shown. In formulas for benzene derivatives, such as

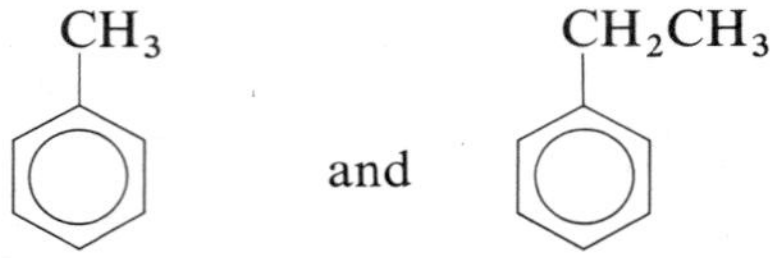

CH₃

and

CH₂CH₃

toluene
methylbenzene

ethylbenzene

it is understood that the substituents replace hydrogen atoms of the ring. The radical formed from benzene by the removal of a hydrogen atom:

is called a **phenyl radical.** The phenyl radical is an example of an **aryl radical,** a group that is produced by the removal of a hydrogen atom from a carbon of the benzene ring of an aromatic compound.

There are three position isomers of any disubstituted benzene derivative, for example:

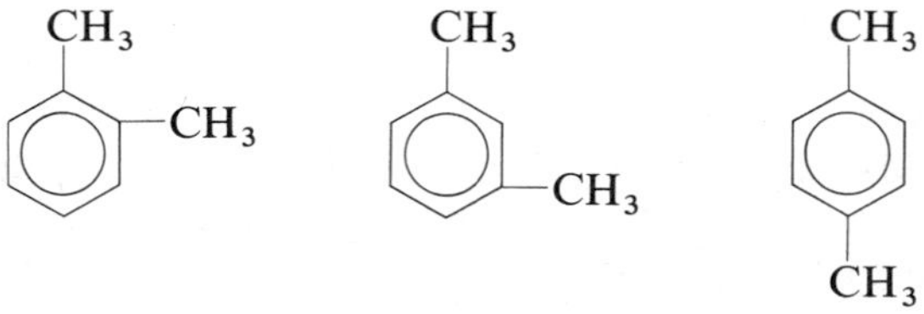

CH₃   CH₃   CH₃

—CH₃   —CH₃

CH₃

o-xylene   *m-xylene*   *p-xylene*

The prefixes *ortho-* (*o-*), *meta-* (*m-*), and *para-* (*p-*) are used to designate the positions of the two substituent groups. Numbers are used when more than two substituents are present on the ring:

$$O_2N-\text{⬡}-NO_2$$
$$NO_2$$

1,3,5-*trinitrobenzene*

Compounds are known in which several rings are fused together. For example, napthalene, $C_{10}H_8$, is a planar molecule that is a resonance hybrid:

There are no hydrogen atoms bonded to the carbon atoms through which the rings are fused.

Aromatic compounds are obtained from processes that utilize the alkanes and cycloalkanes derived from petroleum. Coal is also a source of aromatic compounds. In the preparation of coke for industrial use (principally for the reduction of iron ore), bituminous coal is heated out of contact with air in retorts. Coal tar, a by-product of this process, is a black, viscous material from which various aromatic compounds may be separated.

26.5 Reactions of the Hydrocarbons

Combustion of any hydrocarbon in excess oxygen yields carbon dioxide and water:

$$CH_4(g) + 2O_2(g) \longrightarrow CO_2(g) + 2H_2O(g)$$

The reactions are highly exothermic, a fact that accounts for the use of hydrocarbons as fuels. Carbon compounds exist in oxidation states intermediate between the hydrocarbons and carbon dioxide. These compounds are discussed in sections that follow; they are not generally prepared by direct reaction with elementary oxygen.

The replacement of a hydrogen atom of a hydrocarbon molecule by another atom or group of atoms is called a **substitution reaction.** The alkanes react with chlorine or bromine in the presence of sunlight or ultraviolet light by means of a free-radical chain mechanism (see Section 12.5). The chain is initiated by the light-induced dissociation of a chlorine molecule into atoms:

$$Cl_2 \longrightarrow 2Cl\cdot$$

Propagation of the chain occurs by the reactions

$$Cl\cdot + CH_4 \longrightarrow CH_3\cdot + HCl$$
$$CH_3\cdot + Cl_2 \longrightarrow CH_3Cl + Cl\cdot$$

The chain is terminated by the reactions

$$CH_3 \cdot + Cl \cdot \longrightarrow CH_3Cl$$

$$Cl \cdot + Cl \cdot \longrightarrow Cl_2$$

$$CH_3 \cdot + CH_3 \cdot \longrightarrow CH_3CH_3$$

The reactions that terminate the chain occur at the walls of the container. The energy liberated by the formation of the new bond is transferred to the walls of the container. In a collision between radicals in the gas phase, the united fragments fly apart almost immediately because a third body is not present to carry away the energy released by bond formation. Hence, chain-terminating reactions do not occur with great frequency, and the initial reactants are rapidly consumed.

The overall reaction for the preceding mechanism is

$$CH_4 + Cl_2 \longrightarrow CH_3Cl + HCl$$

In addition to chloromethane, CH_3Cl, the reaction produces dichloromethane, CH_2Cl_2, trichloromethane (chloroform), $CHCl_3$, and tetrachloromethane (carbon tetrachloride), CCl_4. In terms of the chain mechanism, these compounds arise through collisions of chlorine atoms with already chlorinated methane molecules, for example:

$$Cl \cdot + CH_3Cl \longrightarrow CH_2Cl \cdot + HCl$$

$$CH_2Cl \cdot + Cl_2 \longrightarrow CH_2Cl_2 + Cl \cdot$$

Complex mixtures of mono- and poly-substituted isomers are obtained as products of the chlorination of higher alkanes.

Addition reactions are characteristic of unsaturated hydrocarbons. Numerous reagents add to the multiple bonds of alkenes and alkynes. In the addition of Br_2 to ethene, for example, each of the Br atoms of the Br_2 molecule bonds to a C atom of the double bond of ethene, which leaves a single bond between the C atoms:

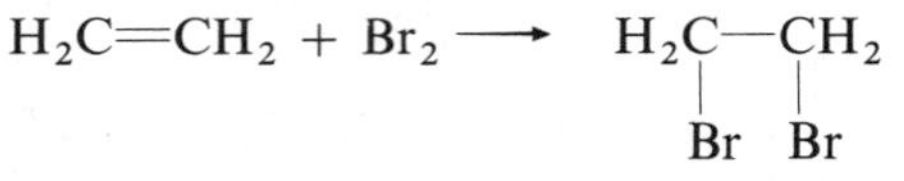

$$H_2C{=}CH_2 + Br_2 \longrightarrow \begin{array}{c} H_2C{-}CH_2 \\ | \quad | \\ Br \quad Br \end{array}$$

1,2-dibromoethane

The Cl_2, Br_2, and H_2 molecules are called **symmetrical reagents.** In an addition reaction involving any one of these reagents, the same type of atom adds to both sides of the multiple bond:

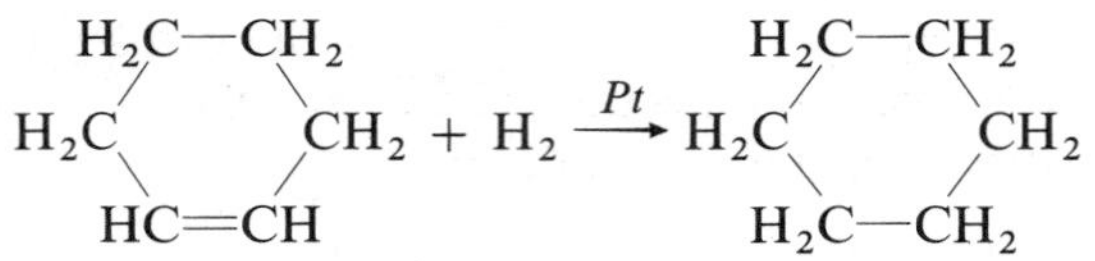

cyclohexane

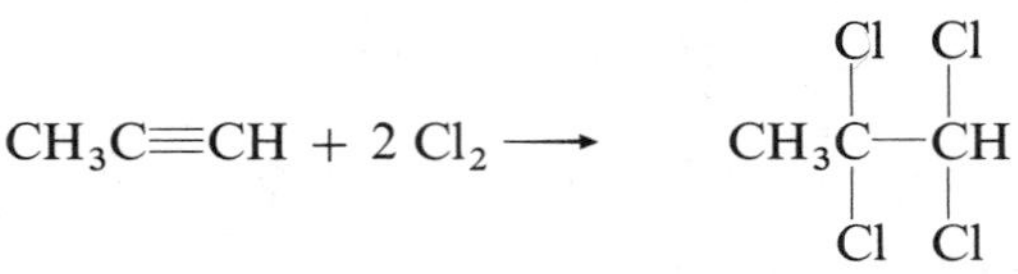

$$CH_3C{\equiv}CH + 2\,Cl_2 \longrightarrow \begin{array}{c} Cl \quad Cl \\ | \quad | \\ CH_3C{-}CH \\ | \quad | \\ Cl \quad Cl \end{array}$$

1,1,2,2-tetrachloropropane

The molecules of an **unsymmetrical reagent** (such as HBr, HCl, and HOH) consist of two different parts. When HCl, for example, adds to a multiple bond, H adds to one side and Cl adds to the other:

$$HC\equiv CH + HCl \longrightarrow CH_2{=}CHCl$$
1-chloroethene

$$CH_2{=}CHCl + HCl \longrightarrow CH_3{-}CHCl_2$$
1,1-dichloroethane

$$CH_3CH_2\overset{\overset{\displaystyle CH_3}{|}}{C}{=}CH_2 + HOH \xrightarrow{H_2SO_4} CH_3CH_2\overset{\overset{\displaystyle CH_3}{|}}{\underset{\underset{\displaystyle OH}{|}}{C}}CH_3$$
2-methyl-2-butanol

$$CH_3CH{=}CH_2 + HBr \longrightarrow CH_3\underset{\underset{\displaystyle Br}{|}}{C}HCH_3$$
2-bromopropane

When unsymmetrical molecules (such as HBr) add to the double bond of an olefin, two isomeric compounds are sometimes produced. Thus, in the last reaction of the previous series, two compounds are obtained:

$$CH_3\underset{\underset{\displaystyle Br}{|}}{C}HCH_3 \qquad CH_3CH_2CH_2Br$$
2-bromopropane   *1-bromopropane*

Over 90% of the product, however, is the 2-bromo-isomer. The principal product of such an addition is that in which the positive part of the addend (a hydrogen atom in these cases) is bonded to the carbon that originally had the larger number of hydrogen atoms (**Markovnikov's rule**).

The initial step of the mechanism of the addition of HBr is thought to be an electrophilic attack by a proton (a Lewis acid) on the π electrons of the double bond (a Lewis base). The product of this step is a positive ion called a **carbocation**; for propene two such ions are possible:

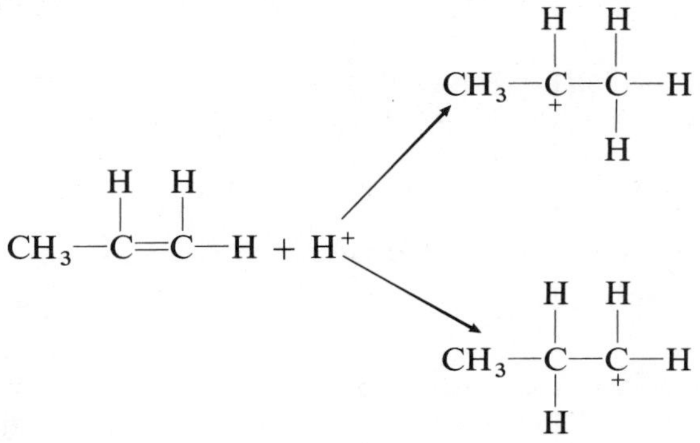

The π electrons are used to bond the incoming proton to one of the carbon atoms of the double bond which leaves the other carbon atom with a positive charge. The stability of carbocations is known to decrease in the order

$$R-\overset{R}{\underset{+}{\underset{|}{C}}}-R \;>\; R-\overset{H}{\underset{+}{\underset{|}{C}}}-R \;>\; R-\overset{H}{\underset{+}{\underset{|}{C}}}-H$$

where R is an alkyl radical. Consequently, in the reaction of propene with HBr, the carbocation produced in larger quantity is the one shown on top.

Carbocations are highly reactive Lewis acids; they exist only transiently and rapidly combine with bromide ion (a Lewis base):

$$CH_3\underset{+}{C}HCH_3 + Br^- \longrightarrow CH_3\overset{}{\underset{\underset{Br}{|}}{C}}HCH_3$$

$$CH_3CH_2CH_2^+ + Br^- \longrightarrow CH_3CH_2CH_2Br$$

Thus, the relative stabilities of the carbocations are responsible for the fact that 2-bromopropane is the principal product of the reaction.

In contrast to unsaturated open-chain hydrocarbons, the principal reactions of benzene are substitutions, not additions:

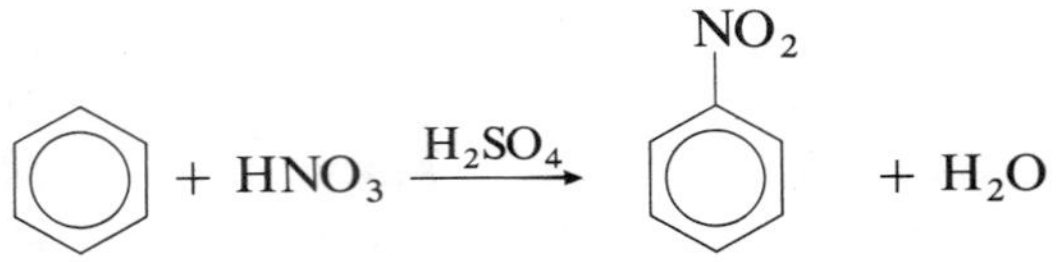

$$\text{benzene} + HNO_3 \xrightarrow{H_2SO_4} \text{nitrobenzene } (NO_2) + H_2O$$

nitrobenzene

$$\text{benzene} + Br_2 \xrightarrow{FeBr_3} \text{bromobenzene } (Br) + HBr$$

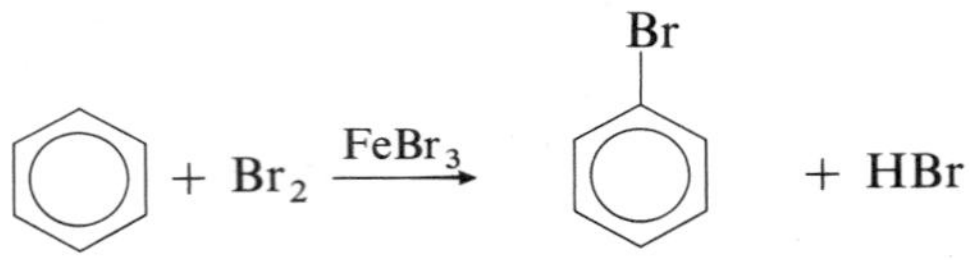

bromobenzene

$$\text{benzene} + CH_3Cl \xrightarrow{AlCl_3} \text{toluene } (CH_3) + HCl$$

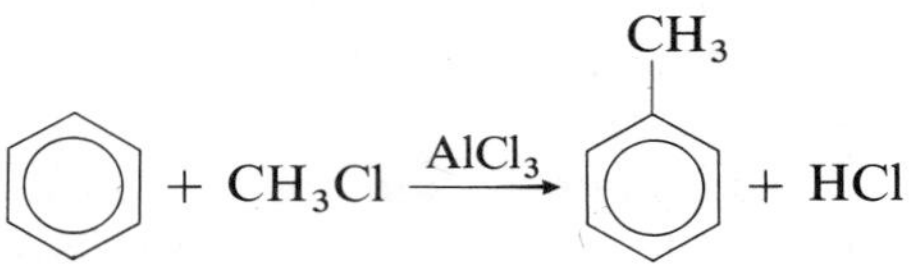

toluene

The last reaction is the Friedel-Crafts synthesis for the preparation of aromatic hydrocarbons.

The catalysts, which are indicated over the arrows in the preceding equations, produce powerful Lewis acids (NO_2^+, Br^+, and CH_3^+) as follows:

$$HNO_3 + 2\,H_2SO_4 \longrightarrow NO_2^+ + H_3O^+ + 2\,HSO_4^-$$

$$Br_2 + FeBr_3 \longrightarrow Br^+ + FeBr_4^-$$

$$CH_3Cl + AlCl_3 \longrightarrow CH_3^+ + AlCl_4^-$$

These cationic Lewis acids, which are short-lived reaction intermediates, are electrophilic, and we shall indicate them as E^+ in the mechanism that follows.

The benzene ring is attacked by the electrophilic group, E^+, which bonds to a carbon atom by means of a pair of π electrons and creates an activated complex with a positive charge:

Notice that the carbon atom to which E is bonded holds two groups in the complex; the remaining five carbon atoms hold only their customary hydrogen atoms. The complex is a resonance hybrid, and the charge is delocalized, thus providing a degree of stability. The π bonding system of benzene is more stable, however, and is restored by the loss of a proton:

The proton is lost to the anion produced in the reaction that generated the cationic electrophile:

$$H^+ + HSO_4^- \longrightarrow H_2SO_4$$

$$H^+ + FeBr_4^- \longrightarrow HBr + FeBr_3$$

$$H^+ + AlCl_4^- \longrightarrow HCl + AlCl_3$$

When a benzene ring holds two groups, three orientations are possible: *ortho* (*o*-), *meta* (*m*-), and *para* (*p*-):

o-chloronitrobenzene   m-chloronitrobenzene   p-chloronitrobenzene

In a reaction in which a second group is introduced into a benzene ring, the orientation that the entering group assumes is directed by the group already present. The chloro group, for example, directs an entering group into the *ortho* and *para* positions. A mixture of *ortho* and *para* isomers is obtained and practically none of the *meta* isomer:

Table 26.4 Directing influence of some functional groups on aromatic substitutions

ortho-para directing:

$$-F, \ -Cl, \ -Br, \ -I, -R, \ -OH, \ -OR, \ -NH_2$$

meta directing:

$$-NO_2, \ -SO_2OH, \ -C\equiv N, \ \overset{\displaystyle O}{\underset{\displaystyle \|}{-C}}-OH, \ \overset{\displaystyle O}{\underset{\displaystyle \|}{-C}}-OR, \ \overset{\displaystyle O}{\underset{\displaystyle \|}{-C}}-R, \ \overset{\displaystyle O}{\underset{\displaystyle \|}{-C}}-NH_2$$

In contrast, the nitro group is a *meta* director. Chlorination of nitrobenzene yields the *meta* isomer almost exclusively:

Groups are either *ortho-para* directing or *meta* directing. A list is given in Table 26.4.

In an aromatic substitution, the attack by the electrophilic group, E^+, occurs at a place on the benzene ring where the electron density is high. The directing influence of a group is believed to result from its effect on the distribution of the electron density of the ring. An *ortho-para* director is believed to build up electron density in these positions. A *meta* director, on the other hand, is believed to drain electron density from the *ortho* and *para* positions, leaving the *meta* position relatively negative.

26.6 Alcohols and Ethers

The alcohols may be considered as derivatives of hydrocarbons in which a hydroxyl group, OH, replaces a hydrogen. The hydroxyl group is one of a number of **functional groups,** which are groups of atoms that give organic compounds bearing them characteristic chemical and physical properties.

The alcohols are named as derivatives of a hydrocarbon consisting of the longest straight chain that contains the OH group. The ending of the parent hydrocarbon is changed from *-ane* to *-anol*, and numbers are used to indicate the positions of substituents, the lowest possible number being assigned to the OH group. Alcohols may be classified as *primary*, *secondary*, or *tertiary*, according to the number of alkyl radicals on the carbon atom holding the OH. Thus,

$$CH_3CH_2CH_2CH_2OH$$

$$\underset{\displaystyle \underset{OH}{|}}{CH_3-CH-CHCH_3} \qquad (CH_3 \ above)$$

$$\underset{\displaystyle \underset{OH}{|}}{CH_3CCH_2CH_3} \qquad (CH_3 \ above)$$

1-butanol	*3-methyl-2-butanol*	*2-methyl-2-butanol*
a primary alcohol	*a secondary alcohol*	*a tertiary alcohol*

Table 26.5 Physical properties of some alcohols

Name	Formula	Melting Point (°C)	Boiling Point (°C)	Solubility in Water (g/100 g H$_2$O, 20°C)
methanol	CH_3OH	-98	65	miscible
ethanol	CH_3CH_2OH	-115	78	miscible
1-propanol	$CH_3(CH_2)_2OH$	-127	97	miscible
1-butanol	$CH_3(CH_2)_3OH$	-90	117	7.9
1-pentanol	$CH_3(CH_2)_4OH$	-79	138	2.4
1-hexanol	$CH_3(CH_2)_5OH$	-52	157	0.6

Alcohols associate through hydrogen bonding:

$$\begin{array}{ccccc} R & & R & & R \\ | & & | & & | \\ O\!-\!H & \cdots & O\!-\!H & \cdots & O\!-\!H \end{array}$$

For this reason, the melting points and boiling points of the alcohols are higher than those of alkanes of corresponding molecular weight (see Table 26.5). The lower alcohols are miscible with water in all proportions because of intermolecular hydrogen bonding:

$$\begin{array}{ccccc} H & & R & & H \\ | & & | & & | \\ O\!-\!H & \cdots & O\!-\!H & \cdots & O\!-\!H \end{array}$$

However, as the size of the alkyl radical increases, the alcohols become more like alkanes in their physical properties, and the higher alcohols are only slightly soluble in water.

Alcohols, like water, are amphiprotic substances. They can act as Brønsted bases (proton acceptors) with very strong acids:

$$H_2SO_4 + ROH \rightleftharpoons ROH_2^+ + HSO_4^-$$

With very strong bases they are Brønsted acids (proton donors):

$$ROH + H^- \longrightarrow OR^- + H_2$$

Alcohols, however, are more weakly acidic and basic than water; consequently, the conjugate acids (ROH_2^+) and conjugate bases (OR^-) are stronger than H_3O^+ and OH^-. The anion OR^-, known as an alkoxide ion, results from the reaction of an alcohol with a reactive metal (compare the reaction of water with sodium):

$$2\,CH_3CH_2OH + 2\,Na \longrightarrow 2\,CH_3CH_2ONa + H_2$$

The first member of the series, methanol (or methyl alcohol), is known as wood alcohol because it can be obtained by the destructive distillation of wood. The principal commercial source of this alcohol is the catalytic hydrogenation of carbon monoxide:

$$CO + 2\,H_2 \longrightarrow CH_3OH$$

The alcohol of alcoholic beverages is ethanol, or ethyl alcohol. It is prepared by the fermentation of starches or sugars. Ethanol is also commercially prepared from ethene. The indirect addition of water to olefins by means of sulfuric acid is a general method of preparing alcohols. The synthesis of ethanol proceeds by the following steps:

$$CH_2{=}CH_2 + H_2SO_4 \longrightarrow CH_3CH_2OSO_2OH$$

$$CH_3CH_2OSO_2OH + H_2O \longrightarrow CH_3CH_2OH + H_2SO_4$$

Alcohols can be prepared from alkyl halides by **displacement,** or **nucleophilic substitution, reactions:**

$$RX + OH^- \longrightarrow ROH + X^-$$

This is an important classification of organic reactions. It includes the reactions of the alkyl halides, as well as the alcohols themselves, with a wide variety of nucleophilic substances. The reactions are usually reversible and are run under conditions that favor the formation of the desired compound; an excess of the nucleophilic reagent may be employed or the product may be removed from the reaction mixture as it forms (for example, by distillation).

A mechanism proposed to account for the reactions of primary and secondary halides or alcohols is the S_N2 mechanism, which can be illustrated by the reaction of hydroxide ion with methyl bromide. The hydroxide ion, a Lewis base, attacks the carbon atom and displaces the bromide ion (also a Lewis base), which takes along the electron pair with which it had been bonded:

activated complex

The attack of the OH^- ion takes place on the side of the carbon atom that is opposite the bromine atom. As the OH^- ion approaches, it begins to form a covalent bond, and the bromide ion begins to break away. The activated complex of the reaction is the form in which both nucleophilic groups are partially bonded to the carbon atom. The reaction causes inversion of the geometric arrangement of the groups attached to the carbon atom, a process that is customarily compared to the inversion of an umbrella in a high wind. The mechanism is given the designation S_N2 because it is the substitution of one nucleophilic group for another, and the rate-determining step (the formation of the activated complex) is bimolecular. The nucleophilic substitution reaction of HBr with CH_3OH proceeds through the formation of $CH_3OH_2^+$ and the displacement of H_2O from this ion by Br^-.

In the case of tertiary alkyl halides, the three alkyl groups bonded to the carbon atom bearing the halogen inhibit the rearward approach of the OH^- ion, and it is thought that tertiary alkyl halides undergo displacement reactions by a different mechanism. The rate-determining step is thought to be the formation of a carbocation:

$$CH_3-\underset{\underset{\displaystyle CH_3}{|}}{\overset{\overset{\displaystyle CH_3}{|}}{C}}-Br \longrightarrow CH_3\overset{\overset{\displaystyle CH_3}{|}}{\underset{\underset{\displaystyle CH_3}{|}}{C^+}} + Br^-$$

The carbocation is a powerful Lewis acid and rapidly reacts with water, which functions as a Lewis base:

$$CH_3-\overset{\overset{\displaystyle CH_3}{|}}{\underset{\underset{\displaystyle CH_3}{|}}{C^+}} + OH_2 \longrightarrow CH_3-\overset{\overset{\displaystyle CH_3}{|}}{\underset{\underset{\displaystyle CH_3}{|}}{C}}-OH_2^+$$

This ionic intermediate rapidly loses a proton to become the tertiary alcohol:

$$CH_3-\overset{\overset{\displaystyle CH_3}{|}}{\underset{\underset{\displaystyle CH_3}{|}}{C}}-OH_2^+ + OH^- \longrightarrow CH_3-\overset{\overset{\displaystyle CH_3}{|}}{\underset{\underset{\displaystyle CH_3}{|}}{C}}-OH + H_2O$$

This mechanism is given the designation S_N1 because it is a nucleophilic substitution in which the rate-determining step is unimolecular. The S_N1 mechanism is favored when the reaction is run in polar solvents (such as water), which aid the first-step ionization of the halide.

Tertiary alcohols also undergo S_N1 reactions. The steps of a typical substitution involving HBr are

$$(CH_3)_3COH + H^+ \rightleftharpoons (CH_3)_3COH_2^+ \qquad \text{(rapid, equilibrium)}$$

$$(CH_3)_3COH_2^+ \longrightarrow (CH_3)_3C^+ + H_2O \qquad \text{(rate determining)}$$

$$(CH_3)_3C^+ + Br^- \longrightarrow (CH_3)_3CBr$$

In all these nucleophilic substitution reactions, appreciable quantities of olefins are formed along with the substitution products; this is particularly true of the reactions of the tertiary compounds. Olefin formation is thought to proceed by the elimination of a proton by a carbocation:

$$H-\overset{\overset{\displaystyle H}{|}}{\underset{\underset{\displaystyle H}{|}}{C}}-\overset{\overset{\displaystyle CH_3}{|}}{\underset{\underset{\displaystyle CH_3}{|}}{C^+}} \longrightarrow H^+ + \overset{\overset{\displaystyle H}{|}}{\underset{\underset{\displaystyle H}{|}}{C}}=\overset{\overset{\displaystyle CH_3}{|}}{\underset{\underset{\displaystyle CH_3}{|}}{C}}$$

Tertiary carbocations are formed more readily than any other type (see Section 26.5). Elimination reactions of primary and secondary alcohols are thought to occur by a mechanism that does not involve the formation of carbocations.

The elimination of one water molecule from one molecule of an alcohol produces an olefin:

$$CH_3CH_2OH \xrightarrow[\textit{above } 150°C]{H_2SO_4} CH_2{=}CH_2 + H_2O$$

Under milder conditions only one molecule of water is removed from two molecules of alcohol, and an ether is produced:

$$CH_3CH_2OH + HOCH_2CH_3 \xrightarrow[130-140°C]{H_2SO_4} CH_3CH_2OCH_2CH_3 + H_2O$$
diethyl ether

Ethers, ROR, may be regarded as derivatives of water, HOH, in which both hydrogen atoms are replaced by alkyl groups, R. The alkyl groups may be alike or different. Ethers are also produced by nucleophilic substitution reactions involving alkoxide ions and alkyl halides:

$$CH_3CH_2O^- + CH_3CH_2CH_2Br \longrightarrow CH_3CH_2OCH_2CH_2CH_3 + Br^-$$
ethyl propyl ether

Unlike alcohols, ethers do not associate by hydrogen bonding. Therefore, the boiling points of ethers are much lower than those of alcohols of corresponding molecular weight:

$$CH_3CH_2OH \qquad CH_3OCH_3$$
ethanol   *dimethyl ether*
boiling point, 78°C   *boiling point*, −24°C

Notice that ethanol and dimethyl ether are isomeric; the type of isomerism displayed by this pair of compounds is known as **functional group isomerism.** The ethers are less reactive than the alcohols.

Polyhydroxy alcohols contain more than one OH group. Examples include 1,2-ethanediol (ethylene glycol):

$$\begin{array}{cc} CH_2 & \!\!\!\!-CH_2 \\ | & | \\ OH & OH \end{array}$$

and 1,2,3-propanetriol (glycerol), which is derived from fats:

$$\begin{array}{ccc} CH_2 & \!\!\!\!-CH- & \!\!\!\!CH_2 \\ | & | & | \\ OH & OH & OH \end{array}$$

Unlike aliphatic alcohols, which do not dissociate in water solution, the aromatic alcohols are weak acids in water:

OH
+ H_2O ⇌ H_3O^+ +   O^-

phenol   *phenoxide ion*

The acidity of phenol (carbolic acid) is attributed to the stability of the phenoxide ion in which the charge is delocalized by the π bonding system of the aromatic ring:

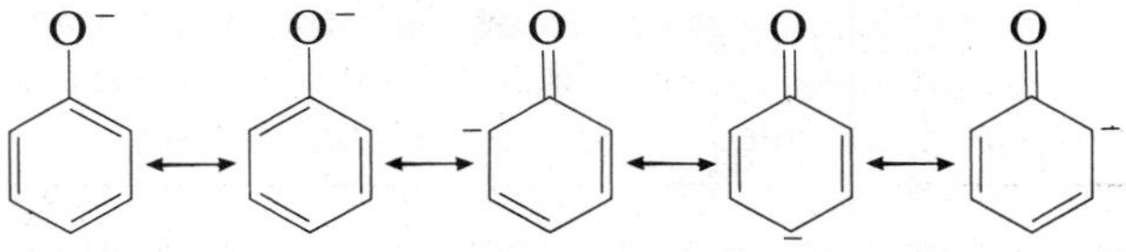

Aromatic halides do not readily undergo displacement reactions. Sodium phenoxide is commercially obtained from chlorobenzene by reaction with NaOH at elevated temperatures and under high pressure:

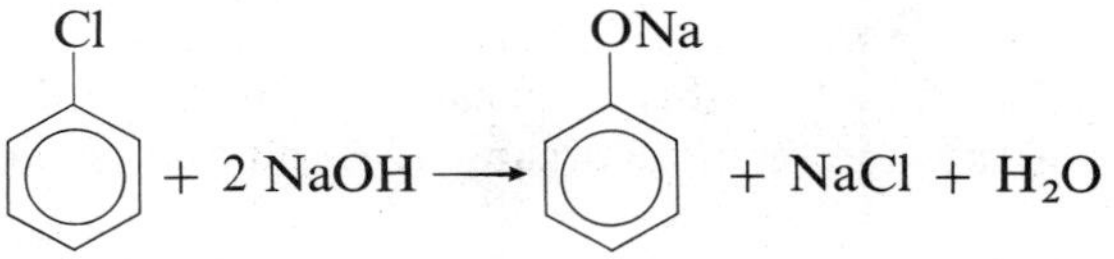

Phenol is derived by acidification of the phenoxide.

26.7 Carbonyl Compounds

A carbon atom can form a double bond with an oxygen atom to produce what is called a **carbonyl group**:

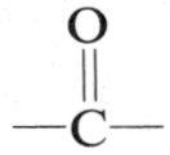

The double bond, like the carbon-carbon double bond, consists of a σ bond and a π bond between the bonded atoms. The carbon atom of the carbonyl group and the atoms bonded to it are coplanar and form bond angles of $120°$, a geometry that is typical of sp^2 hybridized species. The carbonyl double bond differs from the olefinic double bond in that it is markedly polar (oxygen is more electronegative than carbon).

If the carbonyl group is bonded to one hydrogen or to two hydrogens, the compound is an **aldehyde**:

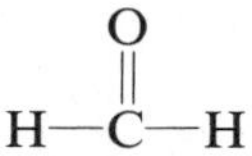

formaldehyde
methanal

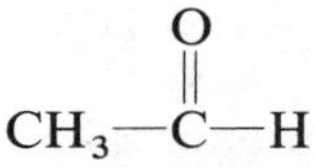

acetaldehyde
ethanal

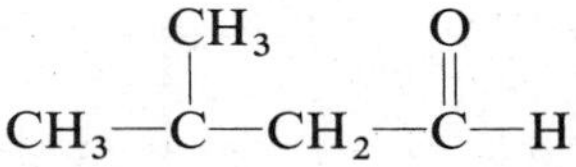

3-methylbutanal

Systematic nomenclature of aldehydes employs the ending- *al*; substituent groups are given numbers based on the assignment of the number 1 (understood) to the carbonyl carbon.

In a **ketone** the carbonyl group is bonded to two alkyl or aryl groups, which may be alike or different:

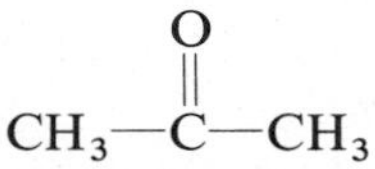

acetone
propanone

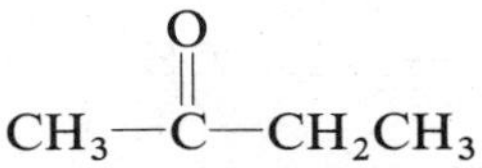

methyl ethyl ketone
butanone

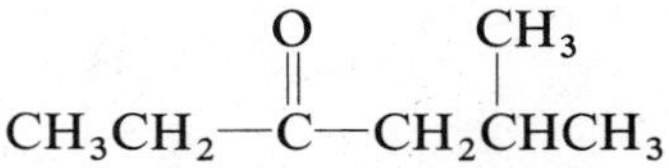

5-methyl-3-hexanone

The ending *-one* is used to designate a ketone. Numbers are employed to indicate the positions of substituents on the longest continuous chain containing the carbonyl group, which is assigned the lowest possible number.

Aldehydes may be prepared by the mild oxidation of primary alcohols. An oxidizing agent such as potassium dichromate in dilute sulfuric acid may be used:

$$3\ RCH_2OH + Cr_2O_7^{2-} + 8\ H^+ \longrightarrow 3\ RC\overset{\displaystyle O}{\overset{\|}{}}{-}H + 2\ Cr^{3+} + 7\ H_2O$$

Equations for a reaction such as this are usually written with the oxidizing agent indicated over the arrow and only the organic reactant and product shown. Thus,

$$CH_3CH_2OH \xrightarrow{Cr_2O_7^{2-}/H^+} CH_3C\overset{\displaystyle O}{\overset{\|}{}}{-}H$$

When an aldehyde is prepared by the oxidation of a primary alcohol, provision must be made to prevent the aldehyde from being destroyed by further oxidation, since aldehydes are easily oxidized to carboxylic acids (see Section 26.8). Many aldehydes may be distilled out of the reaction mixture as they are formed.

Some aldehydes are made commercially by reacting alcohol vapors with air over a copper catalyst at elevated temperatures:

$$2\ CH_3OH + O_2 \xrightarrow{Cu} 2\ H{-}C\overset{\displaystyle O}{\overset{\|}{}}{-}H + 2\ H_2O$$

The oxidation of alcohols may also be conducted by passing the hot alcohol vapor over a heated copper catalyst in the absence of oxygen. This process results in the removal of a molecule of hydrogen from an alcohol molecule and is called a **dehydrogenation**; it avoids the danger of secondary oxidation of the aldehyde product:

$$CH_3CH_2CH_2OH \xrightarrow{Cu} CH_3CH_2C\overset{\displaystyle O}{\overset{\|}{}}{-}H + H_2$$

The oxidation of secondary alcohols produces ketones:

$$CH_3CH_2\overset{\displaystyle OH}{\overset{|}{C}}HCH_3 \xrightarrow{Cr_2O_7^{2-}/H^+} CH_3CH_2\overset{\displaystyle O}{\overset{\|}{C}}CH_3$$

Acetone (propanone) may be prepared by the oxidation of 2-propanol with oxygen or by the dehydrogenation of the alcohol:

$$CH_3\overset{\displaystyle OH}{\overset{|}{C}}HCH_3 \xrightarrow{Cu} CH_3\overset{\displaystyle O}{\overset{\|}{C}}CH_3 + H_2$$

The oxidation of tertiary alcohols results in the destruction of the carbon skeleton of the molecule.

The double bond of the carbonyl group readily undergoes many addition reactions. Reduction of aldehydes or ketones with hydrogen in the presence of a nickel or platinum catalyst yields primary or secondary alcohols:

$$\underset{\textstyle R-\overset{\textstyle O}{\overset{\|}{C}}-H}{} + H_2 \longrightarrow \underset{\textstyle R-\overset{\textstyle OH}{\overset{|}{C}H_2}}{}$$

$$\underset{\textstyle R-\overset{\textstyle O}{\overset{\|}{C}}-R}{} + H_2 \longrightarrow \underset{\textstyle R-\overset{\textstyle OH}{\overset{|}{C}H}-R}{}$$

When an unsymmetrical reagent (such as HCN) adds to the double bond, the positive part of the addend (the hydrogen) adds to the oxygen, and the negative part of the addend (the cyanide group) bonds to the carbon. This mode of addition reflects the polarity of a carbonyl double bond; the oxygen is negative with respect to the carbon:

$$R-\overset{\overset{\textstyle O}{\|}}{C}-R + H-C\equiv N \longrightarrow R-\overset{\overset{\textstyle OH}{|}}{\underset{\underset{\textstyle C\equiv N}{|}}{C}}-R$$

a cyanohydrin

The proposed mechanism for additions of this type consists of the electrophilic attack of the anionic Lewis base:

$$R-\overset{\overset{\textstyle O}{\|}}{C}-R + CN^- \longrightarrow R-\overset{\overset{\textstyle O^-}{|}}{\underset{\underset{\textstyle CN}{|}}{C}}-R$$

followed by combination with a proton:

$$R-\overset{\overset{\textstyle O^-}{|}}{\underset{\underset{\textstyle CN}{|}}{C}}-R + H^+ \longrightarrow R-\overset{\overset{\textstyle OH}{|}}{\underset{\underset{\textstyle CN}{|}}{C}}-R$$

An important addition reaction involves organomagnesium compounds known as **Grignard reagents.** Alkyl halides react with magnesium metal in dry diethyl ether:

$$R-X + Mg \longrightarrow R-Mg-X$$

These reagents are usually given the formula RMgX. The materials, however, possess a high degree of ionic character and may be mixtures of magnesium dialkyls (MgR_2) and magnesium halides (MgX_2). Water must be excluded from a Grignard reaction because these reagents are easily hydrolyzed:

$$R-Mg-X + HOH \longrightarrow R-H + Mg(OH)X$$

Victor Grignard, 1871–1934.
Burndy Library.

The formula Mg(OH)X stands for an equimolar mixture of magnesium hydroxide and magnesium halide.

A Grignard reagent adds to an aldehyde as follows:

$$
\underset{\substack{\| \\ R'-C-H}}{O} + R-Mg-X \longrightarrow \underset{\substack{OMgX \\ | \\ R'-C-H \\ | \\ R}}{}
$$

where the alkyl groups R and R' may be alike or different. The addition compound is readily decomposed by water or by dilute acid.

$$
\underset{\substack{OMgX \\ | \\ R'-C-H \\ | \\ R}}{} + HX \longrightarrow \underset{\substack{OH \\ | \\ R'-C-H \\ | \\ R}}{} + MgX_2
$$

Thus, the ultimate product of the reaction of a Grignard reagent and an aldehyde is a secondary alcohol.

Hydrolysis of an addition compound formed by a Grignard reagent and a ketone yields a tertiary alcohol:

$$
\underset{\substack{\| \\ R'-C-R''}}{O} + RMgX \longrightarrow \underset{\substack{OMgX \\ | \\ R'-C-R'' \\ | \\ R}}{}
$$

$$
\underset{\substack{OMgX \\ | \\ R'-C-R'' \\ | \\ R}}{} + HX \longrightarrow \underset{\substack{OH \\ | \\ R'-C-R'' \\ | \\ R}}{} + MgX_2
$$

The alkyl groups may be alike or different.

In these reactions of Grignard reagents, new carbon-carbon bonds are formed, and the reactions are frequently useful as steps in organic syntheses. The alcohols produced may be oxidized to carbonyl compounds, subjected to displacement reactions, or dehydrated to olefins. Thus, a series of reactions may be employed to synthesize a desired compound from compounds of lower molecular weight.

26.8 Carboxylic Acids and Esters

Oxidation of the aldehyde group yields a **carboxyl group:**

$$
\underset{\substack{\| \\ -C-H}}{O} \longrightarrow \underset{\substack{\| \\ -C-OH}}{O}
$$

Compounds containing the —COOH group are weak acids (**carboxylic acids**); ionization constants for some of these acids are listed in Table 26.6:

Table 26.6   Properties of some carboxylic acids

Acid	Formula	Melting Point (°C)	Boiling Point (°C)	Ionization Constant at 25°C
methanoic (formic)	HCOOH	8	101	1.8×10^{-4}
ethanoic (acetic)	CH_3COOH	17	118	1.8×10^{-5}
propanoic (propionic)	CH_3CH_2COOH	-22	141	1.4×10^{-5}
butanoic (butyric)	$CH_3(CH_2)_2COOH$	-8	164	1.5×10^{-5}
pentanoic (valeric)	$CH_3(CH_2)_3COOH$	-35	187	1.6×10^{-5}
benzoic	C_6H_5COOH	122	249	6.0×10^{-5}

$$R-\overset{\overset{\displaystyle O}{\|}}{C}-OH + H_2O \rightleftharpoons R-\overset{\overset{\displaystyle O}{\|}}{C}-O^- + H_3O^+$$

The charge of the carboxylate anion is delocalized:

$$R-\overset{\overset{\displaystyle O}{\|}}{C}-O^- \longleftrightarrow R-\overset{\overset{\displaystyle O^-}{|}}{C}=O$$

The acids are associated through hydrogen bonding. Lower members of the series form dimers in the vapor state:

$$R-C \overset{O---H-O}{\underset{O-H---O}{\Bigg\langle\quad\Bigg\rangle}} C-R$$

According to systematic nomenclature, the name of a carboxylic acid is derived from the parent hydrocarbon by elision of the final -e, addition of the ending -oic, and addition of the separate word *acid*. The numbers used to designate the positions of substituents on the carbon chain are derived by numbering the chain starting with the carbon of the carboxyl group:

$$CH_3CH_2\overset{\overset{\displaystyle O}{\|}}{C}-OH \qquad\qquad CH_3\overset{\overset{\displaystyle CH_3}{|}}{C}HCH_2\overset{\overset{\displaystyle O}{\|}}{C}-OH$$

propanoic acid   *3-methylbutanoic acid*

Carboxylic acids may be prepared by the oxidation of primary alcohols or aldehydes. Potassium dichromate or potassium permanganate is frequently employed as the oxidizing agent:

$$CH_3CH_2CH_2CH_2OH \xrightarrow{Cr_2O_7^{2-}/H^+} CH_3CH_2CH_2\overset{\overset{\displaystyle O}{\|}}{C}-OH$$

Acetic acid is commercially prepared by the air oxidation of acetaldehyde:

$$2\ CH_3\overset{\displaystyle O}{\overset{\|}{C}}\!\!-\!\!H + O_2 \xrightarrow{\ Mn(C_2H_3O_2)_2\ } 2\ CH_3\overset{\displaystyle O}{\overset{\|}{C}}\!\!-\!\!OH$$

Vigorous oxidation of an alkyl-substituted aromatic compound converts the side chain to a carbonyl group and thus yields benzoic acid:

toluene   benzoic acid

These oxidations do not touch the aromatic ring, an illustration of the stability of this structure.

The oxidation of an unsaturated hydrocarbon yields a variety of products, including carboxylic acids, depending upon conditions. Dilute aqueous permanganate at room temperature oxidizes olefins to glycols:

$$R\!-\!CH\!=\!CH\!-\!R \xrightarrow{\ MnO_4^-\ } R\!-\!\underset{\underset{OH}{|}}{CH}\!-\!\underset{\underset{OH}{|}}{CH}\!-\!R$$

Under more vigorous conditions (more concentrated solutions and heating), the carbon chain is cleaved at the double bond and carboxylic acids are produced:

$$R\!-\!CH\!=\!CH\!-\!R \xrightarrow{\ MnO_4^-\ } R\!-\!\overset{\displaystyle O}{\overset{\|}{C}}\!\!-\!\!OH + HO\!-\!\overset{\displaystyle O}{\overset{\|}{C}}\!\!-\!\!R$$

Alkynes also may be cleaved to yield acids:

$$R\!-\!C\!\equiv\!C\!-\!R \xrightarrow{\ MnO_4^-\ } R\!-\!\overset{\displaystyle O}{\overset{\|}{C}}\!\!-\!\!OH + HO\!-\!\overset{\displaystyle O}{\overset{\|}{C}}\!\!-\!\!R$$

Oxidation of some olefins yields ketones:

$$R\!-\!\underset{\underset{R}{|}}{C}\!=\!\underset{\underset{R}{|}}{C}\!-\!R \xrightarrow{\ MnO_4^-\ } R\!-\!\underset{\underset{R}{|}}{C}\!=\!O + O\!=\!\underset{\underset{R}{|}}{C}\!-\!R$$

Salts of carboxylic acids may be produced by the alkaline hydrolysis of alkyl cyanides; the cyanides are the product of nucleophilic substitution reactions of the CN^- ion with alkyl halides:

$$CH_3CH_2Br + CN^- \longrightarrow CH_3CH_2CN + Br^-$$

$$CH_3CH_2CN + OH^- + H_2O \longrightarrow CH_3CH_2\overset{\displaystyle O}{\overset{\|}{C}}\!\!-\!\!O^- + NH_3$$

Some compounds contain more than one carboxyl group. Examples include

$$\begin{array}{c} COOH \\ | \\ COOH \end{array} \qquad \text{and} \qquad \begin{array}{c} CH_2COOH \\ | \\ HO-C-COOH \\ | \\ CH_2COOH \end{array}$$

oxalic acid   *citric acid*

Carboxylic acids react with alcohols to produce compounds known as **esters**:

$$CH_3\overset{\displaystyle O}{\overset{\|}{C}}{-}OH + CH_3CH_2OH \rightleftharpoons \left[CH_3\overset{\displaystyle OH}{\underset{\displaystyle OCH_2CH_3}{\overset{|}{\underset{|}{C}}}}{-}OH \right] \rightleftharpoons$$

$$CH_3\overset{\displaystyle O}{\overset{\|}{C}}{-}OCH_2CH_3 + H_2O$$

ethyl acetate

Esterification reactions are reversible and proceed to equilibrium; they may be forced in either direction by the appropriate choice of conditions. The name of an ester reflects the alcohol and acid from which it is derived, the ending *-ate* being employed with the base of the name of the acid to give the second portion of the name of the ester. The lower molecular weight esters have pleasant, fruity odors.

The alkaline hydrolysis of an ester produces an alcohol and a salt of a carboxylic acid:

$$R\overset{\displaystyle O}{\overset{\|}{C}}OR' + OH^- \longrightarrow R{-}\overset{\displaystyle O}{\overset{\|}{C}}{-}O^- + R'OH$$

26.9 Amines and Amides

The **amines** may be considered as derivatives of ammonia with one, two, or three hydrogen atoms replaced by alkyl or aryl groups:

$$\begin{array}{c} H \\ | \\ CH_3-N-H \end{array} \qquad \begin{array}{c} CH_3 \\ | \\ CH_3-N-H \end{array} \qquad \begin{array}{c} CH_3 \\ | \\ CH_3-N-CH_3 \end{array}$$

methylamine   *dimethylamine*   *trimethylamine*
a primary amine   *a secondary amine*   *a tertiary amine*

The amines resemble ammonia in that they are weak bases:

$$CH_3NH_2 + H_2O \rightleftharpoons CH_3NH_3^+ + OH^-$$

$$(CH_3)_2NH + H_2O \rightleftharpoons (CH_3)_2NH_2^+ + OH^-$$

$$(CH_3)_3N + H_2O \rightleftharpoons (CH_3)_3NH^+ + OH^-$$

Table 26.7 Properties of some amines				
Amine	Formula	Melting Point (°C)	Boiling Point (°C)	Ionization Constant at 25°C
methylamine	CH_3NH_2	−93	−7	5×10^{-4}
dimethylamine	$(CH_3)_2NH$	−96	7	7.4×10^{-4}
trimethylamine	$(CH_3)_3N$	−124	4	7.4×10^{-5}
ethylamine	$CH_3CH_2NH_2$	−81	17	5.6×10^{-4}
propylamine	$CH_3CH_2CH_2NH_2$	−83	49	4.7×10^{-4}
aniline	$C_6H_5NH_2$	−6	184	4.6×10^{-10}

Ionization constants for some amines are listed in Table 26.7. The amines form ionic salts with acids:

$$CH_3NH_2 + HCl \longrightarrow CH_3NH_3^+ + Cl^-$$

which may be decomposed by hydroxides:

$$CH_3NH_3^+ + OH^- \longrightarrow CH_3NH_2 + H_2O$$

The last reaction is the reverse of the ionization of methylamine in water; it is forced to yield the free amine by employment of an excess of OH^- or by warming.

Amines may be prepared by reacting alkyl halides with ammonia. These reactions are nucleophilic substitutions:

$$NH_3 + CH_3CH_2Br \longrightarrow CH_3CH_2NH_3^+ + Br^-$$

Treatment of the amine salt with hydroxide ion yields the free amine. Mixtures of primary, secondary, and tertiary amines are produced by this type of reaction.

Primary amines may be produced by the catalytic hydrogenation of alkyl cyanides:

$$CH_3CH_2C{\equiv}N + 2H_2 \longrightarrow CH_3CH_2CH_2{-}NH_2$$

Aniline is produced by the reduction of nitrobenzene. Iron and steam, with a trace of hydrochloric acid, serve as the reducing agent:

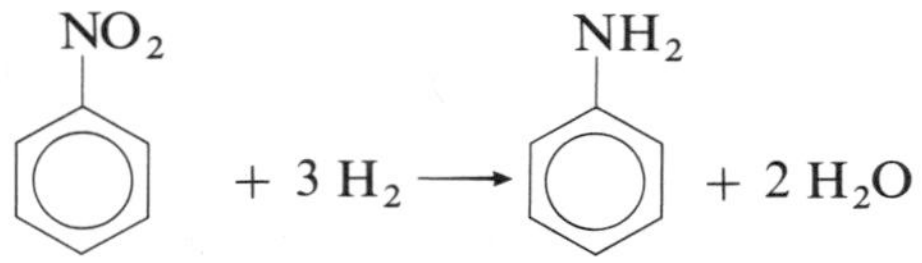

Some compounds that contain more than one amino group are important. Ethylenediamine (1,2-diaminoethane, $H_2N{-}CH_2{-}CH_2{-}NH_2$) is a useful chelating agent (see Section 24.1), and hexamethylenediamine (1,6-diaminohexane, $H_2N(CH_2)_6NH_2$) is used in the manufacture of nylon.

Amides may be produced from ammonia and carboxylic acids. The preparation proceeds through the formation of ammonium salts which eliminate water upon heating:

$$CH_3\overset{\displaystyle O}{\overset{\|}{C}}{-}OH + NH_3 \longrightarrow CH_3\overset{\displaystyle O}{\overset{\|}{C}}{-}O^-NH_4^+$$

$$\underset{\textit{ammonium acetate}}{CH_3\overset{\displaystyle O}{\overset{\|}{C}}{-}O^-NH_4^+} \longrightarrow \underset{\textit{acetamide}}{CH_3\overset{\displaystyle O}{\overset{\|}{C}}{-}NH_2} + H_2O$$

Amides may also be prepared by the nucleophilic substitution reaction of ammonia on an ester:

$$NH_3 + CH_3\overset{\displaystyle O}{\overset{\|}{C}}{-}OCH_2CH_3 \longrightarrow CH_3\overset{\displaystyle O}{\overset{\|}{C}}{-}NH_2 + CH_3CH_2OH$$

Primary and secondary amines react with acids in a similar manner to produce substituted amides, compounds in which alkyl or aryl radicals replace hydrogen atoms of the $-NH_2$ group.

26.10 Polymers

Starch, cellulose, and proteins are examples of natural **polymers**—molecules of high molecular weight that are formed from simpler molecules, called **monomers.** Important polymers, or **macromolecules,** that do not occur in nature have been synthesized; most of these are linear, or chain-type, structures although some are cross-linked.

Many polymers are formed from compounds that contain carbon-carbon double bonds by a process that is called **addition polymerization.** For example, ethene (ethylene) polymerizes upon heating ($100°C$–$400°C$) under high pressure (1000 atm); the product is called polyethylene:

$$\begin{array}{c}\text{H} \quad \text{H} \\ | \quad\; | \\ \text{C}{=}\text{C} \\ | \quad\; | \\ \text{H} \quad \text{H}\end{array} + \begin{array}{c}\text{H} \quad \text{H} \\ | \quad\; | \\ \text{C}{=}\text{C} \\ | \quad\; | \\ \text{H} \quad \text{H}\end{array} \longrightarrow \begin{array}{c}\text{H} \quad \text{H} \quad \text{H} \quad \text{H} \\ | \quad\; | \quad\; | \quad\; | \\ {-}\text{C}{-}\text{C}{-}\text{C}{-}\text{C}{-} \\ | \quad\; | \quad\; | \quad\; | \\ \text{H} \quad \text{H} \quad \text{H} \quad \text{H}\end{array}$$

The preceding equation shows the combination of only two molecules of ethene; the actual polymerization process continues by successive additions of $CH_2{=}CH_2$ molecules until chains of hundreds or thousands of $-CH_2-CH_2-$ units are produced. The product, polyethylene, is a tough, waxy solid.

An addition polymerization is thought to proceed by either a free radical or a carbocation chain mechanism. The free radical type may be initiated by small amounts of hydrogen peroxide or organic peroxides, which are readily split into free radicals (indicated by $Z\cdot$ in the equations that follow):

$$Z_2 \longrightarrow 2Z\cdot$$

A free radical from an initiator combines with one of the π electrons of a molecule of the monomer to form a new free radical:

$$Z\cdot + CH_2{=}CH_2 \longrightarrow Z{-}CH_2{-}CH_2\cdot$$

Chain propagation occurs by the combination of this free radical with another molecule of the monomer:

$$Z{-}CH_2{-}CH_2\cdot + CH_2{=}CH_2 \longrightarrow Z{-}CH_2{-}CH_2{-}CH_2{-}CH_2\cdot$$

and the molecule grows by repeated additions. Chain termination is caused by the combination of two free radicals or by other means, such as the elimination of a hydrogen atom.

A carbocation chain polymerization is initiated by a Lewis acid, such as $AlCl_3$ or BF_3. In the equations that follow, the proton is used to indicate the Lewis acid, or electrophilic, initiator although stronger Lewis acids are more frequently employed. Combination of the initiator with a molecule of the monomer produces a carbocation:

$$H^+ + CH_2{=}CH_2 \longrightarrow CH_3{-}CH_2^+$$

Chain propagation occurs through the combination of the carbocation with both π electrons of a molecule of the monomer:

$$CH_3{-}CH_2^+ + CH_2{=}CH_2 \longrightarrow CH_3{-}CH_2{-}CH_2{-}CH_2^+$$

Chain termination may occur by the elimination of a proton from a long-chain carbocation:

$$CH_3{-}CH_2(CH_2{-}CH_2)_n CH_2{-}CH_2^+ \longrightarrow$$
$$CH_3{-}CH_2(CH_2{-}CH_2)_n CH{=}CH_2 + H^+$$

Important polymers have been made from ethene derivatives. Orlon and acrilan are polymers of acrylonitrile, $CH_2{=}CH{-}C{\equiv}N$; the polymers have the structure

$$-CH_2{-}\underset{\underset{CN}{|}}{CH}\left(CH_2{-}\underset{\underset{CN}{|}}{CH}\right)_n CH_2{-}\underset{\underset{CN}{|}}{CH}-$$

Teflon is a polymer of tetrafluoroethene, $CF_2{=}CF_2$. Vinyl chloride, $CH_2{=}CHCl$, and vinyl acetate:

$$CH_2{=}CH{-}O{-}\overset{\overset{O}{\|}}{C}{-}CH_3$$

are important monomers in the preparation of vinyl plastics. Lucite, or Plexiglas, is a polymer of methyl methacrylate, $CH_2{=}C(CH_3)COOCH_3$.

Natural rubber is a polymer of the diolefin 2-methyl-1,3-butadiene, or isoprene:

$$CH_2{=}\underset{\underset{CH_3}{|}}{C}{-}CH{=}CH_2$$

Such a system of alternating double and single bonds is called a **conjugated system.** Rubber consists of thousands of these units joined in a chain:

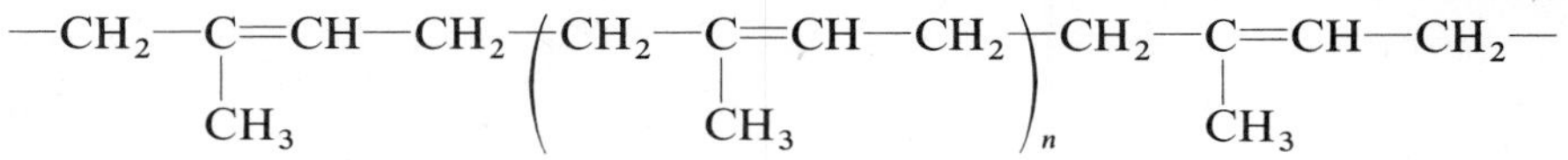

$$-CH_2-\underset{\underset{CH_3}{|}}{C}=CH-CH_2-\left(CH_2-\underset{\underset{CH_3}{|}}{C}=CH-CH_2-\right)_n CH_2-\underset{\underset{CH_3}{|}}{C}=CH-CH_2-$$

Notice that the polymer contains double bonds. Crude rubber is vulcanized by heating with sulfur. It is thought that the sulfur atoms add to some of the double bonds, thereby linking adjacent chains into a complex network. This process adds strength to the final product.

A number of types of synthetic rubber have been made utilizing such monomers as 1,3-butadiene ($CH_2=CH-CH=CH_2$), 2-chloro-1,3-butadiene (chloroprene, $CH_2=C(Cl)-CH=CH_2$), and 2,3-dimethyl-1,3-butadiene ($CH_2=C(CH_3)-C(CH_3)=CH_2$) alone or in combination. The product formed by the polymerization of two different monomers is called a **copolymer;** the molecular structures and properties of such materials depend upon the proportions of the two monomers employed. Buna S rubber is a copolymer of 1,3-butadiene and styrene ($C_6H_5CH=CH_2$). A section of the chain of this material has the structure

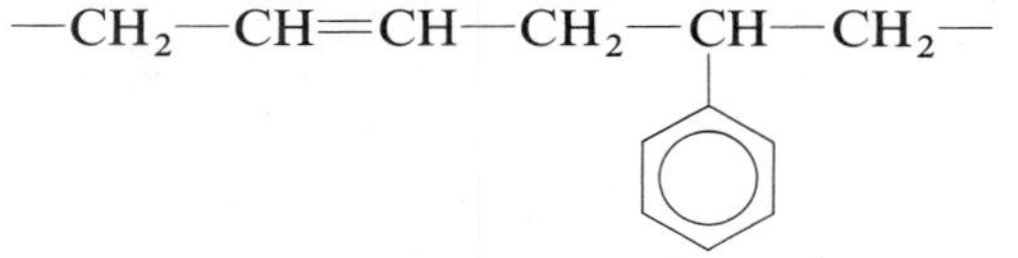

$$-CH_2-CH=CH-CH_2-CH-CH_2-$$

Condensation polymerization occurs between molecules of monomers by the elimination of a small molecule, usually water. Proteins and polysaccharides are condensation polymers. Nylon, like the proteins, is a polyamide. It is formed by the condensation of a diamine [such as hexamethylenediamine, $H_2N(CH_2)_6NH_2$] and a dicarboxylic acid [such as adipic acid, $HOOC(CH_2)_4COOH$]. The polymerization occurs through the elimination of water molecules. The production of a small section of the chain can be illustrated as follows:

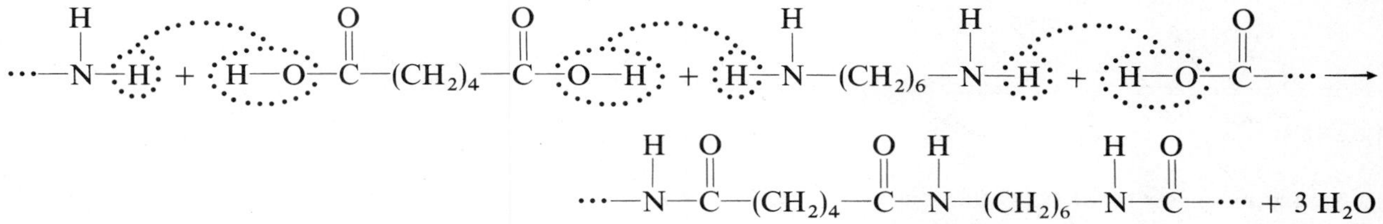

Dacron is a polyester formed by the elimination of water between ethylene glycol ($HOCH_2CH_2OH$) and a dicarboxylic acid (such as terephthalic acid, $p\text{-}HOOCC_6H_4COOH$):

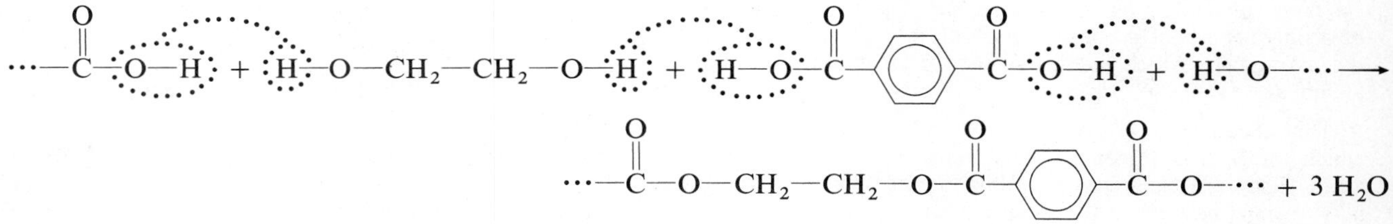

Bakelite is a cross-linked polymer formed from phenol (C_6H_5OH) and formaldehyde ($H_2C{=}O$) by the elimination of water. Notice that the formaldehyde units condense in the *ortho* and *para* positions of phenol:

Summary

The topics that have been discussed in this chapter are

1. The structure, isomerism, physical properties, and nomenclature of the alkanes (the saturated hydrocarbons), the alkenes (hydrocarbons that contain a carbon-carbon double bond), and the alkynes (hydrocarbons that contain a carbon-carbon triple bond.

2. The structure of benzene and its derivates, which are called aromatic compounds. The nomenclature of benzene derivatives.

3. The chemical reactions of the hydrocarbons: substitution and addition reactions.

4. Alcohols (compounds that contain a hydroxyl group), their structure, nomenclature, physical properties, and reactions. Ethers, which may be derived from alcohols.

5. Carbonyl compounds: aldehydes and ketones, their structure, nomenclature, preparation, and reactions.

6. Carboxylic acids (organic acids that contain the carboxyl group), their structure, nomenclature, and preparation.

7. Esters, which are produced from carboxylic acids and alcohols.

8. Amines (organic bases that are derivatives of ammonia), their physical properties and preparation.

9. Amides, which may be produced from a carboxylic acid and ammonia or an amine.

10. Polymers; addition and condensation polymerization.

Key Terms

Some of the more important terms introduced in this chapter are listed below. Definitions for terms not included in this list may be located in the text by the use of the index.

Addition reaction (Sections 26.2 and 26.5) A reaction in which two parts of a reagent add to a multiple bond, one part to each side of the bond.

Addition polymerization (Section 26.10) The formation of a polymer by the successive addition of molecules of the monomer, which is an alkene or a conjugated alkadiene.

Alcohol (Section 26.6) A derivative of an alkane in which the hydroxyl group, —OH, replaces a hydrogen atom. They are classified according to the number of alkyl groups bonded to the carbon atom that holds the

—OH group as a **primary alcohol** (one R group), a **secondary alcohol** (two R groups), or a **tertiary alcohol** (three R groups).

Aldehyde (Section 26.7) A compound that has the general formula

$$
\begin{array}{c} O \\ \parallel \\ R{-}C{-}H \end{array}
$$

where R can be a hydrogen, an alkyl group, or an aryl group.

Alkadiene (Section 26.2) An open-chain hydrocarbon that contains *two* carbon-carbon double bonds; the alkadienes conform to the general formula C_nH_{2n-2}.

Alkane (Section 26.1) An open-chain hydrocarbon in which all carbon-carbon bonds are single bonds; the alkanes conform to the general formula C_nH_{2n+2}.

Alkene (Section 26.2) An open-chain hydrocarbon that contains a carbon-carbon double bond; the alkenes conform to the general formula C_nH_{2n}.

Alkyl halide (Section 26.6) A compound formed from an alkyl radical (R) and a halogen atom (X), RX.

Alkyl radical (Section 26.1): A fragment of an alkane from which a hydrogen atom has been removed.

Alkyne (Section 26.3) An open-chain hydrocarbon that contains a carbon-carbon triple bond; the alkynes conform to the general formula C_nH_{2n-2}.

Amide (Section 26.9) A compound with the general formula

$$\begin{array}{c} O \\ \parallel \\ R-C-NR_2 \end{array}$$

in which R may be a hydrogen atom, an alkyl radical, or an aryl radical, and the three R groups may be alike or different.

Amine (Section 26.9) An organic base that can be considered to be a derivative of ammonia (NH_3) with one hydrogen replaced (RNH_2, a **primary amine**), two hydrogens replaced (R_2NH, a **secondary amine**), or three hydrogens replaced (R_3N, a **tertiary amine**). An R group may be either an alkyl radical or an aryl radical.

Aromatic compound (Section 26.4) Benzene or a derivative of benzene.

Aryl radical (Section 26.4) A radical formed by the removal of a hydrogen atom from a carbon of the benzene ring of an aromatic compound.

Carbocation (Section 26.5) A positive ion formed by a group of atoms that contains a carbon atom that has only six valence electrons.

Carbonyl compound (Section 26.7) A compound that contains a carbonyl group,

$$\begin{array}{c} O \\ \parallel \\ -C- \end{array}$$

Carboxylic acid (Section 26.8) An organic acid that has the general formula

$$\begin{array}{c} O \\ \parallel \\ R-C-OH \end{array}$$

where R can be an alkyl radical or an aryl radical.

Condensation polymerization (Section 26.10) A polymerization in which the monomer molecules combine with the elimination of a small molecule, usually water.

Conjugated double-bond system (Section 26.10) A compound with a carbon chain containing alternating double and single bonds.

Copolymer (Section 26.10) A polymer formed by the polymerization of two different monomers.

Cracking (Section 26.1) An industrial process in which large molecules are broken down into smaller ones.

Cycloalkane (Section 26.1) A hydrocarbon that contains a ring arrangement of carbon atoms and has only carbon-carbon single bonds; the cycloalkanes conform to the general formula C_nH_{2n}.

Cycloalkene (Section 26.2) A hydrocarbon that has a ring arrangement of carbon atoms and contains a carbon-carbon double bond; the cycloalkenes conform to the general formula C_nH_{2n-2}.

Dehydrogenation (Section 26.7) A reaction in which two hydrogen atoms are removed from a molecule to form a double bond.

Displacement reaction (Section 26.6) A reaction in which one group displaces another from an organic molecule. It is also called a **nucleophilic substitution reaction** since it involves the substitution of one nucleophilic substance (a Lewis base) for another. If the rate-determining step is unimolecular, the reaction is called an S_N1 **reaction**; if bimolecular, an S_N2 **reaction**.

Ester of a carboxylic acid (Section 26.8) The product of the reaction of a carboxylic acid

$$\begin{array}{c} O \\ \parallel \\ R-C-OH \end{array}$$

(where R is a hydrogen atom, an alkyl radical, or an aryl radical) and an alcohol or phenol, R'OH (where R' is an alkyl radical or an aryl radical);

$$\begin{array}{c} O \\ \parallel \\ R-C-OR' \end{array}$$

where R and R' may be alike or different.

Ether (Section 26.6) A compound that has the general formula ROR, where the R groups may be alkyl or aryl radicals that are alike or different.

Friedel-Crafts synthesis (Section 26.5) A reaction of an aromatic compound with an alkyl halide in which HCl is eliminated and a new side chain introduced into the benzene ring of the aromatic compound.

Functional group (Section 26.6) An atom or a group of atoms that gives organic compounds bearing them characteristic physical and chemical properties.

Grignard reagent (Section 26.7) An organomagnesium compound, RMgX, where R can be an alkyl or an aryl radical.

Homologous series (Section 26.1) A series of compounds that may be considered to be derived from the first member by the successive addition of a specific number of atoms. The members of a homologous series are called **homologs** and all of them conform to the same general formula.

Ketone (Section 26.7) A carbonyl compound that has the general formula

$$R - \overset{\overset{\displaystyle O}{\|}}{C} - R$$

where the R groups may be alkyl or aryl radicals that are alike or different.

Macromolecule (Section 26.10) A large, high molecular-weight molecule.

Markovnikov's rule (Section 26.5) The principal product of the addition of an acid to the double bond of an alkene is the compound in which the hydrogen atom of the acid bonds to the carbon atom of the double bond that originally had the larger number of hydrogen atoms.

Olefin (Section 26.2) An alkene.

Ortho, meta, and para isomers (Section 26.5) The three isomers of a disubstituted benzene. In the *ortho* isomer, the two substituents are on carbon atoms 1 and 2 of the benzene ring; in the *meta* isomer, 1 and 3; and in the *para* isomer, 1 and 4.

Polymers (Section 26.10) Macromolecules that are formed from simpler molecules called **monomers**.

Saturated hydrocarbon (Section 26.2) A hydrocarbon that contains only carbon-carbon single bonds; an alkane or a cycloalkane.

Stereoisomers (Section 26.2) Compounds that have the same molecular formula and are bonded in the same way but differ in the way the constituent atoms are arranged in space. **Geometric isomers** (Section 26.2), in organic chemistry, are compounds that owe their existence to the restricted rotation about a double bond. **Optical isomers** (discussed in Chapter 24, Section 24.4, and Chapter 27, Section 27.1) are compounds that are mirror images and that are not superimposable.

Structural isomers (Section 26.1) Compounds that have the same molecular formula but differ in the way the atoms are bonded together. Several types exist: **chain isomers** (Section 26.2) are compounds that differ in their arrangement of carbon-carbon chains, **functional-group isomers** (Section 26.6) are compounds that differ in their functional groups (for example, an alcohol and an ether), and **position isomers** (Section 26.2) are compounds that have the same carbon-carbon chains but differ in the position of a substituent (or a multiple bond) in that chain.

Substitution reaction (Section 26.5) A reaction in which an atom or a group of atoms is substituted for another atom or group of atoms in a molecule.

Unsaturated hydrocarbon (Sections 26.2 and 26.3) A hydrocarbon that contains one or more carbon-carbon multiple bonds.

Problems*

Hydrocarbons

26.1 Write structural formulas for the following compounds:
(a) *trans*-2-hexene
(b) 2,4-dimethyl-3-ethylpentane
(c) 2,4-dimethyl-1,4-pentadiene
(d) 3-isopropyl-1-hexyne
(e) 2,2,3-trichlorobutane
(f) *p*-xylene
(g) 2,3,6-trinitrotoluene
(h) naphthalene

26.2 Write structural formulas for:
(a) 2,3-dimethylpentane
(b) 3-ethyl-3-methyl-1-pentene
(c) 2,5-dimethyl-3-hexyne
(d) 1,2,4-tribromobenzene
(e) *p*-dinitrobenzene

26.3 What is incorrect about each of the following names:
(a) 3-pentene
(b) 1,2-dimethylpropane
(c) 2-methyl-2-butyne
(d) 2-methyl-3-butyne
(e) 3,3-dimethyl-2-butene
(f) 2-ethylpentane

26.4 The general formula for the alkanes is C_nH_{2n+2}. What are the general formulas for the (a) alkenes, (b) alkynes, (c) alkadienes, (d) cycloalkanes?

26.5 Write structural formulas for all the mono-, di-, and tri-substituted chloro derivatives of propane.

26.6 Write structural formulas for all the isomers that have the formula C_4H_8.

26.7 Write structual formulas for all the isomers that have the formula $C_4H_8Cl_2$.

26.8 Write structural formulas for all the isomers of (a) bromophenol, (b) dibromophenol, (c) tribromobenzene.

---

* The more difficult problems are marked with asterisks. The appendix contains answers to color-keyed problems.

26.9 *Ortho*-, *meta*-, and *para*-isomers can be identified by determining the number of compounds that can be derived from each isomer when a third substituent is introduced into the ring (Körner's method). If one nitro group is introduced into the benzene ring of each of the following compounds, how many isomeric products would be secured from each?
(a) *o*-dibromobenzene
(b) *m*- dibromobenzene
(c) *p*-dibromobenzene

26.10 Write structural formulas for the ten dichloro substitution isomers of naphthalene.

26.11 *Cis*- and *trans*-isomers are known for certain alkenes. Why do the alkynes not display this type of isomerism?

26.12 For which of the following compounds are *cis*- and *trans*-isomers possible?
(a) 2,3-dimethyl-2-butene
(b) 3-methyl-2-pentene
(c) 2-methyl-2-pentene
(d) $CH(COOH)\!=\!CH(COOH)$

26.13 Define the following: **(a)** olefin, **(b)** homologous series, **(c)** cracking, **(d)** conjugated system, **(e)** addition reaction, **(f)** substitution reaction.

26.14 Draw the resonance forms of **(a)** benzene, **(b)** naphthalene.

26.15 Write equations for the reactions of **(a)** H_2 with *cis*-2-pentene, **(b)** H_2 with 2-pentyne, **(c)** HBr with $(CH_3CH_2)_2C\!=\!CH_2$, **(d)** HBr with $CH_3C\!\equiv\!CH$.

26.16 What is the product(s) of the reaction of bromine with each of the following: **(a)** methane, **(b)** 2-butene, **(c)** 2-butyne, **(d)** benzene (in the presence of $FeBr_3$)?

26.17 Write equations to show how the following can be made by means of addition reactions:
(a) 1,2-dibromoethane from ethene
(b) 1,1-dibromoethane from acetylene
(c) cyclohexanol from cyclohexene

26.18 **(a)** What is Markovnikov's rule? **(b)** How does the mechanism of the addition of HBr to an alkene explain why such an addition follows Markovnikov's rule?

26.19 Compare and contrast the mechanism of a substitution reaction of an alkane with the mechanism of a substitution reaction of an aromatic hydrocarbon.

26.20 What products would be expected from the following: **(a)** nitration of phenol, **(b)** bromination of phenol, **(c)** nitration of nitrobenzene, **(d)** bromination of nitrobenzene, **(e)** nitration of benzoic acid, **(f)** nitration of bromobenzene, **(g)** nitration of toluene, **(h)** reaction of CH_3Cl with toluene in the presence of $AlCl_3$?

Alcohols, Ethers, Carbonyl Compounds, Carboxylic Acids, Esters

26.21 Write structural formulas for
(a) butanal
(b) 2- butanone
(c) methyl ethyl ketone
(d) methyl ethyl ether
(e) methyl ethanoate
(f) sodium benzoate
(g) 3-methylhexanoic acid
(h) 3-methyl-2-hexanone
(i) *o*-bromophenol

26.22 Name each of the following compounds:

(a)
$$CH_3-\underset{\underset{\displaystyle CH_3}{|}}{C}H-CH_2-CH_2-CH_2OH$$

(b)
$$CH_3-\underset{\underset{\displaystyle CH_3}{|}}{C}H-CH_2-\underset{\underset{\displaystyle OH}{|}}{C}H-CH_3$$

(c)
$$CH_3-\overset{\overset{\displaystyle OH}{|}}{\underset{\underset{\displaystyle CH_3}{|}}{C}}-CH_2-CH_2-CH_3$$

(d)
$$CH_3-\underset{\underset{\displaystyle CH_3}{|}}{C}H-\overset{\overset{\displaystyle O}{\|}}{C}-CH_2CH_3$$

(e)
$$CH_3-\underset{\underset{\displaystyle CH_3}{|}}{C}H-O-CH_2CH_3$$

(f)
$$CH_3-\underset{\underset{\displaystyle CH_3}{|}}{C}H-CH_2-CH_2-\overset{\overset{\displaystyle O}{\|}}{C}-H$$

(g)
$$\overset{\overset{\displaystyle O}{\|}}{C}-CH_3 \text{ (phenyl)}$$

(h) cyclohexanone ring structure

(i)
$$\overset{\overset{\displaystyle O}{\|}}{C}-O-CH_3 \text{ (phenyl)}$$

(j)
$$\overset{\overset{\displaystyle O}{\|}}{C}-OH \text{ (phenyl-}NO_2\text{)}$$

26.23 Write structural formulas for all the isomers that have the formula $C_4H_{10}O$.

26.24 Give an example of an isomer of **(a)** ethyl acetate, **(b)** diethyl ether, **(c)** acetone, **(d)** butanal, **(e)** 1-pentanol, **(f)** butanoic acid.

26.25 Why do alcohols have higher boiling points than hydrocarbons of corresponding molecular weight?

26.26 What is the product of the reaction of 1-bromobutane with each of the following: **(a)** Mg, **(b)** NaCN, **(c)** OH^-(aq), **(d)** $NaOCH_2CH_3$?

26.27 What is the product of the reaction of 2-propanol with each of the following: **(a)** Na, **(b)** H_2SO_4 (high temperature), **(c)** H_2SO_4 (moderate temperature), **(d)** propanoic acid, **(e)** $K_2Cr_2O_7$ in acid solution?

26.28 What are the products when the following compounds are oxidized under vigorous conditions:
(a) 2-hexene
(b) 2-methyl-2-butene
(c) 3-hexyne
(d) 1-propanol
(e) 2-propanol
(f) propanal
(g) p-nitrotoluene
What are the products when the following compounds are oxidized using mild conditions:
(h) 2-hexene
(i) 1-propanol

26.29 What are the products when the following compounds are reacted with hydrogen in the presence of a suitable catalyst:
(a) pentanal
(b) 2-pentanone
(c) 2-pentene
(d) 2-pentyne

26.30 State the formulas of all products formed by the reaction of HBr with each of the following. If more than one compound is formed, tell which is produced in the greatest quantity.
(a) CH_3CH_2OH
(b) CH_3CH_2MgBr
(c) $CH_3CH_2COO^-Na^+$

26.31 State the compound formed when each of the following is treated with aqueous NaOH:
(a) $(CH_3)_2CHCOOH$
(b) $(CH_3)_2CHBr$
(c) $CH_3CH_2CH_2CN$
(d) $CH_3CH_2CH_2COOCH_2CH_3$

26.32 What is the final product (obtained by hydrolysis of the initial product) of the reaction of ethyl magnesium bromide with **(a)** methanal, **(b)** butanal, **(c)** butanone?

26.33 What Grignard reagent and what carbonyl compound should be used to prepare the following alcohols:
(a) 3-methyl-2-butanol
(b) 2-methyl-2-butanol
(c) 2-methyl-1-butanol

*26.34 Write equations to show how ethanol may be converted into the following substances (more than one step may be required for a given preparation): **(a)** ethene, **(b)** ethanal, **(c)** diethyl ether, **(d)** ethane, **(e)** acetic acid, **(f)** bromoethane, **(g)** ethyl cyanide, **(h)** 2-hydroxypropanoic acid.

*26.35 Using 1-propanol and 2-propanol as the only organic starting materials, write a series of equations to show how to prepare
(a) 3-hexanol
(b) 2,3-dimethyl-2-butanol
(c) 2-methyl-2-pentanol
(d) 2-methyl-3-pentanol
Use a Grignard reaction as one of the steps in each preparation.

*26.36 Show by a series of reactions how to convert 1-propanol into 2-propanol.

26.37 Interpret the following in terms of the Lewis theory: **(a)** an S_N1 reaction of an alkyl halide, **(b)** an S_N2 reaction of an alcohol, **(c)** the addition of HCN to a ketone.

Amines, Amides

26.38 Draw structural formulas for
(a) diphenylamine
(b) methyl ethyl amine
(c) 1,6-diaminohexane
(d) benzamide
(e) p-nitroaniline

26.39 Write structural formulas for all the isomers that have the formula $C_4H_{11}N$. Name the compounds.

26.40 Given an example of an isomer of
(a) trimethylamine
(b) aminoacetic acid
(c) propionamide

26.41 What are the products of the following reactions:
(a) H_2 and CH_3CN
(b) H_2 and p-nitrotoluene
(c) 1-bromobutane and methylamine
(d) $CH_3COO^-NH_4^+$, heated,
(e) HBr and $CH_3CH_2CO_2^-NH_4^+$
(f) HBr and $(CH_3CH_2)_2NH$
(g) OH^- and CH_3CH_2CN
(h) OH^- and $CH_3CH_2NH_3^+Cl^-$

26.42 When the terms primary, secondary, and tertiary are applied to amines, their meanings are not the same as when they are applied to alcohols. Explain the distinction.

26.43 Write equations to show the reactions of NH_3 with **(a)** 2-bromopropane, **(b)** propanoic acid, **(c)** methyl propanoate.

26.44 Arrange the following compounds in decreasing order of strength as Brønsted acids: CH_3CH_2OH, $CH_3CH_2OH_2^+$, H_2O, H_3O^+, CH_3COOH, $CH_3CH_2NH_2$.

26.45 Arrange the following compounds in decreasing order of strength as Brønsted bases: CH_3CH_2OH, $CH_3CH_2O^-$, OH^-, CH_3COO^-, $CH_3CH_2NH_2$.

26.46 (a) Why is phenol a stronger acid than methanol? **(b)** Why is aniline a weaker base than methylamine?

Polymers

26.47 Describe with examples: **(a)** addition polymers, **(b)** copolymers, **(c)** condensation polymers.

26.48 Compare the function of $AlCl_3$ in a Friedel-Crafts reaction and a polymerization.

26.49 What monomers are used to make **(a)** polyethylene, **(b)** Teflon, **(c)** polystyrene, **(d)** Orlon, **(e)** Nylon, **(f)** rubber, **(g)** Bakelite, **(h)** Dacron?

26.50 Draw the structure of a section of each of the polymers listed in problem 26.49.

BIOCHEMISTRY

CHAPTER
27

Biochemistry is the science that is concerned with the composition and structure of substances that occur in living systems and the chemical reactions that take place in these systems. Impressive progress has been made in this branch of chemistry over the last thirty years. It is currently an area of intense research activity. In this chapter, we will survey some aspects of the field.

The principal component of living material is *water*. It constitutes from about 60% to 90% of all plants and animals. The quantity of inorganic substances present is comparatively small—usually less than 4%; the amount is largest in organisms that have a skeleton. The remainder consists of organic compounds, most of which have complicated structures.

The four most abundant elements of the compounds found in living material are oxygen, carbon, hydrogen, and nitrogen. These four elements account for approximately 96% of the mass of the human body.

27.1 Proteins

Proteins are macromolecules that have molecular weights ranging from about 6000 to over 1,000,000. They are formed from simpler compounds called **α-amino acids.** An α-amino acid is a carboxylic acid that has an amino group, $-NH_2$, bonded to the C atom next to the carboxyl group, $-COOH$. The designation α (alpha) denotes the position of the amino group. The C atom adjacent to the carboxyl group is called the α-carbon atom. The general formula of these compounds is

$$
\begin{array}{c}
\quad\ \ \overset{\displaystyle H}{|}\ \ \ \overset{\displaystyle O}{\|} \\
R-C-C-O-H \\
\quad\ \ \underset{\displaystyle NH_2}{|}
\end{array}
$$

The $-NH_2$ group is basic and will react with acids to form $-NH_3^+$. At low pH, therefore, an amino acid exists as a cation, which will move toward the negative pole in an electric field:

$$
\begin{array}{c}
\quad\ \ \overset{\displaystyle H}{|}\ \ \ \overset{\displaystyle O}{\|} \\
R-C-C-O-H \\
\quad\ \ \underset{\displaystyle NH_3^+}{|}
\end{array}
$$

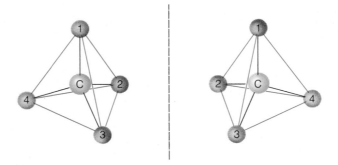

Figure 27.1 Optical isomers of a compound containing carbon atom (dashes represent reflection plane)

The —COOH group is acidic and forms —COO⁻ with bases. At high pH, therefore, amino acids are anions and move toward the positive pole in an electric field:

$$R-\overset{\overset{\displaystyle H}{|}}{\underset{\underset{\displaystyle NH_2}{|}}{C}}-\overset{\overset{\displaystyle O}{\|}}{C}-O^-$$

Internal neutralization, in which the proton from the —COOH group moves to the —NH₂ group, produces a **zwitterion**:

$$R-\overset{\overset{\displaystyle H}{|}}{\underset{\underset{\displaystyle NH_3^+}{|}}{C}}-\overset{\overset{\displaystyle O}{\|}}{C}-O^-$$

The amino acids provide examples of a type of stereoisomerism called **optical isomerism** (see Section 24.4). The simplest optical isomers are compounds that contain chiral carbon atoms—carbon atoms to which four different kinds of groups are bonded (see Figure 27.1). The mirror image (called an **enantiomorph**) of a molecule of this type is not superimposable on the original molecule.

The structures shown in Figure 27.1 are mirror reflections of one another. The dashed line represents the reflection plane. Start with any group (1, 2, 3, or 4) of the structure on the left and trace the shortest distance to the reflection plane. Continue to the right of the reflection plane for the same distance, and the same group will be found in the structure on the right.

A study of this figure reveals that the two structures cannot be superimposed. If groups 1 and 3 of the structures are superimposed, group 4 of one structure coincides with group 2 (not 4) of the other. No matter how the two structures are turned, it is impossible to make all four groups of one structure coincide with the same four groups, respectively, of the other structure. The two structures are isomers.

The physical properties of optical isomers are identical except for their effects on plane-polarized light (see Figure 24.7). The isomers are said to be optically active because they rotate the plane of plane-polarized light in either a clockwise or a counterclockwise direction. The isomer that rotates the plane to the right (clockwise) is said to be *dextro*rotatory, (+) or *d*. The other isomer rotates the plane an equal amount to the left and is said to be *levo*rotatory, (−) or *l*.

With the exception of glycine (for which R equals H), the α-carbon atoms of all α-amino acids have four different kinds of groups bonded to them and are, therefore, chiral. Consequently, optical isomers exist for the α-amino acids. The structures of the isomers are usually diagrammed

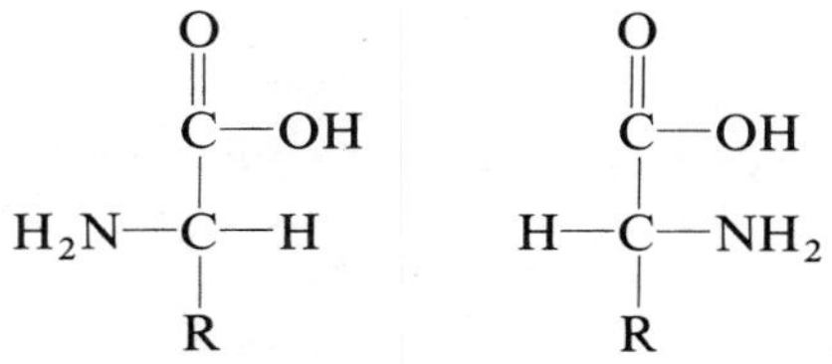

$$
\begin{array}{cc}
\overset{\displaystyle O}{\overset{\displaystyle \|}{C}}\!-\!OH & \overset{\displaystyle O}{\overset{\displaystyle \|}{C}}\!-\!OH \\
H_2N\!-\!\overset{|}{\underset{|}{C}}\!-\!H & H\!-\!\overset{|}{\underset{|}{C}}\!-\!NH_2 \\
R & R \\
\text{L-}\textit{amino acid} & \text{D-}\textit{amino acid}
\end{array}
$$

The diagrams represent structures in which the chiral carbon atoms are in the plane of the paper, the —COOH and —R groups are behind the plane of the paper, and the —H and —NH$_2$ groups are in front of the plane of the paper. With such an arrangement of groups, the L-form is the one in which the amino group is to the left of the chain, and the D-form is the one in which the amino group is to the right. The structures can be related to the diagrams of Figure 27.1. The L-form is the one on the left in the figure if group 1 = —COOH, 2 = —R, 3 = —H, and 4 = —NH$_2$.

All the amino acids derived from proteins are L-amino acids. Only the L-forms can be utilized in the metabolic processes of the body. The sign of rotation should mot be confused with the symbols D- or L- that are used to designate the configuration of the chiral carbon atom. Thus, the L-isomer of alanine (2-aminopropanoic acid) is *dextro*rotatory (+):

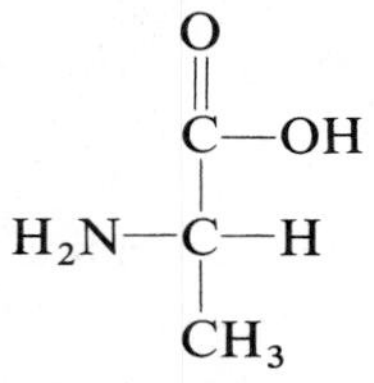

$$
\begin{array}{c}
\overset{\displaystyle O}{\overset{\displaystyle \|}{C}}\!-\!OH \\
H_2N\!-\!\overset{|}{\underset{|}{C}}\!-\!H \\
CH_3 \\
\text{L-}(+)\textit{-alanine}
\end{array}
$$

In protein molecules, many amino acid units are linked together to form long chains. When two amino acids combine, the amino group of one molecule joins, with the elimination of water, to the carboxyl group of the second molecule. A **peptide linkage,**

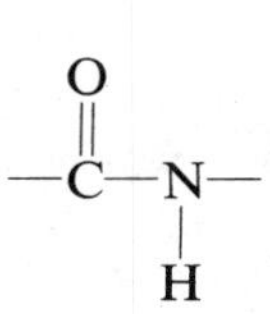

$$
\begin{array}{c}
\overset{\displaystyle O}{\overset{\displaystyle \|}{}} \\
-\overset{}{C}\!-\!\overset{}{\underset{|}{N}}\!- \\
H
\end{array}
$$

is formed:

$$\text{H}-\text{N}-\overset{\overset{\displaystyle H}{|}}{\underset{\underset{\displaystyle H}{|}}{\text{C}}}-\overset{\overset{\displaystyle O}{\|}}{\text{C}}-\text{O}-\text{H} + \text{H}-\text{N}-\overset{\overset{\displaystyle H}{|}}{\underset{\underset{\displaystyle H}{|}}{\text{C}}}-\overset{\overset{\displaystyle O}{\|}}{\text{C}}-\text{O}-\text{H} \longrightarrow$$

$$\text{H}-\text{N}-\overset{\overset{\displaystyle H}{|}}{\underset{\underset{\displaystyle R}{|}}{\text{C}}}-\overset{\overset{\displaystyle O}{\|}}{\text{C}}-\text{N}-\overset{\overset{\displaystyle H}{|}}{\underset{\underset{\displaystyle R}{|}}{\text{C}}}-\overset{\overset{\displaystyle O}{\|}}{\text{C}}-\text{O}-\text{H} + \text{H}_2\text{O}$$

The product, called a **dipeptide,** has a free amino group at one end and a free carboxyl group at the other. It can condense with other amino acids to form long chains called **polypeptides.** Proteins are polypeptides. The hydrolysis of a polypeptide is the reverse of the condensation process and produces amino acids.

The radicals, R, in the preceding formulas vary considerably in their structures. Some of them are simple hydrocarbons, some contain rings (including aromatic rings), some contain additional —NH_2 or —COOH groups, and some incorporate —SH or —OH groups in their structures. Twenty different α-amino acids are regularly found in proteins. A few others are occasionally found as minor constituents.

Proteins contain from about 50 to over 10,000 amino acid residues in a chain. The average molecular weight of an amino acid residue is about 110 to 120. A protein with a molecular weight of 300,000, therefore, contains roughly 300,000/120 or 2500 amino acid residues.

The amino acids in a given protein occur in a definite order. The same type of amino acid may be used many times in the polypeptide chain. Since a protein may contain hundreds or thousands of amino acid residues, a tremendous number of arrangements is possible even though only about 20 different amino acids are used.

The sequence of amino acid residues in a protein is called its **primary structure.** The primary structure of bovine insulin was determined in 1945–52 by Frederick Sanger. A fragment of the structure is shown in Figure 27.2. The complete structure contains 51 amino acid residues in two chains—one of 21 amino acid residues and the other of 30.

The spatial arrangement of the polypeptide chain of a protein is called the **secondary structure** of the protein. The most important arrangement is the **alpha helix.** In an α-helix, the chain is coiled in a form that resembles a spring (see Figure 27.3). There are about 3.6 amino acid residues in each turn. The helix is held together by hydrogen bonds between the loops of the coil. The hydrogen bonds are formed by the H atom of the N—H of one peptide group with the O atom of the C=O group of another peptide group (see Figure 27.3). Each peptide group is joined, by hydrogen bonding, to the third peptide group before it and the third peptide group following it.

Another type of secondary structure is the **pleated sheet** arrangement. Chains are held together in the form of pleated sheets by hydrogen bonds between adjacent chains. This arrangement is not so common as the α-helix arrangement.

The **tertiary structure** of a protein is the arrangement of the secondary structures brought about by interactions between the R groups of the amino acid residues.

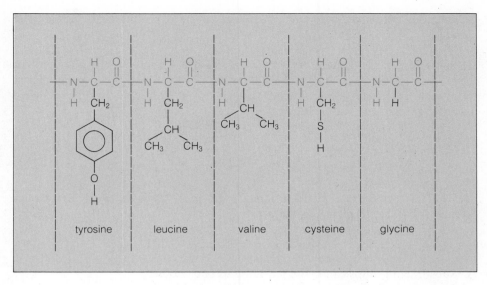

Figure 27.2 Fragment of bovine insulin protein chain. (The polypeptide chain is shown in color. The R radicals are shown in black. The names of the amino acids that make up the chains appear at the bottom.)

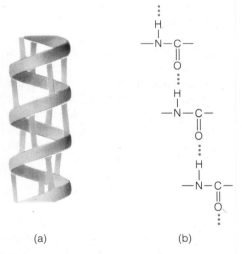

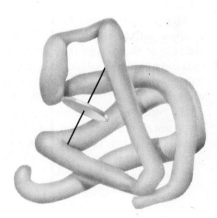

Figure 27.3 (a) Schematic representation for an α-helix. (The colored lines are hydrogen bonds.) (b) Hydrogen bonds between coils of an α-helix.

Figure 27.4 Tertiary structure of a globular protein. (The colored disc is an iron-containing heme group.)

The structure of a globular protein (so called because these proteins are usually approximately spherical in shape) is shown schematically in Figure 27.4.

The interactions that hold the coiled segments of the chain into a compact tertiary form do not involve the peptide linkages of the chain. Instead, these interactions take place between substituents of the R groups of the amino acid residues:

1. Some R radicals contain a —COOH group over and above the one involved in peptide formation. Other R radicals contain a second —NH_2 group. These two functional groups—one attached to one segment of the coil, the other attached to another segment—can react:

$$
\begin{matrix} \text{O} \\ \| \\ -\text{C}-\text{O}-\text{H} \end{matrix} + \text{H}_2\text{N}- \longrightarrow \begin{matrix} \text{O} \\ \| \\ -\text{C}-\text{O}^- \end{matrix} + \text{}^+\text{H}_3\text{N}-
$$

2. By mild oxidation, a disulfide linkage can form between two —SH groups attached to different places on the coil:

$$-\text{S}-\text{H} + \text{H}-\text{S}- \longrightarrow -\text{S}-\text{S}- + 2\,\text{H}^+ + 2e^-$$

3. Substituent groups can hydrogen-bond to one another:

$$
\begin{matrix} & & \text{O}- \\ & & | \\ -\text{O}\cdots\text{H} \\ | \\ \text{H} \end{matrix}
$$

These interactions occur in places where the coils overlap.

The primary, secondary, and tertiary structures describe the arrangement of a single polypeptide chain. Some proteins contain more than one chain. The number of chains and the way in which they are arranged in the complete protein make up what is called the **quaternary structure** of the protein. For example, hemoglobin, the oxygen carrier in blood, is formed from four chains of two different types. Two of the chains each contain 140 amino acid residues. Each of the other two contains 146 amino acid residues. Each of the four chains has an iron-containing heme group (see Section 24.1) and has a tertiary structure similar to that shown in Figure 27.4.

27.2 Carbohydrates

Sugars, starches, and cellulose belong to a group of organic compounds called **carbohydrates.** The name was derived from the fact that most (but not all) compounds of this type have the general formula $C_x(H_2O)_y$, and might, therefore, appear to be hydrates of carbon. Carbohydrates, however, do not contain water. They are hydroxy aldehydes, hydroxy ketones, or substances derived from them. Carbohydrates are the principal substances of which plants are composed. They are an important food for animals—they serve as a source of energy and provide carbon chains for compounds that are synthesized by living organisms.

The simplest carbohydrates are the **monosaccharides,** or simple sugars. The most important monosaccharide is *glucose.* It is the most abundant sugar found in nature, and in animals it is a normal constituent of blood. The two optical isomers of glucose are

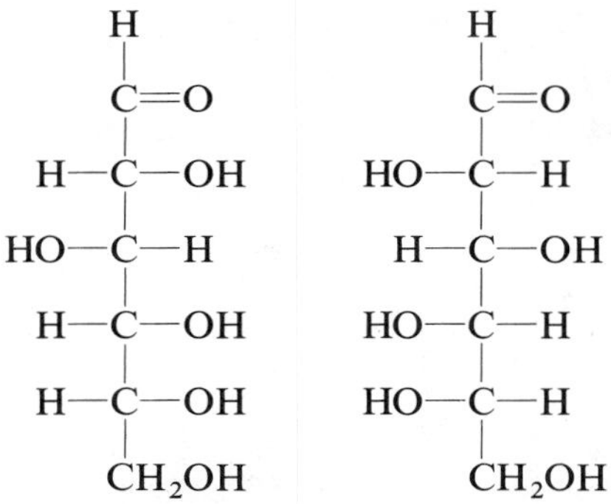

D-(+)-glucose   L-(−)-glucose

There are four chiral carbon atoms in the molecule. The carbon atoms at each end of the molecule do not hold four different groups and are not chiral. Sixteen optical isomers (eight D,L pairs) of this formula exist. Only the preceding pair are called glucose.

D-glucose forms a cyclic molecule by an addition reaction involving the carbonyl group and a hydroxy group:

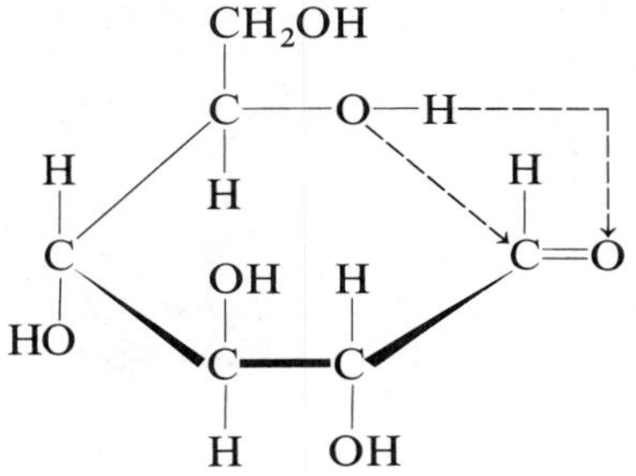

Ring formation produces a new chiral center, and two isomers of D-glucose exist which differ in the orientation of the new OH group:

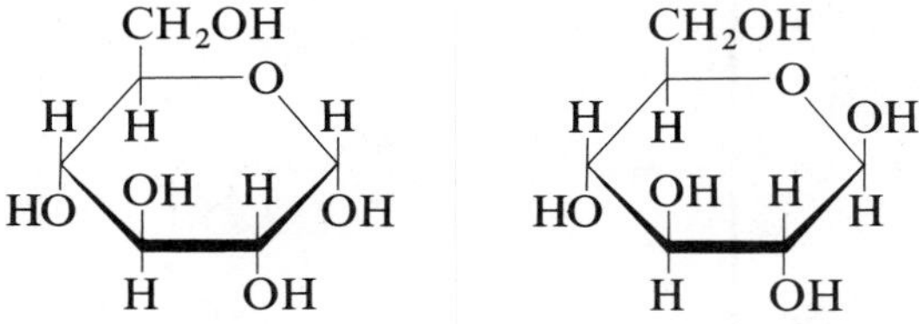

α-D-glucose   β-D-glucose

In these diagrams the carbon atoms of the ring are not shown. The rings are actually puckered, not planar. Notice that in the α form the OH group of the extreme right-hand carbon atom is on the same side of the ring as the OH group of the adjacent carbon atom. These OH groups may be said to be *cis* to each other. In aqueous solution, α and β forms of D-glucose exist in equilibrium, together with a low concentration of the open-chain form.

Fructose is a hydroxy ketone:

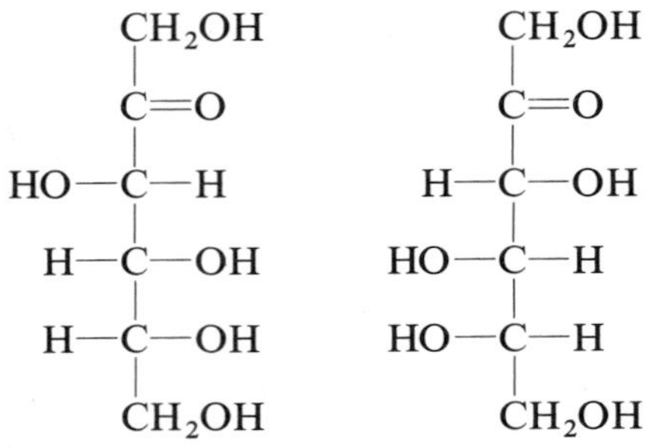

$$
\begin{array}{cc}
\text{CH}_2\text{OH} & \text{CH}_2\text{OH} \\
| & | \\
\text{C}=\text{O} & \text{C}=\text{O} \\
| & | \\
\text{HO}-\text{C}-\text{H} & \text{H}-\text{C}-\text{OH} \\
| & | \\
\text{H}-\text{C}-\text{OH} & \text{HO}-\text{C}-\text{H} \\
| & | \\
\text{H}-\text{C}-\text{OH} & \text{HO}-\text{C}-\text{H} \\
| & | \\
\text{CH}_2\text{OH} & \text{CH}_2\text{OH} \\
\text{D-}(-)\text{-}fructose & \text{L-}(+)\text{-}fructose
\end{array}
$$

There are three chiral carbon atoms in this molecule and eight stereoisomers (or four D,L pairs) of this formula are known. Cyclic forms of D-fructose are known with five- or six-membered rings.

Disaccharides are compsed of two monosaccharide units. *Sucrose*, which is cane sugar, is a disaccharide. Acid hydrolysis of sucrose yields the two simple sugars D-glucose and D-fructose. In the sucrose molecule an α D-glucose unit and the β five-membered ring form of a D-fructose unit are joined through two OH groups by the elimination of a molecule of water:

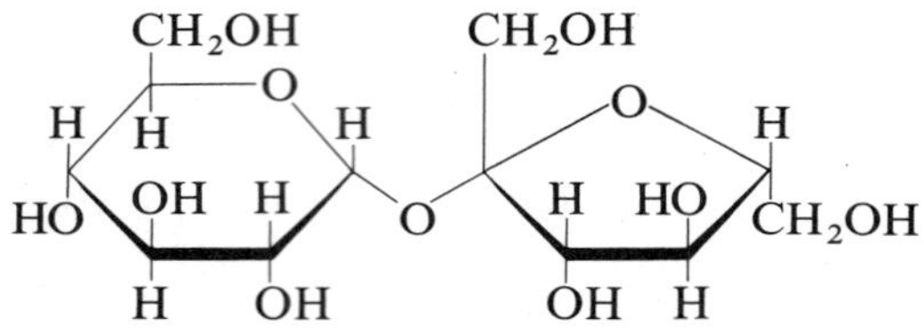

D-*glucose unit*   D-*frucose unit*

Cellulose and the substances that make up starch are **polysaccharides** (carbohydrates that contain many monosaccharide units) and they yield only D-glucose upon hydrolysis. It is estimated that the number of D-glucose units in the molecular structures of these substances may be as high as several thousand. The D-glucose units of cellulose are linked in long chains in β combination:

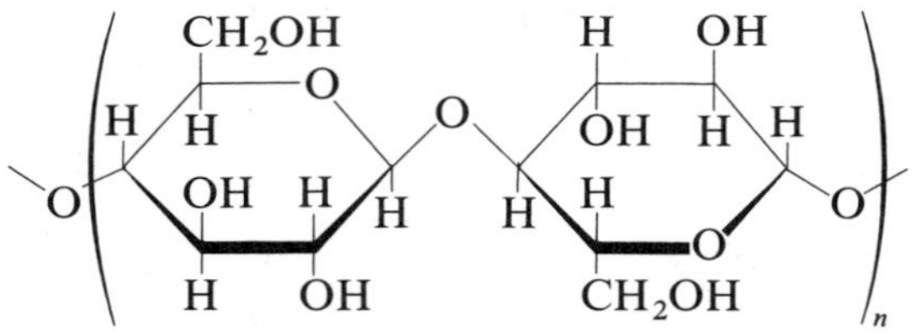

In starch, the D-glucose units are linked in a different manner; there are occasional cross links between long chains of D-glucose units arranged in α combination:

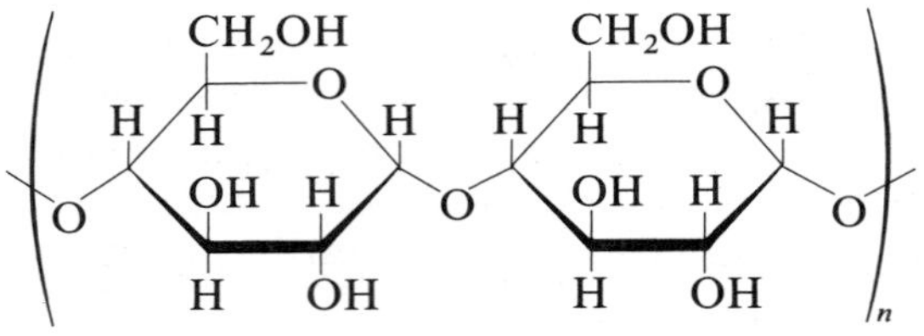

Starch is an important food material, but cellulose cannot be digested by man—an important distinction brought about by the difference in the way that the D-glucose units are linked together.

27.3 Fats and Oils

Fats and oils belong to a larger classification called **lipids.** Lipids are substances that can be dissolved away from biological material by solvents that are nonpolar or slightly polar (such as hydrocarbons, carbon tetrachloride, and diethyl ether). Since the classification is based on solubility, not structure, a wide variety of compounds are lipids. Fats, oils, some vitamins and hormones, and certain constituents of the cell walls are examples of some of the substances classified as lipids.

Fats and oils are esters of fatty acids and glycerol. The **fatty acids** are long straight-chain carboxylic acids (see Table 27.1). Some of them are saturated and some contain one or more double bonds. Almost all the fatty acids isolated from natural sources contain an even number of carbon atoms.

Glycerol is a trihydroxy alcohol, 1,2,3-propanetriol:

$$
\begin{array}{l}
\text{H—O—CH}_2 \\
\quad\quad\quad| \\
\text{H—O—CH} \\
\quad\quad\quad| \\
\text{H—O—CH}_2
\end{array}
$$

The esters formed from 1 mol of glycerol and 3 mol of fatty acid:

$$
\begin{array}{c}
\text{O} \\
\|\\
\text{R—C—O—H}
\end{array}
$$

Table 27.1 Some common fatty acids		
Saturated Acids		
Acid	Formula	Occurrence
lauric	$CH_3(CH_2)_{10}COOH$	coconut oil, palm oil, animal fat
myristic	$CH_3(CH_2)_{12}COOH$	coconut oil, palm oil, animal fat
palmitic	$CH_3(CH_2)_{14}COOH$	animal fat, cottonseed oil, palm oil
stearic	$CH_3(CH_2)_{16}COOH$	animal fat
Unsaturated Acids		
Acid	Formula	Positions of Double Bonds[a]	Occurrence
oleic	$C_{17}H_{33}COOH$	9	corn oil, cottonseed oil, olive oil
linoleic	$C_{17}H_{31}COOH$	9, 12	cottonseed oil, corn oil, linseed oil
linolenic	$C_{17}H_{29}COOH$	9, 12, 15	linseed oil
arachidonic	$C_{19}H_{31}COOH$	5, 8, 11, 14	sardine oil, corn oil, animal fat

[a] All these acids have straight chains. The carbon atom of the COOH group is number 1. The number given in the table indicates the first carbon of the double bond. Number 9, for example, indicates a double bond between carbon atoms 9 and 10.

are called **triglycerides:**

$$
\begin{array}{l}
R-\overset{\overset{\displaystyle O}{\|}}{C}-O-\overset{\overset{\displaystyle H}{|}}{C}-H \\
R'-\overset{\overset{\displaystyle O}{\|}}{C}-O-\overset{|}{C}-H \\
R''-\overset{\overset{\displaystyle O}{\|}}{C}-O-\overset{|}{\underset{\underset{\displaystyle H}{|}}{C}}-H
\end{array}
$$

A natural fat or oil is a mixture of different triglycerides. In fact, a single triglyceride molecule may be formed from three different fatty acids. For this reason, the radicals are indicated in the preceding formula as R, R′, and R″. Palmitic and oleic acids are present in most vegetable and animal fats and oils (see Table 27.1).

Fats are mixtures of triglycerides that are solids at room temperature. Oils are mixtures that are liquids at room temperature. Oils contain a higher percentage of the glyceryl esters of unsaturated fatty acids than fats do. Vegetable oils (such as corn oil and coconut oil) are changed into solids (vegetable shortening and margarine) by reaction with hydrogen. The hydrogen adds to some of the double bonds of the unsaturated triglycerides. In practice, the process is continued until a solid of the desired consistency is obtained. Not all of the double bonds are saturated with hydrogen.

When fats and oils are heated with aqueous solutions of bases, glycerol and the salts of the fatty acids are obtained:

$$
\begin{array}{l}
R-\overset{\overset{\displaystyle O}{\|}}{C}-O-\overset{\overset{\displaystyle H}{|}}{C}-H \\
R'-\overset{\overset{\displaystyle O}{\|}}{C}-O-\overset{|}{C}-H + 3\,OH^- \longrightarrow \\
R''-\overset{\overset{\displaystyle O}{\|}}{C}-O-\overset{|}{\underset{\underset{\displaystyle H}{|}}{C}}-H
\end{array}
\quad
\begin{array}{l}
R-\overset{\overset{\displaystyle O}{\|}}{C}-O^- \\
R'-\overset{\overset{\displaystyle O}{\|}}{C}-O^- \\
R''-\overset{\overset{\displaystyle O}{\|}}{C}-O^-
\end{array}
\;+\;
\begin{array}{l}
H-O-CH_2 \\
H-O-\overset{|}{C}H \\
H-O-CH_2
\end{array}
$$

The process is called **saponification,** which means *soap making*. Soaps are salts of fatty acids. The type of soap obtained depends, of course, upon the base used in the saponification. Sodium soaps, $RCOO^-Na^+$, are the most common.

Fats and oils are used as fuels by the body, and carbon dioxide and water are the products of the oxidation. Fats are also stored as a reserve source of energy. Carbohydrates may be converted into fat and stored also. Fats and oils are also used by the body in the syntheses of constituents of tissues.

27.4 Nucleic Acids

Nucleic acids were first isolated from cell nuclei in the 1860s, and the name was derived from the name of their source. Many different proteins must be constantly

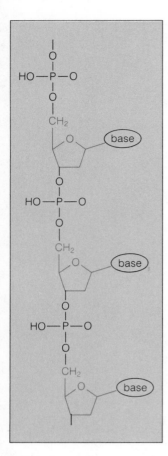

Figure 27.5 Schematic diagram of the primary structure of a nucleic acid. (Three nucleotide units are shown.)

made by an organism to enable it to grow, function, renew itself, and reproduce itself. The nucleic acids direct the syntheses of these proteins. The nucleic acids store, use, and pass on to the next generation the pattern for the reproduction of the organism.

Nucleic acids are high molecular weight, long-chain polymers. Each unit of the chain is formed from three components:

1. a phosphoric acid molecule
2. a five-carbon sugar molecule
3. a molecule of a nitrogen-containing base

A schematic diagram of a fragment of a nucleic acid chain that contains only three units (called **nucleotides**) is shown in Figure 27.5. Actual nucleic acid molecules contain from about 80 to 100 million nucleotides.

There are two types of nucleic acids: **deoxyribonucleic acids** (called **DNA**) and **ribonucleic acids** (called **RNA**). The principal difference between the two is in the sugar molecule employed. DNA contains deoxyribose and RNA contains ribose:

Figure 27.6 Schematic representation of the double helix. (Dots represent hydrogen bonding between complementary bases of the two DNA strands.)

β-D-*deoxyribose*   β-D-*ribose*

The prefix *deoxy-* indicates removal of an oxygen atom. If an atom of oxygen is removed from the number *2* carbon atom of ribose, the structure of deoxyribose is obtained.

In the chain of a nucleic acid, carbon atom number *3* of one sugar molecule and carbon atom number *5* of the next sugar molecule are joined by ester linkages to a molecule of phosphoric acid (see Figure 27.5). One of four different nitrogen base radicals replaces the OH group of carbon atom number *1* of each sugar molecule.

The secondary structure of DNA is a **double helix** (see Figure 27.6). Two chains of DNA are coiled together in such a way that the bases are in the interior of the coils. The structure is held together by hydrogen bonds between the bases of one chain and the bases of the other chain.

The structures of the four different base radicals found in DNA are shown in Figure 27.7. The formulas are drawn in pairs with hydrogen bonds indicated between the pairs. Notice that **adenine** (abbreviated **A**) and **thymine** (abbreviated **T**) complement each other. The positions of the atoms make it possible for two strong hydrogen bonds to form between A on one chain and T on the other chain of the double helix. **Guanine (G)** and **cytosine (C)** complement each other in the same way. Three strong hydrogen bonds are formed between this pair of bases.

In any sample of DNA there is the same amount of A as there is T and there is the same amount of G as there is C. The bases attached to one DNA chain complement the bases attached to the other DNA chain. If there is an A on chain 1, T is found on chain 2 opposite it. If T is found on chain 1, A will be opposite it on chain 2. The same type of base pairing occurs between G and C. Notice that the two hydrogen-bonded pairs shown in Figure 27.7 are about the same length.

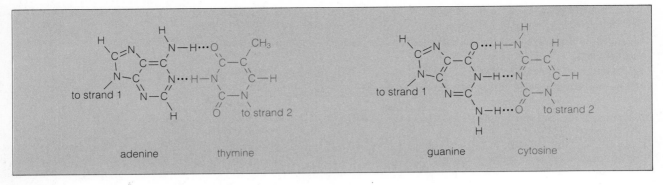

adenine   thymine   guanine   cytosine

Figure 27.7 The four base radicals found in DNA. (The formulas are drawn in complementary pairs.)

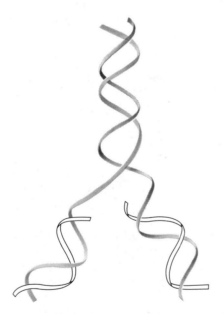

Figure 27.8 Replication of DNA double helix. (The strands unwind and each serves as a template for the synthesis of a new complementary strand.)

Consequently, they hold the two chains of the double helix a uniform distance apart.

When a cell divides, the two chains of a DNA double helix come apart. Each chain serves as a template, or pattern, for the synthesis of a new, complementary chain. Two identical double helices result from the process—each one containing one of the chains of the original double helix (see Figure 27.8).

Nucleotides from the surrounding solution form the new chains. The base of a nucleotide pairs with the complementary base of the DNA chain through the formation of hydrogen bonds. The nucleotides line up, therefore, in a way that is dictated by the order in which the bases occur in the DNA chain. The new chains formed from the nucleotides are complementary to the original DNA chains.

The order in which the bases occur in a DNA molecule constitutes the information used for the synthesis of proteins. Two types of RNA are used for these processes: messenger RNA and transfer RNA. RNA employs the same bases that DNA does with the exception of thymine. In RNA, **uracil (U)** is used in place of thymine (T). The structure of uracil is the same as that of thymine except that a H atom replaces the $-CH_3$ group. U, therefore, complements A in the same way that T does (see Figure 27.7).

An RNA chain is synthesized from nucleotides in the nucleus of the cell in such a way that the order of bases in the RNA chain complements the order of a section of the DNA chain. The DNA chain, therefore, acts as a template. Suppose, for example, that the order of bases of a small portion of the DNA chain is

C—C—A—A—C—G

The order of bases in the RNA chain that is made by using this part of the DNA chain as a pattern is

G—G—U—U—G—C

since G complements C and U complements A.

The chain formed in this way is called **messenger RNA.** A messenger RNA chain may contain about 500 bases in an order dictated by a *section* of a DNA chain, which contains altogether more than 100 million bases. The information in the section of the DNA chain (the order of bases) is said to be **transcribed** by the messenger RNA. This information is the message that the messenger carries from the cell nucleus to the ribosomes, where the protein is made.

The order of bases directs the order in which different amino acids are assembled in the formation of the protein. Each amino acid of the protein is specified by a set of three base groups (a base triplet) of the messenger RNA chain. The sequence

G—G—U—U—G—C

pertains to two amino acids. The code G—G—U specifies the amino acid glycine. The code U—G—C specifies the amino acid cysteine. The portion of the messenger RNA chain that contains the three bases of a code is called a **codon.** The preceding sequence described two codons.

A second type of RNA, called **transfer RNA,** is used to decipher the code and build the protein. Transfer RNA chains are the smallest nucleic acid chains. They consist of approximately 80 nucleotides and may contain a few bases other than the customary four.

A particular transfer RNA molecule bonds with a specific amino acid and carries it to the messenger RNA where it is inserted into the protein being made at the position directed by the code of the messenger RNA. The transfer RNA is said to **translate** the code. Each transfer RNA contains a base triplet (called an **anticodon**) that complements the triplet of a codon of messenger RNA. The anticodon of a particular transfer RNA corresponds to the specific amino acid that it carries.

Suppose, for example, that the codon of messenger RNA specifies the code GGU. The anticodon CCA that corresponds to this codon is found on transfer RNA that carries the amino acid glycine.

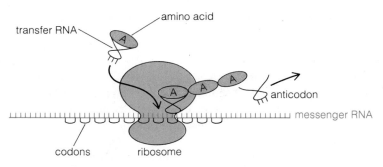

Figure 27.9 Synthesis of a protein by transfer RNA on messenger RNA.

The synthesis of a protein is thought to occur in steps (see Figure 27.9). Transfer RNA carrying an amino acid becomes attached to the messenger RNA by codon-anticodon pairing. A second transfer RNA becomes attached to the next codon of the messenger RNA chain. A peptide linkage is formed between the two amino acids of the two transfer RNA's. The first transfer RNA is then released and leaves behind its amino acid now bound into a dipeptide. A third transfer RNA, of a type specified by the next codon, becomes attached to the messenger RNA. As soon as a peptide bond connects the dipeptide to the amino acid of the third transfer RNA, the second transfer RNA is released. The process stops when a codon that does not correspond to any amino acid is encountered. The synthesis of more than one protein is directed by a single messenger RNA.

Specific enzymes catalyze every step of the entire process including the formation of the messenger RNA and the bonding of an amino acid to the transfer RNA. Formation of the polypeptides is rapid. It is estimated that hundreds of peptide bonds are formed in 1 min.

Any change in the DNA of the cell will result in a change in the messenger RNA produced from it and, consequently, an alteration in the resulting protein. Such changes are called **mutations.** The changes may prove to be beneficial, harmful, or trivial. They are passed on from generation to generation. Mutations are responsible for genetic diseases, such as Tay-Sachs disease, sickle-cell anemia, hemophilia, and Huntington's chorea.

27.5 Enzymes

Enzymes are biological catalysts. The term is derived from a Greek word that means *in leaven* since the action of yeast in fermentation was the first enzymatic action identified. Enzymes are involved in every reaction that occurs in a cell.

A cell contains over 1000 enzymes. A large number is required because enzymes are **specific** in their action. Some are so specific that they catalyze only one reaction of one particular compound. Enzymes are known that will hydrolyze only one definite type of peptide bond—one that involves a particular kind of amino acid residue. The enzymes that are least specific are probably those that hydrolyze most ester linkages.

Some enzymes are **stereochemically specific.** Such an enzyme might act on only the L form of an amino acid and not the D form, or an enzyme might be active only toward β linkages between glucose units of a carbohydrate and not α linkages.

Enzymes are amazingly efficient. The **turnover number** of an enzyme is the number of moles of reactant transformed in 1 min by the action of the enzyme. Turnover numbers vary from about 10,000 mol/min to 5,000,000 mol/min. Since these values are so high, only a small amount of each enzyme is required by a cell.

The action of enzymes permits the cell reactions to occur rapidly at the relatively low temperature of the body and at a pH near 7. When comparable reactions are run in the laboratory without the aid of enzymes, they require higher temperatures and the addition of acids or bases.

The **substrate** of an enzyme is the compound upon which it acts. If E is used to designate the enzyme, S the substrate, and P the product of the reaction, the overall process can be indicated as follows:

$$E + S \rightleftharpoons ES \longrightarrow E + P$$

ES, formed from the enzyme and the substrate, may be considered to be the activated complex for the reaction. The catalytic effect is brought about by a significant lowering of the energy of activation for the reaction (see Section 12.7).

Enzymes are surface catalysts. They are globular proteins and are much larger than the substrate upon which they act. Reaction occurs at an **active site** on the surface of the enzyme. The substrate is held to the active site by hydrogen bonding, ionic attraction, or some other type of chemical interaction.

The contour of an active site fits the shape of the substrate exactly, or more probably, the active site changes its conformation to fit the substrate molecule after it is attached. In the process, bonds within the substrate are stretched and weakened. In this way, the reaction is facilitated. After the products of the reaction leave, the active site is free to bind another substrate molecule.

Enzymes are specific in their action because they function by means of active sites that are tailored to fit a specific substrate or type of substrate. The stereochemical specificity of some enzymes is a result of the conformation of their active sites—the L form of a molecule might fit and the D form might not.

Even though the reaction occurs on a relatively small part of the surface—the active site—the rest of the enzyme molecule is important. It probably helps to form the active site into the required conformation and to hold it that way until after the reaction has occurred.

Most enzymes are **conjugated proteins,** that is, they contain nonprotein parts together with protein parts. If the nonprotein part is not easily removed, it is called a **prosthetic group.** If it is easily removed, it is called a **coenzyme.** Neither the coenzyme nor the protein part from which it is removed is active when the two are separated. Prosthetic groups and coenzymes are important because they help to form the active sites of enzymes. Many coenzymes are vitamins or compounds derived from vitamins. Some enzymes require the presence of metal ions (such as Co^{2+}, Fe^{2+}, Zn^{2+}, Cu^{2+}, and Mg^{2+}) to function.

The action of enzymes can be inhibited, or retarded, in several ways. An **inhibitor** is a substance that combines with an enzyme at the active site and prevents the enzyme from functioning. In **competitive inhibition,** the enzyme, E, and the inhibitor, I, combine reversibly:

$$E + I \rightleftharpoons EI$$

The combination of the enzyme and substrate is also reversible:

$$E + S \rightleftharpoons ES$$

The concentrations of S and I determine the amounts of ES and EI formed. Since the desired reaction depends upon the formation of ES, the reaction can be slowed down or stopped altogether if the concentration of I is relatively high.

The product of a reaction may serve as an inhibitor. **Product inhibition** is important since it serves to control the extent of reactions that occur in a cell. As the concentration of the product of some reactions increases, product inhibition gradually slows the reaction down and prevents an undesirable accumulation of product in the cell.

If a process occurs in a series of steps:

$$A \longrightarrow B \longrightarrow C \longrightarrow D$$

control of the entire sequence can be achieved if a product of the last step (D) inhibits a previous step (say, A → B). Such inhibition is called **feedback inhibition**. It too controls the amounts of materials produced in a cell.

If an inhibitor combines with an enzyme in an irreversible manner, the effect is called **noncompetitive inhibition**:

$$E + I \longrightarrow EI$$

The substrate cannot displace this type of inhibitor from the active site. Many poisons work by inhibiting enzyme reactions that are necessary for life processes.

Some drugs are competitive inhibitors. Antibiotics probably work by inhibiting enzyme reactions that are required by microorganisms. Sulfanilamide is believed to act as an inhibitor of an enzyme for which the substrate is *p*-aminobenzoic acid:

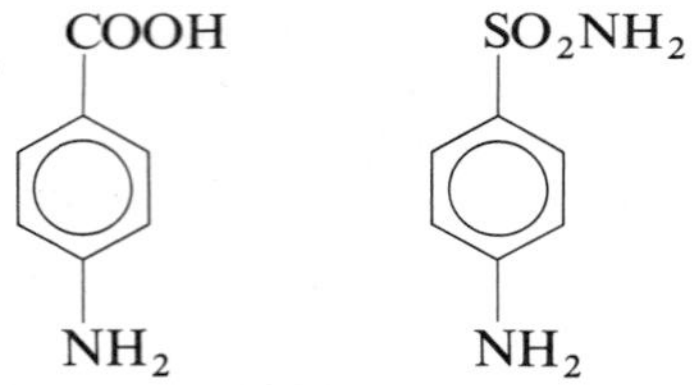

p-aminobenzoic acid   *sulfanilamide*

Notice the similarity between the structures of the two compounds, which compete for the active site. Certain microorganisms utilize the enzymatic reaction of *p*-aminobenzoic acid. Inhibition of the reaction by sulfanilamide causes the death of these microorganisms.

Many insecticides and herbicides are effective because they are enzymatic inhibitors.

27.6 Metabolism

Metabolism is a general term that refers to all the reactions that occur in a living organism. Metabolic reactions are divided into two groups:

1. Catabolic processes are those in which substances are broken down into simpler substances. These reactions provide the energy required by the organism. They also produce simple compounds from which larger ones are made.

2. Anabolic processes are those in which large molecules are synthesized from smaller ones. These processes require energy and produce substances needed for the reproduction and survival of the organism.

The primary source of energy for all life is the light from the sun. In **photosynthesis,** the chlorophyll of plants absorbs sunlight, which is used as a source of energy to bring about the manufacture of carbohydrates from CO_2 and H_2O. In this way, energy of the sun is stored as chemical energy in the plants. Humans obtain the energy required to sustain life from their food, which comes from plants or from animals that have been fed on plants.

A large amount of energy is liberated by the oxidation of glucose and other substances derived from food. If 1 mol of glucose is oxidized to CO_2 and H_2O, about 2870 kJ of energy is liberated. In the body, the oxidation is conducted gradually, in a controlled manner, through a sequence of many steps. Each step is catalyzed by an enzyme and involves a comparatively small energy effect. Enzymes cannot function at temperatures much higher than body temperature.

A part of the energy from the oxidation of glucose is dissipated as heat, which maintains body temperature. About 44% of the total energy available, however, is captured and stored. The energy is stored in high-energy molecules. The most important molecule of this type is **adenosine triphosphate,** or **ATP:**

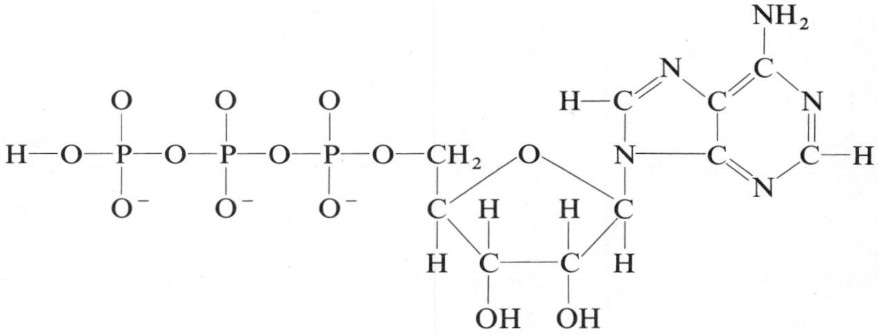

ATP is prepared from **adenosine diphosphate, ADP.** The structures of the two compounds are the same except that ATP has three phosphate groups and ADP has two:

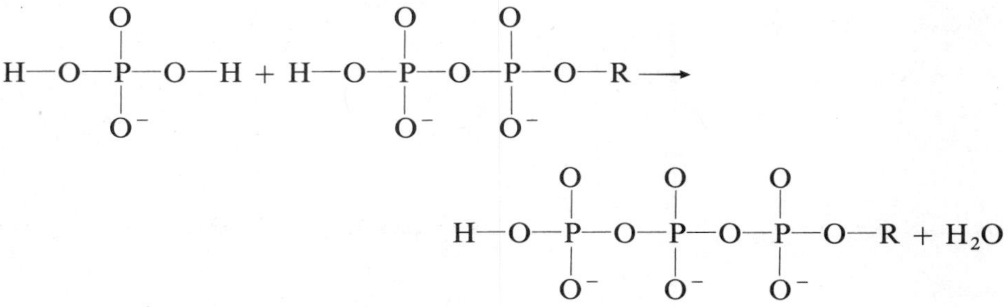

In this equation, R is used to indicate the organic part of the molecules.

The free energy change for the preparation of ATP from ADP is about $+33$ kJ/mol—which means that energy is required and the reaction will not occur by itself. In the cell, the reaction is **coupled** with another reaction that liberates more energy than the preparation of ATP requires. The overall change in free energy for the coupled reactions is negative. The following figures are typical:

cell reaction	$\Delta G = -50\,\text{kJ}$
preparation of ATP	$\Delta G = +33\,\text{kJ}$ (stored)
coupled reactions	$\Delta G = -17\,\text{kJ}$ (lost as heat)

The energy stored in ATP is released by the reverse of the preceding reaction. When ATP loses a phosphate group, ADP is produced and the free energy change is about $-33\,\text{kJ}$. The energy released is used, by means of reaction coupling, to drive reactions in which components of the organism are synthesized. All reactions of this type (anabolic reactions) require energy. The energy of ATP hydrolysis is also used to produce muscle contractions, but the mechanism of the process is not known.

The oxidation of glucose in the body takes place in many steps. Several of the steps lead to the production of one or more molecules of ATP. In all, the oxidation of 1 mol of glucose produces 38 mol of ATP. The process, therefore, stores $38 \times 33\,\text{kJ} = 1254\,\text{kJ}$ of energy. Since the total energy liberated by the oxidation of 1 mol of glucose is 2870 kJ,

$$(1254\,\text{kJ}/2870\,\text{kJ})100 = 44\%$$

of the energy liberated is stored.

Summary

The topics that have been discussed in this chapter are

1. Amino acids and the macromolecules derived from them, proteins. The primary, secondary, tertiary, and quaternary structures of proteins.

2. The structures of carbohydrates—which are hydroxy aldehydes, hydroxy ketones, and substances derived from them.

3. The esters of fatty acids and glycerol, fats and oils.

4. Nucleic acids, their structures and role in the biological synthesis of proteins.

5. Enzymes, which are biological catalysts, and how they function; enzyme inhibition.

6. Some considerations involving energy changes in metabolic reactions.

Key Terms

Some of the more important terms introduced in this chapter are listed below. Definitions for terms not included in this list may be located in the text by use of the index.

Active site (Section 27.5) The position on the surface of an enzyme at which a reaction occurs.

α-amino acid (Section 27.1) A carboxylic acid with an amino group ($-NH_2$) on the carbon atom adjacent to the carboxyl group (the alpha, α, position).

Anabolic process (Section 27.6) A biological process in which large molecules are synthesized from smaller ones.

Anticodon (Section 27.4) The portion of a transfer RNA chain that contains a sequence of three bases that complement the base triplet of a messenger RNA codon. A given anticodon corresponds to a specific amino acid that the particular transfer RNA carries.

Carbohydrate (Section 27.2) A hydroxy aldehyde, a hydroxy ketone, or a substance derived from them.

Catabolic process (Section 27.6) A biological process in which substances are broken down into simpler substances.

Chiral carbon atom (Section 27.1) A carbon atom that has four different groups bonded to it. Molecules that have a carbon atom of this type are **chiral** (their mirror images are not superimposable on them).

Codon (Section 27.4) A protion of a messenger RNA chain that contains a sequence of three bases that constitute a code.

Coupled reactions (Section 27.6) A pair of biological reactions such that one drives the other since the first reaction liberates more energy than the second requires.

Deoxyribonucleic acid, DNA (Section 27.4) A nucleic acid composed of nucleotides that contain deoxyribose units as the sugar component.

Disaccharide (Section 27.2) A carbohydrate molecule that is composed of two monosaccharide units.

Double helix (Section 27.4) The secondary structure of DNA, a coil made from two chains of DNA.

Enzyme (Section 27.5) A biological catalyst.

Enzyme inhibition (Section 27.5) A process in which the action of an enzyme is prevented or retarded. In **competitive inhibition,** an inhibitor combines reversibly with the active site and hence competes with the substrate. In **noncompetitive inhibition,** an inhibitor combines in an irreversible manner with the active site. In **product inhibition,** a product of the enzyme-catalyzed reaction acts as an inhibitor and controls the reaction rate.

Fats and oils (Section 27.3) Mixtures of esters of fatty acids and glycerol (1,2,3-propanetriol). Fats are solids at room temperature; oils are liquids.

Fatty acids (Section 27.3) Long, straight-chain carboxylic acids; some are saturated and some contain one or more carbon-carbon double bonds.

Lipids (Section 27.3) Substances that can be dissolved away from biological material by solvents that are nonpolar or slightly polar. The classification is large and includes fats, oils, some vitamins and hormones, and certain constituents of the walls of cells.

Messenger RNA (Section 27.4) an RNA chain synthesized in the nucleus of a cell in such a way that the order of bases complements the order in a section of the DNA chain. Messenger RNA moves from the cell nucleus to the ribosomes where it directs, through its sequence of codons, the synthesis of a protein from amino acids.

Metabolism (Section 27.5) All the reactions that occur in a living organism.

Monosaccharide (Section 27.2) A simple sugar; a molecule obtained from the hydrolysis of a polymeric carbohydrate.

Mutation (Section 27.4) A change in the DNA of a cell that ultimately results in an alteration of the proteins synthesized by that cell.

Nucleic acid (Section 27.4) A high molecular weight, long-chain polymer formed from nucleotides.

Nucleotide (Section 27.4) A unit of a nucleic acid formed from a phosphoric acid molecule, a five-carbon sugar molecule, and a molecule of a nitrogen-containing base.

Peptide linkage (Section 27.1) The —CO—NH— linkage formed when two amino acids combine. The amino group of one molecule joins, with the elimination of water, to the carboxyl group of another molecule.

Polypeptide (Section 27.1) A protein; a macromolecule formed by the condensation of many amino acid molecules.

Polysaccharide (Section 27.2) A carbohydrate molecule that contains many monosaccharide units; examples are starch and cellulose, both of which yield D-glucose upon hydrolysis.

Primary structure of a protein (Section 27.1) The sequence of amino acids in the protein chain.

Protein (Section 27.1) A macromolecule formed from amino acids; a polypeptide.

Quaternary structure of a protein (Section 27.1) The number of protein chains and the way their tertiary structures are arranged in the complete protein.

Ribonucleic acid, RNA (Section 27.4) A nucleic acid composed of nucleotides that contain ribose units as the sugar component.

Saponification (Section 27.3) The process in which a triglyceride or a mixture of triglycerides is heated with an aqueous solution of a base to yield glycerol and the salts of fatty acids (**soaps**).

Secondary structure of a protein (Section 27.1) The spatial arrangement of a protein chain, which includes the α-helix (a coiled arrangement similar to that of a spring) and the pleated sheet.

Substrate (Section 27.5) A substance that undergoes reaction at the active site of an enzyme.

Tertiary structure of a protein (Section 27.1) The arrangement of secondary structures of proteins brought about by interactions between groups (other than the peptide linkage) in the protein chain.

Transfer RNA (Section 27.4) The smallest type of nucleic acid found in the cell. A particular transfer RNA carries a specific amino acid and has an anticodon that corresponds to this amino acid.

Triglyceride (Section 27.3) A fat or oil; an ester formed from 1 mol of glycerol and 3 mol of fatty acids (which may be alike or different).

Turnover number (Section 27.5) The number of moles of reactant transformed per unit time by the action of an enzyme.

Zwitterion (Section 27.1) A form of an amino acid brought about by internal neutralization; the proton of the —COOH group moves to the —NH$_2$ group to form —COO$^-$ and —NH$_2^+$ groups.

Problems*

Amino Acids, Proteins

27.1 Alanine is 2-aminopropanoic acid. Write the structure of the polypeptide formed from three molecules of alanine.

27.2 Write a chemical equation for the reaction of valine, 2-amino-3-methylbutanoic acid, **(a)** with H$^+$(aq) in water solution, **(b)** with OH$^-$(aq) in water solution, **(c)** in which a zwitter ion is prepared.

27.3 Do optical isomers of glycine (2-amino-acetic acid) exist? Why or why not?

27.4 What is meant by the primary, secondary, tertiary, and quaternary structures of a protein?

27.5 What types of interactions hold a protein into its secondary structure? tertiary structure?

27.6 Approximately how many amino acid residues are present in a protein with a molecular weight of 1,000,000?

*27.7 Consider two different amino acids, which we will call A and B. If we prepare a dipeptide from these amino acids, how many arrangements (AA, AB, etc.) are possible? If we prepare a polypeptide containing three amino acid residues, how many arrangements of these two amino acids (AAA, AAB, etc.) are possible? If x is the number of different amino acids used, and y is the number of amino acid residues in a polypeptide, what is the formula for the number of possible polypeptides?

27.8 What is the difference in meaning between the symbols D and (+) as applied to amino acids?

27.9 Give examples of each of the following types of isomers: **(a)** chain, **(b)** position, **(c)** functional group, **(d)** geometric, **(e)** optical.

27.10 For which of the following are optical isomers possible:
(a) 2-hydroxypropanoic acid,
(b) hydroxyacetic acid,
(c) 2-bromo-2-chloropropane,
(d) 1-bromo-1-chloroethane,
(e) the cyanohydrin prepared from butanone?

Carbohydrates

27.11 What is a carbohydrate?

27.12 What are a monosaccharide, a disaccharide, and a polysaccharide?

27.13 Glyceraldehyde is 2,3-dihydroxypropanal. Draw the structure of the compound. How many chiral carbon atoms does the molecule contain? How many optical isomers of glyceraldehyde should exist?

27.14 The number of optical isomers of a compound is 2^n, where n is the number of chiral carbon atoms in the compound. How many optical isomers of 2,3,4-trihydroxybutanal should exist? How many optical isomers of 1,3,4-trihydroxy-2-butanone should exist?

Fats and Oils

27.15 Write an equation to show how a soap can be made from a trigylceride.

27.16 (a) Draw the structure of the triglyceride made from 3 mol of linolenic acid (see Table 27.1) and 1 mol of glycerol. **(b)** Write a chemical equation for the hydrogenation of this compound.

27.17 (a) Draw the structure of the triglyceride made from 3 mol of lauric acid (see Table 27.1) and 1 mol of glycerol. **(b)** Write a chemical equation for the combustion of this compound in oxygen.

27.18 What is the chemical difference between a fat and an oil?

27.19 If oleic acid, linoleic acid, and linolenic acid were hydrogenated completely, what saturated fatty acid would result? (See Table 27.1.)

Nucleic Acids

27.20 What is the difference between an α-helix and a double helix?

27.21 Are the two chains of a DNA double helix identical? Explain.

27.22 Describe the process by which a DNA double helix is duplicated in cell division.

27.23 What is a mutation?

---

* The more difficult problems are marked with asterisks. The appendix contains answers to color-keyed problems.

27.24 Assume that the average molecular weight of an RNA nucleotide is 340, and the average molecular weight of an amino acid residue found in a protein is 120. What must the minimum molecular weight of a messenger RNA molecule be if it is used to synthesize a protein that has a molecular weight of 90,000? The molecular weight is a minimum since some messenger RNA molecules are used to synthesize more than one protein.

27.25 The anticodon of a transfer RNA molecule that carries the amino acid serine is AGA. **(a)** What is the corresponding codon of messenger RNA? **(b)** What sequence of bases on the DNA chain (strand 1) would lead to the production of this codon on a messenger RNA molecule? **(c)** What sequence of bases on strand 2 of the DNA double helix would occur opposite the portion described in part **(b)**?

Enzymes and Metabolism

27.26 What is the turnover number of an enzyme?

27.27 Compare the action of competitive and noncompetitive inhibitors.

27.28 Describe how product inhibition and feedback inhibition work. Why are they important to the operation of the cell?

27.29 By what mechanism are enzymes thought to function? Why are they specific in their action?

27.30 What is the difference between catabolic processes and anabolic processes?

27.31 What are coupled reactions?

27.32 What purpose do high energy molecules serve? How do they function?

INTERNATIONAL SYSTEM OF UNITS (SI)

SI Base Units		
Measurement	Unit	Symbol
length	meter	m
mass	kilogram	kg
time	second	s
electric current	ampere	A
temperature[a]	kelvin	K
amount of substance	mole	mol
luminous intensity	candela	cd

[a]Temperature may also be expressed in degrees Celsius (symbol °C).

SI Supplementary Units		
	SI Unit	
Measurement	Name	Symbol
plane angle	radian	rad
solid angle	steradian	sr

SI Prefixes			
Factor		Prefix	Symbol
10^{12}	1 000 000 000 000	tera-	T-
10^{9}	1 000 000 000	giga-	G-
10^{6}	1 000 000	mega-	M-
10^{3}	1 000	kilo-	k-
10^{2}	100	hecto-	h-
10^{1}	10	deka-	da-
10^{-1}	0.1	deci-	d-
10^{-2}	0.01	centi-	c-
10^{-3}	0.001	milli-	m-
10^{-6}	0.000 001	micro-	μ-
10^{-9}	0.000 000 001	nano-	n-
10^{-12}	0.000 000 000 001	pico-	p-
10^{-15}	0.000 000 000 000 001	femto-	f-
10^{-18}	0.000 000 000 000 000 001	atto-	a-

VALUES OF SOME CONSTANTS AND CONVERSION FACTORS

Physical Constants

Constant	Symbol	Value
Avogardo's number	N	$6.02205 \times 10^{23}/\text{mol}$
Bohr radius	a_0	5.29177×10^{-11} m
electron rest mass	m	9.109535×10^{-28} g 5.485803×10^{-4} u
electronic charge, unit charge	e	1.60219×10^{-19} C
Faraday	F	9.64846×10^{4} C
ideal gas constant	R	8.2057×10^{-2} liter·atm/(K·mol) 8.31441 J/(K·mol)
molar volume, ideal gas at STP	—	22.4138 liter
neutron rest mass	—	1.674954×10^{-24} g 1.008665 u
Planck's constant	h	6.62618×10^{-34} J·s
proton rest mass	—	1.672649×10^{-24} g 1.007276 u
speed of light in a vacuum	c	2.997925×10^{8} m/s

Conversion Factors

Unit	Abbreviation	Definition or Equivalent
ångstrom	Å	10^{-10} m $= 10^{-8}$ cm $= 0.1$ nm $= 100$ pm
atmosphere	atm	1.01325×10^{5} Pa (or N/m^2) 760 torr (or mm of mercury)
calorie	cal	4.1840 J
coulomb	C	A·s
curie	Ci	$3.7 \times 10^{10}/\text{s}$
electron volt	eV	1.6022×10^{-19} J
electrostatic unit	esu	3.33564×10^{-10} C
erg	erg	10^{-7} J
joule	J	N·m $=$ kg·m^2/s^2 $=$ V·C 10^7 erg
Kelvin temperature scale	K	K $=$ °C $+ 273.15$ Triple point of water (0.01°C) is 273.16 K Freezing point of water (0°C) is 273.15 K
newton	N	kg·m/s^2
pascal	Pa	N/m^2 $=$ kg/(s^2·m)
torr (or mm of mercury)	torr	1.31579×10^{-3} atm 1.33322×10^{2} Pa
unified atomic mass unit	u	1.660566×10^{-24} g 931.502 MeV
Volt	V	J/C $=$ kg·m^2/(A·s^2)

APPENDIX C

NOTES ON MATHEMATICAL OPERATIONS

C.1 Exponents

An **exponent** is a superscript added to a **base.** It indicates a mathematical operation that is to be performed on the base. In the expression a^n the exponent is n and the base is a. The following types of exponents are frequently encountered.

1. Exponent is a positive integer. In the expression a^n the exponent n is the number of times that the base a is to be taken as a factor in the expansion. Therefore, $(n - 1)$ is the number of times the base is to be multiplied by itself. Hence,

$$a^4 = a \times a \times a \times a$$

2. Exponent is a negative integer. The expression a^{-n} is the reciprocal of a^n. For example,

$$a^{-4} = \frac{1}{a^4} = \frac{1}{a \times a \times a \times a}$$

3. Exponent is a fraction of the type 1/n. The value of n is the index of a root of the base. Thus,

$$a^{1/2} = \sqrt{a}$$
$$a^{1/3} = \sqrt[3]{a}$$

4. Exponent is a fraction of the type m/n. This exponent indicates two operations (those of parts 1 and 3). Hence, $a^{m/n}$ is $\sqrt[n]{a^m}$, and

$$a^{3/2} = \sqrt{a^3} = \sqrt{a \times a \times a}$$

5. Exponent is zero. Provided the base is not zero, the value of the expression is unity. Thus,

$$a^0 = 1 \qquad (a \neq 0)$$

Some properties of exponents are summarized in the following equations.

1. $a^m a^n = a^{m+n}$ Thus, $a^4 a^2 = a^6$
2. $(a^m)^n = a^{mn}$ Thus, $(a^4)^2 = a^8$
3. $(ab)^n = a^n b^n$ Thus, $(ab)^3 = a^3 b^3$
4. $a^m/a^n = a^{m-n}$ Thus, $a^5/a^2 = a^3$; $a^2/a^5 = a^{-3} = 1/a^3$
5. $a^n/a^n = 1$ Thus, $a^3/a^3 = 1$

C.2 Scientific Notation

Very large and very small numbers are frequently encountered in scientific studies. The velocity of light in a vacuum, for example, is 29,979,000,000 cm/s, and the distance between the centers of two hydrogen atoms in an H_2 molecule is 0.0000000075 cm. **Scientific notation** is used to simplify the handling of cumbersome values such as these. When scientific notation is employed, the value is expressed in the form

$$a \times 10^n$$

where a, the decimal part, is a number with one digit to the left of the decimal point and all others to the right, and n, the exponent of 10, is a positive or negative integer or zero.

A number can be converted into this form by moving the decimal point until there is only one nonzero digit to the left of it. For each place the decimal point is moved to the *left*, n is *increased* by one. For each place the decimal point is moved to the *right*, n is *decreased* by one. For example,

$$29{,}979{,}000{,}000 \text{ cm/s} = 2.9979 \times 10^{10} \text{ cm/s}$$
$$0.0000000075 \text{ cm} = 7.5 \times 10^{-9} \text{ cm}$$

Mathematical operations involving numbers expressed in this manner are carried out in the following ways.

1. Multiplication. The decimal parts are multiplied and the exponents of 10 are added algebraically:

$$(3.0 \times 10^5)(2.0 \times 10^2) = (3.0 \times 2.0) \times 10^{5+2}$$
$$= 6.0 \times 10^7$$
$$(4.0 \times 10^7)(5.0 \times 10^{-3}) = (4.0 \times 5.0) \times 10^{7+(-3)}$$
$$= 20 \times 10^4$$
$$= 2.0 \times 10^5$$

2. Division. The decimal parts are divided, and the exponent of 10 in the denominator is algebraically subtracted from the exponent of 10 in the numerator:

$$\frac{6.89 \times 10^{-7}}{3.36 \times 10^3} = \left(\frac{6.89}{3.36}\right) \times 10^{(-7)-(+3)}$$
$$= 2.05 \times 10^{-10}$$

3. Addition and subtraction. The numbers must all be expressed with the same power of 10. The answer, which has this same power of 10, is found by adding or subtracting the decimal parts:

$$(6.25 \times 10^3) + (3.0 \times 10^2) = (6.25 \times 10^3) + (0.30 \times 10^3)$$
$$= 6.55 \times 10^3$$

4. Taking a root. When a square root is taken, the number is written in such a way that the exponent of 10 is perfectly divisible by 2. The answer is obtained by taking the square root of the decimal part and dividing the power of 10 by 2:

$$\sqrt{2.21 \times 10^{-7}} = \sqrt{22.1 \times 10^{-8}}$$
$$= 4.70 \times 10^{-4}$$

When a cube root is taken, the cube root of the decimal part is obtained and the exponent of 10 is divided by 3:

$$\sqrt[3]{1.86 \times 10^8} = \sqrt[3]{186 \times 10^6}$$
$$= 5.71 \times 10^2$$

5. Raising to a power. When a number is squared, the decimal part is squared and the exponent of 10 is multiplied by 2:

$$(1.36 \times 10^4)^2 = (1.36)^2 \times 10^{2(+4)}$$
$$= 1.85 \times 10^8$$

When a number is cubed, the decimal part is cubed and the exponent of 10 is multiplied by 3:

$$(2.06 \times 10^{-5})^3 = (2.06)^3 \times 10^{3(-5)}$$
$$= 8.74 \times 10^{-15}$$

In general,

$$(a \times 10^n)^p = a^p \times 10^{p(n)}$$

C.3 Logarithms

The logarithm of a number is the power to which a base must be raised in order to secure the number. Common logarithms (abbreviated log) employ the base 10. If

$$a = 10^n$$
$$\log a = n$$

and, therefore,

$$\log 1000 = \log 10^3 = 3$$
$$\log 0.01 = \log 10^{-2} = -2$$

Appendix C   Notes on Mathematical Operations

The logarithm of a number that is merely 10 raised to a power is the exponent of 10. The logarithm of a number such as 3.540, however, cannot be determined by inspection. The logarithms of numbers from 1 to 10 can be obtained from the table of logarithms found in Appendix D. Decimal points are omitted from this table. Each of the numbers listed is assumed to have a decimal point following the first digit. Each of the logarithms should have a decimal point preceding the value listed. The logarithm of 3.540 is 0.5490.

The logarithm of a number greater than 10 or less than 1 can be obtained in the following way. The number is expressed in scientific notation. For example, consider

$$3.540 \times 10^{12}$$

Since logarithms are exponents and since $a^m a^n = a^{m+n}$, the logarithm of this value is obtained by adding the logarithm of 3.540 to the logarithm of 10^{12}. Hence,

$$\begin{aligned} \log (3.540 \times 10^{12}) &= \log 3.540 + \log 10^{12} \\ &= 0.5490 + 12 \\ &= 12.5490 \end{aligned}$$

Another example is

$$\begin{aligned} \log (2.00 \times 10^{-5}) &= \log 2.00 + \log 10^{-5} \\ &= 0.301 + (-5) \\ &= -4.699 \end{aligned}$$

Notice that the number of digits that follow the decimal point in the recorded logarithm is equal to the number of significant figures in the original value.

Sometimes it is necessary to find an **antilogarithm,** the number that corresponds to a given logarithm. In such instances the procedure used to find a logarithm is reversed. The given logarithm is written in two parts: a decimal fraction (called a **mantissa**) and a positive or negative whole number (called a **characteristic**). For example,

$$\text{antilog} (3.740) = \text{antilog} (0.740 + 3)$$

The antilogarithm of the mantissa (0.740) is obtained by finding the number corresponding to this logarithm in the table (it is 5.50), and the antilogarithm of the characteristic (3) is merely 10^3. Therefore,

$$\text{antilog} (3.740) = 5.50 \times 10^3$$

or

$$3.740 = \log (5.50 \times 10^3)$$

All the mantissas recorded in a table of logarithms are *positive*. This fact must be taken into account when an antilogarithm of a negative number is found. For example, to take the antilogarithm of -3.158, we must write the number in such a way that the mantissa is positive. Thus,

$$\text{antilog} \, (-3.158) = \text{antilog} \, (0.842 - 4)$$
$$= 6.95 \times 10^{-4}$$

or

$$-3.158 = \log \, (6.95 \times 10^{-4})$$

Since logarithms are exponents, mathematical operations involving logarithms follow the rules for the use of exponents. When each of the following operations is performed, the logarithms of the values involved are found, the logarithms are treated as indicated, and the antilogarithm of the result is secured as the final answer.

1. *Multiplication.* $\qquad \log \, (ab) = \log a + \log b$

2. *Division.* $\qquad\qquad \log \, (a/b) = \log a - \log b$

3. *Extraction of a root.* $\quad \log \, (a^{1/n}) = \dfrac{1}{n} \log a$

4. *Raising to a power.* $\quad \log \, (a^n) = n \log a$

Logarithms that employ the base 10 are called common logarithms. Natural logarithms (abbreviated ln) employ the base e, where

$$e = 2.71828 \ldots$$

The relation between natural logarithms and common logarithms is

$$\ln a = 2.303 \log a$$

Thus, to find the natural logarithm of 6.040, we multiply the common logarithm of 6.040 by 2.303:

$$\ln 6.040 = 2.303 \log 6.040$$
$$= 2.303(0.7810)$$
$$= 1.7986$$

Logarithms can be used to evaluate an expression of the type e^n, where e is the base of natural logarithms. Since

$$\ln a = 2.303 \log a$$
$$\log a = \frac{\ln a}{2.303}$$

and since

$$\ln e^n = n$$
$$\log e^n = \frac{n}{2.303}$$

Therefore,

$$e^n = \text{antilog} \frac{n}{2.303}$$

For example, the value of $e^{2.209}$ can be found in the following way:

$$e^{2.209} = \text{antilog} \frac{2.209}{2.303} = \text{antilog } 0.9590$$

$$= 9.100$$

C.4 Quadratic Equations

An algebraic equation in the form

$$ax^2 + bx + c = 0$$

is called a **quadratic equation** in one variable. An equation of this type has two solutions given by the **quadratic formula**

$$x = \frac{-b \pm \sqrt{b^2 - 4ac}}{2a}$$

When the quadratic formula is used to find the answer to a chemical problem, two solutions are obtained. One of them, however, must be discarded because it represents a physical impossibility.

Assume that x in the equation

$$x^2 + 0.50\,x - 0.15 = 0$$

is the number of moles of a gas that dissociate under a given set of conditions. The values of the coefficients are: $a = 1$, $b = 0.50$, and $c = -0.15$; and

$$x = \frac{-0.50 \pm \sqrt{(0.50)^2 - 4(1)(-0.15)}}{2(1)}$$

$$x = \frac{-0.50 \pm 0.92}{2}$$

$$x = +0.21, -0.71$$

The value -0.71 is discarded because a negative amount of a substance is physically impossible.

	0	1	2	3	4	5	6	7	8	9
10	0000	0043	0086	0128	0170	0212	0253	0294	0334	0374
11	0414	0453	0492	0531	0569	0607	0645	0682	0719	0755
12	0792	0828	0864	0899	0934	0969	1004	1038	1072	1106
13	1139	1173	1206	1239	1271	1303	1335	1367	1399	1430
14	1461	1492	1523	1553	1584	1614	1644	1673	1703	1732
15	1761	1790	1818	1847	1875	1903	1931	1959	1987	2014
16	2041	2068	2095	2122	2148	2175	2201	2227	2253	2279
17	2304	2330	2355	2380	2405	2430	2455	2480	2504	2529
18	2553	2577	2601	2625	2648	2672	2695	2718	2742	2765
19	2788	2810	2833	2856	2878	2900	2923	2945	2967	2989
20	3010	3032	3054	3075	3096	3118	3139	3160	3181	3201
21	3222	3243	3263	3284	3304	3324	3345	3365	3385	3404
22	3424	3444	3464	3483	3502	3522	3541	3560	3579	3598
23	3617	3636	3655	3674	3692	3711	3729	3747	3766	3784
24	3802	3820	3838	3856	3874	3892	3909	3927	3945	3962
25	3979	3997	4014	4031	4048	4065	4082	4099	4116	4133
26	4150	4166	4183	4200	4216	4232	4249	4265	4281	4298
27	4314	4330	4346	4362	4378	4393	4409	4425	4440	4456
28	4472	4487	4502	4518	4533	4548	4564	4579	4594	4609
29	4624	4639	4654	4669	4683	4698	4713	4728	4742	4757
30	4771	4786	4800	4814	4829	4843	4857	4871	4886	4900
31	4914	4928	4942	4955	4969	4983	4997	5011	5024	5038
32	5051	5065	5079	5092	5105	5119	5132	5145	5159	5172
33	5185	5198	5211	5224	5237	5250	5263	5276	5289	5302
34	5315	5328	5340	5353	5366	5378	5391	5403	5416	5428
35	5441	5453	5465	5478	5490	5502	5514	5527	5539	5551
36	5563	5575	5587	5599	5611	5623	5635	5647	5658	5670
37	5682	5694	5705	5717	5729	5740	5752	5763	5775	5786
38	5798	5809	5821	5832	5843	5855	5866	5877	5888	5899
39	5911	5922	5933	5944	5955	5966	5977	5988	5999	6010
40	6021	6031	6042	6053	6064	6075	6085	6096	6107	6117
41	6128	6138	6149	6160	6170	6180	6191	6201	6212	6222
42	6232	6243	6253	6263	6274	6284	6294	6304	6314	6325
43	6335	6345	6355	6365	6375	6385	6395	6405	6415	6425
44	6435	6444	6454	6464	6474	6484	6493	6503	6513	6522
45	6532	6542	6551	6561	6571	6580	6590	6599	6609	6618
46	6628	6637	6646	6656	6665	6675	6684	6693	6702	6712
47	6721	6730	6739	6749	6758	6767	6776	6785	6794	6803
48	6812	6821	6830	6839	6848	6857	6866	6875	6884	6893
49	6902	6911	6920	6928	6937	6946	6955	6964	6972	6981

	0	1	2	3	4	5	6	7	8	9
50	6990	6998	7007	7016	7024	7033	7042	7050	7059	7067
51	7076	7084	7093	7101	7110	7118	7126	7135	7143	7152
52	7160	7168	7177	7185	7193	7202	7210	7218	7226	7235
53	7243	7251	7259	7267	7275	7284	7292	7300	7308	7316
54	7324	7332	7340	7348	7356	7364	7372	7380	7388	7396
55	7404	7412	7419	7427	7435	7443	7451	7459	7466	7474
56	7482	7490	7497	7505	7513	7520	7528	7536	7543	7551
57	7559	7566	7574	7582	7589	7597	7604	7612	7619	7627
58	7634	7642	7649	7657	7664	7672	7679	7686	7694	7701
59	7709	7716	7723	7731	7738	7745	7752	7760	7767	7774
60	7782	7789	7796	7803	7810	7818	7825	7832	7839	7846
61	7853	7860	7868	7875	7882	7889	7896	7903	7910	7917
62	7924	7931	7938	7945	7952	7959	7966	7973	7980	7987
63	7993	8000	8007	8014	8021	8028	8035	8041	8048	8055
64	8062	8069	8075	8082	8089	8096	8102	8109	8116	8122
65	8129	8136	8142	8149	8156	8162	8169	8176	8182	8189
66	8195	8202	8209	8215	8222	8228	8235	8241	8248	8254
67	8261	8267	8274	8280	8287	8293	8299	8306	8312	8319
68	8325	8331	8338	8344	8351	8357	8363	8370	8376	8382
69	8388	8395	8401	8407	8414	8420	8426	8432	8439	8445
70	8451	8457	8463	8470	8476	8482	8488	8494	8500	8506
71	8513	8519	8525	8531	8537	8543	8549	8555	8561	8567
72	8573	8579	8585	8591	8597	8603	8609	8615	8621	8627
73	8633	8639	8645	8651	8657	8663	8669	8675	8681	8686
74	8692	8698	8704	8710	8716	8722	8727	8733	8739	8745
75	8751	8756	8762	8768	8774	8779	8785	8791	8797	8802
76	8808	8814	8820	8825	8831	8837	8842	8848	8854	8859
77	8865	8871	8876	8882	8887	8893	8899	8904	8910	8915
78	8921	8927	8932	8938	8943	8949	8954	8960	8965	8971
79	8976	8982	8987	8993	8998	9004	9009	9015	9020	9025
80	9031	9036	9042	9047	9053	9058	9063	9069	9074	9079
81	9085	9090	9096	9101	9106	9112	9117	9122	9128	9133
82	9138	9143	9149	9154	9159	9165	9170	9175	9180	9186
83	9191	9196	9201	9206	9212	9217	9222	9227	9232	9238
84	9243	9248	9253	9258	9263	9269	9274	9279	9284	9289
85	9294	9299	9304	9309	9315	9320	9325	9330	9335	9340
86	9345	9350	9355	9360	9365	9370	9375	9380	9385	9390
87	9395	9400	9405	9410	9415	9420	9425	9430	9435	9440
88	9445	9450	9455	9460	9465	9469	9474	9479	9484	9489
89	9494	9499	9504	9509	9513	9518	9523	9528	9533	9538
90	9542	9547	9552	9557	9562	9566	9571	9576	9581	9586
91	9590	9595	9600	9605	9609	9614	9619	9624	9628	9633
92	9638	9643	9647	9652	9657	9661	9666	9671	9675	9680
93	9685	9689	9694	9699	9703	9708	9713	9717	9722	9727
94	9731	9736	9741	9745	9750	9754	9759	9763	9768	9773
95	9777	9782	9786	9791	9795	9800	9805	9809	9814	9818
96	9823	9827	9832	9836	9841	9845	9850	9854	9859	9863
97	9868	9872	9877	9881	9886	9890	9894	9899	9903	9908
98	9912	9917	9921	9926	9930	9934	9939	9943	9948	9952
99	9956	9961	9965	9969	9974	9978	9983	9987	9991	9996

STANDARD ELECTRODE POTENTIALS AT 25°C

Acid Solution	
Half Reaction	$\mathscr{E}°$ (volts)
$Li^+ + e^- \rightleftharpoons Li$	-3.045
$K^+ + e^- \rightleftharpoons K$	-2.925
$Rb^+ + e^- \rightleftharpoons Rb$	-2.925
$Cs^+ + e^- \rightleftharpoons Cs$	-2.923
$Ra^{2+} + 2e^- \rightleftharpoons Ra$	-2.916
$Ba^{2+} + 2e^- \rightleftharpoons Ba$	-2.906
$Sr^{2+} + 2e^- \rightleftharpoons Sr$	-2.888
$Ca^{2+} + 2e^- \rightleftharpoons Ca$	-2.866
$Na^+ + e^- \rightleftharpoons Na$	-2.714
$Ce^{3+} + 3e^- \rightleftharpoons Ce$	-2.483
$Mg^{2+} + 2e^- \rightleftharpoons Mg$	-2.363
$Be^{2+} + 2e^- \rightleftharpoons Be$	-1.847
$Al^{3+} + 3e^- \rightleftharpoons Al$	-1.662
$Mn^{2+} + 2e^- \rightleftharpoons Mn$	-1.180
$Zn^{2+} + 2e^- \rightleftharpoons Zn$	-0.7628
$Cr^{3+} + 3e^- \rightleftharpoons Cr$	-0.744
$Ga^{3+} + 3e^- \rightleftharpoons Ga$	-0.529
$Fe^{2+} + 2e^- \rightleftharpoons Fe$	-0.4402
$Cr^{3+} + e^- \rightleftharpoons Cr^{2+}$	-0.408
$Cd^{2+} + 2e^- \rightleftharpoons Cd$	-0.4029
$PbSO_4 + 2e^- \rightleftharpoons Pb + SO_4^{2-}$	-0.3588
$Tl^+ + e^- \rightleftharpoons Tl$	-0.3363
$Co^{2+} + 2e^- \rightleftharpoons Co$	-0.277
$H_3PO_4 + 2H^+ + 2e^- \rightleftharpoons H_3PO_3 + H_2O$	-0.276
$Ni^{2+} + 2e^- \rightleftharpoons Ni$	-0.250
$Sn^{2+} + 2e^- \rightleftharpoons Sn$	-0.136
$Pb^{2+} + 2e^- \rightleftharpoons Pb$	-0.126
$2H^+ + 2e^- \rightleftharpoons H_2$	0.0000
$S + 2H^+ + 2e^- \rightleftharpoons H_2S$	$+0.142$
$Sn^{4+} + 2e^- \rightleftharpoons Sn^{2+}$	$+0.15$
$SO_4^{2-} + 4H^+ + 2e^- \rightleftharpoons H_2SO_3 + H_2O$	$+0.172$
$AgCl + e^- \rightleftharpoons Ag + Cl^-$	$+0.2222$
$Cu^{2+} + 2e^- \rightleftharpoons Cu$	$+0.337$
$H_2SO_3 + 4H^+ + 4e^- \rightleftharpoons S + 3H_2O$	$+0.450$
$Cu^+ + e^- \rightleftharpoons Cu$	$+0.521$
$I_2 + 2e^- \rightleftharpoons 2I^-$	$+0.5355$
$MnO_4^- + e^- \rightleftharpoons MnO_4^{2-}$	$+0.564$
$O_2 + 2H^+ + 2e^- \rightleftharpoons H_2O_2$	$+0.6824$
$Fe^{3+} + e^- \rightleftharpoons Fe^{2+}$	$+0.771$

(continued)

Half Reaction	$\mathscr{E}°$ (volts)
$Hg_2^{2+} + 2e^- \rightleftharpoons 2\,Hg$	$+0.788$
$Ag^+ + e^- \rightleftharpoons Ag$	$+0.7991$
$2\,NO_3^- + 4\,H^+ + 2e^- \rightleftharpoons N_2O_4 + 2\,H_2O$	$+0.803$
$Hg^{2+} + 2e^- \rightleftharpoons Hg$	$+0.854$
$2\,Hg^{2+} + 2e^- \rightleftharpoons Hg_2^{2+}$	$+0.920$
$NO_3^- + 4\,H^+ + 3e^- \rightleftharpoons NO + 2\,H_2O$	$+0.96$
$Br_2 + 2e^- \rightleftharpoons 2\,Br^-$	$+1.0652$
$O_2 + 4\,H^+ + 4e^- \rightleftharpoons 2\,H_2O$	$+1.229$
$MnO_2 + 4\,H^+ + 2e^- \rightleftharpoons Mn^{2+} + 2\,H_2O$	$+1.23$
$Tl^{3+} + 2e^- \rightleftharpoons Tl^+$	$+1.25$
$Cr_2O_7^{2-} + 14\,H^+ + 6e^- \rightleftharpoons 2\,Cr^{3+} + 7\,H_2O$	$+1.33$
$Cl_2 + 2e^- \rightleftharpoons 2\,Cl^-$	$+1.3595$
$Au^{3+} + 2e^- \rightleftharpoons Au^+$	$+1.402$
$PbO_2 + 4\,H^+ + 2e^- \rightleftharpoons Pb^{2+} + 2\,H_2O$	$+1.455$
$Au^{3+} + 3e^- \rightleftharpoons Au$	$+1.498$
$Mn^{3+} + e^- \rightleftharpoons Mn^{2+}$	$+1.51$
$MnO_4^- + 8\,H^+ + 5e^- \rightleftharpoons Mn^{2+} + 4\,H_2O$	$+1.51$
$Ce^{4+} + e^- \rightleftharpoons Ce^{3+}$	$+1.61$
$2\,HOCl + 2\,H^+ + 2e^- \rightleftharpoons Cl_2 + 2\,H_2O$	$+1.63$
$PbO_2 + SO_4^{2-} + 4\,H^+ + 2e^- \rightleftharpoons PbSO_4 + 2\,H_2O$	$+1.682$
$Au^+ + e^- \rightleftharpoons Au$	$+1.691$
$MnO_4^- + 4\,H^+ + 3e^- \rightleftharpoons MnO_2 + 2\,H_2O$	$+1.695$
$H_2O_2 + 2\,H^+ + 2e^- \rightleftharpoons 2\,H_2O$	$+1.776$
$Co^{3+} + e^- \rightleftharpoons Co^{2+}$	$+1.808$
$S_2O_8^{2-} + 2e^- \rightleftharpoons 2\,SO_4^{2-}$	$+2.01$
$O_3 + 2\,H^+ + 2e^- \rightleftharpoons O_2 + H_2O$	$+2.07$
$F_2 + 2e^- \rightleftharpoons 2\,F^-$	$+2.87$

Alkaline Solution

Half Reaction	$\mathscr{E}°$ (volts)
$Al(OH)_4^- + 3e^- \rightleftharpoons Al + 4\,OH^-$	-2.33
$Zn(OH)_4^{2-} + 2e^- \rightleftharpoons Zn + 4\,OH^-$	-1.215
$Fe(OH)_2 + 2e^- \rightleftharpoons Fe + 2\,OH^-$	-0.877
$2\,H_2O + 2e^- \rightleftharpoons H_2 + 2\,OH^-$	-0.82806
$Cd(OH)_2 + 2e^- \rightleftharpoons Cd + 2\,OH^-$	-0.809
$S + 2e^- \rightleftharpoons S^{2-}$	-0.447
$CrO_4^{2-} + 4\,H_2O + 3e^- \rightleftharpoons Cr(OH)_3 + 5\,OH^-$	-0.13
$NO_3^- + H_2O + 2e^- \rightleftharpoons NO_2^- + 2\,OH^-$	$+0.01$
$O_2 + 2\,H_2O + 4e^- \rightleftharpoons 4\,OH^-$	$+0.401$
$NiO_2 + 2\,H_2O + 2e^- \rightleftharpoons Ni(OH)_2 + 2\,OH^-$	$+0.490$
$HO_2^- + H_2O + 2e^- \rightleftharpoons 3\,OH^-$	$+0.878$

Data from A. J. de Bethune and N. A. Swendeman Loud, "Table of Electrode Potentials and Temperature Coefficients," pp. 414–424 in *Encyclopedia of Electrochemistry* (C. A. Hampel, editor), Van Nostrand Reinhold, New York, 1964, and from A. J. de Bethune and N. A. Swendeman Loud, *Standard Aqueous Electrode Potentials and Temperature Coefficients*, 19 pp., C. A. Hampel, publisher, Skokie, Illinois, 1964.

F.1 Ionization Constants

Monoprotic Acids		
acetic	$HC_2H_3O_2 \rightleftharpoons H^+ + C_2H_3O_2^-$	1.8×10^{-5}
benzoic	$HC_7H_5O_2 \rightleftharpoons H^+ + C_7H_5O_2^-$	6.0×10^{-5}
chlorous	$HClO_2 \rightleftharpoons H^+ + ClO_2^-$	1.1×10^{-2}
cyanic	$HOCN \rightleftharpoons H^+ + OCN^-$	1.2×10^{-4}
formic	$HCHO_2 \rightleftharpoons H^+ + CHO_2^-$	1.8×10^{-4}
hydrazoic	$HN_3 \rightleftharpoons H^+ + N_3^-$	1.9×10^{-5}
hydrocyanic	$HCN \rightleftharpoons H^+ + CN^-$	4.0×10^{-10}
hydrofluoric	$HF \rightleftharpoons H^+ + F^-$	6.7×10^{-4}
hypobromous	$HOBr \rightleftharpoons H^+ + OBr^-$	2.1×10^{-9}
hypochlorous	$HOCl \rightleftharpoons H^+ + OCl^-$	3.2×10^{-8}
nitrous	$HNO_2 \rightleftharpoons H^+ + NO_2^-$	4.5×10^{-4}

Polyprotic Acids		
arsenic	$H_3AsO_4 \rightleftharpoons H^+ + H_2AsO_4^-$	$K_{a1} = 2.5 \times 10^{-4}$
	$H_2AsO_4^- \rightleftharpoons H^+ + HAsO_4^{2-}$	$K_{a2} = 5.6 \times 10^{-8}$
	$HAsO_4^{2-} \rightleftharpoons H^+ + AsO_4^{3-}$	$K_{a3} = 3 \times 10^{-13}$
carbonic	$CO_2 + H_2O \rightleftharpoons H^+ + HCO_3^-$	$K_{a1} = 4.2 \times 10^{-7}$
	$HCO_3^- \rightleftharpoons H^+ + CO_3^{2-}$	$K_{a2} = 4.8 \times 10^{-11}$
hydrosulfuric	$H_2S \rightleftharpoons H^+ + HS^-$	$K_{a1} = 1.1 \times 10^{-7}$
	$HS^- \rightleftharpoons H^+ + S^{2-}$	$K_{a2} = 1.0 \times 10^{-14}$
oxalic	$H_2C_2O_4 \rightleftharpoons H^+ + HC_2O_4^-$	$K_{a1} = 5.9 \times 10^{-2}$
	$HC_2O_4^- \rightleftharpoons H^+ + C_2O_4^{2-}$	$K_{a2} = 6.4 \times 10^{-5}$
phosphoric	$H_3PO_4 \rightleftharpoons H^+ + H_2PO_4^-$	$K_{a1} = 7.5 \times 10^{-3}$
	$H_2PO_4^- \rightleftharpoons H^+ + HPO_4^{2-}$	$K_{a2} = 6.2 \times 10^{-8}$
	$HPO_4^{2-} \rightleftharpoons H^+ + PO_4^{3-}$	$K_{a3} = 1 \times 10^{-12}$
phosphorous (diprotic)	$H_3PO_3 \rightleftharpoons H^+ + H_2PO_3^-$	$K_{a1} = 1.6 \times 10^{-2}$
	$H_2PO_3^- \rightleftharpoons H^+ + H_2PO_3^{2-}$	$K_{a2} = 7 \times 10^{-7}$
sulfuric	$H_2SO_4 \rightleftharpoons H^+ + HSO_4^-$	strong
	$HSO_4^- \rightleftharpoons H^+ + SO_4^{2-}$	$K_{a2} = 1.3 \times 10^{-2}$
sulfurous	$SO_2 + H_2O \rightleftharpoons H^+ + HSO_3^-$	$K_{a1} = 1.3 \times 10^{-2}$
	$HSO_3^- \rightleftharpoons H^+ + SO_3^{2-}$	$K_{a2} = 5.6 \times 10^{-8}$

Bases		
ammonia	$NH_3 + H_2O \rightleftharpoons NH_4^+ + OH^-$	1.8×10^{-5}
aniline	$C_6H_5NH_2 + H_2O \rightleftharpoons C_6H_5NH_3^+ + OH^-$	4.6×10^{-10}
dimethylamine	$(CH_3)_2NH + H_2O \rightleftharpoons (CH_3)_2NH_2^+ + OH^-$	7.4×10^{-4}
hydrazine	$N_2H_4 + H_2O \rightleftharpoons N_2H_5^+ + OH^-$	9.8×10^{-7}
methylamine	$CH_3NH_2 + H_2O \rightleftharpoons CH_3NH_3^+ + OH^-$	5.0×10^{-4}
pyridine	$C_5H_5N + H_2O \rightleftharpoons C_5H_5NH^+ + OH^-$	1.5×10^{-9}
trimethylamine	$(CH_3)_3N + H_2O \rightleftharpoons (CH_3)_3NH^+ + OH^-$	7.4×10^{-5}

F.2 Solubility Products

Bromides	
$PbBr_2$	4.6×10^{-6}
Hg_2Br_2	1.3×10^{-22}
$AgBr$	5.0×10^{-13}

Carbonates	
$BaCO_3$	1.6×10^{-9}
$CdCO_3$	5.2×10^{-12}
$CaCO_3$	4.7×10^{-9}
$CuCO_3$	2.5×10^{-10}
$FeCO_3$	2.1×10^{-11}
$PbCO_3$	1.5×10^{-15}
$MgCO_3$	1×10^{-5}
$MnCO_3$	8.8×10^{-11}
Hg_2CO_3	9.0×10^{-17}
$NiCO_3$	1.4×10^{-7}
Ag_2CO_3	8.2×10^{-12}
$SrCO_3$	7×10^{-10}
$ZnCO_3$	2×10^{-10}

Chlorides	
$PbCl_2$	1.6×10^{-5}
Hg_2Cl_2	1.1×10^{-18}
$AgCl$	1.7×10^{-10}

Chromates	
$BaCrO_4$	8.5×10^{-11}
$PbCrO_4$	2×10^{-16}
Hg_2CrO_4	2×10^{-9}
Ag_2CrO_4	1.9×10^{-12}
$SrCrO_4$	3.6×10^{-5}

Fluorides	
BaF_2	2.4×10^{-5}
CaF_2	3.9×10^{-11}
PbF_2	4×10^{-8}
MgF_2	8×10^{-8}
SrF_2	7.9×10^{-10}

Hydroxides	
$Al(OH)_3$	5×10^{-33}
$Ba(OH)_2$	5.0×10^{-3}
$Cd(OH)_2$	2.0×10^{-14}
$Ca(OH)_2$	1.3×10^{-6}
$Cr(OH)_3$	6.7×10^{-31}
$Co(OH)_2$	2.5×10^{-16}
$Co(OH)_3$	2.5×10^{-43}
$Cu(OH)_2$	1.6×10^{-19}

Hydroxides (continued)	
$Fe(OH)_2$	1.8×10^{-15}
$Fe(OH)_3$	6×10^{-38}
$Pb(OH)_2$	4.2×10^{-15}
$Mg(OH)_2$	8.9×10^{-12}
$Mn(OH)_2$	2×10^{-13}
$Hg(OH)_2$ (HgO)	3×10^{-26}
$Ni(OH)_2$	1.6×10^{-16}
$AgOH$ (Ag_2O)	2.0×10^{-8}
$Sr(OH)_2$	3.2×10^{-4}
$Sn(OH)_2$	3×10^{-27}
$Zn(OH)_2$	4.5×10^{-17}

Iodides	
PbI_2	8.3×10^{-9}
Hg_2I_2	4.5×10^{-29}
AgI	8.5×10^{-17}

Oxalates	
BaC_2O_4	1.5×10^{-8}
CaC_2O_4	1.3×10^{-9}
PbC_2O_4	8.3×10^{-12}
MgC_2O_4	8.6×10^{-5}
$Ag_2C_2O_4$	1.1×10^{-11}
SrC_2O_4	5.6×10^{-8}

Phosphates	
$Ba_3(PO_4)_2$	6×10^{-39}
$Ca_3(PO_4)_2$	1.3×10^{-32}
$Pb_3(PO_4)_2$	1×10^{-54}
Ag_3PO_4	1.8×10^{-18}
$Sr_3(PO_4)_2$	1×10^{-31}

Sulfates	
$BaSO_4$	1.5×10^{-9}
$CaSO_4$	2.4×10^{-5}
$PbSO_4$	1.3×10^{-8}
Ag_2SO_4	1.2×10^{-5}
$SrSO_4$	7.6×10^{-7}

Sulfides	
Bi_2S_3	1.6×10^{-72}
CdS	1.0×10^{-28}
CoS	5×10^{-22}
CuS	8×10^{-37}
FeS	4×10^{-19}
PbS	7×10^{-29}
MnS	7×10^{-16}

Sulfides (continued)	
HgS	$1.6 \times 10^{-}$
NiS	3×10^{-2}
Ag_2S	5.5×10^{-51}
SnS	1×10^{-26}
ZnS	2.5×10^{-22}

Miscellaneous	
$NaHCO_3$	1.2×10^{-3}
$KClO_4$	8.9×10^{-3}
$K_2[PtCl_6]$	1.4×10^{-6}
$AgC_2H_3O_2$	2.3×10^{-3}
$AgCN$	1.6×10^{-14}
$AgCNS$	1.0×10^{-12}

F.3 Instability Constants

AlF_6^{3-}	1.4×10^{-20}
$Al(OH)_4^-$	1.3×10^{-34}
$Al(OH)^{2+}$	7.1×10^{-10}
$Cd(NH_3)_4^{2+}$	7.5×10^{-8}
$Cd(CN)_4^{2-}$	1.4×10^{-19}
$Cr(OH)^{2+}$	5×10^{-11}
$Co(NH_3)_6^{2+}$	1.3×10^{-5}
$Co(NH_3)_6^{3+}$	2.2×10^{-34}
$Cu(NH_3)_2^+$	1.4×10^{-11}
$Cu(NH_3)_4^{2+}$	4.7×10^{-15}
$Cu(CN)_2^-$	1×10^{-16}
$Cu(OH)^+$	1×10^{-8}
$Fe(CN)_6^{4-}$	1×10^{-35}
$Fe(CN)_6^{3-}$	1×10^{-42}
$Pb(OH)^+$	1.5×10^{-8}
$HgBr_4^{2-}$	2.3×10^{-22}
$HgCl_4^{2-}$	1.1×10^{-16}
$Hg(CN)_4^{2-}$	4×10^{-42}
HgI_4^{2-}	5.3×10^{-31}
$Ni(NH_3)_4^{2+}$	1×10^{-8}
$Ni(NH_3)_6^{2+}$	1.8×10^{-9}
$Ag(NH_3)_2^+$	6.0×10^{-8}
$Ag(CN)_2^-$	1.8×10^{-19}
$Ag(S_2O_3)_2^{3-}$	5×10^{-14}
$Ag(S_2O_3)_3^{5-}$	9.9×10^{-15}
$Zn(NH_3)_4^{2+}$	3.4×10^{-10}
$Zn(CN)_4^{2+}$	1.2×10^{-18}
$Zn(OH)_4^{2-}$	3.6×10^{-16}
$Zn(OH)^+$	4.1×10^{-5}

Chapter 1

1.5 (a) 137.0, **(b)** 10.00, **(c)** 0.900, **(d)** 5.0, **(e)** 112, **(f)** 0.00210

1.7 (a) 1×10^5 cm, **(b)** 1×10^{-6} kg, **(c)** 1.0×10^7 ns, **(d)** $1.00 \times 10^{-16} \mu m$

1.9 (a) 1×10^{-1} nm, **(b)** 1×10^2 pm, **(c)** 0.099 nm, 99 pm

1.11 1.21 km

1.13 $0.954 m^3$

1.16 (a) 1.00 m is 1.09 yards (9% increase), **(b)** 1.00 liter is 1.06 quarts (6% increase), **(c)** 1.00 kg is 2.20 pounds (10% increase)

1.18 (a) 200 g Pt, **(b)** 1.25×10^3 g alloy

*1.21 (a) 2000 g alloy, **(b)** 300 g Cu and 60 g Ni left over

1.23 89 km/hr

1.25 (a) 464.5 m/s, **(b)** 4.01×10^7 m, **(c)** 6.38×10^6 m

1.27 $0.0570 cm^3$

1.29 6.4×10^{12} g Au

1.31 (a) $0.858 g/cm^3$, **(b)** floats

*1.33 4.881×10^{27} g

Chapter 2

2.8 4.480×10^{-23} g

2.10 (a) 4.6135 mol Pt and 0.52024 mol Ir, **(b)** 2.7783×10^{24} atoms Pt and 3.1329×10^{23} atoms Ir

*2.12 They would extend 6.022×10^{18} km, which is over 4×10^{10} times the distance

2.14 (a) $H_3B_3S_6$, **(b)** $Na_2S_2O_4$, **(c)** V_3S_4, **(d)** $Na_8P_8O_{24}$, **(e)** C_4H_8

2.16 $LiBF_4$

2.18 C_2H_6N

2.20 $C_9H_8O_4$

2.22 $C_6O_7H_8$

2.24 (a) 0.210 mol C and 0.350 mol H, **(b)** C_3H_5, **(c)** 2.87 g

2.26 (a) 0.389 mol C, 0.389 mol H, and 0.0555 mol N, **(b)** 4.67 g C, 0.39 g H, and 0.78 g N, **(c)** 1.77 g O, **(d)** 0.111 mol O, **(e)** $C_7H_7NO_2$

2.28 $x = 3$

2.30 34.4% Ni

2.33 629 g Zn

2.35 0.6334 g Xe and 0.3666 g F

*2.38 83.9% C, 12.0% H, and 4.1% O

2.41 (a) $V_2O_5 + 2H_2 \longrightarrow V_2O_3 + 2H_2O$, **(b)** $2B_2O_3 + 7C \longrightarrow B_4C + 6CO$, **(c)** $4Bi + 3O_2 \longrightarrow 2Bi_2O_3$, **(d)** $CaC_2 + 2H_2O \longrightarrow Ca(OH)_2 + H_2C_2$, **(e)** $Ba(NO_3)_2 + H_2SO_4 \longrightarrow BaSO_4 + 2HNO_3$

2.44 138.1 g H_3PO_4

2.47 3.26 g HI

2.49 73.7% Na_2SO_3

2.50 4.99 g NH_4SCN

2.53 1.29 g SF_4

2.55 (a) 0.2956 g Ti, **(b)** 3.807 g $TiCl_3$, **(c)** 78.80% yield

*2.57 45.0% BaO

2.60 221 g H_2SO_4

2.63 16.0 ml H_3PO_4 solution

2.66 0.842 g CaO

Chapter 3

3.1 37°C

3.4 20°C

3.5 1.36 kJ/°C

3.7 139 kJ

3.9 144 J

3.11 26.7°C

3.13 873 kJ/mol

3.15 2.25 kJ/°C

3. $C_2H_6(l) + \frac{15}{2}O_2(g) \longrightarrow 6CO_2(g) + 3H_2O(l)$
$\Delta H = -3268$ kJ

3.20 19.42 kJ

3.22 +76.3 kJ

3.24 $+50.0$ kJ

3.27 -149.9 kJ

3.30 -1081.6 kJ

*__3.32__ -71.4 kJ

3.34 (a) $Ag(s) + \frac{1}{2}Cl_2(g) \longrightarrow AgCl(s)$ $\Delta H_f^\circ = -127$ kJ, **(b)** $\frac{1}{2}N_2(g) + O_2(g) \longrightarrow NO_2(g)$ $\Delta H_f^\circ = +33.8$ kJ, **(c)** $Ca(s) + C(graphite) + \frac{3}{2}O_2(g) \longrightarrow CaCO_3(s)$ $\Delta H_f^\circ = -1206.9$ kJ

3.36 -1125.2 kJ

3.40 (a) $CH_3OH(l) + \frac{3}{2}O_2(g) \longrightarrow CO_2(g) + 2\,H_2O(l)$, **(b)** -764.1 kJ

3.42 (a) $N_2H_4(l) + O_2(g) \longrightarrow N_2(g) + 2\,H_2O(l)$ $\Delta H = -622.4$ kJ, **(b)** $+50.6$ kJ/mol

3.45 -351.5 kJ/mol

3.48 $+96$ kJ/mol

3.51 132 kJ/mol

3.54 -23 kJ

3.58 (a) -150 kJ, **(b)** -159 kJ

Chapter 4

4.2 (a) H^+, smaller mass, **(b)** Ne^{2+}, higher charge

*__4.3__ **(a)** 1.82×10^{14} g/cm^3, **(b)** 1.32×10^{18} g, I couldn't

*__4.6__ 2.7 cm

4.8 (a) 56 protons and 82 neutrons in the nucleus and 56 electrons outside it, **(b)** $^{209}_{83}Bi$

4.15 99.75% $^{51}_{23}V$ and 0.25% $^{50}_{23}V$

4.18 69.72

4.22 1.13×10^{15}/s

4.23 (a) 4.29×10^{14}/s, 2.84×10^{-19} J, **(b)** 7.50×10^{14}/s, 4.97×10^{-19} J

4.25 (a) 333 m, 5.97×10^{-28} J, **(b)** 250 nm, 7.96×10^{-16} J

4.26 251 photons

*__4.29__ **(a)** 4.97×10^{-19} J, **(b)** 3.59×10^{-19} J, **(c)** 5.41×10^{14}/s, 555 nm

4.31 97.24 nm

4.33 from $n = 6$ to $n = 2$

*__4.37__ **(a)** 2.961×10^{16}/s, **(b)** 10.80 nm, **(c)** 72.93 nm and 54.02 nm

*__4.42__ 28, Ni

*__4.46__ **(a)** 0.0243 nm, **(b)** 3.51×10^{-36} m

4.50 (a) 18, **(b)** impossible, **(c)** 10, **(d)** 2, **(e)** 2, **(f)** impossible, **(g)** 6

4.52 (a) 9, **(b)** 17, **(c)** 8

4.56 (a) Cl, **(b)** Mn, **(c)** Ca, **(d)** Zn, **(e)** Kr

4.59 (a) $1s^2\,2s^2\,2p^6\,3s^2\,3p^6\,3d^{10}\,4s^2\,4p^6\,5s^2$, **(b)** $1s^2\,2s^2\,2p^6\,3s^2\,3p^6\,3d^{10}\,4s^2\,4p^6\,4d^{10}\,5s^2\,5p^2$, **(c)** $1s^2\,2s^2\,2p^6\,3s^2\,3p^6\,3d^{10}\,4s^2\,4p^6\,4d^{10}\,4f^6\,5s^2\,5p^6\,6s^2$, **(d)** $1s^2\,2s^2\,2p^6\,3s^2\,3p^6\,3d^8\,4s^2$, **(e)** $1s^2\,2s^2\,2p^6\,3s^2\,3p^6$

$3d^{10}\,4s^2\,4p^6\,4d^{10}\,4f^{14}\,5s^2\,5p^6\,5d^1\,6s^2$, **(f)** $1s^2\,2s^2\,2p^6$ $3p^6\,3d^{10}\,4s^2\,4p^6\,4d^{10}\,4f^{14}\,5s^2\,5p^6\,5d^{10}\,6s^2$

4.60 (a) 0, **(b)** 2, **(c)** 6, **(d)** 2, **(e)** 1, **(f)** 0

Chapter 5

5.4 (a) Si, **(b)** Sn, **(c)** Ga, **(d)** Al, **(e)** Mg, **(f)** Si, **(g)** Si

5.11 (a) Sr, **(b)** Sn, **(c)** Sb, **(d)** Sb, **(e)** Se, **(f)** S, **(g)** Ar

5.18 (a) Na, **(b)** Ar, **(c)** Na, **(d)** Cl, **(e)** Ar, **(f)** Na, Mg, Al

5.19 -824 kJ/mol

5.22 -603 kJ/mol

5.27 (a) $1s^2\,2s^2\,2p^6$, **(b)** $1s^2\,2s^2\,2p^6\,3s^2\,3p^6\,3d^4$, **(c)** $1s^2\,2s^2\,2p^6\,3s^2\,3p^6\,3d^7$, **(d)** $1s^2\,2s^2\,2p^6\,3s^2\,3p^6\,3d^{10}\,4s^2\,4p^6\,4d^8$, **(e)** $1s^2\,2s^2\,2p^6\,3s^2\,3p^6\,3d^{10}\,4s^2\,4p^6\,4d^{10}$, **(f)** $1s^2\,2s^2\,2p^6\,3s^2\,3p^6\,3d^{10}\,4s^2\,4p^6\,4d^{10}\,5s^2\,5p^6$

5.28 (a) 0, 4, 3, 2, 0, 0, **(b)** diamagnetic: Mg^{2+}, Ag^+, I^-; paramagnetic: Cr^{2+}, Co^{2+}, Pd^{2+}

5.31 (a) Kr: Se^{2-}, Br^-, Rb^+, Sr^{2+}, Y^{3+}, **(b)** Zn^{2+}: Cu^+, Ga^{3+}, **(c)** Zn: Ga^+, Ge^{2+}, As^{3+}, **(d)** O^{2-}: N^{3-}, F^-, Na^+, Mg^{2+}, Al^{3+}, **(e)** Ca^{2+}: P^{3-}, S^{2-}, Cl^-, K^+, Sc^{3+}

5.36 (a) Cs^+, **(b)** S^{2-}, **(c)** S^{2-}, **(d)** Cr^{2+}, **(e)** Ag, **(f)** Ag^+

5.38 (a) Cr_2O_3, **(b)** $Ca_3(PO_4)_2$, **(c)** $Ag_2Cr_2O_7$, **(d)** $Mg(ClO_3)_2$, **(e)** $Ni(NO_3)_2$, **(f)** $ZnCO_3$

5.40 (a) manganese(II) sulfate, **(b)** magnesium phosphate, **(c)** lead(II) carbonate, **(d)** mercury(II) chloride, **(e)** sodium peroxide, **(f)** aluminum sulfate

Chapter 6

6.3 (a) $\ddot{\text{F}}\!-\!C\!\equiv\!N\!:$, $^{(2+)}:F\!\equiv\!C\!-\!\ddot{N}\!:^{(2-)}$; the atoms of the first structure have no formal charges

(b) , ; the second structure violates the adjacent charge rule

6.4 (a) $H\!-\!C\!\equiv\!N\!:$ **(b)** $H\!-\!\ddot{\underset{..}{S}}\!:$ **(c)** $H\!-\!\overset{H}{\underset{H}{Si}}\!-\!H$

(d) $H\!-\!\ddot{\underset{..}{I}}\!:$ **(e)** $H\!-\!\overset{..}{\underset{H}{P}}\!-\!H$

6.7 (a) **(b)**

(c) **(d)**

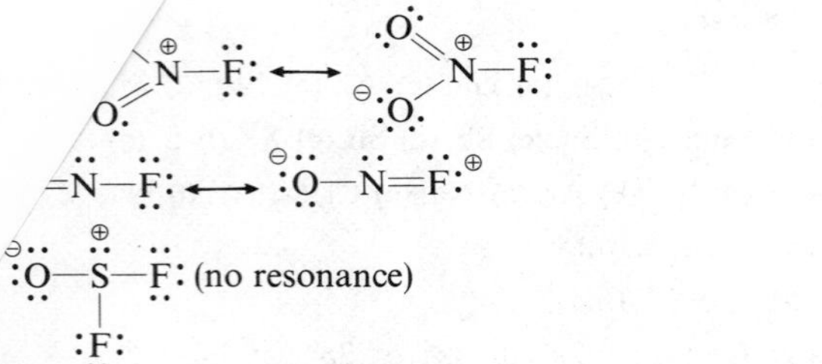

(d) 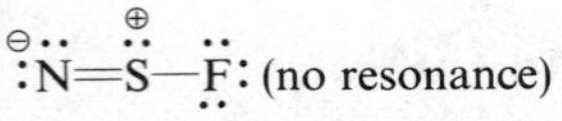 InCl$_2^+$ (2, 2, 0) linear, **(e)** 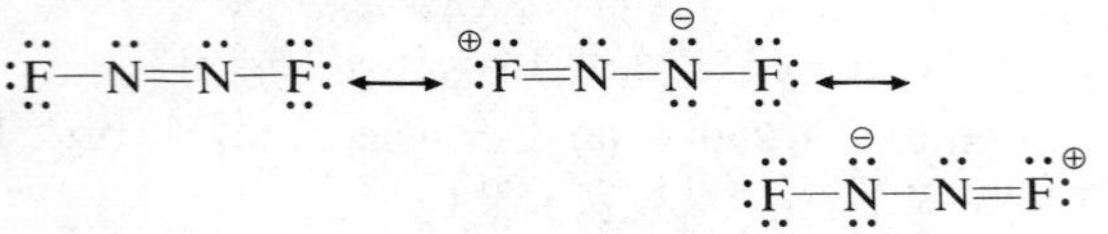 BeF$_3^-$ (3, 3, 0) triangular planar, **(f)** GeF$_2$ (3, 2, 1) angular, **(g)** AsF$_4^-$ (5, 4, 1) irregular tetrahedral, **(h)** XeF$_2$ (5, 2, 3) linear, **(i)** AsH$_4^-$ (4, 4, 0) tetrahedral

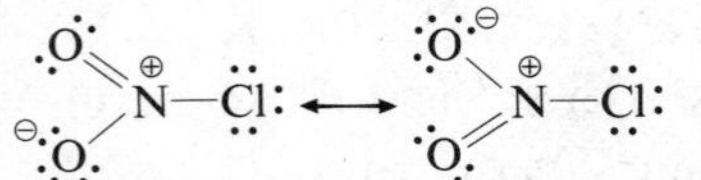

7.8 (a) d^2sp^3, **(b)** d^2sp^3, **(c)** sp^3, **(d)** sp, **(e)** sp^2, **(f)** sp^2, **(g)** dsp^3, **(h)** dsp^3, **(i)** sp^3

7.11 (a)

(left column, top)

$:N-F: \longleftrightarrow :O-N-F:$ (resonance structures with O)

$=N-F: \longleftrightarrow :O-N=F:$

$:O-S-F:$ (no resonance)
$:F:$

(d) $:N=S-F:$ (no resonance)

(e) $:F-N=N-F: \longleftrightarrow :F=N-N-F: \longleftrightarrow$

$:F-N-N=F:$

6.13 $:N=P=N-H \longleftrightarrow :N\equiv P-N-H$

6.17 (two resonance structures of $N-Cl$ with two O)

6.19 $H-N=N=N: \longleftrightarrow H-N-N\equiv N:$

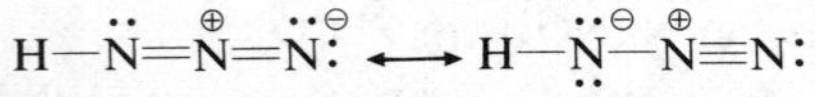

6.24 (a) CuCl$_2$, **(b)** MgSe, **(c)** LiI, **(d)** PbCl$_4$, **(e)** CdI$_2$, **(f)** Al$_2$S$_3$, **(g)** SnI$_2$

6.26 5.86%

6.31 Electronegativity difference given in parentheses. **(a)** ionic (2.0), **(b)** covalent—very low polarity (0), **(c)** covalent—moderate polarity (0.8), **(d)** covalent—low polarity (0.4), **(e)** covalent—low polarity (0.4), **(f)** covalent—very low polarity (0), **(g)** covalent—low polarity (0.2), **(h)** covalent—moderate polarity (1.0), **(i)** ionic (2.1)

6.34 (a) BrF$_5$, **(b)** S$_2$Cl$_2$, **(c)** P$_4$N$_4$, **(d)** TeF$_6$, **(e)** S$_5$N$_2$

6.36 (a) phosphorus pentachloride, **(b)** diiodine pentoxide, **(c)** silicon tetrafluoride, **(d)** sulfur trioxide, **(e)** tetrasulfur tetranitride

Chapter 7

7.3 Numbers in parentheses are: total number of electron pairs, number of bonding pairs, and number of non-bonding pairs. **(a)** BH$_4^-$ (4, 4, 0) tetrahedral, **(b)** XeF$_5^+$ (6, 5, 1) square pyramidal, **(c)** BeCl$_2$ (2, 2, 0) linear, **(d)** SbF$_2^+$ (3, 2, 1) angular, **(e)** SnCl$_3^-$ (4, 3, 1) trigonal pyramidal, **(f)** AsH$_3$ (4, 3, 1) trigonal pyramidal, **(g)** TeF$_4$ (5, 4, 1) irregular tetrahedral, **(h)** IF$_3$ (5, 3, 2) T-shaped, **(i)** SiF$_5^-$ (5, 5, 0) trigonal bipyramidal

7.4 (a) sp^3, **(b)** d^2sp^3, **(c)** sp, **(d)** sp^2, **(e)** sp^3, **(f)** sp^3, **(g)** dsp^3, **(h)** dsp^3, **(i)** dsp^3

7.7 Numbers in parentheses are: total number of electron pairs, number of bonding pairs, and number of nonbonding pairs. **(a)** BiCl$_5^{2-}$ (6, 5, 1) square pyramidal, **(b)** SeF$_5^-$ (6, 5, 1) square pyramidal, **(c)** ClF$_2^+$ (4, 2, 2) angular,

7.11
(a) 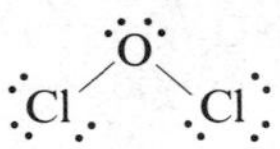 Cl–O–Cl — *angular*
(b) O–As–O with O — *trigonal pyramidal*
(c) 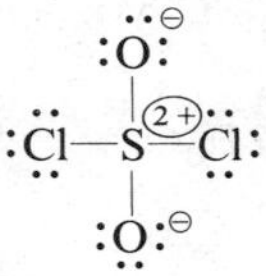 Cl–S²⁺–Cl with O — *tetrahedral*
(d) Cl–S⁺–Cl — *trigonal pyramidal*
(e) 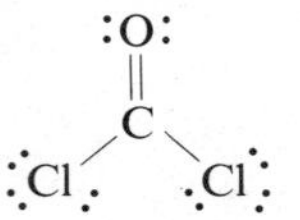 Cl–C(=O)–Cl — *triangular planar*
(f) S=C=S — *linear*

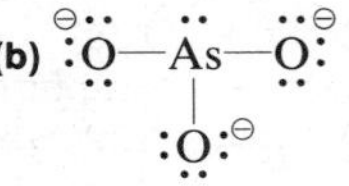

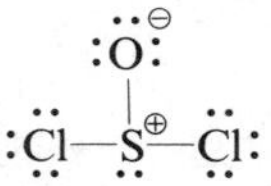

7.15 (a) 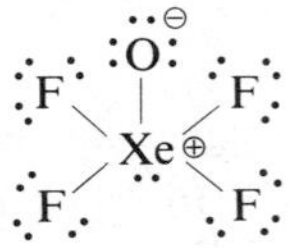 XeF$_4$ (Xe$^+$ with 4 F) — *square pyramidal*
(b) I$^+$ with 5 OH — *octahedral*
(c) 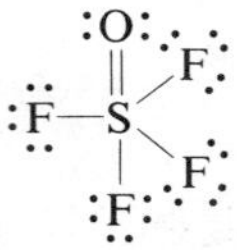 F–S(=O) with 3 F — *trigonal bipyramidal*
(d) Xe^{2+} with OH groups and O — *octahedral*

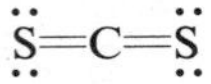

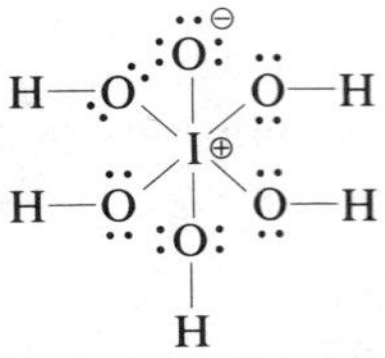

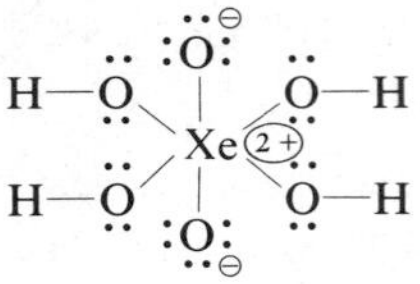

7.19

	H$_2$	H$_2^+$	HHe	He$_2$	He$_2^+$
σ^*1s			↑	↑↓	↑
$\sigma1s$	↑↓	↑	↑↓	↑↓	↑↓
bond order:	1	0.5	0.5	0	0.5

He$_2$ does not exist

7.22 (a)

	C$_2$	C$_2^{2-}$
σ^*2p	——	——
π^*2p	—— ——	—— ——
$\sigma2p$	——	↑↓
$\pi2p$	↑↓ ↑↓	↑↓ ↑↓
σ^*2s	↑↓	↑↓
$\sigma2s$	↑↓	↑↓

(b) bond order: C$_2$, 2; C$_2^{2-}$, 3; **(c)** C$_2^{2-}$ is isoelectronic with N$_2$ and CO.

7.25 (resonance structures of O$_3$ and SO$_2$)

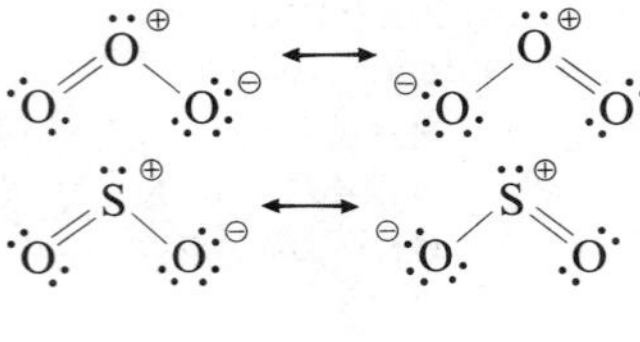

Resonance accounts for the bond distances in O_3. Since there are no d orbitals in the valence level of the O atom, $p\pi$–$d\pi$ interactions are impossible in O_3. There are d orbitals in the valence level of the S atom, and the "extra" shortening of the bonds in SO_2 is caused by $p\pi$–$d\pi$ interactions.

Chapter 8

8.3 (a) 750 ml, (b) 150 ml, (c) 1.50×10^3 ml

8.5 1.00×10^3 balloons

8.7 (a) 4.04 liter, (b) 79°C, (c) -156°C

8.10 (a) 2.34 atm, (b) 1.22×10^3°C, (c) -49°C

8.12 205 ml

8.14 346°C

8.17 5.97 liter

8.19 0.474 atm

8.21 34.1 g/mol

8.24 30.0 liter CH_4, 45.0 liter O_2, 30.0 liter NH_3, 90.0 liter H_2O

8.26 $4\,NH_3(g) + 5\,O_2(g) \longrightarrow 4\,NO(g) + 6\,H_2O(g)$, 40.0 liter NO

8.28 0.50 liter Cl_2, 1.50 liter N_2, 9.00 liter HCl

8.31 5.71 g/liter

8.34 (a) 8.0×10^{-5} g/m³, (b) 1.2×10^{-6} mol SO_2/m³, (c) 2.7×10^{-8} atm, (d) 2.7×10^{-6} %

8.36 (a) $10\,Na(s) + 6\,N_2O(g) + 2\,NH_3(1) \longrightarrow 4\,NaN_3(s) + 6\,NaOH(s) + N_2(g)$, (b) 6.20×10^3 ml N_2O, (c) 1.03×10^3 ml N_2

8.38 (a) $CaH_2(s) + 2\,H_2O(1) \longrightarrow 2\,H_2(g) + Ca(OH)_2(s)$, (b) 4.70 g CaH_2

8.40 (a) $Al_4C_3(s) + 12\,H_2O(l) \longrightarrow 3\,CH_4(g) + 4\,Al(OH)_3(s)$, (b) 1.00 liter CH_4

8.42 0.507 liter gas

8.45 (a) 0.125 mol hydrocarbon, 0.8125 mol O_2, 0.500 mol CO_2, 0.625 mol H_2O, (b) 2 (hydrocarbon), 13 O_2, 8 (CO_2), 10 (H_2O), (c) $2\,C_4H_{10}(g) + 13\,O_2(g) \longrightarrow 8\,CO_2(g) + 10\,H_2O(l)$

8.47 $p_{N_2O} = 0.194$ atm, $p_{N_2} = 0.306$ atm

8.49 (a) $X_{CO} = 0.250$, $X_{CO_2} = 0.750$, (b) 0.350 mol, (c) 2.45 g CO, 11.55 g CO_2

8.51 91.1 ml

*__8.54__ (a) 1.38 mol, (b) $(1 - x)$ mol N_2O_4 undissociated, $2x$ mol NO_2 produced; therefore, 0.62 mol N_2O_4 and 0.76 mol NO_2, (c) $X_{N_2O_4} = 0.45$, $X_{NO_2} = 0.55$, (d) $P_{N_2O_4} = 0.45$ atm, $p_{NO_2} = 0.55$ atm

8.58 279 m/s at 100K, 624 m/s at 500K

8.60 CO effuses 1.25 times faster than CO_2

8.62 (a) 1.19 g/liter, (b) 58.0 g/mol

8.64 0.900 g/liter

8.68 (a) 0.514 liter, 0.0321 liter, (b) the first caused principally by the effect of intermolecular forces, the second

caused principally by the effect of molecular vol⟨...⟩ (c) 0.611, 1.31

8.70 (a) 22.41 atm, (b) 21.94 atm, (c) the ideal gas ⟨...⟩ value deviates by 2% from the van der Waals resu⟨...⟩

8.72 (a) 30.62 atm, (b) 30.48 atm, (c) the ideal gas la⟨...⟩ value deviates by 0.5% from the van der Waals result⟨...⟩ Better agreement because the temperature is higher.

8.74 (a) 1.77×10^{-26} liter/molecule, (b) 0.0478%

Chapter 9

9.2 (a) OF_2 is angular, BeF_2 is linear, (b) PF_3 is trigonal pyramidal, BF_3 is triangular planar, (c) SF_4 is irregular tetrahedral, SnF_4 is tetrahedral

9.5 PF_3 is trigonal pyramidal with a nonbonding electron pair on the P atom at the apex of the pyramid; PF_5 is trigonal bipyramidal and does not have a nonbonding pair.

9.7 Each molecule is linear: $S{=}C{=}O$, $O{=}C{=}O$, $S{=}C{=}S$. Both double bonds are the same in CO_2 and in CS_2, and in these molecules, the bond polarities cancel (the centers of positive and negative charge both fall in the center of the molecule). In SCO, the two bonds are not identical ($S{=}C$ is virtually nonpolar), bond polarities do not cancel and the molecule has a dipole.

9.9 The London forces increase in the same order; I_2 is the largest molecule, has the strongest London forces, and has the highest melting point.

9.12 The HF_2^- ion consists of an HF molecule and a F^- ion held together by hydrogen bonding: $F{-}H \cdots F^-$

9.14 The anion of an acid salt (HSO_4^-, for example) contains a H atom with which it can hydrogen bond with H_2O molecules.

9.16 Since the molecules are similar, the London forces of each are similar. The H atoms and electron pairs of the $-NH_2$ groups, however, can enter into hydrogen bonding. Since $H_2NCH_2CH_2NH_2$ has *two* $-NH_2$ groups, hydrogen bonding is more extensive in this compound.

9.20 The boiling point is the temperature at which the vapor pressure of a liquid equals the pressure of the surroundings. The temperature at which the vapor pressure of a liquid equals 1 atm is called the normal boiling point. At 0.50 atm, the boiling points are: diethyl ether, 15°C; ethyl alcohol, 60°C; water, 80.3°C.

9.24 (a) at -1°C: vapor $\rightarrow$ solid at 5.5×10^{-3} atm, solid $\rightarrow$ liquid at 33 atm, (b) at 50°C: vapor $\rightarrow$ liquid at 0.1 atm, (c) at -50°C: vapor $\rightarrow$ solid at 4×10^{-5} atm

9.26 (a) at -60°C: vapor $\rightarrow$ solid at 4.0 atm, (b) at 0°C: vapor $\rightarrow$ liquid at 34 atm

9.28 Increasing the pressure on a solid-liquid equilibrium system will cause the equilibrium to shift to the more dense form. If solid is the more dense form, the substance will freeze (curve slants to the right); if liquid is the more dense form, the substance will melt (curve slants to the left).

9.31 (a) network, (b) metallic, (c) London, (d) ionic, (e) London, (f) London and dipole-dipole

negativity difference is larger for
dipole-dipole forces are stronger.
rF is larger than that of ClF; London
er. **(b)** BrCl. Stronger London forces in
.pole forces exist between BrCl molecules
en Cl_2 molecules. **(c)** CsBr. Ionic forces are
.n London or dipole-dipole forces. **(d)** Cs.
onding is stronger than London attractions.
.amond). Network bonding is stronger than
. forces.

.45 g/cm^3

8 1 atom/unit cell, simple cubic

J.40 55.8

9.42 316 pm

9.45 a cube 2.169 cm on an edge

9.46 128 pm

9.49 348 pm

9.51 70.8 pm

9.53 14.9°, 30.9°

9.55 (a) one Cs^+ ion, one Cl^- ion, **(b)** 412 pm, **(c)** 357 pm

***9.58 (a)** 385 pm, **(b)** 6.98 g/cm^3

9.61 NiS, BaO, CaS, NaBr, AgCl, KCl

***9.63 (a)** 0.5%, **(b)** 8.241 g/cm^3, **(c)** 8.235 g/cm^3

***9.65 (a)** 58.41 g/mol, **(b)** 0.08%

Chapter 10

10.2 (a) CH_3OH, **(b)** NaCl, **(c)** CH_3F

10.5 (a) Fe^{3+}, **(b)** Li^+, **(c)** F^-, **(d)** Sn^{2+}, **(e)** Al^{3+}, **(f)** Mg^{2+}

10.8 -47 kJ/mol

10.10 -785 kJ/mol; energy required to separate water molecules from each other and energy released when the water molecules hydrate the ions

10.12 4.81×10^{-2} mol CO_2, 2.12 g CO_2

***10.15** $m = 100 M/(100x - xy)$

10.16 CH_3OH, 0.294; H_2O, 0.706

10.19 71.4% $C_6H_{12}O_6$

10.21 7.01 g KOH

10.23 (a) 50.6 g conc HBr, **(b)** 33.7 ml conc HBr

10.25 (a) 5.51 M, **(b)** 6.93 m

10.27 3.35 M HCl

10.29 25.5 ml conc H_3PO_4

10.31 0.316 M HNO_3

10.35 0.264 atm

10.36 0.514

10.39 70 g/mol

10.42 (a) 0.434 atm, **(b)** negative, **(c)** evolved, **(d)** maximum-boiling

10.45 333 g $C_2H_4(OH)_2$

10.47 $-1.75°C$

10.49 122 g/mol

***10.52 (a)** N_2, 0.800 atm; O_2, 0.200 atm, **(b)** N_2, 1.49×10^{-5}; O_2, 7.97×10^{-6}, **(c)** $-0.0024°C$

10.54 27 g $C_3H_5(OH)_3$

10.56 62.0 g/mol

10.59 2.91 atm

10.61 6.70×10^4 g/mol

10.63 (a) 0.0106 atm, **(b)** 11.0 cm

10.65 27.1 atm

10.67 2.67

10.69 $-0.242°C$

Chapter 11

11.1 (a) $ZnS(s) + 2H^+ + 2Cl^- \longrightarrow H_2S(g) + Zn^{2+} + 2Cl^-$,
(b) $2Na^+ + CO_3^{2-} + Sr^{2+} + 2C_2H_3O_2^- \longrightarrow SrCO_3(s) + 2Na^+ + 2C_2H_3O_2^-$,
(c) $Sn^{2+} + 2Cl^- + 2NH_4^+ + SO_4^{2-} \longrightarrow$ N.R.,
(d) $Mg^{2+} + 2NO_3^- + Ba^{2+} + 2OH^- \longrightarrow Mg(OH)_2(s) + Ba^{2+} + 2NO_3^-$, **(e)** $3Na^+ + PO_4^{3-} + 3H^+ + 3Br^- \longrightarrow H_3PO_4 + 3Na^+ + 3Br^-$

11.6 (a) $6+$, **(b)** $6+$, **(c)** $5+$, **(d)** $1+$, **(e)** $4+$, **(f)** $2-$, **(g)** $3+$, **(h)** $5+$

11.8 (a) $3+$, **(b)** $3+$, **(c)** $4+$, **(d)** $5+$, **(e)** $4+$, **(f)** $5+$, **(g)** $3+$, **(h)** $5+$

11.10 Oxidized (reducing agent): **(a)** Zn, **(b)** $SbCl_3$, **(c)** Mg, **(d)** NO, **(e)** H_2
Reduced (oxidizing agent): **(a)** Cl_2, **(b)** $ReCl_5$, **(c)** $CuCl_2$, **(d)** O_2, **(e)** WO_3

11.12 (a) $2H_2O + 4MnO_4^- + 3ClO_2^- \longrightarrow 4MnO_2 + 3ClO_4^- + 4OH^-$,
(b) $8H^+ + Cr_2O_7^{2-} + 3H_2S \longrightarrow 2Cr^{3+} + 3S + 7H_2O$,
(c) $6H_2O + P_4 + 10HOCl \longrightarrow 4H_3PO_4 + 10Cl^- + 10H^+$,
(d) $3Cu + 8H^+ + 2NO_3^- \longrightarrow 3Cu^{2+} + 2NO + 4H_2O$,
(e) $PbO_2 + 4HI \longrightarrow PbI_2 + I_2 + 2H_2O$

11.15 (a) $6H_2O + 4AsH_3 + 24Ag^+ \longrightarrow As_4O_6 + 24Ag + 24H^+$,
(b) $14H^+ + 2Mn^{2+} + 5BiO_3^- \longrightarrow 2MnO_4^- + 5Bi^{3+} + 7H_2O$,
(c) $4H^+ + 2NO + 4NO_3^- \longrightarrow 3N_2O_4 + 2H_2O$,
(d) $11H^+ + 2MnO_4^- + 5HCN + 5I^- \longrightarrow 2Mn^{2+} + 5ICN + 8H_2O$,
(e) $12H^+ + 3Zn + 2H_2MoO_4 \longrightarrow 3Zn^{2+} + 2Mo^{3+} + 8H_2O$

11.17 (a) $6H^+ + ClO_3^- + 6I^- \longrightarrow Cl^- + 3I_2 + 3H_2O$,
(b) $10H^+ + 4Zn + NO_3^- \longrightarrow 4Zn^{2+} + NH_4^+ + 3H_2O$
(c) $3H_3AsO_3 + BrO_3^- \longrightarrow 3H_3AsO_4 + Br^-$,
(d) $2H_2SeO_3 + H_2S \longrightarrow 2Se + HSO_4^- + H^+ + 2H_2O$,
(e) $4H_2O + 2ReO_2 + 3Cl_2 \longrightarrow 2HReO_4 + 6Cl^- + 6H^+$

11.21 (a) $8OH^- + S^{2-} + 4I_2 \longrightarrow SO_4^{2-} + 4H_2O + 8I^-$,

(b) $H_2O + 3CN^- + 2MnO_4^- \longrightarrow 3CNO^- + 2MnO_2 + 2OH^-$,

(c) $2H_2O + 4Au + 8CN^- + O_2 \longrightarrow 4Au(CN)_2^- + 4OH^-$,

(d) $H_2O + Si + 2OH^- \longrightarrow SiO_3^{2-} + 2H_2$,

(e) $4OH^- + 2Cr(OH)_3 + 3BrO^- \longrightarrow 2CrO_4^{2-} + 3Br^- + 5H_2O$

11.23 (a) $OH^- + 5HClO_2 \longrightarrow 4ClO_2 + Cl^- + 3H_2O$,
(b) $8OH^- + 8MnO_4^- + I^- \longrightarrow 8MnO_4^{2-} + IO_4^- + 4H_2O$,
(c) $4OH^- + 2H_2O + P_4 \longrightarrow 2HPO_3^{2-} + 2PH_3$,
(d) $OH^- + SbH_3 + 3H_2O \longrightarrow Sb(OH)_4^- + 3H_2$,
(e) $CO(NH_2)_2 + 3OBr^- \longrightarrow CO_2 + N_2 + 3Br^- + 2H_2O$

11.26 (a) $PbS + 4H_2O_2 \longrightarrow PbSO_4 + 4H_2O$,
(b) $4OH^- + 2Cr(OH)_3 + 3H_2O_2 \longrightarrow 2CrO_4^{2-} + 8H_2O$,
(c) $6H^+ + 2MnO_4^- + 5H_2O_2 \longrightarrow 2Mn^{2+} + 5O_2 + 8H_2O$,
(d) $Ag_2O + H_2O_2 \longrightarrow 2Ag + O_2 + H_2O$

11.32 (a) bromic acid, **(b)** potassium bromate, **(c)** hydrobromic acid, **(d)** sodium nitrite, **(e)** potassium hydrogen sulfate, **(f)** potassium sulfite, **(g)** sodium hydrogen carbonate, **(h)** sodium dihydrogen phosphate, **(i)** copper(II) nitrate

11.35 (a) Cl_2O_7, **(b)** N_2O_3, **(c)** SO_2, **(d)** B_2O_3, **(e)** Al_2O_3, **(f)** ZnO, K_2O

11.38 $0.3858\ M$

11.40 $41.28\%\ Mg(OH)_2$

11.42 $69.5\%\ KHC_8H_4O_4$

11.44 (a) 0.0448 g NaCl, **(b)** 0.895% NaCl

11.46 (a) $3N_2H_4 + 2BrO_3^- \longrightarrow 3N_2 + 2Br^- + 6H_2O$, **(b)** $24.0\%\ N_2H_4$

11.49 (a) $1/4$, **(b)** $1/6$, **(c)** $1/5$, **(d)** $1/6$, **(e)** $1/2$, **(f)** $1/4$

11.51 $6.00\ M$ HCl, $3.00\ M\ H_2SO_4$, $2.00\ M\ H_3PO_4$

11.53 $0.409\ N$

11.55 (a) 90.0 g, **(b)** one

11.57 (a) $0.1200\ N\ K_2Cr_2O_7$, **(b)** $0.0750\ N$ KMnO$_4$, **(c)** $0.0150\ M$ KMnO$_4$

Chapter 12

12.2 (a) rate of formation of $C = k[A][B]$, **(b)** 0.016 liter/(mol·s)

12.4 (a) $k = 0.100$ mol/(liter·s), rate $= 0.100$ mol/(liter·s), **(b)** $k = 0.100$/s, rate $= 0.0050$ mol/(liter·s), **(c)** $k = 0.100$ liter/(mol·s), rate $= 0.00025$ mol/(liter·s)

12.6 (a) $3/16$ of the original rate, **(b)** zero, **(c)** $2/27$ of the original rate, **(d)** 4 times the original rate, **(e)** 8 times the original rate

12.12 step 1: $ICl + H_2 \longrightarrow HCl + HI$; step 2: $HI + ICl \longrightarrow HCl + I_2$

*12.14 $[A^*] = (k_1/k_2)[A]$; therefore, rate $= (k_1k_3/k_2)[A]$

12.18 chain initiating: $Cl_2 \longrightarrow 2Cl$; chain propagating: $Cl + CH_4 \longrightarrow CH_3 + HCl$, $CH_3 + Cl_2 \longrightarrow CH_3Cl + Cl$; chain terminating: $2Cl \longrightarrow Cl_2$, $2CH_3 \longrightarrow C_2H_6$, $CH_3 + Cl \longrightarrow CH_3Cl$

12.21 266 kJ/mol

12.24 7.9×10^5 liter/(mol·s)

12.26 4.7×10^{-3} liter/(mol·s)

12.28 668 K

12.30 52.3 kJ/mol

Chapter 13

13.2 (a) left, **(b)** left, **(c)** left, **(d)** left, **(e)** left

13.3 endothermic

13.6 (a) decrease, **(b)** increase, **(c)** increase, **(d)** no effect, **(e)** decrease, **(f)** decrease, **(g)** no effect, **(h)** no effect, **(i)** no effect

13.8 61

13.10 (a) $[Cl_2] = 0.050$ mol/liter, $[PCl_5] = 0.60$ mol/liter, **(b)** 0.042 mol/liter

13.12 7.6×10^{-4} mol/liter

13.14 5.5×10^{-4} mol/liter

13.16 $[H_2] = [I_2] = 0.319$ mol/liter, $[HI] = 2.362$ mol/liter

13.18 0.025 mol/liter

13.20 (a) $[CO] = [H_2O] = 0.0335$ mol/liter, $[CO_2] = [H_2] = 0.0665$ mol/liter, **(b)** 3.94, **(c)** 3.94

13.22 (a) 2.19×10^{-10} mol/liter, **(b)** 6.71×10^{-9} atm

13.24 (a) 0.563 mol^2/liter^2, **(b)** $K = 1.78$ liter^2/mol^2, $K_p = 8.79 \times 10^{-4}$/atm^2

13.26 0.024 atm

13.28 $p_{PCl_3} = p_{Cl_2} = 0.75$ atm, initial pressure of PCl$_5$ = 1.00 atm, 75% dissociated

Chapter 14

14.4 Solvent system: acid (NH$_4$Cl) reacts with base (NaNH$_2$) to form solvent ($2NH_3$) and salt (NaCl). Brønsted: acid$_1$ (NH$_4^+$) reacts with base$_2$ (NH$_2^-$) to form conjugate base$_1$ (NH$_3$) and acid$_2$ (NH$_3$)—NH$_3$ is amphiprotic. Lewis: a nucleophilic displacement of NH$_3$ by NH$_2^-$ on NH$_4^+$.

14.6 (a) $H_2AsO_3^-$, **(b)** $HAsO_4^{2-}$, **(c)** NO_2^-, **(d)** S^{2-}

14.8 (a) H_3AsO_4, **(b)** PH_4^+, **(c)** $HC_2H_3O_2$, **(d)** HS^-, **(e)** HPO_4^{2-}

14.11 (a) $HOCl + OH^- \rightleftharpoons H_2O + OCl^-$, **(b)** $H_2O + NH_2^- \rightleftharpoons NH_3 + OH^-$, **(c)** $HCO_3^- + OH^- \rightleftharpoons H_2O + CO_3^{2-}$, **(d)** $NH_3 + H^- \rightleftharpoons H_2 + NH_2^-$

14.16 (a) H_3O^+, H_3PO_4, HCN, H_2O, NH_3, **(b)** NH_2^-, OH^-, CN^-, $H_2PO_4^-$, H_2O

14.17 (a) yes, **(b)** no, **(c)** no, **(d)** yes

14.20 HBr is a strong acid, H$_2$S is a weak acid, and AsH$_3$ is not acidic. The acid strength of the hydrides of the elements of a period increases from left to right (with increasing electronegativity).

14.22 (a) H_3PO_4, **(b)** H_3AsO_4, **(c)** H_2SO_4, **(d)** H_2CO_3, **(e)** HBr

14.23 (a) P^{3-}, (b) NH_3, (c) SiO_3^{2-}, (d) NO_2^-, (e) F^-

14.26 (a) acid: CS_2, base: SH^-; (b) acid: H^+, base: $Fe(CO)_5$; (c) acid: Ag^+, base: NH_3; (d) acid: H^+, base: $Mn(CO)_5^-$; (e) acid: HF, base F^-

*__14.28__ (a) nucleophilic: NH_2^- displaced by H^-, (b) electrophilic: NO_2^+ displaced by H^+, (c) nucleophilic: H_2O displaced by Cl^-, (d) nucleophilic: GeS displaced by Ge, (e) nucleophilic: OH^- displaced by O^{2-}, (f) electrophilic: Br^+ displaced by $FeBr_3$, (g) electrophilic: CH_3^+ displaced by $AlCl_3$, (h) nucleophilic: I^- displaced by OH^-

Chapter 15

15.1 1.5×10^{-4}

15.4 2.3×10^{-5}

15.5 7.3×10^{-7}

15.7 $4.4 \times 10^{-4}\ M$

15.9 (a) $1.8 \times 10^{-3}\ M$, (b) 0.72%

15.12 $0.18\ M$

15.15 $1.0 \times 10^{-3}\ M$

15.16 $[H^+] = 1.5 \times 10^{-5}\ M$, $[N_3^-] = 0.13\ M$, $[HN_3] = 0.10\ M$

15.19 $[NH_4^+] = 2.7 \times 10^{-5}\ M$, $[OH^-] = 0.10\ M$, $[NH_3] = 0.15\ M$

15.21 (a) $[H^+] = 0.020\ M$, $[OH^-] = 5.0 \times 10^{-13}\ M$, (b) $[OH^-] = 0.040\ M$, $[H^+] = 2.5 \times 10^{-13}\ M$

15.23 (a) 4.14, (b) 1.08, (c) 10.52, (d) 12.62

15.25 (a) $3.5 \times 10^{-14}\ M$, (b) $4.8 \times 10^{-5}\ M$, (c) $1.3 \times 10^{-2}\ M$, (d) $4.6 \times 10^{-13}\ M$

15.27 3.1×10^{-3} mol $HClO_2$

15.29 7.3×10^{-6}

15.32 6.57–8.35

15.35 (a) 3.80, (b) 0.46%

15.36 5.5×10^{-5}

15.38 $0.70\ M$

15.40 4.85

15.41 $0.045\ M$

15.44 $[NH_4^+]/[NH_3] = 0.56$

15.46 (a) $[H^+] = [HCO_3^-] = 1.2 \times 10^{-4}\ M$, $[CO_3^{2-}] = 4.8 \times 10^{-11}\ M$, $[CO_2] = 0.034\ M$, (b) 3.92

15.48 (a) $4.9 \times 10^{-21}\ M$, (b) $7.4 \times 10^{-8}\ M$

15.50 $1.9 \times 10^{-3}\ M$

15.51 8.18

15.52 2.82

15.55 $0.59\ M$

15.57 4.3×10^{-7}

*__15.59__ 5.82

15.61 (a) 3.92, (b) 8.46, (c) 12.16

15.63 6.2×10^{-6}

Chapter 16

16.1 2.0×10^{-14}

16.4 $2.1 \times 10^{-4}\ M$

16.5 1.6×10^{-5} mol $CuCO_3$/liter lower than 1.3×10^{-4} mol Ag_2CO_3/liter

16.8 3.3×10^{-13} mol $Ni(OH)_2$

16.10 $[Na^+] = 0.16\ M$, $[Cl^-] = 0.30\ M$, $[Ba^{2+}] = 0.070\ M$, $[C_2O_4^{2-}] = 2.1 \times 10^{-7}\ M$

16.12 $9.5 \times 10^{-4}\ M$

16.14 $1.2\ M$

16.17 $0.18\ M$

16.19 ion product (1.8×10^{-22}) smaller than K_{sp}, no

16.22 $0.23\ M$ (minimum)

16.24 $2.3 \times 10^{-7}\ M$

16.26 $0.086\ M$ (minimum)

16.28 (a) 2.4×10^{-2} mol AgCl/liter, (b) 1.4×10^{-3} mol AgBr/liter, (c) 1.9×10^{-5} mol AgI/liter

Chapter 17

17.3 (a) -1364.3 kJ/mol, (b) -1366.8 kJ, (c) -277.9 kJ/mol

17.6 -5459.55 kJ

17.8 -66.58 kJ

17.10 -184.79 kJ/mol

17.14 (a) $+103.0$ J/K, (b) -587.6 kJ

17.16 $+267.4$ kJ, -207.0 kJ, $H_2S(g)$ and $H_2O(l)$

17.18 For BF_3, $\Delta G° = +12.09$ kJ; for BCl_3, $\Delta G° = -266.19$ kJ; BCl_3 will hydrolyze at $25°C$, BF_3 will not.

17.21 -48.39 kJ, yes

17.23 (a) -101.01 kJ, (b) -120.6 J/K, (c) -120.6 J/K

17.25 (a) $+38.3$ kJ, no, (b) -11.8 kJ, yes

17.28 239.7 K, $-33.4°C$

17.30 $+18.25$ kJ/mol

17.32 208 J/(K · mol)

17.34 2.438 J/(K · mol), diamond

17.35 -1095.02 kJ

17.36 5.12×10^{-4}

17.37 0.443

17.39 2.07

17.41 $+38.5$ kJ

17.43 $+27$ kJ

Chapter 18

18.7 0.570 g Ni

18.9 (a) $Pb^{2+} + 2\,H_2O \longrightarrow PbO_2(s) + 4\,H^+ + 2e^-$, (b) 0.558 g PbO_2, (c) 46.6 min

18.11 28.6 min

18.13 (a) 780 C, **(b)** 0.867 A

18.15 333 g C

18.17 9.22 liter Cl_2

18.20 (a) 5.00 liter Cl_2, **(b)** 0.893 M

18.22 5.77 hr

18.25 (a) $+2.227$ V, **(b)** $Mg + Sn^{2+} \longrightarrow Mg^{2+} + Sn$, **(c)** the Sn^{2+}/Sn electrode

18.27 (a) $Pt(s)|I_2(s)|I^-(aq)\|Cl^-(aq)|Cl_2(g)|Pt(s)$, **(b)** $+0.8240$ V, **(c)** the Cl_2/Cl^- electrode

18.29 -1.789 V

18.32 (a) $PbSO_4$; Cd^{2+}; Cr^{3+}; **(b)** Au^+; PbO_2, SO_4^{2-}, H^+; $HOCl$, H^+; Ce^{4+}; **(c)** Ga; Cr; Zn; **(d)** Au^+; Cl^-; Cr^{3+}; Tl^+

18.34 (a) no, **(b)** $H_2O_2 + 2Ag^+ \longrightarrow 2Ag + O_2 + 2H^+$, **(c)** $Ag^+ + Fe^{2+} \longrightarrow Ag + Fe^{3+}$, **(d)** no, **(e)** $H_2SO_3 + 2H_2S \longrightarrow 3S + 3H_2O$

18.37 (a) In^+ will disproportionate, **(b)** In^{3+}, **(c)** yes, In^{3+}, **(d)** $3In^+ \longrightarrow 2In + In^{3+}$, $6H^+ + 2In \longrightarrow 2In^{3+} + 3H_2$, $2In + 3Cl_2 \longrightarrow 2In^{3+} + 6Cl^-$

18.39 (a) no, **(b)** yes, **(c)** no

18.40 (a) -1.208 V, **(b)** no, **(c)** yes, Ti^{2+}

18.43 (a) $+1.397$ V, **(b)** yes, **(c)** no

18.45 (a) $Pb(s)|PbSO_4(s)|SO_4^{2-}(aq)\|Pb^{2+}(aq)|Pb(s)$, **(b)** $Pb^{2+} + SO_4^{2-} \longrightarrow PbSO_4$, **(c)** $+0.233$ V, **(d)** -45.0 kJ

18.47 **(a)** $Pt(s)|H_2(g)|OH^-(aq)\|H^+(aq)|H_2(g)|Pt(s)$, **(b)** $+0.828$ V, -79.9 kJ

18.49 (a) $+0.2943$ V, **(b)** -56.79 kJ, **(c)** -121.7 J/K

18.52 (a) -509.5 kJ, **(b)** -43.5 kJ/mol

18.54 (a) $+125.2$ kJ/mol, **(b)** $+1.76$ V, **(c)** *theoretically*, H_2O_2, Co^{3+}, $S_2O_8^{2-}$, O_3, and F_2

18.55 7.1×10^3

18.57 1.0×10^{-14}

18.59 -0.284 V

18.61 (a) $+2.157$ V, **(b)** $Mg + Ni^{2+} \longrightarrow Mg^{2+} + Ni$, **(c)** the Ni^{2+}/Ni electrode

18.64 2.04 M

18.66 (a) $\mathscr{E}° = -0.13$ V, no, **(b)** $\mathscr{E} = +0.05$ V, yes

18.68 (a) increased by 0.00891 V,   decreased by 0.00891 V

18.69 (a) $+0.010$ V, **(b)** $[Sn^{2+}] = 1.37\ M$, $[Pb^{2+}] = 0.63\ M$

18.70 (a) $+0.136$ V, **(b)** $H_2 + 2H^+(5.00\ M) \longrightarrow H_2 + 2H^+(0.0250\ M)$, **(c)** $+0.175$ V

Chapter 19

19.3 (a) $2Na(s) + H_2(g) \longrightarrow 2NaH(s)$, **(b)** $Ca(s) + H_2(g) \longrightarrow CaH_2(s)$,

(c) $H_2(g) + Cl_2(g) \longrightarrow 2HCl(g)$, **(d)** $N_2(g) + 3H_2(g) \rightleftharpoons 2NH_3(g)$, **(e)** $Cu_2O(s) + H_2(g) \longrightarrow 2Cu(s) + H_2O(g)$, **(f)** $CO(g) + 2H_2(g) \longrightarrow CH_3OH(l)$, **(g)** $WO_3(s) + 3H_2(g) \longrightarrow W(s) + 3H_2O(g)$

19.5 The low melting point, boiling point, and critical temperature reflect the weak nature of the London forces between H_2 molecules. The forces are weak because the electron cloud of the molecule is relatively small.

19.7 (a) 0.503 g H_2, **(b)** 0.958 g H_2

19.11 -805.8 kJ/mol

19.14 (a) $CaF_2(s) + H_2SO_4(l) \longrightarrow 2HF(g) + CaSO_4(s)$, $HF(l) + KF(s) \longrightarrow KHF_2(s)$, $2KHF_2(l) \xrightarrow[\text{heat}]{\text{electrolysis}} H_2(g) + F_2(g) + 2KF(l)$, **(b)** $2NaCl(l) \xrightarrow[\text{heat}]{\text{electrolysis}} 2Na(l) + Cl_2(g)$, **(c)** $Cl_2(g) + 2Br^-(aq) \longrightarrow 2Cl^-(aq) + Br_2(l)$, **(d)** $2IO_3^-(aq) + 5HSO_3^-(aq) \longrightarrow I_2(s) + 5SO_4^{2-}(aq) + 3H^+(aq) + H_2O$, **(e)** $PBr_3(l) + 3H_2O \longrightarrow 3HBr(g) + H_3PO_3(aq)$

19.16 (a) $SiO_2(s) + 6HF(aq) \longrightarrow 2H^+(aq) + SiF_6^{2-}(aq) + 2H_2O$, $SiO_2(s) + 4HF(aq) \longrightarrow SiF_4(g) + 2H_2O$, **(b)** $CO_3^{2-}(aq) + 2HF(aq) \longrightarrow 2F^-(aq) + CO_2(g) + H_2O$, **(c)** $KF(s) + HF(l) \longrightarrow KHF_2(l)$, **(d)** $CaO(s) + 2HF(aq) \longrightarrow CaF_2(s) + H_2O$

19.19 (a) $FeCl_2$, **(b)** $RbCl$, **(c)** BeF_2

19.22 (a) HCl, **(b)** Cl_2, **(c)** HF, **(d)** F_2, **(e)** Cl_2, **(f)** Cl_2

19.24 28.2 F

19.26 1324 g Cl_2

Chapter 20

20.2 (a) $2HgO(s) \xrightarrow{\text{heat}} 2Hg(l) + O_2(g)$, **(b)** $2O_2^{2-}(aq) + 2H_2O \longrightarrow 4OH^-(aq) + O_2(g)$, **(c)** $2NaNO_3(l) \xrightarrow{\text{heat}} 2NaNO_2(l) + O_2(g)$, **(d)** $2KClO_3(s) \xrightarrow[\text{MnO}_2]{\text{heat}} 2KCl(s) + O_2(g)$, **(e)** $2H_2O \xrightarrow{\text{electrolysis}} 2H_2(g) + O_2(g)$

20.4 (a) $2C_4H_{10} + 13O_2 \longrightarrow 8CO_2 + 10H_2O$, **(b)** $C_5H_{12}S + 9O_2 \longrightarrow 5CO_2 + 6H_2O + SO_2$, **(c)** $2C_3H_8O + 9O_2 \longrightarrow 6CO_2 + 8H_2O$, **(d)** $2ZnS + 3O_2 \longrightarrow 2ZnO + 2SO_2$, **(e)** $2PbS + 3O_2 \longrightarrow 2PbO + 2SO_2$

20.5 (a) $CH_4(g) + O_2(g) \longrightarrow C(s) + 2H_2O(g)$, **(b)** $2CH_4(g) + 3O_2(g) \longrightarrow 2CO(g) + 4H_2O(g)$, **(c)** $CH_4(g) + 2O_2(g) \longrightarrow CO_2(g) + 2H_2O(g)$

20.9 (a) $PbS + 4H_2O_2 \longrightarrow PbSO_4 + 4H_2O$, **(b)** $4OH^- + 2Cr(OH)_3 + 3H_2O_2 \longrightarrow 2CrO_4^{2-} + 8H_2O$, **(c)** $6H^+ + 2MnO_4^- + 5H_2O_2 \longrightarrow 2Mn^{2+} + 5O_2 + 8H_2O$, **(d)** $Ag_2O + H_2O_2 \longrightarrow 2Ag + O_2 + H_2O$

20.11 (a) -285.9 kJ,
(b) -333.3 kJ, they are more exothermic

*__20.12__ $+299.5$ kJ/mol

20.15 (a) $S(s) + O_2(g) \longrightarrow SO_2(g)$,
(b) $S(s) + S^{2-}(aq) \longrightarrow S_2^{2-}(aq)$,
(c) $S(s) + SO_3^{2-}(aq) \longrightarrow S_2O_3^{2-}(aq)$,
(d) $S(s) + Fe(s) \longrightarrow FeS(s)$,
(e) $S(s) + 2F_2(g) \longrightarrow SF_4(g)$,
$S(s) + 3F_2(g) \longrightarrow SF_6(g)$,
(f) $2S(s) + Cl_2(g) \longrightarrow S_2Cl_2(l)$,
(g) $S(s) + 4H^+(aq) + 4NO_3^-(aq) \longrightarrow SO_2(g) + 4NO_2(g) + 2H_2O$

20.17 (a) $Fe(s) + S(s) \longrightarrow FeS(s)$,
$FeS(s) + 2H^+(aq) \longrightarrow Fe^{2+}(aq) + H_2S(g)$,
(b) $S(s) + O_2(g) \longrightarrow SO_2(g)$,
$SO_2(g) + H_2O \longrightarrow H_2SO_3(aq)$,
(c) $H_2SO_3(aq) + 2OH^-(aq) \longrightarrow SO_3^{2-}(aq) + H_2O$,
$S(s) + SO_3^{2-}(aq) \longrightarrow S_2O_3^{2-}(aq)$,
(d) $2SO_2(g) + O_2(g) \xrightarrow{V_2O_5} SO_3(g)$,
$SO_3(g) + H_2O \longrightarrow H_2SO_4(l)$,
$H^+(aq) + HSO_4^-(aq) + OH^-(aq) \longrightarrow H_2O + HSO_4^-(aq)$,
(e) $SO_3(g) + H_2SO_4(l) \longrightarrow H_2S_2O_7(l)$

20.20 (a) $CH_3C(NH_2)S + H_2O \longrightarrow CH_3C(NH_2)O + H_2S$,
(b) $SO_2 + H_2O \longrightarrow H_2SO_3$,
(c) $SO_3 + H_2O \longrightarrow H_2SO_4$,
(d) $Al_2S_3 + 6H_2O \longrightarrow 2Al(OH)_3 + 3H_2S$,
(e) $SeO_3 + H_2O \longrightarrow H_2SeO_4$,
(f) $TeO_3 + 3H_2O \longrightarrow H_6TeO_6$,
(g) $H_2SO_5 + H_2O \longrightarrow H_2SO_4 + H_2O_2$,
(h) $H_2S_2O_7 + H_2O \longrightarrow 2H_2SO_4$

20.22 (a) $SO_3^{2-}(aq) + 2H^+(aq) \longrightarrow SO_2(g) + H_2O$,
(b) $S^{2-}(aq) + 2H^+(aq) \longrightarrow H_2S(g)$,
(c) $S_2O_3^{2-}(aq) + 2H^+(aq) \longrightarrow S(s) + SO_2(g) + H_2O$

*__20.24__ SO_2 in acid solution goes to SO_4^{2-} and $S_4O_6^{2-}$, SO_3^{2-} in alkaline solution goes to SO_4^{2-} and $S_2O_3^{2-}$

20.25 The S atom in SF_4 has four bonding electron pairs and one nonbonding electron pair (a total of five electron pairs) in its valence level. The O atom is limited to four electron pairs since its valence level has only four orbitals.

20.27 (a) acidification produces $H_2S(g)$, **(b)** acidification produces $SO_2(g)$, **(c)** $BaSO_4(s)$ precipitates when a solution of $BaCl_2$ is added, **(d)** acidification produces $S(s)$ and $SO_2(g)$

20.29 (a) angular, **(b)** angular, **(c)** triangular planar, **(d)** trigonal pyramidal, **(e)** tetrahedral, **(f)** tetrahedral, **(g)** irregular tetrahedral, **(h)** octahedral, **(i)** octahedral

Chapter 21

21.4 At STP, the density of pure N_2 is 1.250 g/liter, the density of "N_2 from air" is 1.257 g/liter

21.7 (a) $P_4O_6(s) + 6H_2O \longrightarrow 4H_3PO_3(aq)$,
$P_4O_6(s) + 8OH^-(aq) \longrightarrow 4HPO_3^{2-}(aq) + 2H_2O$,
$P_4O_6(s) + 4OH^-(aq) + 2H_2O \longrightarrow 4H_2PO_3^-(aq)$,
$P_4O_6(s) + H^+(aq) \longrightarrow$ N.R.,

(b) $Sb_4O_6(s) + H_2O \longrightarrow$ N.R., $Sb_4O_6(s) + 4OH^-(aq) \longrightarrow 4SbO_2^-(aq) + H_2O$, $Sb_4O_6(s) + 4H^+(aq) + 4Cl^-(aq) \longrightarrow 4SbOCl(s) + 2H_2O$,
(c) $Bi_2O_3(s) + H_2O$ N.R., $Bi_2O_3(s) + OH^-(aq) \longrightarrow$ N.R., $Bi_2O_3(s) + 6H^+(aq) \longrightarrow 2Bi^{3+}(aq) + 3H_2O$

21.11 (a) $2NH_3(aq) + Ag^+(aq) \longrightarrow Ag(NH_3)_2^+(aq)$,
(b) $NH_3(aq) + H^+(aq) \longrightarrow NH_4^+(aq)$,
(c) $NH_3(aq) + H_2O + CO_2(aq) \longrightarrow NH_4^+(aq) + HCO_3^-(aq)$,
(d) $4NH_3(g) + 3O_2(g) \longrightarrow 2N_2(g) + 6H_2O(g)$,
(e) $4NH_3(g) + 5O_2(g) \longrightarrow 4NO(g) + 6H_2O(g)$,
(f) $NH_3(g) + HCl(g) \longrightarrow NH_4Cl(s)$,
(g) $2NH_3(g) + 2V(s) \longrightarrow 2VN(s) + 3H_2(g)$,
(h) $2NH_3(l) + 2Na(s) \longrightarrow 2Na^+ + 2NH_2^- + H_2(g)$

21.13 (a) $3Cu(s) + 8H^+(aq) + 2NO_3^-(aq) \longrightarrow 3Cu^{2+}(aq) + 2NO(g) + 4H_2O$,
(b) $4Zn(s) + 9H^+(aq) + NO_3^-(aq) \longrightarrow 4Zn^{2+}(aq) + NH_3(g) + 3H_2O$,
(c) $4HNO_3(l) + P_4O_{10}(s) \longrightarrow 4HPO_3(s) + 2N_2O_5(s)$,
(d) $H^+(aq) + NH_3(aq) \longrightarrow NH_4^+(aq)$,
(e) $2H^+(aq) + Ca(OH)_2(s) \longrightarrow Ca^{2+}(aq) + 2H_2O$

21.15 (a) $N_2O(g) + H_2O$, **(b)** $N_2(g) + H_2O$,
(c) $PbO(s) + NO_2(g) + O_2(g)$, **(d)** $NaNO_2(s) + O_2(g)$,
(e) $Na(s) + N_2(g)$

21.16 (a) $2NH_3(aq) + OCl^-(aq) \longrightarrow N_2H_4(l) + Cl^-(aq) + H_2O$,
(b) $2NO_2(g) + 5H_2(g) + 2H^+(aq) \longrightarrow 2NH_3OH^+(aq) + 2H_2O$,
(c) $NH_4F + 2HF(l) \longrightarrow NF_3(g) + 3H_2(g)$

21.19 (a) $Li_3N(s) + 3H_2O \longrightarrow 3Li^+(aq) + 3OH^-(aq) + NH_3(g)$,
(b) $2AlN(s) + 3H_2O \longrightarrow Al_2O_3(s) + 2NH_3(g)$,
(c) $Ca_3P_2(s) + 6H_2O \longrightarrow 3Ca^{2+}(aq) + 6OH^-(aq) + 2PH_3(g)$,
(d) $CaNCN(s) + 3H_2O \longrightarrow CaCO_3(s) + 2NH_3(g)$,
(e) $NCl_3(l) + 3H_2O \longrightarrow NH_3(g) + 3HOCl(aq)$,
(f) $PCl_3(l) + 3H_2O \longrightarrow H_3PO_3(aq) + 3H^+(aq) + 3Cl^-(aq)$
(g) $PCl_5(s) + H_2O \longrightarrow POCl_3(l) + 2H^+(aq) + 2Cl^-(aq)$, $PCl_5(s) + 4H_2O \longrightarrow H_3PO_4(aq) + 5H^+(aq) + 5Cl^-(aq)$,
(h) $H_4P_2O_7(l) + H_2O \longrightarrow 2H_3PO_4(aq)$,
(i) $3NO_2(g) + H_2O \longrightarrow 2H^+(aq) + 2NO_3^-(aq) + NO(g)$

21.20 (a) N_2O, **(b)** N_2O_3, **(c)** N_2O_5, **(d)** P_4O_{10}, **(e)** P_4O_{10},
(f) P_4O_6, **(g)** As_2O_5

21.22 (a) $H_2NNH_2(l) + H^+(aq) \longrightarrow H_2NNH_3^+(aq)$,
$H_2NNH_2(l) + 2H^+(aq) \longrightarrow H_3NNH_3^{2+}(aq)$,
(b) $H_2NOH(s) + H^+(aq) \longrightarrow H_3NOH^+(aq)$,
(c) $NO_2^-(aq) + H^+(aq) \longrightarrow HNO_2(aq)$,
(d) $NH_3(aq) + H^+(aq) \longrightarrow NH_4^+(aq)$,
(e) $(NH_4)_2CO_3(s) + 2H^+(aq) \longrightarrow 2NH_4^+(aq) + CO_2(g) + H_2O$

21.23 (a) $4As(s) + 3O_2(g) \longrightarrow As_4O_6(s)$,
(b) $P_4(s) + 3O_2(g) \longrightarrow P_4O_6(s)$, $P_4(s) + 5O_2(g) \longrightarrow P_4O_{10}(s)$,
(c) $2PCl_3(l) + O_2(g) \longrightarrow 2POCl_3(l)$,
(d) $2NO(g) + O_2(g) \longrightarrow 2NO_2(g)$,
(e) $2Sb_2S_3(s) + 9O_2(g) \longrightarrow Sb_4O_6(s) + 6SO_2(g)$

21.24 (a) $Na_3As(s) + 3H_2O \longrightarrow AsH_3(g) + 3Na^+(aq) +$
$3OH^-(aq)$,
(b) $As_2O_5(s) + 3H_2O \longrightarrow 2H_3AsO_4(aq)$,
(c) $Mg_3Bi_2(s) + 6H_2O \longrightarrow 2BiH_3(g) + 3Mg(OH)_2(s)$,
(d) $As_4O_6(s) + 6H_2O \longrightarrow 4H_3AsO_3(aq)$,
(e) $SbCl_3(s) + H_2O \longrightarrow SbOCl(s) + 2H^+(aq) + 2Cl^-(aq)$,
(f) $PH_4I(s) \longrightarrow PH_3(g) + H^+(aq) + I^-(aq)$

21.26 (a) trigonal pyramidal, **(b)** angular, **(c)** tetrahedral,
(d) octahedral, **(e)** trigonal bipyramidal, **(f)** octahedral

21.28 (a) $NH_4^+(aq) + OH^-(aq) \longrightarrow NH_3(g) + H_2O$,
(b) $P_4(s) + 3OH^-(aq) + 3H_2O \longrightarrow PH_3(g) +$
$3H_2PO_2^-(aq)$,
(c) $As_4O_6(s) + 12OH^-(aq) \longrightarrow 4AsO_3^{3-}(aq) + 6H_2O$,
(d) $Sb_2O_5(s) + 2Na^+(aq) + 2OH^-(aq) + 5H_2O \longrightarrow$
$2NaSb(OH)_6(s)$

Chapter 22

22.5 (a) hydrogen cyanide, **(b)** hydrocyanic acid,
(c) potassium cyanide, **(d)** potassium cyanate,
(e) potassium thiocyanate, **(f)** pentacarbonyliron(0),
see Section 24.3,
(g) sodium hydrogen carbonate, **(h)** sodium carbonate

22.6 (a) $CO(g) + Cl_2(g) \longrightarrow COCl_2(g)$,
(b) $CO(g) + S(s) \longrightarrow COS(g)$,
(c) $2CO(g) + O_2(g) \longrightarrow 2CO_2(g)$,
(d) $CO(g) + FeO(s) \longrightarrow Fe(l) + CO_2(g)$,
(e) $4CO(g) + Ni(s) \longrightarrow Ni(CO)_4(g)$

22.10 $2e^- + H_2O + OCN^- \longrightarrow CN^- + 2OH^-$,
$2e^- + H_2O + PbO \longrightarrow Pb + 2OH^-$,
$PbO + CN^- \longrightarrow Pb + OCN^-$

22.13 (a) $B_2O_3 + 3Mg \longrightarrow 2B + 3MgO$,
(b) $2BBr_3 + 3H_2 \longrightarrow 2B + 6HBr$, **(c)** $2B + N_2 \longrightarrow$
$2BN$,
(d) $2B + Mg \longrightarrow MgB_2$, **(e)** $BF_3 + F^- \longrightarrow BF_4^-$
(f) $B_2O_3 + 3H_2O \longrightarrow 2B(OH)_3$,
(g) $B(OH)_3 + OH^- \longrightarrow B(OH)_4^-$,
(h) $B(OH)_3 \longrightarrow HBO_2 + H_2O$, **(i)** $2B(OH)_3 \longrightarrow$
$B_2O_3 + 3H_2O$,
(j) $2LiH + B_2H_6 \longrightarrow 2LiBH_4$

22.15 B_4H_{10}

22.17 (a) $Xe + F_2 \longrightarrow XeF_2$, **(b)** $Xe + 2F_2 \longrightarrow XeF_4$,
(c) $Xe + 3F_2 \longrightarrow XeF_6$, **(d)** $XeF_6 + H_2O \longrightarrow XeOF_4$
$+ 2HF$, **(e)** $XeF_6 + 3H_2O \longrightarrow XeO_3 + 6HF$

22.19 (a) $5XeO_3 + 6Mn^{2+} + 9H_2O \longrightarrow 5Xe +$
$6MnO_4^- + 18H^+$,
(b) $0.120 M$, $3.59 g XeO_3$

22.20 The London forces of He are weaker than those of
H_2. Whereas both He and H_2 have two electrons, the
electron cloud of He is smaller and less polarizable than
that of H_2.

Chapter 23

****23.7** 2.4 cm

23.9 (a) Pt, Au, and Ag are relatively unreactive elements.
(b) $BaSO_4$, $SrSO_4$, and $PbSO_4$ are very insoluble in

water—the least soluble of all sulfates. **(c)** The principal
anions found in sea water are CO_3^{2-}, HCO_3^-, SO_4^{2-}, Cl^-,
and Br^-. The Na^+ and Mg^{2+} salts of all of these anions
are very soluble in water. **(d)** The sulfides of Pb, Ni, and
Bi are very insoluble in water.

23.16 (a) $UO_3 + 2Al \longrightarrow U + Al_2O_3$, **(b)** $3V_2O_5 +$
$2Al \longrightarrow 6V + 5Al_2O_3$, **(c)** $Ta_2O_5 + 10Na \longrightarrow 2Ta +$
$5Na_2O$, **(d)** $ThO_2 + 2Ca \longrightarrow Th + 2CaO$,
(e) $WO_3 + 2Al \longrightarrow W + Al_2O_3$

23.18 $CaCO_3(s) \xrightarrow{heat} CaO(s) + CO_2(g)$,
$CaO(s) + H_2O \longrightarrow Ca^{2+}(aq) + 2OH^-(aq)$,
$Mg^{2+}(aq) + 2OH^-(aq) \longrightarrow Mg(OH)_2(s)$,
$Mg(OH)_2(s) + 2H^+(aq) + 2Cl^-(aq) \longrightarrow Mg^{2+}(aq) +$
$2Cl^-(aq) + 2H_2O$,

$MgCl_2(l) \xrightarrow[heat]{electrolysis} Mg(l) + Cl_2(g)$

23.20 (a) Leaching of the carbonate ore by dilute sulfuric
acid gives a solution of $CuSO_4$, from which the Cu is
obtained by electrolysis. **(b)** The CuS ore is concentrated
by flotation and smelted to Cu_2S (matte). The Cu_2S is
reduced by blowing air through the molten material and
impure, blister Cu is obtained. This impure Cu is refined
by electrolysis. The impure Cu is made the anode of the
cell and pure Cu plates out on the cathode.

23.23 (a) $PbCO_3(s) \longrightarrow PbO(s) + CO_2(g)$,
$PbO(s) + C(s) \longrightarrow Pb(l) + CO(g)$,
(b) $2PbS(s) + 3O_2(g) \longrightarrow 2PbO(s) + 2SO_2(g)$,
$PbS(s) + 2O_2(g) \longrightarrow PbSO_4(s)$,
$PbS(s) + 2PbO(s) \longrightarrow 3Pb(l) + SO_2(g)$,
$PbS(s) + PbSO_4(s) \longrightarrow 2Pb(l) + 2SO_2(g)$

23.24 (a) $2Na + H_2 \longrightarrow 2NaH$, **(b)** $Na + N_2 \longrightarrow$ N.R.,
(c) $2Na + O_2 \longrightarrow Na_2O_2$, **(d)** $2Na + Cl_2 \longrightarrow 2NaCl$,
(e) $2Na + S \longrightarrow Na_2S$, **(f)** $12Na + P_4 \longrightarrow 4Na_3P$,
(g) $2Na + 2C \longrightarrow Na_2C_2$, **(h)** $2Na + 2H_2O \longrightarrow$
$2NaOH + H_2$,
(i) $2Na + 2NH_3 \longrightarrow 2NaNH_2 + H_2$

23.25 Since Cs is the largest I A metal atom, Cs is the atom
from which it is most easy to remove one electron (smallest
first ionization energy). The Li^+ ion, however, is the
smallest I A metal cation and more energy is released
when Li^+ is hydrated in water solution than when any
other I A cation is hydrated. Standard electrode potentials
refer to processes that occur in water solution. The relatively large amount of energy released by the hydration
of the Li^+ ion makes up for the relatively large amount
of energy required for the ionization of Li.

23.27 (a) $Ca + H_2 \longrightarrow CaH_2$, **(b)** $3Ca + N_2 \longrightarrow Ca_3N_2$,
(c) $2Ca + O_2 \longrightarrow 2CaO$, **(d)** $Ca + Cl_2 \longrightarrow CaCl_2$,
(e) $Ca + S \longrightarrow CaS$, **(f)** $6Ca + P_4 \longrightarrow 2Ca_3P_2$,
(g) $Ca + 2C \longrightarrow CaC_2$, **(h)** $Ca + 2H_2O \longrightarrow Ca(OH)_2 +$
H_2, **(i)** $Ca + 2NH_3 \longrightarrow Ca(NH_2)_2 + H_2$

23.30 (a) $2Al + 3Cl_2 \longrightarrow 2AlCl_3$,
(b) $4Al + 3O_2 \longrightarrow 2Al_2O_3$,
(c) $2Al + 3S \longrightarrow Al_2S_3$,
(d) $2Al + N_2 \longrightarrow 2AlN$,
(e) $2Al + 2OH^- + 6H_2O \longrightarrow 2Al(OH)_4^- + 3H_2$,
(f) $2Al + 6H^+ \longrightarrow 2Al^{3+} + 3H_2$

23.33 The Al^{3+} ion has a high charge and hydrolyzes readily. The S^{2-} ion also hydrolyzes. Al_2S_3 is completely hydrolyzed in water.

$$2 Al^{3+}(aq) + 3 S^{2-}(aq) + 6 H_2O \longrightarrow$$
$$2 Al(OH)_3(s) + 3 H_2S(g)$$

23.34 (a) $Sn + 2 Cl_2 \longrightarrow SnCl_4$, **(b)** $Sn + O_2 \longrightarrow SnO_2$,
(c) $Sn + 2 S \longrightarrow SnS_2$, **(d)** $Sn + 2 H^+ \longrightarrow Sn^{2+} + H_2$,
(e) $Sn + OH^- + 2 H_2O \longrightarrow Sn(OH)_3^- + H_2$, **(f)** $3 Sn + 4 H^+ + 4 NO_3^- \longrightarrow 3 SnO_2 + 4 NO + 2 H_2O$

23.35 (a) $PbO_2(s) + 2 H_2O + 2 OH^-(aq) \longrightarrow Pb(OH)_6^{2-}(aq)$,
(b) $3 PbO_2(s) \longrightarrow Pb_3O_4(s) + O_2(g)$,
(c) $2 PbO_2(s) \longrightarrow 2 PbO(s) + O_2(g)$,
(d) $PbO(s) + H_2O + OH^-(aq) \longrightarrow Pb(OH)_3^-(aq)$,
(e) $SnS(s) + S^{2-}(aq) \longrightarrow$ N.R.,
(f) $SnS_2(s) + S^{2-}(aq) \longrightarrow SnS_3^{2-}(aq)$,
(g) $PbCl_2(s) + Cl^-(aq) \longrightarrow PbCl_3^-(aq)$,
(h) $SnF_4(s) + 2 F^-(aq) \longrightarrow SnF_6^{2-}(aq)$

23.36 (a) $2 Fe + 3 Cl_2 \longrightarrow 2 FeCl_3$, $2 Cr + 3 Cl_2 \longrightarrow 2 CrCl_3$, $Zn + Cl_2 \longrightarrow ZnCl_2$,
(b) $3 Fe + 2 O_2 \longrightarrow Fe_3O_4$, $4 Cr + 3 O_2 \longrightarrow 2 Cr_2O_3$, $2 Zn + O_2 \longrightarrow 2 ZnO$,
(c) $Fe + S \longrightarrow FeS$, $Cr + S \longrightarrow CrS$, $Zn + S \longrightarrow ZnS$,
(d) $Fe + N_2 \longrightarrow$ N.R., $2 Cr + N_2 \longrightarrow 2 CrN$, $Zn + N_2 \longrightarrow$ N.R.,
(e) $3 Fe + 4 H_2O \longrightarrow Fe_3O_4 + 4 H_2$, $2 Cr + 3 H_2O \longrightarrow Cr_2O_3 + 3 H_2$, $Zn + H_2O \longrightarrow ZnO + H_2$,
(f) $Fe + 2 H^+ \longrightarrow Fe^{2+} + H_2$, $Cr + 2 H^+ \longrightarrow Cr^{2+} + H_2$, $Zn + 2 H^+ \longrightarrow Zn^{2+} + H_2$,
(g) $Fe + OH^- \longrightarrow$ N.R., $2 Cr + 6 OH^- + 6 H_2O \longrightarrow 2 Cr(OH)_6^{3-} + 3 H_2$, $Zn + 2 OH^- + 2 H_2O \longrightarrow Zn(OH)_4^{2-} + H_2$

23.39 $CrO + 2 H^+ \longrightarrow Cr^{2+} + H_2O$,
$CrO + OH^- \longrightarrow$ N.R.,
$Cr_2O_3 + 6 H^+ \longrightarrow 2 Cr^3 + 3 H_2O$,
$Cr_2O_3 + 2 OH^- + 3 H_2O \longrightarrow 2 Cr(OH)_4^-$,
$CrO_3 + H_2O \longrightarrow H_2CrO_4$,
$CrO_3 + 2 OH^- \longrightarrow CrO_4^{2-} + H_2O$

*****23.41 (a)** $+0.0592$ V, **(b)** $+0.770$ V

23.42 (a) $2 La + 3 Cl_2 \longrightarrow 2 LaCl_3$,
(b) $4 La + 3 O_2 \longrightarrow 2 La_2O_3$,
(c) $2 La + 3 S \longrightarrow La_2S_3$,
(d) $2 La + N_2 \longrightarrow 2 LaN$,
(e) $2 La + 6 H_2O \longrightarrow 2 La(OH)_3 + 3 H_2$,
(f) $2 La + 6 H^+ \longrightarrow 2 La^{3+} + 3 H_2$

Chapter 24

24.2 (a) $3+$, **(b)** $3+$, **(c)** 0, **(d)** $3+$, **(e)** $2+$

24.4 If the reaction is written $Ag^+ + 2 NH_3 \longrightarrow Ag(NH_3)_2^+$, Ag^+ is functioning as a Lewis acid and NH_3 as a Lewis base. For the reaction written $Ag(H_2O)_2^+ + 2 NH_3 \longrightarrow Ag(NH_3)_2^+ + 2 H_2O$, NH_3 (a Lewis base) is displacing H_2O (another Lewis base) from a complex with Lewis acid (Ag^+)—the reaction is a base (nucleophilic) displacement.

*****24.5** 1.4×10^{-34}

24.7 (a) $K_2[Rh(H_2O)Cl_5]$, **(b)** $[Co(NH_3)_4(SO_4)]NO_3$,
(c) $Na_3[ReO_2(CN)_4]$, **(d)** $[Co(NH_3)_2(en)_2]Cl_2$,
(e) $K_4[Ni(CN)_4]$, **(f)** $K_4[Ni(CN)_4]$,
(g) $[Cu(NH_3)_4]_3[CrCl_6]_2$

24.9 (a) potassium tetracyanonickelate(0),
(b) potassium tetracyanonickelate(II),
(c) ammonium pentachloroaquoferrate(III),
(d) tetraamminecopper(II) tetrachloroplatinate(II),
(e) nitritopentaammineiridium(III) chloride,
(f) hexaamminecobalt(III) tetracyanonickelate(II)

24.11 (a) $KFe[Fe(CN)_6]$, **(b)** $Fe[Fe(CN)_6]$,
(c) $Cu_2[Fe(CN)_6]$, **(d)** $K_2Fe[Fe(CN)_6]$

24.13 (a) $[Co(NH_3)_5(SO_4)]NO_3$,
(b) $[Mn(CO)_5(NCS)]$,
(c) $[Pt(NH_3)_3Cl][Pt(NH_3)Cl_3]$,
(d) $[Co(en)_2(H_2O)Br]Br_2 \cdot H_2O$

24.15 Let $a = NH_3$. There is one stereoisomer of $[Pta_4][PtCl_6]$ and one of $[Pta_3Cl][PtaCl_5]$. For $[Pta_3Cl_3][PtaCl_3]$ there are two:

and

For $[Pta_4Cl_2][PtCl_4]$ there are two:

and

24.17 Let $a = NH_3$ and py = pyridine.

24.18 The dipole moment of the *trans* isomer is zero.

24.20 $[Ni(CN)_4]^{2-}$ has no unpaired electrons, $[NiCl_4]^{2-}$ has two unpaired electrons

24.23 **(a)** The pairing energy must be greater than 250 kJ/mol and less than 460 kJ/mol. Actually, P is 335 kJ/mol. **(b)** Since $C_2O_4^{2-}$ induces a smaller Δ_o than H_2O (see the spectrochemical series), $[Mn(C_2O_4)_3]^{3-}$ should be a high-spin complex.

24.24 $[Co(NH_3)_6]^{2+}$ is a d^7 high-spin octahedral complex, $[Co(NH_3)_6]^{3+}$ is a d^6 low-spin octahedral complex. See Figure 24.16 for splitting diagrams.

*24.27 $[Ti(H_2O)_6]^{3+}$ is a d^1 complex. Light is *absorbed* by an electron transition from a t_{2g} orbital to an e_g orbital (Δ_o must be supplied). Red and violet light is *not* absorbed, which produces the color of the complex. If a larger Δ_o is induced, the *absorbance* takes place at shorter wavelengths (more energetic radiation, toward the violet end) and *transmittance* takes place at longer wavelengths (at the red end). If Δ_o is increased gradually, the transmitted color changes from red to yellow to orange to green, and so forth.

Chapter 25

25.3 **(a)** $^{218}_{85}At \longrightarrow ^{214}_{83}Bi + ^4_2He,$
(b) $^{198}_{79}Au \longrightarrow ^{198}_{80}Hg + ^{0}_{-1}e,$ **(c)** $^{25}_{13}Al \longrightarrow ^{25}_{12}Mg + ^0_1e,$
(d) $^{108}_{47}Ag + ^{0}_{-1}e \longrightarrow ^{108}_{46}Pd$

25.5 **(a)** $^{221}_{87}Fr \longrightarrow ^{217}_{85}At + ^4_2He,$
(b) $^{66}_{29}Cu \longrightarrow ^{66}_{30}Zn + ^{0}_{-1}e,$ **(c)** $^{18}_9F \longrightarrow ^{18}_8O + ^0_1e,$
(d) $^{133}_{56}Ba + ^{0}_{-1}e \longrightarrow ^{133}_{55}Cs$

25.7 2.6 MeV

25.9 0.22 MeV

25.11 19.99999 u

25.13 3.52 MeV

25.15 30.98066 u

25.17 Since the mass difference (0.00088 u) is less than 0.00110 u, the process will occur by electron capture.

25.19 50.9448 u

25.21 0.00365 g

25.23 288 days

25.25 **(a)** 6.23×10^{-3}/min, **(b)** 111 min

25.27 256 min

25.29 $1/10^{645}$

25.31 **(a)** 1.6×10^{17} atoms, **(b)** 1.4×10^{-5} g

25.33 $^{228}_{88}Ra,$ $^{228}_{89}Ac,$ $^{228}_{90}Th,$ $^{224}_{88}Ra,$ $^{220}_{86}Rn,$ $^{216}_{84}Po,$ $^{216}_{85}At,$ $^{212}_{83}Bi,$ $^{212}_{84}Po,$ $^{208}_{82}Pb$

25.35 **(a)** $^{82}_{35}Br + ^1_0n \longrightarrow ^{83}_{35}Br + \gamma,$
(b) $^{10}_5B + ^1_0n \longrightarrow ^7_3Li + ^4_2He,$
(c) $^{35}_{17}Cl + ^1_0n \longrightarrow ^{35}_{16}S + ^1_1p,$
(d) $^7_3Li + ^1_1p \longrightarrow ^7_4Be \longrightarrow ^1_0n,$
(e) $^{130}_{52}Te + ^2_1H \longrightarrow ^{130}_{53}I + 2^1_0n,$
(f) $^{43}_{20}Ca + ^4_2He \longrightarrow ^{46}_{21}Sc + ^1_1p,$
(g) $^{237}_{93}Np + ^4_2He \longrightarrow ^{240}_{95}Am + ^1_0n,$
(h) $^{238}_{92}U + ^{22}_{10}Ne \longrightarrow ^{256}_{102}No + 4^1_0n$

25.38 $^{249}_{98}Cf + ^{15}_7N \longrightarrow ^{260}_{105}? + 4^1_0n$

25.39 7.5915 MeV/nucleon

25.41 **(a)** 5.5 MeV, **(b)** 0.20%

Chapter 26

26.2 **(a)** $CH_3-CH-CH-CH_2-CH_3,$ with CH_3 CH_3 substituents,

(b) $CH_2{=}CH-\underset{CH_2-CH_3}{\overset{CH_3}{\underset{|}{\overset{|}{C}}}}-CH_2-CH_3,$

(c) $CH_3-\underset{CH_3}{\overset{|}{CH}}-C{\equiv}C-\underset{CH_3}{\overset{|}{CH}}-CH_3,$

(d) benzene ring with Br, Br, Br; **(e)** benzene ring with NO_2, NO_2

26.3 **(a)** The lowest possible number should be used—should be 2-pentene, **(b)** The longest chain consists of 4 atoms—should be 2-methylbutane, **(c)** Compound does not exist—C number 2 is pentavalent, **(d)** The lowest possible numbers should be used—should be 3-methyl-1-butyne, **(e)** Compound does not exist—C number 3 is pentavalent, **(f)** The longest chain consists of 6 atoms—should be 3-methylhexane

26.5 *monosubstituted*: 1- and 2-chloropropane, *disubstituted*: 1,1-; 2,2-; 1,2-; and 1,3-dichloropropane, *trisubstituted*: 1,1,1-; 1,1,2-; 1,1,3-; 1,2,2-; and 1,2,3-trichloropropane

26.7 *dichlorobutanes*: 1,1-; 2,2-; 1,2- (*d, l* pair); 1,3-(*d, l* pair); 1,4-; 2,3- (*d, l* pair and an optically inactive form), *dichloro-2-methylpropanes*: 1,1-; 1,2-; 1,3-

26.9 **(a)** two (2,3- and 3,4-dibromonitrobenzene), **(b)** three (2,6-; 2,4-; and 3,5-dibromonitrobenzene), **(c)** one (2,5-dibromonitrobenzene)

26.11 the *four* atoms of the $Z-C{\equiv}C-Z$ grouping are linear

26.12 **(b)** and **(d)**

26.16 **(a)** bromomethane, dibromomethane, tribromomethane, and tetrabromomethane, **(b)** 2,3-dibromobutane, **(c)** 2,2,3,3-tetrabromobutane, **(d)** bromobenzene

26.20 **(a)** o- and p-nitrophenol, **(b)** o- and p-bromophenol, **(c)** m-dinitrobenzene, **(d)** m-bromonitrobenzene, **(e)** m-nitrobenzoic acid, **(f)** o- and p-bromonitrobenzene, **(g)** o- and p-nitrotoluene, **(h)** o- and p-xylene

26.22 **(a)** 4-methyl-1-pentanol, **(b)** 4-methyl-2-pentanol, **(c)** 2-methyl-2-pentanol, **(d)** 2-methyl-3-pentanone, **(e)** ethyl isopropyl ether, **(f)** 4-methylpentanal, **(g)** methyl phenyl ketone, **(h)** cyclohexanone, **(i)** methyl benzoate, **(j)** m-nitrobenzoic acid

26.23 1-butanol; 2-butanol (*d, l* pair); 2-methyl-1-propanol; 2-methyl-2-propanol; methyl *n*-propyl ether; methyl isopropyl ether; diethyl ether

26.26 (a) *n*-butyl magnesium bromide, (b) *n*-butyl cyanide, (c) 1-butanol, (d) ethyl *n*-butyl ether

26.28 (a) acetic acid and butanoic acid, (b) acetic acid and acetone, (c) propanoic acid, (d) propanoic acid, (e) acetone, (f) propanoic acid, (g) *p*-nitrobenzoic acid; (a) 2,3-hexanediol, (b) propanal

26.30 (a) CH_3CH_2Br, (b) CH_3CH_3 and $MgBr_2$, (c) CH_3COOH and $NaBr$

26.32 (a) 1-propanol, (b) 3-hexanol, (c) 3-methyl-3-pentanol

***26.35** Two Grignard reagents may be prepared by converting the two alcohols to bromo compounds and then reacting these bromo compounds with Mg in dry ether: $CH_3CH_2CH_2MgBr$ and $(CH_3)_2CHMgBr$. In addition, propanal may be obtained by the mild oxidation of 1-propanol and acetone may be obtained by the oxidation of 2-propanol. The desired products are obtained by hydrolyzing the product of the reaction of:
(a) $CH_3CH_2CH_2MgBr$ and CH_3CH_2CHO,
(b) $(CH_3)_2CHMgBr$ and $(CH_3)_2CO$, $CH_3CH_2CH_2MgBr$ and $(CH_3)_2CO$,
(d) $(CH_3)_2CHMgBr$ and CH_3CH_2CHO

***26.36** Dehydration of 1-propanol to propene (with H_2SO_4), addition of HBr to propene to produce 2-bromopropane, alkaline hydrolysis of the bromide to give 2-propanol

26.39 1-aminobutane, 2-aminobutane, 2-methyl-1-aminopropane, 2-methyl-2-aminopropane, 2-methyl-2-aminopropane, methyl *n*-propyl amine, methyl isopropyl amine, diethyl amine, dimethyl ethyl amine

26.41 (a) ethyl amine, (b) *p*-methylaniline, (c) methyl *n*-butyl amine, (d) acetamide, (e) propanoic acid, (f) diethyl-ammonium bromide, (g) propanoic acid, (h) ethyl amine

26.43 the products are: (a) $(CH_3)_2CHNH_3^+Br^-$, (b) $CH_3CH_2COO^-NH_4^+$, (c) $CH_3CH_2CONH_2$ and CH_3OH

26.45 $CH_3CH_2O^-$, OH^-, $CH_3CH_2NH_2$, CH_3COO^-, CH_3CH_2OH

26.49 $CH_2{=}CH_2$, (b) $CF_2{=}CF_2$, (c) $C_6H_5CH{=}CH_2$, (d) $CH_2{=}CHCN$, (e) $H_2N(CH_2)_6NH_2$ and $HOOC(CH_2)_4COOH$, (f) $CH_2{=}C(CH_3)CH{=}CH_2$, (g) C_6H_5OH and $HCHO$, (h) $HOCH_2CH_2OH$ and *p*-$HOOCC_6H_4COOH$

Chapter 27

27.1 $H_2N-\overset{\displaystyle CH_3}{\underset{}{CH}}-\overset{\displaystyle O}{\overset{\|}{C}}-NH-\overset{\displaystyle CH_3}{\underset{}{CH}}-\overset{\displaystyle O}{\overset{\|}{C}}-NH-\overset{\displaystyle CH_3}{\underset{}{CH}}-\overset{\displaystyle O}{\overset{\|}{C}}-OH$

27.3 Since neither carbon atom in glycine is chiral, optical isomers of glycine do not exist.

27.6 approximately 8333

***27.7** 4, 8, x^y

27.10 (a), (d), and (e)

27.13 $CH_2(OH)CH(OH)CHO$, carbon atom number 2 is chiral, two optical isomers exist (a *d, l* pair)

27.14 $CH_2(OH)CH(OH)CH(OH)CHO$, 2 chiral carbon atoms (numbers 2 and 3), 4 optical isomers (2 *d, l* pairs); $CH_2(OH)CH(OH)C(O)CH_2OH$, one chiral carbon atom (number 3), 2 optical isomers (one *d, l* pair)

27.19 stearic acid

27.24 765,000

27.25 (a) U—C—U, (b) A—G—A, (c) T—C—T

INDEX

A reference to an *illustration* is marked with an *i* following the page number, one to a *definition* is marked with a *d*, and one to a *table* is marked with a *t*.

A reference to an *illustration* is marked with an *i* following the page number, one to a *definition* is marked with a *d*, and one to a *table* is marked with a *t*.

A reference to an *illustration* is marked with an *i* following the page number, one to a *definition* is marked with a *d*, and one to a *table* is marked with a *t*.

A reference to an *illustration* is marked with an *i* following the page number, one to a *definition* is marked with a *d*, and one to a *table* is marked with a *t*.